EUROPEAN PUMPS & PUMPING

The practical reference book on pumps & ancillary equipment and buyers guide to European manufacturers and suppliers

Published by

CHW Roles & Associates Ltd
PO Box 25, Sunbury-on-Thames
Middlesex TW16 5QB
United Kingdom
Tel/Fax: 081-783 0088

© 1994 The entire content of this publication is protected by copyright, full details of which are available from the Publishers. No part of this publication may be reproduced, stored in a retrieval system, or transmitted in any form or by any means - electronic, mechanical, photocopying, recording or otherwise - without prior permission of the copyright owner.

Whilst every care is taken in the compilation of European Pumps & Pumping the Publishers accept no responsibility for any errors or omissions which may have occurred

Printed in England   ISBN 0 907485 07 3

RELIABILITY AND SAVINGS WITH AHLSTROM PUMPS

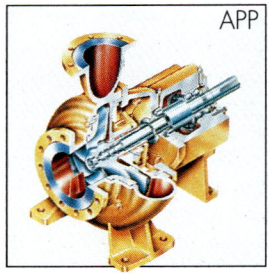
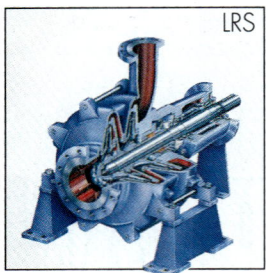
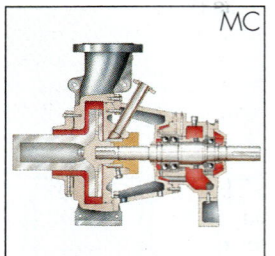
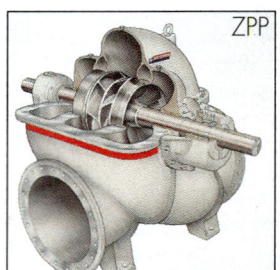
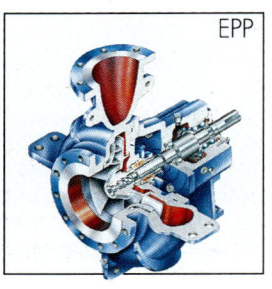
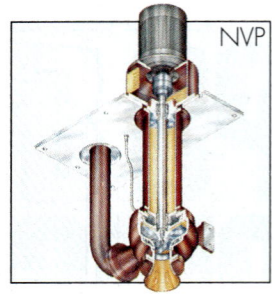

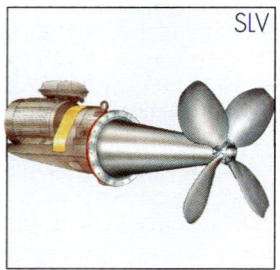

Ahlstrom has been manufacturing pumps for more than 100 years. Our basic objective has always been continuous development of pumps and pumping solutions for our customers. The close cooperation with our customers has created us better capabilities to improve pumping reliability and achieve greater savings.

Ahlstrom has successfully developed several pump innovations which contribute to greater reliability. Such are for example the ROTOKEY self-locking impeller mounting and the dynamic seal without any wearing parts and external sealing water. The AHLSTAR process pump can also be equipped with a degassing impeller thus enabling pumping of liquids with gas content up to 50%.

Ahlstrom's comprehensive pump selection with various impeller alternatives for the process pumps ensure that you will always get the optimum solution for each application. Considerable energy savings can be achieved with the right pump type combined with good efficiency.

The high quality of Ahlstrom pumps originates from our own foundry and the precise FMS machining centers. The Ahlstrom Pump Factories have received the Quality System Certificate ISO 9002.

AHLSTROM MACHINERY, Pump Industry: Finland: Tel. +358 5 2291 111, Fax +358 5 263958, **Austria:** Tel. +43 7242 60692, Fax +43 7242 60711, **France:** Tel. +33 88 532 100, Fax +33 88 532 115, **Germany, Netherlands:** Tel. +49 70 21 5009-0, Fax +49 70 21 81085, **Great Britain:** Tel. +44 61 705 1631, Fax +44 61 797 1494, **Norway, Denmark, Iceland:** Tel. +47 32 837890, Fax +47 32 830805, **Russia:** Tel. +7 812 535 1543, Fax +7 812 534 8872, **Spain:** Tel. +34 3 414 0000, Fax +34 3 201 9816, **Sweden:** Tel. +46 11 282 200, Fax +46 11 282 253

Foreword

By Mr L M Teasdale
Technical Director, British Pump Manufacturers Association

The universality of pumping arises from its central position in an advanced economy. As systems for distribution of fluids have extended and manufacturing processes have demanded ever higher temperatures and pressures, and man's continuing desire to recover more of the Earth's surface from the twin barriers of water and drought, so dependency on the pump and on pumped systems has increased. The present importance of the pump is clear when it is realised that in developed, developing and underdeveloped countries, all rely upon pumping to provide the facility for moving liquids from where they are to where they are wanted.

This widespread application places upon engineers a great responsibility to understand the means of designing effective pumps, selecting that most appropriate for the task; operating it sensibly; and ensuring continuous reliable functioning. It also creates difficulties in equipping engineers to achieve these aims.

By its broad range of cover, pumping is a subject important to all branches of engineering but is not the specialisation of any one. It is a fringe subject to all despite its central importance.

It follows that given the combination of broad fields of application and man's ingenuity, there is likely to be a wide range of invented solutions. And it is so. The number of pump types is legion with most having some unique feature exploitable in an application even if it is sometimes difficult to see it. Some may have manufacturing advantages and others operating benefits and it is this range of options that makes selection complex, but which ensures that there is a best solution in every case.

Books such as this set out to aid those engaged in pumps and pumping, whether as specifiers or as users, to make the best choices and to use the machines wisely.

It is often forgotten by users of pumps that their liquid will have particular characteristics and that these may affect a pump's performance. Similarly, the system designer can make the mistake of optimising his system from the viewpoint of its function leaving the pump designer to fit a pump. It has been known for the pump designer to produce a complex design with attractive features but 'delicate', requiring tender care. Each is guilty of a 'blinkered' outlook, and the most successful installation is one where all the interests are balanced to give a well-matched combination.

"European Pumps & Pumping" is structured in such a way to help the serious reader to build his or her knowledge of the subject. It takes the non-specialist pump engineer through the basic language of pumps and on to the broad categories available. Installation problems are discussed as is the control of the system - an often overlooked aspect. The book then moves into the ancillary, but still important, areas of shaft sealing and drivers. Finally, it deals with pumping economics and gives a rapid selection procedure.

"European Pumps & Pumping" is not just a summary of the subject, but a detailed presentation of the factors to be taken into account whatever is the reader's role in the pumped system. It is both a text book and a reference book.

How to use this book

For detailed pumping problems, "European Pumps & Pumping" is best used as a reference book. The detailed contents (pages 5 - 10) and the reference index (pages 390 - 396) will simplify finding the appropriate section.

When planning new or modifying existing pumping installations, cost is often the decisive factor in determining the best method of approach. Simpler installations intended for short service life do not require such extensive investigation work. For larger plant with long service life, alternative solutions must be examined, particularly on account of the often considerable energy costs. Chapter 13 deals with these decisive factors and is important. As a text book though, "European Pumps & Pumping" can be read from cover to cover to obtain a comprehensive picture of the various aspects of liquid transportation. Individual chapters can of course be studied separately.

Considerable coverage has been given to the question of pump choice. The first steps in choosing a pump are to define:

- The transportation job
- The operating conditions: - operational time, environment, control and regulation requirements
- Liquid properties
- Pump flow and differential head

The next step is to find a suitable pump or a number of alternatives. Apart from the information given in Chapter 3 — Pumps, the general sections of Chapter 14 — Pump selection, provide two alternative "short cuts":-

Direct selection table - Section 14.2

Rapid selection table - Section 14.3

The direct selection table is based on current pump designations which are associated with a particular pump use or industry, e.g. deep-well pumps, oil cargo pumps etc.

The rapid selection table is designed to aid pump selection according to the basic 'technical' features. By narrowing the selection to specific values for flow, differential head, viscosity, particle content, etc., it is possible, using the table, to obtain suggestions for suitable pump types. Both tables link up with Chapter 16 — Pump classification guide and guide to manufacturers and suppliers, listing the names of suitable manufacturers and suppliers for the categories of pumps recommended. The names and addresses of these companies are also listed alphabetically by country.

The pump manufacturers and suppliers guide (Section 16.3) is cross-referenced to Chapter 3 — Pumps, enabling the various pump categories to be compared and potential manufacturers and suppliers to be identified. A guide to manufacturers and suppliers of ancillary products such as couplings, motors and packing materials is also provided, (Section 16.4). More detailed information on the pump alternatives selected, materials, couplings, etc., is supplied in the appropriate chapters.

Artus™ colour coded plastic shims

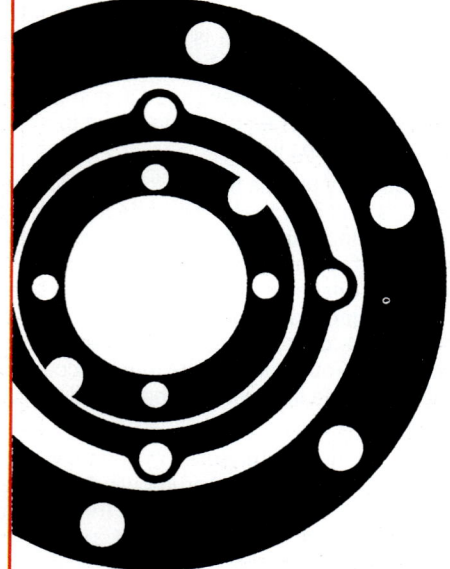

We are one of the most successful suppliers of plastic shims and gaskets to the pump industry. Our Artus shim stock is colour coded for different thicknesses ensuring the right stock (0.012mm to 1.50mm) is selected.

They are non-rusting, have a high resistance to chemicals, have a significant reduction in weight against steel and perform better at low temperatures. Artus also costs less than metal, although we do also offer metals in sheet form and finished components in a thickness range from 0.012mm to 0.787mm. Our technical support department can work with you to ensure your requirement is both designed and manufactured to precision, giving the ultimate combination in service. So to keep your pumps pumping, call Hughes Wynne on :

0932 569700

Hughes Wynne
L I M I T E D

68 Guildford Street, Chertsey, Surrey GB KT16 9BB. England
Fax 0932 569652

Worth pumping up and down about

CHEMCON

HIGH PERFORMANCE WITH PRECISION . . . .
Europe's leading range of pumps

CHEMCON PTFE pump and filter units from Production Techniques are the outright winners in the race for improved filtration performance and increased cost efficiency. The fully submersible PFU 30000 sets the pace by combining pump and filter in a single compact PTFE housing, ensuring long life in corrosive environments. Fluid lifespans too are greatly increased as a result of its unbeatable filtration qualities.

The pneumatic bellows CBP 500 pump is designed for continuous operation over long periods and can be used for pumping corrosive liquids. The pump consists of two convoluted PTFE bellows and end caps with the main pump housing manufacturered in polypropylene. The unit's simplicity of construction and operation makes it suitable both for plant and laboratory work.

Both the PFU 30000 and the CBP 500 are pneumatically controlled by an external valve system incorporated into a control box with easy access for maintenance, keeping down time and production loss to a minimum.

To find out more about why CHEMCON PTFE pump and filter units, are the driving force in cost-efficient filtration and recirculation contact Production Techniques now!

Telephone 0252 616575.
A WINNING COMBINATION
Production Techniques Ltd
13 Kings Road, Fleet, Hants GU13 9AU
Tel: 0252 616575 · Fax: 0252 615818

CLASAL

DOUBLE ACTING PISTON-PUMPS
1 till 50 m³/h

* quality
* self priming up to 8,5 m
* long life
* since 1928

APPLICATIONS :
  watering
  marine
  agriculture
  water supply

we're searching agents in several countries

CLASAL NV
Ruddervoordestraat 82
B 8210 ZEDELGEM - BELGIUM
TEL 050/278253 FAX 050/275494

Contents

Chapter 1 — The properties of liquids — 11

1.1 Explanation of terms — 11
1.1.1 Introduction — 11
1.1.2 Changes of state — 11
1.1.3 Viscosity — 13
1.1.4 Density and relative density — 14
1.1.5 Compressibility — 14
1.1.6 pH value — 15
1.1.7 Hazards — 16

1.2 Water — 19
1.2.1 Demineralized water — 19
1.2.2 Fresh water — 20
1.2.3 Brackish water — 22
1.2.4 Sea water — 22

1.3 Oils — 24
1.3.1 General — 24
1.3.2 Viscosity — 24

1.4 Liquid solid mixtures — 26
1.4.1 General — 26
1.4.2 Sewage — 28
1.4.3 Sludge — 29
1.4.4 Pulps — 29

1.5 Liquid gas mixtures — 30

1.6 Table of liquid properties — 30
1.6.1 General — 30
1.6.2 Liquid table — 32

Chapter 2 — Liquid flow — 49

2.1 Basic equations — 50
2.1.1 Explanation of terms — 50
2.1.2 Continuity equation — 50
2.1.3 Bernoulli's equation — 50
2.1.4 Momentum equation — 51
2.1.5 Energy equation — 52

2.2 Pipe flow losses — 53
2.2.1 Reynolds Number — 53
2.2.2 Pressure losses in straight pipes — 53
2.2.3 Pressure losses in fittings — 54
2.2.4 Hydraulic diameter — 54
2.2.5 Total losses in a pipe system — 55

2.3 Liquid solid mixtures — 56
2.3.1 General — 56
2.3.2 Homogeneous mixtures — 56
2.3.3 Heterogeneous mixtures — 57
2.3.4 Paper pulp — 59

2.4 Pressure loss diagrams — 60
2.4.1 General — 60
2.4.2 Water — 63
2.4.3 Oil — 66
2.4.4 Paper pulp — 70

2.5 Flow measurement — 72
2.5.1 General — 72
2.5.2 Instruments for pipe systems — 73
2.5.3 Instruments for open channels — 75
2.5.4 Other instruments — 76
2.5.5 Measurement standards — 77

Chapter 3 — Pumps — 79

3.1 Introduction — 81

3.2 Rotodynamic pumps — 82
3.2.1 Rotodynamic pump theory — 82
3.2.2 Rotodynamic pump curves — 87
3.2.3 Rotodynamic pump classification — 91
3.2.4 Types of rotodynamic pump — 92

005 Standard water pumps — 93
010 Double suction axially-split pumps — 93
015 Heating, water and sanitation pumps — 94
020 Automatic supply packages — 94
025 Standard pumps to prEN 733 — 95
030 Agricultural pumps without driver — 96
035 Fixed irrigation pumps — 96
040 Machine tool coolant pumps — 96
045 Marine pumps — 96
050 Electrically driven submersible pumps — 97
055 Vertical wet pit pumps — 98
060 Vertical dry pit pumps — 99
065 Portable self-priming pumps and submersible pumps not electrically driven — 99
070 Deep well pumps with ejector — 100
075 Submersible deep well pumps — 100
080 Wash water pump packages — 101
085 Multi-stage segmental pumps > 300m — 101
090 Multi-stage segmental pumps < 300m — 102
095 Standard pumps to ISO 2858, ISO 3069, ISO 3661 — 102
100 End suction pumps to ASME/ANSI B73.1 — 103
105 Inline pumps to ASME/ANSI B73.2 — 103
110 End suction pumps to API 610 — 104
115 Double suction pumps to API 610 — 104
120 Magnetic drive pumps — 104
125 Canned motor pumps — 105
130 Non-metallic pumps — 106
135 Pulp pumps — 106
140 Pumps for solids > 10mm — 107
145 Pumps for solids < 10mm — 107
150 Hygienic quality pumps — 107
155 High speed single stage pumps — 108
160 High speed multi-pump packages — 108
165 Multi-stage axially-split pumps — 109
170 Multi-stage radially-split pumps — 109
175 Vertical multi-stage pumps — 110
180 Non-clogging pumps with standard motor — 111
185 Submersible non-clogging pumps — 111
190 Other non-clogging pumps — 112
195 Mixed flow pumps — 112
200 Axial propeller pumps — 112
205 Power recovery turbines — 112

3.3 Liquid ring and peripheral pumps	**113**
3.3.1 Liquid ring pumps	113
3.3.2 Peripheral pumps	113
210 Liquid ring pumps	113
215 Peripheral pumps	114
3.4 Positive displacement pumps	**114**
3.4.1 Rotary positive displacement pump theory	115
3.4.2 Reciprocating positive displacement pump theory	116
3.4.3 Positive displacement pump curves	119
3.4.4 Classification of positive displacement pumps	120
3.4.5 Types of positive displacement pump	120
220 External gear pumps	121
225 Internal gear pumps	122
230 Screw pumps with 2 or more screws	123
235 Progressing cavity pump	123
240 Lobe pumps and rotary piston pumps	124
245 Rigid vane and flexible vane pumps	126
250 Peristaltic hose pump	127
255 Rotary peristaltic pump	127
260 Piston pumps for hydraulic power - radial and axial pistons	127
265 Portable plunger pump packages for high pressure descaling/cleaning	128
270 Plunger pumps - horizontal and vertical	128
275 Piston pumps	130
280 Diaphragm pumps	131
285 Metering pumps	131
290 Jet pumps	132
295 Air lift pumps	134
300 Barrel emptying pumps	134
305 Rigid screw pumps	135
310 Pitot tube pumps	135
3.5 Other pump types	**136**
3.5.1 Theory	136
3.5.2 Classification	136
315 Miscellaneous types	136
3.6 Suction performance	**137**
3.6.1 Cavitation	137
3.6.2 Net Positive Suction Head	138
3.6.3 Permissible suction lift	138
2.6.4 Cavitation effects on pump operation	140
3.6.5 Self priming	142
3.6.6 The effect of dissolved and entrained gases	142
3.6.7 Examples	143
3.7 Standards	**144**
3.7.1 General	144
3.7.2 Issuing authorities	144
3.7.3 Safety	145
3.7.4 Reliability and operational life	146
3.7.5 Dimensions and performance	146
3.7.6 Vibration and noise	147
3.7.7 Forces and moments on connections	147
3.7.8 Components	148
3.7.9 Testing	148
Chapter 4 — Pumps and piping systems	**151**
4.1 System curves	**152**
4.1.1 Pump nominal duty point	152
4.1.2 Single pipe system	152
4.1.3 Variable system curves	153
4.1.4 Branched pipe systems	154
4.1.5 Viscous and non-Newtonian liquids	156
4.2 Valve pressure drop	**156**
4.2.1 General	156
4.2.2 Isolating valves	156
4.2.3 Non-return valves	156
4.2.4 Control valves	156
4.3 Multiple pump systems	**158**
4.3.1 Series pump operation	158
4.3.2 Parallel pump operation	159
4.3.3 Pressure boosting	160
4.3.4 Pressure maintenance	161
4.4 Pump hydraulic specification data	**161**
4.4.1 Hydraulic tolerances	161
4.4.2 Pump Q and H at maximum efficiency	162
4.4.3 Minimum pump flow	163
4.5 Water hammer in pump installations	**164**
4.5.1 Hydraulic gradient	164
4.5.2 Causes of water hammer	165
4.5.3 Pump behaviour after power loss	165
4.5.4 Protection against water hammer	166
4.6 Pressure pulsations in piping systems	**168**
4.6.1 Rotodynamic pumps	168
4.6.2 Positive displacement pumps	169
4.7 Bibliography	**169**
Chapter 5 — Flow regulation	**171**
5.1 Introduction	**173**
5.2 Variable flow requirements	**173**
5.3 Flow regulation	**174**
5.3.1 Methods	174
5.3.2 Control signals	174
5.4 On-off control of constant speed pump	**174**
5.4.1 Principle and application	174
5.4.2 Costs	175
5.4.3 Problems when starting/loading	175
5.4.4 Problems when stopping/unloading	175
5.4.5 Operational sequences	176
5.4.6 Storage volumes	177
5.4.7 Power consumption	178
5.4.8 Example	178
5.5 Pole changing induction motor	**179**
5.5.1 Principle and application	179
5.5.2 Costs	180
5.5.3 Problems when starting	180
5.5.4 Problems when stopping	180
5.5.5 Storage volumes	180
5.5.6 Power consumption	180
5.5.7 Examples	180
5.6 Multi-speed gearbox	**180**
5.6.1 Principle and application	180
5.6.2 Costs	181
5.6.3 Problems when starting	181
5.6.4 Problems when stopping	181
5.6.5 Storage volumes	181
5.6.6 Power consumption	181

5.6.7 Example	181
5.7 Throttling by control valve	**181**
5.7.1 Principle and application	181
5.7.2 Costs	182
5.7.3 Sizing control valve	182
5.7.4 Power consumption	182
5.7.5 Example	183
5.8 By-pass return	**183**
5.8.1 Principle and application	183
5.8.2 Costs	184
5.8.3 Sizing by-pass	184
5.8.4 Overflow by-pass return	184
5.8.5 Power consumption	184
5.8.6 Example	184
5.9 Infinitely variable speed	**185**
5.9.1 Principle and application	185
5.9.2 Costs	188
5.9.3 Conversion of pump curves to various speeds	189
5.9.4 Efficiencies of various methods	189
5.9.5 Power consumption	190
5.9.6 Regulation system : schematics	191
5.9.7 Schematics for various systems	192
5.10 Factors affecting choice of flow regulation method	**194**
5.10.1 Direct flow regulation	195
5.10.2 Speed regulation	195

Chapter 6 — Pump materials — 199

6.1 Introduction	**200**
6.2 Typical materials	**200**
6.2.1 Grey cast iron	201
6.2.2 Spheroidal and nodular cast iron	202
6.2.3 Low alloy steel	202
6.2.4 Alloyed cast iron	202
6.2.5 11-13% Cr steel	204
6.2.6 Stainless steel	204
6.2.7 Super stainless steel	205
6.2.8 Copper alloys	206
6.2.9 Aluminium alloys	206
6.2.10 Nickel alloys	206
6.2.11 Other metallic materials	206
6.2.12 Non-metallic materials	207
6.2.13 Coatings	209
6.3 Material strength and integrity	**209**
6.4 Corrosion and erosion	**211**
6.4.1 Liquid corrosion of metals	211
6.4.2 Corrosion rate	211
6.4.3 Types of corrosion	211
6.4.4 Corrosion testing	213
6.5 Abrasive resistant materials	**214**
6.5.1 Pump selection	214
6.5.2 Hard metallic materials	214
6.5.3 Rubber cladding	214
6.6 Materials resistant to cavitation damage	**214**
6.7 Material selection	**215**
6.7.1 Basic information	215
6.7.2 Material combinations	215
6.7.3 Corrosion resistance	216
6.7.4 Liquid effect on materials unknown	217
6.8 Bibliography	**217**

Chapter 7 — Shaft seals — 219

7.1 Introduction	**220**
7.2 Methods of sealing : Rotary shafts	**220**
7.2.1 Non-contacting seal	220
7.2.2 Lip seal	220
7.2.3 Hover seal	220
7.2.4 Auxiliary pump	220
7.2.5 Soft packing	221
7.2.6 Mechanical seal	221
7.3 Process liquid seals for rotary shafts	**221**
7.3.1 Soft packing	222
7.3.2 Mechanical seal	225
7.4 Methods of sealing : Reciprocating rods	**231**
7.4.1 Lip seal	231
7.4.2 Soft packing	232
7.5 Soft packing process liquid seals for reciprocating rods	**232**
7.5.1 Operating principles	232
7.5.2 Design variations	232
7.5.3 Packing material	232
7.5.4 External systems	232
7.5.5 Maintenance	233
7.5.6 Trouble-shooting	233
7.6 Selection of process liquid sealing	**233**
7.6.1 Process liquid	233
7.6.2 Size, speed and pressure	233
7.6.3 Local environment	233
7.6.4 Cost	234
7.6.5 Standardisation	234
7.7 Bibliography	**234**

Chapter 8 — Shaft couplings — 235

8.1 Introduction	**236**
8.2 Types of coupling	**236**
8.3 Misalignment	**237**
8.4 Forces and moments	**238**
8.5 Service factors	**238**
8.6 Speed	**239**
8.7 Size and weight	**239**
8.8 Environment	**239**
8.9 Installation and disassembly	**240**
8.10 Service life	**240**
8.11 Shaft alignment	**242**
8.11.1 General	242
8.11.2 Methods of alignment	242
8.11.3 Determining shim thickness	243
8.11.4 Graphical method of determining shim thickness	244
8.12 Choice of coupling	**244**
8.12.1 Costs	244

    8.12.2 Factors influencing choice   244

8.13 Guards   246

8.14 Bibliography   246

Chapter 9 — Electric motors for pumps   247

9.1 General electrical technology, basic principles   248
    9.1.1 Electrical units   248
    9.1.2 Electrical systems   248
    9.1.3 Active, apparent and reactive power   249
    9.1.4 Phase compensation   249
    9.1.5 Speed   249
    9.1.6 Torque   250
    9.1.7 Voltage   250
    9.1.8 Starting   251
    9.1.9 Efficiency   251

9.2 Regulations and standards   251
    9.2.1 Controlling authorities   251
    9.2.2 Physical protection   253
    9.2.3 Cooling categories   253
    9.2.4 Mounting arrangements   254
    9.2.5 Terminal markings and direction of rotation   255
    9.2.6 Rated power, centre height and outline dimensions   255
    9.2.7 Temperature classification   255
    9.2.8 Potentially explosive atmospheres   256
    9.2.9 Certification   256

9.3 Motor types   256
    9.3.1 Constant speed AC motors   256
    9.3.2 Variable speed AC motors   257
    9.3.3 Variable speed DC motors   258

9.4 Motor starters   259
    9.4.1 Direct-on-line   260
    9.4.2 Star-Delta   260
    9.4.3 Soft start   260
    9.4.4 Resistance   260

9.5 Noise   260

9.6 Maintenance   261

9.7 Example   261
    Special Note : Engines   261
    Special Note : Turbines   262
    Special Note : Power recovery turbines   262

Chapter 10 — Ancillary equipment   263

10.1 Baseplates, skids and trailers   264

10.2 Belt drives   265

10.3 Gearboxes   265

10.4 Relief valves   266

10.5 Accumulators   267

10.6 Pulsation dampeners   267

10.7 Instrumentation   269

Chapter 11 — Quality assurance and testing   271

11.1 Introduction   272
    11.1.1 Physical properties   272
    11.1.2 Heat treatment   272
    11.1.3 Chemical composition   273
    11.1.4 Corrosion resistance   273
    11.1.5 Non destructive testing   273
    11.1.6 Repairs   273
    11.1.7 Welding   273
    11.1.8 Inspection   273
    11.1.9 Assembly   273
    11.1.10 Packaging   274
    11.1.11 Pressure testing   274
    11.1.12 Running tests   274
    11.1.13 Painting   275
    11.1.14 Purchased equipment   275
    11.1.15 Functional testing   275
    11.1.16 Witnessing   275
    11.1.17 Clarification of specifications   275
    11.1.18 Certification   276
    11.1.19 Documentation   277

11.2 Mass produced standard pumps   277

11.3 Pumps built to purchaser's specification   277

11.4 Guide-lines for documentation   279
    11.4.1 Rotodynamic pumps   279
    11.4.2 Positive displacement pumps   280

11.5 Bibliography   280

Chapter 12 — Installation and maintenance   283

12.1 Installation   284

12.2 Foundations   284

12.3 Tanks and sumps   285
    12.3.1 General   285
    12.3.2 Submersible pumps   286

12.4 Pipe systems for pumps   287

12.5 Care and maintenance   290
    12.5.1 General considerations   290
    12.5.2 Preventative maintenance   292
    12.5.3 Stocking spare parts   293
    12.5.4 Trouble-shooting guide   294

Chapter 13 Pump economics   295

13.1 Economic optimisation   296
    13.1.1 Introduction   296
    13.1.2 New and existing plant   296

13.2 Economic assessment criteria   297
    13.2.1 Investment calculation - new plant   297
    13.2.2 Investment calculation - existing plant   298
    13.2.3 Estimated profits and service life   298
    13.2.4 Energy costs   299

13.3 Important system characteristics   300
    13.3.1 Pumping efficiency factor   300
    13.3.2 Demand variations   301
    13.3.3 Availability   302
    13.3.4 Hydraulic power   303

13.4 Partial optimisation   304

13.4.1 Economic pipe diameter — 304
13.4.2 Component efficiency — 305
13.4.3 Existing plant — 305

Chapter 14 — Pump selection — 307

14.1 General operating conditions — 308

14.2 Selection of pump according to pump designation
Direct selection table — 309

14.3 Selection of pump according to duty and capabilities
Rapid selection table — 311

14.4 Selection of pump according to hydraulic performance — 314
14.4.1 Pumps for low viscosity liquids — 314
14.4.2 Pumps for viscous liquids — 314
14.4.3 Pumps for highly viscous liquids — 316

14.5 Pumps for liquid solid mixtures — 316
14.5.1 Pumping non-abrasive solids — 317
14.5.2 Pumping abrasive solids — 317
14.5.3 "Gentle" pumping — 317
14.5.4 Pumping waste water, sewage — 318

14.6 Check list for pump purchase specification — 319

14.7 Purchasing — 322

Chapter 15 — Case studies — 323

15.1 Pumping fresh water to a high reservoir — 324
15.2 A small sewage pumping station — 325
15.3 A fresh water booster station — 327
15.4 Circulation pump for domestic central heating — 328
15.5 Adjustable jet pump — 329
15.6 Modernisation of water supply for industrial use — 329
15.7 Contractors' use of electric powered submersible sump pumps — 330
15.8 Engine driven self-priming pumps — 332
15.9 Land reclamation pump — 333
15.10 Cargo pump for tankers — 333
15.10.1 Large tankers for crude oil — 333
15.10.2 Smaller tankers for finished products — 334
15.10.3 Tankers for LNG — 334
15.11 Liquid detergent manufacture - pump installation — 335
15.12 Positive displacement pumps - variable discharge conditions — 335
15.13 Energy recovery turbines — 336
15.14 Economic aspects of energy utilisation — 337
15.15 Choosing non-return valves — 340

Chapter 16 — Pump classification guide and guide to manufacturers and suppliers — 343

16.1 Introduction — 344
16.2 Pump classification guide — 344
16.3 Pump manufacturers and suppliers guide — 348
16.4 Manufacturers and suppliers of ancillary products — 366
16.5 Names and addresses of European manufacturers and suppliers — 368
16.6 Trade Names — 381

Chapter 17 — Units and conversions — 382

17.1 SI, The International System of Units — 383
17.2 Conversion factors for SI units — 384
17.2.1 Plane angle — 384
17.2.2 Length — 384
17.2.3 Area — 384
17.2.4 Volume — 385
17.2.5 Time — 385
17.2.6 Linear velocity — 385
17.2.7 Linear acceleration — 385
17.2.8 Angular velocity — 385
17.2.9 Angular acceleration — 385
17.2.10 Mass — 385
17.2.11 Density — 385
17.2.12 Force — 386
17.2.13 Torque — 386
17.2.14 Pressure, stress — 386
17.2.15 Dynamic viscosity — 386
17.2.16 Kinematic viscosity — 386
17.2.17 Energy — 387
17.2.18 Power — 387
17.2.19 Flow — 387
17.2.20 Temperature — 387
17.3 Other conversion factors — 388
17.3.1 Hardness — 388
17.3.2 Material toughness — 389
17.3.3 Compressibility — 389
17.4 Normal quantities and units used within pump technology — 389

Chapter 18 — Reference index — 390

Chapter 19 — Useful pump terms translated — 397

Index of advertisers — 400

1 The properties of liquids

The purpose of Chapter 1 is to supply the necessary information about those properties of liquids which must be known for pump selection. In the case of hazards, guidance is given regarding legislation and sources of useful information.

The chapter starts with explanations of the meanings of the terms in which various parameters of liquids are defined, explained and set out in examples. The most common liquids, water, oils and certain liquid solid mixtures, are treated in greater depth, in separate sections, than other liquids. The liquids table, Section 1.5, contains data on some 400 liquids, in the form of direct data and by reference to supplementary diagrams for more detailed information.

Contents

1.1 Explanation of Terms
- 1.1.1 Introduction
- 1.1.2 Changes of state
- 1.1.3 Viscosity
- 1.1.4 Density and relative density
- 1.1.5 Compressibility
- 1.1.6 pH value
- 1.1.7 Hazards
    - 1.1.7.1 Health hazards
    - 1.1.7.2 Physical hazards
    - 1.1.7.3 Environmental hazards
    - 1.1.7.4 Installation hazard assessment
    - 1.1.7.5 Information sources

1.2 Water
- 1.2.1 Demineralized water
- 1.2.2 Fresh water
- 1.2.3 Brackish water
- 1.2.4 Sea water

1.3 Oils
- 1.3.1 General
- 1.3.2 Viscosity

1.4 Liquid solid mixtures
- 1.4.1 General
- 1.4.2 Sewage
- 1.4.3 Sludge
- 1.4.4 Pulps

1.5 Liquid gas mixtures

1.6 Table of liquid properties
- 1.6.1 General
- 1.6.2 Liquid table

1.1 Explanation of terms

1.1.1 Introduction

Liquids, together with solids and gases, are the forms in which substances occur in nature. We speak of the solid state, the liquid state and the gaseous state. The three physical states are sometimes called phases. Liquids and gases can be combined in the general group called fluids. Fluids differ from solids in that they will readily change shape to suit the container.

A solid body subjected to a small shear force undergoes a small elastic deformation and returns to it's original shape when the force is removed. When subjected to larger shear force the shape may be permanently changed due to plastic deformation.

A fluid when subjected to an arbitrarily small shear force undergoes a continuous deformation. This happens regardless of the inertia of the fluid. For a fluid the magnitude of the shear force and the speed of deformation are directly related. In a solid body it is the deformation itself which is related to the shear force.

A fluid may be either a liquid or a gas. A gas differs from a liquid in that it will expand to completely fill the container. A gas at conditions very close to boiling point or in contact with the liquid state is usually called a vapour. Fluids are compressible; gases being much more compressible than liquids.

A substance can exist in all three states. A typical example of this is ice, water and steam. When ice is heated at constant pressure, the ice converts to water at the melting point and to steam at the boiling point. If the steam pressure is increased at constant temperature, the steam converts to water at the saturation (vapour) pressure.

Solid particles can be suspended or mixed in a liquid. Such a combination, liquid plus particles, is called a suspension. When the particles distribute themselves evenly through the liquid, we speak of a homogeneous mixture. When concentration gradients occur, we speak of a heterogeneous mixture.

The word solution refers to an otherwise pure liquid in which a solid, another liquid or gas has been dissolved.

Two liquids which are not soluble in each other, can be mixed by mechanical action. Such a mixture is called an emulsion. Emulsions can be very difficult to separate.

Liquids generally display greatly varying properties. For the purposes of pumping the following characteristics of liquids should generally be known:-

- changes of state,
- viscosity,
- density,
- compressibility,
- pH value,
- hazards.

1.1.2 Changes of state

Melting point
SI Unit, °C

The melting point is the temperature at which a substance changes from a solid to a liquid state and also solidifies from a liquid to a solid. The melting point in most substances is pressure dependent only to a very limited degree. In those cases where the pressure dependence has to be taken into account, the boundary between solid and liquid state is shown by the melting curve in a pressure-temperature diagram.

Boiling point
SI Unit, °C

The boiling point is the temperature at which a liquid converts to vapour or gas at a particular local pressure. The boiling point is usually stated at a standardised atmospheric pressure, 101.325 kPa (760 mm Hg). The boiling point of water at this pressure is 100 °C. The boiling point of all liquids is heavily dependent upon pressure.

Vapour pressure
SI Unit, Pa, kPa, MPa
Preferred Unit, bar

All liquids have a tendency to evaporate. Vapour or gas accumulates above the free surface of a liquid by reason of interchange of molecules. The partial pressure of the vapour rises to a point at which as many molecules are being returned to the liquid as are leaving it. At this equilibrium state the partial pressure of the liquid is called the vapour or saturation pressure. Vapour pressure depends on temperature alone and increases with the temperature of the liquid. At a particular temperature, the equilibrium pressure above the surface of a liquid can never be less than the vapour pressure. Any attempt to lower the pressure of the vapour (by means of a vacuum pump for example) immediately results in increased evaporation, i.e. the liquid boils. If the pressure in a liquid decreases locally to the vapour pressure at the actual temperature, vapour bubbles are generated in the liquid. In a pump installation, the formation of vapour bubbles (cavitation) can cause serious mechanical damage and can seriously diminish the performance of the installation. Various liquids display widely differing values of vapour pressure as a function of temperature. Some substances in a solid state can also change directly to the gaseous state without passing through the liquid state. As an example, a complete pressure-temperature diagram, showing the phases for water is illustrated below. At the triple point, all three states may exist simultaneously. In practice the only liquid which exhibits the solid-gas transformation is carbon dioxide.

NOTE: The SI unit of pressure is very small. In most practical applications it may be necessary to use Pa or kPa for suction pressures and MPa for discharge pressures. The pump industry is concerned that the change in prefix could confuse operators and lead to potentially dangerous situations. The MPa is a very large unit; small numbers representing large pressures. The bar is a much more suitable unit and does not require prefixes.

Whenever a pressure value is quoted it must be qualified as absolute or gauge.

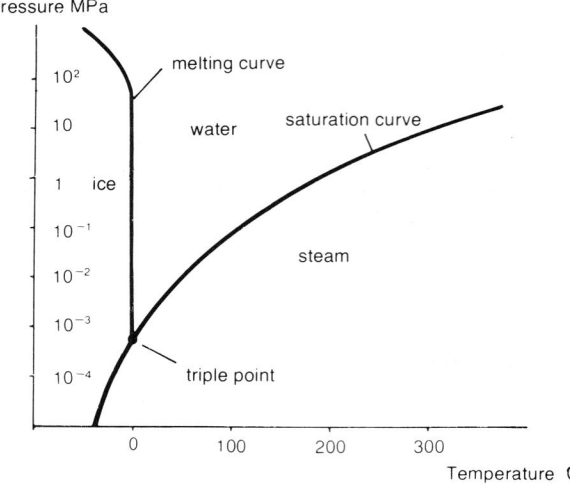

Figure 1.1 Phase diagram for water

Comments

Melting point (freezing or solidifying point) must be known when assessing the pumpability of a liquid and the risk of blockage, freezing or solidifying.

Information regarding the vapour pressure is required for calculating the permitted suction lift and $NPSH_a$. Vapour pressure is a critical factor in the choice of pump type, speed and shaft sealing requirements.

1.1.3 Viscosity

SI Units
Absolute viscosity   Ns/m²
Kinematic viscosity   m²/s

Other Units
Absolute viscosity   cP,
Kinematic viscosity   cSt

Viscosity (the ability to flow) is a property of liquids treated under the heading of rheology. The word rheology derives from the Greek "rheos" meaning flow.

Between two layers of liquid flowing at different speeds, a tangential resistance, a shear stress, is developed because of molecular effects. We say that the shear stress is caused by the internal friction of the liquid, or conversely that the liquid transmits shear forces by reason of its internal friction.

A liquid in motion is continuously deformed by the effects of the shear forces. The magnitude of the stress depends on the rate of shear deformation and the sluggishness of the liquid, i.e. the viscosity.

Viscosity is defined for flow in layers, laminar flow, by Newton's Law of Viscosity.

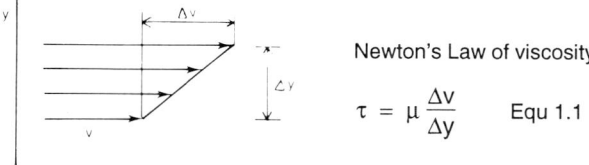

Newton's Law of viscosity

$$\tau = \mu \frac{\Delta v}{\Delta y} \quad \text{Equ 1.1}$$

Figure 1.2 Definition of viscosity

Newton's Law of Viscosity

τ = shear stress   (N/m²)
μ = dynamic viscosity   (kg/ms)
Δv = change in viscosity   (m/s)
Δy = distance between layers   (m)

Absolute viscosity is stated in the SI system as the unit

1 kg/ms = 1 Ns/m²

Other units are:

1 Poise = 1P = 0.1 kg/ms

or

1 centipoise = 1 cP = 0.01 P = 0.001 kg/ms

In viscous flow equations the dynamic viscosity divided by the density of the liquid is given the symbol ν. This parameter is called kinematic viscosity.

$$\nu = \frac{\mu}{\rho} \quad \text{Equ 1.2}$$

where

ν = kinematic viscosity   (m²/s)
μ = dynamic viscosity   (kg/ms)
ρ = density   (kg/m³)

The SI unit for kinematic viscosity is 1 m²/s.

Sometimes, for convenience, the following units are used:
    1 Stoke = 1 St = 0.0001 m²/s
or more usually
    1 cSt = 0.01 St = 0.000001 m²/s = 1 mm²/s

Water at 20 °C and 0.1 MPa has a kinematic viscosity of 1 cSt.

Newtonian liquids
A liquid which follows Newton's Law of Viscosity in laminar flow and has constant viscosity, regardless of shear rate and time, is known as a Newtonian liquid.

Examples of Newtonian liquids are: water, aqueous solutions, low molecular liquids, oils and oil distillates. Black liquor, liquid resin and resinous sebacic acid also behave like Newtonian liquids.

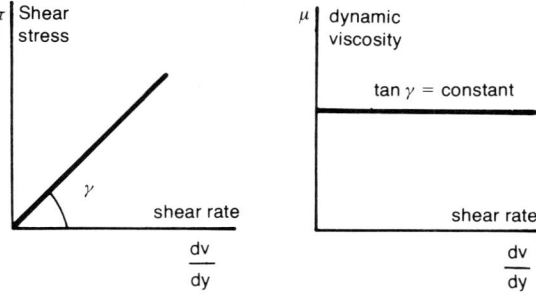

Figure 1.3 Newtonian liquid

Non-Newtonian liquids
Liquids which do not fulfil the requirements for Newtonian liquids are called non-Newtonian liquids. Most high molecular liquids, suspensions and emulsions display non-Newtonian properties. Non-Newtonian liquids usually fall into three main groups:

i) Non-time-dependant
    pseudo-plastic
    dilatant
    plastic

ii) Time dependant
    thixotropic
    rheopectic
    irreversible

iii) Visco-elastic

Liquids in Group i) are not effected by the length of time of the flow process. The shear stress in the case of laminar flow at a given temperature is determined entirely by the shear rate. In analogy with Newtonian liquids, we may say

$$\tau = \mu_1 \frac{dv}{dy} \quad \text{Equ 1.3}$$

where μ_1 = apparent dynamic viscosity

In the case of pseudo-plastic liquids, the apparent viscosity decreases with increasing shear rate.

Examples: high molecular solutions, rubber, latex, certain molten materials, mayonnaise.

In the case of dilatant liquids the apparent viscosity increases with increasing shear rate.

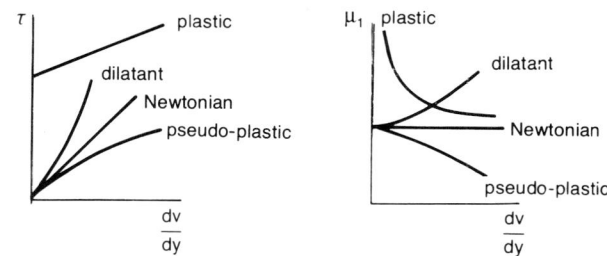

Figure 1.4 Non-time-dependent non-Newtonian liquids

Examples: base in oil paints, suspensions with high concentrations of small particles: cement, lime, sand, starch.

Plastic liquids require a certain minimum shear stress - yield stress - in order to initiate flow. The apparent viscosity decreases from an infinitely high value as the shear rate increases.

Examples: toothpaste, ointments, grease, margarine, paper pulp, printers ink, emulsions.

Liquids in Group ii), time-dependent, the apparent viscosity is affected not only by the shear rate but also by the length of time during which the flow continues.

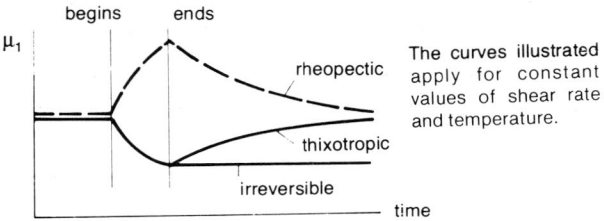

Figure 1.5 Illustrations of time-dependent non-Newtonian liquids

In the case of thixotropic liquids, μ_1 diminishes when flow commences. When the flow ceases, the liquid returns to its original viscosity after a certain period of time.

Examples: paint, gelatinous foodstuffs.

Rheopectic liquids display increasing viscosity under mechanical influence and resume their original viscosity when the influence ceases.

Examples: certain suspensions of gypsum.

Irreversible liquids do not resume their original viscosity at all or perhaps only after a very long time after the removal of the influence. These liquids must be pumped carefully.

Examples: cheese coagulates, yoghurt, marmalade.

Group iii), visco-elastic liquids, contains liquids exhibiting both elastic and viscous properties. Visco-elastic liquids undergo both elastic and viscous deformations. When the flow ceases there is a certain reversal of elastic deformation.

Examples: asphalt, liquid nylon, rubber, polymer solutions.

Comments

- Dynamic viscosity is temperature-dependent. Increasing temperature results in diminishing viscosity. Certain liquids can be pumped easier when heated.

- Viscosity is also an indication of the lubricating qualities of the liquid. At low values, less than 1 cSt, additional lubrication may be necessary.

- Temperature dependence on density has also to be considered in kinematic viscosity. At pressures of 10 MPa a certain pressure dependence can be observed in the viscosity.

- Loss coefficients in pipe flows are dependent upon Reynold's number which is in turn dependent upon dynamic viscosity.

- The performance of rotodynamic pumps depends upon Reynold's number. Standard pump data is always for water and this has to be corrected when pumping other liquids. Positive displacement pumps are better for high viscosities but there is no specific value to aid selection. The viscosity effects are dependant upon the pump size. For rotodynamic pumps, increased viscosity reduces efficiency and increases NPSHr. The effects of viscosity on positive displacement pumps is dependant upon the pump type as well as size.

- Viscosity is defined as the ratio of shear stress and shear rate in laminar flow. In the case of turbulent flow, this ratio is affected by exchange of momentum between the layers caused by the random motion of the liquid particles.

- In the past viscosity has been specified by the results of proprietary test methods such as Engler, Saybolt and Redwood. These tests measured the time taken for an oil sample to flow through an orifice. In Chapter 18, Units and Conversions, the necessary basis is given for conversion to SI units.

Penetration

For some non-Newtonian liquids, for example grease and asphalt, determination of viscosity does not provide sufficient information about flow properties and supplementary information regarding consistency is required. This information is obtained from determination of penetration. For greases, penetration is a measure of consistency. That is the depth, in tenths of a millimetre, to which a cone forces its way down into a test receptacle containing a grease sample heated to 25 °C. The penetration depends on whether the consistency is changed by stirring, shaking and so on. Hard greases have low penetration figures whilst soft greases have high penetration. Manufacturers of greases state the penetration figure for each quality, e.g. 240 - 325. Two lubricant greases, furthermore, may have the same penetration figure but may still have different flow capabilities depending upon the viscosity of their respective component oils. The oil grades and thickeners should also be specified.

Determination of penetration for asphalt is carried out at 25 °C in the same way, in principle, as with lubricant grease, but using a loaded pin instead of a cone. The values measured are used to classify asphalt, typical penetration figures being 10 - 50. See BS 2000 Part 50, 1985, Cone penetration of lubricating grease - which is identical to the Institute of Petroleum standard IP 50/84.

1.1.4 Density and relative density

Density

Unit, kg/m³, symbol ρ

The density of a liquid is the relationship between the mass of the liquid and its volume. The density of liquids changes slightly with temperature and very slightly with pressure unless the liquid is very compressible. The density of a liquid is to a very slight extent affected by the quantity of dissolved gases. Since the ability to dissolve gases is temperature and pressure dependent, there is an indirect dependence on these parameters. The effect of dissolved gases on density can generally be neglected however.

Relative density

Unit, non-dimensional, symbol d

Relative density, formerly called specific gravity, is the ratio of a liquid's density to water at standard conditions, atmospheric pressure, 101.325 kPa and 4 °C.

Comments

Information about the density of a liquid, or it's relative density, is required when converting pressure to head. Rotodynamic pumps calculate power using differential head and specific gravity. Positive displacement pumps calculate power using differential pressure therefore density or specific gravity has no influence.

1.1.5 Compressibility

All liquids are compressible to some extent. This compressibility can generally be neglected when considering rotodynamic pumps. However when considering reciprocating positive displacement pumps its value must be quantified. It may be noted for example that water is about 100 times more elastic than steel and about 0.012 times as elastic as air. Compressibility is very temperature dependant and slightly pressure dependant. Any values used must relate to the operating conditions. Classically, compressibility is expressed in terms of the bulk modulus defined by the relationship

$$\text{Compressibility} = \frac{1}{K} \qquad \text{Equ 1.4}$$

$$K = \rho \cdot \frac{\Delta p}{\Delta \rho} = -V \frac{\Delta p}{\Delta V} \qquad \text{Equ 1.5}$$

where

K	=	bulk modulus of the liquid	(N/m²)
p	=	liquid pressure	(Pa)
ρ	=	density of the liquid	(kg/m³)
V	=	volume of liquid	(m³)

Δ = change of magnitude

The change in volume due to a change in pressure can be calculated directly from the definition

$$\Delta V = -\frac{V \Delta p}{K} \quad \text{Equ 1.6}$$

where the minus sign indicates that the volume decreases with increasing pressure.

Densities and specific volumes of water under a wide range of pressures and temperatures are readily available in good steam tables. However compressibility data for other liquids can be difficult to obtain. Some data is available in the form of "contraction per unit volume per unit pressure increase" which is designated by β. In the cases where definitive data is not available the compressibility can be approximated as a multiple of water compressibility.

Comments

In the case of acoustics however, compressibility is of critical importance. The acoustic velocity, or wave speed, is directly related to the bulk modulus and compressibility. If acoustic resonance occurs in the pipework the acoustic velocity must be known to effect a successful cure. Acoustic resonance can be a very serious problem creating destructive piping vibrations and large pressure pulsations. In suction pipework the resonance can create cavitation conditions leading to pump damage and reduced performance. The acoustic velocity calculated from the bulk modulus applies to pure clean liquid. If the liquid contains gas bubbles or solid particles the acoustic velocity will be greatly reduced from the theoretical value. Testing may be the only approach to find the true value. See Figure 1.6 for indicative values of bulk modulus and acoustic velocity.

1.1.6 pH value

The concentration of hydrogen ions in an aqueous solution is a measure of the acidity of the solution and is expressed as the pH value.

The definition of the pH value is

$$pH = \log_{10}\frac{1}{H^+} \quad \text{Equ 1.7}$$

where H^+ is the concentration of hydrogen ions (mol/l).

If, for example, $H^+ = 10^{-4}$ mol/l, then the pH value = 4.

The hydrogen exponent is another name for the pH value.

The numerical value of pH varies between the limits 0 and 14. Acid solutions have pH values between 0 and 6.5, neutral solutions pH = 6.5 to 7.5 and the alkaline or base solutions pH = 7.5 to 14. Acid solutions turn blue litmus red and alkaline solutions turn red litmus blue.

All aqueous solutions reacting as acids contain a surfeit of hydrogen ions H+. All aqueous solutions reacting as alkalis have a surfeit of hydroxide ions OH-. The product of the hydrogen ion concentration and the hydroxide ion concentration is always a constant at any fixed temperature.

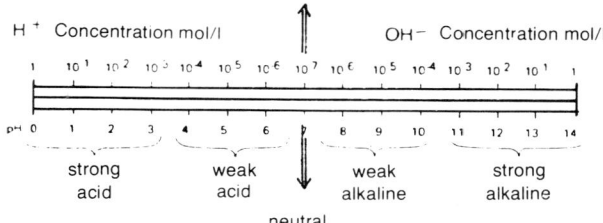

Figure 1.7 Ion concentration, pH value, degree of acidity and alkalinity

At 22 °C for example the ion product is:

$H^+ \times OH^- = 10^{-14}$ mol/l

When the hydrogen ion concentration and the hydroxide ion concentration are equal, i.e.

$H^+ = OH^- = 10^{-7}$ then

pH = 7 and the solution reacts neutrally.

Pure water is a very poor electrical conductor. Water only acquires good conductivity in solutions of so-called electrolytes such as salts, acids or bases. When dissolved in water, these substances split into two components with opposing electrical charges, by means of which an electric current may be carried. This act of splitting is called electrolytic dissociation. The charged particles so formed are called ions. Distinction is made between the positively charged cation particles and the negatively charged anion particles. The positively charged particles are symbolised by + and the negatively charged particles by -. Metals and hydrogen in acids produce cations by electrolytic dissociation. The acid remnants and the alkaline hydroxyl groups on the other hand generate negatively-charged anions.

For liquids with pH values of less than 4, the hydrogen ion concentration is a powerful factor affecting the speed of the dissolving process for most metals. The positively-charged hydrogen ions relinquish their charge at the surface of the metal to the metal atoms and break down their solid connections, whilst they themselves, robbed of their ion construction, return to the atomic state and unite into molecules leaving the metal surface in the form of gas bubbles. The dissolving process is faster, therefore, when the concentration of hydrogen ions is greater, i.e. the lower the pH value of the liquid.

Within the pH range of about 4 to 9, the oxygen present in the liquid, from air for example, has an influence on the corrosion process, say, for iron. The discharged hydrogen does not

Liquid	Bulk modulus N/m²	Wave propagation speed m/s	Estimated wave propagation speed in pipe	
			Steel	PVC
Acetone	$0.8 \cdot 10^9$	1000	850	280
Petrol	$1.0 \cdot 10^9$	1170	1000	320
Ethyleneglycol	$2.7 \cdot 10^9$	1560	1150	300
Glycerol	$4.5 \cdot 10^9$	1890	1400	380
Chloroform	$1.0 \cdot 10^9$	820	700	230
Mercury	$24 \cdot 10^9$	1330	430	-
Methanol	$0.8 \cdot 10^9$	1000	850	280
Oil	$1.5 \cdot 10^9$	1300	1000	300
Turpentine	$1.4 \cdot 10^9$	1270	1000	300
Water, dist	$2.2 \cdot 10^9$	1480	1100	300
Water, sea	$2.4 \cdot 10^9$	1530	1100	300

Figure 1.6 Elastic properties of some liquids at 25 °C and 0.1 MPa (1bar)

leave in the form of gas bubbles, as is the case with pH values in the region of 0 to 4, but combines immediately with the oxygen to produce water. At the same time the iron is oxidised to rust. The effect of pH on non-metallic materials must be considered on a case by case basis.

Although pH is an important factor in the attack of metals, this is by no means the only factor which needs to be considered. The concentration (as distinct from pH) of the chemical, particularly of acids and alkalis, has an important bearing on the degree of corrosion or attack. Dissolved oxygen plays an important part in corrosion of metal components. There is now a whole range of organic chemical compounds and biological compounds which will give rise to chemical attack, and many of these are completely unrelated to pH. Even very small concentrations of some modern chemicals can create serious hazards both from the point of view of chemical attack, leakage and toxic hazards.

1.1.7 Hazards

The word "hazard" is in common use in the English language and it must be defined here to show the context in which it used hereafter;

hazard - a physical situation with a potential for human injury, damage to property, damage to the environment or some combination of these.

It can be seen from the definition that three distinct types of effect are considered but in some cases one hazard may lead to others. Fire, for example, can be a serious health hazard. A hazardous substance is defined as:-

a substance which, by virtue of it's chemical properties, constitutes a hazard.

1.1.7.1 Health hazards

The pump user must consider the effects of the liquid, and it's vapour, on the health of the operators and employees. Most countries have legislation limiting the exposure of employees to substances judged to be hazardous. If the liquid to be pumped is listed in local regulations the pump manufacturer must be informed. The type of health hazard must be specified.

Another health hazard, sometimes not recognised as such, is noise. Some countries have regulations stipulating the acceptable noise levels and exposure times. Some pump types are inherently noisy. Large equipment, in general, is noisy. Noise levels can be attenuated by fitting acoustic enclosures however these tend to drastically diminish the maintainability of the equipment by hindering access. In some instances, costly acoustic enclosures have been removed at site and scrapped in order to achieve acceptable access. One easy solution to this hazard is to declare certain areas as "Ear Protection Zones".

1.1.7.2 Physical hazards

Physical hazards include fire and explosions as well as corrosion. The degree of risk attached to the hazard is dependant upon the properties of the liquid and the vapour; if the liquid evaporates quickly at ambient temperature and whether the vapour is lighter or heavier than air. A light vapour can disperse easily in an open installation whereas a heavy vapour can become trapped in sumps, depressions and pipe trenches. The upper and lower limits of flammability are important when considering allowable concentrations.

The pumping equipment itself may pose a physical hazard. Within the EEC, the Machinery Directive, 89/392/EEC, and the amending directive 91/368/EEC, which came into force on 1st January 1993, place the responsibility for safety on the machine designer. The machine must be designed to be safe in all aspects; installation, commissioning, operation and maintenance. If the designer is unable to devise a completely safe machine the areas of concern must be documented and recommended precautions communicated to the user. Because this is a legal requirement in all EEC countries the machine designer may not be relieved of the obligations by a third party.

1.1.7.3 Environmental hazards

As a race we are becoming more aware of the limitations of our environment. The Earth's resources and waste-disposal capabilities are finite. Stricter limitations will be imposed gradually on the amount of pollutant which can be released while the list of pollutants will become longer. The pump user must be aware of the full consequences of leakage of liquid/vapour from the pump and installation. The environment can be considered in two separate identities, local and global. If the site is surrounded or close to a town what risk is likely to the population, structures or habitat in the event of a failure? In the global sense, what are the likely cumulative effects of product leakage ?

1.1.7.4 Installation hazard assessment

The user and system designer are in full possession of all the relevant available facts regarding the liquid and the installation. Any assumptions made should be passed to the pump manufacturer and identified as such. The user must assess the risks attached to all the possible hazards and decide what, if any, leakage is acceptable. Liquid properties reviewed during the assessment should include :-

autoignition point	the temperature above which a substance will start to burn without an ignition source being necessary,
flash point	the lowest temperature at which a liquid gives off sufficient vapour to burn if an ignition source is present,
atmospheric boiling point	the temperature at which the liquid boils at atmospheric pressure, 101.325 kPa,
vapour specific gravity	specific gravity is the ratio of a vapour's density to air at standard conditions, atmospheric pressure, 101.325 kPa and 4 °C,
surface tension	the surface tension at process temperature will indicate how difficult the liquid is to seal.

The nature of the hazards will also dictate the type of pipe connections to be used; screwed, flat-face flanged, raised-face flanged, ring-type joints. Process upset conditions must be considered as part of the assessment. Upset conditions which last for more than one or two hours may have a significant impact on pump and ancillary equipment selection.

The physical location of the pump, indoor or outdoor, will decide the behaviour of the leakage once outside the pump. Will any vapour cloud quickly disperse on a breeze which always blows over the un-manned site or will a manned enclosed pump house gradually build up a dangerous concentration of vapour? Only the user can assess these questions and specify the necessary precautions.

It is the responsibility of the user to define exactly what the pump is intended to do. It is the responsibility of the pump manufacturer to supply equipment to meet the required performance.

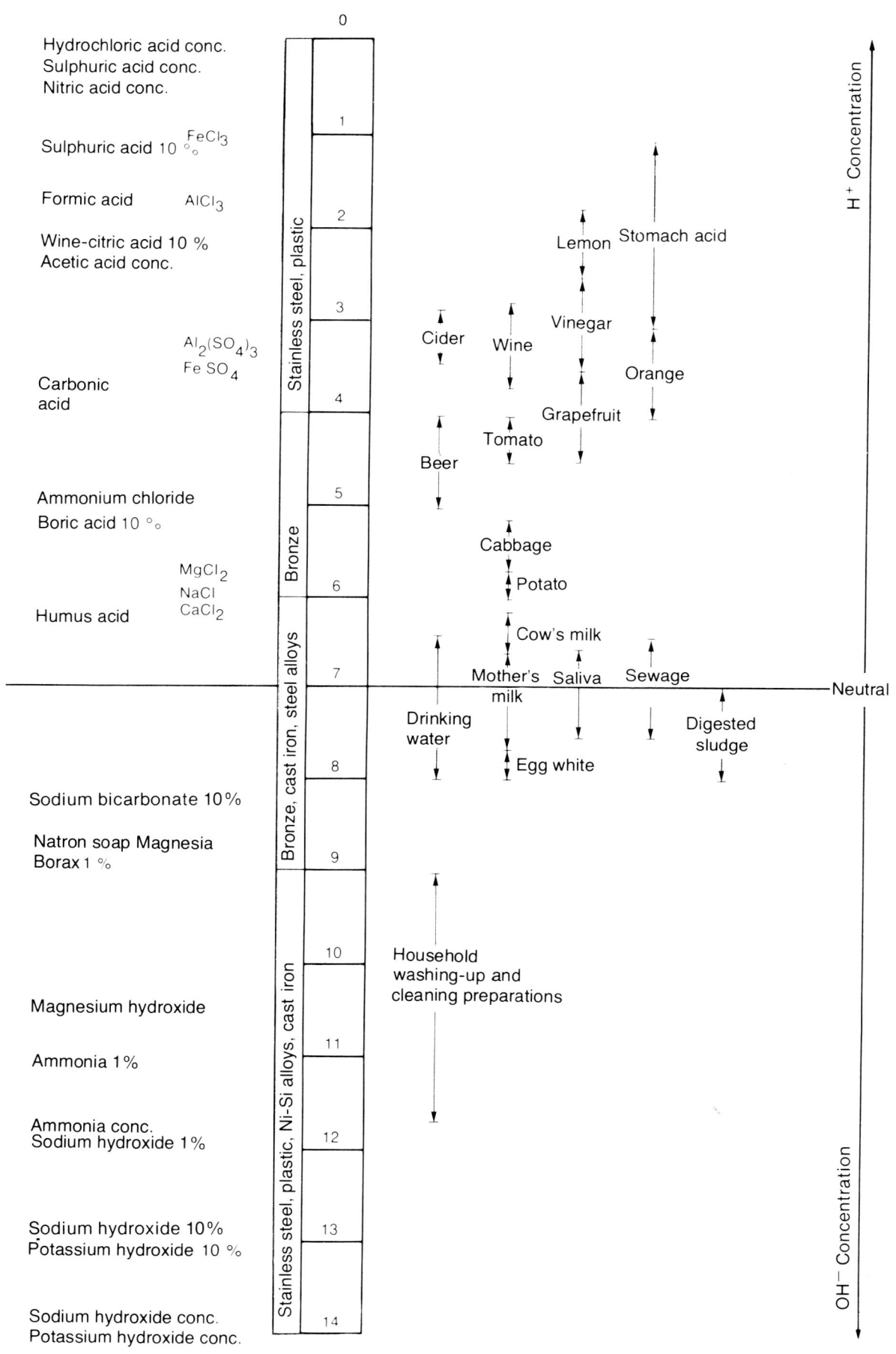

Figure 1.8 Illustration of pH values. The materials quoted can be regarded as a first introduction to the choice of materials. See also Chapter 6 "Materials"

1.1.7.5 Information sources

Basic information on plant and system design, together with risk assessment, can be obtained from various sources. The British Institution of Chemical Engineers publishes several good texts:

Process Plant Design and Operation — Doug Scott and Frank Crawley

Nomenclature for Hazard and Risk Assessment in the Process Industries — David Jones.

The Institution of Chemical Engineers, Davis Building, 165-171 Railway Terrace, Rugby, CV21 3HQ, United Kingdom, Tel 0788 578214, Fax 0788 560833.

Information regarding trade names and manufacturers of dangerous chemicals is available from the Chemical Industries Association, Kings Buildings, Dean Stanley Street, Smith Square, London, SW 1P, United kingdom, Tel 071-735 3001, Tlx 916672 and the Fire Protection Association, 140 Aldersgate, London, EC1A 4HY, United Kingdom, Tel 071-606 3757.

British Standards BS 5908 - Code of practice for fire precautions in the chemical and allied industries - summarises the statutory requirements, storage and movement of materials, process plant piping, ventilation, fire protection etc. This publication also lists a wide range of reference literature including: Home Office publications, American Petroleum Institute, American Society of Mechanical Engineers, US Bureau of Mines and publications of Oil Companies European Organisation for Environmental and Health Protection.

BS 5908 classifies substances according to the operating temperature and flash point temperature, agreeing with the Institute of Petroleum:

	Flash point °C	Operating temp °C
Class I	< 21	
Class II(1)	21 to 55	below flash point
Class II(2)	21 to 55	at or above flash point
Class III(1)	55 to 100	below flash point
Class III(2)	55 to 100	at or above flash point

The Institute of Petroleum is located at 61 New Cavendish Street, London, W1M 8AR, U.K., Tel 071-636 1004, Fax 071-255 1472.

American practice is outlined in ANSI B31.3. ANSI can be contacted at American National Standards Institute, Inc., 1430 Broadway, New York, New York 10018, USA, Tel 212 354 3473.

In the UK, the Control of Substances Hazardous to Health (COSHH) Regulations 1988, state "The exposure of employees to substances hazardous to health should be prevented, or, where this is not reasonably practicable, adequately controlled." COSHH covers the following significant areas:

Regulation 6   assessment of exposure,

Regulation 7   control of exposure,

Regulation 8 & 9 use and maintenance of control measures,

Regulation 10   monitoring exposure.

Separate regulations cover the hazards of carcinogenic substances, lead and asbestos. COSHH does not apply to underground mining installations or to the hazards posed by micro-organisms.

Advice is provided by the Health and Safety Executive (HSE), publication EH40, on the exposure limits for a wide range of chemicals. The Approved Code of Practice "Control of Carcinogenic Substances" L5, 1990, deals with practical guidance on the COSHH requirements. Lead is covered by the Control of Lead at Work Regulations 1980, Appendix 1, Approved Code of Practice 2, 1985. Asbestos is covered by HSE The Control of Asbestos at Work Regulations 1987, SI 1987 No.2115.

Airborne dust and fumes can be a personnel hazard, see HSE Dust: general principles of protection, Guidance Note EH44, 1991, and EH54 1990 on fumes.

The Health and Safety Executive produces a wide range of literature on all aspects of safety. The HSE can be found at Baynards House, 1 Chepstow Place, Westbourne Grove, London, W2 4TF, U.K., Tel 071-221 0870, Fax 071-221 9178.

A very useful source of information is "European Agreement concerning the International Carriage of Dangerous Goods by Road (ADR)". In the UK this document is issued by the Department of Transport as is its companion volume concerning transportation by rail (RID). These documents are published and recognized in; Austria, Belgium, Czechoslovakia, Denmark, Finland, France, Germany, Greece, Hungary, Italy, Luxembourg, the Netherlands, Norway, Poland, Portugal, Spain, Sweden and Switzerland. The Department of Transport is at 2 Marsham Street, London, SW1P 3EB, U.K. Tel 071-276 3000, Fax 071-276 0818.

Conditions are laid down in the Petroleum (Consolidation) Act 1928 and in particular The Highly Flammable Liquid and Liquefied Gases Regulations SI (Statutory Instrument) 1972 No. 917, governing the storage, handling and conveyance of gases and liquids constituting a fire hazard.

According to the above regulations: -
a highly flammable liquid means any liquid, liquid solution, emulsion or suspension which, when tested in the manner specified in Schedule 1 to the Regulations, gives off a flammable vapour at a temperature of less than 32 °C and when tested in the manner specified in Schedule 2 to the Regulations, supports combustion;

a dangerous concentration of vapours means a concentration greater than the lower flammable limit of the vapours.

Fire hazardous liquids are listed in the Petroleum (inflammable Liquids) Order SI 1968 No 570 under Part 1 of the Schedule and any solution or mixture containing any of those substances specified in Part 1 of the Schedule, if it gives off an inflammable vapour at a temperature below 22.8 °C.

Note: The Fire Hazard categories specified in the Liquid Table 1.5.2 are classified according to the following:

CLASS 1   Liquids having a flash point below + 21 °C.

CLASS 2a Liquids with a flash point above + 21 °C but not exceeding + 30 °C.

CLASS 2b Liquids with a flash point above + 30 °C but not exceeding + 60 °C.

CLASS 3   Motor fuel and heating oil with a flash point in excess of + 60 °C.

Below 32 °C (73 °F)

Abel Apparatus: Statutory Method - BS 2000: Part 33:1982 (IP 33/59(78))

-18 °C to 71.5 °C (0 °F to 160 °F)

Abel Apparatus: Non-Statutory Method - BS 2000: Part 170:1982 (= IP 170/75(81))

-7 °C to 371 °C

Pensky-Martens Apparatus - BS 6664: Part 5:1990 (= IP 34/67 and = ISO 2719)

American literature can be checked for additional information. ANSI/NFPA 325M, Fire Hazard Properties of Flammable Liquids, Gases and Volatile Solids, categorises substances by health, flammability and chemical reactivity. ANSI/NFPA 704 Identification of the Fire Hazards of Materials, may be useful. The National Fire Protection Association is located at 1 Batterymarch Park, PO Box 9101, Quincy, MA 02269-9101, USA. ACGHI Documentation of threshold limit values for substances in workroom air is available from American Conference of

Government Industrial Hygienists, PO Box 1937, Cincinnati, Ohio 45201, USA.

Gas constituting a fire hazard is not classified according to a flash point, the criteria being the way in which the gas can be ignited and the flame propagated. An explosive atmosphere occurs if the percentage concentration by volume of an explosive gas or vapour in air is such that the air will burn when ignited and flame propagation occurs. The upper and lower explosive limits are referred to as the Explosive Range or Flammability Limits.

Regulations governing such installations so as to reduce the risk of explosion caused by electrical equipment are laid down and certified by National bodies, European and International agreements. This aspect of hazards and safety is fully covered in Chapter 9 but some basic information is given here. BS 5345, Code of practice for selection, installation, and maintenance of electrical apparatus for use in potentially explosive atmospheres (other than mining applications or explosive processing and manufacture), gives guidance and references for all aspects of electrical installations.

EN 50 014 to EN 50 020, plus EN 50 028 and EN 50 039 are the relevant European standards issued by European Committee for Standardization and adopted by all EEC member countries. The International Electrotechnical Committee has issued standard IEC 79, in approximately 17 sections for world use.

Accredited test houses check for standard compliance and issue certificates for suitable equipment. The UK, Germany and USA have the most popular test houses, see Chapter 9.

1.2 Water

1.2.1 Demineralized water

Demineralized water is chemically pure water and can be produced by various methods :

multiple distillation,
multi-bed ion exchange,
reverse osmosis,
electro-dialysis.

Chemical impurities and dissolved gasses are removed which cause the water to become chemically active. At reasonable temperatures, glass and austenitic stainless steels are completely corrosion resistant. Cast iron and bronze are both attacked resulting in contamination of the water. At higher temperatures austenitic stainless steels suffer from intergranular corrosion.

The hydraulic properties of demineralized water can be determined with very great accuracy. The viscosity and vapour (saturation) pressures given below conform to the international

Temperature °C	Dynamic viscosity μ kg/m.s			
	0.1 MPa	1 MPa	10 MPa	100 MPa
0	$1.793 \cdot 10^{-3}$	$1.789 \cdot 10^{-3}$	$1.768 \cdot 10^{-3}$	$1.651 \cdot 10^{-3}$
25	$0.891 \cdot 10^{-3}$	$0.891 \cdot 10^{-3}$	$0.889 \cdot 10^{-3}$	$0.891 \cdot 10^{-3}$
50	$0.547 \cdot 10^{-3}$	$0.547 \cdot 10^{-3}$	$0.549 \cdot 10^{-3}$	$0.568 \cdot 10^{-3}$
75	$0.377 \cdot 10^{-3}$	$0.378 \cdot 10^{-3}$	$0.380 \cdot 10^{-3}$	$0.404 \cdot 10^{-3}$
100	-	$0.282 \cdot 10^{-3}$	$0.285 \cdot 10^{-3}$	$0.300 \cdot 10^{-3}$
150	-	$0.182 \cdot 10^{-3}$	$0.185 \cdot 10^{-3}$	$0.207 \cdot 10^{-3}$
200	-	-	$0.136 \cdot 10^{-3}$	$0.156 \cdot 10^{-3}$
250	-	-	$0.108 \cdot 10^{-3}$	$0.128 \cdot 10^{-3}$
300	-	-	$0.087 \cdot 10^{-3}$	$0.110 \cdot 10^{-3}$
350	-	-	-	$0.096 \cdot 10^{-3}$

Figure 1.9 Dynamic viscosity of chemically pure water at various temperatures and pressures

standard values accepted at the ICPS-8 (Eighth International Conference on Properties of Steam, 1975). The values stated for density agree with ICPS-6 (1964).

Temperature °C	Density ρ kg/m³			
	0.1 MPa	1 MPa	10 MPa	100 MPa
0	999.8	1000.2	1004.7	1045.5
50	988.0	988.5	992.4	1027.4
100	-	958.6	962.8	1000.0
150	-	917.1	922.2	965.3
200	-	-	871.1	924.2
250	-	-	806.0	876.7
300	-	-	715.4	823.2
350	-	-	-	762.5

Figure 1.10 Density of chemically pure water at various temperatures and pressures

Temperature °C	Dynamic Viscosity μ kg/m.s	Kinematic Viscosity ν m²/s	Density kg/m³	Remarks
-25	$5.842 \cdot 10^{-3}$	$5.889 \cdot 10^{-6}$	992.1	
-20	$4.342 \cdot 10^{-3}$	$4.365 \cdot 10^{-6}$	994.7	Super-
-15	$3.342 \cdot 10^{-3}$	$3.353 \cdot 10^{-6}$	996.7	cooled
-10	$2.650 \cdot 10^{-3}$	$2.655 \cdot 10^{-6}$	998.3	water
-5	$2.156 \cdot 10^{-3}$	$2.158 \cdot 10^{-6}$	999.3	
0	$1.793 \cdot 10^{-3}$	$1.793 \cdot 10^{-6}$	999.8	
5	$1.518 \cdot 10^{-3}$	$1.518 \cdot 10^{-6}$	1000.0	
10	$1.306 \cdot 10^{-3}$	$1.306 \cdot 10^{-6}$	999.8	
15	$1.137 \cdot 10^{-3}$	$1.138 \cdot 10^{-6}$	999.2	
20	$1.003 \cdot 10^{-3}$	$1.005 \cdot 10^{-6}$	998.3	NB: All
25	$0.891 \cdot 10^{-3}$	$0.895 \cdot 10^{-6}$	997.2	values are
30	$0.798 \cdot 10^{-3}$	$0.801 \cdot 10^{-6}$	995.7	for a
35	$0.720 \cdot 10^{-3}$	$0.724 \cdot 10^{-6}$	994.1	pressure of
40	$0.654 \cdot 10^{-3}$	$0.659 \cdot 10^{-6}$	992.3	0.1 MPa
50	$0.547 \cdot 10^{-3}$	$0.554 \cdot 10^{-6}$	988.0	(1bar).
60	$0.466 \cdot 10^{-3}$	$0.474 \cdot 10^{-6}$	983.2	
70	$0.403 \cdot 10^{-3}$	$0.412 \cdot 10^{-6}$	997.7	
80	$0.354 \cdot 10^{-3}$	$0.364 \cdot 10^{-6}$	971.6	
90	$0.315 \cdot 10^{-3}$	$0.326 \cdot 10^{-6}$	965.2	
95	$0.298 \cdot 10^{-3}$	$0.310 \cdot 10^{-6}$	961.7	
99	$0.285 \cdot 10^{-3}$	$0.297 \cdot 10^{-6}$	958.9	

Figure 1.11 Dynamic viscosity, kinematic viscosity and density for chemically pure water in the temperature range -25° to 99 °C and at a pressure of 0.1 MPa (1 bar)

Temperature °C	Vapour pressure MPa	Vapour pressure, metres, column of water
0	0.00061	0.06
10	0.00127	0.13
20	0.00234	0.24
30	0.00424	0.43
40	0.00738	0.76
50	0.01234	1.27
60	0.01992	2.07
70	0.03116	3.27
80	0.04736	4.97
90	0.07011	7.41
100	0.10133	10.8
110	0.14327	15.4
120	0.19854	21.5
150	0.47600	52.9
200	1.5549	183
250	3.9776	508
300	8.5927	1229
350	16.535	2810

Figure 1.12 Vapour (saturation) pressure for chemically pure water

Various characteristics of chemically pure water:
- Melting point, ice to water, 0 °C at 100 kPa
- Latent heat of fusion, ice to water, 335 kJ/kg
- Specific heat 4.182 kJ/kg K at 100 kPa and 20 °C

- Boiling point 100 °C at 100 kPa
- Latent heat to steam 2260 kJ/kg at 100 °C
- Coefficient of expansion $\gamma = 0.000207$ per deg °C at 100 kPa and 20 °C
- Bulk modulus $K = 2.2 \times 10^{-9}$ N/m^2 at 100 kPa and 20 °C

1.2.2 Fresh water

General

Fresh water is the water derived from rivers, streams and wells. Generally containing less than 1% salt, sodium chloride, it can be 'hard' or 'soft'. Hard water is high in calcium and/or magnesium salts and results in poor lather when using soap. Dissolved gasses are always present, usually oxygen and carbon dioxide. The oxygen content should be about 5 mg/l. During treatment for drinking water chlorides are added. Unless stated otherwise, fresh water is considered to be clean; this is taken as meaning no measurable concentration of solids over 10 microns. Untreated water should have micro-organisms 0.5 to 1.0 microns.

Fresh water has enormous usage, among other things as raw water for drinking purposes, as cooling and process water within various industries and for irrigation.

For the purposes of pumping installations, the most interesting aspect is the corrosive action of water on the commonly used construction materials, generally steel, grey cast iron and bronze. The characteristics of water which contribute to corrosion are:
- pH value,
- hardness and carbonic acid content,
- content of various chemicals, primarily salts,
- acidity.

Apart from these, the temperature of the water and the following factors will influence installation corrosion and wear:
- velocity of flow; normally several m/s in pipes and 10-40 m/s in a centrifugal pump,
- NPSH available and cavitation,
- content of solid bodies, e.g. sand and sludge from various sources.

Not all materials are suitable for higher velocity service. Not all materials have equal cavitation resistance. If the absolute pressure of fresh water is reduced to approximately 0.8 bara the dissolved gasses will start to evolve. If this occurs in a pump suction cavitation symptoms and failures may occur. Small quantities of solids can cause high wear rates and accelerated corrosion in pumps designed and selected for clean water.

Consideration should be given to the rate of corrosion. Popular pump types can be supplied with an internal corrosion allowance providing an estimated component life. It would be very difficult here to make general rules when even an insignificant quantity of salt may magnify an attack of corrosion. Water containing chlorides can be especially troublesome. The corrosive action on steel, for example, is increased by a factor of 8 at 50 mg/litre of Cl_2, which means that steel is not practicable for this purpose.

pH value and choice of material

Natural waters usually have pH values of between 4 and 8. They are divided into two main groups according to their acid content.
- Ground water from deep down: this contains very little acid and it is thus the hydrogen ion concentration which is the decisive factor in the aggressiveness of the water. It should be pointed out that iron is attacked noticeably at pH values of 6 to 7 in this low acid water.
- Surface water, which is acidic: here the pH value is no absolute measure of aggressiveness although it is important to know what it is.
- Drinking water may be low acid ground water or high acid surface water which has, furthermore, been treated chemically and filtered. Waterworks supply the necessary information in this respect. Special conditions apply for lime-containing water, which is dealt with in the following section.

Since pumps used for conveying fresh water are for the most part constructed of cast-iron, some general aspects are given below about the use of this material with particular reference to the pH value of the water.

- Grey cast iron may be used without any real problems within the pH range 7 to 10. If chlorides are present, it may be that cast iron is not adequate.
- Grey cast iron can often be used within the pH range 5 to 7, but the effects of those factors arising from the lower pH values can be great. Within this pH range, cast iron is superior to steel as regards resistance to corrosion. The high carbon content of grey cast-iron (3 to 4 %) means that at moderate speeds of flow the graphite, together with corrosion products, can build up an anti-corrosive film, so-called graphitization.

For pH values at which cast-iron is not resistive, bronze, steel or stainless steel has to be used. Aluminium bronze is particularly suitable for fresh water, although nickel aluminium bronze may be preferred for longer life. (See also Chapter 6, Materials). Steel can be used successfully under the right conditions. High velocity areas can be coated. 11-13Cr steel is popular for boiler feed applications although stainless steel is sometimes used to protect against poor water quality control. It should be emphasised that the pH value of water can always be adjusted by suitable chemical treatment before pumping. Care must be taken to ensure any chemical treatment is performed correctly and that additional corrosion problems are not introduced.

When pumping corrosive liquids, particularly liquids supporting electrolytic cells, the correct choice of material combinations for adjacent components is critical. Materials for shafts, sleeves, bushes and wear rings must also be considered. Material properties at clearances can be varied by coating or plating one or both components.

Hardness of water

The degree of hardness of water depends upon the presence of impurities, mainly calcium (Ca) and magnesium (Mg) in the form of carbonates, although non-carbonates, for example sulphates, nitrates and chlorides, also have an effect. The quantities in which these impurities occur are a measure of the hardness of the water. Hardness can be expressed in terms of the specific substance such as calcium hardness (Ca-H), magnesium hardness (Mg-H) and so on. The total hardness comprises the sum of the individual hardnesses. Distinction can also be made between permanent and temporary hardness. Temporary hardness consists of alkali ions bound to carbonates, and permanent hardness of alkali ions bound to non-carbonates. Temporary hardness is so called because it disappears with heating. Distribution under the various headings is according to the chemically equivalent content of alkali ions. Since the various alkaline metals have different atomic weights, the unit for hardness - 1 milli-equivalent per litre (meq/l) is defined as:

$$1 \text{ meq/l} = \frac{1 \text{ m mol/l}}{\text{chemical valency}} \qquad \text{Equ 1.8}$$

The hardness unit 1 meq/litre corresponds to the following ion contents in mg/l:

1 meq calcium hardness   =   20.04 mg/l Ca++

1 meq magnesium hardness =   12.16 mg/l Mg++

1 meq strontium hardness =   43.82 mg/l Sr++

1 meq barium hardness   =   68.68 mg/l Ba++

In the so-called English degree of hardness, Eng°, is expressed as the equivalent content of calcium carbonate ($CaCO_3$) per Imperial gallon of water, thus:

$$1 \text{ Eng°} = 14.2 \text{ mg } CaCO_3/l \qquad \text{Equ 1.9}$$

For the various oxides of mineral alkaline metals:

1 Eng° = 8.0 mg CaO/l
1 Eng° = 5.7 mg MgO/l
1 Eng° = 14.78 mg SrO/l
1 Eng° = 21.88 mg BaO/l

and the relation in terms of meq/l is:
1 Eng° = 0.285 (meq/l)
1 meq/l = 3.5 (Eng°)

It is usual in the UK to state the hardness of water in English hardness degrees. A common classification is:

    3   very soft
    3 - 6   soft
    6 - 10   medium hard
    10 - 15   rather hard
    15 - 24   hard
       - 24   very hard

Other units of concentration are used in other countries; for conversion of various hardness units, see figure 1.13.

Units	Alkali ions meq/l	German hardness degrees °dH	English hardness degrees	French hardness degrees	ppm CACO$_3$ (USA)
1 meq/l alkali ions	1.00	2.8	3.5	5.0	50.5
1 German hardness degrees °dH	0.356	1.0	1.25	1.78	17.8
1 English hardness degree	0.285	0.80	1.0	1.43	14.3
1 French hardness degree	0.20	0.56	0.70	1.00	10.0
1 ppm CACO$_3$ (USA)	0.02	0.056	0.07	0.10	1.0

Figure 1.13 Conversion factors for various degrees of hardness

Example: 1°dH = 1.78 French hardness degrees

1 Eng° = 0.80 German hardness degrees °dH

Soft water in general is more suitable for the majority of tasks, typically for household use, than hard water. When washing, a high bicarbonate content in the water is damaging because soaps consisting of a mixture of sodium stearate and palmitate generate insoluble calcium salts of organic acids with calcium bicarbonate. Hard water always contains calcium carbonates whose solubility diminishes with increase of temperature. That is why calcium carbonate is deposited as fur and scale in boilers, heat exchangers and other heating vessels. In order to combat the deposition of scale, which can lead to local overheating in steam boilers, soft water has to be used as a source of supply. Water intended for such applications has to be softened, or dehardened.

Carbonic acid and carbonate equilibrium

Precipitation seeping down through the ground, to become ground water, absorbs carbon dioxide (CO_2) from the air in the soil, generated there by the oxidisation of organic material or by the action of various acids on limestone. Carbon dioxide and water from carbonic acid H_2CO_3 which converts the carbonates $CaCO_3$ (limestone) and $MgCO_3$, both difficult to dissolve in water, into soluble bicarbonates $Ca(HCO_3)_2$ and $Mg(HCO_3)_2$. The latter contains some CO_2 from the original carbonate (bonded carbonic acid), and some CO_2 from the carbonic acid which converted the carbonate into bicarbonate (semi-bonded carbonic acid) (see figure 1.14). In order to keep the bicarbonate dissolved, a certain extra amount of CO_2 is required (free attached carbonic acid).

If there is enough carbonate in the ground and if all the CO_2 is used up in the conversion of this to bicarbonate and in keeping the bicarbonate in solution, then the water is in a state of equilibrium as regards carbonate-carbonic acid. Thus a special condition of equilibrium arises between lime and the attached free carbonic acid. If the free carbonic acid content is less than that required for equilibrium, lime is separated. If the carbonic acid content increases the lime is re-dissolved. If, on the other hand, there is a surplus of CO_2, this is called "free surplus carbonic acid" or "aggressive carbonic acid". It is this part of the carbonic acid content which usually causes corrosion.

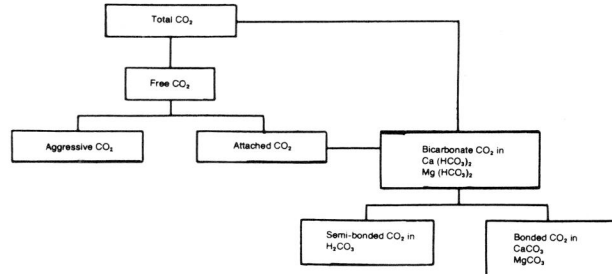

Figure 1.14 Various states for carbon dioxide (CO_2) in water

Water without free carbonic acid, if oxygen is present at the same time, will easily generate carbon-containing protective films which prevent corrosion on exposed pump surfaces, vessels and pipes. If there is free "aggressive carbonic acid" in the water, the build-up of the protective layer is hampered, particularly if there is a lack of oxygen. If this kind of water comes into contact with lime, concrete, treated lime or soda, the lime is dissolved until its chemical equilibrium is restored. Free carbonic acid is thus aggressive to lime. Iron, steel and light metals are attacked by all free content of carbonic acid, so that the lower the value of pH, the greater is the speed of corrosive breakdown. Ground water often contains bicarbonates with "attached carbonic acid" and perhaps also "aggressive carbonic acid". Water containing aggressive carbonic acid can be aerated and/or dosed with a suitable alkali (lime, soda, sodium hydrate) to bond the carbonic acid and to increase the pH value. At the same time, however, it must be observed that the so-called lime-saturated pH value should not be overstepped. If that happened, troublesome precipitations might result.

For practical purposes, the pH for natural carbonated water depends almost entirely on the relationship between the "bonded" and the "free carbonic acid" and, according to Klut, is determined by the relationship:

$$pH = 6.82 + \log\left(\frac{\text{bonded } CO_2}{\text{free } CO_2}\right) \qquad \text{Equ 1.10}$$

If the equilibrium value pH is set in relation to the carbonate hardness, then relationships are obtained as shown in figures 1.15 and 1.16. Using figure 1.16 distinction can also be made between "attached" and "aggressive" carbonic acid.

Carbonate precipitation can cause trouble in certain assemblies in pumps, for example shaft seals and water-lubricated plain bearings.

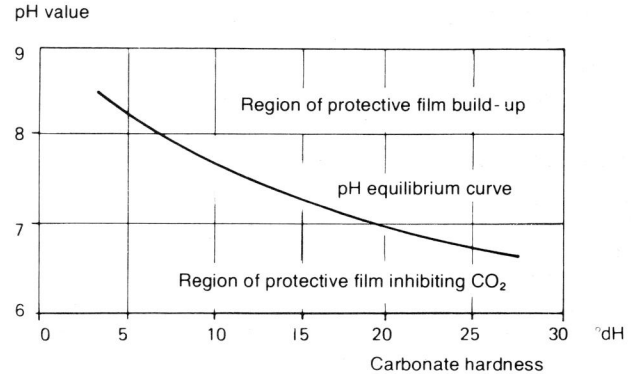

Figure 1.15 Condition of equilibrium for the creation of carbon content protective films in acidic waters

Figures 1.15 and 1.16 can be used to measure the aggressivity of natural water when the pH value and the carbonate hardness are known. Oxygen-rich carbonated water is aggressive only when the pH value is less than the values along the equilibrium curve. In other cases, a lime protective film is built up. If the pH value is considerably below the values on the equilibrium curve, the effect of other pH reducing substances may be suspected. Corrosion is then caused by these.

In the case of waters having low oxygen content there will be no build-up of protective film. All free content of carbonic acid adds to the aggressivity of the water. The corrosion rate increases with decreasing pH value.

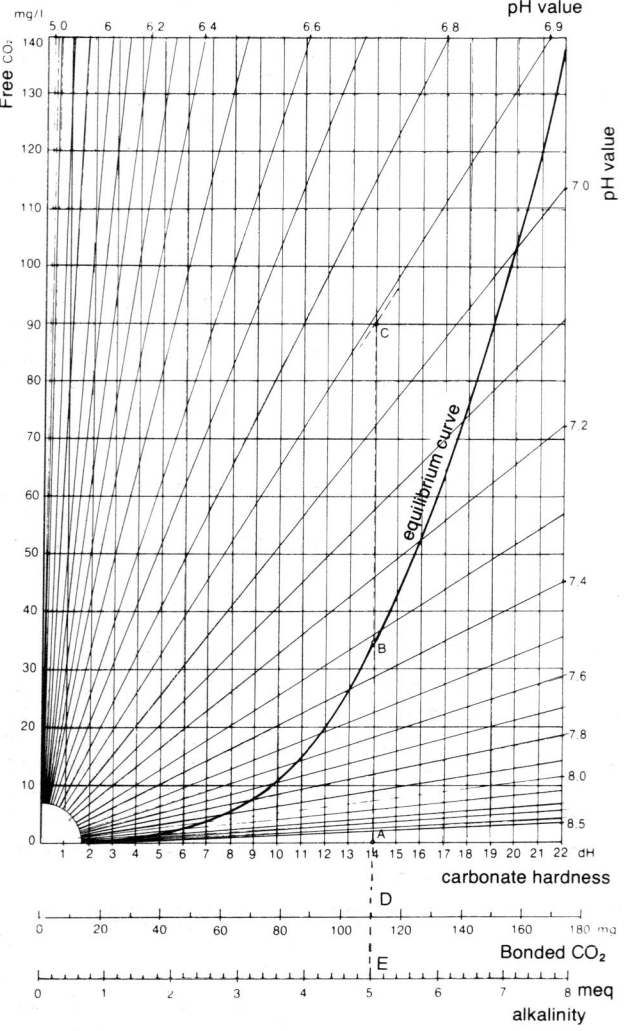

Example:
The following values are obtained after chemical analysis of raw water:
Carbonate hardness	14 °dH	(A)
Free carbonic acid	90 mg/l	(AC)
According to the nomogram:		
Free attached carbonic acid	35 mg/l	(AB)
Free aggressive carbonic acid	55 mg/l	(BC)
Bonded carbonic acid	110 mg/l	(D)
pH value	6.91	(C)
Alkalinity	5.0 meq	(E)

Figure 1.16 Carbonate hardness relative to carbonic acid content

The so very important oxygen content of raw water can vary greatly. In both spring water and ground water emanating from upper strata, the oxygen content is almost always sufficient to build up the natural lime rust-protective layer for carbonate hardnesses of 6 °dH. Soft surface water, because of lack of lime, cannot build up lime rust-protective layers and is, therefore, always more or less aggressive.

1.2.3 Brackish water

Brackish water occurs at the mouths of rivers where fresh water and sea water mix. The salt and chlorides content are diluted to approximately 1 to 2.5% and 4000 ppm respectively, giving a pH range of 6 to 9. Because of the turbulent flow regime brackish water contains suspended solids, typically silt and sand. Particle sizes would range from 1 to 200 microns. The percentage of solids, and the size distribution, could only be quantified by sampling.

The river water may be contaminated with industrial waste such as nitrogen compounds or caustic. Ammonia, sodium carbonate and sodium hydroxide are some of the most popular wastes. If nitrogen compounds are present, the oxygen content may be reduced due to the activity of bacteria and fungi. Small micro-organisms, as in fresh water, may be present.

Cast iron, steel and 11-13Cr steel will generally not be suitable. The presence of ammonia or caustic will cause stress corrosion cracking problems with brass. The addition of 0.25% to 4% tin to copper alloys improves corrosion resistance and resistance to stress corrosion cracking. Aluminium bronzes, with 10% aluminium, and nickel aluminium bronzes, with 4% nickel, are generally corrosion resistant. Nickel aluminium bronze has proved superior where mud or silt deposits accumulate. Austenitic stainless steels, such as AISI 316, have given adequate service on brackish water with 0.2% ammonia compounds up to 35 °C. The solids content must be quantified to allow proper pump selection.

1.2.4 Sea water

Sea water is pumped continuously in two important processes; crude oil secondary recovery and desalination. Secondary recovery may require pressures up to 350 barg. Desalination, by reverse osmosis, requires only 70 barg. These two processes account for the recent developments in pump materials.

General data

Sea water contains a mixture of inorganic salts. Cooking salt, sodium chloride NaCl, forms about 75% of the salt content. The salt content at great depths varies insignificantly between the oceans of the world. At a depth of 500 m the salt content is about 3.5%. At the surface, however, the salinity is affected considerably by the climatic and other factors. In the Northern Baltic, the salinity is almost 0%, and in the Red Sea it is about 4%. Evaporation, rain, polar ice, rivers and industrial/chemical effluents are all factors which affect the properties of sea water.

The temperature at the surface of the sea varies from about -2 °C in the Arctic Ocean and all around Antarctica to about +37 °C in the Persian Gulf. The sea bed temperature is generally between + 2 °C to + 5 °C. At temperatures between 10 and 15 °C the oxygen content varies between 8 and 10 mg/l. Chlorides are present up to 25000 ppm. Unless polluted there is no hydrogen sulphide. Sea water properties vary throughout the year and care must be taken to establish the full range. At one location temperatures ranged from 3.3 to 27 °C, pH 7.9 to 8.4, whilst oxygen varied from 72 to 100% saturation. North Sea properties have been measured as:

density	1020 kg/m^3
freezing point	-2 °C
boiling point	100 °C
temperature	2 to 4 °C,
chlorides	12.8 to 17.4 g/l,
oxygen	6.5 to 8.4 mg/l,
pH	8.0 to 8.3.

The physical properties of sea water vary little from chemically pure water.

The qualities of sea water and corrosion

Operational experience has shown, especially in the field of ships' pumps, that the quality of sea water is of great significance as regards resistance to corrosion. There are examples of identical ships where the one vessel has had serious problems with its sea water pumps whilst the pumps in a sister ship have remained intact. This phenomenon has usually been explained by the fact that the ships have been plying in different types of sea water. By and large, those ships engaged in coastal and river traffic are attacked more by corrosion than are those which sail the open seas. As regards coast-based industries and power stations, the quality of the water may vary greatly, depending upon where the installation is located. In such cases however, the quality of water to be dealt with is usually already known at the design stage, enabling the choice of pump materials to be made accordingly. This is not usually possible in the case of marine pump installations. Coastal installations and others subjected to saline atmospheres require special attention to external construction, particularly where dissimilar metals are used, to reduce the effects of galvanic corrosion. The variations in the properties of sea water and the effects these variations have on corrosion rates can be summarised as follows:

- The temperature of sea water is of relatively great importance with regard to corrosion rates. In general, higher temperatures increase the corrosion risks.
- Solid contaminants damage the protective-film on the material, thus giving corrosion a chance to attack. The rate of mechanical wear can of course, also be increased.
- Chemical contaminants of various types can also lead to increased rate of corrosion.
- Tests have shown that oxygen contents over 1 mg/l will promote stress corrosion cracking. If pre-treatment is considered, oxygen reduction is of prime importance. The pH value of sea water is normally about 8.
- The dosing of additives may perhaps be regarded as a deliberate chemical contamination of sea waters. Properly conducted, such dosing should not bring any problems from the corrosion point of view. Too strong a dose of sodium hypochlorite, sometimes used to inhibit algae, can on the other hand cause serious pitting corrosion.

Summarising, it can be confirmed that clean and cold sea water with normal oxygen content gives the least corrosion problems.

Brief note on the choice of material

Care must be taken with material combinations when pumping sea water. An electrolytic cell is set up between dissimilar metals due to the high conductivity of the sea water. The less noble material should have a much larger surface area than the noble material. The problem is exacerbated by changes in the electrolytic potential caused by high flow velocities.

Steel is not very good in moving sea water; corrosion of steel increases by a factor of 4 from static to 4.5 m/s. When suitable, cast iron castings with stainless steel impellers last longer overall than nickel aluminium bronze impellers. Ni-resist and ductile ni-resist have been used for some applications; ductile ni-resist is difficult to cast; both are difficult to weld repair.

Copper alloys give good service lives in sea water and are generally easy to cast and machine. Nickel aluminium bronze can be difficult for some machining operations, like tapping. Special tools may be required. Nickel aluminium bronze and magnesium aluminium bronze are better than aluminium bronze and gunmetal. Gunmetal casings with nickel aluminium impellers have been popular. Aluminium bronze is resistant to stress corrosion cracking provided metallurgy and heat treatment are correct. Nickel aluminium bronze has exceptional resistance to cavitation damage compared to some stainless steels; BS 1400 AB2 = 0.025mm, AISI 321 = 0.305 mm in 3% NaCl. Welding of nickel aluminium bronze is possible, but heat treatment after is necessary to reduce corrosion possibilities. Aluminium and nickel aluminium bronze are limited to 18 to 25 m/s to eliminate erosion problems. Aluminium bronze can be used in combination with stainless steel and titanium without problems. Sea water piping 90/10 Cu Ni has adequate service life.

Stainless steels are the most popular materials for pumping sea water. Care must be taken when reviewing test results and when specifying materials. The chemical composition, and therefore corrosion resistance, of austenitic stainless steels varies with material production; cast AISI 316 is different to wrought AISI 316.

Ferritic steels suffer from local corrosion and pitting in sea water. Stainless steels without molybdenum are generally poor. AISI 304 is generally regarded as the lowest quality austenitic stainless steel. 304 is not used for sea water. AISI 316 has proved successful in tests up to 90 °C on synthetic sea water although when tested at 20 °C with bubbled chlorine gas pitting occurred. The general concensus is that cast 316L castings with 316 shafts is the minimum working combination. Cast 316L has 10 to 15% ferrite and a higher chromium content than wrought 316L which has no ferrite structure. 316L castings are as good as Alloy 20 for pitting and crevice corrosion resistance, have good stress corrosion cracking resistance and are easier to cast and weld repair. The ferrite content improves corrosion resistance and aids weldability. 316 shafts have to be designed carefully to avoid sharp corners with attendant stress concentrations resulting in stress corrosion cracking. Pitting can be a problem in the stuffing box/mechanical seal area. 316 shafts and impellers have been replaced by monel at extra cost but with no real economic benefit.

Research over a number of years tended to show that molybdenum played an important part in the corrosion resistance. An early theory suggested the Cr + Mo percentage should be over 30%. This was followed by an Uddeholm equation for Pitting Resistance Equivalent (PRE).

$$PRE = (Cr\ weight\%) + (3.3 \times Mo\ weight\%) \quad Equ\ 1.11$$

Uddeholm thought that PRE should equal 28 minimum. More research and testing. Later, nitrogen was found to have an important role in the corrosion resistance. A new equation was proposed.

$$PRE_N = (Cr\ weight\%) + (3.3 \times Mo\ weight\%)$$
$$+16 \times N_2\ weight\%) \quad Equ\ 1.12$$

Larsen of Ingersoll-Rand suggested the PRE_N value be set at 35. Some users have requested PRE_N values of over 40. Nitrogen strengthened steels had been available since the early 70's. Now more exotic varieties were compounded. Another benefit provided by nitrogen hardened stainless steels was improved cavitation damage resistance. The table in Figure 1.17 shows the range of chromium and nickel contents of current popular materials.

	Composition %	
Designation	Cr	Ni
AISI 316	16-18	10-14
ACI CF-8M	18-21	9-12
AISI 316L	16-18	10-14
ACI CF-3M	17-21	9-13
Avesta 254SMO	20	18
Avesta 254SLX	20	25
ACI CN-7M	19-22	27.5-30.5
Firth Vickers Rex 734	21.6	9
Armco Nitronic 50	20.5-23.5	11.5-13.5
Mather & Platt Zeron 25	24-26	5.5-7.5
Mather & Platt Zeron 100	24-26	6-8
Bonar Langley Ferralium 255	24-27	4.5-6.5
Bonar Langley Ferralium 288	26-29	6-9

Figure 1.17 Materials for sea water

Some exotic stainless steels have a low tensile strength combined with high ductility. Higher pressure pumps require much thicker casings resulting in heavy pumps. More costly, stronger materials may be better resulting from thinner casings and smaller corrosion allowances. Materials, such as titanium, may cost less than proprietary stainless steels.

NOTE: Water, more exactly brine, extracted from wells in oil-fields may have up to 30% salt content, mostly sodium chloride. Oxygen content can be zero but hydrogen sulphide is often present. The pH value is normally low, about 4, and the brine reacts as a weak acid causing severe corrosion.

Stainless steels containing 25% Cr and 25% Ni are typically used for these applications.

See also Chapter 6 "Pump materials".

1.3 Oils

1.3.1 General

Oils are classified according to their origins into mineral oils, animal and vegetable oils, but in the context of pumps they can all be treated the same. Along with mineral oils, we can also reckon such petroleum products as solvents, petrol, kerosene and the like, which should be considered when pumping oil stocks. When pumping oil, the upper and lower operating temperature limits, viscosity, cloud point, lowest flow (pour point), solidifying temperature and the vapour pressure should all be established. The flow capability of oils follows Newton's Law. In common with water, they have constant viscosity independent of shear rate and time. The viscosity is temperature dependent, oil flowing more easily when heated. The viscosity falls as the temperature rises. In order to assess the needs of a pump installation, the viscosity-temperature relationship must be known and the way the oil behaves with variations in operational temperatures must also be clarified.

Cloud point, pour point and solidifying temperatures:

Mineral oils transform gradually from the liquid to the solid state, as opposed to other liquids (water, for example, which has an exact freezing point). When the oil is chilled, it goes cloudy at a certain temperature because of precipitation of paraffin crystals, i.e. generation of wax. This temperature is called the cloud point (cold filter plugging point). The pour point is reached if the temperature is further reduced. A few degrees below this temperature, the oil changes to a completely solid form, the solidifying point. Because of waxing, it is considered that mineral oils can only be handled by pumping at a temperature of at least 10 °C above the pour point. When considering pumping high viscosity oils with solidifying temperatures near to or above the ambient temperature, (for outdoor installations for example), notice must be taken of the pour point and the installation must be designed so that the pipes and pump can be heated. Pipes which are not heated must be able to be emptied in order to prevent stoppages building up if pumping is interrupted in ambient temperatures below the lowest flow temperature. Low sulphurous heating oils have higher pour point temperatures than high sulphurous oils.

Complex petroleum products, petrol for example, have a vapour pressure range which is dependent upon the most easily flowing components. This property affects the calculation of the NPSH available for the pump. Mineral oils are classed as dangerous liquids in fire hazard Class 3. Light distillates such as petrol and photogene are in fire hazard Class 1, see "Liquid Table" 1.6.2.

1.3.2 Viscosity

Burner oils

Low sulphur oils (max 0.8% by mass - EEL Council Directive 25/716/EEL) are covered by classes C, and C2 have viscosities of between 1 - 2 cSt at 37 °C and together with Class D oils can be stored and handled at ambient temperatures likely to be encountered in the UK. Class E - H are residual blended oils for atomising burners and normally require preheating before atomisation. The four lines on the chart show average viscosity/temperature relationships for fuel of Classes E to H at the maximum viscosity allowed by the specifications. The approximate viscosity/temperature relationship for any petroleum fuel within these classes, for which viscosity at one temperature is known, can be determined by drawing through the known viscosity/temperature intersection a line parallel to those shown. From this line can be read the temperature required for any desired viscosity e.g. that specified by burner manufacturers for proper atomisation. As the temperature is

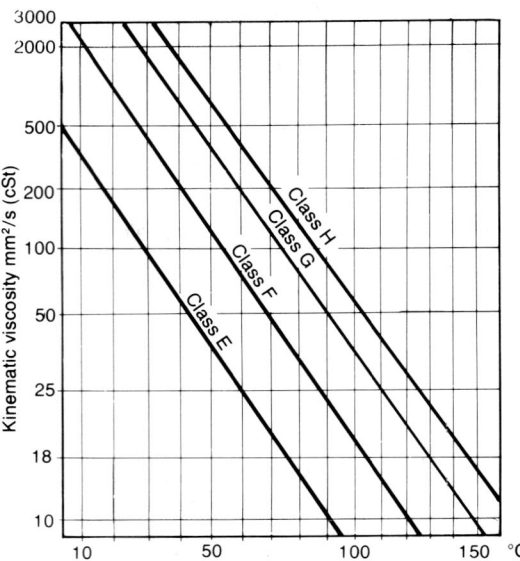

Figure 1.18 Kinematic viscosity/temperature chart (from BS 2869:1970)

lowered towards the pour point (lowest flow temperature) there is an increasing upward deviation from the viscosity indicated on the chart. This deviation is of such a magnitude that in no case shall the chart be used within 15 °C of the pour point, information on which can be obtained from the suppliers.

Class of fuel	Min. temperature for storage	Min. temperature for outflow from storage and handling
E	7 °C	7 °C
F	20 °C	27 °C
G	32 °C	38 °C
H	(special purpose fuels)	

Engine oils

The SAE system was devised in America and laid down by the Society for Automotive Engineers (SAE) in 1926. The society has recently changed its name to "The Engineering Society for Advancing Mobility Land Sea Air and Space" but retains its well known abbreviation of SAE. Lower numbers in the SAE series mean thinner oils and the letter W after the number indicates that the oil is suitable for use in winter. The SAE system is accepted and used internationally.

Below the cloud point (cold filter plugging point) temperature, there is no rectilinear relationship because of waxing.

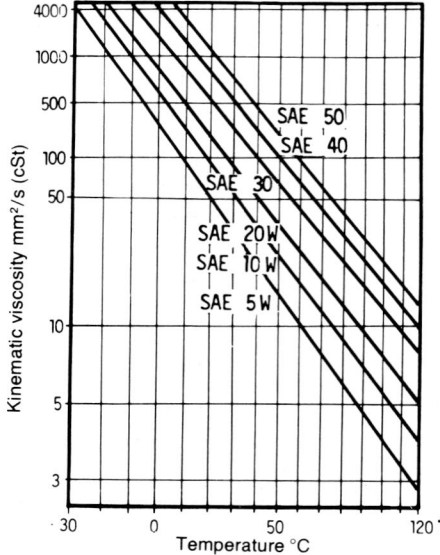

Figure 1.19 Kinematic viscosity for engine oils, SAE 5W to SAE 50

Pour point -20 °C to -30 °C (lower for certain special qualities).

Lowest recommended handling temperature: 10 °C to 15 °C above pour point.

Gearbox oils

Below the cloud point (cold filter plugging point) temperature, there is no rectilinear relationship, because of waxing.

Pour point -20 °C to -30 °C (lower for certain special qualities).

Lowest recommended handling temperature: 10 °C to 15 °C above pour point.

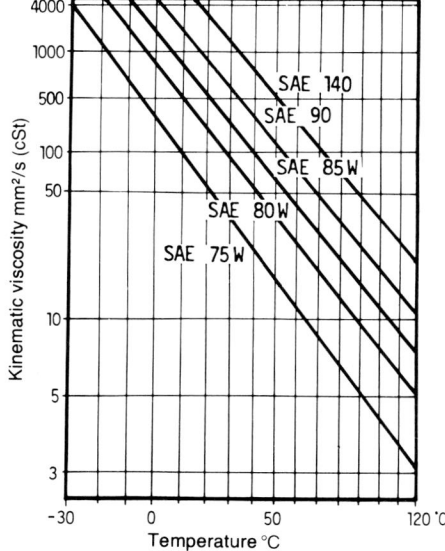

Figure 1.20 Kinematic viscosity for gearbox oils, SAE 75W to SAE 140

Industrial oils

The ISO (International Organisation for Standardisation) has developed a system of viscosity classification for lubricants for industrial use which came into use effectively from 1977. The system consists of 18 viscosity categories stated in centiStokes at 40 °C. Each class of viscosity is identified by an ISO VG (viscosity grade) number which in general coincides with the mean value in accordance with Figure 1.20.

ISO Viscosity Class	Kinematic viscosity at 40 °C Mean value	Minimum	cSt (mm²/s) Maximum
ISO VG 2	2.2	1.98	2.42
ISO VG 3	3.2	2.88	3.52
ISO VG 5	4.6	4.14	5.06
ISO VG 7	6.8	6.12	7.48
ISO VG 10	10	9.00	11.0
ISO VG 15	15	13.5	16.5
ISO VG 22	22	19.8	24.2
ISO VG 32	32	28.8	35.2
ISO VG 46	46	41.4	50.6
ISO VG 68	68	61.2	74.8
ISO VG 100	100	90.0	110.0
ISO VG 150	150	135	165
ISO VG 220	220	198	242
ISO VG 320	320	288	352
ISO VG 460	460	414	506
ISO VG 680	680	612	748
ISO VG 1000	1000	900	1100
ISO VG 1500	1500	1350	1650

Figure 1.21 Viscosity classes in accordance with ISO 3448/BS 4231:1982.

The classification system has special advantages:
- the ISO VG number gives information on the viscosity of the oil.
- ISO 3448 is fully supported by the leading national standardisation organisations such as ASTM, DIN, BSI, JIS and is expected therefore to achieve international application. Thus it will be easier to compare the viscosity of the oils with that specified by the machine manufacturers.
- ISO 3448 is directly comparable with BS 4231: 1982 for classifications at 40 °C.

Pour point 20 °C to 50 °C (lower values for certain special qualities).

Lowest recommended handling temperature: 10 °C to 15 °C above pour point.

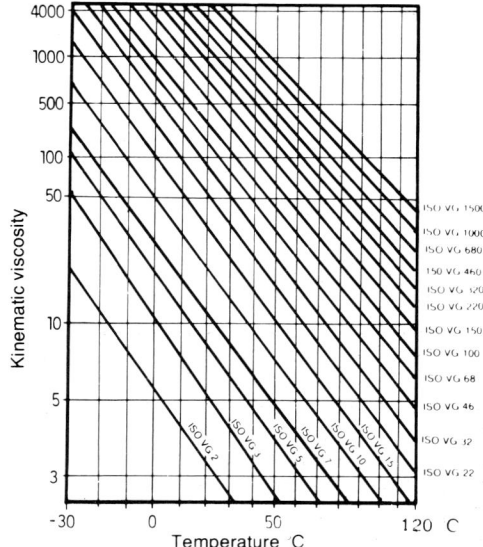

Figure 1.22 Kinematic viscosity for industrial oils in accordance with ISO 3448 and BS 4231: 1982.

Comments

The scales in the viscosity diagrams are themselves not linear but are adjusted so that the viscosity relationships are linear. The diagram may be used to construct the viscosity relationship for other oils if the associated viscosity temperature values are known for two points. These relationships will also be linear.

Liquid	Kinematic viscosity ν mm²/s (cSt) at a given temperature	
Aviation fuel	0.7/0 °C	0.6/38 °C
Castor oil	300/38 °C	42/54 °C
Corn oil	29/38 °C	9/100 °C
Groundnut oil	42/38 °C	23/54 °C
Kerosene	2.0/20 °C	1.6/38 °C
Linseed oil	31/38 °C	19/54 °C
Olive oil	43/38 °C	24/54 °C
Petrol	0.7/20 °C	0.6/38 °C
Soya oil	35/38 °C	19/54 °C
Turpentine	2.1/20 °C	2.0/38 °C
Valtran	37/38 °C	22/54 °C

Figure 1.23 Kinematic viscosity for various oils

Other liquid data

Liquid	Density kg/m³ at 20 °C	Region of boiling °C at 0.1 MPa	Vapour pressure kPa at 20 °C	Fire* hazard class
Aviation fuel	720	40-150	25-35	1
Burner oils —				
BS 2869 Class D	840			3
Class E	940			3
Class F	950			3
Castor oil	960		(0)	
Corn oil	920		(0)	
Engine oil SAE				
5W-50	900		(0)	3
Gearbox oil SAE				
75W-140	910		(0)	3
Groundnut oil	910		(0)	
Industrial oil ISO				
VG2-1500	880-905		(0)	3
Kerosene	780	170-250	0.1	2b
Linseed oil	930		(0)	
Liquid resin	970		(0)	
Olive oil	910		(0)	
Petrol	730	40-180	25-70	1
Soya oil	940		(0)	
Turpentine	860		(0)	2b
Valtran	920		(0)	

Figure 1.24 Some liquid properties for various oils.
*Refers to classification in the "Liquid Table" 1.6.2 - See section 1.1.7 for reference to other classifications.

1.4 Liquid solid mixtures
1.4.1 General

Particles of inorganic and organic solid matter are to be found, more or less finely distributed, suspended in liquids either as contaminants or for the purpose of transportation. Liquid solid mixtures belong to the non-Newtonian liquids. The characteristics of the mixture depend upon:
- the pure liquid properties,
- the size of the solid particles,
- the density of the solid particles,
- shape, hardness and abrasiveness of the solid particles,
- the deformability of the solids,
- the friability of the solids,
- and also the concentration of particles in the liquid.

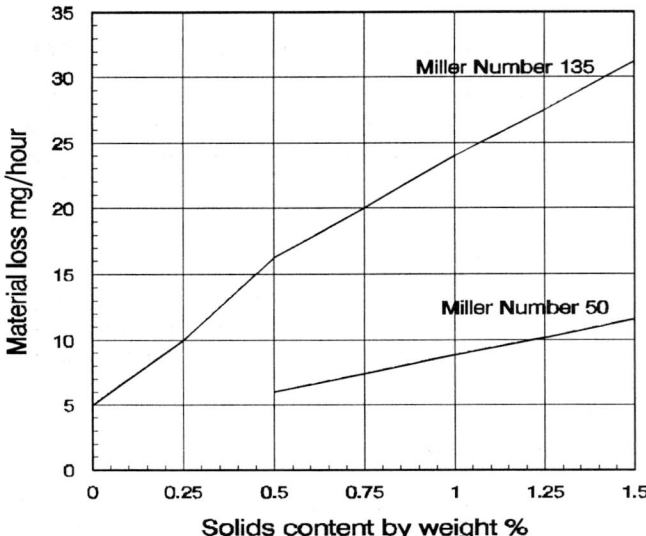

Figure 1.25 Valve material loss

These parameters are of decisive importance in the choice of the pump. It is usual in pump catalogues to state the suitability of pumps for various kinds of liquids, e.g. clean, no solid contamination, slightly contaminated, contaminated, slurry, mud, transport of pulps, solids, etc. without closer definition. However, all applications for liquid solid mixtures should be discussed in detail with pump manufacturers. Small quantities of solids can have a serious effect on wear. Vetter, Thiel and Störk presented test results in their paper 'Reciprocating pump valve design' for valve wear with low concentrations of 15 μm quartz with a Miller Number of 135, see Figure 1.25. These test results can be taken as representative of fine sand. A concentration by weight of only 0.25% doubled the wear rate. A concentration of 1% solids with a Miller Number of 50 in the same valves would increase valve wear by 80%.

In the quick-choice table, Chapter 12 "Choice of Pump", the following parameters are used for the particle content of the liquid:
Sizes <0.1, 0.1 to 1.0, 1.0 to 10 and 10 to 100 mm
Deformable
Abrasive
Concentration approx. <1%, >1%.

At the head of the table there are 62 different types of pump, marked "especially suitable", "suitable" or "suitable but with reservations" for these eight particle content parameters.

Liquid properties

The physical properties of the carrier liquid influence the properties of the liquid solid mixture. Solids in the liquid tend to increase the viscosity. The clean liquid viscosity is necessary to evaluate the mixture viscosity. Also the abrasiveness of a mixture is effected by the liquid viscosity. Sand in oil is much less abrasive than sand in water. Solids in liquid tend to increase the relative density. Liquid relative density is required to calculate the mixture relative density.

Size of particle

Particles less than 1 μm occur in clean fresh water at very low concentrations. Solids in chemical process applications are usually small, up to 1 mm. Solids transported in liquids can vary over a wide range, dependant upon the working pressure and the choice of pump, up to 25 mm is not uncommon. Larger hard solids are encountered with dredge pumps, quarrying and building site applications. A standard requirement is to pass a 100 mm sphere. Large soft solids, such as fruit, fish and sewage waste up to 140 mm are pumped on a regular basis. Examples of particle sizes and approximate pump capabilities are shown in Figure 1.26.

Particle density

The particle density is important for power consumption. Energy is expended moving the particles and the liquid. The power required for rotodynamic pumps is a function of the mixture density which is dependant upon the solids density.

Shape, hardness and abrasiveness

The shape of manufactured or processed solids can be defined. Naturally occurring solids may be difficult to describe. The hardness of solids can be measured and should be specified.

Because of the difficulties caused by shape, shape and hardness can be combined to give abrasiveness. Abrasiveness is quantified by testing representative samples of the mixture. 'Representative' is critically important. Extrapolating data from one mixture to another is unreliable. A popular test method is the Miller Number; originally intended for reciprocating piston pumps and now standardised as ASTM G75. The test measures the weight loss of a reciprocating metal block, 28% chrome iron, due to the effect of a 50% concentration by weight in distilled water. The Miller Number is calculated from the test results to produce a linear abrasive scale; a Miller Number of 200 produces twice as much wear as a Miller Number of 100 all other conditions being equal. Miller Num-

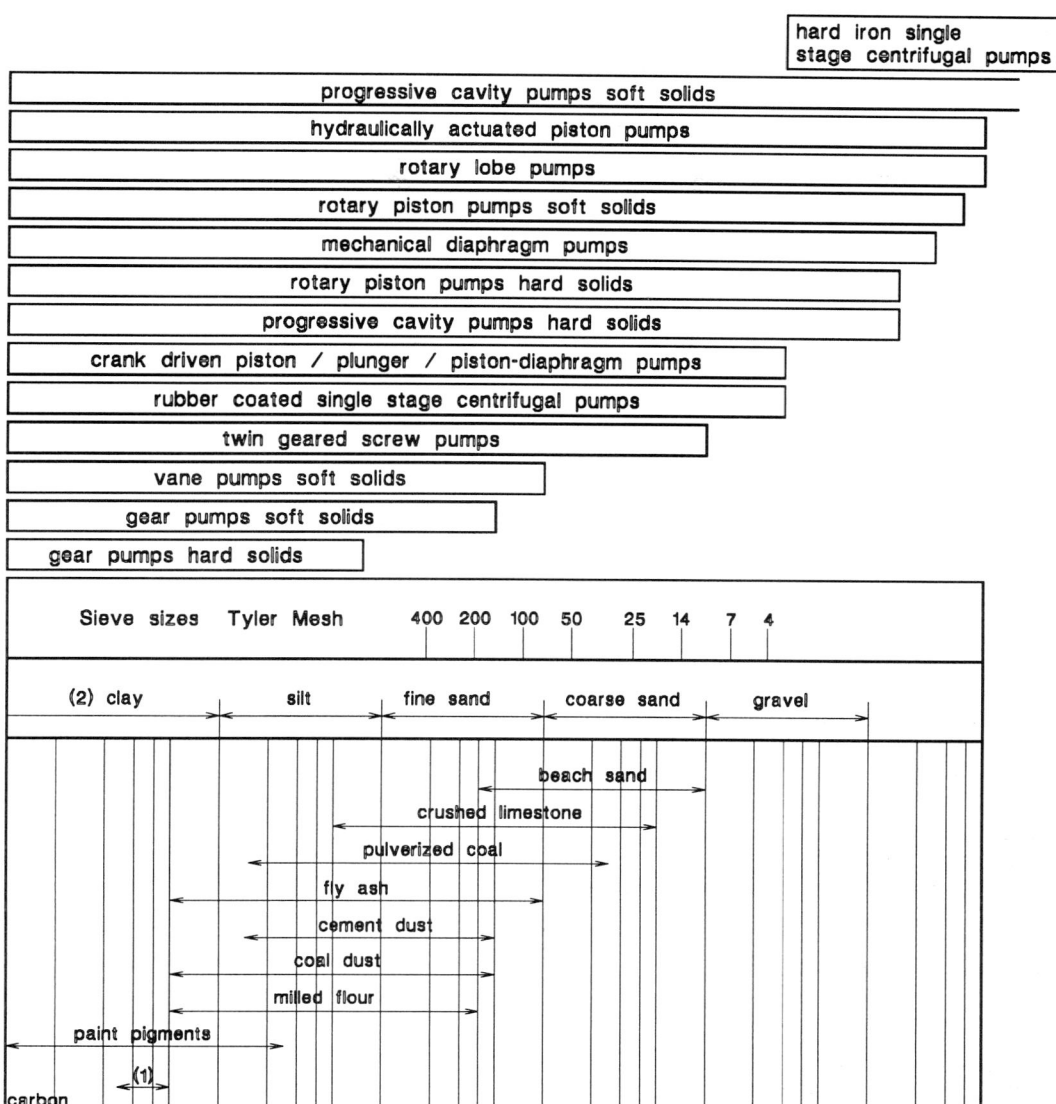

Figure 1.26 Particle sizes and pump capabilities

(1) micro-organisms in fresh water
(2) designation of the International Society for Soil Science

bers less than 50 are considered non abrasive; over 50 consideration should be given to wear problems. Figure 1.27 indicates typical values of Miller Number.

Solid	Minimum values	Maximum values
Bauxite	9	134
Coal	6	57
Copper Ore	20	135
Iron Ore	28	157
Limestone	22	46
Magnetite	64	134
Mud, drilling	10	10
Snad	51	246
Sewage (digested)	15	15
Sewage (raw)	25	25
Shale	53	59
Tailings	24	644

Table 1.27 Typical Miller Number Values

Deformable particles

Particles of organic matter, soft, fibrous particles which can be compressed, deformed or may be subject to damage. For example, dispersions of plastic or natural rubber in water - latex. Latex can coagulate under pressure and therefore requires careful pumping. Other soft solids; fruit, vegetables, fish; can be pumped with minimum damage by correct pump selection.

Friable particles

Friable means easily crushed or easily broken. Hard solids, like some coals, can be crushed or broken when trapped by certain pump components. Friable solids can be handled by gear pumps, some screw pumps and reciprocating pumps with valves.

Solids concentration

Concentration refers to the quantity of particles in suspension (dry substance = DS content) and is expressed as a percentage by weight or volume. Equations 1.13, 1.14, 1.15 and 1.16 show the volumetric relationships between liquid, particles and mixture. Equations 1.17, 1.18, 1.19 and 1.20 show the mass relationships.

$$\rho_M = \rho_L + \frac{C_V(\rho_s - \rho_L)}{100} \quad \text{Equ 1.13}$$

$$C_V = \frac{100(\rho_M - \rho_L)}{(\rho_S - \rho_L)} \quad \text{Equ 1.14}$$

$$\rho_L = \frac{\left(\rho_M - \frac{C_V \rho_S}{100}\right)}{\left(1 - \frac{C_V}{100}\right)} \quad \text{Equ 1.15}$$

$$\rho_S = \frac{100}{C_V}\left(\rho_M - \rho_L + \frac{C_V \rho_L}{100}\right) \quad \text{Equ 1.16}$$

$$\rho_M = \frac{1}{\left(\frac{C_M}{100\rho_S} + \frac{1}{\rho_L} - \frac{C_M}{100\rho_L}\right)} \quad \text{Equ 1.17}$$

$$C_M = \frac{\left(1 - \frac{\rho_M}{\rho_L}\right)}{\left(\frac{\rho_M}{100\rho_S} - \frac{\rho_M}{100\rho_L}\right)} \quad \text{Equ 1.18}$$

$$\rho_L = \frac{\left(1 - \frac{C_M}{100}\right)}{\left(\frac{1}{\rho_M} - \frac{C_M}{100\rho_S}\right)} \quad \text{Equ 1.19}$$

$$\frac{1}{\rho_S} = \frac{100}{C_M}\left(\frac{1}{\rho_M} - \frac{1}{\rho_L} + \frac{C_M}{100\rho_L}\right) \quad \text{Equ 1.20}$$

Where:

- C_M = percentage of solids in mixture by weight
- C_V = percentage of solids in mixture by volume
- ρ_L = density of liquid kg/m³
- ρ_M = density of mixture kg/m³
- ρ_S = density of solids kg/m³

The relationship between C_M and C_V is given by:

$$\frac{\rho_L C_M}{\rho_S} + \frac{C_M C_V}{50} + \frac{C_V \rho_S}{\rho_L} - \frac{C_M C_V \rho_S}{100\rho_L} -$$

$$- \frac{C_M C_V \rho_L}{100\rho_S} - C_M - C_V = 0 \quad \text{Equ 1.21}$$

Figure 1.28 shows a nomogram for estimating the mixture density when the carrier liquid is water.

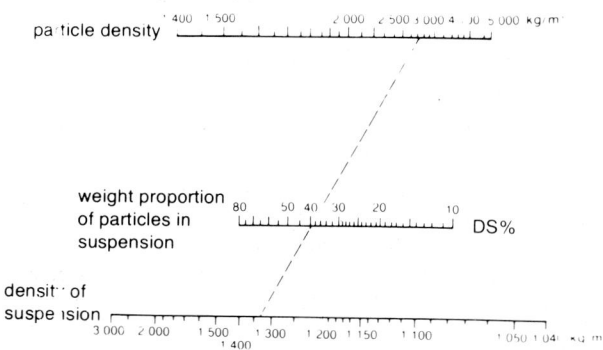

Figure 1.28 Mixture density nomogram for water

The example shows that a suspension having a particle content of 40% by weight and a particle density of 2,700 has a total density of about 1,340.

Figure 1.29 indicates typical values of solids concentration found in some pump applications.

Pumped mixture	Solids concentration by weight %
Polluted municipal sewage water	0.1
Paper pulp in rotodynamic pumps	0.5 to 6
In dense pulp pumps	6.0 to 20
Centrifuged sewage	up to 35
Copper concentrate pipeline	up to 45
Coal pipeline traditional	13 to 50
Activated sewage sludge	20 to 50
Carbonation slurry	50 to 60
Limestone slurry	50 to 60
Iron ore slurry	20 to 65
Coal water fuel	up to 75
Sand	20 to 75
Clay, raw cement sludge	50 to 75
Mine backfill slime	up to 78

Figure 1.29 Examples of solids content in pumped mixtures

It is not possible to generalise the effect of solids concentration on pump selection, performance or component life. When solids are known to be present the pump selection should be fully discussed with manufacturers. Figure 1.26 indicates the maximum solid size handling capabilities of various pump types ignoring concentration. Small particles, up to 40 μm, can be pumped at 65% by weight in water and look like watery toothpaste. Solids concentrations of 80% look like toothpaste. Small solids can be pumped at high concentrations. Large solids must be pumped at low concentrations. Also large solids must be handled by large pumps. Depending upon the pump type maximum hard solid size would be 4% to 20% of the nozzle size, 10 to 50% for soft solids.

One important aspect of solids handling has not been mentioned, velocity. Abrasive solids, and large solids, are pumped at much lower velocities than clean liquids. Experience has shown that abrasive wear in pumps is related to flow velocity raised to a power between 2.5 and 5 depending upon the component. If the velocity is increased by 10% the wear rate will increase by a minimum of 27% up to 61%. The velocity-wear relationship can be used with the Miller Number to estimate the effects of duty changes. Consider a piston pump operating at 50 rpm on drilling mud and having a valve seat life of 5000 hours. If the pump speed was increased to 75 rpm the predicted seat life would be 660 to 1800 hours. If the pump was operated at 50 rpm on mine tailings the predicted life would be 80 to 2000 hours. If the pump operated at 40 rpm on tailings then predicted seat life would be 130 to 6300 hours.

1.4.2 Sewage

Sewage is a generic term for:-

Soil sewage: discharge from water-closets, urinals, slop sinks including waste water from domestic baths etc. and even industrial effluent.

Surface water: rainfall and storm water.

Drain water: drainage from building sites, fields, leaks in broken pipes etc.

Waste water from households, businesses, hotels, offices, restaurants serves as a carrier for the contaminants from water closets and other sources of waste in the sewage system. The size of the waste thereafter is limited in principle to the area of the intake connected to the pipeline. It is not easy to define the limits of length of a soft and flexible objects. These objects, such as sheets of plastic, nylon stockings etc., pass unchecked down the water closet and into the public sewer.

It is forbidden to discharge, into any open drain or sewer, material which is likely to cause damage, create a hazard or which could have a prejudicial effect upon the treatment of their contents (i.e. waste matter which should be dealt with by commercial waste disposal companies). Such waste could be dirty engine oil, volatile liquids producing flammable or toxic vapour or biological waste. The legal regulations applying to local sewage works are enforced by the Local Water Authorities supervised by government bodies such as River Authorities and Environmental Protection agencies.

There is a steady increase in the quantity of unwanted or forbidden objects collected at rain water and sewage grills. Especially difficult, and containing a high content of textile matter amongst others, is the sewage from densely populated city centres. The heart of the city with its restaurants, offices and department stores presents a far greater sewage problem than do the surrounding dormitory suburbs.

The demand on pumps and other means of transporting the daily sewage should be assessed according to the contaminants (permitted or not) which are contained in the sewage. The larger contaminants can be classified as deformable according to the quick-choice table in Chapter 12 "Choice of Pump" - thus requiring a non-clogging (chokeless) design. Different types of industry produces varying types of effluent which often contains substances which can cause damage to the sewage treatment works and the recipient water installation.

In the UK the law regarding the discharge of industrial waste into sewers is restricted by the Public Health (Drainage of Premises) Act 1937, amended in 1961. The responsibility for administering and enforcing this legislation is in the hands of the Water Authorities. Provisions for the restriction of pollution of tidal waters and estuaries appear in sections 2-5 of the Rivers (Prevention of Pollution) Act 1951 and amended by the 1961 Act. The Clean Rivers (Estuaries and Tidal Waters) Act 1960 also applies with limitations specified in the Schedule to the Act 1960. The control of Pollution Act 1974 sections 43 and 44 extends legislation with respect to specific pollutants. Limiting values for the discharge of substances which are toxic, corrosive, explosive and/or damaging to pump and system materials etc. are the subject of consent which must be obtained from the local Water Authority. (Many of these liquids appear in the Liquid tables in Section 1.6.2). Cooling water, surface water and drain water display properties as in Section 1.2. Solid precipitations from the circulating water can accumulate in cooling tower installations in the form of lime furring, sludge and concentrations of minerals giving rise to corrosion, hot spots and blockage in pumps and piping.

1.4.3 Sludge

Sludge is the name given to the residue which forms when inorganic and organic particles are separated when a liquid is cleaned. The following mechanical and chemical cleaning methods are used:

Sedimentation-settling: the particles fall to the bottom in a separator under their own weight;

Flotation: the particles are made to float by the injection of small air bubbles;

Centrifuging: the particles are separated by centrifugal force;

Filtering: the liquid is passed through a filter which allows the liquid to pass through whilst trapping the particles, the filter may consist of a grill, porous material or one or more beds of filtering media;

Precipitation-flocculation: by the addition of chemical reagents, the particles form insoluble combinations -flocs- which can be separated by sedimentation or flotation.

When pumping, the DS content-value - pumpability- of the sludge, is of interest as is also the size and hardness of the particles. Since sludge always occurs in the treatment of water and in industrial processes, there can be no general definition. The values obtained in the treatment of sludge in municipal sewage works can be used however as a guide.

DS-content after various treatment stages:

- Before thickening DS-content %

Chemical sludge (scum) after precipitation or flotation	0.5 to 1
Mechanical sludge, sedimentation or flotation	2 to 3

- After thickening

Mechanically agitated sludge	6 to 10
Activated sludge	2 to 3
Bio-filtration	4 to 8
Mechanical and activated sludge	5 to 8

- After straining in the centrifuge, filter screen press, vacuum filtering etc.   > 30

In the bottom, or compression zone, of sedimentation basins, sand traps, oil tanks and other containers of stationary liquids, the bottom layer, during sedimentation is subjected to mechanical pressure due to the weight of sludge lying above it, which causes liquid to be compressed out of the bottom layer thus increasing its density. Other names given to sludge are sediment and slurry.

1.4.4 Pulps

General

Pulps are used in the cellulose and paper industries. It is a name for fibre suspensions in water. Pulps belong to the non-Newtonian liquids; group 1, the time-independent, plastic liquids.

The density of a pulp can be calculated from equations 1.13 and 1.17 with reservation for possible air content. Often the affects of the fibres and air cancel each other out. The density then coincides with that of the water.

Pulps exhibit widely varying pH values, depending on the production method and the bleaching process. Pumps made of acid-resistant steel are usually needed.

Pulp quality

The pumping characteristics of pulp depends, in the first instance, upon the raw material used, the additives and the method of production. Examples of raw materials used are coniferous (soft) wood (fibre length 3 to 4 mm), deciduous (hard) wood (fibre length 1 to 1.5 mm) and rag (fibre length 25 to 30 mm). Basically the methods of production can be divided into chemical and mechanical methods. The chemical method is most common. Sulphate and sulphite pulps are produced both bleached and unbleached. The most important pulp to be produced mechanically is groundwood pulp.

Aside from the concentration - the DS content - the flow characteristics of the pulp are also effected by the fibre ratio - length/diameter and the degree of pulverisation.

Air content in pulp

Pulp differs from many other suspensions in that it consists of three phases; water, solid-fibre and air.

Air in the pulp occurs either in the form of bubbles or in a combination. Air occurs as bubbles either in a free form or attached to the fibre. Air in the combined form occurs in solution in the water or absorbed in the fibre. The content of air in the pulp depends upon; quality, concentration, additives, pulverisation, temperature and time; and also the pulp handling process. Groundwood pulp contains more air than sulphate or sulphite pulp. Better sizing increases the air content considerably as does pulverising.

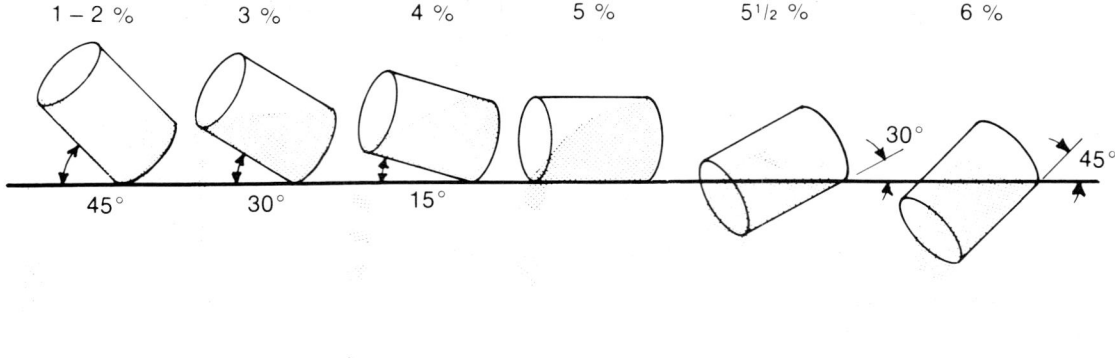

Figure 1.30 Illustration of fluidity of paper pulp. The percentage figures given indicate the DS-content

An increase in temperature increases the content of free air whilst the ability of air to dissolve in water decreases. Generally speaking, the air content of pulp reduces in storage. The air content of pulp increases quickly to a certain level specific to each type of pulp when mixed with air. This happens typically when the pulp is allowed to fall freely into a vessel or cistern. Air can also be taken up by the pulp at shaft seals in pumps if there is a vacuum.

Pulp properties which complicate pumping.
- Content of solid particles: The pulp fibres build up a network, the density of which increases with increased concentration. It is relatively difficult to get a high concentration pulp moving, since when high energy pulses are transmitted locally in the network, it is probable that the pulp will break up and movement will take place only locally.
- Air content: Air in the form of bubbles in the pulp is very inconvenient from the point of view of pumping. An air content of only 1% to 2% is enough to change the pumping characteristics.
- Pressure drop: For flow at low speeds the pressure drop for pulp is much higher than for water. In general, pulp which is hot or which contains additives is easier to pump than pure, cold pulp. It can be somewhat of a problem to get pulp moving again in a pipe system after a shut-down, especially in the case of high-concentration pulp. This is because the starting resistance, "stiction", is greater than the pressure drop at low flow speeds. Dilution with water in the pump can assist restarting.
- Blockage characteristics: Flocculation of the pulp, sticks, twigs and synthetic fibre, can cause blockage in the pump. A total blockage of both pump and pipe is not an unknown occurrence.
- The tendency to thicken in reducing sections: Rapidly reducing conical sections on the suction side of the pump, e.g. reduction in the suction pipe from 400 mm to 150 mm, can cause flocculation at concentrations of only 3 to 4%.

1.5 Liquid gas mixtures

Interest in pumping liquid-gas mixtures centres around:
    small oil and gas wells,
    boiler feed pumps,
    aircraft fuel systems,
    chemical, pharmaceutical and food processing.

Small oil and gas wells, especially off-shore, where it would not be economic to lay two pipelines; one for crude oil and one for natural gas or fit pumps and compressors with associated drivers and controls. Boiler feed pumps are generally high speed multi-stage centrifugal pumps operating with small NPSH margins. Process upsets initiating cavitation, can cause severe costly damage, resulting in unscheduled shutdowns. Aircraft fuel systems can be prone to air entrainment due to manoeuvring. Loss of fuel pressure and reduced flow rates can be disastrous. Chemical, pharmaceutical and food processing operations saw the handling of gas and liquids together as a process simplification resulting in cost savings.

As with all pumping processes the pump manufacturer requires all the operating data to be able to assess the application. Detailed information regarding the nature of the gas, or liquid vapour, is essential. Also, details of the gas flow; steady state controlled percentage or variations from bubbles to slugs.

Users hoped it would be possible to use centrifugal pumps to handle the mixture without pre-treatment. Centrifugal pumps were the first choice equipment because of reliability and familiarity. However tests indicated that 2 to 4% gas content seriously effected pump performance and increasing gas content to 15% caused pumps to lose their prime. Development since 1989 has not significantly improved gas handling characteristics. Current configurations consist of an efficient in-line separator feeding liquid to a centrifugal pump and the gas to a compressor; both discharging into a single pipeline. This arrangement is reported to cope with 40% gas by volume. Flow controls on both machines are necessary for gas content variations.

Progressive cavity pumps and vane pumps are both reported as running satisfactorily with liquid gas mixtures; neither suitable for 100% gas. It may be possible for diaphragm pumps or hydraulically operated piston pumps to be useful in this service.

The greatest success to date is with specially adapted geared twin screw pumps. Pumps are available which can handle gas volume fraction from 0 to 97%; dry running on 100% gas is not yet possible. Pumps are available for differential pressures of 48 bar; 69 bar in certain applications. Discharge pressures up to 153 barg are possible. Mechanical seal problems are avoided by fitting double seals with a lubricating/cooling barrier liquid.

Positive displacement pumps have the advantage in being able to cope with varying differential pressures, due to downstream friction loss changes, without substantial flow changes.

1.6 Table of liquid properties

1.6.1 General

The liquid table gives necessary data for pump applications. The characteristics of liquids are not given to that degree of accuracy which, in special cases, may be required in order to specify other process equipment. Note that the commonest liquids; water, oils and certain suspensions; are dealt with in more detail in dedicated sections, 1.2, 1.3 and 1.4.

Compressibility and wave speed are not included in the liquid tables. These characteristics are stated for a number of liquids in Section 1.1.5.

The liquid table contains the following data:

- Substance: the liquid properties are arranged in alphabetical order according to the name of the substance. The trade names with synonyms are given in Figure 1.31. The letter D indicates a dangerous (flammable or otherwise hazardous) liquid within the terms of the Petroleum (Inflammable liquids and Other Dangerous Substances) Order 1947 - Amended 1968 SI 570. The letter T indicates a high level of toxicity.
- Chemical formula: for identification of substance. In practice a substance can unfortunately not be assigned to a single formula because of contamination. Liquids containing contaminants which are insoluble or solid have the worst pumping characteristics, often causing troublesome wear on pumps and shaft seals.
- Concentration in H_2O: the concentration of the substance in water expressed as percentage by weight.
- Solubility in H_2O: attention is drawn in the table to various diagrams referring to the solubility of the substance in water. The solubility is expressed as a concentration in percentage by weight at varying temperatures. When pumping, precipitation often occurs when the concentration exceeds that stated by the solubility curve for the given temperature. Conversely, there will be precipitation if the temperature drops below the stated solubility temperature.
- Viscosity: is stated in mm^2/s (cSt) at +20 °C. Viscosity below $5 mm^2/s$ indicates that the liquid is easy flowing like water. Viscosity below $1 mm^2/s$ indicates that the liquid has poor lubrication qualities and poor frictional dampening properties for acoustic and pressure waves. The liquid table also refers to diagrams showing dependence of viscosity on temperature. The viscosity will also be used for the determination of frictional losses in the pipework. This depends primarily on whether the flow is laminar or turbulent, see Chapter 2. In the laminar region, the pipe frictional losses are proportional to the viscosity, whereas its effect can be neglected in the turbulent region for the internal roughnesses of pipes which occur in practice.

  Certain liquids are non-Newtonian and require special care when determining the equivalent (apparent) viscosity.
- Vapour pressure (absolute): is stated in kiloPascals (kPa) at +20 °C. See curves in Figure 1.34 which show the temperature dependence of vapour pressure. Note that boiling point at atmospheric pressure (101.325 kPa 760 mm Hg) gives another point on the vapour pressure curve. The curves in Figure 1.34 are set out schematically and may deviate from the exact values. This is especially applicable to aqueous solutions - curve 13. All references to this curve apply to chemicals in solution in water. Liquid pressure near to the vapour pressure at the actual liquid temperature can cause cavitation in a pump or pipe.
- Density: is stated in kg/m^3 at +25 °C. Reference is made to various figures for other temperatures and concentrations. The density is used among other things for conversion of pressure to head and for the calculation of the power required for the pump.
- pH region: is an expression of acidity or alkalinity and is grouped within pH regions 0 - 4, 4 - 6, 6 - 9 and 9 - 14. This group classification assists in the choice of pump material, the pH regions coinciding with commonly used pump material resistivity to corrosion.
- Melting point: note that many contaminants gradually begin to degrade at temperatures immediately above the melting point which can have a harmful effect on the function of the pump.
- Fire hazard class is stated in accordance with section 1.1.7.

NB!

Where there is no information in the liquid table, this means that the actual liquid property is not known and it does not mean that the information omitted is of no general interest in pumping.

TRADE NAMES AND SYNONYMS

Designation	Entered in table as
Acetic ether	Ethylacetate
Alcohol	Ethyl alcohol
Alum	Aluminium sulphate
Ammonium hydroxide	Ammoniac
Benzol	Benzene
Black liquor	Sulphate liquor
Bromite	Silver bromide
Butanol	Butyl alcohol
Carbitol	Ethyldiglycol
Caustic soda	Sodium hydroxide
Cellosolve	Sodium chloride
Cellosolve acetate	Ethylglycolacetate
Chlorhydric acid	Hydrochloric acid
Chloroacetic acid	Monochloroacetic acid
Chlorobenzene	Monochlorobenzene
Chrome alum	Potassium chrome sulphate
Chrome oxide	Chromic acid
Cooking salt	Sodium chloride
Copper vitriol	Copper sulphate
Cyancalium	Potassium cyanide
Dichloromethane	Methylene chloride
Diethylether	Ether
Diphonylether	Diphenyl oxide
Ethanol	Ethyl alcohol
Ferrichloride	Iron (III) chloride
Ferrosulphate	Iron (II) sulphate
Fixerbad	Sodium tiosulphate
Formaline	Formaldehyde
Fumaric acid	Malein acid
Glauber salt	Sodium sulphate
Hartshorn salt	Ammonium carbonate
Hydrofluoric acid	Hydrofluoric acid
Methanol	Methyl alcohol
Methylaldehyde	Formaldehyde
Naphthalene chloride	Chloronaphthalone
Natron saltpetre	Sodium nitrate
Oleic acid	Fatty acid
Palmitic acid	Fatty acid
Perhydrol	Hydrogen peroxide
Phenol	Carbolic acid
Potash	Potassium carbonate
Potash nitrate	Potassium nitrate
Potash salts	Potassium hydroxide
Pulping liquor	Calcium bisulphite liquor
Radical vinegar	Conc. Acetic acid
Salammoniac	Ammonium chloride
Saltlake	Sodium chloride
Slaked lime	Calcium hydroxide
Soda	Sodium carbonate
Soda lye	Sodium hydroxide
Sodium borate	Borax
Styrol	Styrene
Sublimate	Chloride of mercury
Sulphite liquor	Calcium bisulphite
Tannin	Tannic acid
Trichloroethanyl	Chloral
Urea	Carbamide
Water glass	Sodium silicate

Figure 1.31 Trade names and synonyms

1.6.2 Liquid Table

Substance		Chemical formula	Viscosity mm²/s at 20°C	Vapour pressure kPa at 20°C	Vapour pressure at various temps. See fig 1.34 curve no	Conc. % in H₂O	Density at 25°C	Density at various temps. See Fig no	Density at various conc. See Fig no	pH region	Boiling point °C at 101.3 kPa (760 mm Hg)	Melting point °C	Fire hazard class - See Section 1.1.7	Solubility in H₂O at various temps. See Fig no	Comments
Abietinic acid		C₂₀H₃₀O₂					780/20°C					172			
Acetaldehyde	D	CH₃CHO	<1	100	10		1050	1.35		0-4	21	-125	1		
Acetic acid	D	CH₃COOH	1.2	3.5	19	5-100	1050	1.35	1.37	0-4	119		2b		E-mod and speed of sound - Figure 1.6
Acetic acid anhydride	D	(CH₃CO)₂O			19	50	1060			0-4	136	-73	1		
Acetone	D	CH₃COCH₃	<1	30	19		1080	1.35		0-4	57	-95	1		
Acetyl chloride		CH₃COCl			20		790				51	-112	2b		
Alkyd solution in paint naphtha			Fig 1.32	0.9	14		1100				150		1		Non-Newtonian - Figure 1.32
Alkyd solution in xylene	D		Fig 1.32				900				139		2a		Non-Newtonian - Figure 1.32
Alkyd tixotrope sol. in paint naphta			Fig 1.32				1000				150		2b		Non-Newtonian - Figure 1.32
Allyl alcohol	D	CH₂CHCH₂OH		2.3			900				97		1		
Ally chloride	D	CH₂CHCH₂CL					850				45				
Aluminium		Al					940								
Aluminium chloride		AlCl₃	<1				2700					660			
Aluminium nitrate		Al(NO₃)₃	<5	2.2	13	5	1030		1.38	0-4				1.41	
Aluminium sulphate		Al₂(SO₄)₃		2.2		10	1090			0-4				1.41	
Ammonium alum		NH₄Al(SO₄)₂		2.2	13	10	1050		1.38	0-4				1.41	
Ammonium bromide		NH₄Br				10	1110		1.36	0-4					
Ammonium carbonate		(NH₄)₂CO₃			13	5	1050		1.38	4-9					
Ammonium chloride		NH₄Cl		2.2	13	10	1030		1.38	4-6					
				1.8	13	10-50	1070			4-6					
Ammonium fluoride		NH₄F		2.3		26	1030		1.38	4-6					
Ammonium hydroxide		NH₄OH		2.2	1	6	1060	1.35		9-14					
			<5	9.3		14	960			9-14					
			<5	34.5		10	910			9-14					
Ammonium nitrate		NH₄NO₃	<5	2.2	13	25	1040			6-9					
				0.5	13	10-20	1230		1.38	6-9					
					13	60	1040			6-9					
Ammonium oxalate		(COONH₄)₂H₂O			13	30	1040			6-9					
Ammonium perchlorate		NH₄ClO₄				10	1060			4-6					
Ammonium persulphate		(NH₄)₂S₂O₈		2.2		0-conc	1060		1.38	4-6				1.41	
Ammonium sulphate		(NH₄)₂SO₄		0.7		10	1280			4-6					
Amyl acetate		CH₃COOC₅H₁₁	<5	0.7	20	50	880				149	-78	2b		
Amyl alcohol		C₅H₁₁OH	6				810				133	-78	2b		
Amyl chloride		C₅H₁₁Cl					890				109		1		
Amyl polycaptan		C₅H₁₁SH	6	0.1	24		850				100		3		
Aniline	D	C₆H₅NH₂				5	1020			0-4	184	-6	3		
Aniline hydrochloride		C₆H₅NH₂·HCl				20	1090			0-4					
Antimony		Sb					1250				226	630			
Anthracene oil		C₁₄H₁₀					2500		1.37						
Arsenic acid		H₃AsO₄				10	1070			0-4			2b		
Asphalt solution in naphta			1500-7500	25			900				150		1		Non-Newtonian Figures - 1.6, 1.23, 1.24
Aviation fuel			<1				720				40				
Barium chloride		BaCl₂		2.2	13	10	1090		1.39	6-9					
Beer				1.8		26	1280			4-6					
Benzene	T	C₆H₆	<1	13	13		880	1.35			80	5.5	1	1.41	
Benzene sulphonic acid		C₆H₅SO₃H			16		1010				100	525			
Benzoic acid		C₆H₅COOH					1270				250	122			

32

1.6.2 Liquid Table

Substance		Chemical formula	Viscosity mm²/s at 20 °C	Vapour pressure kPa at 20 °C	Vapour pressure at various temps. See fig 1.34 curve no	Conc. % in H₂O	Density at 25 °C	Density at various temps. See Fig no	Density at various conc. See Fig no	pH/ region	Boiling point °C at 101.3 kPa (760 mm Hg)	Melting point °C	Fire hazard class - See Section 1.1.7	Solubility in H₂O at various temps. See Fig no	Comments
Benzyl alcohol		C₇H₈O	5	<0.1			1050	1.35			205	70	3		
Biphenyl		C₆H₅C₆H₅					1990				256	741			
Borax		Na₂B₄O₇		2.2	13	3.5	2370			4 - 6					
					13	10	1030			4 - 6					
Borax acid		H₃BO₃			14	50									
Bromine		Br₂	<1/0 °C	12			3120/20 °C	1.35			59	-7.2			See Section 1.3 and Figure 1.6
Burner oils	D		Fig 1.18				840/20 °C						3		See Section 1.3 and Figure 1.6
	E		Fig 1.18				940/20 °C						3		See Section 1.3 and Figure 1.6
	F		Fig 1.18				950/20 °C						3		
Butane		C₄H₁₀					600/0 °C				- 0.5	138			
Butyl acetate	D	CH₃COOC₄H₉	<1	2	8	50	880	1.35			125	-77	3		
Butyl alcohol	D	C₄H₉OH	<5	0.9	19		810	1.35			118	-90	3		
Butyl glycol		C₆H₁₄O₂	<5	<0.1	18		900	1.35			171		2a		
Butyric acid		C₃H₇-COOH			21		960	1.35			164	6.5	3		
Calcium bisulphate		Ca(HSO₃)₂				25	1040			4 - 6					
Calcium chloride		CaCl₂			13		1060	1.38	1.36						
Calcium hydroxide		Ca(OH)₂			13	5			1.38	9 - 14					
Calcium nitrate		Ca(NO₃)₂			13		990								
Camphor		C₁₀H₁₆O					1320				209	176			
Carbamide		(H₂N)CO(NH₂)				50	1115					132			
Carbolic acid	D	C₆H₅OH	11		13		1070	1.35		4 - 6	43		1		
Carbon disulphide		CS₂	<1	48	12		1262	1.35			46	-112			
Carbon tetrachloride	T	CCl₄	<1	20	15		1600	1.35			77	-23			
Castor oil			Fig 1.23	(0)			960								
Chloral		CCl₃CHO					1520				98	-57			
Chloramine		CH₃C₆H₄SO₂NCl											183			
Chloride of lime		Ca(ClO)₂ 4H₂O				0.1	2350								
Chloride of mercury	T	HgCl₂			13		1000			4 - 6					E-mod and speed of sound - Figure 1.6
	D						7150								
Chlorine	D	Cl₂			3		1200	1.35			- 35	-103			
Chlornaphthalene		C₁₀H₇-Cl	2.5	31			1490				263	-25			
Chloroform	D	CHCl₃	<1		14		1280	1.35			61	2			
Chlorosulphonic acid		HClSO₃	<5		21		2700				158	-80			
Chromic acid	D	H₂CrO₄	<5			10 - 50				0 - 4					
Citric acid		C₆H₈O₇·H₂O	<5		13		1550			0 - 4		153			
Copper (II) chloride		CuCl₂2H₂O				10	3390			0 - 4		620			
							1090								
Copper cyanide		CuCN			13		2920			0 - 4				1.41	
Copper nitrate		Cu(NO₃)₂6H₂O			13		2070					114		1.41	
		Cu(NO₃)₂3H₂O					2320								
							1090								
Copper sulphate	D	CuSO₄		2.2	13	10	3610			0 - 4				1.41	
					13	18	1160			0 - 4					
							1210								
Corn oil			Fig 1.23				1040								
Cresol	D	C₆H₄(CH₃)OH	71	0.3	25		950	1.35		0 - 4	191	30	1		
Cyclohexanol	D	C₆H₅OH	<5	0.7			950				161		2b		
Cyclohexanone	D	C₆H₁₀O									156				
Diacetone alcohol			150	0.1	23		940				170	-43	2b		
Dichloroacetic acid	D	CHCl₂-COOH	<5				1550	1.35		0 - 4	192	-11			
Dichloroethane	D	CH₂Cl-CH₂Cl		31.2	14		1180	1.35			59		1		
Dichloroethylene	D	C₂H₂Cl₂		32			1250	1.35			60	-81	1		

33

1.6.2 Liquid Table

Substance		Chemical formula	Viscosity mm²/s at 20 °C	Vapour pressure kPa at 20 °C	Vapour pressure kPa at various temps. See fig 1.34 curve no	Conc. % in H₂O	Density at 25 °C	Density at various temps. See Fig no	Density at various conc. See Fig no	pH/ region	Boiling point °C at 101.3 kPa (760 mm Hg)	Melting point °C	Fire hazard class- See Section 1.1.7	Solubility in H₂O at various temps. See Fig no	Comments
Dichlorohydrin		(CH₂Cl₂)₂CHOH					1360			0 - 4	176				
Dichlorophenoxyacetic acid		Cl₂C₆H₃O CH₂COOH													
Diesel fuel			Fig 1.18				830				210		3		
Diethyl carbonate		(C₂H₅O)₂CO	82		19		980				128	- 48	2b		
Dioctylphthalate							980						1		
Dioxan	D	O₂(CH₂)₄	2	3.9			1040	1.35			101	12	3		
Dipentene							850				180	- 70	3		
Diphenyloxide		C₆H₅OC₆H₄	4	0.03			1070				253	27	3		
Engine oil SAE 5W-50			Fig 1.19				900/20 °C								See section 1.3 and Figure 1.6
Ether	D	C₄H₁₀O	<1	74			710	1.35			35		3		
Etherdiethylene		C₂H₂O		150			870				11	- 113	1		
Ethyl acetate	D	CH₃COOC₂H₅	<1	13.8	16		900	1.35			77		1		
Ethyl alcohol		C₂H₅OH	<5	8			790	1.35			78		1		E-mod and speed of sound - Figure 1.6
Ethyldiglycol		HO(CH₂)₄ O₂C₂H₅	<5	0.02			990				202		3		
Ethyl chloride		C₂H₅Cl	<1	133	9		890	1.35			12	- 139	1		
Ethyl glycol	D	C₄H₁₀O₂	2.3	0.5			930				135		2b		E-mod and speed of sound - Figure 1.6
Ethyl glycol acetate		C₆H₃O₃	<5	0.2			970				150		2b		
Ethylene diamine	D	H₂NCH₂ CH₂NH₂	1.5	1.4			900			9 - 14	117	11			
Ethylene glycol	D	C₂H₆O₂	18	<0.13	24		1110	1.35			198		3		
Fluorine	T	F									- 188	- 223			
Formaldehyde	D	HCHO	1.5	5.3	5		820		1.35		- 20				
Formic acid	D	HCOOH			18	20 - 50	1220	1.35			101				
Freon		CF₂Cl₂					1470				- 29	- 155			
Fruit juices					13										
Furfural	D	C₅H₄O₂	1.4	0.3	22		1160	1.35			162		3		
Furfurole	D	C₅H₆O₂		0.13	23		1160	1.35			171		3		
Gallic acid		C₆H₂(OH)₃COOH					1700								
Glycerine		CH₂OHCHOH CH₂OH	1200				1260	1.35		0 - 4	290				See Figures 1.23, 1.24
Ground nut oil			Fig 1.23				910								
Hexane		C₆H₁₄	<1	30	14		660		1.35	0 - 4	69	- 90	1		
Hydrobromic acid		HBr · H₂O	<1			50	1780		1.37	0 - 4		- 89			
Hydrochloric acid	D	HCl		0.6	13	20	1520		1.37	0 - 4					
				1.3	13	36	1100			0 - 4	110				
Hydrofluoric acid	T	HF		1.3		1	1180	1.35	1.37	0 - 4	20	- 83			
						10	990								
				2.2		40	1005 1030								
				1		75	1130 1240								
Hydrogen		H₂			21		1460	1.35			- 253	- 259			
Hydrogen peroxide		H₂O₂		2.2	21	10	1040				151	- 89			
Hydrogen sulphide	D	H₂S					950/-50 °C	1.35			- 60	- 86			
Hydriodic acid		HI·H₂O					1700				127				
Industrial oils as per ISO - Cutting oil			Fig 1.22	(0)			900/20 °C						3		See Section 1.3 and Figure 1.6
Gear oil			Fig 1.20	(0)			905/20 °C						3		

34

1.6.2 Liquid Table

Substance		Chemical formula	Viscosity mm²/s at 20 °C	Vapour pressure kPa at 20 °C	Vapour pressure at various temps. See fig 1.34 curve no	Conc. % in H₂O	Density at 25 °C	Density at various temps. See Fig no	Density at various conc. See Fig no	pH region	Boiling point °C at 101.3 kPa (760 mm Hg)	Melting point °C	Fire hazard class See Section 1.1.7	Solubility in H₂O at various temps. See Fig no	Comments
Hydraulic oils			Fig 1.22	(0)			880/20 °C								
Turbine oil			Fig 1.22	(0)			885/20 °C								
Iodine	D	I₂					4940				184	114	3		
Iodoform		CHI₃					4010					120	3		
Iron (II) chloride		FeCl₂	1	2.2	13	10 - 50	1085	1.35		0 - 4				1.41	
Iron (III) chloride		FeCl₃	15	0.7	13	50	1550			0 - 4					
Iron nitrate		Fe(NO₃)₃		2.2		10	1080			0 - 4					
				1.8		25	1230			0 - 4					
Iron (II) sulphate		FeSO₄·7H₂O			13		1050								
Iron (III) sulphate		Fe₂(SO₄)₃		1.2	13	30 - 50	800			0 - 4					
Isobutylalcohol		(CH₃)₂CHCH₂OH	<5		16		780				108	-89	2a		
Isopropylalcohol		CH₃CHOHCH₃	<5		17		1190				82		1		
Isopropylnitrate											100				
Kerosene			<5	0.1			780				170				See Figures 1.23, 1.24
Lacquer			200 - 500		10		900				150		2b		Non-Newtonian, see Section 1.4
Lactic acid		H₆C₃O₃				1020				6 - 9			2b		Non-Newtonian, see Section 1.4
Latex			1000 - 4000			50	1050				100				
Lead		Pb		(0)			11350					327			
Lead acetate		(CH₃COO)₂Pb·3H₂O				10				6 - 9				1.41	
Lead nitrate		Pb(NO₃)₂				30				0 - 4				1.41	
Linseed oil			20 - 300			930	970								
Liquid resin			Fig 1.23				910/20 °C								See Section 1.3 - Figure 1.6
Lubricating oil SAE 75W-140			Fig 1.20												
Magnesium carbonate		MgCO₃				10 - 50	2960			4 - 6					
Magnesium chloride		MgCl₂·6H₂O		2.2	13	10	2320			4 - 6		708			
				2.2	13	15	1080			4 - 6					
				2.2	13	25	1130			4 - 6					
				2.2	13	30	1150			4 - 6					
							1280								
Magnesuim sulphate		MgSO₄·7H₂O		2.2	13	10	1680			4 - 6					
				2.2	13	20	1100			4 - 6					
Maleic		(HCCOOH)₂					1300			0 - 4		130			
Manganese chloride		MnCl₂			13	50	1590			4 - 6		650			
Manganese chloride hydrate		MnCl₂·4H₂O				10	1300								
Manganese sulphate		MnSO₄·7H₂O				30	2980			4 - 6		58			
							1060								
							2010								
							2090								
							1220								
Mercury	T	Hg					13600				357	-39			E-mod and speed of sound - Figure 1.6
Methyl acetate		CH₃CO₂CH₃	<1	13	12		930	1.35			58		1		
Methyl alcohol	D	CH₃OH	<1	72.4	14		790	1.35			65		1		E-mod and speed of sound - Figure 1.6
Methylene chloride	D	CH₂Cl₂	<1	15	11		1320	1.35			40		1		
Methylethylketone		C₄H₈O			16		810				80		2b		
Methyl glycol		C₃H₈O₂	1.6	1.7			970				125		1		
Methylchloride		CH₃Cl					1790				-24				
Milk, fresh					13		1020			4 - 6					
Milk, sour					13		1020								Non-Newtonian, see Section 1.4
Molasses			100 - 50000		20		1110	1.35			132		2a		
Monochlorobenzene		C₆H₅Cl	1	1.6	24		1410			0 - 4	189	63			
Monochloracetic acid		CH₂ClCO₂H	2.2 (50 °C)		13										Non-Newtonian, see Section 1.4
Mustard			approx 5000												

1.6.2 Liquid Table

Substance		Chemical formula	Viscosity mm²/s at 20 °C	Vapour pressure kPa at 20 °C	Vapour pressure at various temps. See fig 1.34 curve no	Conc. % in H₂O	Density at 25 °C	Density at various temps. See Fig no	Density at various conc. See Fig no	pH/region	Boiling point °C at 101.3 kPa (760 mm Hg)	Melting point °C	Fire hazard class- See Section 1.1.7	Solubility in H₂O at various temps. See Fig no	Comments
Naphthalene		C₁₀H₈			27		1150				210	80			
Naphthalene sulphonic acid		C₁₀H₇·SO₃H					1450					102			
Nickel nitrate		Ni(NO₃)₂ 6H₂O			13	10	1030			0 - 4		57			
Nickel sulphate		NiSO₄ 7H₂O			13	10	2050			4 - 6	137				
					13	10	1050								
					13	10	1950								
Nitric acid	D	HNO₃			13	10	1060			4 - 6					
					16	1	1500		1.37	0 - 4	83	-42			
				2	16	5	1004			0 - 4					
				1.5	16	10	1030			0 - 4					
					16	20	1050			0 - 4					
					16	40	1120			0 - 4					
					16	65	1250			0 - 4					
					16	80	1480			0 - 4					
Nitro benzine	D	C₆H₅NO₂	2	0.2			1452	1.35			211				
Nonane		C₉H₂₀	1	0.7			1200	1.35			151	-54	3		
							720						2b		
Octane		C₈H₁₈	<1	2			700	1.35			126		1		
Octanol		C₆H₁₃CHOH CH₃	11				830			0 - 4	194	-15			
Oleum		H₂SO₄ + 13%SO₃					1910								
Olive oil			Fig 1.23				910								
Oxalic acid		C₂H₂O₄ 2H₂O		(0)	13	10 - 50	1650			0 - 4		101			
					13	0.5	1000			0 - 4					
					13	10	1020			0 - 4					
					13	15	1030			0 - 4					
Ozone		O₃									-112	-192			
Paraffin							770				240	45 - 55			
Pentane		C₅H₁₂	<1	67	11		620				36	-130	1		
Peracetic acid		CH₃COOOH					1230				105				
Petrol			<1				730				40				
Phenoldicarbonate							1130					80			
Phenolphthalein							1280				312	261			
Phosphoric acid	D	C₂₀H₁₄O₄		3	13	10	1840			0 - 4					
		H₃PO₄		1.3	13	20	1050			0 - 4					
					13	40	1120			0 - 4					
					13	70	1370			0 - 4					
					13	85	1530			0 - 4	181				
Phthalic acid		C₆H₄(CO₂H)₂					1690					206			
Phthalic anhydride		C₆H₄(CO)₂O					1600					130			
Potassium bi-carbonate		KHCO₃			13	10	1530			6 - 9					
Potassium bi-chromate		K₂Cr₂O₇			13	25	2170			0 - 4		214			
Potassium bi-sulphate		KHSO₄					2680								
							1050								
Potassium bromide		KBr			13	5	2320		1.39	0 - 4	1435	730			
Potassium carbonate		K₂CO₃		2.2			1035			9 - 14		891		1.40	
							2750								
							2420								
Potassium chlorate	D	KClO₃			13	20	1190			6 - 9		356		1.40	
Potassium chloride		KCl			13	20	2320		1.35	6 - 9		776		1.40	
						10 - 30	1980								
							1130								
Potassium chromate	D	K₂CrO₄					2730					968		1.40	

1.6.2 Liquid Table

Substance		Chemical formula	Viscosity mm²/s at 20°C	Vapour pressure kPa at 20°C	Vapour pressure at various temps. See fig 1.34 curve no	Conc. % in H₂O	Density at 25°C	Density at various temps. See Fig no	Density at various conc. See Fig no	pH/region	Boiling point °C at 101.3 kPa (760 mm Hg)	Melting point °C	Fire hazard class - See Section 1.1.7	Solubility in H₂O at various temps. See Fig no	Comments
Potassium chrome sulphate	D	KCr(SO₄) 12H₂O			13		1830								
Potassium cyanate	D	KOCN					2060								
Potassium cyanide	T	KCN			13		1520			9 - 14	635				
Potassium fluoride	T	KF 2H₂O			13	5 - 10	2450								
Potassium hydroxide	D	KOH		2.2	13	10 - 90	2040		1.36	9 - 14	156	41		1.40	
				2.2	13	30	1290			9 - 14	360				
				2.2	13	50	1510			9 - 14					
Potassium iodide		KI					3130								
Potassium nitrate		KNO₃		2.2		10	2110		1.39						
Potassium oxalate		K₂C₂O₄ H₂O				15	1080	1.38							
Potassium perchlorate		KClO₄					2130			0 - 4					
Potassium permanganate		KMnO₄					1170								
				2.2	13	20	2520			0 - 4					
Potassium persulphate		K₂S₂O₈					2700								
							1040			0 - 4					
						1 - conc	2480			0 - 4					
Potassium silicate		K₂SiO₃			13		2660					976			
Potassium sulphate		K₂SO₄			2		580	1.35				-190			
Propane		C₃H₈	2.8				800	1.35			-42	-126	1		
Propanol		C₃H₇OH			21		990				97	-21	1		
Propionic acid		C₂H₅COOH				25	1030				141				
Pyridine	D	C₅H₅N		3.6	19		980				116	-42			
Pryogallol	D	C₆H₃(OH)₃					1460				309				
							1030								
Quinine		C₂₀H₂₄N₂O₂					1440					57			
Salicylic acid		C₆H₄OHCO₂H	1			conc	1000			0 - 4	211	159			
Seawater - 4% NaCl				2.3	13		1020			6 - 9	100				See Section 1.2.4 - Figure 1.6
Sebacic acid		C₈H₁₆(COOH)₂					915				240	55			
Silver bromide		AgBr					6470				432				
Silver chloride		AgCl					5560				455				
Silver nitrate	D	AgNO₃			13		4350		1.39		212				
Size			500 - 3000							6 - 9	100				Non-Newtonian, see Section 1.4
Sodium acetate		NaC₂H₃O₂ 3H₂O			13	7	1450			6 - 9	123	58		1.40	
Sodium arsenate	T	Na₂HAsO₄ 7H₂O			13		1050								
Sodium bi-carbonate		NaHCO₃			13	5	1880			9 - 14					
							2160								
Sodium bi-sulphate		NaHSO₄ H₂O			13	10	1040			0 - 4		58			
Sodium bi-sulphite		NaHSO₃			13	10	2100								
							1080								
Sodium bromide		NaBr 2H₂O		2.2	13	10	1480			4 - 6					
							1100								
Sodium carbonate		Na₂CO₃			13	5 - 10	2180			9 - 14		850		1.40	
				2.2	13	10 - 50	2530		1.39	9 - 14				1.40	
Sodium chlorate		NaClO₃		2.2	13	10	1150			4 - 6		250		1.40	
						28	2480								
Sodium chloride		NaCl			13		1410		1.39		1413	801		1.40	
				2.2	13	1 - conc	2170			6 - 9					

1.6.2 Liquid Table

Substance		Chemical formula	Viscosity mm²/s at 20 °C	Vapour pressure kPa at 20 °C	Vapour pressure at various temps. See fig 1.34 curve no	Conc. % in H₂O	Density at 25 °C	Density at various temps. See Fig no	Density at various conc. See Fig no	pH/ region	Boiling point °C at 101.3 kPa (760 mm Hg)	Melting point °C	Fire hazard class- See Section 1.1.7	Solubility in H₂O at various temps. See Fig no	Comments
Sodium di-chromate		Na₂Cr₂O₇·2H₂O					2520					988			
Sodium fluoride	T	NaF			13		2550			4-6				1.40	
Sodium fluorsilicate		Na₂SiF₆				5	1050							1.40	
Sodium hydrogen difluoride		NaHF₂					2680								
Sodium hydrosulphate		NaHSO₄					2080								
		NaHSO₄·H₂O					2440								
Sodium hydroxide	D	NaOH		2.2	13	10-70	2100				1390	58		1.40	
				2.2	13	30	2130		1.36	9-14		318			
				2.2	13	50	1330			9-14					
				2.3	13	5	1530			9-14					
Sodium hypochlorite	D	NaOCl			13		1020			6-9					
Sodium nitrate		NaNO₃					2260		1.38	6-9		307		1.40	
Sodium perchlorate		NaClO₄·H₂O				4	1030			4-6				1.40	
Sodium peroxide		Na₂O₂				10	2020			6-9		130			
							1070								
Sodium phosphate primary		NaH₂PO₄·12H₂O				10	2810					60			
							1110								
Sodium phosphate secondary		Na₂HPO₄·12H₂O		2.2		10	1910			0-4					
							1070								
Sodium phosphate tertiary		Na₃PO₄·12H₂O		2.3		50	1520								
							1250								
				2.2		conc	1620			9-14					
Sodium silicate		Na₂SiO₃			13		1110		1.38			1088		1.40	
Sodium sulphate		Na₂SO₄·10H₂O			13	1	2400			4-6		32			
				2.2		5	1460			4-6					
				2.2			1000								
Sodium sulphide		Na₂S·9H₂O		2.2	13	20	1020			6-9		40			
							1420								
Sodium sulphite		Na₂SO₃·7H₂O			13		1070			4-6					
Sodium thiosulphate		Na₂S₂O₃·5H₂O		2.2	13	1-conc	1530					-70			
					13		1730								
Soya oil			Fig 1.23	(0)			920				380				
Stearic acid		CH₃(CH₂)₁₆COOH	<1				840				145				
Styrene		C₆H₈	Fig 1.33	0.95	21		910						2b		
Sulphate liquor 20% DS*			Fig 1.33		13		1090								*DS - Dry substance, content by weight - see equations 1.13 to 1.20
40% DS*			Fig 1.33				1195								
50% DS*			Fig 1.33				1250								
60% DS*							1305								
Sulphur		S			5		2060	1.35			445	113			
Sulphur dioxide		SO₂·H₂O		2.1	13	10	1380	1.35		0-4	-10	-73			
Sulphuric acid	D	H₂SO₄			13	15	1030		1.37	0-4	338				
Sulphurous acid	D	H₂SO₃			13	20	1840		1.37	0-4					
					13	30	1070		1.37	0-4					
					20	40	1100			0-4					
					20	50	1140			0-4					
					20	60	1220			0-4					
					20	70	1300			0-4					
					20	80	1400			0-4					
					20	90	1500			0-4					
					20		1610			0-4					
					20		1730								
							1820								
Sweetened juice				0.1	13										

1.6.2 Liquid Table

Substance	Chemical formula	Viscosity mm²/s at 20°C	Vapour pressure kPa at 20°C	Vapour pressure at various temps. See fig 1.34 curve no	Conc. % in H₂O	Density at 25°C	Density at various temps. See Fig no	Density at various conc. See Fig no	pH region	Boiling point °C at 101.3 kPa (760 mm Hg)	Melting point °C	Fire hazard class- See Section 1.1.7	Solubility in H₂O at various temps. See Fig no	Comments
Tannic acid	C₇₆H₅₂O₄₆				5 - 50	1035			0 - 4					
Tartaric acid	C₄H₆O₆				10 - 50				0 - 4					
Tetrachloroethane	CHCl₂CHCl₂	1.1	0.8	13		1590	1.35		0 - 4	146				
Tetrachloroethylene D	C₂Cl₄	<1	2.4	20		1620	1.35			121		3		
Tetralin	C₁₀H₁₂		0.1	19		970				207				
Tin	Sn					5750				2260	232			
Tin (II) chloride	SnCl₂ 2H₂O					2710					38			
Tin (IV) chloride	SnCl₄	<5	8			3950				652	246	1		
Tincture of iodine						905				78				
Thioglycollic acid	CH₂SHCO₂H					1330				120				
Tuolene D	C₆H₅CO₃	<1	3.5	18		860	1.35			110	-17	1		
Trichloroacetic acid D	CCl₃CO₂H					1620				198	56			
Trichloroethylene D	C₂HCl₃	<1	7.5	16		1460	1.35		0 - 4	87		2b		
Turpentine						860				156				E-mod and speed of sound - Figure 1.6
Unslaked lime	CaO	Fig 1.23			10	3250			9 - 14	2850	2580			
Valtran			(0)			920								
Vinylacetate	C₄H₆O₂		15.5			930			0 - 4	73		1		
Vinegar						990			6 - 9					See Section 1.2 - Figure 1.6
Water		1	2.3	13		997				100				
Xylene D	C₆H₄(CH₃)₂	<1	1			860	1.35			139		2a		
Zinc	Zn					7140					419			
Zinc chloride	ZnCl₂		2.2	13	5	2910			4 - 6		283			
			2.2	13	20	1030			4 - 6					
			2.2	13	30	1150			4 - 6					
			2.2	13	40	1220			4 - 6					
			2.2	13	60	1420			4 - 6					
Zinc cyanide	Zn(CN)₂					1750								
Zinc sulphate	ZnSO₄ 7H₂O			13		1850					100		1.41	
						1960								

1.6.3 Supplementary diagrams

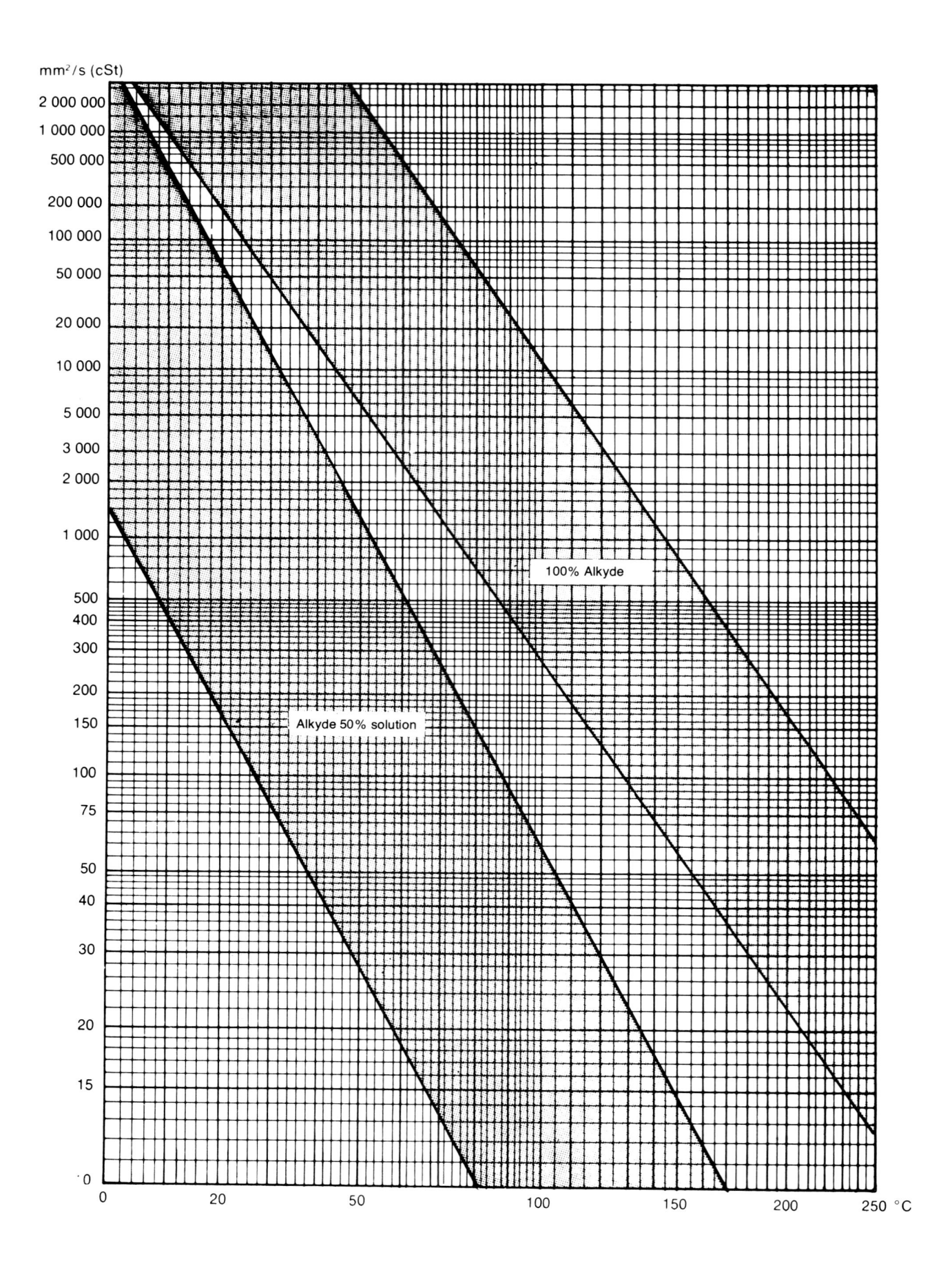

Figure 1.32 Viscosity of alkydes at various temperatures

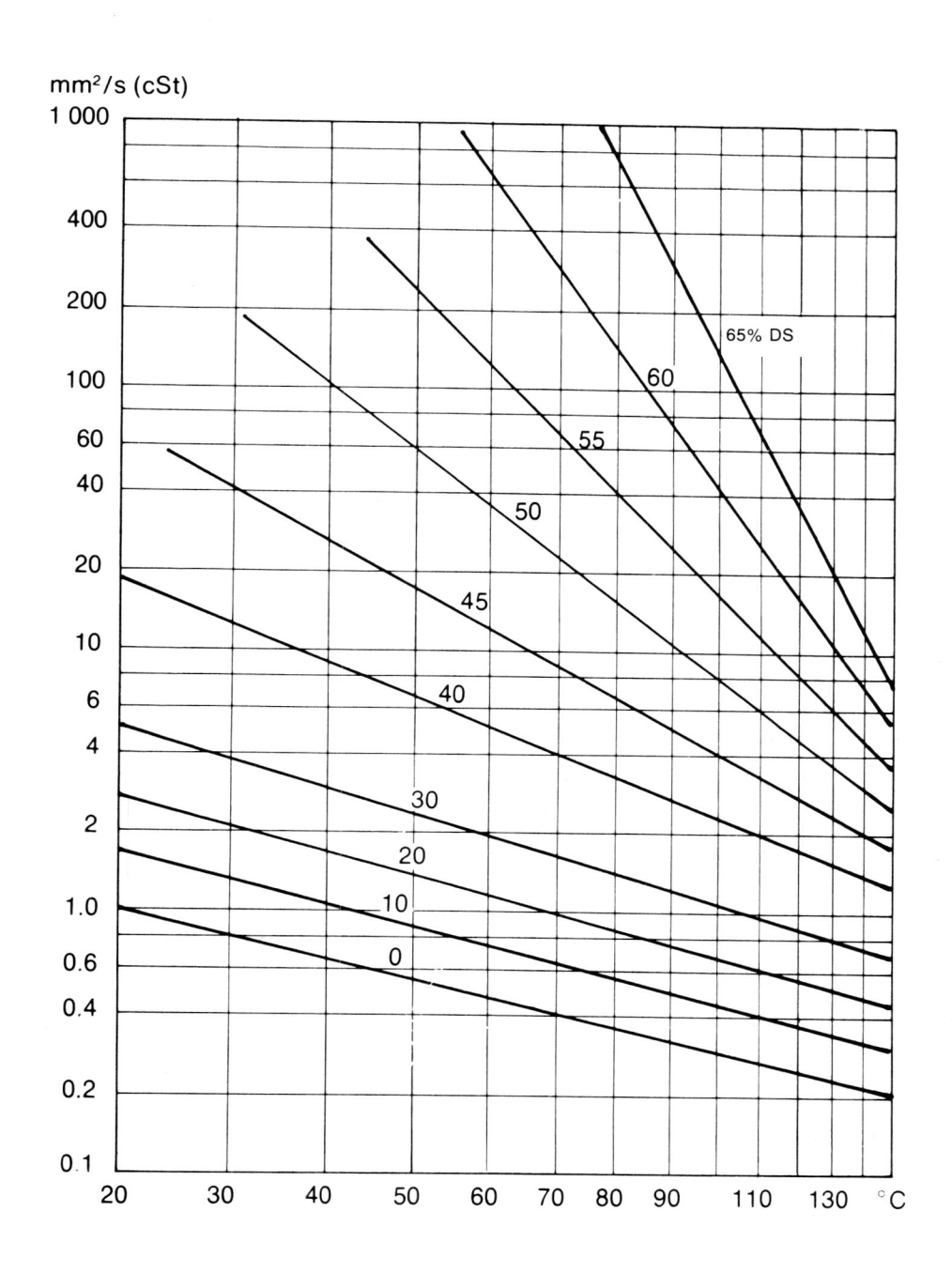

Figure 1.33 Viscosity temperature diagram for black liquor

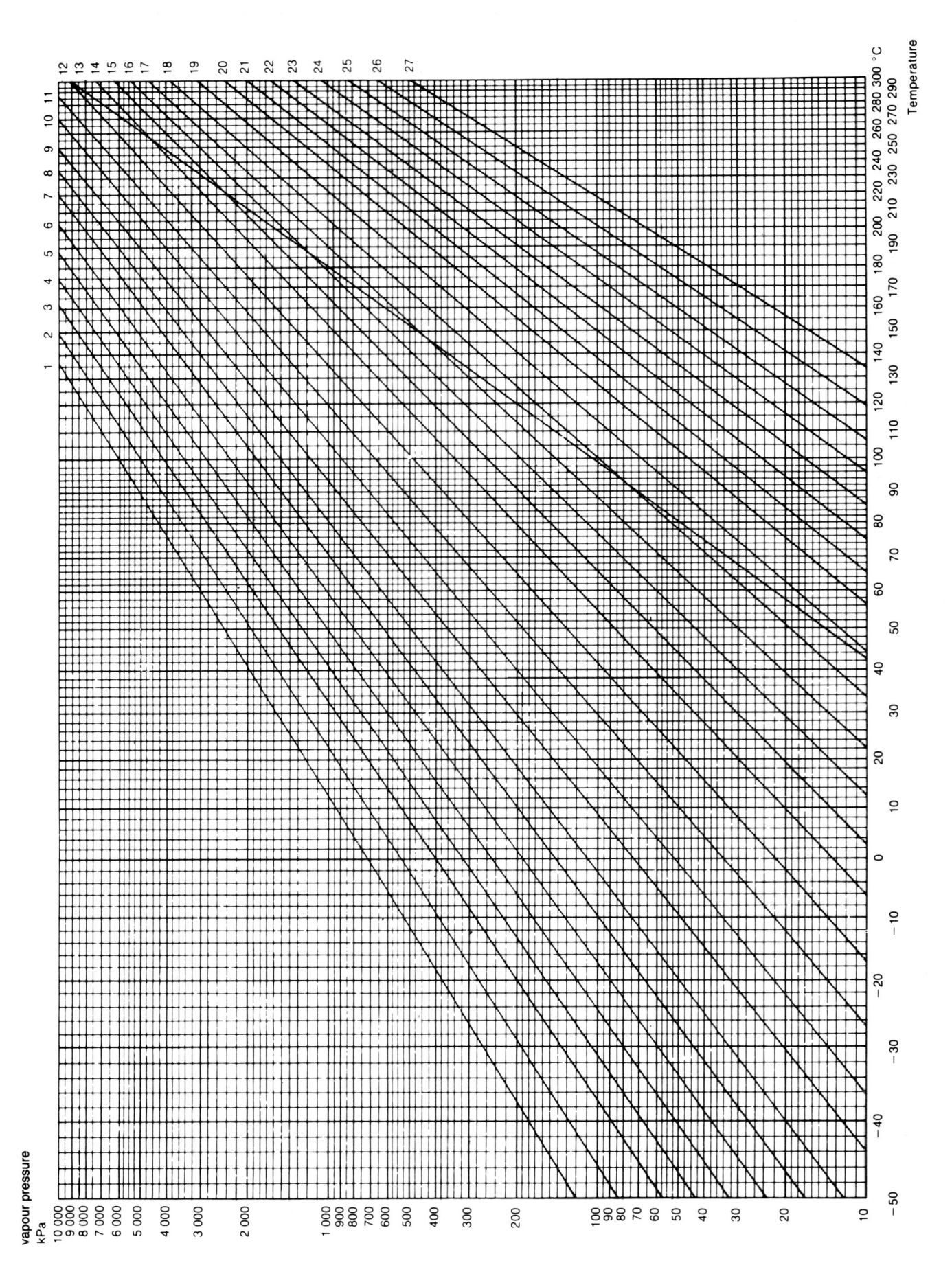

Figure 1.34 Vapour pressure diagram

Substance	Temperature °C									
	-100	-75	-50	-25	0	20	50	100	150	200
Acetaldehyde						783				
Acetic acid						1049	1018	960	896	827
Acetone	920	893	868	840	812	791	756			
Ammonia			695		636	609	561	458		
Aniline					1039	1022	996	951		
Benzene					900	879	847	793	731	661
Benzylalcohol					1061	1045	1022			
Bromine					3188	3120				
Butane	698	676	652	627	601	579	542	468	296	
Butanol					825	810				
Butyric acid					977	958	927			774
Carbon bi-sulphide	1432		1362		1292	1262				
Carbon dioxide			1154	1070	925	772				
Chlorine	1717		1595		1469	1411	1314	1111	810	
Chlorobenzene					1128	1106	1074	1019	960	896
Chloroform			1618		1526	1490	1433			
Cyclohexane						779	750	700	645	578
Cyclohexanone					963	952	925			
Cyclohexene				842	830	811	780	732		
Dichloroacetic acid						1552				
Dichloroethane					1207	1176				
Dichloroethylene						1250				
Dioxane						1030	1010			
Ethanol					806	789	763	716	649	557
Ether (Diethylether)	842	816	790	764	736	714	676	611	518	
Ethyl acetate					924	901	864	797	721	621
Ethylene glycol					1128	1112				
Ethyl chloride					919	892	846			
Formic acid						1220	1184			
Furfurol					1181	1160	1128			
Glycerine					1273	1261	1242	1209		
Hexane		742	721	700	678	659	631	580	520	438
Hydrochloric acid	1235		1076		920					
Hydrofluoric acid	1660		1123		1002	987				
Hydrogen sulphide	1170		980		870					
Iodine	5060		5010		4960	4940			3780	
Methanol					810	792	765	714	650	553
Methyl acetate					959	934	894	822	734	610
Methylene chloride					1362	1326				
Methylethylketone					826	803				
Nitrobenzene					1223	1203	1174			
Nonane			769	751	733	718	694	653	609	
Octane		860	757	738	719	703	678	635	588	532
Octanol					842	829				
Pentane	737	715	693	670	646	626	596	533	460	
Phenol					1092	1071	1050			
Propane	646	619	590	560	528	501	450			
Propanol						804	779	733	674	592
Styrol						910				
Sulphur	2100		2080		2070	2060			1780	1740
Sulphur dioxide			1557		1435	1383	1296	1118	768	
Tetrachloroethane					1626	1593				
Tetrachloroethylene					1656	1621				
Toluole					885	868	839	739	737	672
Trichlorethylene						1463				
Xylene					881	864	838	795		678

Figure 1.35 Density for liquids at various temperatures

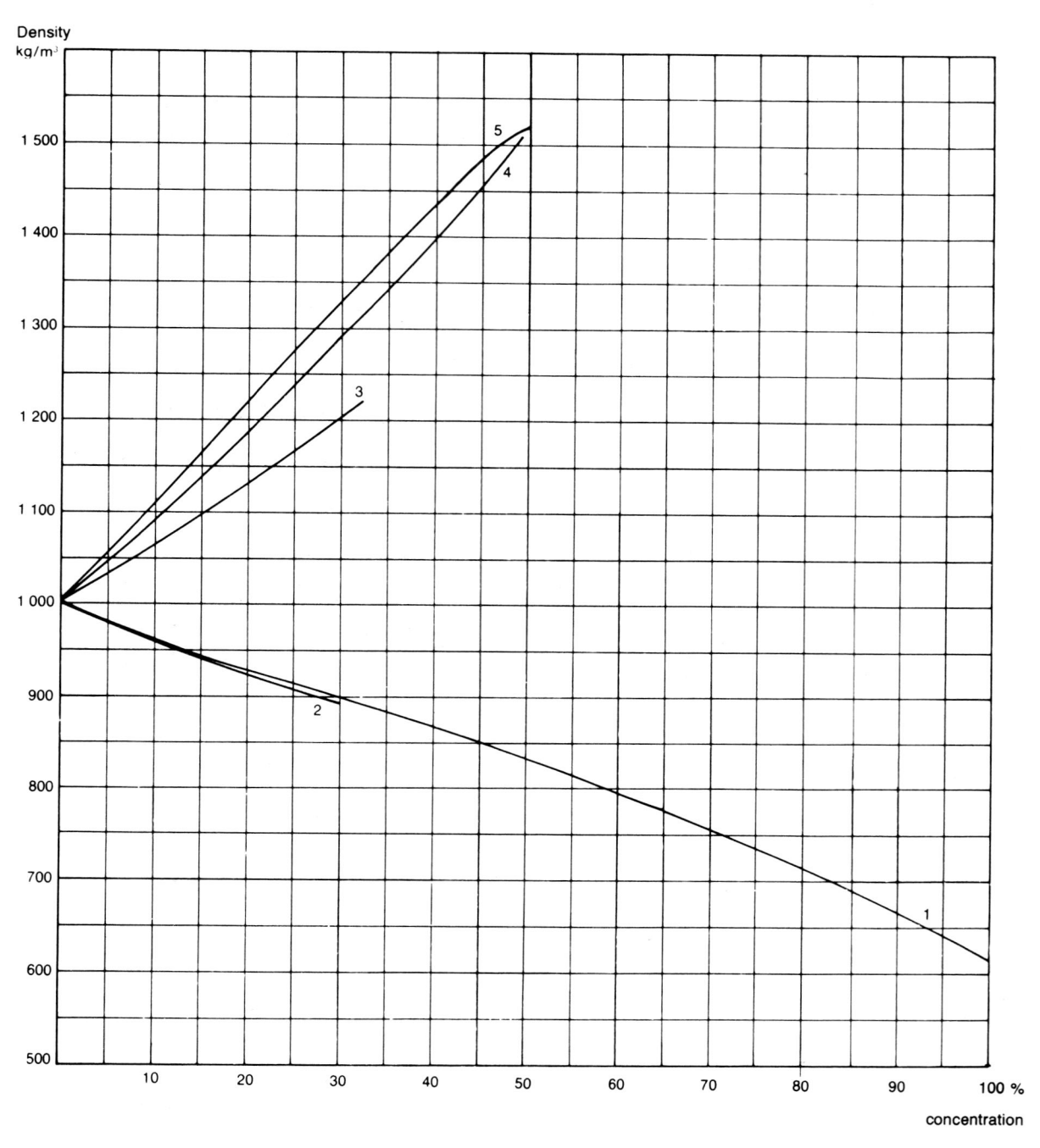

1. Ammonia solution   at 15°C
2. Ammonium hydroxide solution   20°C
3. Calcium hydroxide solution   20°C
4. Potassium hydroxide solution   15°C
5. Sodium hydroxide solution   15°C

Figure 1.36 Density against concentration for various bases

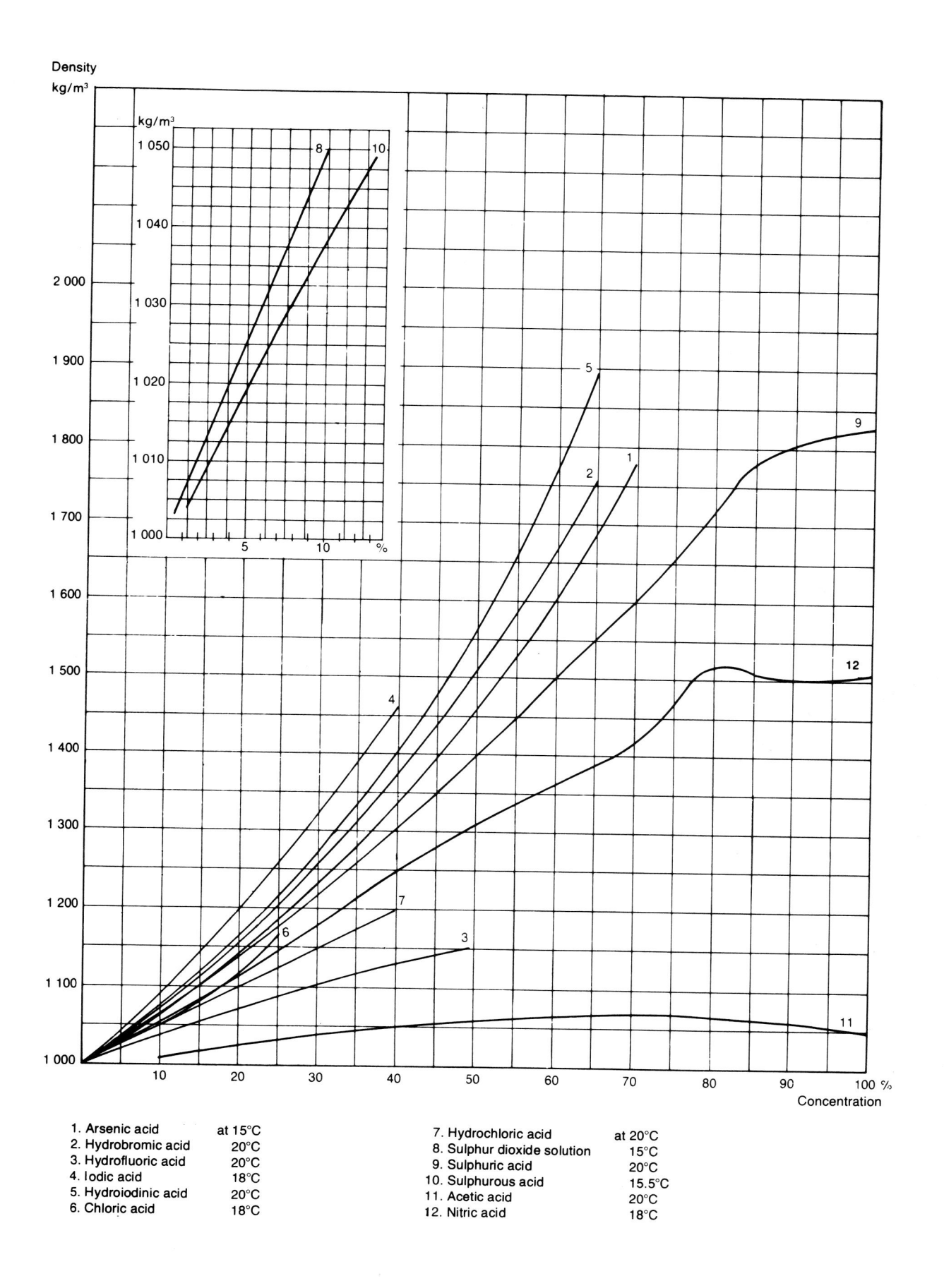

1. Arsenic acid	at 15°C	7. Hydrochloric acid	at 20°C
2. Hydrobromic acid	20°C	8. Sulphur dioxide solution	15°C
3. Hydrofluoric acid	20°C	9. Sulphuric acid	20°C
4. Iodic acid	18°C	10. Sulphurous acid	15.5°C
5. Hydroiodinic acid	20°C	11. Acetic acid	20°C
6. Chloric acid	18°C	12. Nitric acid	18°C

Figure 1.37 Density against concentration of various acids

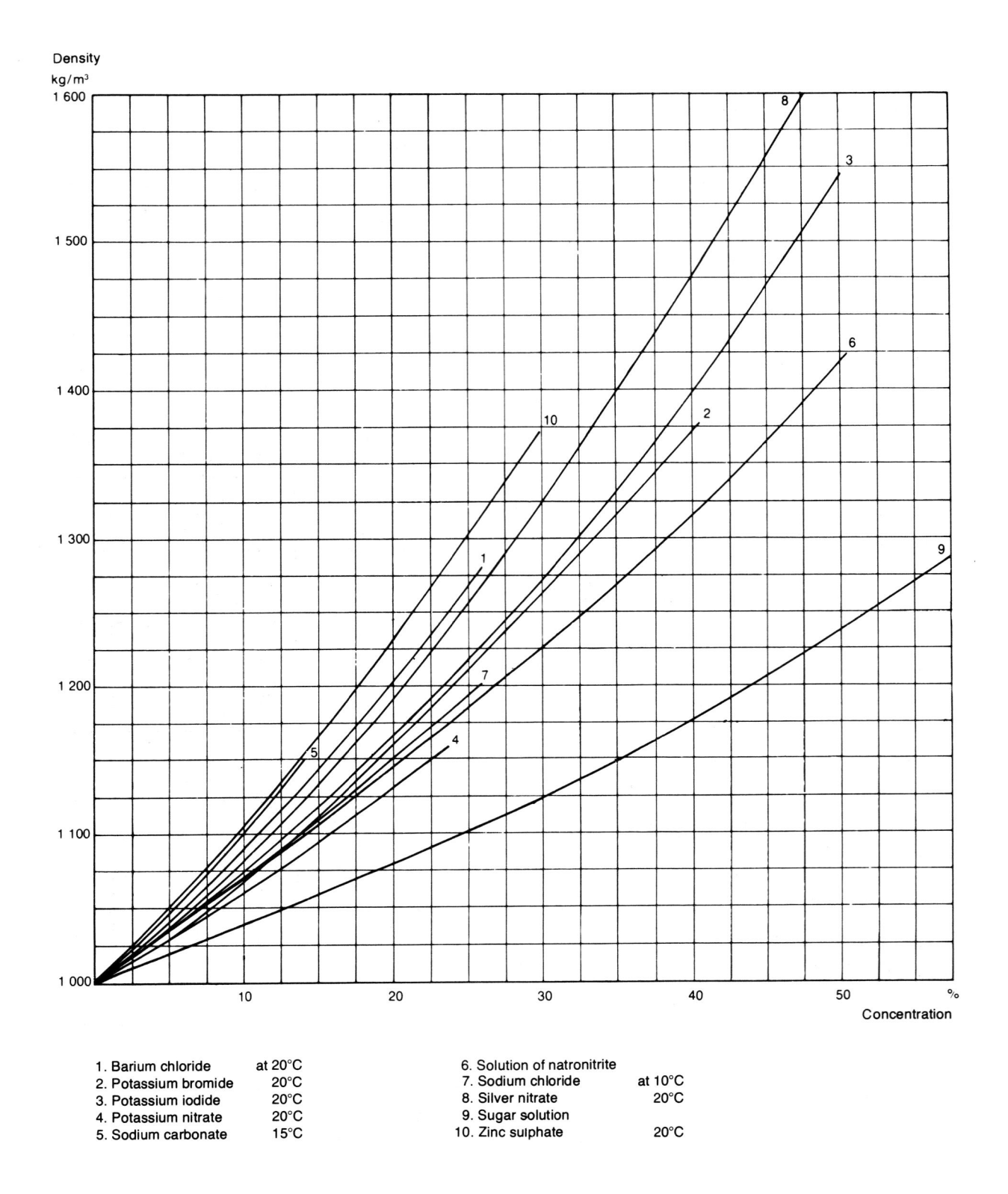

1. Barium chloride   at 20°C
2. Potassium bromide   20°C
3. Potassium iodide   20°C
4. Potassium nitrate   20°C
5. Sodium carbonate   15°C
6. Solution of natronitrite
7. Sodium chloride   at 10°C
8. Silver nitrate   20°C
9. Sugar solution
10. Zinc sulphate   20°C

Figure 1.38 Density against concentration for various salts

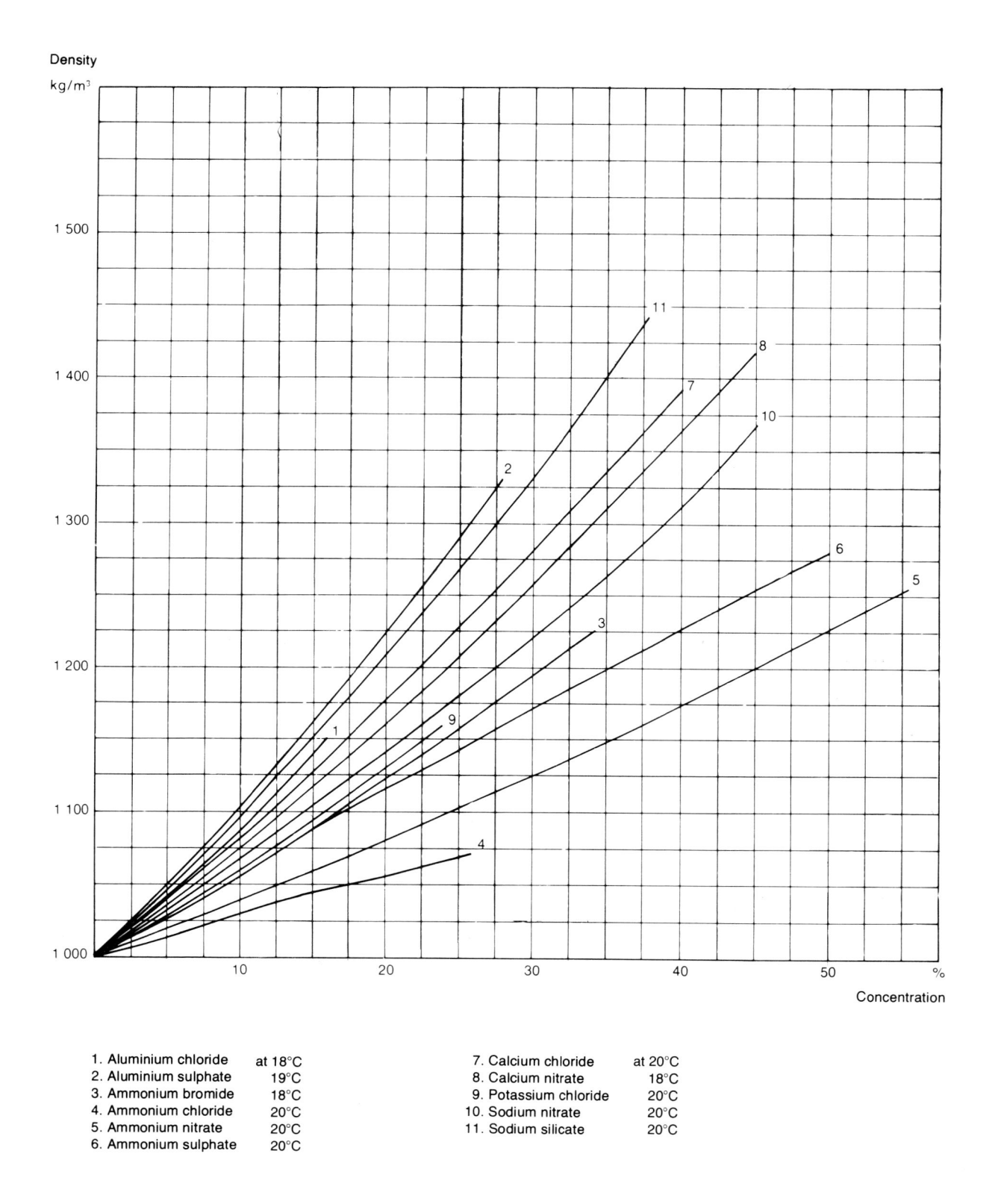

Figure 1.39 Density against concentration for various salts

1.	Aluminium chloride	at 18°C	7.	Calcium chloride	at 20°C
2.	Aluminium sulphate	19°C	8.	Calcium nitrate	18°C
3.	Ammonium bromide	18°C	9.	Potassium chloride	20°C
4.	Ammonium chloride	20°C	10.	Sodium nitrate	20°C
5.	Ammonium nitrate	20°C	11.	Sodium silicate	20°C
6.	Ammonium sulphate	20°C			

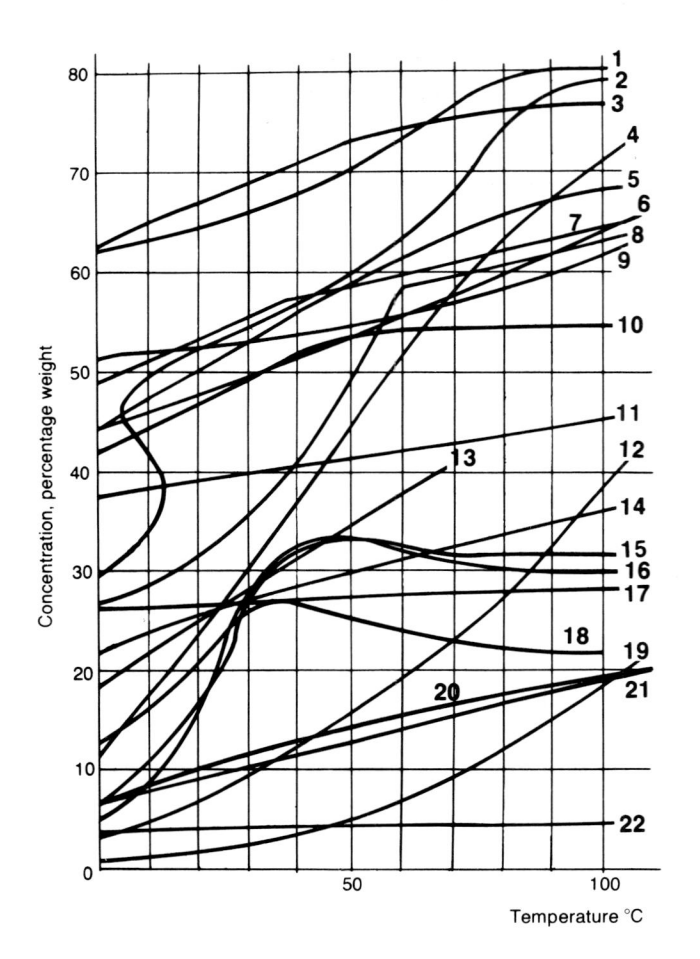

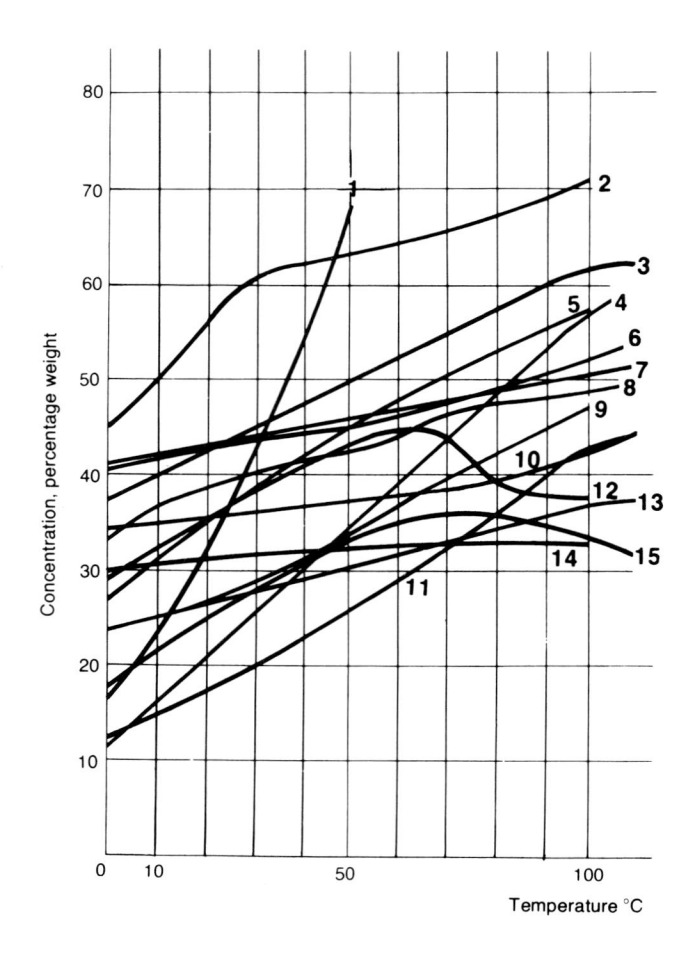

	Curve no.
Potassium bicarbonate	13
Potassium hydroxide	7
Potassium carbonate	9
Potassium chlorate	12
Potassium chloride	14
Potassium chromate	11
Potassium nitrate	4
Potassium perchlorate	19
Potassium sulphate	21
Sodium acetate	8
Sodium bicarbonate	20
Sodium bromide	10
Sodium dichromate	1
Sodium fluoride	22
Sodium hydroxide	2
Sodium carbonate	15
Sodium chlorate	5
Sodium chloride	17
Sodium nitrate	6
Sodium perchlorate	3
Sodium sulphate	16
Sodium sulphite	18

Figure 1.40 Solubility of salts in water at various temperatures

	Curve No.
Aluminium chloride	14
Aluminium nitrate	3
Aluminium sulphate	9
Ammonium perchlorate	4
Ammonium sulphate	7
Barium chloride	13
Lead acetate	1
Lead nitrate	5
Iron (II) chloride	8
Copper (II) chloride	6
Copper (II) nitrate	2
Copper (II) sulphate	11
Magnesium chloride	10
Magnesium sulphate	15
Zinc sulphate	12

Figure 1.41 Solubility of salts in water at various temperatures

2 Liquid flow

The purpose of Chapter 2, is to lay the basis for determination of flow characteristics in the pipe system of a pump installation.

Equations are presented which are fundamental to all flow calculations.

The process of calculation and the necessary data for quantifying flow losses are reviewed for Newtonian liquids. Suspensions, solid particles suspended in water, have special flow characteristics and are dealt with in a special section.

Pre-computed pressure losses for water, oils and paper pulp in pipes are presented in diagram form. The diagrams for water and oils can also be used for other Newtonian liquids with similar viscosities. The chapter ends with a review of flow rate measuring devices.

Contents

2.1 Basic equations
    2.1.1 Explanation of terms
    2.1.2 Continuity equation
    2.1.3 Bernoulli's equation
    2.1.4 Momentum equation
    2.1.5 Energy equation

2.2 Pipe flow losses
    2.2.1 Reynolds Number
    2.2.2 Pressure losses in straight pipes
    2.2.3 Pressure losses in fittings
    2.2.4 Hydraulic diameter
    2.2.5 Total losses in a pipe system

2.3 Liquid solid mixtures
    2.3.1 General
    2.3.2 Homogeneous suspensions
    2.3.3 Heterogeneous suspensions
    2.3.4 Paper pulp

2.4 Pressure loss diagrams
    2.4.1 General
    2.4.2 Water
    2.4.3 Oil
    2.4.4 Paper pulp

2.5 Flow measurement
    2.5.1 General
    2.5.2 Instruments for pipe systems
    2.5.3 Instruments for open channels
    2.5.4 Other instruments
    2.5.5 Measurement standards

2.1 Basic equations

2.1.1 Explanation of concepts

The processes of flow occurring in nature are usually very complex and difficult to deal with. In a good many technical applications, however, perfectly acceptable results can be obtained from calculations based upon simplified considerations. Some of the concepts and conditions germane to this concept are discussed below.

A flow process is steady if all flow parameters, pressure, velocity, etc. at any specific point in the flow field are independent of time. According to this definition practically all flow processes are unsteady. In practice, many flow processes can, with sufficient accuracy, be treated as steady using suitable mean time values. The basic equations presented in later sections are valid for steady flow.

Extra care should be taken with certain pumping installations where steady state conditions may not exist. Low head, high flow centrifugal pumps with wide impeller tip widths produce velocity variations across the flow stream. Some positive displacement pumps, with a small number of pumping elements, produce cyclic flow and pressure variations. The amplitude of these variations is dependant upon the piping system as well as the pump construction.

Flow is generally three-dimensional, i.e. the flow parameters vary with all three co-ordinates defining some point in space. In a good many practical cases of flow the number of dimensions studied may be reduced whilst still maintaining accuracy. A common example is pipe flow, where the parameters are assumed to vary in only one direction, i.e. the longitudinal dimension of the pipe. One-dimensional flow assumes that flow parameters can be described by mean values across the flow through-section. In principle, different mean values should be applied when studying continuity, momentum and energy. In the case of pipe flow, the mean velocity in the pipe is defined as the volume flow divided by the cross-sectional area of the pipe ($v = Q/A$). This mean velocity can as a rule be used with adequate accuracy for most purposes.

An important property of a flowing medium is its density and the changes in density which occur during flow. A gas is compressed, i.e. the density increases, when the pressure somewhere in the flow increases. This type of flow is called compressible flow. In a liquid system density variations are very small under great changes of pressure. Liquid flow can thus, with sufficient accuracy, be treated as incompressible. This is also the case with gases at low flow velocities when the pressure changes are small.

A streamline is a curve to which the velocity vector is tangential at any point. In the case of steady flow, the streamlines remain unchanged with time and describe the path of a liquid particle passing through the flow field. The streamlines through all points on a closed curve in the flow field constitute a flow tube. No mass will pass through the circumferential boundary surface of a flow tube. The flow tube is thus reminiscent of an ordinary pipe. In the case of a pipe, however, there are strong frictional effects associated with the wall of the pipe which do not necessarily occur in a flow tube.

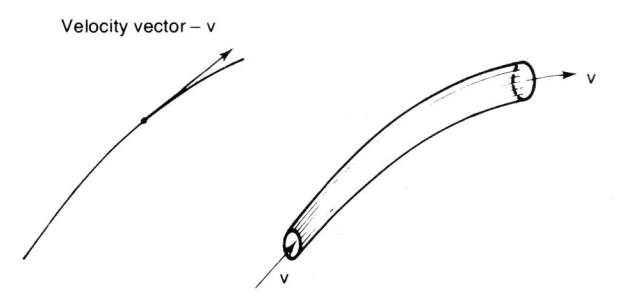

Figure 2.1 Streamline and flow tube

2.1.2 The continuity equation

The continuity equation is a statement for the condition that mass is not created or destroyed during a flow process.

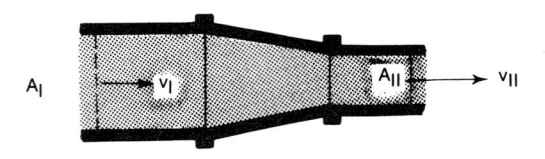

Figure 2.2 Example of one-dimensional flow

Assuming that the flow is steady, the mass flow $\dot{m}$ must be of equal magnitude everywhere along the pipe or the flow tube. In the case of one-dimensional flow in figure 2.2,

$$\dot{m} = \rho_I \cdot v_I \cdot A_I = \rho_{II} \cdot v_{II} \cdot A_{II} \qquad \text{Equ 2.1}$$

or for an incompressible liquid flow,

$$Q = v_I \cdot A_I = v_{II} \cdot A_{II} \qquad \text{Equ 2.2}$$

where
- Q = volume flow (m^3/s)
- v = Q/A = flow velocity (m/s)
- A = cross-sectional area (m^2)

When the cross-sectional area in a pipe reduces, then, according to the continuity equation, the flow velocity increases. If the area is halved, the velocity doubles and so on.

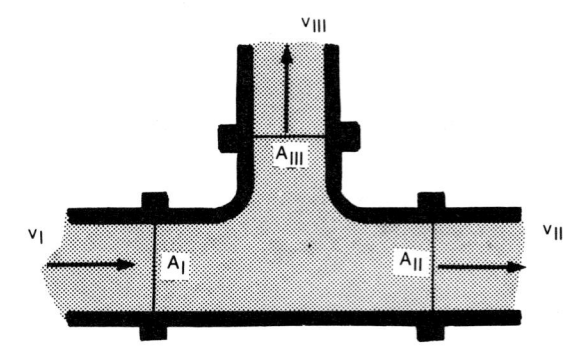

Figure 2.3 Branching

Since there is no increase in mass at the point of branching, the mass flow entering will equal the total mass discharging per unit time. Using the symbols of figure 2.3,

$$Q_I = Q_{II} + Q_{III}$$

or

$$v_I \cdot A_I = v_{II} \cdot A_{II} + v_{III} \cdot A_{III} \qquad \text{Equ 2.3}$$

2.1.3 Bernoulli's equation

Bernoulli's equation is an equation of motion for incompressible flow based on Newton's Second Law of Motion adapted for fluids and Euler's Equation of Motion. The equation expresses the relationship between:
- static pressure,
- potential energy,
- kinetic energy and
- friction losses.

Bernoulli's equation, for steady, one-dimensional, constant temperature and incompressible flow between stations I and II, becomes:

$$\frac{p_I}{\rho} + \frac{v_I^2}{2} + gz_I = \frac{p_{II}}{\rho} + \frac{v_{II}^2}{2} + gz_{II} + \frac{\Delta p_f}{\rho} \qquad \text{Equ 2.4}$$

where
- p = static pressure (gauge) (Pa)

ρ = density (kg/m³)
v = velocity (m/s)
g = gravitational acceleration (m/s²)
z = height above datum (m)
Δp_f = frictional pressure loss (Pa)

Equation 2.4 expresses the various energy terms per unit mass of fluid, J/kg.

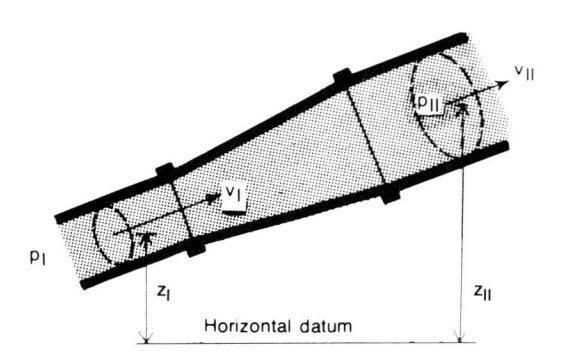

Figure 2.4 Symbols used in Bernoulli's equation

In rotodynamic pump technology it is practical to express the Bernoulli equation in terms of "head" i.e. metres of liquid column. If all terms in Bernoulli's equation are divided through by g, we get:

$$\frac{p_I}{w} + \frac{v_I^2}{2g} + z_I = \frac{p_{II}}{w} + \frac{v_{II}^2}{2g} + z_{II} + \frac{\Delta p_f}{w} \quad \text{Equ 2.5}$$

where
w = specific weight (N/m³)
w = ρg

The various terms are then called:

$\frac{p}{\rho g}$ or $\frac{p}{w}$ = static head (m)

$\frac{v^2}{2g}$ = velocity head (m)

z = potential head (m)

$\frac{\Delta p_f}{w} = h_f$ = head loss (m)

Since all the terms in Equation 2.5 refer to height they are easy to illustrate graphically.

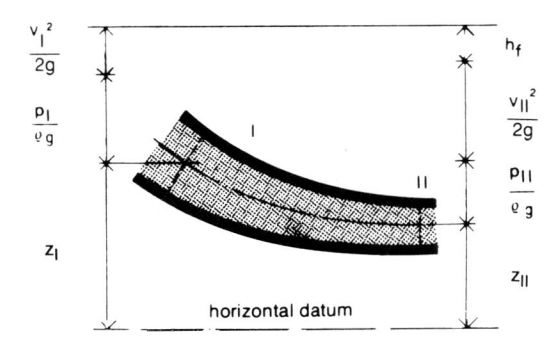

Figure 2.5 Graphic illustration of Bernoulli's equation

Total head is sometimes designated by H. The relationship between pressure and head is p/w = h. The three head terms can be converted to equivalent pressure terms. Static pressure plus velocity, or dynamic pressure is called total pressure or stagnation pressure.

$$\text{total pressure} = p + \frac{\rho v^2}{2} = wh + \frac{wv^2}{2} \quad \text{Equ 2.6}$$

Total and static pressure are measured in different ways. A pressure tap at right angles to the direction of flow will sense the static pressure. Ahead of a Pitot tube the velocity is reduced to zero and the static pressure rises to stagnation pressure.

As a rule, it is easy to measure the static pressure, whereas the total pressure can easily be affected by measurement errors. A Prandtl tube or Pitot-static tube measures both total and static pressure. The differential pressure, the dynamic pressure, $p_{dyn} = wv^2/2g$ is obtained directly.

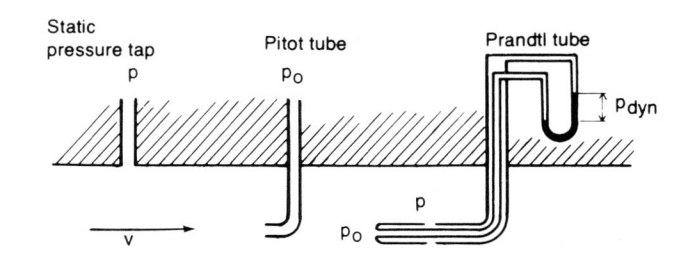

Figure 2.6 Measuring of static, total and dynamic pressure

2.1.4 The momentum equation

The product of mass times flow velocity for a liquid particle is called its momentum. The momentum equation for steady flow reads:

The resultant F of all external forces on a control volume is equal to the changes in momentum for the outflowing and inflowing mass per unit time, or,

$$F = \dot{m} \left(\bar{v}_2 - \bar{v}_1 \right) \quad \text{Equ 2.7}$$

where
$\dot{m}$ = mass flow   dm/dt   (kg/s)
$\bar{v}$ = velocity vector   (m/s)

The momentum equation is illustrated in the following example. The problem is to determine the force required to hold a pipe bend in position.

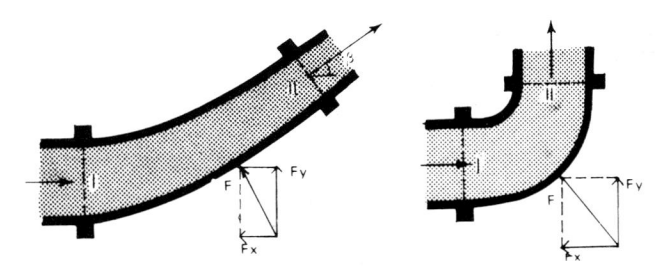

Figure 2.7 Pipe bends, arbitrary (left) and 90° bend

The control volume is defined by the interior of the pipe bend. Then, for the arbitrary bend in the x axis:

$-F_x + p_I \cdot A_I - p_{amb} \cdot A_I - p_{II} \cdot A_{II} \cdot \cos \beta + p_{amb} \cdot A_{II} \cdot \cos \beta = \dot{m} (v_{II} \cdot \cos \beta - v_I)$

or:

$F_x = \dot{m} \cdot v_I + \left(p_I - p_{amb}\right) \cdot A_I - [\dot{m} \cdot v_{II} + \left(p_{II} - p_{amb}\right) \cdot A_{II}] \cdot \cos \beta$

Similarly, in the y axis:

$F_y = \left[\dot{m} \cdot v_{II} + \left(p_{II} - p_{amb}\right) \cdot A_{II}\right] \cdot \sin \beta$

For a 90° bend $\beta = 90°$, therefore:

$F_x = \dot{m} \cdot v_I + \left(p_I - p_{amb}\right) \cdot A_I$

$$F_y = \dot{m} \cdot v_{II} + \left(p_{II} - p_{amb}\right) \cdot A_{II}$$
$$F = \sqrt{F_x^2 + F_y^2} \qquad \text{Equ 2.8}$$

Note that it is a gauge pressure and not the absolute pressure in the pipe which determines the magnitude of the force. Note also that the forces can be determined without detailed knowledge of the internal flow process through the bend. The momentum equation applies regardless of whether the process has losses or not.

It is often the case that the terms m and v are small compared with the others. For a 90° degree bend with A_I being equal to A_{II}.

$$F = \sqrt{2} \cdot \left(p - p_{amb}\right) \cdot A \qquad \text{Equ 2.9}$$

where
- p = pressure in pipe (absolute) (Pa)
- p_{amb} = ambient pressure (absolute) (Pa)
- A = pipe cross-sectional area (m²)
- F = fixing force (N)

The fixing force has an angle of 135 degrees to the incoming direction of flow.

2.1.5 The energy equation

The energy equation is an extension of the energy principle which states that energy cannot be created or destroyed but can only be converted to other forms. The energy equation for steady state one-dimensional pump flow is:

$$W_{in} = \left(h_{II} - h_I\right) + \frac{\left(c_{II}^2 - c_I^2\right)}{2} + g\left(z_{II} - z_I\right) + Q_{out} \qquad \text{Equ 2.10}$$

The term h, specific enthalpy, is made up from two components; u, ($u = Tc_p$) internal energy and pv, which represents the mechanical energy due to pressure, often called flow work. The energy equation can be rewritten:

$$W_{in} = \left(T_{II}c_{p_{II}} - T_I c_{p_I}\right) + \left(p_{II}v_{II} - p_I v_I\right) + \frac{\left(c_{II}^2 - c_I^2\right)}{2} + g\left(z_{II} - z_I\right) + Q_{out} \qquad \text{Equ 2.11}$$

where
- T = temperature (°C)
- c_p = specific heat (J/kg/°C)
- p = pressure (Pa)
- v = specific volume (m³/kg)
- c = velocity (m/s)
- z = height above datum (m)
- g = gravitational acceleration (m/s²)
- W_{in} = work input to the shaft (J/kg)
- Q_{out} = heat rejected (J/kg)

Note in these equations the velocity is represented by c not v. Engineers frequently use the same symbol for different quantities. The context is important and should be explanatory. All the terms in the energy equation are calculated per kg of medium flowing. The terms $c^2/2$ and gz respectively represent the kinetic and potential energy of the medium per kg. The effect of a pump on the energy equation is shown diagrammatically in figure 2.8.

Mechanical work is input through the rotating shaft, all the variables change and some heat is rejected from the pump body. Notice particularly that a temperature rise has occurred. The change from suction conditions to discharge conditions usually takes place very quickly within the pump so the ideal process is isentropic, constant entropy, not constant temperature. The isentropic temperature rise is small for water, of the order of 0.4 °C per 100 bar pressure rise at 50 °C rising to 2.6 °C per 100 bar rise at 250 °C. Pump inefficiencies will generally appear as increased temperature rise and heat rejection. Noise and vibration will take their energy as portions of heat rejection. In equations 2.10 and 2.11, g is the only constant, for the location, not universally. Even c_p is a variable, from less than 4.0 to over 5.0 for water.

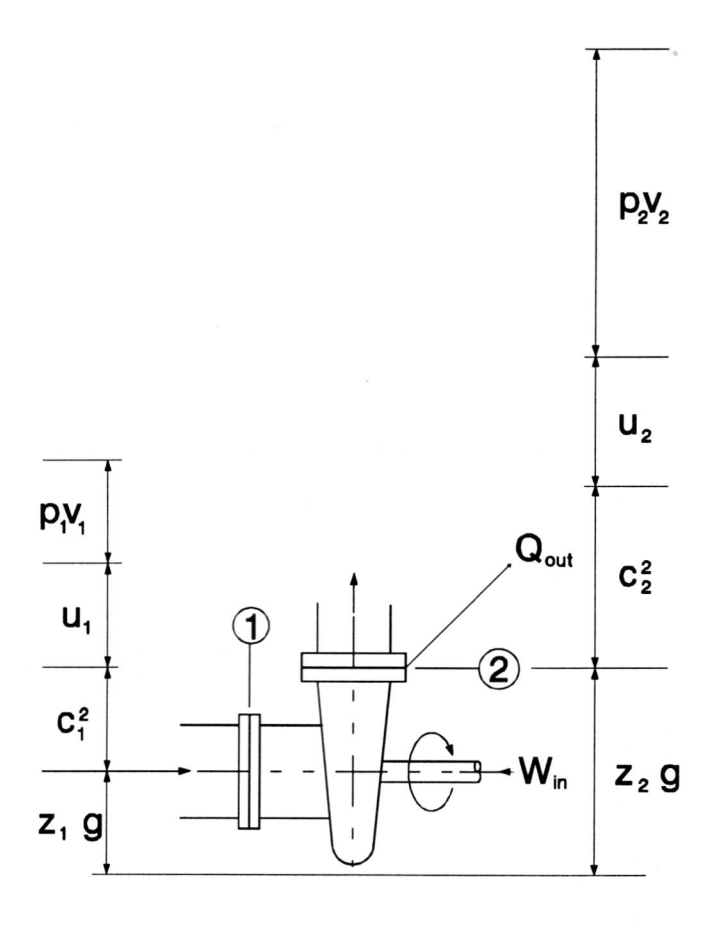

Figure 2.8 Illustration of the energy equation

Rotodynamic pump designers take the centre-line of the first stage impeller eye as the datum for all calculations and parameters such as $NPSH_r$. Most positive displacement pump designers take the centre-line of the inlet connection as the datum. A system designer will not know precisely where the impeller eye or inlet connection will be. The system designer should base all the system calculations on the foundation level where the pump will be located or the liquid surface level for submerged pumps.

The quality of any calculated result is dependant upon the accuracy of the input data. Pumps always operate in the pressurised liquid region so that saturated liquid data will not be accurate. Steam tables, such as the U.K. Steam Tables, based on the 1967 IFC Formulation, contain tabulated data for pressurised water up to 1000 bara. However this data is aimed at the power generating industry and is tailored to their needs. Inconsistencies exist in the data for low temperatures at pressures close to atmospheric. This data should be examined carefully before use and 'conditioned' if thought appropriate. Unconditional use of raw data can result in errors of up to 6% when compared to results obtained from calculations in sections 3.2 and 3.4. Computerised steam tables alleviate some of the inconsistencies.

In some cases it will be possible to ignore the effects of changes in velocity, specific volume and height, between inlet and outlet, without loss of accuracy. Pump differential head or pressure is the important criterion. No assumptions should be made on pumps which operate at low differentials, below 50m.

2.2 Pipe flow losses

2.2.1 Reynolds Number

Flow losses occur because of the effect of internal friction. Shear stresses arise as soon as velocity gradients exist.

$$\tau = \mu \cdot \frac{dv}{dy} \quad (N/m^2) \qquad \text{Equ 2.12}$$

where

- τ = shear stress (N/m²)
- μ = viscosity (Ns/m²)
- v = flow velocity (m/s)
- y = thickness of flow stream (m)

The work done by the shear force generates heat and adds to the internal energy of the liquid. The increase in internal energy means that the static pressure becomes a little less hence the name pressure loss or pressure drop. The level to which the liquid temperature can rise due to friction heating is dependant upon the heat transfer mechanism to the surroundings.

The flow losses are great wherever the shear stress is great, i.e. where the velocity gradients are great. High velocity gradients occur in the boundary layer, at flow around sharp corners, in strong vortex flows and so on. Viscous flow in pipes is characterised by Reynolds Number.

$$Re = \frac{v \cdot d}{\nu} \qquad \text{Equ 2.13}$$

where

- Re = Reynolds Number (dimensionless)
- v = flow velocity (m/s)
- d = pipe diameter (m)
- ν = kinematic viscosity (m²/s)

The following relationship exists between kinematic viscosity, ν, and dynamic viscosity, μ.

$$\nu = \frac{\mu}{\rho} \qquad \text{Equ 2.14}$$

where

- ρ = liquid density (kg/m³)

In the case of low Reynolds Numbers, Re <2300, the flow is laminar. The liquid flows in layers with different velocities which do not mix with each other. At Re > 2300 the flow is turbulent, that is to say the liquid particles do not flow in a straight line in the direction of motion of the main flow.

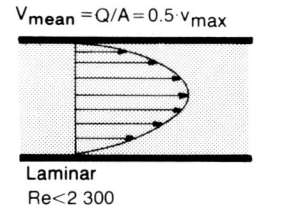

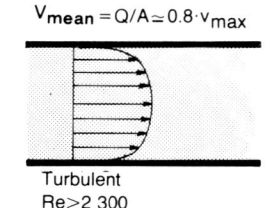

Laminar Re<2 300 — $V_{mean} = Q/A = 0.5 \cdot v_{max}$

Turbulent Re>2 300 — $V_{mean} = Q/A \cong 0.8 \cdot v_{max}$

Figure 2.9 Laminar and turbulent flow

Since the velocity gradients differ, the shear stress and flow losses will also differ between the two cases. The critical value of Re Number has been stated as Re$_{cr} \cong 2300$. It should be noted however, that the flow may continue to be laminar at considerably higher values of Re if the flow is very well protected from disturbances and entrance conditions to the section are the optimum. Low values of Re Number occur at low flow velocities, in small pipe diameters or at high viscosity.

Liquid	Kinematic Viscosity ν m²/s	Transition to turbulent flow (for the pipe diameters given) V_{cr} m/s			
		d = 0.001m	d = 0.01 m	d = 0.1 m	d = 1 m
Water, 20 °C	$1 \cdot 10^{-6}$	2.3	0.23	0.023	0.0023
SAE 5W 60 °C	$1 \cdot 10^{-5}$	23	2.3	0.23	0.023
Burner oil Class E (BS 2869: 1970) 30 °C	$1 \cdot 10^{-4}$	230	23	2.3	0.23
SAE 50 18 °C	$1 \cdot 10^{-3}$	2300	230	23	2.3

Figure 2.10 Transition to turbulent flow in pipes

It can be seen from the examples that laminar flow is less usual in practice for liquids with viscosity similar to that of water, but is quite common for oils.

2.2.2 Head losses in straight pipes

The following formula applies to head losses for flow in a straight pipe.

$$h_f = \lambda \cdot \frac{l}{d} \cdot \frac{v^2}{2g} \quad (m) \qquad \text{Equ 2.15}$$

where

- λ = loss coefficient for straight pipes (dimensionless)
- l = length of pipe (m)
- d = pipe diameter (m)
- v = Q/A = flow velocity (m/s)

The loss coefficient is dependent upon Reynolds Number and upon the internal roughness of the pipe. This is shown in Figure 2.11.

Pipe and pipe material	Condition of pipe	Factor k Calculation values mm
Steel pipe	Seamless, new	0.03 - 0.05
	Welded, new	0.03 - 0.10
	in use, rusted	0.15 - 0.5
	Seamless, galv.	0.1 - 0.16
	Welded, galv.	0.1 - 0.2
	Heavy asphalt coating	0.5
	Mature water mains	1.2
	Stainless, acid resist.	0.045
Cast iron pipe	Bitumen coated	0.1-0.3
	Main conduit	0.1
	Other conduits	0.2-0.3
	in use, rusted	0.4-2.0
Copper pipe Brass, glass, Al	Diameter <200 mm	0.01
	Technically smooth	0.002
Concrete pipe	Water main	0.1-0.2
	Waste, drains, new	0.2-0.5
	Waste, drains, old	1.0
Asbestos cement pipe Earthenware pipe	Newly laid	0.05
	Newly laid drains	0.2
	Older drains	1.0
	Mature foul sewer	3.0
PE, PVC	Dia. <200> mm	0.01
	Dia. 200 mm	0.05
GRP	All dimensions	0.1
Wood	New	0.2-1.0

Figure 2.12 Examples of approximate roughness in pipes

The roughness of the pipe can be assessed from the values in figure 2.12.

The loss coefficient can also be calculated from the following formulae:

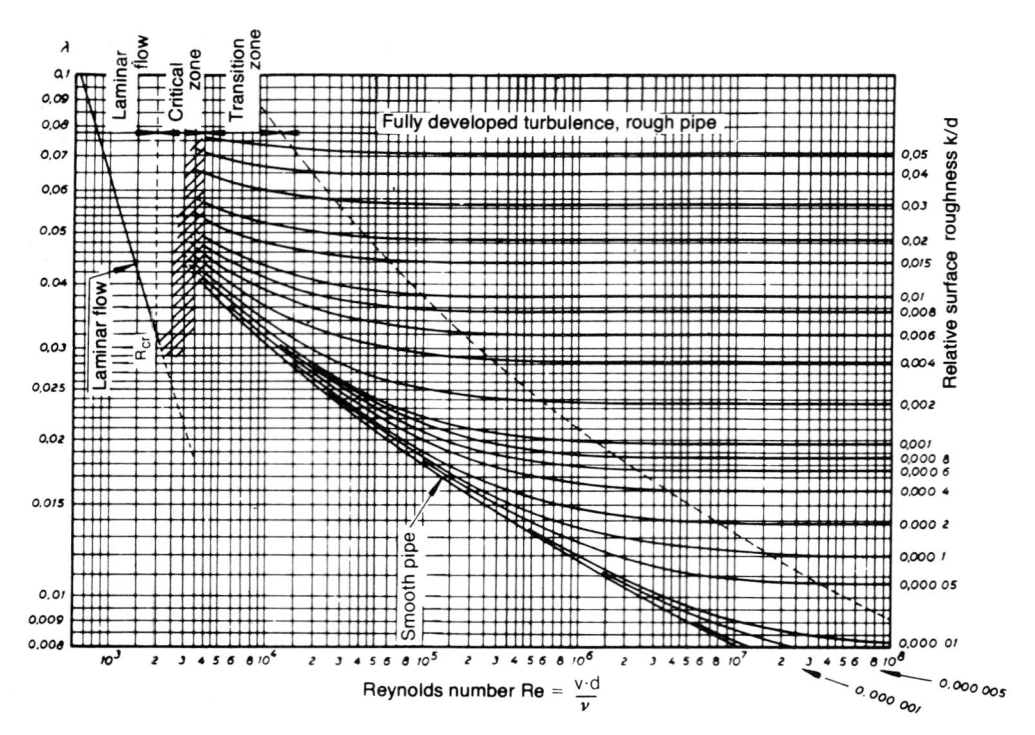

Figure 2.11 Loss coefficient for straight pipes

Laminar flow Re < 2300

$$\lambda = \frac{64}{Re} \qquad \text{Equ 2.16}$$

Turbulent flow Re > 4000

$$\frac{1}{\sqrt{\lambda}} = -2 \cdot \log_{10}\left(\frac{2.51}{Re\sqrt{\lambda}} + \frac{K/d}{3.71}\right) \qquad \text{Equ 2.17}$$

Equation 2.17 is sometimes called the Colebrook-White equation; Colebrook in France and Prandtl-Colebrook in Germany. In the region Re = 2300 to Re = 4000 laminar and turbulent flows may alternate in various sections of the pipe. The resultant loss coefficient will then assume values between those obtained in formulae 2.16 and 2.17.

The expression given in 2.17 is somewhat inconvenient for manual calculation. The following formula, proposed by Miller, can be used for turbulent flow:

$$\lambda = \frac{0.25}{\left[\log_{10}\left(\frac{K}{3.7d} + \frac{5.74}{Re^{0.9}}\right)\right]^2} \qquad \text{Equ 2.18}$$

Other formulae which often occur in literature for the calculation of straight pipe losses in the turbulent region are the Hazen-Williams and Manning formulae. These, however, do not provide any additional information to the equations presented above and have, therefore, been omitted.

2.2.3 Head losses in fittings

Head losses in bends, valves, etc. can be calculated with the help of this formula:

$$h_f = \zeta \cdot \frac{v^2}{2g} \quad (m) \qquad \text{Equ 2.19}$$

where

ζ = loss coefficient of fitting (dimensionless)
v = flow velocity (m/s)

The magnitude of the head losses are in principle influenced, as in the case of straight pipe flow, by surface roughness and by Reynolds Number. Examples of approximate values of loss coefficients for losses in fittings are illustrated in figure 2.13. All values are those for conditions of normal roughness and at high Reynolds Number, i.e. at fully developed turbulent flow. In cases where the velocity changes it must always be the highest velocity which is used for calculating h_f in accordance with Equation 2.19. In the case of varying diameters, the velocity to be applied is the one resulting from the smallest diameter, and in the case of T-pieces, the velocity related to the total volume flow is applicable.

An alternative method of expressing the magnitude of head losses in fittings is the use of the concept of equivalent pipe length, l_{eq}, i.e. the equivalent length, (leq) of straight pipe giving rise to the same losses in pressure as the fitting, at the same flow velocity. By comparing Equations 2.15 and 2.19 we get:

$$l_{eq} = \frac{\zeta}{\lambda} \cdot d \qquad \text{Equ 2.20}$$

where

ζ = loss coefficient of fitting (dimensionless)
λ = loss coefficient for straight pipes (dimensionless)

It can be seen from Equation 2.20 that the equivalent pipe length is a function of λ, i.e. the surface roughness of a straight pipe and Reynolds Number. It is therefore difficult to quantify l_{eq}, as a specific property for, say, a pipe bend and is thus of an approximate character. For λ = 0.02, the equivalent pipe length of a short radius 90° bend (r = d) is l_{eq} = 20d.

For additional information and data on losses see Internal Flow Systems by D. S. Miller and Crane Co. (USA) Technical Papers 409 and 410.

2.2.4 Hydraulic diameter

For the case of flow in a partially filled pipe, in non-cylindrical ducts or in open channels the head losses can be calculated in principle in the same way as described in earlier sections. Instead, however, of using the diameter of the pipe, the hydraulic diameter, d_h, must be used in these cases.

$$d_h = \frac{4 \cdot A}{P} \qquad \text{Equ 2.21}$$

where

d_h = hydraulic diameter (m)
A = cross-sectional area flowing (m2)
P = wetted perimeter (m)

For a completely full cylindrical pipe

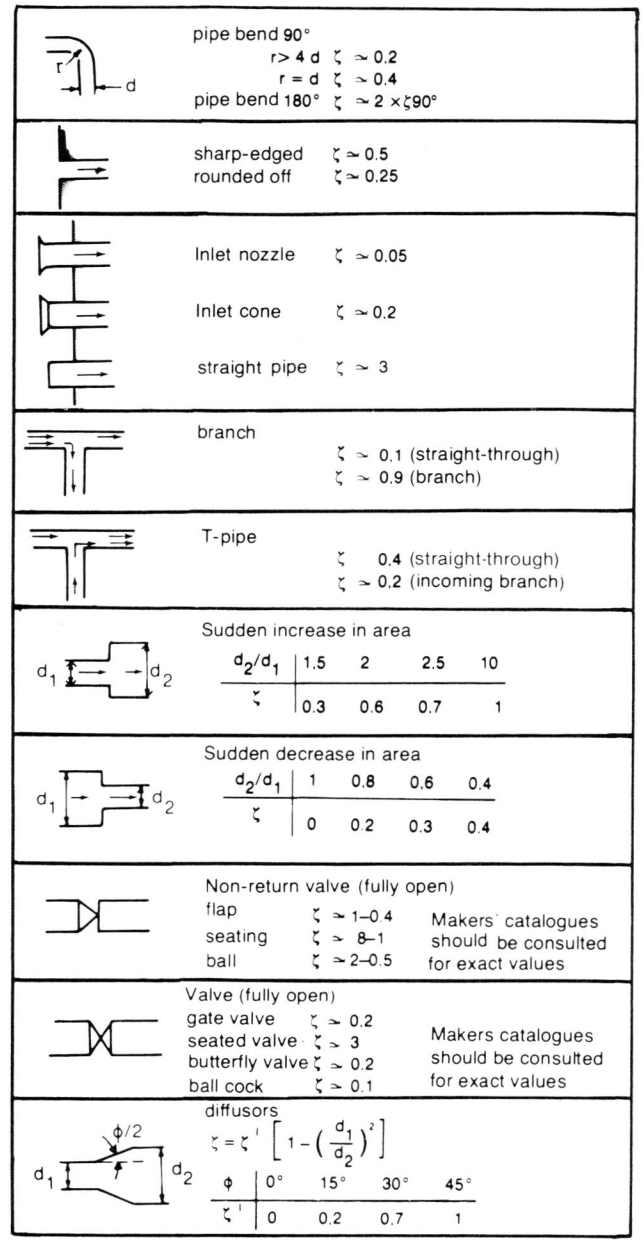

Fitting	Loss coefficient
pipe bend 90°, r > 4d	ζ ≈ 0.2
pipe bend 90°, r = d	ζ ≈ 0.4
pipe bend 180°	ζ ≈ 2 × ζ90°
sharp-edged	ζ ≈ 0.5
rounded off	ζ ≈ 0.25
Inlet nozzle	ζ ≈ 0.05
Inlet cone	ζ ≈ 0.2
straight pipe	ζ ≈ 3
branch (straight-through)	ζ ≈ 0.1
branch (branch)	ζ ≈ 0.9
T-pipe (straight-through)	ζ ≈ 0.4
T-pipe (incoming branch)	ζ ≈ 0.2

Sudden increase in area

d_2/d_1	1.5	2	2.5	10
ζ	0.3	0.6	0.7	1

Sudden decrease in area

d_2/d_1	1	0.8	0.6	0.4
ζ	0	0.2	0.3	0.4

Non-return valve (fully open)
- flap ζ ≈ 1–0.4
- seating ζ ≈ 8–1
- ball ζ ≈ 2–0.5

Makers' catalogues should be consulted for exact values

Valve (fully open)
- gate valve ζ ≈ 0.2
- seated valve ζ ≈ 3
- butterfly valve ζ ≈ 0.2
- ball cock ζ ≈ 0.1

Makers catalogues should be consulted for exact values

diffusors $\zeta = \zeta' \left[1 - \left(\frac{d_1}{d_2}\right)^2\right]$

φ	0°	15°	30°	45°
ζ'	0	0.2	0.7	1

Figure 2.13 Approximate values for loss coefficients for head losses in fittings. The values for flow meters are stated in Section 2.5

$$d_h = \frac{4 \frac{\pi \cdot d^2}{4}}{\pi \cdot d} = d$$

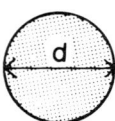

i.e. in this case the hydraulic and geometric diameters coincide.

For a half-filled cylindrical pipe,

$$d_h = \frac{4 \cdot \frac{\pi \cdot d^2}{2 \cdot 4}}{\frac{\pi \cdot d}{2}} = d$$

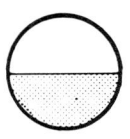

For a half-filled rectangular section,

$$d_h = \frac{4 \cdot b \cdot \frac{h}{2}}{b + 2 \cdot \frac{h}{2}} = \frac{2 \cdot b \cdot h}{b + h}$$

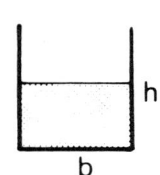

With Reynolds Number

$$Re = \frac{v \cdot d_h}{\nu} \qquad \text{Equ 2.22}$$

and relative roughness k/d_h, Figure 2.11 and formulae 2.16 and 2.18 apply with unchanged numerical values. The value of Re_{cr} will change as the section proportions change and will have to be evaluated by experiment or computer simulation.

Pipes flowing entirely due to gravitational head can operate completely filled. The degree of fullness depends, among other things, on the slope of the pipe and the operating conditions. For a pipe the fullness is defined by the ratio d^1/d where d^1 is the depth of liquid in the pipe and d is the diameter of the pipe. Figure 2.14 illustrates how various flow parameters change with the degree of fullness of the pipe. Both flow velocity and hydraulic diameter have their greatest value at a filling coefficient of just below 1.

Q = volume flow
A = area
V = velocity
d = geometric diameter
d_h = hydraulic diameter

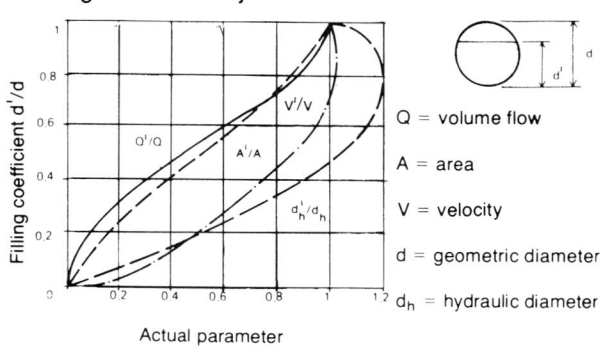

Figure 2.14 Flow in a partially-filled pipe

For gravity flow pipes carrying solids and liquids care must be taken with regard to sedimentation; the avoidance of which can place restrictions on the viable configuration and operation of the pipe.

2.2.5 Total head losses in the pipe system

The system total losses consist of the sum of the straight pipe head losses, the head losses in fittings and valves etc. and static losses due to changes in elevation.

$$h_f = h_f \text{ (straight pipe)} + h_f \text{ (fittings and valves)} + \Delta z \qquad \text{Equ. 2.23}$$

With the usual symbols head loss

$$h_f = \lambda \cdot \frac{1}{d} \cdot \frac{v^2}{2g} + \Sigma\zeta \cdot \frac{v^2}{2g} + \Delta z$$

$$= \left(\lambda \cdot \frac{1}{d} + \Sigma\zeta\right) \cdot \frac{v^2}{2g} + \Delta z$$

$$= \left(\lambda \cdot \frac{1}{d} + \Sigma\zeta\right) \cdot \frac{Q^2}{\left(\frac{\pi \cdot d^2}{4}\right)^2} \cdot \frac{1}{2g} + \Delta z \qquad \text{Equ 2.24}$$

where
- h_f = head loss (m)
- λ = loss coefficient for straight pipe
- ζ = loss coefficient for fittings, valves etc.
- $\Sigma\zeta$ = sum of all loss coefficients
- l = length of pipe (m)
- d = diameter of pipe (m)
- Q = volume flow (m³/s)
- v = flow velocity (m/s)
- Δz = ($z_{II} - z_I$) (m)

Another useful method for calculating head losses in fittings was proposed by William B. Hooper in the Chemical Engineering magazine; August 1981 and November 1988; The Two-K method.

The flow losses in both suction and discharge systems must be evaluated. For variable flow systems a Q-h_f curve should be produced for both systems; the suction system curve should also show NPSHA (Net Positive Suction Head Available). NPSHA is the difference between the liquid pressure and its vapour pressure. It is best for the system designer to use the top of the pump foundations as the datum level; the pump manufacturer can easily adjust the data to satisfy the pump configuration. Centrifugal pumps use the centre of the impeller eye as the datum whereas reciprocating pumps use the centre of the suction connection.

In long systems, or systems where the flow velocity is high; these terms are of necessity vague; the possibility of surge or water-hammer due to transients or fast acting valves must be considered. Very high peak pressures can be generated and the system/pump must be either designed to cope or protected from these effects. Relief valves and bursting discs can be fitted for protection, standard relief valves can be damaged in the event. Pulsation dampeners and air vessels may be fitted to reduce the magnitude of the pressure pulses generated. See Chapter 4 for more information.

Some positive displacement pumps produce uneven flows. Reciprocating pumps, due to the speed variation of the pistons or plungers, produce a cyclic flow variation in the suction and discharge systems. The flow variation in the suction system produces addition pressure losses which must be considered. The Hydraulic Institute (USA) and VDMA (Germany) have both produced data to allow this phenomenon to be evaluated.

In general, it is better to have more NPSHA than less. A wider range of pump types will be available if there is plenty NPSHA. Low NPSHA usually means bigger pumps running slowly.

Conversely high suction pressure can also cause problems; increased thrust, bearing cooling problems and reduced efficiency. The suction pressure should be considered relative to the discharge pressure. If the discharge pressure is greater than 16 barg and the suction pressure is greater than 20% of the discharge consult some manufacturers about possible problems and cost/size implications.

Comments

The treatment of pipe flow losses in Section 2.2 applies to Newtonian liquids of which the commonest is water. Non-Newtonian liquids require special attention regarding the determination of pressure losses. A special type of flow loss is caused by control valves in the pipe system. The control valve pressure losses are dealt with from a sizing point of view in Chapter 4 "Pump and piping systems" and from the regulation viewpoint in Chapter 5 "Flow regulation".

2.3 Liquid solid mixtures

2.3.1 General

Liquid solid mixtures, solid particles in liquid, behave as non-Newtonian liquids. There are many parameters which affect the flow characteristics of liquid solid mixtures. All pumping of solids, i.e. transportation of solid materials in liquid, is characterised by a certain degree of uncertainty. It is normally recommended that testing is carried out, with representative solids, to evaluate critical velocity, head loss and wear characteristic before the design is finalised.

Some of the main aspects of the behaviour of liquid solid mixtures are set out below, along with guide-line values for the estimation of the more important parameters for certain liquid solid mixtures.

2.3.2 Homogeneous mixtures

When the solid particles distribute themselves evenly throughout the cross-section of the pipe, without any concentration gradients, we speak of a homogeneous mixture or suspension. In practice, homogeneous mixtures are considered to occur when the particle size is less than about 50μm (0.050 mm).

Homogeneous mixtures with sufficiently high particle content behave as time-independent, plastic, non-Newtonian liquids. These kinds of liquid to some degree follow Binghams Law, see also Section 1.4.

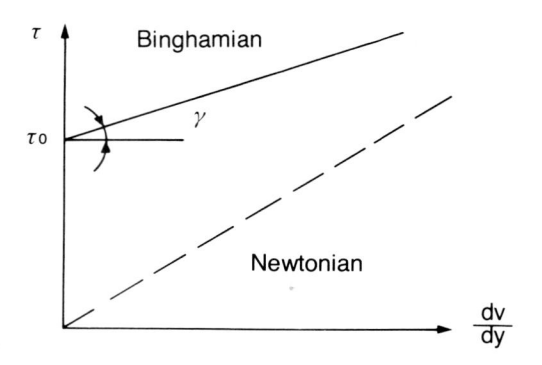

Figure 2.15 Homogeneous suspensions with flow characteristics as per Bingham

The pressure drop for laminar flow in straight piping measured in metres head of mixture is:

$$h_f = \frac{32}{\rho g} \cdot \frac{l}{d} \cdot \left(\frac{\tau_0}{6} + \frac{\tan \gamma \cdot v}{d} \right) \qquad \text{Equ 2.25}$$

where
h_f	=	loss of head	(m)
ρ	=	density of liquid solid mixture	(kg/m^3)
g	=	acceleration due to gravity	(m/s^2)
l	=	length of pipe	(m)
d	=	diameter of pipe	(m)
τ_0	=	viscous shear stress at boundary	(N/m^2)
$\tan \gamma$	=	plastic dynamic viscosity	(Ns/m^2)
v	=	velocity of flow	(m/s)

The viscous shear stress at boundary τ_0 and plastic viscosity, Tan γ, can be determined with the aid of a viscometer or by measuring the pressure drop in a test loop. Plastic viscosity, Tan γ, is also called apparent viscosity in some texts.

The loss of head for turbulent flow in straight pipe measured in metres of mixture is:

$$h_f = \lambda \cdot \frac{l}{d} \cdot \frac{v^2}{2g} \qquad \text{Equ 2.26}$$

i.e. exactly the same formula as for Newtonian liquids. The loss coefficient must be determined using the apparent viscosity.

Expressed in metres liquid column, the loss of head for a homogeneous liquid solid mixture is the same as for pure water in turbulent flow.

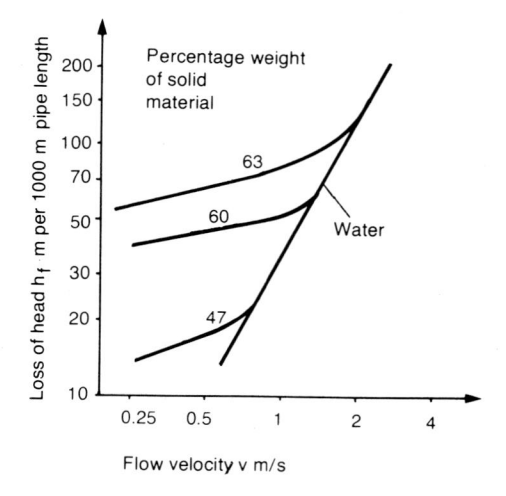

Figure 2.16 Examples of loss of head in pipe flow for a homogeneous liquid solid mixture (ρ_p = 2600 kg/m^3, d = 50 mm)

The transition from laminar to turbulent flow occurs, as for water, at Re ≅ 2300 but with the difference that the apparent viscosity should now be used for Reynolds Number. The velocity of flow at transition V_{cr} for Re = 2300 is:

$$V_{cr} \cong \frac{1150 \cdot \tan \gamma}{\rho \cdot d} \left(1 + \sqrt{1 + \frac{\tau_0 \cdot \rho \cdot d^2}{3600 \cdot \tan^2 \gamma}}\right) \quad \text{Equ 2.27}$$

where
- V_{cr} = critical velocity of flow at transition (m/s)
- $\tan \gamma$ = plastic viscosity (Ns/m²)
- ρ = density of liquid solid mixture (kg/m³)
- τ_0 = shear stress at boundary (N/m²)
- d = diameter of pipe (m)

Sedimentation in the pipe is avoided completely if the flow velocity in the pipe is greater than V_{cr}. Thus V_{cr} is of extraordinary importance when sizing pump installations for homogeneous mixtures in pipes. In short or carefully monitored pipes, $v < V_{cr}$ can sometimes be accepted.

There is an optimum solids concentration for the transportation of solid material by hydraulic methods. If the concentration is too low, the liquid solid mixture loses its plastic characteristic and with it the possibility of making use of the advantageous 'plug' flow. Besides which an unnecessarily large amount of liquid is pumped per kilogram of solid material if the liquid is discarded. If the particle content is too high the transportation must be carried out at increased velocity due to increased transitional velocity, V_{vc}, so as to avoid sedimentation.

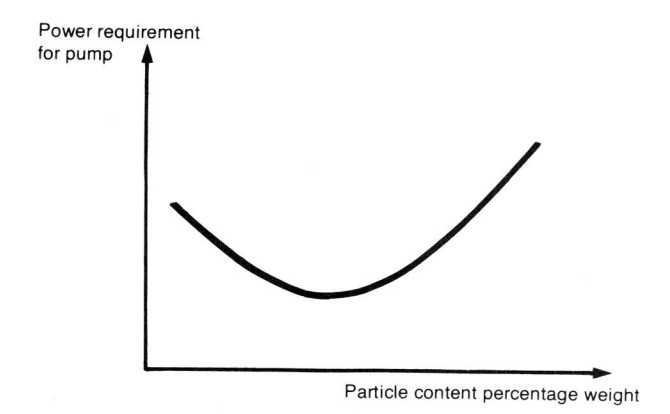

Figure 2.17 Optimum particle content for homogeneous liquid solid mixtures for the transportation of a given quantity of solid material per unit time

If, for example, the particle concentration in the liquid solid mixture in Figure 2.16 were to increase from 60% to 63%, the velocity of flow would have to increase from about 1.7 to 2.2 m/s. Thereby increasing the transportation capacity by a factor of 1.4, the loss of head by about 1.7 and the pump power consumption, with pipe losses only, by a factor of about 2.3. When the transportation capacity is constant but using a smaller pipe diameter the power requirement of the pump increases 1.7 times. The particle content is thus a critical economic parameter.

2.3.3 Heterogeneous mixtures

Heterogeneous mixtures or suspensions, that is to say liquid solid mixtures with concentration gradients across the pipe section, are considered in practice to occur for particle sizes of more than about 50 μm (0.050 mm). The main difference compared with homogeneous liquid solid mixtures is that the liquid and the particles in a heterogeneous mixture retain their identity. The flow is a two - phase flow. The viscosity of a heterogeneous mixture in principle is the same as that for the pure liquid.

The solid particles are conveyed suspended in the liquid if the particles are small and the velocity high, or by jumps if the particles are large or the speed low.

Particle size mm	Normal means of transportation
0.05 to 0.2	suspended
0.20 to 1.5	combined suspended
> 1.50	jumps

Figure 2.18 Normal means of transportation for various particle sizes

For all practical purposes, particle sizes in the order of 0.05 to 0.2 mm are transported suspended in the liquid. Liquid solid mixtures with particle sizes down towards the lower limit, 0.05 to 0.1 mm, comprise the boundary with the homogeneous mixtures. The most effective way of minimizing the risk of sedimentation and blockage of the pipe is to maintain the pipe flow in the turbulent regime. The pressure losses, expressed as loss of head in metres mixture, can then be calculated in the same way as for the liquid.

Liquid solid mixtures of particles with sizes up towards the upper limit, 0.1 to 0.2 mm, are on the boundary with the next group where the combined means of transportation prevails. The particle region 0.05 to 0.2 mm is not so well understood as other size ranges.

Mixtures with particles of 0.2 mm have been the subject of comprehensive pressure loss measurements; Durand, Condolios et al. Figure 2.19 shows an example of a loss diagram for a horizontal pipe.

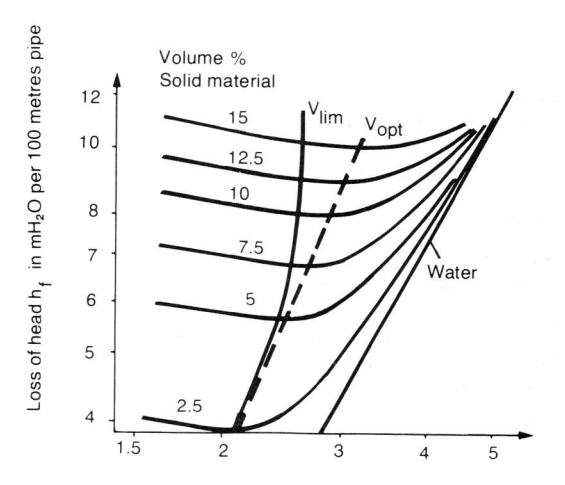

Figure 2.19 Example of loss of head in heterogeneous suspension, sand ρ_p - 2650 kg/m³, d_p mean ≅ 0.4 mm, d = 200 mm

The experimental results shown cover the region:
- d = pipe diameter — 40 to 600 mm
- d_p = particle size — 0.2 to 25 mm
- K_{vol} = volumetric concentration — 2.5 to 15%
- ρ_p = particle density liquid — 1500 to 5000 kg/m³ water

The loss of head expressed in metres of water (m H₂O) is greater than for pure water and exhibits a minimum at the optimum flow velocity v_{opt}. The loss of head for liquid solid mixtures of $d_p > 0.2$ mm flowing in horizontal pipes can be assessed using the formula:

$$h_f = \lambda \cdot \frac{l}{d} \cdot \frac{v^2}{2g} (1 + k_f) \quad \text{Equ 2.28}$$

where
- h_f = loss of head (m H₂O)
- λ = loss coefficient for water
- d = pipe diameter (m)
- v = velocity of flow (m/s)
- k_f = supplementary loss factor over and above that of pure water

The supplementary losses above those for pure water can be assessed from:

$$k_f = 85 \cdot K_{vol} \left[\frac{g \cdot d}{v^2} \cdot \left(\frac{\rho p}{\rho v} - 1 \right) \cdot k_p \right]^{3/2} \quad \text{Equ 2.29}$$

where
- K_{vol} = proportion of solid material by volume
- ρ_p = particle density (kg/m³)
- ρ_v = liquid density (kg/m³)
- k_p = factor for particle characteristics

The particle factor is assessable from:

$$k_p = \frac{v_s}{\sqrt{\frac{4}{3} \cdot \left(\frac{\rho p}{\rho v} - 1 \right) \cdot g \cdot d_p}} \quad \text{Equ 2.30}$$

where
- v_s = sedimentation velocity for a single particle in still water (m/s)
- d_p = particle diameter (m)

Some text books use the dimensionless "particle resistance coefficient" C_D instead of the particle factor k_p. The relationship between them is given by:

$$k_p = \frac{1}{\sqrt{C_D}} \quad \text{Equ 2.31}$$

The optimum flow velocity v_{opt}, i.e. the flow velocity for minimum loss of head at any given volumetric concentration, can be assessed from Equation 2.32.

$$v_{opt} = 3.5 \cdot K_{vol}^{1/3} \left[g \cdot d \cdot \left(\frac{\rho p}{\rho v} - 1 \right) \cdot k_p \right]^{1/2} \quad \text{Equ 2.32}$$

For particles >0.2 mm and density ρp = 2650 kg/m³, the text books give:

$$v_{opt} = 2.9 \cdot K_{vol}^{0.15} \cdot \sqrt{g \cdot d} \quad \text{Equ 2.33}$$

Another velocity of interest is the "limiting" velocity, below which the particles are in continuous contact with the bottom of the pipe. The limiting velocity can be evaluated, at testing, from:

$$v_{lim} = F_{lim} \sqrt{2g \cdot d \left(\frac{\rho p}{\rho v} - 1 \right)} \quad \text{Equ 2.34}$$

where the factor F_{lim} depends upon the volumetric concentration and size of particles and is determined experimentally.

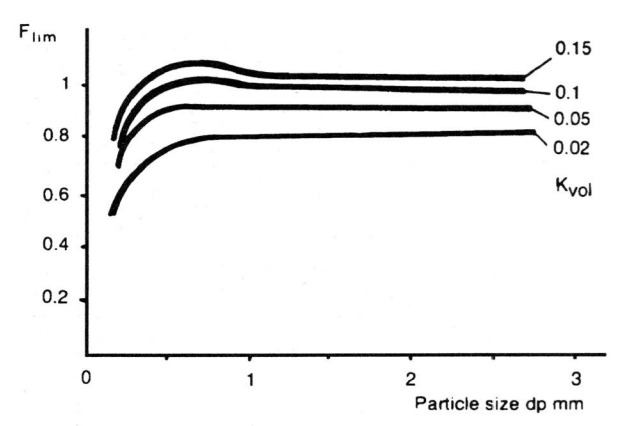

Figure 2.20 Factor F_{lim} for the determination of the limiting velocity v_{lim}

Liquid solid mixtures with particle contents of different sizes are dealt with as follows:
- Small particles which form a liquid solid mixture with the pure liquid at the velocity used are considered to be a part of the carrier liquid. The liquid and the small particles (d_p less than perhaps 0.1 mm, certainly less than 0.05 mm) constitute a new liquid with high density and viscosity values.
- The larger particles together with the carrier liquid form a heterogeneous mixture. Where there is a variation of size within the group of larger particles, a weighted mean value is used to determine the particle factor k_p.

Small particles increase the density of the carrier liquid and reduce the ratio of densities of transported particles and carrier liquid. Because of this both the supplementary losses and the limiting velocity for minimum loss and sedimentation are reduced. Additions of small particles can thus be economically attractive in the transportation of larger particles.

The sedimentation velocity v_s used to calculate the particle factor k_p should be quantified experimentally. The affect of the density of the liquid and its viscosity, together with the geometrical shape of the particle all affect the particle factor. The sedimentation velocity for sand in water is illustrated in figure 2.21.

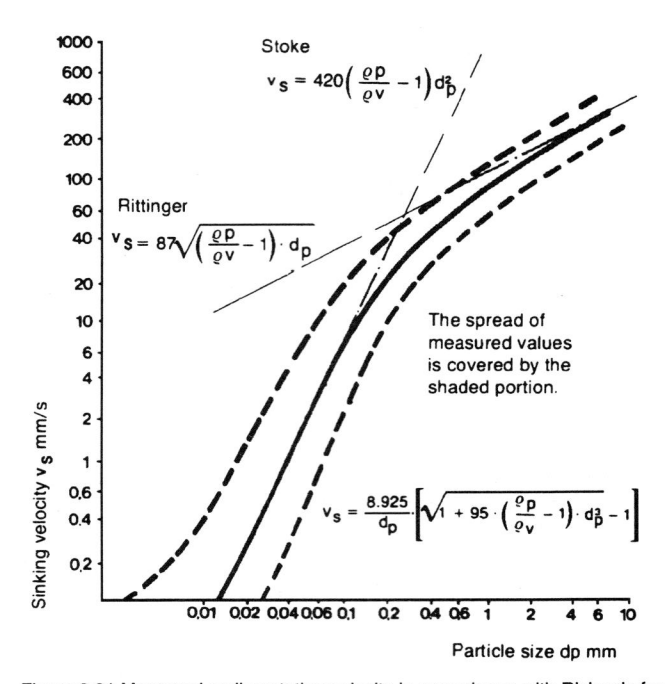

Figure 2.21 Measured sedimentation velocity in accordance with Richards for quartz sand (ρp = 2650 kg/m³) in water (10 °C, v = 1.3 × 10⁻⁶ m²/s) compared with various theoretical expressions

It is extremely difficult to give general guide-line values for the most suitable flow velocity for the hydraulic transportation of solid materials. Ash, however, recommends the following values:

Solids	Mean velocity m/s
waste from ore dressing plant	1.5 - 2.1
dust from precipitators	1.5 - 1.8
fly ash	1.8 - 2.1
fine sand	2.4 - 3.0
normal sand with fine	3.4
coarse sand without fine	3.7 - 4.0
shingle, max. 13 mm	4.3

These values are based on pipe diameters between 100 and 200 mm. Note the effect of the pipe diameter on the limiting velocity in Equation 2.34.

Special note

The formulae presented for the evaluation of properties for liquid solid mixtures are the result of extensive research. Further research is being conducted and more accurate expressions may be available at the time of printing. However all published data is approximate. The equations are derived by finding the "best fit" to test results. It is very unlikely that any of the test results would closely approximate a particular actual operating condition and there may be no way of evaluating how far the "best fit" equation is from the closest suitable test results. In practice, it will be the size distribution/content of the solids which determines the flow properties. When solids; copper/iron ore, coal; are crushed variable quantities and sizes of fines are also produced. Model tests on representative crushed samples are essential to confirm calculated results. Wear tests, such as Miller Number, ASTM G75, should also be undertaken to assess what precautions, if any, are required against wear.

Discussions with pump manufacturers, of several pump types, should be held as soon as approximate figures are available for solid sizes and distribution, flows and pressures. The discussions will focus on:

- efficiency; how efficient will the pump be and how will the efficiency be affected by changing operating conditions and wear,
- wear; which components will wear, predicted life of components and accessibility for maintenance,
- adjustment; how can the pump be adjusted, at site, to optimize efficiency and component life for changing operating conditions,

    multi-stage centrifugal pumps can be de-staged or have impeller diameters modified; reciprocating pumps can have different diameter pistons or plungers fitted,

- NPSHR; some pumps require a lot of NPSHA and this can radically effect the design of suction systems.

2.3.4 Paper pulp*

General

The behaviour of pulp suspensions when flowing through a pipe is unusual from several aspects. The loss of head for chemical pulps goes through a maximum and a minimum value when the flow velocity increases from zero, and the head loss is finally less than the value for pure water.

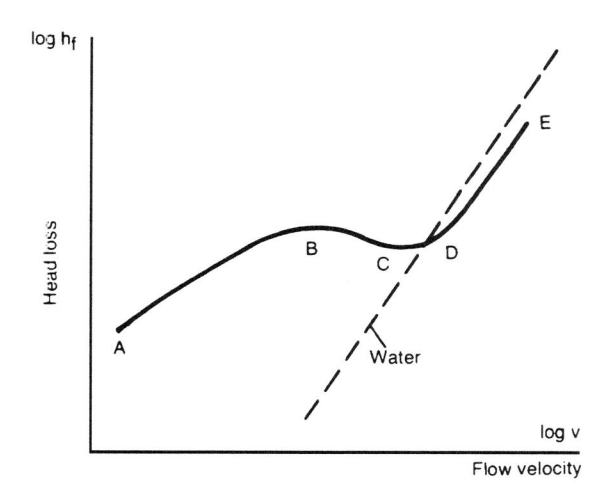

Figure 2.22 Loss of head for chemical pulp in pipe flow

For velocities up to point D, after the intersection of the two curves, the suspension flows like a plug surrounded by a thin boundary layer in which the whole of the viscous flow occurs. The character of the boundary layer varies according to the position of the working point on the curve:

A to B   The surface of the plug is disturbed by intermittent contact with the pipe wall.

* Section 2.3.4 is based upon results presented in "Friction losses in the pumping of paper pulp" published formerly by Scanpump Ltd, (now part of ABS Pumps Ltd, Industrial & Process Division)

B to C   The boundary layer consists mainly of pure water in laminar flow.

C to E   The flow in the boundary layer is turbulent.

Plug flow can be regarded as ended when the stress at the edge of the plug is equal to the ultimate shear stress of the network of fibres which form the plug. When the velocity increases beyond this point the plug breaks up and finally the whole cross section flows turbulently.

The friction loss curves for mechanical pulps do not usually have any maximum or minimum. The flow characteristics are modified by the relatively low strength of the network and the high proportion of non-fibrous solid material.

Calculation of pipe flow losses

The following symbols are used:

h_f = loss of head (m)
l = pipe length (m)
v = flow-velocity (m/s)
k = surface roughness of pipe (m)
d = pipe diameter (m)
ρ = density of suspension (kg/m^3)
μ = dynamic viscosity (Ns/m^2)
τ = shear stress (N/m^2)
DS = proportion by weight of fibre (absolutely dry) (%)
F_k = length/diameter ratio of fibre

Pipe flow losses can be determined with the help of a number of diagrams reproduced below.

The non-dimensional flow parameter S is defined as:

$$S = \left(\frac{V^5 \cdot \rho^2 \cdot \eta}{\tau_D^3} \cdot d\right)^{1/6} \qquad \text{Equ 2.35}$$

and similarly the non-dimensional loss parameter F:

$$F = \frac{\rho \cdot g \cdot h_f \cdot d}{4 \cdot \tau_D \cdot l} \qquad \text{Equ 2.36}$$

The shear stress τ_D is a measure of the strength of the plug of plaited fibres which occurs at most flow velocities in practical operations.

$$\tau_D = \tau_D^1 \cdot f_F \cdot f_R \cdot f_M \cdot f_D \qquad \text{Equ 2.37}$$

If the pulp never dries out, $f_D = 1$. If it has dried out and is absorbed again, $f_D = 0.75$.

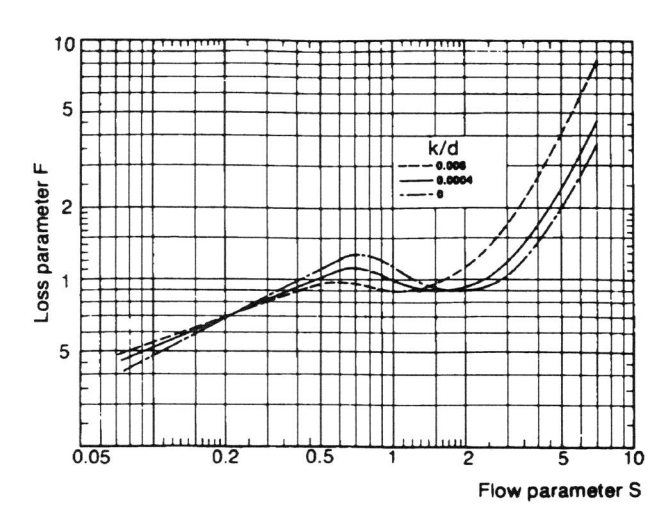

Figure 2.23 Pipe flow characteristics of chemical pulp

The diagrams, especially figure 2.23, apply to chemical pulps. The different characteristics of mechanical pulps (fibres, networks, etc.) give the loss curve a different shape and generally increased losses. Correction factors should therefore be used. Approximate values can be obtained, however, using the above method provided that:

- the loss parameter F never assumes a value lower than the maximum point in figure 2.23 for the section of the curve S >0.6.
- $f_k = 1$
- a safety factor of 1.2 should be used.

The guide-line values for head losses in bends and fittings are given in Section 2.4.4 in the form of equivalent pipe lengths.

It is typical of pulp suspensions, as for all plastic non-Newtonian liquids, that a certain yield shear stress (stiction) has to be overcome before a flow can be made to start up at all. This stiction can be reduced considerably by dilution.

Example

A chemical pulp, 281 m³/h, has to be pumped through a 70 m long stainless steel pipe diameter 300 mm. The pulp has never been dried and has a fineness grading 600 C.s.f., the temperature is 30 °C and the concentration 2.7% absolutely dry. The average fibre length is 2.4 mm and mean diameter 40 μm. Determine the loss of head through the pipe.

Fibre ratio $F_K = 2.4/0.040 = 60 \rightarrow f_F = 0.89$

Stainless steel pipe k/d = 0.045/300 = 1.5 x 10⁻⁴ → $f_R = 1.11$

Fineness grading 600 Csf → $f_M = 0.98$

Undried chemical pulp → $f_D = 1$

DS content 2.7 % → $\tau^1{}_D = 27$ N/m²

Shear stress τ_D → 27 · 0.89 · 1.11 · 0.98 · 1 = 26.1 N/m²

Water at 30 °C → μ = 0.8 · 10⁻³ Ns/m², ρ ~ 1000 kg/m³

Flow velocity

$$v = \frac{Q}{\frac{\pi \cdot d^2}{4}} = \frac{281 \cdot 4}{3600 \cdot \pi \cdot 0.3^2} = 1.1 \text{ m/s}$$

The flow parameter

$$S = \frac{1.1^5 \cdot 1000^2 \cdot 0.8 \cdot 10^{-3}}{26.1^3 \cdot 0.300} = 0.79$$

Read off loss parameter F = 1.05 (figure 2.23)

$$h_f = \frac{4 \cdot \tau_D \cdot l}{\rho \cdot g \cdot d} \cdot F = \frac{4 \cdot 26.1 \cdot 70}{100 \cdot 9.81 \cdot 0.3} \cdot 1.05$$

$h_f = 2.4$ m

2.4 Pressure losses - nomograms & diagrams

2.4.1 General

Pre-computed pressure losses in pipes for a number of commonly applied cases of flow are set out in this section in the form of diagrams. These diagrams and nomograms can be used to calculate the losses in a system, i.e. flow losses in the pipes, valves, bends, etc. This method is quick and provides acceptable results if carried out accurately. A number of diagrams for the determination of the associated values of velocity, frictional losses, diameter and volume flow are presented. There are also diagrams for determining the flow velocity in pipes in relation to the volume flow, mass flow and diameter together with diagrams for the determination of head loss (m) in bends fittings etc., related to the velocity of flow.

It is most practical when sizing pipes to be able, right from the outset, to assess the approximate economic flow velocity. By this means, the final economic sizing can be carried out rapidly, since fewer alternatives have to be tested. Figure 2.25

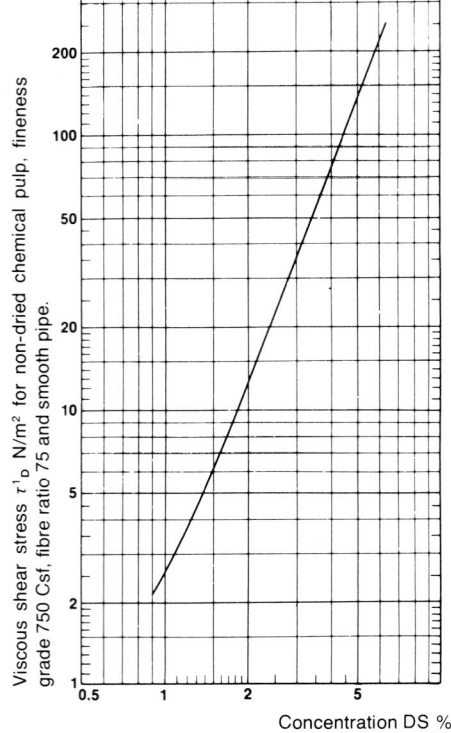

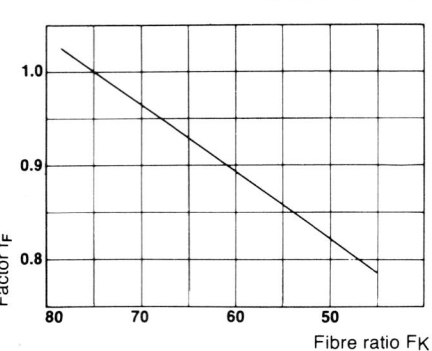

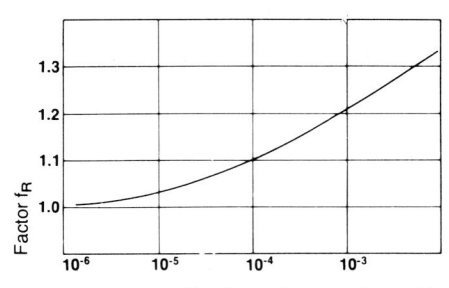

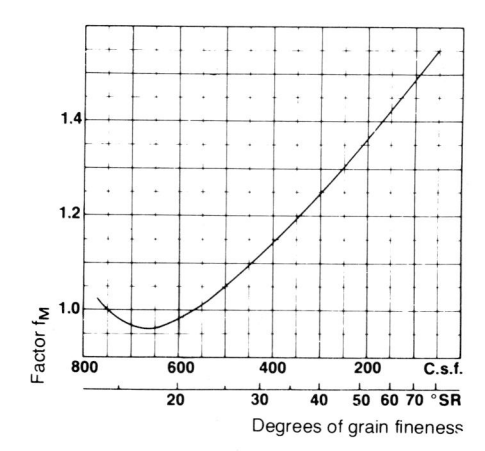

Figure 2.24 Properties of chemical pulps

gives guide-lines for normal flow velocities in some common cases.

Liquid Application	m/s
Dirty water	
Mine water lines	1.75 to 2.5
Hot water	
Condensate	1.0 to 1.5
District heating distribution	1.0 to 2.5
District heating service main	0.75 to 2.0
Domestic heating circulation	0.3 to 0.6
Feed water lines	1.2 to 4.3
Crude oil	
Oil pipelines	0.25 to 1.8
Lube oil	
Drain line (1)	0.15 to 0.75
Feed lines	1.0 to 2.3
Pump discharge line	0.6 to 1.8
Pump suction line (2)	0.45 to 1.5
Pump suction line (3)	0.3 to 1.0
Petrol	
Pump discharge	1.0 to 1.4
Pump suction	0.45 to 0.75
Refrigerant	
Condenser to receiver	≤ 1.0
Receiver to system (4)	≤ 1.5
Sewage	
Pump discharge, centrif	≥ 2.4
Pump suction, centrif	≥ 1.5
Warm water	
Economiser tubes	0.75 to 1.5
Water	
Pump discharge, centrif	8 to 15
Pump discharge, recip	3.0 to 4.5
Concrete suction bays	≤ 0.3
Pump suction headers	≤ 1.0
Pump suction, centrif	0.45 to 1.5
Pump suction, recip	≤ 0.6
Consumer supply	1.0 to 1.4
Service mains	0.6 to 1.5
Trunk main	1.0 to 3.0

(1) 1:50 fall minimum, half full, (2) flooded suction, (3) suction lift, (4) 0.115 bar / 25m maximum

Figure 2.25 Guide-line flow velocities for general applications

High flow velocities give small pipe dimensions = low plant procurement costs, but cause greater pump power consumption through greater frictional losses = high running costs.

Quite apart from economy, other factors can effect the choice of flow velocity:
noise in central heating equipment,
erosion in copper hot water pipes,
static electricity for liquids with low flash points,
risk of cavitation in suction lines,
risk of cavitation with sudden changes in velocity,
sedimentation with liquid solid mixtures, i.e. critical velocity,
vibration of pipework,
surge and pressure pulsation problems.

Special attention should be given to pump suction systems. The best systems are short, straight and one pipe size larger, at least, than the pump suction connection. Any bends should be more than five pipe diameters from the pump connection. If the NPSH margin is small it is critically important to evaluate all system losses for all flow rates. If the suction system causes pump cavitation, operational problems will reduce the maintenance interval. Typical problems include:
loss of pump capacity,
high rotating/reciprocating parts wear,
sealing problems,
fatigue failures,
discharge pipework noise and vibration,
pipe connection leaks.

Remember thermodynamic data is for pure substance, no contaminants. Fresh water will contain dissolved gases. Gases will evolve from solution at a much higher pressure than the vapour pressure of the water; typically at 8m absolute when cold. Gas evolution causes the same problems, and looks exactly the same, as cavitation. For liquids with dissolved gases, the NPSHA must be based on the gas evolution pressure, not liquid vapour pressure. A pump cannot distinguish between released gas and saturated vapour, the effects and the damage are the same.

Thermodynamic data must be comprehensive if used for pump applications. Data on compressed liquid state must be available as well as saturation line data.

The higher velocities, listed in figure 2.25, demand better pipework designs and high quality manufacture and assembly. Poorly fitted slip-on flanges, badly aligned, will cause extra turbulence and losses. Bent pipe is preferable to fabricated bends. In most cases, large radius should be used in place of short. The minimum of weld neck flanges, or connectors which use locating seal rings, such as Grayloc's, will preserve good flow patterns.

Losses in bends, valves etc., can be expressed as equivalent straight pipe lengths by using the values in figure 2.26. The values for valves are mean values and vary according to the design.

By adding the equivalent length of pipe and the straight length, the total pipe length is obtained enabling the flow losses, expressed as head (m), to be calculated from the diagrams.

Obstacle type	Equivalent pipe length	
	d = 10 - 50 mm	d = 80 - 400 mm
Gate valve	15 - 10 x d	10 - 5 x d
Check valve (with flap)	200 - 150 x d	150 - 100 x d
Angle valve (seating)	400 - 200 x d	150 - 100 x d
Stop valve (straight seating)	1000 - 500 x d	500 - 400 x d
Pipe bend 45°	10 x d	10 x d
Pipe bend 90°	30 - 20 x d	15 - 10 x d
Tank outflow (rounded)	10 - 5 x d	10 - 5 x d
Tank outflow (sharp edged)	40 - 30 x d	40 - 30 x d
T-branch through side outflow	50 - 40 x d	40 x d

Figure 2.26 Conversion factors for equivalent pipe lengths, (guide-line values)

As pointed out in Section 2.2.3, the equivalent pipe length depends upon the straight pipe loss coefficient A, i.e. pipe roughness, Reynolds Number etc. The numerical values in figure 2.26, therefore, must be regarded as approximate guide-line values.

Alternatively, losses in bends, valves etc., can be separately calculated by converting the values for ζ obtained from figure 2.13 into metres head (column of liquid) at specific flow velocities with the help of the diagrams. Note that the stated value of ζ is for turbulent flow. In the case of laminar flow conversion to equivalent pipe length is the handiest way to calculate these losses.

	Content			
2.28	Flow velocity in pipes for volume, m³/h and l/min, and pipe diameter			
2.29	Losses in bends and fittings etc., (m)			
	Frictional losses in straight pipe			
	Liquid	Temp °C	Factor k roughness mm	Type of pipe
2.30	Water	10	0.01 - 0.05	steel, Cu, PE, PVC
2.31	Water	10	0.1	steel, cast iron, GRP
2.32	Water	10	0.2	steel, cast iron, concrete, wood
2.33	Water	10	0.5	steel, cast iron, concrete
2.34	Water	10	1.0	concrete, earthenware
2.35	Water	10	2.0	cast iron
2.36	Oil	-	0.03	steel d = 20 & 25 mm
2.37	Oil	-	0.03	steel d = 32 & 40 mm
2.38	Oil	-	0.03	steel d = 50 & 60 mm
2.39	Oil	-	0.03	steel d = 80 & 100 mm
2.40	Oil	-	0.03	steel d = 125 & 150 mm
2.41	Oil	-	0.03	steel d = 175 & 200 mm
2.42	Oil	-	0.03	steel d = 250 & 300 mm
2.43	Oil	-	0.03	steel d = 350 & 400 mm
2.44	Normal pulps	40	0.045	stainless steel fittings
2.45	Normal pulps	40	0.045	stainless steel 80 to 600 mm

Figure 2.27 Index of diagrams for calculating pressure losses in pipe lines

The applicability of the diagrams is wider than the impression given by figure 2.27. The diagrams for water can be used for all water-like Newtonian liquids, i.e. for liquids with corresponding viscosity. Note, however, that if the pressure loss is stated in Pa (Pascal) the density must also coincide with that of water.

Fig. 2.28 Flow velocity in pipes in for volume flow and diameter. Interpolate between the lines for other diameters

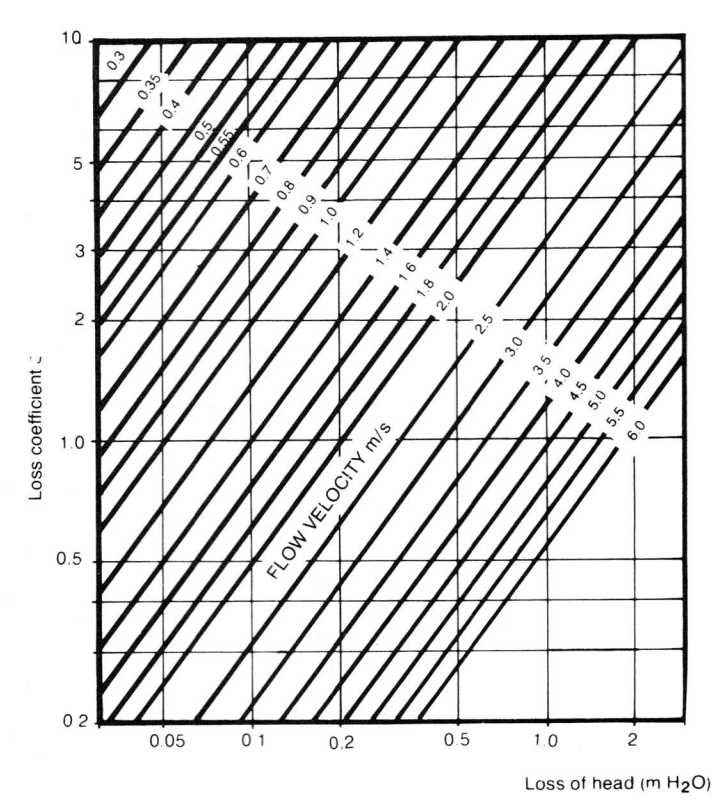

Figure 2.29 Relation between loss coefficient and loss of head in bends and fittings

The diagrams for oils can be used for all Newtonian liquids with higher viscosities than water, under the same conditions.

Example:
Calculate the total loss of head in 3 x 90° short radius pipe bends in a 100 mm diameter pipe for a water flow of 700 l/m.
From Figure 2.28, velocity of flow = 1.5 m/s
From Figure 2.13, ζ = 3 x 0.4 = 1.2
From Figure 2.29, loss of head = 0.14 m

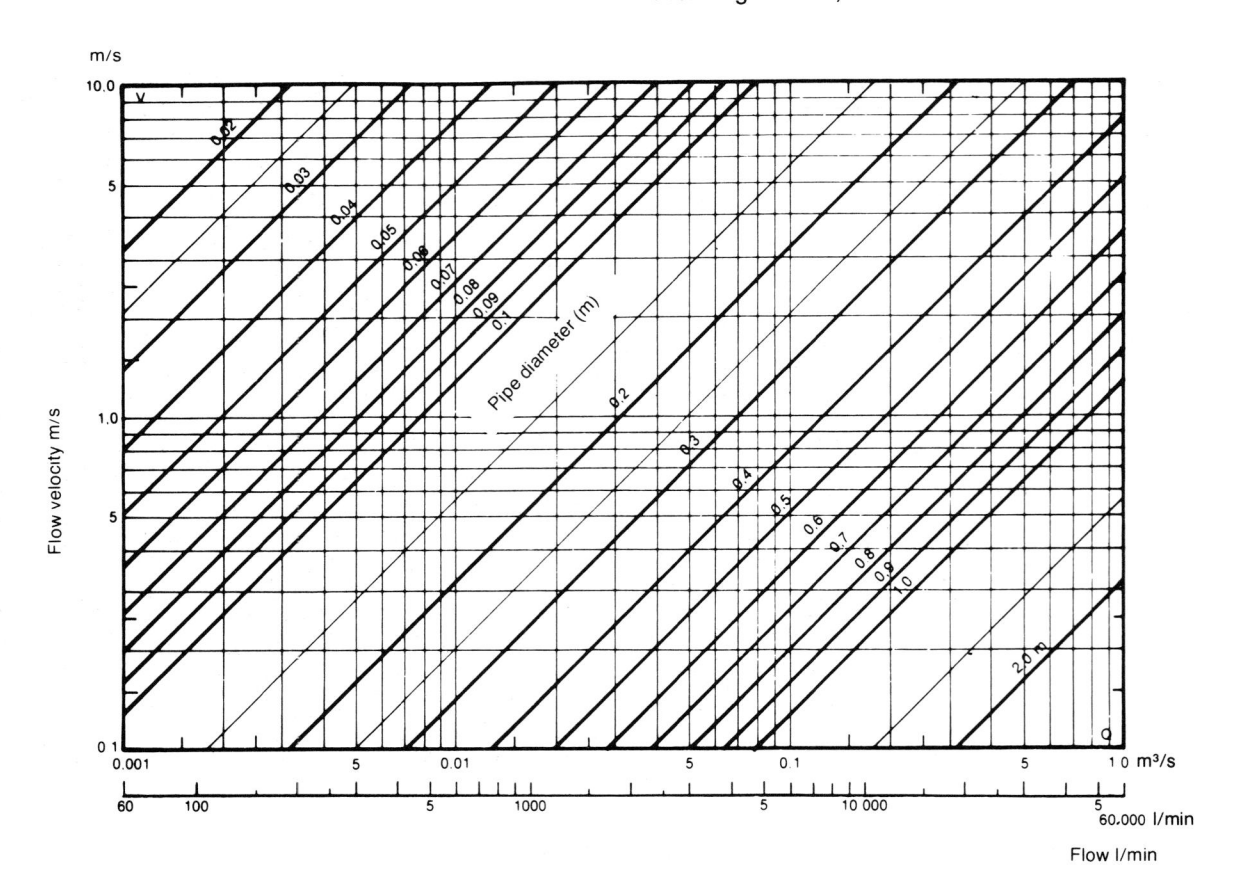

2.4.2 Water

The diagrams also apply to other Newtonian liquids having similar viscosity to that of water when the pressure loss is expressed as head (metres of liquid column).

Example:

A flow of 10 l/s is required through a 120 mm diameter pipe, 200 m long. Calculate the flow velocity and the loss of head. From the diagram, the velocity is 0.9 m/s and loss of head 200 x 7/1000 = 1.4 m.

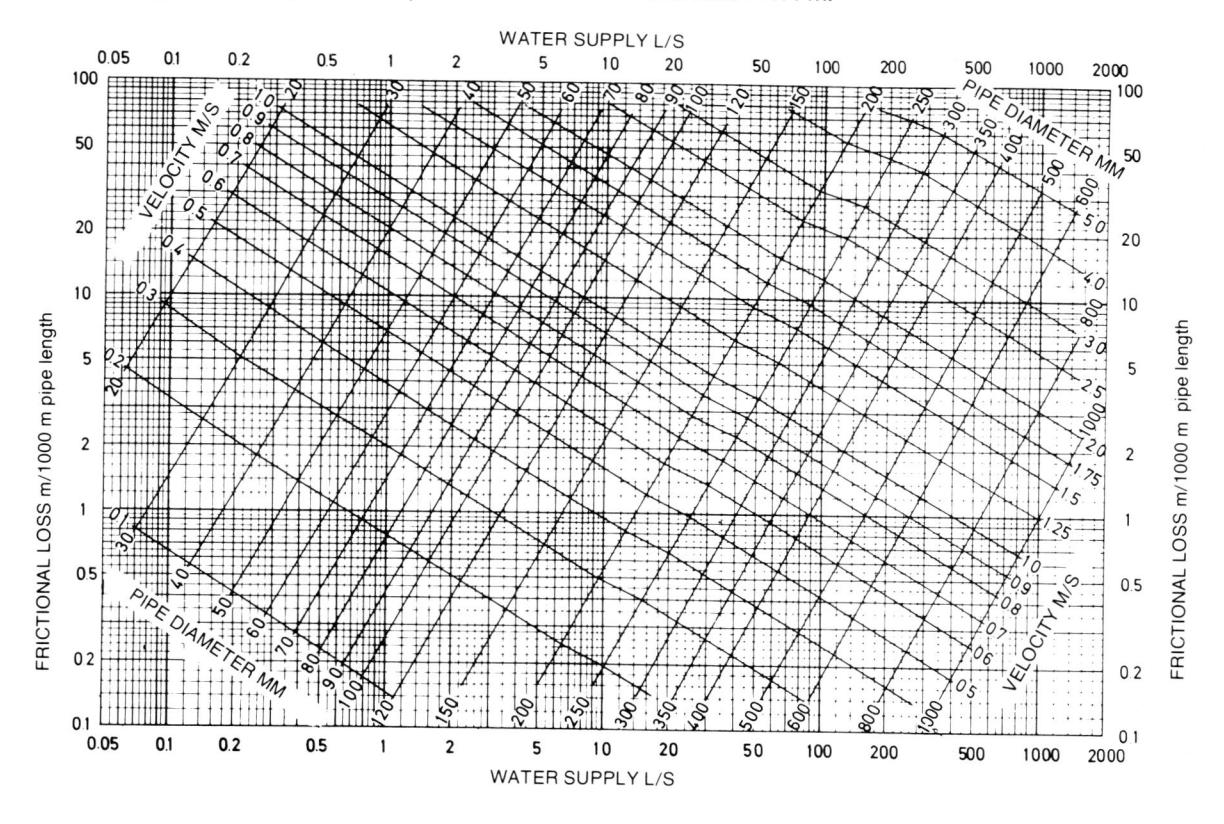

Figure 2.30 Frictional loss diagram for water at +10 °C. Pipe type: steel, Cu, PE, PVC. k = 0.01 mm for diameter ≤ 200 mm and k = 0.05 mm for diameter > 200 mm

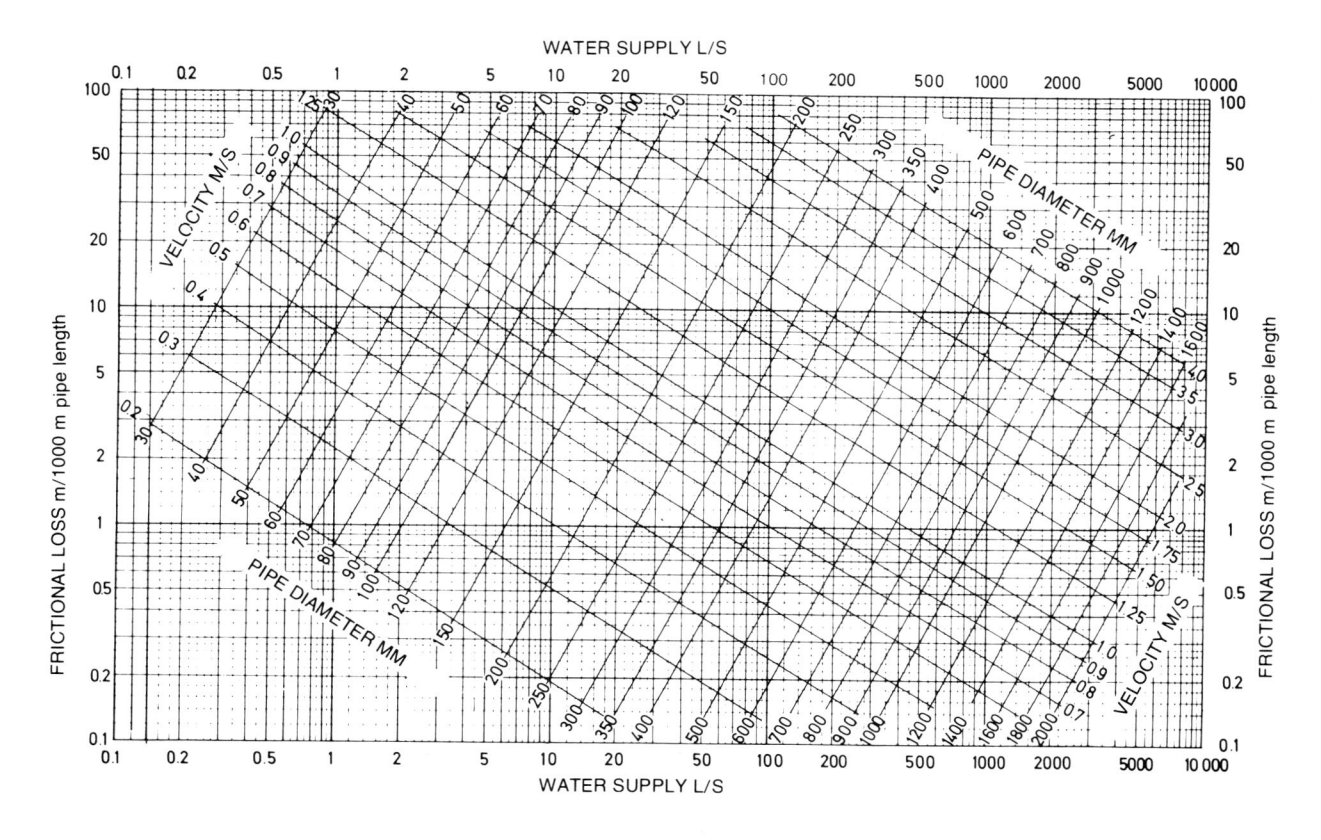

Figure 2.31 Frictional loss diagram for water at +10°C, k = 0.1 mm. Pipe type: steel, cast iron, GRP

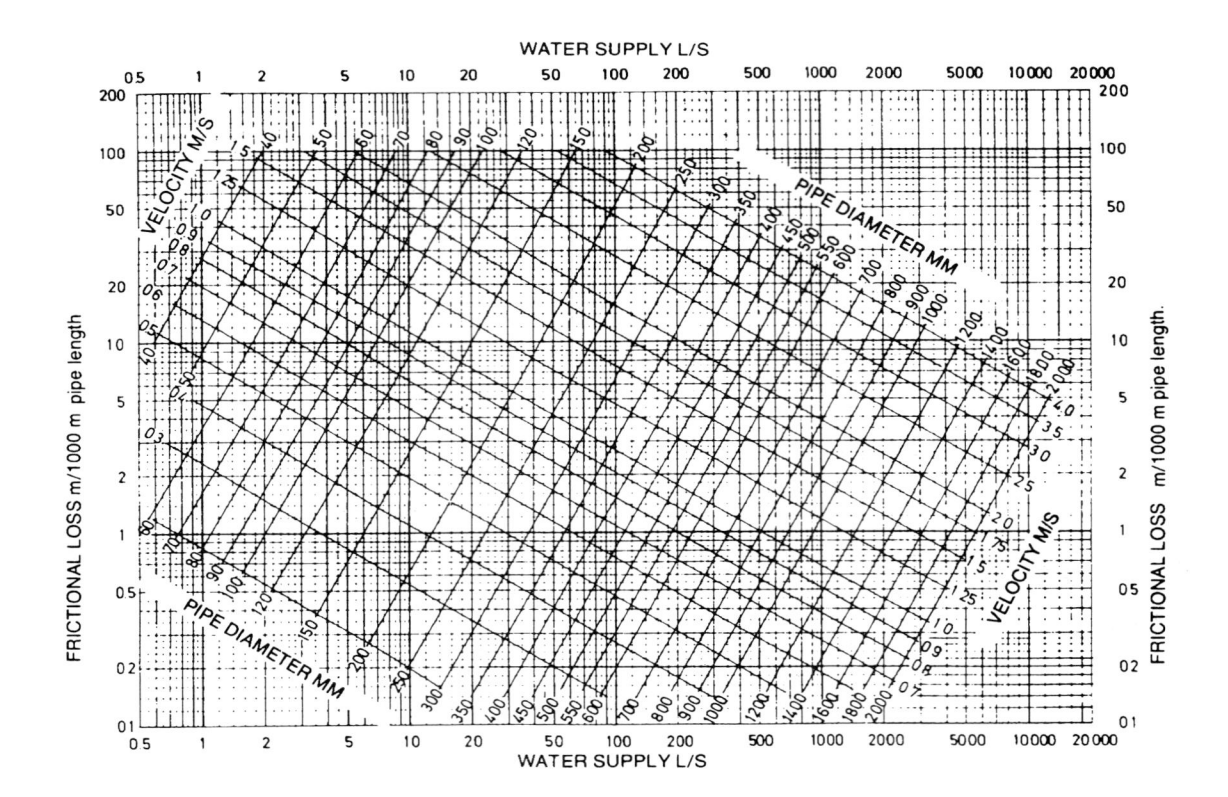

Figure 2.32 Frictional loss diagram for water at + 10 °C, k = 0.2 mm. Pipe type: steel, cast iron, concrete, wood

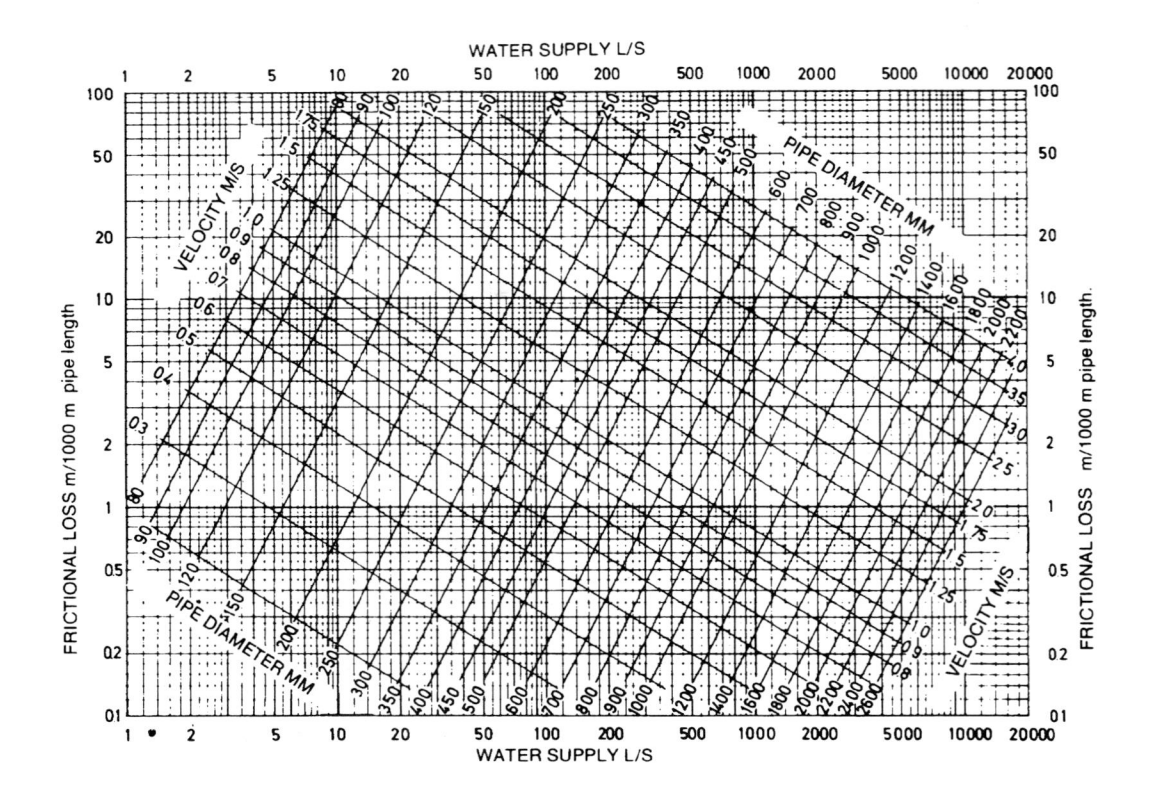

Figure 2.33 Frictional loss diagram for water at + 10 °C, k = 0.5 mm. Pipe type: steel, cast iron, concrete

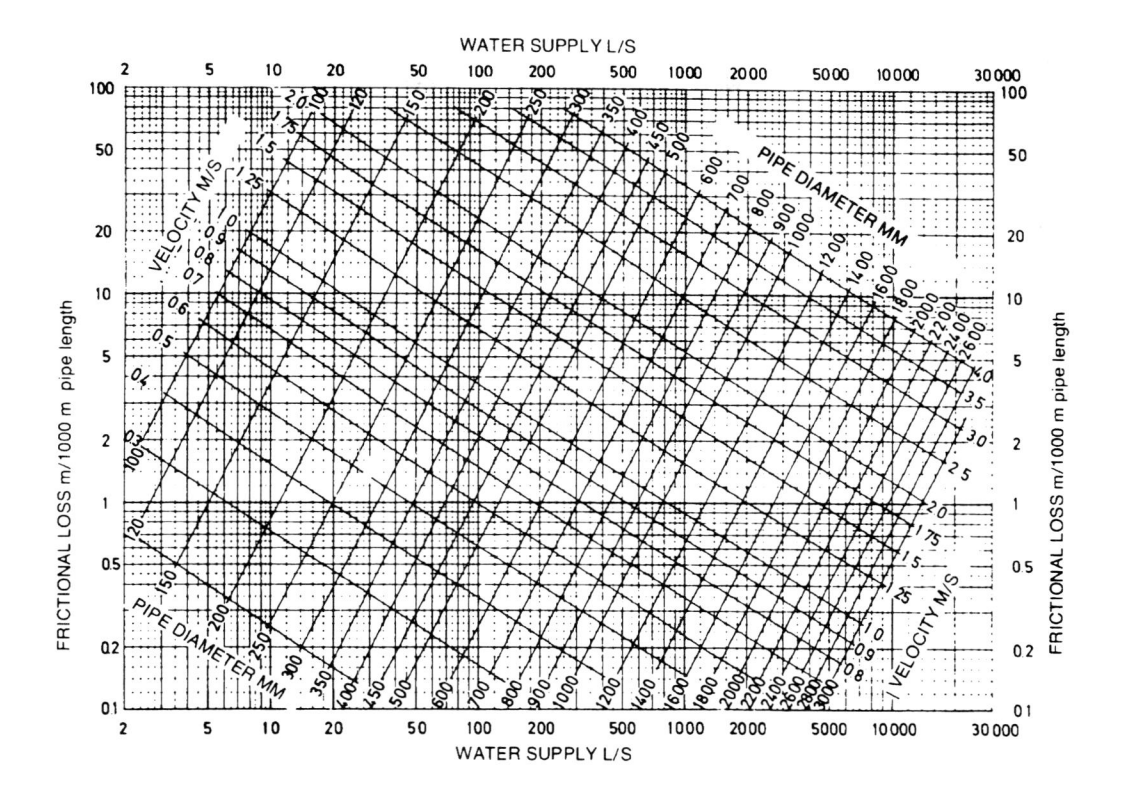

Figure 2.34 Frictional loss diagram for water at + 10 °C, k = 1.0 mm. Pipe type: cast iron, concrete, earthenware

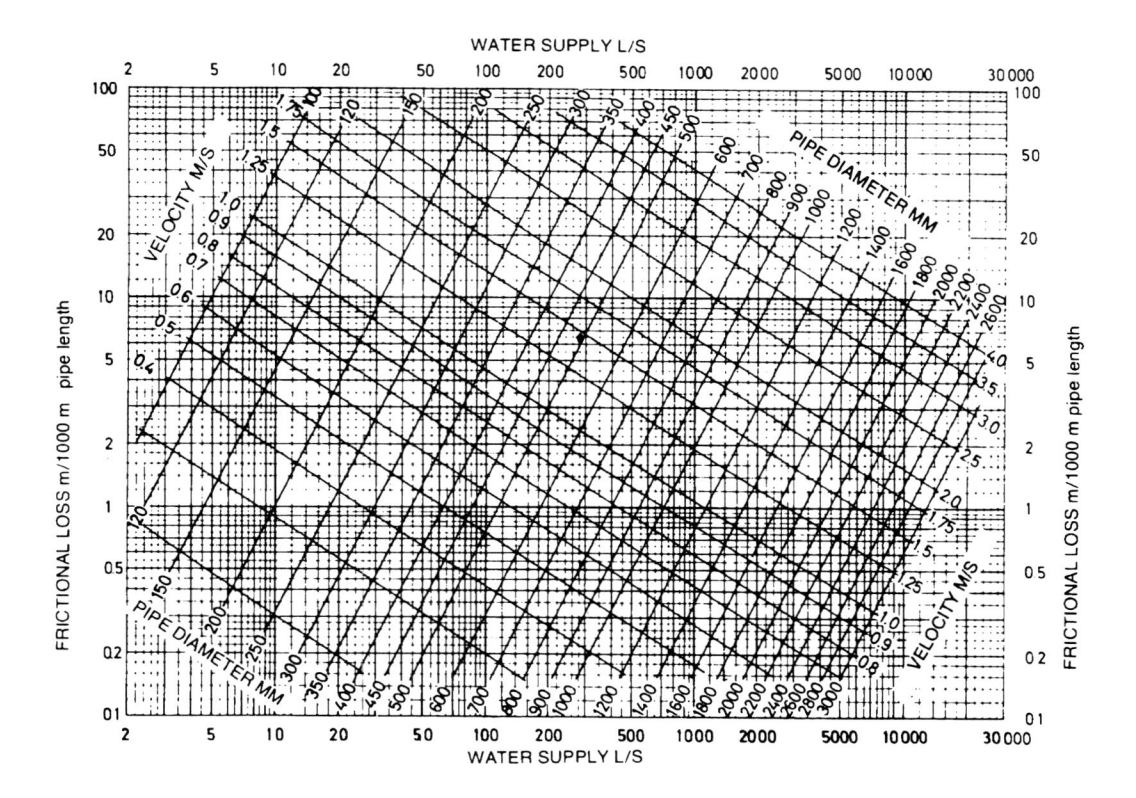

Figure 2.35 Frictional loss diagram for water at + 10 °C, k = 2.0 mm. Pipe type: cast iron

2.4.3 Oil

The diagrams apply equally for other Newtonian liquids at the stated viscosity.

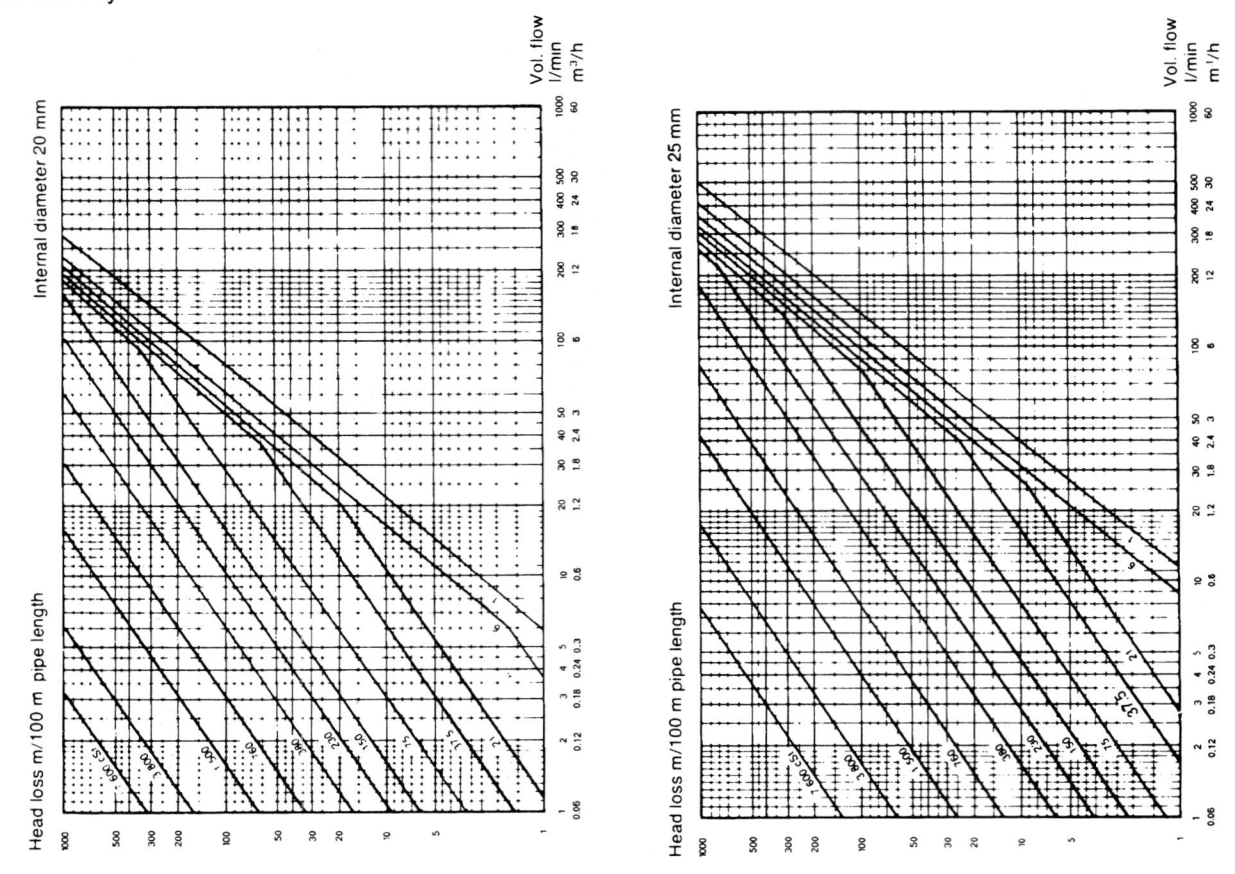

Figure 2.36 Frictional loss diagrams for oil at various viscosities, k = 0.03 mm, steel pipe

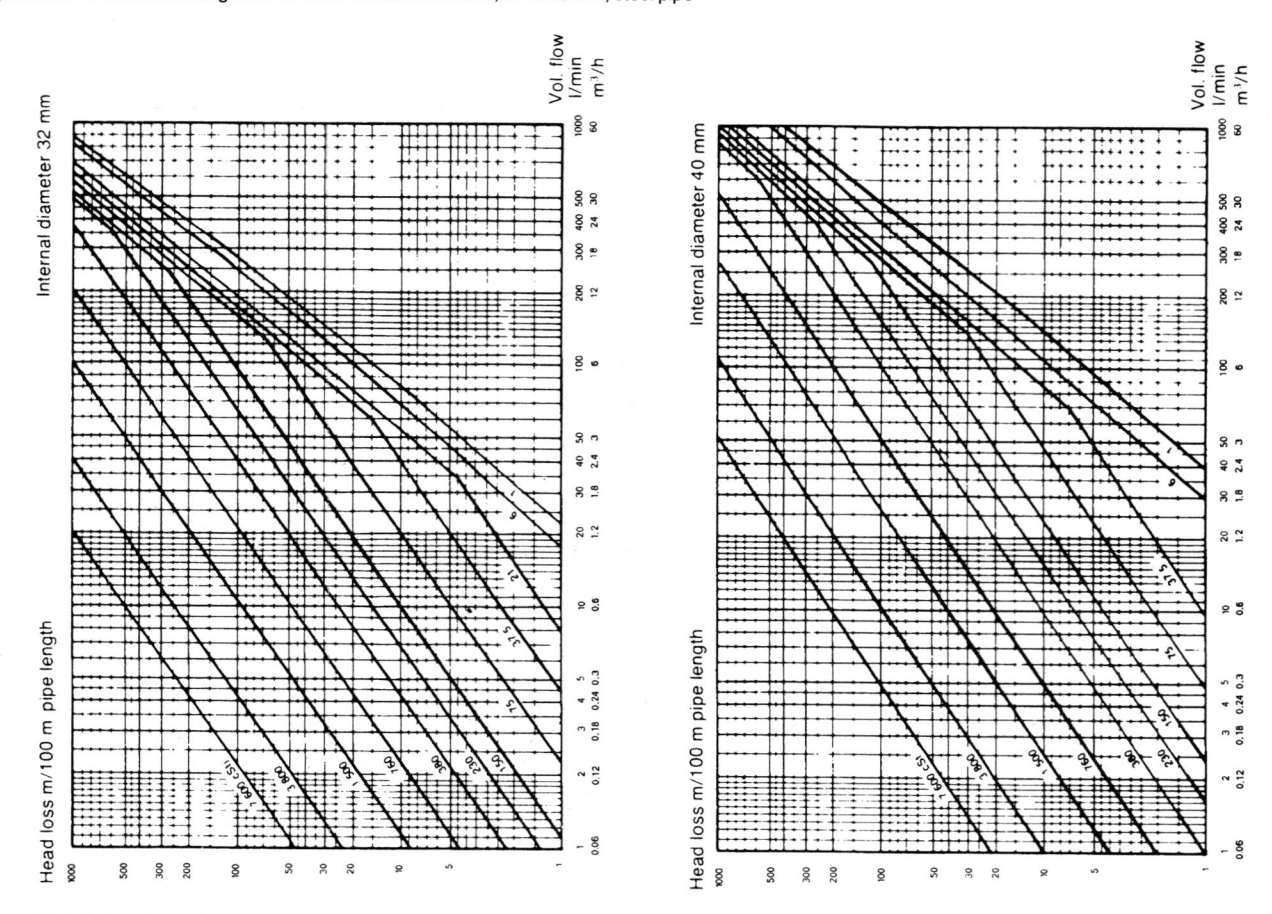

Figure 2.37 Frictional loss diagrams for oil at various viscosities, k = 0.03 mm, steel pipe

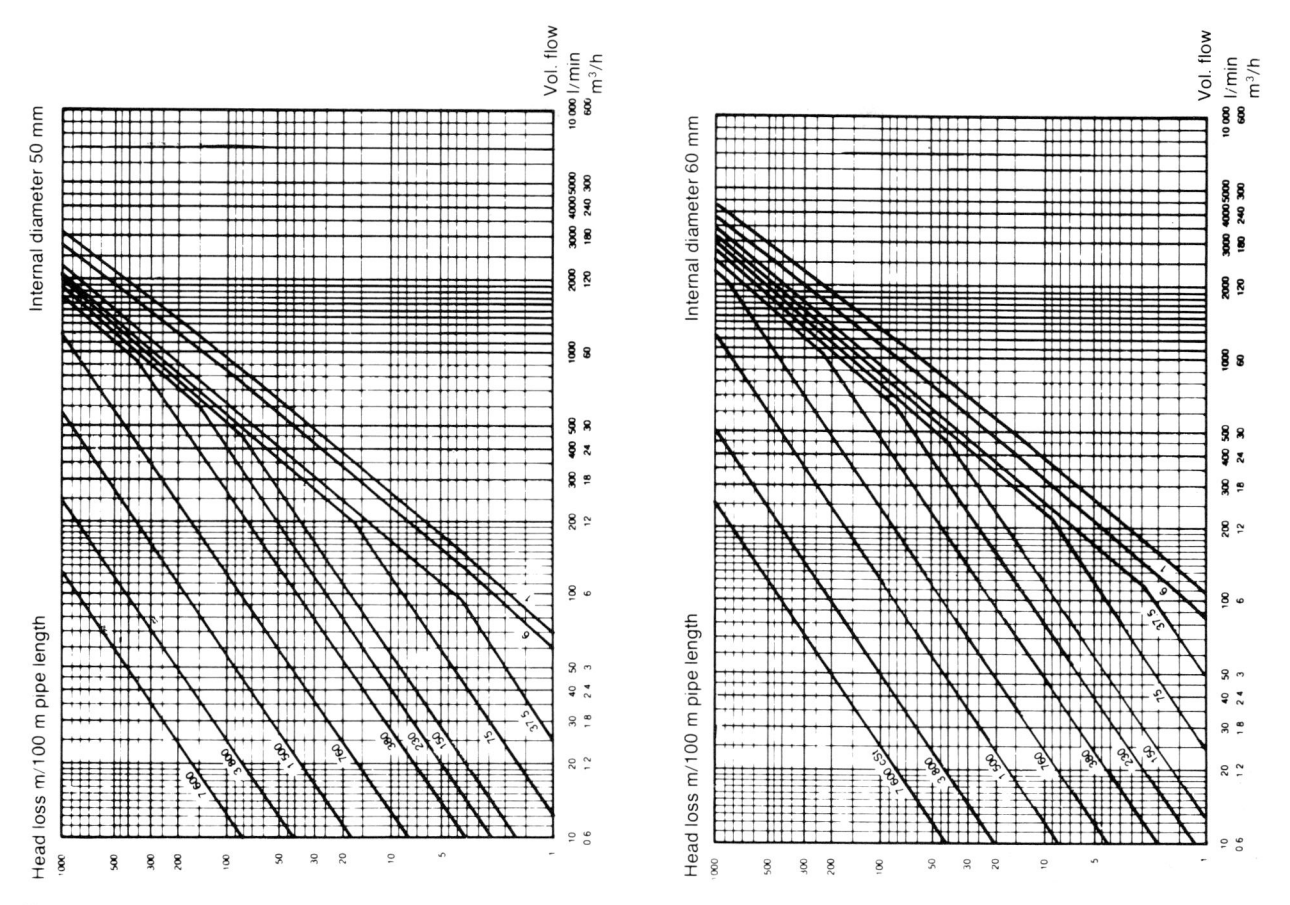

Figure 2.38 Frictional loss diagrams for oil at various viscosities, k = 0.03 mm, steel pipe

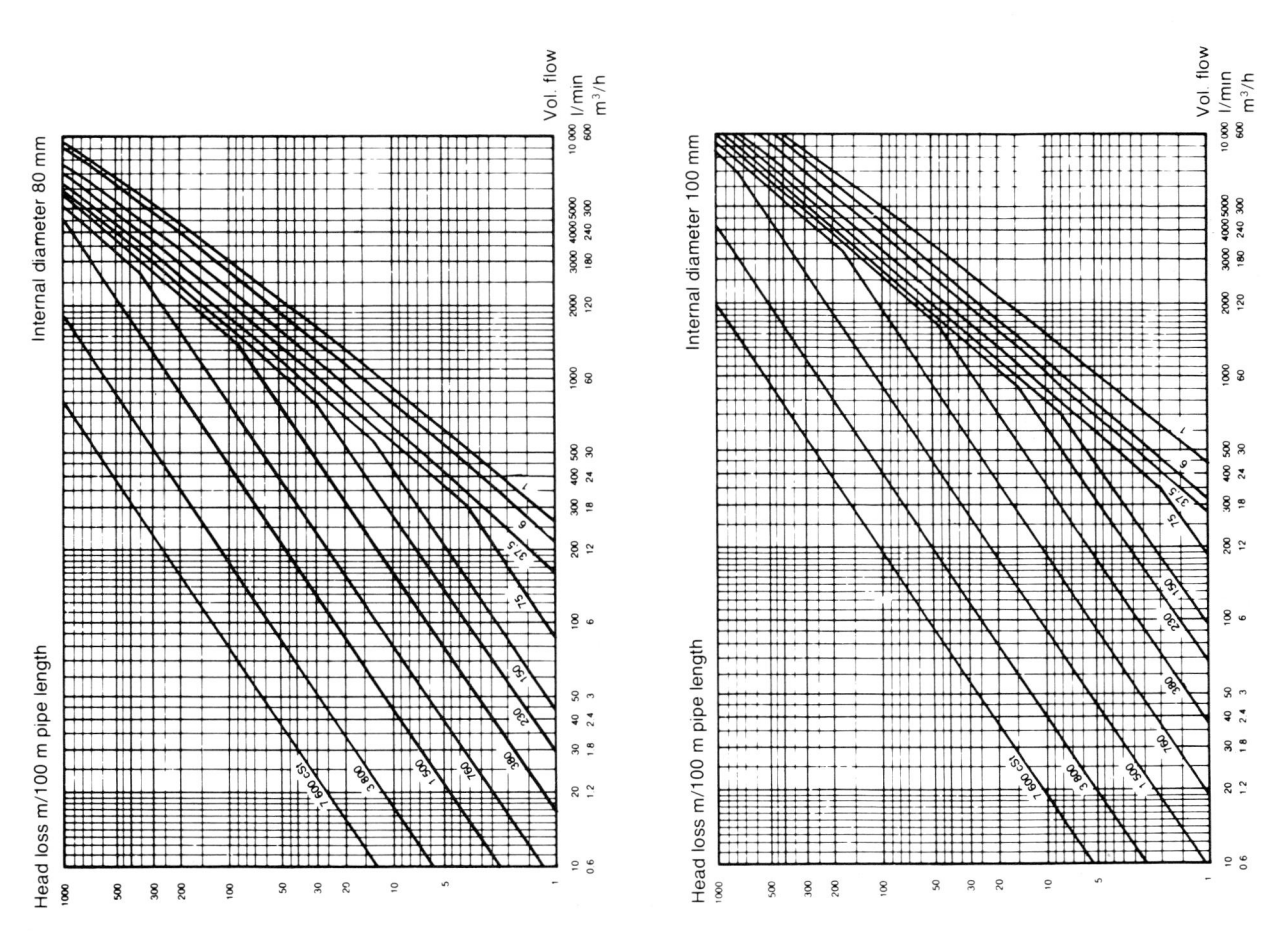

Figure 2.39 Frictional loss diagrams for oil at various viscosities, k = 0.03 mm, steel pipe

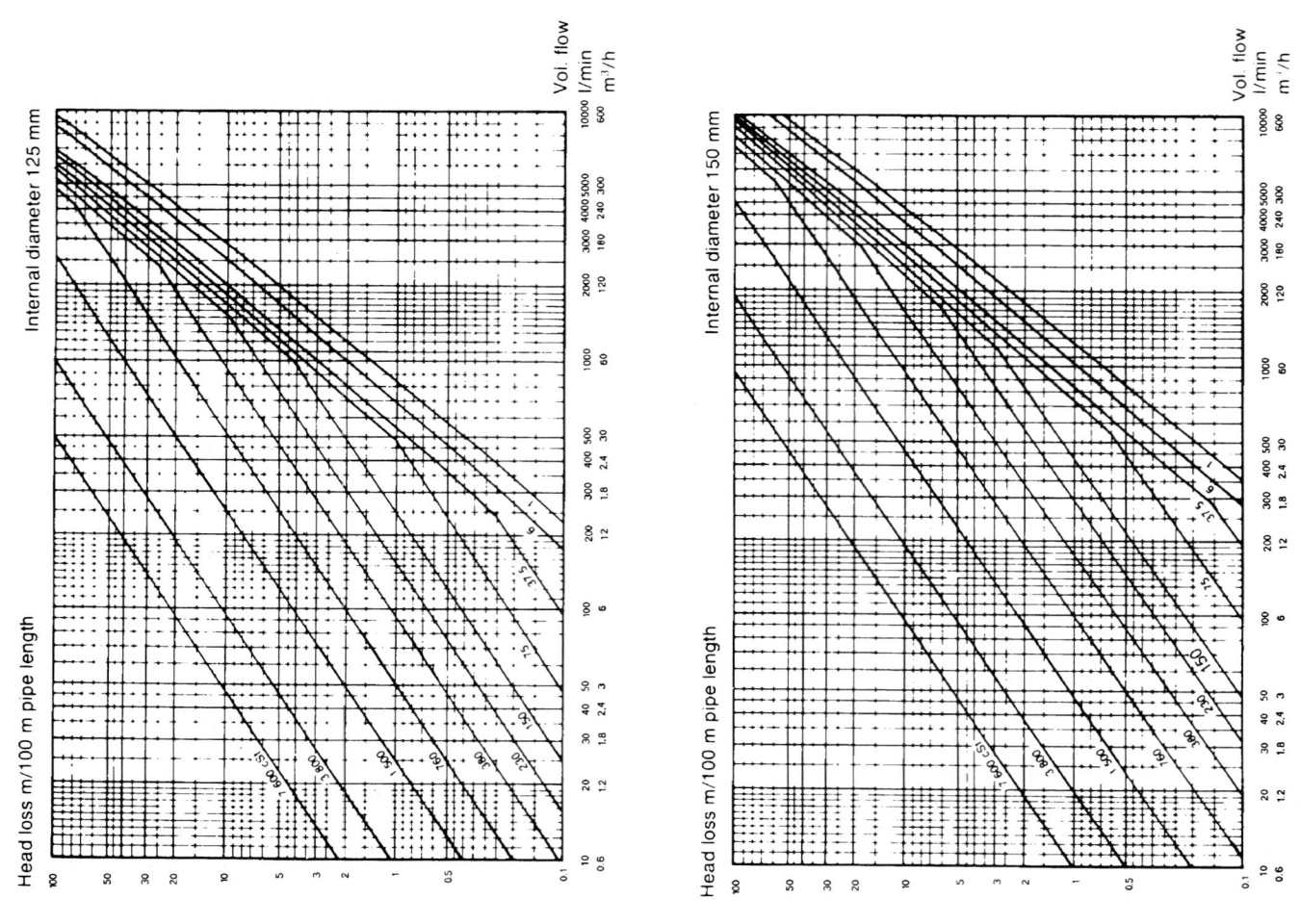

Figure 2.40 Frictional loss diagrams for oil at various viscosities, k = 0.03 mm, steel pipe

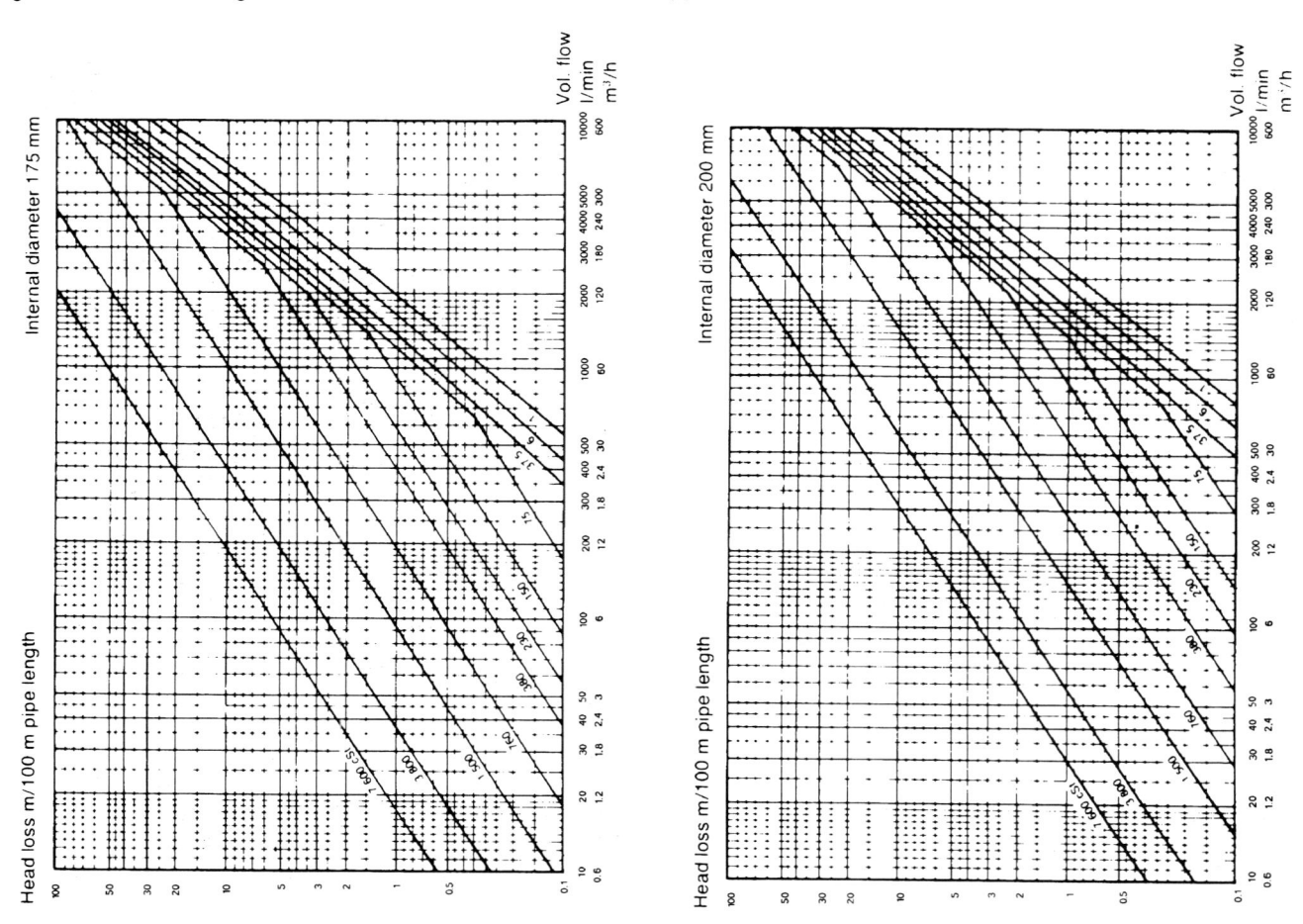

Figure 2.41 Frictional loss diagrams for oil at various viscosities, k = 0.03 mm, steel pipe

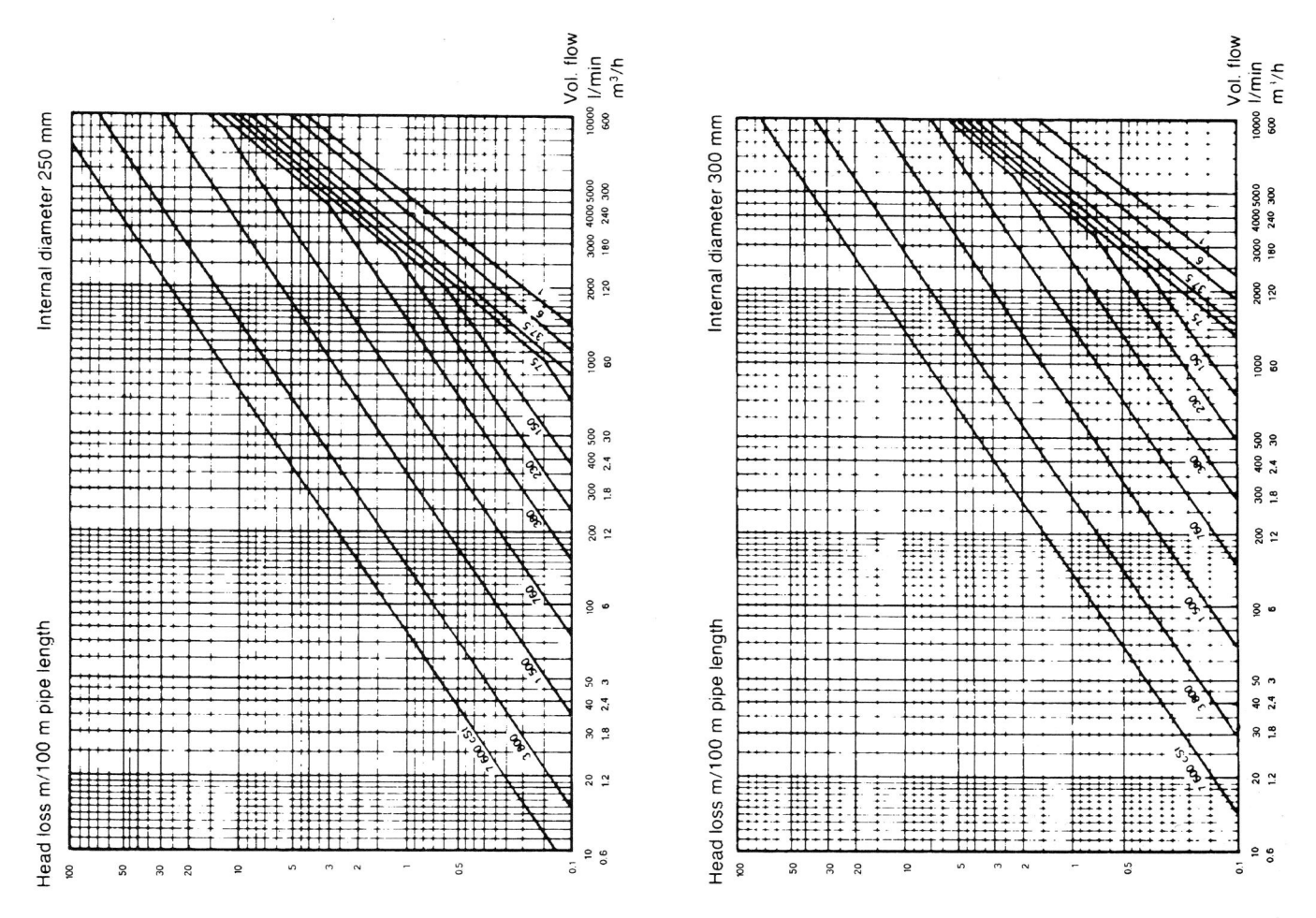

Figure 2.42 Frictional loss diagrams for oil at various viscosities, k = 0.03 mm, steel pipe

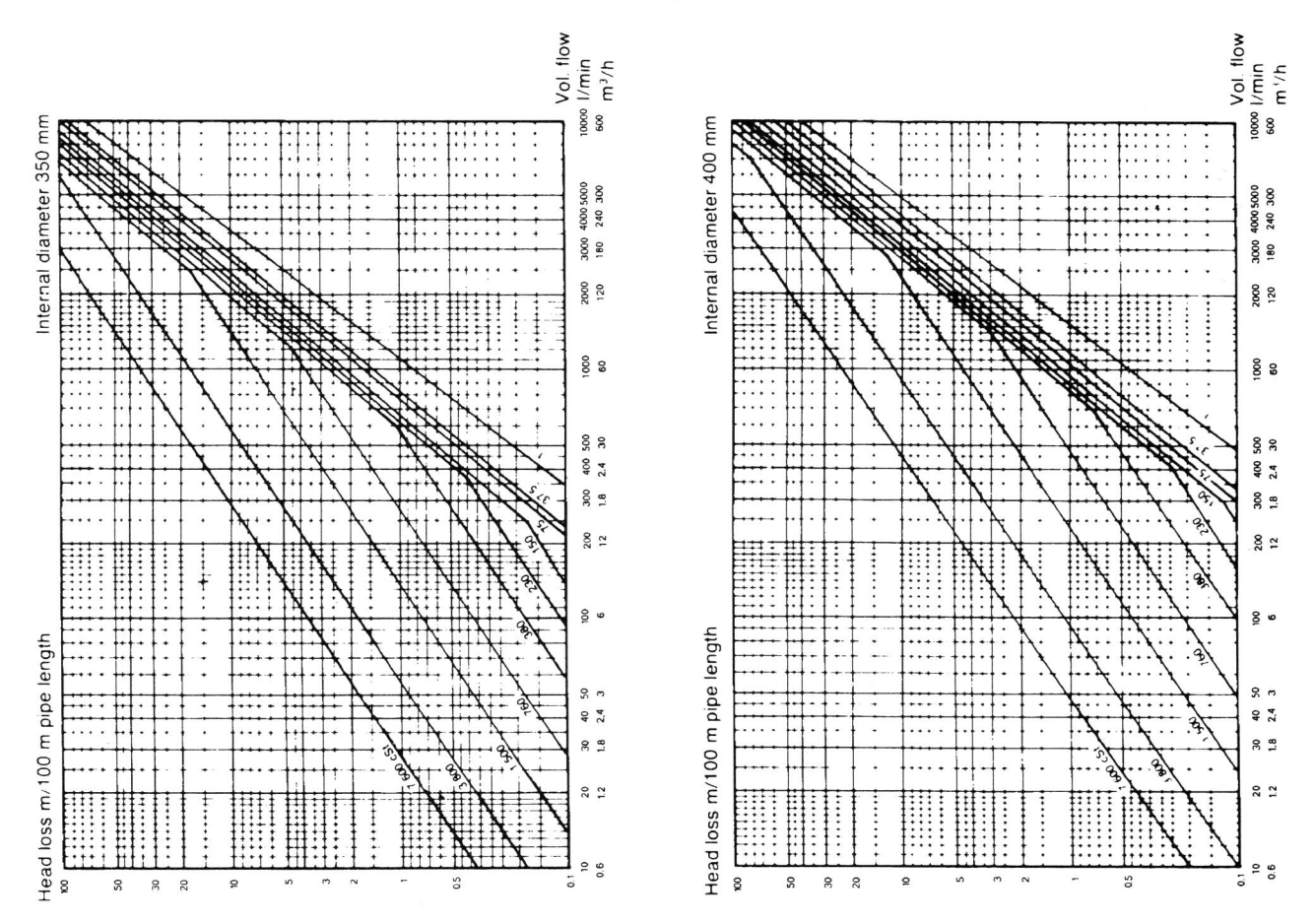

Figure 2.43 Frictional loss diagrams for oil at various viscosities, k = 0.03 mm, steel pipe

2.4.4 Paper pulp

The diagrams* apply for paper pulp of normal quality in stainless steel pipe:

Surface roughness of pipe	0.045 mm
Undried pulp	
Fibre ratio	75
Fineness grade	750 C.s.f.
Temperature	40 °C

The diagrammatic values are computed in accordance with the method set out in Section 2.3.4. If the actual pulp differs from normal pulp, the frictional losses may be assessed using the formula:

$$h_f = f_F \cdot f_M \cdot f_D \cdot h_{fN} \qquad \text{Equ 2.38}$$

h_{fN} = loss of head for normal pulp (m)
f = factors in accordance with Section 2.3.4.

It is not possible, on the other hand, to make corrections for the roughness of the pipe in this way.

The guide-line values for losses in bends and fittings are expressed by means of an equivalent pipe length - l_{eq} i.e. the extra length of straight pipe which would result in a corresponding pressure loss.

$$l_{eq} = L_e \cdot d \qquad \text{Equ 2.39}$$

l_{eq} = equivalent length of pipe (m)
L_e = factor in accordance with fig 2.44
d = diameter of pipe (m)

Since the equivalent pipe length is related to straight pipe flow it is in principle only valid for normal pulp and cannot be applied to other liquids in other pipes.

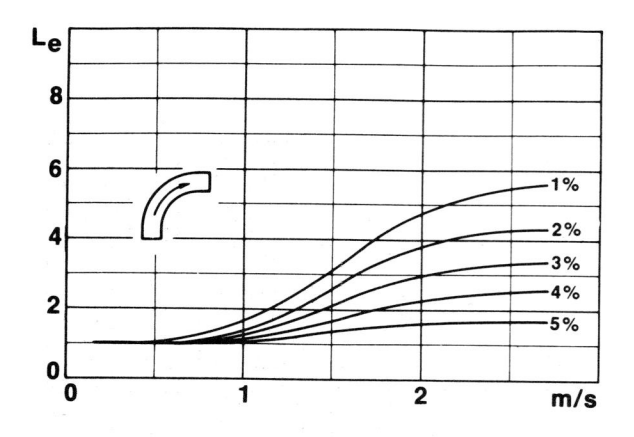

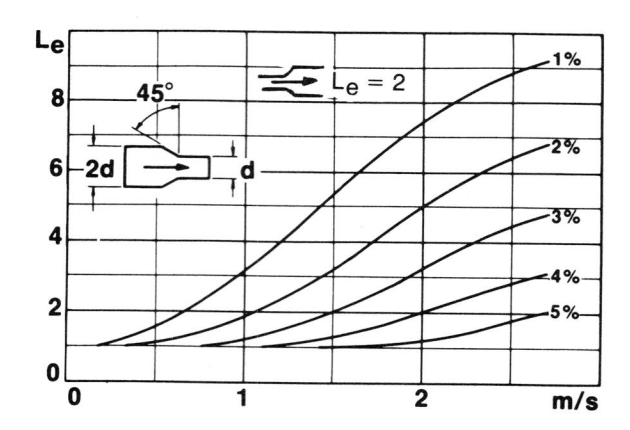

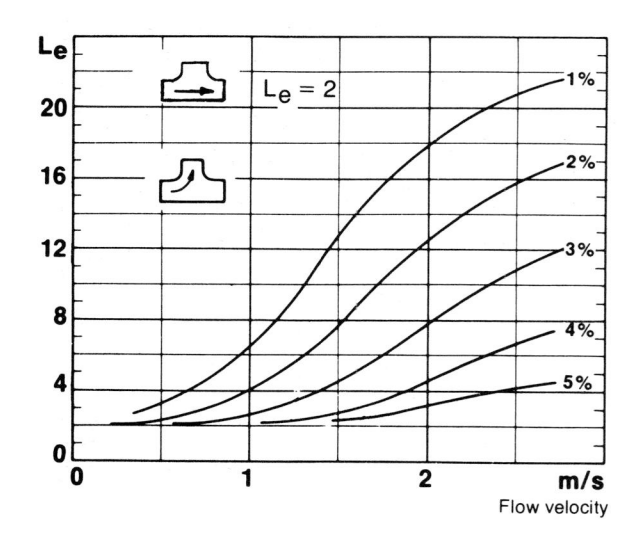

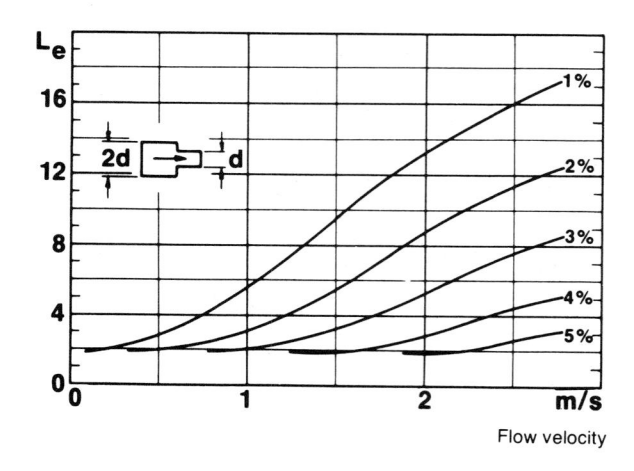

Figure 2.44 Factor for calculating equivalent pipe lengths for normal pulp in stainless steel pipe

* These diagrams are published by permission of ABS Pumps Ltd, Industrial & Process Division — (Scanpump - J & S)

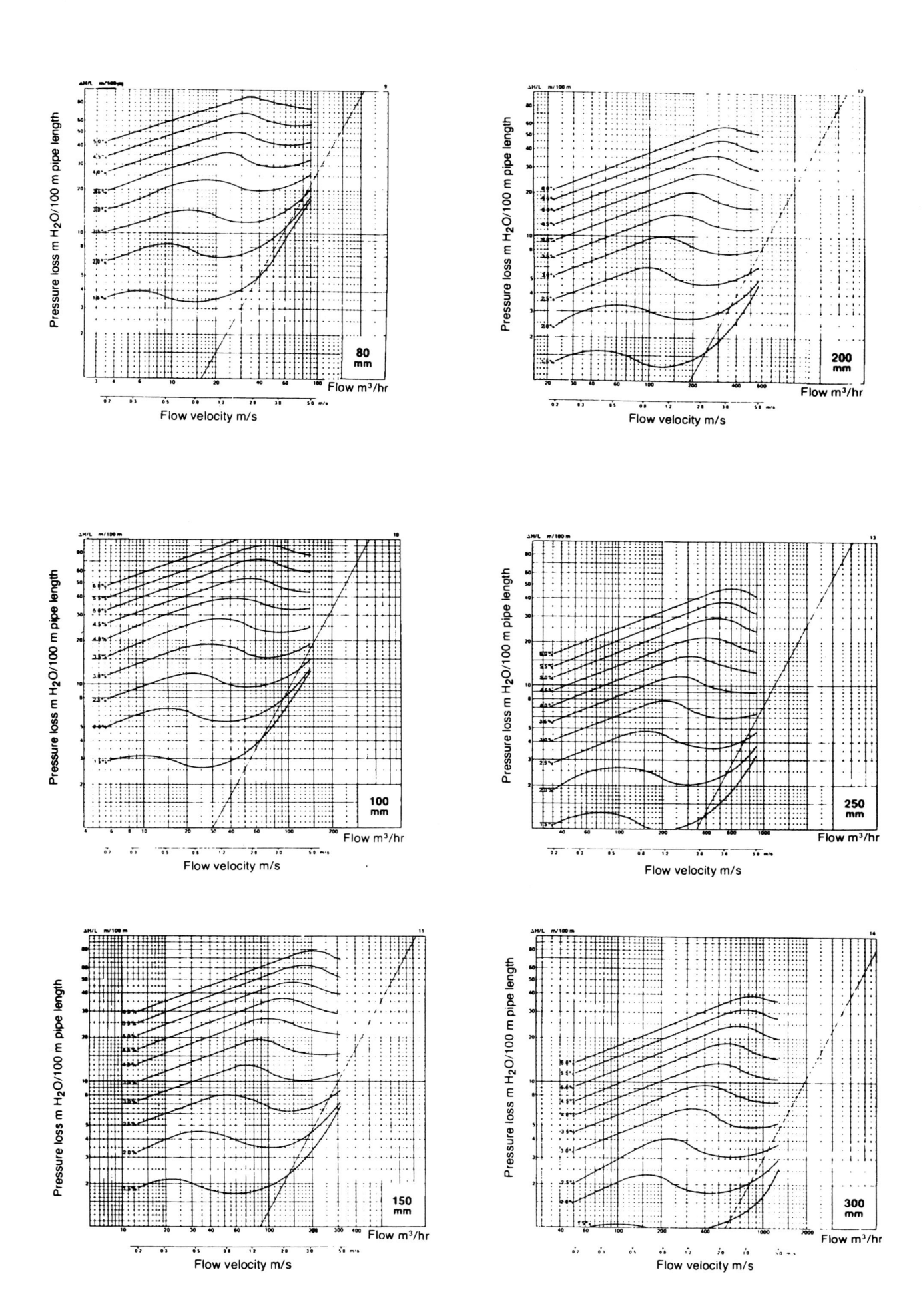

Figure 2.45 Frictional loss diagrams, normal pulp in stainless steel pipe

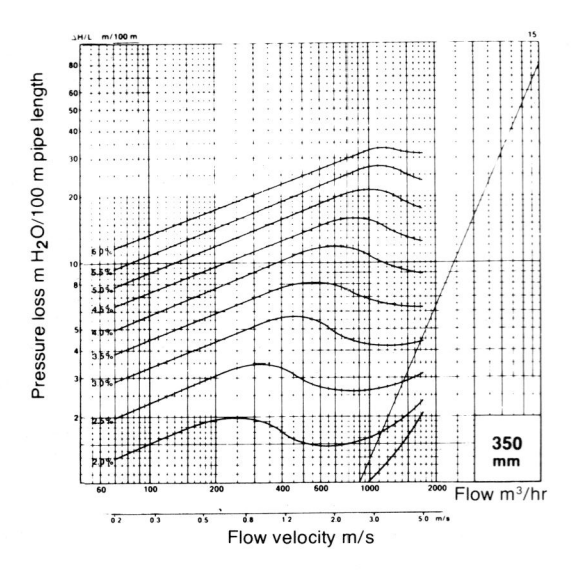

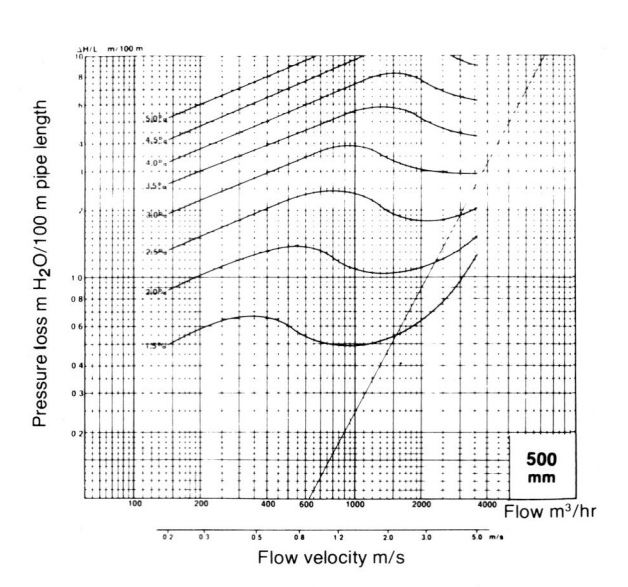

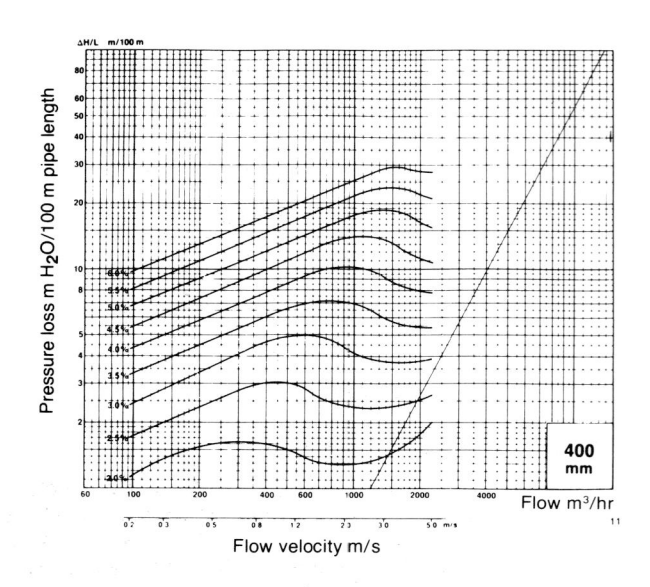

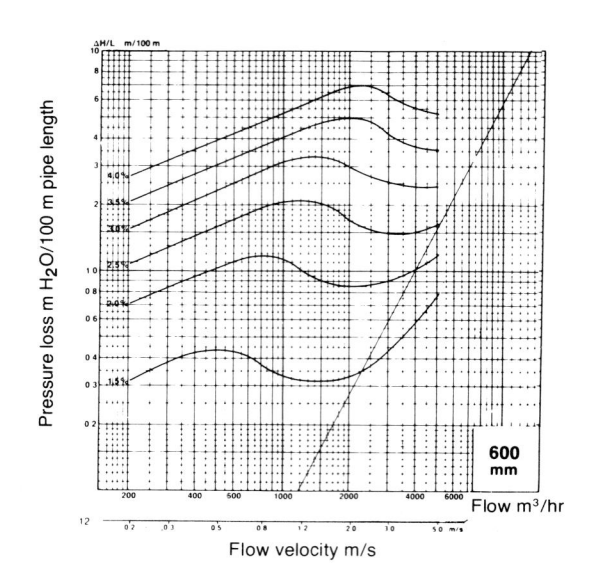

Figure 2.45 Frictional loss diagrams, normal pulp in stainless steel pipe

2.5 Flow measurement

2.5.1 General

Measurement requirements

The purpose of flow measurement may be:

- pump performance monitoring,
- pump control,
- leakage detection,
- to obtain sizing data,
- to obtain costing data,
- to obtain productivity data.

The measuring installation may be:

- permanent,
- temporary.

The measurement requirements during testing may be:

- continuous,
- periodic,
- intermittent.

Measurement data may be needed:

- currently,
- continuously,
- cumulatively.

Methods of measurement

When choosing a suitable method of measurement, account must first be taken of the category of liquid which is to be measured:

- pure liquid compounds free of contamination,
- contaminated liquids with a content of dissolved or undissolved; liquids, solids or gases,
- liquids having special characteristics, such as high viscosity, hot, reactive, corrosive, explosive or toxic

Pure liquid compounds with no solid particle content or gas bubbles can be measured using most of the measurement equipment commercially available. The exceptions are extremely pure liquids with measuring processes which make use of contaminants normally occurring in liquids. For example, distilled water flow cannot be measured with laser or electromagnetic flow meters.

When measuring liquids contaminated with solid particles, equipment with narrow passages and sludge trapping cavities are unsuitable. When measuring liquids with special characteristics the equipment must be selected so as to obtain compatible performance and material combinations having due regard to the particular properties of the liquid. Laser, electromagnetic, ultrasonic and turbine devices are more suitable than others, for hazardous liquids, because of the absence of external pipe connections.

Measuring range

Instruments should be selected so that the main area of operation falls within 20% to 60% of their full-scale range.

Too large a scale may mean that a good part of the measurement occurs below 20% capacity, with resultant errors in measurement.

Too small a scale may cause the capacity of the meter to be exceeded, which will also give incorrect readings. In order to avoid measuring in unsuitable ranges, the measuring equipment in some cases may be fitted with several scales in order to minimize measurement errors.

Accuracy of measurement

Accuracy is usually stated in percentage error of full scale and applies in ideal conditions as regards installation of the equipment according to the manufacturers instructions.

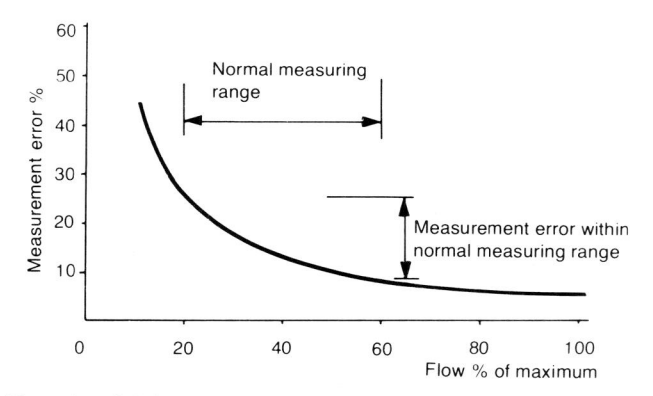

Figure 2.46 Relative measurement error with actual flow for a meter with a measurement error of ± 5% of full scale deflection. The absolute error is assumed to be ± 5% of maximum value within the complete measuring range

It can be seen from figure 2.46 that in the region below 10% of maximum value, the measurement accuracy is obviously poor. Within the normal measurement range, the errors extend from ± 8 to 25% of the actual measurement value, despite the claim that the meter has a measurement error within ± 5% at full-scale.

Straight pipe length

Most flow meters require a well developed velocity profile ahead of the measuring instrument. Requirements are often placed on the minimum length of straight pipe in order to maintain the guaranteed accuracy. The necessary straight pipe lengths, in many cases, may be difficult or expensive to arrange. However, pump manufacturers will not be pleased if the quality of their products are impugned by dubiously installed flow meters.

Flow losses

Different measuring instruments give rise to considerably varying flow losses in the installation. Loss of head associated with flow measurement can be responsible for undesirable additions to the pump energy requirement and power consumption.

2.5.2 Instruments for pipe systems

Electromagnetic flow meters

A special section of pipe is enclosed by electromagnets which create a magnetic field through the flowing cross-section. As the liquid moves through the magnetic field, a weak electrical potential difference appears at right angles to the flow which is measured by electrodes in the otherwise electrically insulated bore of the pipe. This voltage is directly proportional to the flow velocity and, since the section is full, also to the flow itself. When used in contaminated water, the electrodes can be fitted with scrapers or vibrated ultrasonically to keep them clean of deposits. Both AC and DC versions are available.

In the case of the electromagnetic flow meter, the output signal is directly proportional to the flow without any need for linearisation. This means that it can be switched simply between different display scales. It can even be arranged so that it switches automatically to a more sensitive scale as soon as the flow drops below a certain limit, for example, 20% of full scale. The electromagnetic flow meter does not recognize the direction of flow through the meter head. This can cause trouble if, for example, the meter is fitted to a pump line from a pump station which uses reverse flushing with product to clean out the pump. The reverse flush quantity will then have been measured three times.

The process liquid must be conductive to some extent; liquids as low as 0.008 µS/cm have been measured successfully. Because the device is electrical, certification is advisable for hazardous areas.

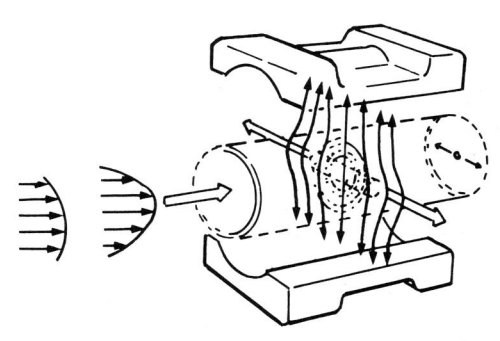

Figure 2.47 Electromagnetic flow meter

Positive displacement flow meters

Most positive displacement pumps, when driven as a "motor", i.e. with reversed direction of flow, work like turbines. There are many positive types of mechanical flow meter, as for positive displacement pumps, adapted to various liquid characteristics and capacities. Some types, such as rotary piston flow meters, rely on the process liquid viscosity for lubrication and sealing, minimum viscosity of the order of 20 cP. Flow rates as low as 0.002 l/m can be measured and working pressures over 500 bar can be accommodated.

Orifice plate and nozzle

Measuring flanges and orifices are dimensioned in accordance with ISO R541. In addition, the ISO standard provides information about installation requirements and the calculation process for conversion of the pressure differential into flow. The guaranteed degree of accuracy is also specified.

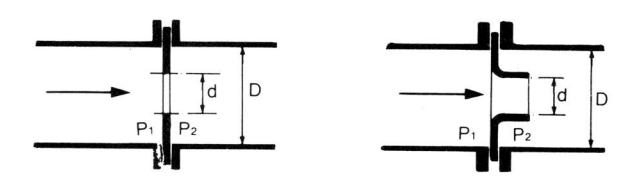

Figure 2.48 Orifice plate and nozzle. The flow is proportional to the square root of the pressure differential

The range of measurement is limited at the lower end by increasing measurement errors at low Reynolds Numbers and at the top end by sharply rising head losses. Orifice plates and nozzles are simple devices with no moving parts. Problems can arise with the small bore instrument pipework connecting the manometer.

Normal straight pipe lengths are of the order of 20 × pipe diameter upstream of the orifice or nozzle and 10 × pipe diameter downstream.

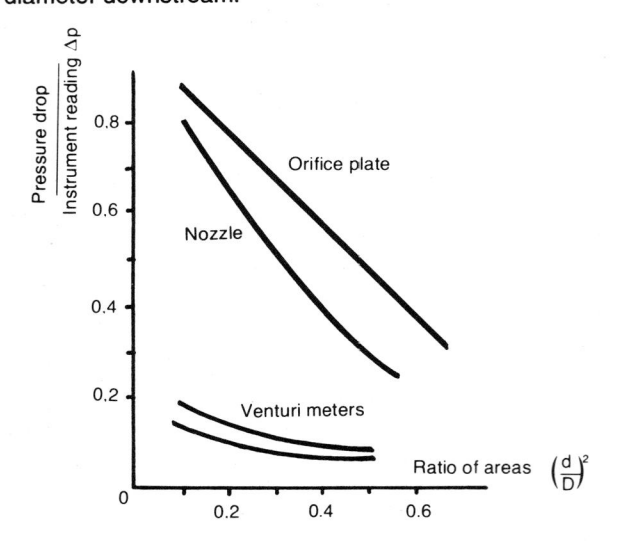

Diameter ratio d/D	Loss coefficient ζ		
	Orifice	Nozzle	Venturi
0.4	85	30	6
0.5	30	10	2
0.6	12	3	1
0.7	5	1	0.5
0.8	2	0.5	0.2

Figure 2.49 The pressure drop for a differential pressure flow meter expressed as a proportion of the instrument reading and as a loss coefficient.

As can be seen from figure 2.49, the nozzle has lower pressure losses than the orifice plate. From this point of view, the best differential pressure flow meter is the venturi meter.

Venturi meters

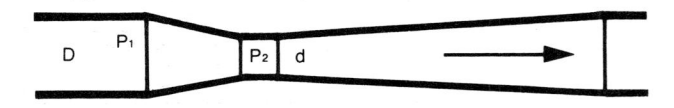

Figure 2.50 Venturi meter, long version

Venturi meters are available in long and short versions. The lowest flow losses occur with the long version. Venturi meters are standardised in ISO R781 as regards dimensions, installation, calculation procedure and degree of accuracy.

Turbine meters

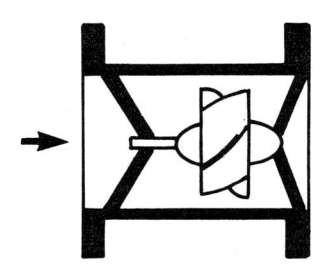

Figure 2.51 Simpler turbine meter

A turbine meter consists essentially of a turbine impeller driven by the flow of liquid. Either the number of revolutions is measured, as for summation measurements, or the rpm (pulses per unit time) in volume flow measurements. Turbine meters are available in both axial flow and Pelton Wheel versions. Axial flow meters can be very large whereas Pelton Wheel meters are particularly suitable for very small flows with high accuracy. Axial flow turbine meters can have aerofoil profile blades and specially shaped stators to obtain very high degrees of accuracy and are often used for calibration purposes. Simpler axial turbine meters, the so-called Woltmann meter, are not shaped for highest accuracy, however, but for general sensitivity.

Ultrasonic meters

Flow measurement using ultrasonic sound waves can be performed in two different ways and require different liquid characteristics to work correctly.

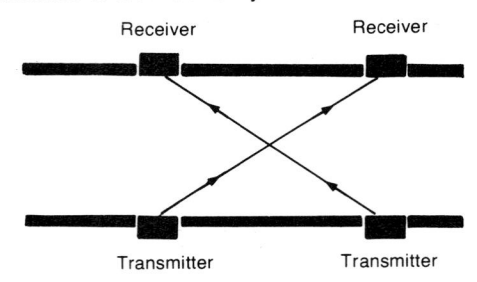

Figure 2.52 Flow measurement, ultrasonic

The first method requires the liquid to have some impurities, like solids or gas bubbles, to reflect the sound waves. The injected ultrasonic signal is reflected and scattered by impurities. The receiver detects the returning signals and uses Doppler techniques to convert the frequency shift into liquid velocity. The frequency shift is directly proportional to flow.

Pure clean liquids can be measured by a different method. Two transmitters inject sound waves, one in the direction of flow, and the other against the direction of flow. The difference in transit time between transmitter and receiver for these two signals is a measure of the velocity of flow and therefore of the flow itself. The difference in transit time can be converted to frequency, this being directly proportional to the flow. The frequency difference is independent of the speed of sound in the liquid.

Swirl flow meters

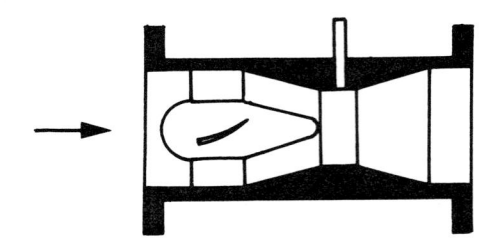

Figure 2.53 Swirl flow meter

A set of fixed vanes impart a tangential component of velocity, rotational component, at the same time as creating a secondary flow. The rotating secondary flow gives rise to pressure pulsations which are detected by a fixed sensor in the smallest section. The number of pulsations per unit time is a measure of the flow. This meter contains no moving parts.

Vortex shedding flow meters

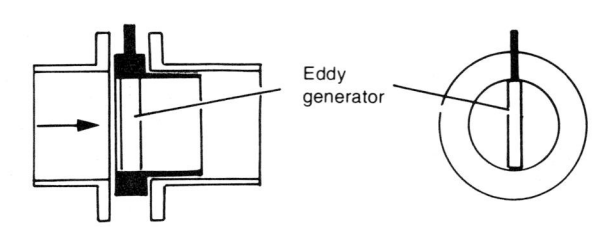

Figure 2.54 Vortex shedding flow meter

When liquids flow round specially shaped bodies, there is a periodic release of vortices trailing behind the bodies. Since the boundary layer release changes from one side of the body to the other, the pressure distribution also changes around the body. The number of changes per unit time is proportional to the flow. This type flow meter can contain no moving parts depending upon how the vortices are counted.

The vortices can be counted by a thermistor detecting the cooling or heating effect changes caused by the different flow patterns. As the vortices are shed alternately from each side the pressure differential across the body changes direction. The direction of the pressure gradient can be detected by a moving sphere or disc. The vortices can be counted by a thin vane, downstream of the body, deflecting from side to side. Lastly, the vortices can be counted by ultrasonic sound waves.

Variable orifice flow meters

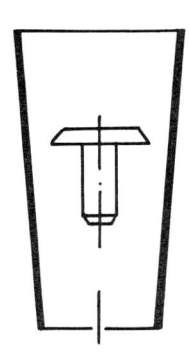

Figure 2.55 Variable orifice flow meter.

A spring loaded restriction, shaped like a poppet valve and mounted in a fixed orifice, moves to adjust the flow area in response to the liquid velocity. The flow can be read directly from the position of the restriction. Variable aperture flow meters are available in various sizes and flows over 1400 m³/h can be accommodated.

Calorimetric flow meters

Calorimetric flow meters work on the principle of applying a known quantity of heat to the flowing liquid and measuring the consequential temperature rise. This technique is most useful in small flow meters, up to 10 l/min.

2.5.3 Instruments for open channels

Measurement flumes

Measurement flumes are based on the idea that when the flow of water is so constrained a specific relationship will exist between flow and water level upstream of the flume, or, in some designs, between flow and difference in water level along the length of the flume. Flow measurement is thus converted to a level measurement.

The commonest forms of measurement flume are probably the Parshall flume and the Palmer-Bowlus flume. The relationship, water level to flow, is determined empirically and therefore each flume has to be an exact copy of the master flume from which the calibration curve is derived. The degree of accuracy is stated in various sources as being in the order of ±3% to ±5% for precision made flumes and measurement devices.

Weirs and notches

Weirs consist of a wall, of various proportions, the full width of the channel. Theoretical expressions can be derived, from Bernoulli's equation, for the relationship between the head above the weir to the flow rate. Theoretical expressions have proved very inaccurate and calibration of the finished installation is essential. Empirical equations are available for designing weirs but calibration is still imperative after construction for accurate measurements.

Francis's Formula

$$Q = 1.84(L - 0.2H)H^{1.5} \qquad \text{Equ 2.40}$$

Method	Liquid	Measurement range	Error	Straight pipe requirements	Head loss	Comments
Electromagnetic	(1)	≤ 40:1	1% FSD or 2% of reading	5D	very low	liquid must have electrical conductivity
Positive displacement	(2)	≤ 100:1	± 0.25% possible	short	moderate	various types; vane, lobe, rotary piston, screw
Orifice plate	(3)	4:1	2% of reading	25D	large	easily manufactured those to ISO R541 need no calibration
Nozzle	(3)	6:1	2% of reading	25D	large	those to ISO R541 need no calibration
Venturi meter	(3)	8:1	2% of reading	25D	small	those to ISO R781 need no calibration
Axial turbine	(4)	20:1	0.25% FSD or 1% of reading	25D	small	
Pelton Wheel	(4)	≤ 200:1	± 0.5%	short	small	0.01 l/m to 1200 m³/h
Ultrasonic Doppler	(5)	10:1	1% FSD or 2% of reading	20D	very small	
Ultrasonic timing	clean	10:1	1% FSD or 2% of reading	20D	very small	
Swirl	clean	10:1	1.5% FSD or 2.5% of reading	15D	moderate	
Vortex shedding	(6)	15:1	1.5% FSD or 2.5% of reading	15D	moderate	
Variable orifice	(7)	10:1	1.5% FSD	short	moderate	
Calorimetric	clean	100:1	± 10% of reading	short	moderate	low flows
Coriolis effect	(1)		≤ 1%	short	moderate	costly

FSD = full scale deflection
(1) liquid can contain solid particles and gas bubbles
(2) clean liquid is best, solids cause wear, some viscosity desirable for lubrication and to reduce slip
(3) solids damage orifice and reduce accuracy, clean liquid preferred, range limited by Reynolds Number
(4) clean liquid, range limited by Reynolds Number
(5) liquid must contain impurities
(6) slightly contaminated possible, limited by Reynolds Number
(7) clean liquids, high viscosity possible

Figure 2.56 Typical properties of flow meters in pipes. The characteristics quoted vary, of course, in the case of special types

Bazin's Formula

$$Q = \left(0.4046 + \frac{0.003607}{H}\right)\sqrt{2g}\, LH^{1.5} \quad \text{Equ 2.41}$$

where
- L = length of weir (m)
- H = height of liquid (m)
- g = gravitational acceleration (m/s^2)

Notches consist of an accurately shaped sharp-edged "partial orifice" over which the liquid flows. A notch can be rectangular or triangular in shape. The flow rate is directly related to the height (H) of the upper liquid surface above the notch. The coefficient of discharge, Cd, must be evaluated for the specific installation.

Rectangular notch

$$Q = \frac{2}{3} C_d \sqrt{2g}\, LH^{1.5} \quad \text{Equ 2.42}$$

Vee notch

$$Q = \frac{8}{15} C_d \sqrt{2g}\, \tan\frac{\theta}{2}\, H^{2.5} \quad \text{Equ 2.43}$$

where
- L = length of weir (m)
- H = height of liquid (m)
- g = gravitational acceleration (m/s^2)
- C_d = coefficient of discharge
- θ = angle of vee notch (degree)

The Sutro-weir is not an overfall weir in the strict sense of the word. It is, however, a simple arrangement for flow measurement of heavily contaminated liquids. The Sutro-weir does not have sludge pockets and gives a linear relationship between upstream liquid level and flow.

Vanes

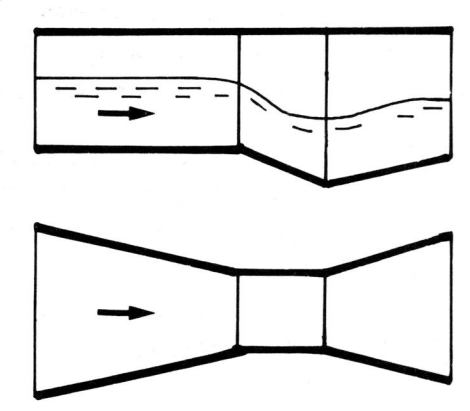

Figure 2.57 Examples of measurement flumes

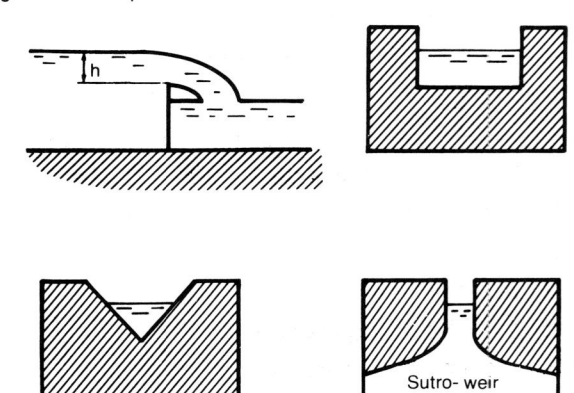

Figure 2.58 Different types of weir and notch for flow measurement

A very accurate flow measurement can be carried out using a measurement vane.

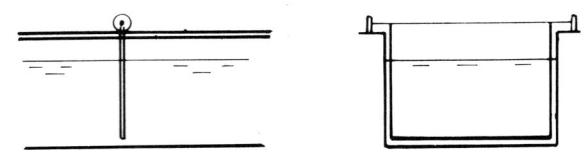

Figure 2.59 Flow measurement using vane

The vane, which is shaped to fit the channel, moves with the liquid flow. The time is measured to travel between two predetermined positions.

Method of measurement	Liquid	Range of measurement	Degree of accuracy	Remarks
Flume	Clean to heavily contaminated	10:1	5% of max.	Several shapes available. Low head loss
Weir Notch	Clean to slightly contanimated	10:1	2% of max.	Several shapes available. Moderate loss of head. Risk of sludging
Vane	Clean	High ~ 20:1	0.5% of max.	

Figure 2.60 Typical data for open channel flow meters

2.5.4 Other instruments

Tanks

Measuring the time taken to fill or empty a tank of known volume is a well tried method of flow measurement. As an alternative to volume measurement, the tank can be weighed before and after filling. Properly performed, the weighing method can offer the most accurate of all known methods and is therefore often used for calibration. Both volume and weighing measurement methods are ideal when the flow is not constant, from reciprocating pumps for example.

When the weighing method is used and the results subsequently converted to volume, care should be exercised during the conversion. Ensure all parties concerned are using the same, agreed, liquid data. Even for water.

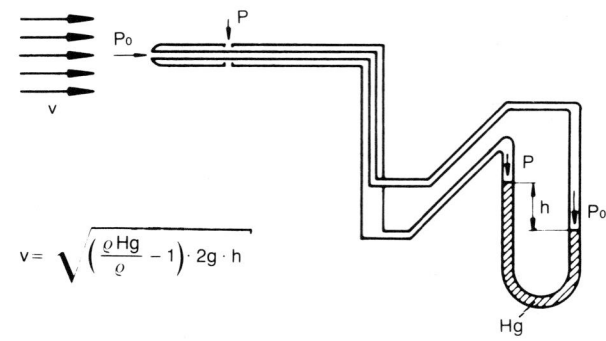

$$v = \sqrt{\left(\frac{\varrho Hg}{\varrho} - 1\right) \cdot 2g \cdot h}$$

Figure 2.61 Pitot-static tube with U-tube manometer

Pitot-static tube

Velocity of flow can be determined by the use of the Pitot-static tube. Because of the velocity profile in pipe-lines and ducts, the velocity at a particular point cannot be converted to a flow. By measuring the dynamic pressure at a number of points the flow can be summated with greater accuracy. Various means of measurement with several outlet pressure points distributed around the pipe or duct section are available commercially.

Current meter

A current meter is a device used in large channels, canals and rivers, to measure the current speed at a particular position. Some instruments are mounted on traversing gantries so that a velocity distribution across the section can be averaged. The device converts the linear motion of the liquid into rotary motion which can be measured. Two designs are available. Figure 2.62 shows a propeller version, the rotary speed of which varies linearly with the velocity of flow. The other version consists of a simple Pelton Wheel device which uses hollow cones as buckets and is mounted with the axis of rotation vertical.

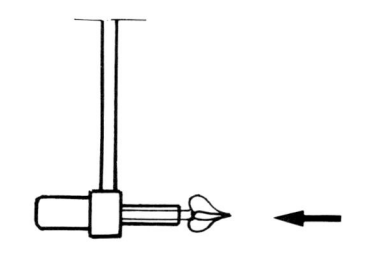

Figure 2.62 Propeller current meter

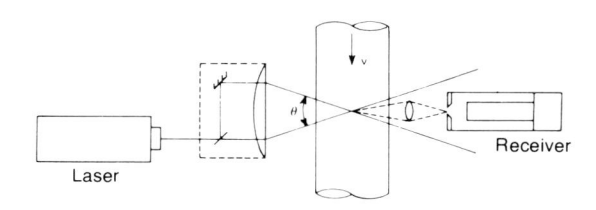

Figure 2.63 Flow measurement using lasers

Laser-doppler

A laser beam is divided into two beams in a prism. The beams are focused together again in the volume to be measured in the flow field, where an interference pattern is generated. A particle crossing the interference band reflects light pulses to a receiver. The frequency of the reflected light is a measure of the velocity of flow. The measuring equipment is very accurate but costly and its use is limited to the laboratory.

Figure 2.64 Simple probe for hotwire anemometer

Hot wire anemometer

The hot wire anemometer is used to measure flow velocity. The dissipation of heat from the hot wire is a measure of the velocity component at right angles to the wire. Small dimensions, the length and diameter of element are of the order of 1 mm and 0.01 mm respectively, and fast response makes it possible to draw up such niceties as degree of turbulence etc. As an alternative to wire, thin film may also be used. The meter is delicate and limited to laboratory use.

Various arrangements for checking flow

A number of methods are set out below, which are intended rather for checking flow than for accurate flow measurements.

- Pipe bend
  Pressure outlets are placed perpendicular to the direction of through flow at the inner and outer bend radii respectively. The error can be limited to ±10 % for a calibrated bend.
- Sight glass
  The position of a pivoted body in the flow (flap, paddle, etc.) can be observed through a viewing glass.
- Pump curve
  The flow can be assessed from pressure measurements and via the known H-Q curve of the pump. This method must be used with caution. A knowledge of how the pump performance has changed, since new, is very important.
- Free jet
  The outlet velocity of a free jet can be estimated by measuring the length of throw, and with the velocity the flow.
- Floats
  The time taken for a float, or other trace substance, to travel a given length gives a good idea of the flow in an open channel of known section.
- Tracers
  Trace substances can be used in pipe systems. An example is salt dosing in pulses. The salt cloud is traced by means of electrodes which detect the changed electrical conductive properties of the water.
- Concentration measurements
  With a known constant dosing, the concentration in the liquid is dependent on the flow. Solutions of salt in water have been used with good results.
- Pump speed
  For speed regulated pumps, the measurement of pump speed provides a means of checking the flow. If both the pump and system curve are known, this method can provide reasonable accuracy. Again wear in the pump and it's effect on pump performance must be considered. Positive displacement pumps produce a relatively constant output over a wide pressure range. System response is not so important in these situations.

2.5.5 Measurement standards

Methods of measurement of liquid flow are dealt with in various standards.

Measurement in pipe and duct systems

BS 1042 Pt 1 S1.1 Orifice plates, nozzles & venturis ≡ ISO 5167

BS 1042 Pt 1 S1.2 Square edged orifices

BS 1042 Pt 1 S1.4 Guide to the use of devices in S1.1 & S1.2

BS 1042 Pt 2 S2.1 Pitot-static tubes ≡ ISO 3966

BS 1042 Pt 2 S2.2 Measurement of velocity ≡ ISO 7145

BS 1042 Pt 2 S2.3 Pitot-static & current meters for swirling conditions ≡ ISO 7194

BS 5857 Pt 1 S1.1 Using tracers, general ≡ ISO 2975/1

BS 5857 Pt 1 S1.2 Constant rate injection ≡ ISO 2975/2

BS 5857 Pt 1 S1.3 Using radioactive tracers ≡ ISO 2975/3

BS 5857 Pt 1 S1.4 Transit time ≡ ISO 2975/6

BS 5857 Pt 1 S1.5 Using radioactive tracers ≡ ISO 2975/7

BS 5875 Glossary of terms ≡ ISO 4006

Measurement in open channels

BS 3680 Pt 1 Glossary of terms, see also ISO 772

BS 3680 Pt 2 Dilution methods

BS 3680 Pt 2A Constant rate injection

BS 3680 Pt 2C Radioactive tracers ≡ ISO 555/3

BS 3680 Pt 3A Stream flow measurement ≡ ISO 748

BS 3680 Pt 3D Moving boat method ≡ ISO 4369

BS 3680 Pt 4A Thin plate weirs = ISO 1438/1

BS 3680 Pt 4B Long base weirs

BS 3680 Pt 4C Flumes

BS 3680 Pt 8 Measuring instruments

BS 3680 Pt 9 Water level instruments

MARINE HYDRAULICS

Forward Industries

Marine Hydraulic Equipment goes Forward

Forward Industries specialise in manufacturing high pressure hydraulic pumps, motors, cylinders and actuators for Marine applications

- Marine hydraulic systems design for stabilisers and steering gears.
- Bulk carrier valve actuators, power units and control systems.
- VSG pumps and motors
- Dynapad piston pumps
- H & P and Benmar actuators
- Powerline Marine cylinders and actuators

FORWARD INDUSTRIES LIMITED
P.O.Box 8, South Marston Park, Swindon,
Wiltshire SN3 4RA
Telephone : 0793 823241   Fax : 0793 828474

An OMI International PLC Company

LIQUID DYNAMICS

HURSTFIELD INDUSTRIAL ESTATE,
STOCKPORT, U.K. SK6 7BB

TEL: 061 442 6222
FAX: 061 443 1486

Smoothflow Equipment

PROBLEM?

REDUCE:	Installation/Maintenance/Operation Costs.
PREVENT:	Acceleration Head Loss, Pump Wear, Cavitation Damage, Pipe Vibration, Flowmeter Inaccuracy, etc.
ALLEVIATE:	Pressure Surges in the System.

SOLUTION:

INSTALL: Liquid Dynamics **PULSATION PREVENTERS**.

The combination of Large Ports, Liquid-inside-the-Bag configuration and absence of internal poppet-valves (giving un-obstructed ports) ensures the most efficient operation.

Materials:	All machinable alloys/plastics, all mouldable elastomers + PTFE etc.
Pressures:	From Partial Vacuum to over 2000 Bar.
Volumes:	From a few mL to 4000 Litres and more.
Specifications:	ASME VIII, BS 5500, Stoomwezen, etc.
Programs:	Pulsation Prediction & Analysis.

TYPICAL STAINLESS/PTFE SHOCKGUARD PREVENTER

3 Pumps

This chapter comprises the most extensive chapter in "European Pumps & Pumping". After a general "Introduction" four main categories of pumps are classified and treated. Each main category can be studied independently. A main category generally contains the following sections: basics (theory), performance curves, design classification and a description of different design types. This classification into categories recurs in both the list of contents as well as in those sections dealing with classification by design. This classification by categories is applied consistently throughout the rest of the chapters in the book. The description of over sixty different types of pumps is relatively brief, but should, however, be of benefit both to the layman as well as to the specialist in this field.

The sections entitled "Suction performance" and "Standards" are intended to be applicable to all sorts of pumps, even if rotodynamic pumps have tended to dominate due to their wide range of application. "Suction performance" comprises not only the now classic concepts of cavitation, suction lift and NPSH, but also the evacuation of suction lines and the effect of gas dissolved or entrained in liquid. Standardisation for pumps is undergoing rapid development and those which are applicable up to and including 1992 are given. The latest developments of new standards, or new editions in committee, are mentioned.

As well as the list of contents of this chapter, the Reference index and the chapter "Pump selection" provide quick access to current data on a particular pump type.

Contents

3.1 Introduction

3.2 Rotodynamic pumps

    3.2.1 Rotodynamic pump theory
    3.2.1 Rotodynamic pump curves
    3.2.3 Rotodynamic pump classification
    3.2.4 Types of rotodynamic pump
        005 Standard water pumps
        010 Double suction axially-split pumps
        015 Heating, water and sanitation pumps
        020 Automatic supply packages
        025 Standard pumps to prEN 733
        030 Agricultural pumps without driver
        035 Fixed irrigation pumps
        040 Machine tool coolant pumps
        045 Marine pumps
        050 Electrically driven submersible pumps
        055 Vertical wet pit pumps
        060 Vertical dry pit pumps
        065 Portable self-priming pumps and submersible pumps not electrically driven
        070 Deep well pumps with ejector
        075 Submersible deep well pumps
        080 Wash water pump packages
        085 Multi-stage segmental pumps <300m
        090 Multi-stage segmental pumps >300m
        095 Standard pumps to ISO 2858, ISO 3069, ISO 3661
        100 End suction pumps to ASME/ANSI B73.1
        105 Inline pumps to ASME/ANSI B73.2
        110 End suction pumps to API 610
        115 Double suction pumps to API 610
        120 Magnetic drive pumps
        125 Canned motor pumps
        130 Non-metallic pumps
        135 Pulp pumps
        140 Pumps for solids >10mm
        145 Pumps for solids <10mm
        150 Hygienic quality pumps
        155 High speed single stage pumps
        160 High speed multi-pump packages
        165 Multi-stage axially-split pumps

- 170 Multi-stage radially-split pumps
- 175 Vertical multi-stage pumps
- 180 Non-clogging pumps with standard motor
- 185 Submersible non-clogging pumps
- 190 Other non-clogging pumps
- 195 Mixed flow pumps
- 200 Axial propeller pumps
- 205 Power recovery turbines

3.3 Liquid ring and peripheral pumps
- 3.3.1 Liquid ring pumps
- 3.3.2 Peripheral pumps
  - 210 Liquid ring pumps
  - 215 Peripheral pumps

3.4 Positive displacement pumps
- 3.4.1 Rotary positive displacement pump theory
- 3.4.2 Reciprocating positive displacement pump theory
- 3.4.3 Positive displacement pump curves
- 3.4.4 Classification of positive displacement pumps
- 3.4.5 Types of positive displacement pump
  - 220 External gear pumps
  - 225 Internal gear pumps
  - 230 Screw pumps with 2 or more screws
  - 235 Progressing cavity pump
  - 240 Lobe pumps and rotary piston pumps
  - 245 Rigid vane and flexible vane pumps
  - 250 Peristaltic hose pump
  - 255 Rotary peristaltic pump
  - 260 Piston pumps for hydraulic power - radial and axial pistons
  - 265 Portable plunger pump packages for high pressure descaling/cleaning
  - 270 Plunger pumps - horizontal and vertical
  - 275 Piston pumps
  - 280 Diaphragm pumps
  - 285 Metering pumps
  - 290 Jet pumps
  - 295 Air lift pumps
  - 300 Barrel emptying pumps
  - 305 Rigid screw pumps
  - 310 Pitot tube pumps

3.5 Other pump types
- 3.5.1 Theory
- 3.5.2 Classification
  - 315 Miscellaneous types

3.6 Suction performance
- 3.6.1 Cavitation
- 3.6.2 Net Positive Suction Head
- 3.6.3 Permissible suction lift
- 3.6.4 Cavitation effects on pump operation
- 3.6.5 Self priming
- 3.6.6 The effect of dissolved and entrained gases
- 3.6.7 Examples

3.7 Standards
- 3.7.1 General
- 3.7.2 Issuing authorities
- 3.7.3 Safety
- 3.7.4 Reliability and operational life
- 3.7.5 Dimensions and performance
- 3.7.6 Vibration and noise
- 3.7.7 Forces and moments on connections
- 3.7.8 Components
- 3.7.9 Testing

3.1 Introduction

Use is made of pumps everywhere there is a need for fluid transportation. Sometimes it is clean liquid which is pumped, sometimes the liquid is used as a medium for carrying thermal energy or solid suspended material etc. Due to widely varying operational requirements there are a large number of different pump designs available. A classification into four main categories has been made in an attempt to provide a clear picture of the types of pumps in use today.

- Rotodynamic pumps
- Liquid ring pumps
- Positive displacement pumps
- Other pumps

The construction of rotodynamic pumps is characterised by one or more impellers equipped with vanes which rotate in a pump casing. Those forces which come into being when liquid flows around the vanes are utilised in the conversion of energy. The rotodynamic pumps are called radial pumps or axial pumps depending on the main direction in which the fluid flows through the impellers. Alternatively, the terms "centrifugal pumps" for radial pumps, and "propeller" for axial pumps are used. Of course, intermediate forms between purely radial and purely axial pumps are also found and as such, are termed "mixed flow" pumps. Figure 3.1 shows the most usual and most important of the rotodynamic pumps.

Radial pump / Centrifugal pump

Axial pump / Propeller pump

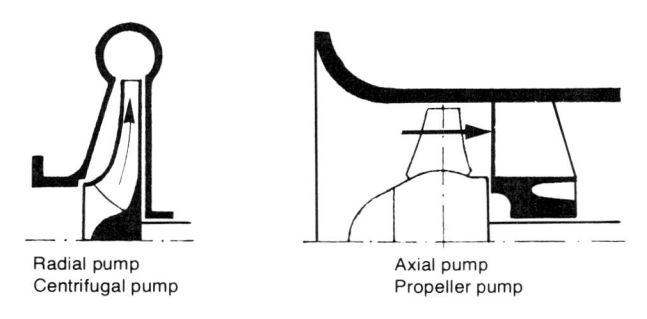

Figure 3.1 Examples of rotodynamic pumps

Certain aspects of liquid ring pumps are related to positive displacement pumps. A liquid ring is formed along the inner wall of the pump casing when the eccentrically placed impeller rotates. During a revolution of the impeller the partial volumes in the gaps between the vanes inside the liquid ring are first increased (inlet) and then decreased (outlet).

A variant of liquid ring pumps has the impeller placed centrally and brings about the varying partial volumes trapped during the revolution by allowing the liquid ring to partially wander out into channels in the side walls of the pump casing. The depth of these side channels varies during the revolution and hence one of the alternative names "side channel" pumps. Liquid ring pumps are self-priming.

Eccentric

Side channel

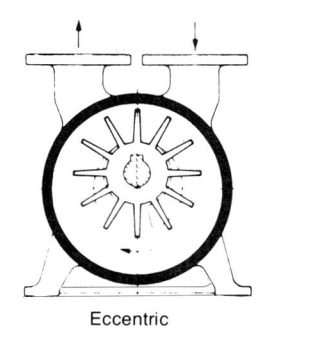

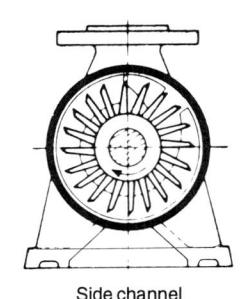

Figure 3.2 Examples of liquid ring pumps

Positive displacement pumps operate with enclosed liquid volumes which are forced forward in the direction of pumping, or squeezed and expelled into the pump outlet. The oldest and perhaps the most typical positive displacement pump is the piston pump. A certain volume of liquid is pressed out of the cylinder for each stroke of the piston. The plunger pump is a variant of the piston pump. The piston pump belongs to the sub-category among the positive displacement pumps which operate by means of a reciprocating motion. The other sub-category comprises rotating positive displacement pumps. These include, for example, gear pumps where the liquid is transported in gaps between the gear teeth and where the difference in pressure is maintained by means of, among others, the seal afforded by the contact surface of the meshing gears. Other examples of rotating displacement pumps are screw pumps, vane pumps, hose pumps, etc.

Reciprocal motion (plunger pump)

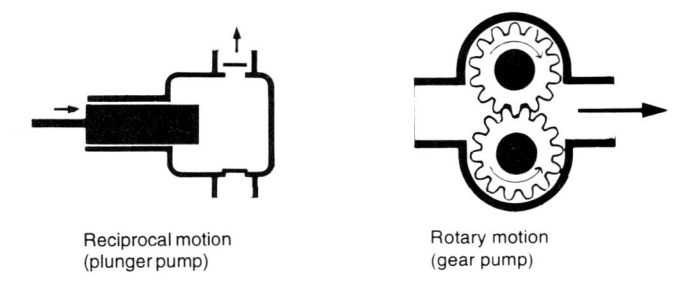

Rotary motion (gear pump)

Figure 3.3 Examples of positive displacement pumps

Jet pumps and Pitot tube pumps can be cited as examples of the main category termed "Other pumps". The static pressure in the nozzle of the jet pump falls when the speed of the motive fluid increases. The pumped liquid is sucked in and mixed with the motive fluid. The overall speed of the fluid flow is reduced and the static pressure is increased in the jet pump venturi (diffuser). Because jet pumps can handle liquids and gases they are useful for priming other pumps. Jet pumps powered by steam or compressed air are commonly called ejectors.

Jet pump

Pitot tube pump

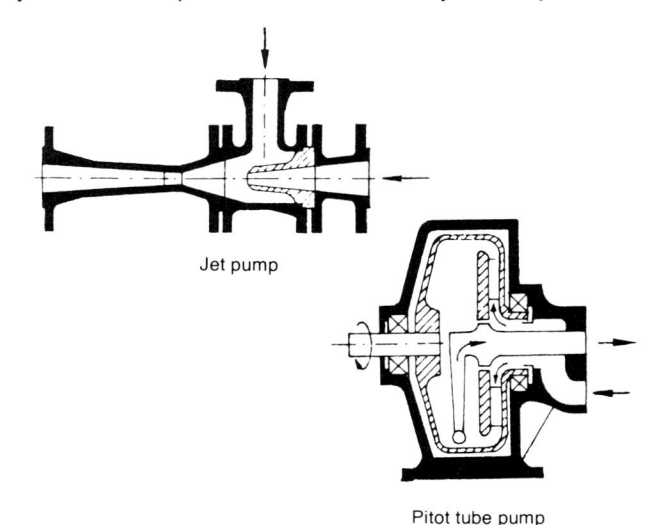

Figure 3.4 Examples of pumps from the main category "Other pumps"

The rotating casing of the Pitot tube pumps accelerates the incoming liquid to a high speed. A high static pressure is present at the stationary mouth of the Pitot tube and is brought about both by the effect of centrifugal forces and by the stagnation pressure of the moving liquid. Pressure increases of up to approximately 100 bar can be achieved in this matter.

The performance of pumps and above all rotodynamic pumps, is usually shown in the form of a curve. The main interest is the relationship between the volume of liquid transported per unit of time, the so-called volume flow or flow rate or just "flow" and the pressure increase or head rise of the pump. The different main categories have, in this respect, drastically different characteristics, see Figure 3.5.

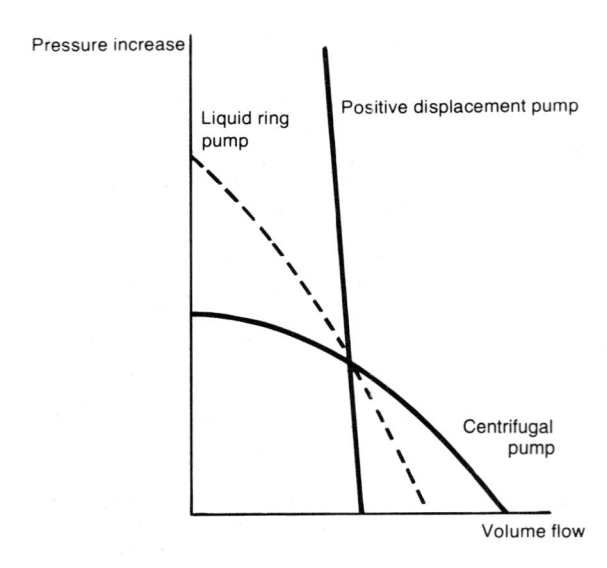

Figure 3.5 Pump curves for different pump types

The volume flow in positive displacement pumps is virtually independent of the pressure difference, i.e. the performance curve is almost a vertical line. The performance curve for the centrifugal pump shows increasing pressure rise with decreasing volume flow. Liquid ring pumps have a curve lying between these two groups of pumps, see the special section for pump curves for the category of "Other pumps". The flow in a pump is determined by the relationship of the specified pump pressure increase to the back-pressure, i.e. the resistance in the system, which is required in order to drive forward the flow.

If the liquid pressure somewhere in a pump is reduced to the vapour pressure of the liquid, the liquid evaporates. So-called "cavitation" occurs when the vapour bubbles return to liquid form in areas of higher pressure, this phenomenon causes intense local pressure spikes. Cavitation is not desirable since it reduces both pump performance and the operating life of the pump. The concept NPSH is employed to describe the risk of cavitation and states the necessary pressure or head measured above the vapour pressure of the liquid such that the effects of cavitation are limited or eliminated. For further details see Section 3.6.

3.2 Rotodynamic pumps

3.2.1 Rotodynamic pump theory

3.2.1.1 Differential head

In the past pumps were used almost exclusively for transporting water from a lower to higher level, e.g. in a mine, water distribution or in an irrigation installation. Then it was practical to use the concept of delivery head (delivery height) as a criterion of the pumps performance ability. Despite the fact that today pumps are used to a large extent for purposes where the alteration in the elevation of the medium is of subsidiary importance, this concept of head still remains. It is useful because the ideal pump differential head is related to the speed of the vanes, velocity head $c^2/2$. In this section the pump differential head will always be shown as ΔH and called differential head. In some newer publications it may be shown as H_{t2-1} indicating the difference in total heads between suction and discharge. In pumping circles it is standard practice only to refer to "head" on the understanding that everyone knows that it is really differential head.

The process which takes place when a liquid is passing through the pump is isentropic, i.e. the exchange of heat between the pumped liquid and the environment is so small that it can be neglected and is more than compensated by extra heat generated by inefficiencies. An equal amount of mass flows through per unit of time in and out through the pump, the external leakage normally being negligible in comparison with the mass flow through the pump. Some pump designs have internal recirculation for thrust balancing which complicates the theory a little. Such a process is described by the energy equation for steady flow reckoned per unit of mass of pumped liquid and using the terms as given in figure 3.6*.

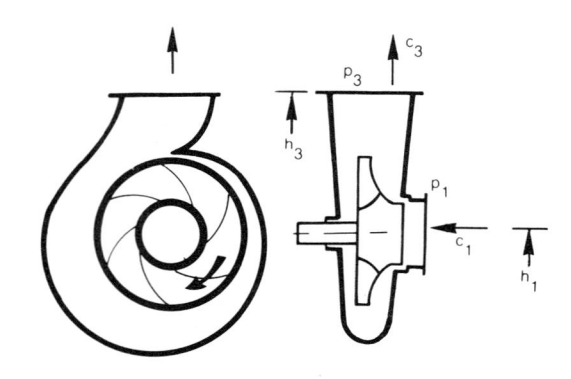

Figure 3.6 Pump terms

$$l_t = u_3 - u_1 + \frac{p_3 - p_1}{\rho} + \frac{c_3^2 - c_1^2}{2} + g(h_3 - h_1) \quad \text{Equ 3.1}$$

Put into words, the energy equation states the following: The mechanical work l_t, which is supplied via the pump shaft, causes; an increase of the internal energy (u), an increase in the static pressure (p/ρg), an increase in kinetic energy ($c^2/2$) and an increase in potential energy (gh); in the liquid.

It is desirable that as large as possible a part of the shaft work supplied results in an increase of static pressure in the liquid. In principle kinetic and potential energy can be completely converted into static pressure. It is more difficult to make use of the increase in the internal energy. It corresponds to an unnecessary and undesirable increase in temperature of the liquid and is considered as a loss. It is against this background that the pump's differential head H is defined as that useful part of the change in state of the liquid measured in metres of liquid column.

$$H = \frac{p_3 - p_1}{\rho g} + \frac{c_3^2 - c_1^2}{2g} + h_3 - h_1 \quad \text{Equ 3.2}$$

where
- p = static pressure (N/m²)
- c = absolute velocity (m/s)
- h = potential head (m)
- ρ = density (kg/m³)
- g = gravitational acceleration (m/s²)

Note that the potential head of the inlet flow is taken at the centre-line of the impeller.

3.2.1.2 Losses and efficiency

Varying liquid properties can change the proportions of the power flow. Increased viscosity, for example, leads to higher impeller friction and blade losses. The shaft power P supplied is defined as the product of rotary moments and angular velocity at the pump's shaft coupling. The pump efficiency is designated by η.

$$\eta = \frac{P_u}{P} \quad \text{Equ 3.3}$$

where
- η = pump efficiency
- P_u = useful work output (W)
- P = power input (W)

* Absolute flow velocity is designated by c in this chapter in line with current practice for rotodynamic machinery.

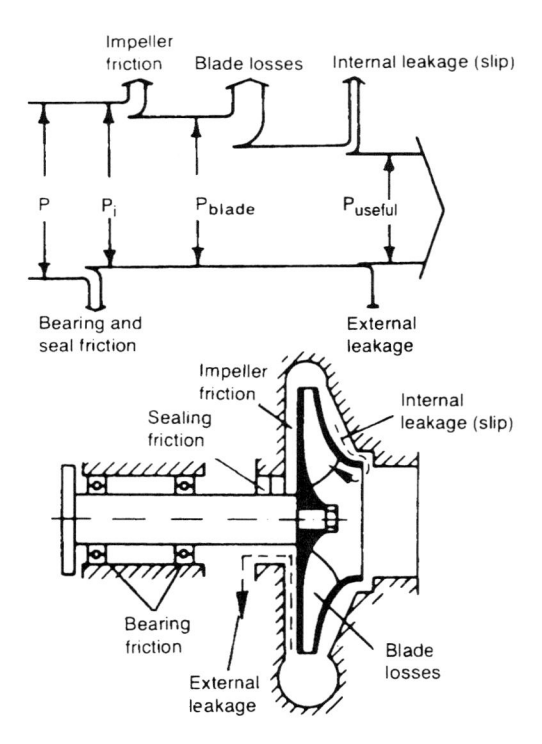

Figure 3.7 Power flow through a pump

The useful power (P_u) is thus the shaft power which remains from that supplied after all losses have been overcome. The useful power is applied to the pumped liquid and causes it to change state. The useful power is associated with the previously mentioned mechanical work according to the following relationship.

$$P = \dot{m} \cdot l_t \qquad \text{Equ 3.4}$$

where $\dot{m}$ denotes the mass flow through the pump.

$$P_u = \dot{m} \cdot g \cdot \Delta H = \rho \cdot Q \cdot g \cdot \Delta H \qquad \text{Equ 3.5}$$

where
$\dot{m}$	=	mass flow	(kg/s)
ρ	=	density	(kg/m³)
Q	=	volume flow	(m³/s)
g	=	gravitational acceleration	(m/s²)
ΔH	=	differential head	(m)

The pump efficiency can be shown to be the product of other efficiencies concerned with specific losses within the pump.

$$\eta = \eta_m \cdot \eta_i \cdot \eta_c \cdot \eta_v \qquad \text{Equ 3.6}$$

A part of the shaft power is used in overcoming friction resistance in bearings and seals, Pjm. This loss of power does not generally affect the liquid but is transformed into heat which is given off to the environment. The pump mechanical efficiency is given by:

$$\eta_m = \frac{P - P_{jm}}{P} \qquad \text{Equ 3.7}$$

The impeller efficiency, η_i, is a measure of how much of the power available, ($P - P_{jm}$), is converted to kinetic energy in the liquid and how much power is lost in blade losses and disc friction.

The casing efficiency, η_c, is a measure of how much kinetic energy, produced by the impeller, is recovered as static pressure rise compared to losses which appear as increased internal energy, temperature increase.

The volumetric efficiency, η_v, evaluates how much flow is delivered from the pump discharge compared to the flow through the impeller. External leakage must go through the impeller, and be worked on, before it can escape from the seal. Internal leakage, slip, allows a small quantity of liquid to recirculate through the impeller, reducing the effective pump flow.

The individual losses mentioned above are not constant but vary depending upon where the pump operates on its curve. At low flows/high heads, bearing friction can increase due to higher axial thrust and radial loads. The power conversion to kinetic energy in the impeller is influenced by the velocity triangles, see figure 3.8. For a constant speed pump, the velocity triangles can only be correct for one flow rate, Best Efficiency Point (BEP). At other flows mismatches in the blade and liquid angles occur causing additional losses. How the casing efficiency varies is dependant upon the type of casing design. Kinetic energy can be recovered as static pressure rise in two types of diffuser; vaned and vaneless. A vaned diffuser will suffer the same velocity triangle problems as the impeller with off-design flows. The vaneless diffuser or volute does not have blade angles to cause extra losses but does operate with flow paths away from the optimum in the specially shaped passages, again increasing losses. Internal leakage is effected by the pump differential head and the measures employed, by the manufacturer, to reduce it, such as wear rings, and wear in the pump.

A very important efficiency to consider, in today's energy conscious climate, is the overall efficiency, η_{gr}, often called the "wire-to-water" efficiency. This should be evaluated from the actual power consumed by the pump driver compared to the pump useful power. Overall efficiency takes into account any drive train systems between the driver and the pump and is the only valid efficiency to be used in "Life Cycle Costs". Extra care should be exercised with electric motors controlled by variable frequency invertors. Invertors can inject harmonics into the mains supply which may cause serious errors in electrical power measurement.

3.2.1.3 Euler's equation

Figure 3.8 shows a pump impeller with associated velocity triangle diagrams. The velocities u_1 and u_2 are the peripheral velocities of the impeller blade at inlet and outlet. The outlet peripheral velocity is commonly called the tip speed. Liquid enters the impeller at a velocity c_1 at a radius r_1. The fluid leaves the impeller at a radius r_2 with a velocity c_2. The design of the blade forces the relative velocity, i.e. the velocity which is experienced by an observer travelling with the movement of the blade, to alter its magnitude and direction from w_1 to w_2. Due to this the absolute velocity c_2 at the impeller outlet will also deviate from the flow velocity c_1 at entry. The tangential component of absolute velocity is designated by c_u, i.e. in the direction of the peripheral speed.

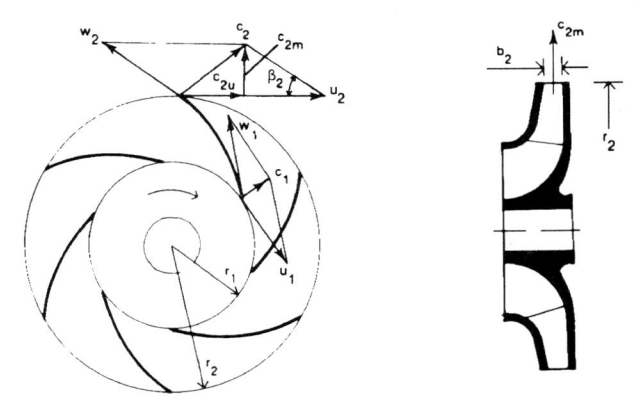

Figure 3.8 Impeller with velocity triangles at inlet and outlet

By applying the law of conservation of momentum from fluid mechanics to the flow through the impeller, we obtain in the tangential direction the following

$$M_t = \dot{m}(r_2 \cdot C_{2u} - c_{1u}) \qquad \text{Equ 3.8}$$

Here M_t, is the rotary moment, which the impeller must impart to the fluid, in order that the flow according to figure 3.8 may be produced. The flow velocities in equation 3.8 are intended to be representative mean values for the liquid which is flowing through the impeller.

During the time Δt the pump shaft rotates through the angle $\Delta\phi$ at the same time as the mass Δm passes in and an equal mass Δm passes out through the impeller tip. If both sides of equation 3.8 are multiplied by $\Delta\phi$ and at the same time $\dot{m} = \Delta m/\Delta t$ is introduced, then

$$M_t \cdot \Delta\phi = \frac{\Delta\phi}{\Delta t} \cdot \Delta m (r_2 \cdot c_{2u} - r_1 \cdot c_{1u}) \qquad \text{Equ 3.9}$$

but $M_t \cdot \Delta\phi/\Delta m$ is the work per unit mass which the rotary moment carries out whilst the impeller rotates through angle $\Delta\phi$. This work per unit mass is called blade work and is designated I_b. Furthermore, the quotient $\Delta\phi/\Delta t$ is equal to the constant angular velocity ω of the pump impeller.

Thus

$$I_b = \omega (r_2 \cdot c_{2u} - \phi \cdot c_{1u}) \qquad \text{Equ 3.10}$$

That part of the blade work which corresponds to the blade losses results in an increase of the internal energy of the liquid whilst the rest ($\eta_h \cdot I_b$) provides a useful change of the state of the liquid ($g \cdot H$). If then the peripheral speed $u = r_2 \cdot \omega$ is applied, we finally obtain

$$\frac{g \cdot \Delta H}{\eta_h} = u_2 \cdot c_{2u} - u_1 \cdot c_{1u} \qquad \text{Equ 3.11}$$

where

g =	gravitational acceleration	(m/s^2)
ΔH =	differential head	(m)
η_h =	hydraulic efficiency	
u =	peripheral velocity	(m/s)
c_u =	tangential component of the absolute velocity	(m/s)

Equation 3.11 is Euler's equation as it is normally written for pumps.

3.2.1.4 Pump curves

In accordance with Euler's equation differential head is dependent upon the size and direction of the velocity vectors and the size of the hydraulic losses. Both these factors are affected by, among others, the flow rate Q, which passes through the pump.

In many cases the tangential component of the absolute velocity entering the impeller is small, i.e. $c_{1u} \approx 0$. In such cases Euler's equation is simplified to

$$\frac{g \cdot \Delta H}{\eta_h} = u_2 \cdot c_{2u} \qquad \text{Equ 3.12}$$

If the flow through the pump were free of losses ($\eta_h = 1$), the ideal differential head, ΔH_i, would be given by the following equation

$$\Delta H_i = \frac{u_2 \cdot c_{2u}}{g} \qquad \text{Equ 3.13}$$

ΔH_i is often called the Euler head. With reference to figure 3.8

$$\Delta H_i = \frac{u_2}{g}\left(u_2 - \frac{c_{2m}}{\text{Tan}\beta_2}\right)$$

and

$$c_{2m} = \frac{Q}{2\pi \cdot r_2 \cdot b_2}$$

or

$$\Delta H_i = \frac{u_2}{g}\left(u_2 - \frac{Q}{2\pi \cdot r_2 \cdot b_2 \cdot \text{Tan}\beta_2}\right) \qquad \text{Equ 3.14}$$

The angle β_2, in the velocity triangle is somewhat smaller than the blade angle at the impeller blade tip. This angular difference is called the deviation angle and is due to the blade's inability to completely control the relative flow of a real liquid as opposed to an ideal liquid with zero viscosity and density. The size of the deviation angle is primarily dependent upon the number of blades.

For a given pump, operating at a certain constant speed, the ideal differential head H_i decreases linearly according to equation 3.14 with increasing flow Q. The actual differential head H differs from H_i due to the hydraulic losses h_f. These, as in all other cases of flow, are dependent upon the inlet flow direction towards the body around which the liquid flows. In the case of pumps the inlet flow direction, towards for example the blades, varies with the flow rate. A certain inlet flow angle gives the most favourable flow and thus the smallest losses. Both higher and lower values of Q result in an increase of h_f. By subtracting h_f from H_i, the actual H-Q curve of the pump at constant speed is obtained, see figure 3.9. The shape of the H-Q curve varies from one pump type to another and between different designs of the same pump types. Detail design particulars, such as; blade profiles, blade angles, number of blades, area ratios, diffuser type; all effect the curve shape.

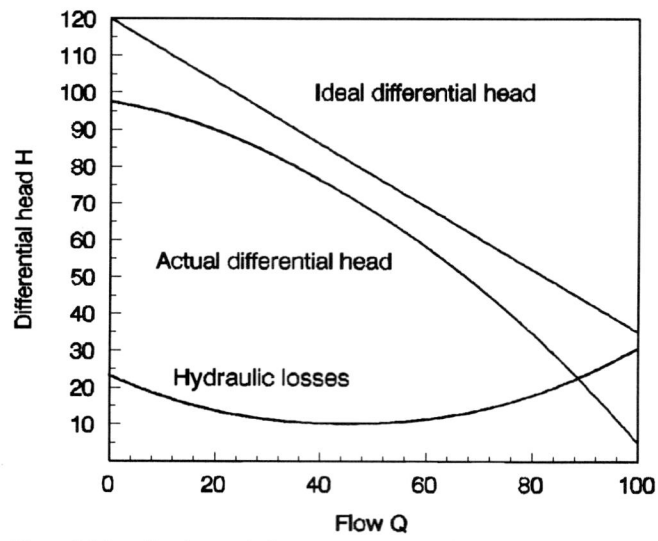

Figure 3.9 Centrifugal pump H-Q curve

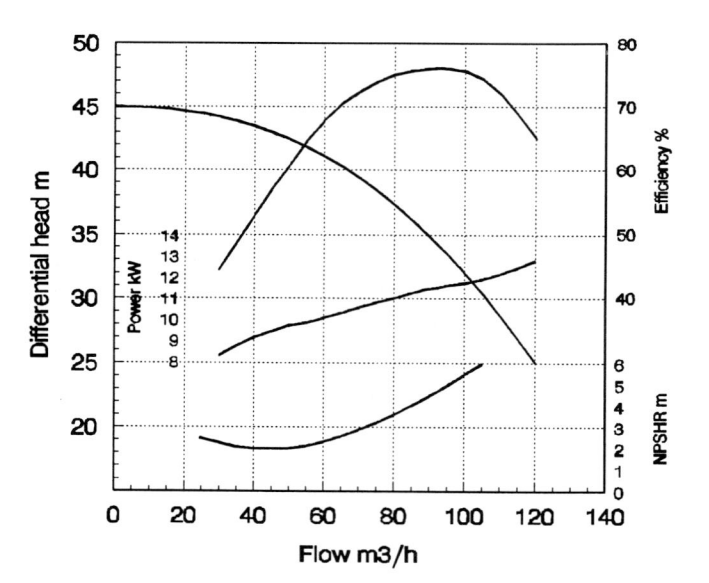

Figure 3.10 Constant speed, fixed diameter pump curve

In addition to the pump's H-Q curve, the required shaft power P, the pump's efficiency η and NPSHR, as shown in figure 3.10 are also usually given as a function of the flow in a pump characteristic curve. The shaft power is based on water, i.e. specific gravity = 1.0. Variations in specific gravity will affect the power, and the discharge pressure even though the differential head is constant. The pump can, in principle, operate at any point whatsoever along the H-Q curve. The position of the operating point in an actual case is determined by the characteristics of the system, suction plus discharge, to which the pump is connected. The art of good pump selection is to operate as close as possible to the Best Efficiency Point (BEP); also some users specify the duty point must be to the left of BEP.

3.2.1.5 Specific speed

An important parameter which is frequently used to describe different pump types and to characterize a duty point is specific speed N_s, defined by the following relationship.

$$N_s = n \cdot \frac{Q_{BEP}^{0.5}}{H_{BEP}^{0.75}} \qquad \text{Equ 3.15}$$

where
- N_s = specific speed
- n = shaft speed (rpm)
- Q_{BEP} = pump flow at best efficiency point (m³/h)
- H_{BEP} = pump ΔH at best efficiency point (m)

Note that N_s is not dimensionless and therefore has different numerical values in different systems of units. N_s can be defined in words as the mechanical speed of a pump identical with the one under consideration, and which with identical velocity vector diagrams gives a flow of 1 m³/h at a differential head of 1 m. All identically shaped pumps have therefore the same specific speed independent of their size.

The specific speed is used, among others, for characterising the shape of rotodynamic pump impellers. Drawing on experience gained in this field, it is generally possible to estimate which impeller shape will, under normal circumstances, give the best results for a given speed, flow and differential head. Figure 3.11 illustrates this relationship. The axial flow pump has shown itself to be the most suitable type in the case of, for example, a large flow and a small differential head, i.e. in the case of high specific speed. Axial flow pumps are therefore said to have high specific speeds.

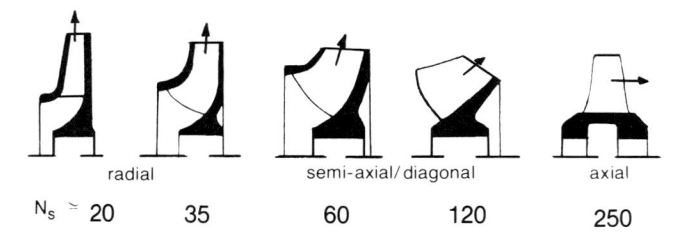

	radial		semi-axial/diagonal		axial
N_s ≈	20	35	60	120	250

Figure 3.11 Pump impellers having different specific speeds.

There is another form of specific speed in popular use, Suction specific speed, N_{ss}. Suction specific speed is evaluated using the same equation as specific speed but $NPSH_R$ is used in place of differential head. N_{ss} is an indication of how good the suction performance of a pump is. Some users specify a limiting range, 175 to 215, of N_{ss} in an attempt to limit cavitation problems.

Cavitation, especially when operating well away from BEP, is a very complex phenomenon and specifying any N_{ss} value will not necessarily solve or help the problem. Other problems associated with rotodynamic pumps operating well away from the BEP are surge and internal recirculation. Both of these problems are associated with low flows. Manufacturers take great care to avoid surge and recirculation problems by specifying minimum flows for pumps. It is up to the user to correctly assess the full operating regime and communicate this to the manufacturer.

A very interesting and educational paper was presented by R. Palgrave, the chief engineer of Ingersoll-Rand Gateshead, now Ingersoll-Dresser Pumps, on the relationship and effects of both specific speed parameters. The paper, presented to the Institution of Mechanical Engineers in September 1984 on "Deaerators and Boiler Feed Pumps", indicates the alternatives available to the pump designer in response to changing operating conditions and customer restraints. The paper shows pump performance as a 3D surface, which connects flow, head and NPSH, rather than a conventional 2D curve see figure 3.12.

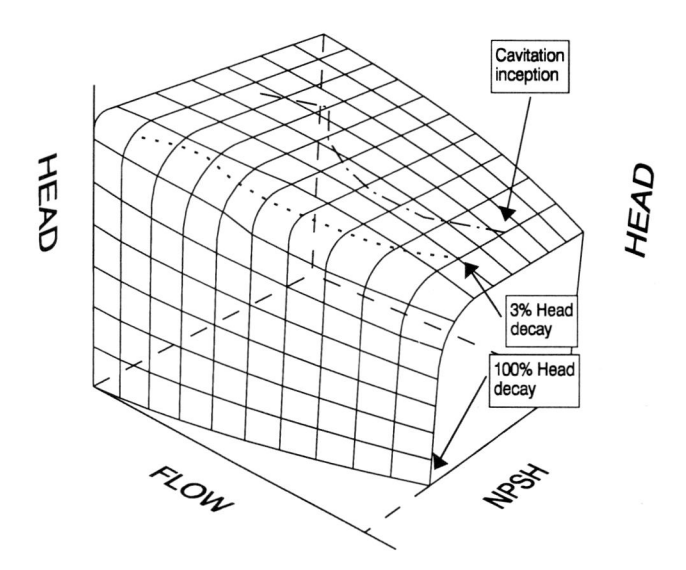

Figure 3.12 Pump performance surface

3.2.1.6 Similarity

Under certain conditions, the performance of a pump can be recalculated in a specially simple manner so as to be valid for all sizes of pump at different operational speeds. These conditions are as follows.

- The pumps compared should be identical or geometrically similar.
- The pumps compared should operate at similar operational points, i.e. with geometrically similar velocity diagrams.
- Any difference in efficiency should be negligible.

In a comparison between two pumps I and II, which fulfil these conditions, the following relationships between their respective speeds nI and nII, dimensions expressed as the impeller diameters D_I and D_{II}, and their performance, will apply.

- For the flows

$$\frac{Q_I}{Q_{II}} = \frac{n_I \cdot D_I}{n_{II} \cdot D_{II}}$$

- For the differential heads

$$\frac{H_I}{H_{II}} = \frac{n_I^2 \cdot D_I^2}{n_{II}^2 \cdot D_{II}^2}$$

- For the power requirements

$$\frac{P_I}{P_{II}} = \frac{n_I^3 \cdot D_I^3}{n_{II}^3 \cdot D_{II}^3}$$

The affinity laws which apply to a pump with a fixed impeller diameter, form a special case of the general similarity relationships. When $D_I = D_{II}$, then

$Q \propto n$

$H \propto n^2$

$P \propto n^3$

When the principles of similarity are used in practice, the variation in efficiency between the pumps compared is ignored except when calculating power. Similar pumps have the same

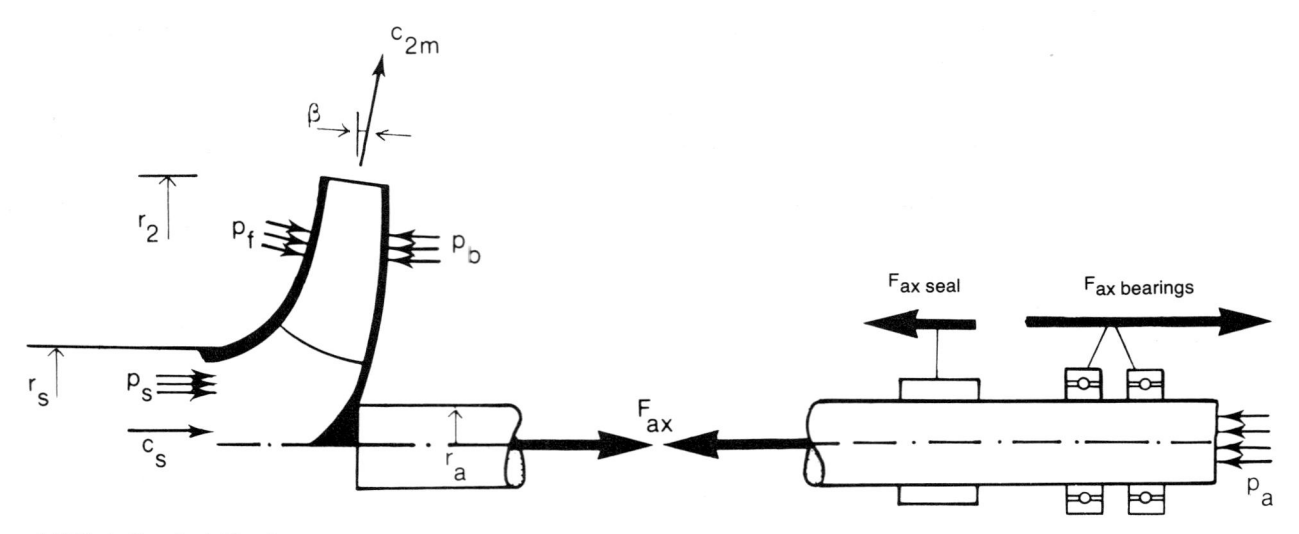

Figure 3.13 Illustration of axial thrust

blade angles. Geometrically similar velocity diagrams imply the same flow angles. The inlet flow direction, which is so important for the hydraulic losses, is therefore unaltered in the cases compared. Against this background a constant hydraulic efficiency would seem to be a reasonable approximation. In practice the efficiency will improve slightly as speed increases. In the case of very large pumps which are tested by means of model tests in the laboratory, a somewhat higher efficiency can be expected for the full-scale pump than that which is measured during the model test. Parasitic losses tend to increase as size reduces.

3.2.1.7 Axial thrust and radial forces

The pump impeller works best if the impeller tip width is aligned centrally with the throat of the diffuser. The pump shaft is adjusted during assembly to accomplish this. The unbalanced axial thrust (axial force) on the pump shaft must be absorbed by a thrust bearing or balanced out with a special arrangement. A special balance disc or drum, however, always requires a certain amount of driving power and internal leakage and therefore reduces the overall efficiency of the pump. It is thus advantageous if the axial force can be limited directly by means of the pump impeller design. This is not always possible and some large pumps have a balance disc/drum to reduce the axial thrust and a thrust bearing to locate the shaft. Balance discs and drums are always designed to leave a residual thrust which must always be in the same direction to avoid thrust bearing problems.

The application of the momentum equation, equation 2.7, to a control volume, precisely that which encloses the impeller, gives us the following, using the terms as shown in figure 3.12.

$$F_{ax} = \int_{r_a}^{r_s} p_b \cdot 2\pi \cdot r \cdot dr + \int_{r_s}^{r_2} p_b \cdot 2\pi \cdot r \cdot dr -$$

$$- \int_{r_s}^{r_2} p_f \cdot 2\pi \cdot r \cdot dr -$$

$$- p_s \cdot \pi \cdot r_s^2 + \dot{m} \cdot c_{2m} \cdot \sin\beta - \dot{m} \cdot c_s \qquad \text{Equ. 3.18}$$

If the pressure on the front and back face of the impeller for $r < r_s$ is assumed to be equal ($p_b = p_f$) and if the angle β is small (Sine $\beta = 0$) the above expression can be simplified to

$$F_{ax} = \int_{r_a}^{r_s} p_b \cdot 2\pi \cdot r \cdot dr - p_s \cdot \pi \cdot r_s^2 - \dot{m} \cdot c_s \qquad \text{Equ. 3.19}$$

A normal procedure in order to limit F_{ax} is to provide the back wall of the impeller with an extra sealing edge at the same diameter as r_s, back wear ring, and at the same time to bore holes through the wall close to the shaft for pressure equalisation. Then

$$F_{ax} = -p_s \cdot \pi \cdot r_a^2 - \dot{m} \cdot c_s \qquad \text{Equ. 3.20}$$

As well as the axial force from the impeller, the shaft is also subjected to forces from seals and bearings as well as the atmospheric pressure at the end of the shaft. The bearing force is

$$F_{ax\,bearing} = (p_a - p_s) \cdot \pi \cdot r_a^2 - \dot{m} \cdot c_s + F_{shaft\,seal}$$

Equ. 3.21

In general the forces generated by seals and atmospheric pressure on the open shaft end are very small. Another way of limiting the axial thrust is to provide the impeller with blades on the reverse side. This causes the rotation of the liquid on the reverse side to be greater than that achieved by friction on the front side. The increased rotation therefore causes lower pressure for $r < r_2$ and also therefore a reduction in the actual

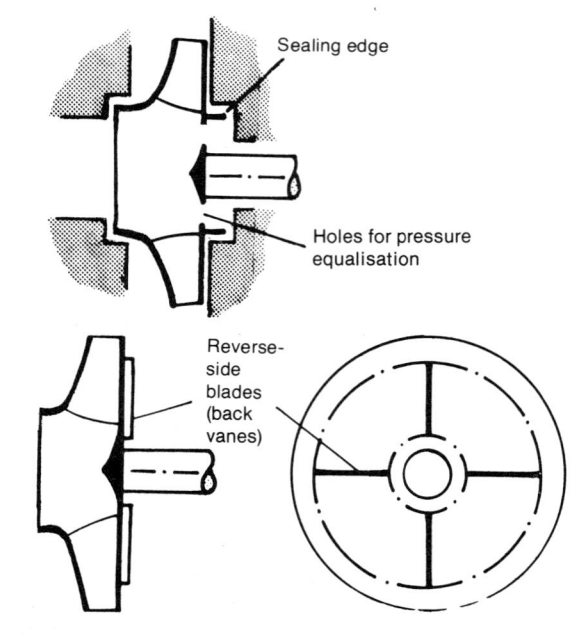

Figure 3.14 Methods for limiting the axial thrust

force. In the case of multi-stage pumps or double suction impellers low axial thrust can be obtained by having impellers facing in opposite directions.

Considerable radial forces occur during partial loading in all pumps equipped with volute casings. The design of such pumps seeks to achieve an even pressure distribution around the impeller. This condition determines the shape of the volute casing. During part load, however, the pressure around the periphery of the impeller will vary. The radial forces on the impeller are taken up by the bearing via the shaft. The diameter of the shaft and the distance from the bearings determines the deflection.

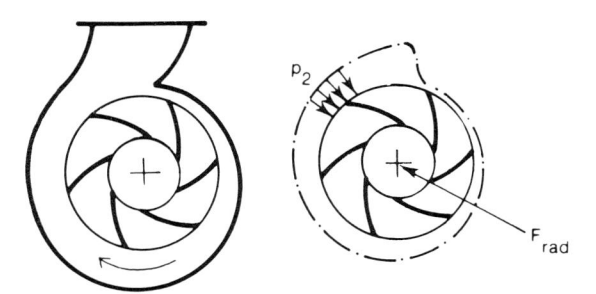

Figure 3.15 Illustration of radial forces on the impeller in the case of reduced flow ($Q < Q_{BEP}$)

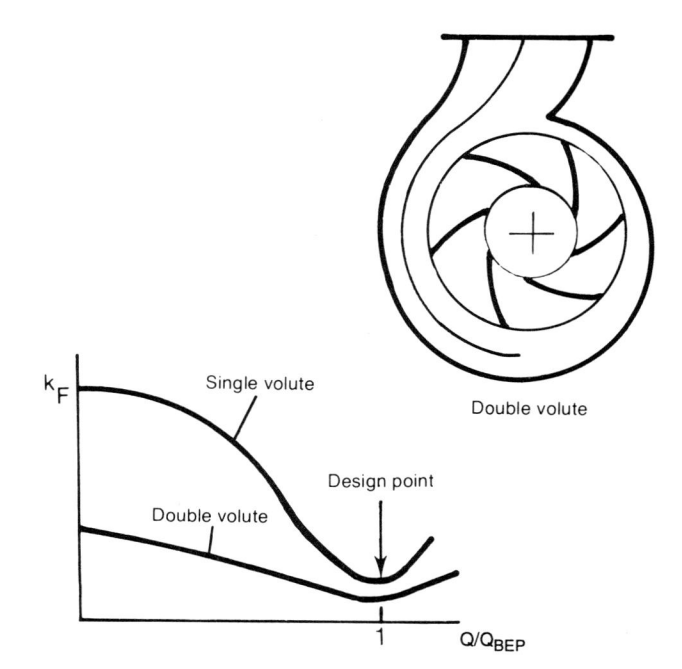

Figure 3.16 Examples of the effect of the position of the operating point and the type of the volute casing on the parameter k_F

Since the pressure varies around the periphery of the impeller the flow in the individual blade passages will also vary as the impeller rotates. In simple theory the liquid flow is considered as constant velocity across the peripheral section of a blade passage. In reality the flow may be in two or three distinct zones with different velocity profiles.

The size of the radial force can be estimated by means of the following formula:

$$F_{rad} = k_F \cdot D_2 \cdot b_2 \cdot \rho \cdot g \cdot H \qquad \text{Equ. 3.22}$$

where

F_{rad}	=	radial force	(N)
k_F	=	empirical constant	(dimensionless)
D_2	=	diameter of impeller	(m)
b_2	=	tip width	(m)
ρ	=	density of liquid	(kg/m^3)
g	=	gravitational acceleration	(m/s^2)
H	=	differential head	(m)

The parameter k_F assumes different values for different designs of pumps and also varies greatly with the volume flow through the pump.

For a centrifugal pump with a single volute k_F can have a value up to 0.4 at zero discharge (Q=0).

An effective method of limiting the size of the radial force is to provide a double volute, see figure 3.16.

The radial force is greatest at Q= 0, shut-off head. The radial force causes a deflection of the pump shaft and subjects the shaft to rotary fatigue. The condition that the maximum shaft deflection at the zero discharge should be less than 0.05 mm at the shaft seal is often used as guiding value in deciding the dimensions of the shaft.

The size of the axial thrust and radial force is a determining factor in the design and construction of the pump (clearances, bearings, bearing arrangement, etc.). They are also often the primary cause of breakdowns. Note that both the axial thrust and the radial force increase with partial load and are greatest at zero discharge. Pumps with front and back wear rings on each impeller have much stiffer shafts than simple theory predicts. Each wear ring acts as a hydrodynamic bearing, to some extent, and provides support for the shaft. Large pumps, and high differential head pumps, often have restrictions regarding low flow operation. The manufacturer may have stated the pump should not be operated continuously at less than 25% BEP flow. Low flow operation is restricted to reduce heat build up in the pump; it obviously helps with axial thrust and radial load.

3.2.2 Rotodynamic pump curves

Different ways of presenting data

In addition to the relationship between head and flow (H-Q) the complete data set for the rotodynamic pump also comprises power input, efficiency and NPSH$_r$. Figure 3.17 shows an example of a complete pump characteristic. This diagram also shows performance for different diameters of impeller, i.e. the different performances which can be attained for one and the

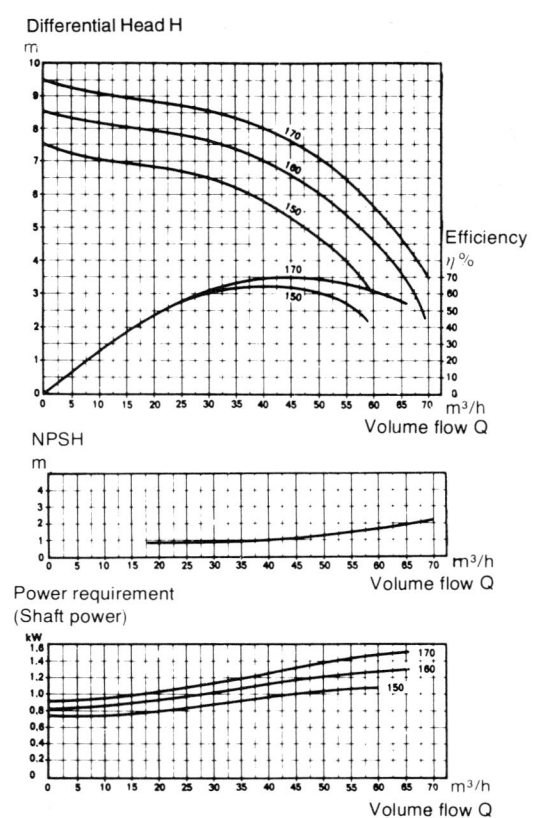

Figure 3.17 Example of performance curve for a centrifugal pump with a speed of 1450 rpm on water. The figures 150-170 refer to the impeller diameter in mm

same pump by machining the impeller. With moderate adjustments of the impeller diameter the NPSH-curve is not affected and the efficiency curve is only affected to an insignificant degree.

Rotodynamic pumps are usually designed to be driven directly by standard squirrel cage electric motors. This philosophy means that most pumps run at one speed out of a choice of eight.

Number of poles	50 Hz	60 Hz
2 pole	2900 rpm	3500 rpm
4 pole	1450 rpm	1750 rpm
6 pole	960 rpm	1160 rpm
8 pole	730 rpm	880 rpm

Figure 3.18 Approximate full load speeds of electric motors

Motors are available with more than eight poles but these are not popular and are costly. The actual full load speed of the motor is dependant upon its size. Big motors are more efficient and have less slip and therefore run closer to the synchronous speed. From figure 3.18 it can be seen that there is a speed ratio of about two between two pole motors and four pole motors. For a pump designed at 2900 rpm to run at 1450 rpm it would produce 50% flow and 25% differential head. Pump manufacturers needed a cost effective mechanism to increase the operating range of a pump without changing motors or speeds with gearboxes or vee-belts. Reducing the impeller diameter, from the maximum possible, is a simple and very effective solution. The impeller can be reduced to about 70% on most designs.

In the absence of pump performance data for different diameters then the modification in performance in the case of reduction in the diameter from D_1 to D_2 according to figure 3.19 can be estimated according to the following rule

$$\frac{H_2}{H_1} = \frac{D_2^2}{D_1^2} \quad \text{and} \quad \frac{Q_2}{Q_1} = \frac{D_2}{D_1} \qquad \text{Equ 3.23}$$

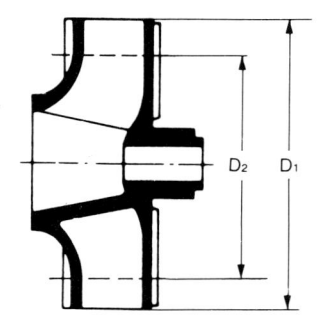

Figure 3.19 Impeller with diameter reduced from D_1 to D_2

A new approximate H-Q curve is obtained by recalculating sufficient points to produce a smooth curve.

The efficiency is sometimes indicated by iso-efficiency lines on the H-Q diagram as shown in figure 3.20.

When a pump is designed to be capable of accepting different types and sizes of driver the pump characteristic will show the efficiency of the pump alone. If the characteristic shows an overall efficiency, including driver and any transmission losses, it will be stated clearly.

In the case of axial or mixed flow pumps, different characteristics can be achieved through adjustment of the angular position of the impeller blades, see figure 3.21. Similarly the performance can be modified by adjusting stationary inlet guide vanes, see figure 3.22.

If the pump speed is varied then performance changes can also be expressed as a series of curves in the H-Q and the P-Q diagrams. An existing pump curve with H-Q and P-Q diagrams is recalculated from, for example, a maximum speed to some other speed by recalculating 4 or 5 points on the

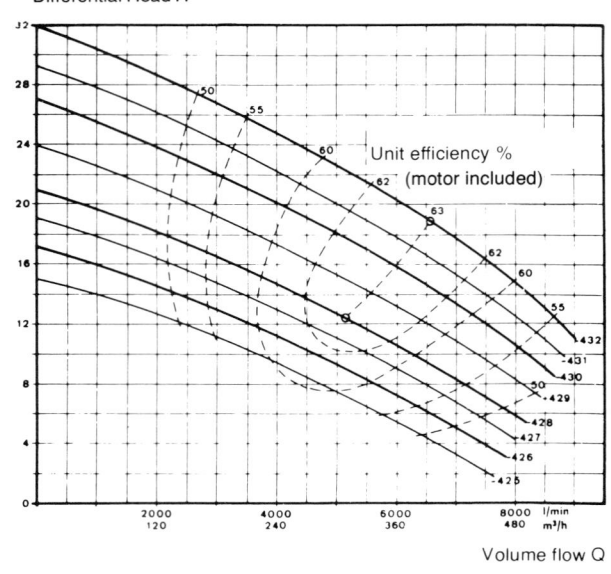

Figure 3.20 H-Q diagram with iso-efficiency lines. The figures on the extreme right of each H-Q curve are the different impeller diameters

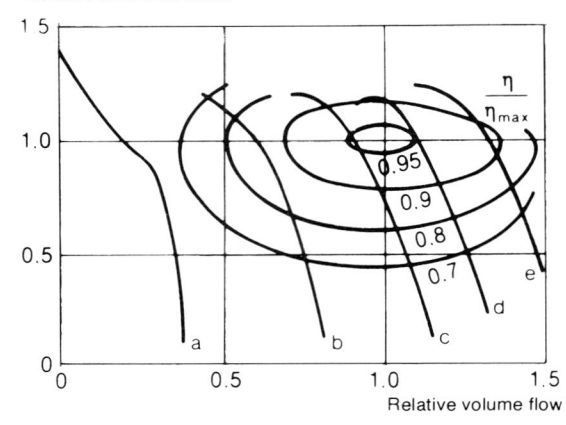

Figure 3.21 Pump curves for an axial flow pump with adjustable impeller blades. The curves a to e designate different blade settings. The performance is expressed relative to Best Efficiency Point (ηmax)

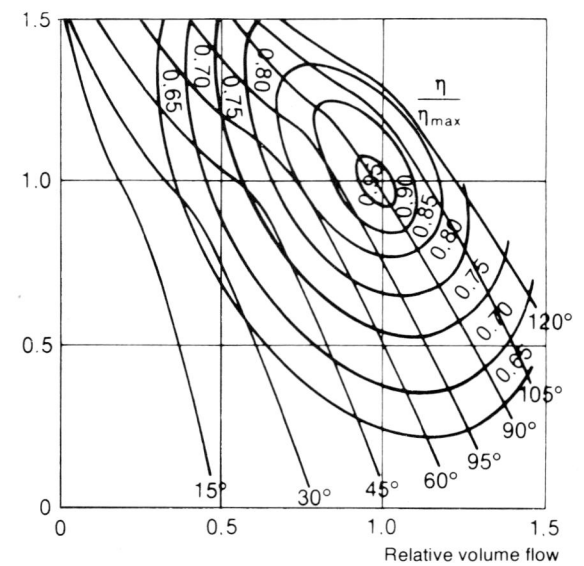

Figure 3.22 Pump curves for mixed flow pump with adjustable inlet guide vanes. The angular values relate to the angle setting of the guide vanes.

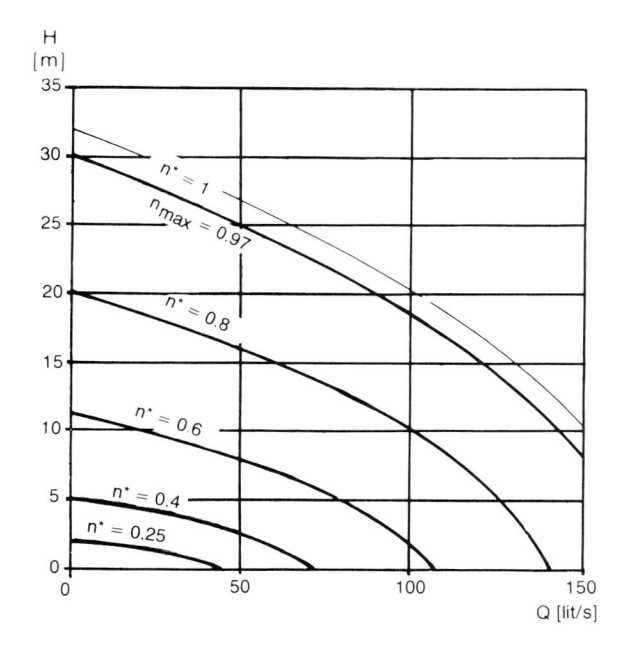

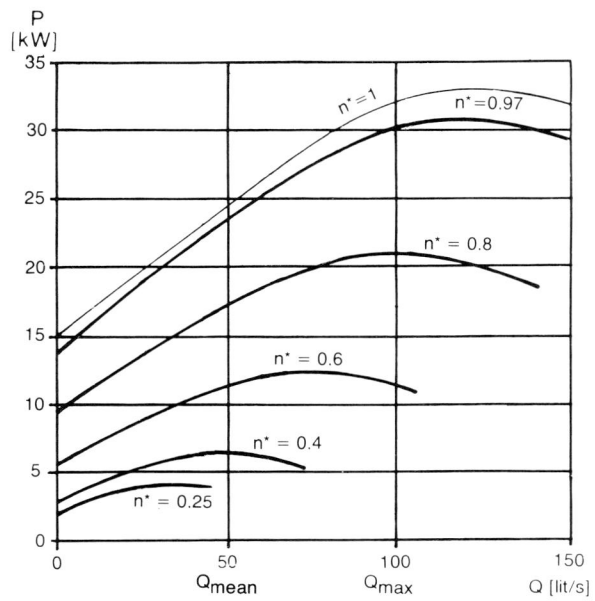

Figure 3.23 Example of performance curves for a variable speed pump. The power requirement relates to a pump with fluid coupling. The speed is expressed relative to the full speed of a direct driven pump. Due to slip in the hydraulic coupling at the full load n*$_{max}$ = 0.97, which gives somewhat lower performance than for corresponding direct drive pumps

maximum curves one after the other to the new speed, with the help of the affinity laws and the transmission efficiencies for the actual speed converter. Figure 3.23 shows an example of such recalculated pump performances. A more detailed treatment of speed regulated pumps is given in Chapter 5.

The pump curves described hitherto have related to one and the same size pump. In order to be able to display data for many sizes of pump simultaneously, range charts are used. Logarithmic scales may be used on either or both axes to show the individual performance range with some accuracy, see figure 3.24. Diagrams of this type are used both in order to be able to find a suitable size of pump quickly, and to systematise classification into different sizes. Due to the use of different pump impeller diameters or blade angles, one pump size covers a "window" in the diagram. Each pump covers a flow range of approximately 2:1 and a head range of 1:6. In order to cover varying needs for volume flow and differential head, 20 to 50 different sizes of pump can be required in a range to provide good coverage. It is very useful to indicate the BEP as a point on the maximum and minimum impeller curves. This shows very quickly how good a selection will be.

Good range design for a specific type of pump is very complicated and many factors must be considered. Problems arise because small pumps cost more to manufacture than large pumps. Good range coverage is achieved with centrifugal pumps when pumps are spaced on a geometric power scale. The step size should lie between 1.2 and 1.75. Experience has shown that because of the small pump cost problem, small pump design should not be compromised by interchangeability considerations but designed specifically for that size.

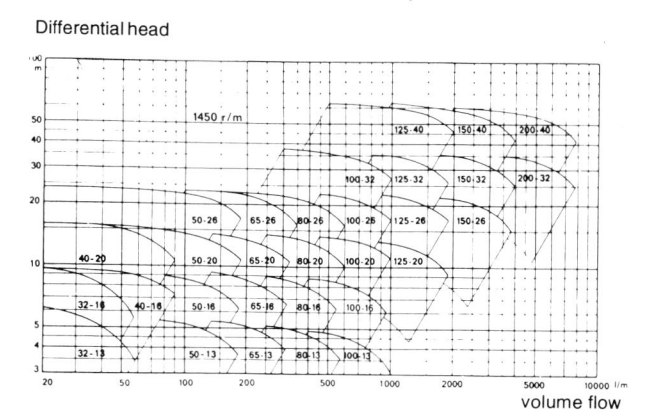

Figure 3.24 Range chart for small single stage pumps. The first figure of each designation relates to the discharge dimension in mm and the second figure to the maximum impeller diameter in cm

A rotodynamic pump's H-Q curve is stated as being stable or unstable depending upon whether the differential head is constantly rising or not when flow decreases, figure 3.25. The differential head when Q = 0 is called "shut-off head" or "closed valve head" and in the case of a centrifugal pump with stable pump curve is the highest pressure the pump can produce. The unstable part of the curve can cause difficulties due to the fact that the point of interception with the system curve is not clearly definable in some cases. Unstable pump curves are therefore undesirable and are usually avoided when the friction losses in the system are small and when a number of pumps are operating in parallel.

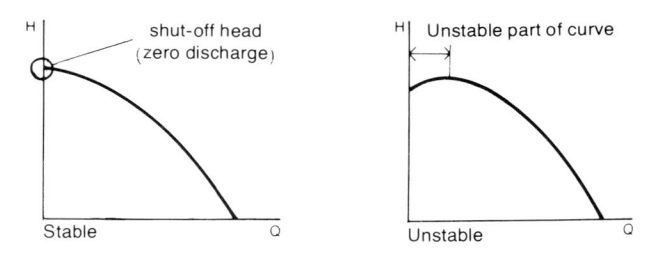

Figure 3.25 Stable and unstable pump curves

Dependent upon the slope of the H-Q curve, a distinction is made in theoretical considerations between steep and flat curves. As a measure of this slope the equation H_{max}/H_0 is evaluated for the differential head at optimum (BEP) and closed valve, as shown in figure 3.26.

When drawing a H-Q curve the choice of scales can cause the same pump curve to appear to be flat or steep. Choice of an operating point to the right or left of the flow at best efficiency, Q_0 in figure 3.26, decides in practice if the curve together with a piping system will function as a steep or a flat pump curve.

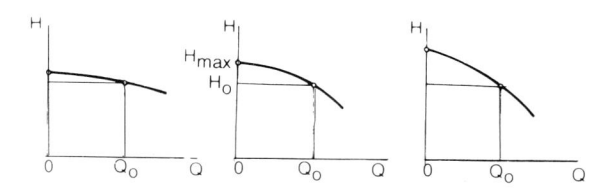

Figure 3.26 Flat and steep pump curves for centrifugal pumps

Parallel pump operation can be very difficult, especially when the system curve is mostly static pressure and there is little pipe friction, such as boiler feed pumps, and experienced users specify a minimum value, percentage head, for the slope of the H-Q curve between the duty point and closed valve.

The different varieties of rotodynamic pump; centrifugal pumps, mixed flow pumps, axial pumps; have performance curves which differ greatly in appearance as shown in figure 3.27. An increased specific speed gives an increasingly steep H-Q curve whilst the power curve alters from rising to decreasing with the flow.

The shape of the power curve is, together with variations in the flow, the determining factor for the size of the driver. In the case of axial pumps, see figure 3.27, the greatest power requirement is when the flow is zero, which can mean that the pump's starting conditions must be adapted to this fact. In the case of rotodynamic pumps the power requirements which are given in the pump catalogues, unless otherwise stated, refer to liquids comparable with cold water, i.e. with a density of 1000 kg/m³ and a viscosity of 1 cP approximately.

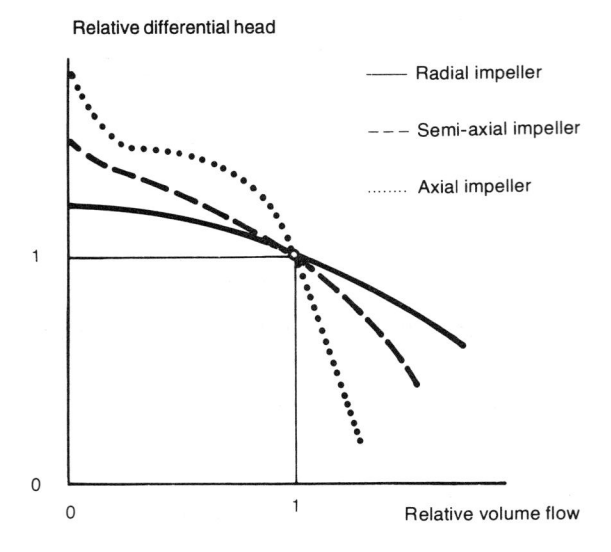

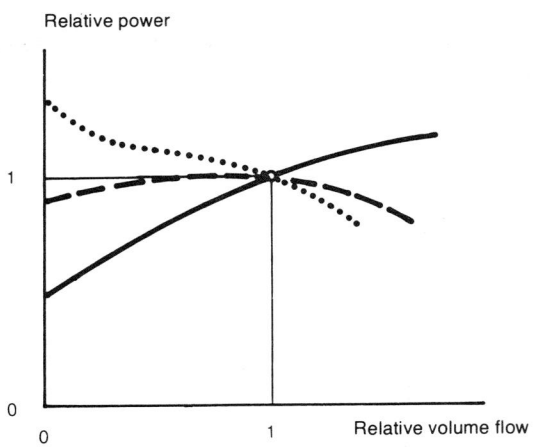

Figure 3.27 Rotodynamic pump curves relative to the BEP

In the case of liquids having different densities and viscosities from that of water, the use of the unit, "m" for differential head can cause misunderstandings. The unit "m" stands for "meters of liquid column". If reference is made to meters of water column it is usually written "m H$_2$0" so as to avoid confusion. The reason for this confusion is that rotodynamic pumps give the same differential head in meters liquid column (m), irrespective of the density of the liquid. The pump's power requirement, and discharge pressure, is on the other hand proportional to the density. For densities different from that of water the power requirements and discharge pressure quoted in the purchase order always relate to the liquid stated. When pumping liquids which differ from water, the manufacturer should state how the pump will perform, on test, with water. The standards for acceptance testing of rotodynamic pump performances is covered by various current national and international standards, for further details see Section 3.7.9 and chapter 11. The manufacturers' guarantees relate to an installation designed and built to recognized standards and sound engineering practice. An unsuitable design of, above all, the inlet pipe in an installation can reduce the pump performance and cause severe operational problems.

A rotodynamic pump's performance falls off with wear of the pump parts. This wear occurs both at the internal and external seals and at clearances, affecting the internal and external leakage, figure 3.28, and at the blades, decreasing the blade work, figure 3.29. It is very difficult to give a limit for the extent

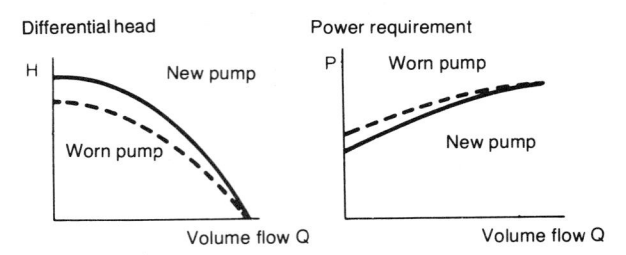

Figure 3.28 The effect on performance of wear in seals and clearances

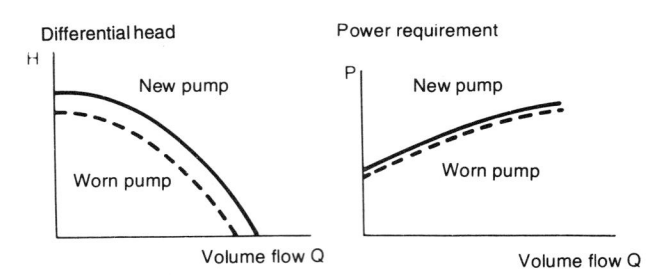

Figure 3.29 The effect on performance of wear on pump blades

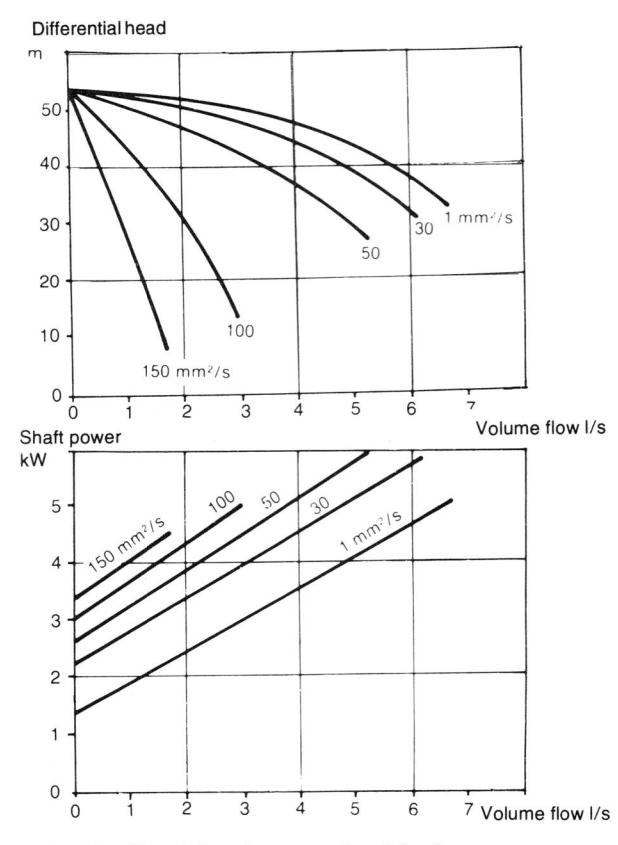

Figure 3.30 The effect of viscosity on a small centrifugal pump
N_s = 11, discharge connection 50 mm. Max efficiency falls from approx 50% to approx 5%

of wear but reductions in delivery head of the order of 10% during the pumping of clean liquids are, however, possible.

The performance of rotodynamic pumps falls off rapidly with increasing viscosity of the pumped liquid. This reduction expresses itself in such a manner that the H-Q curve falls but shut-off head is retained. The power requirement rises considerably primarily due to the increase of impeller friction. Figure 3.30 shows typical examples of the reduction in performance for smaller centrifugal pumps. Figures 3.31 and 3.32 show correction factors, in USgpm and feet, for large and small centrifugal pumps. These graphs are based on testing carried out by Stepanoff of Ingersoll-Rand, and are reproduced by kind permission of Ingersoll-Dresser Pumps. For further details see Chapter 14 on the choice of pumps.

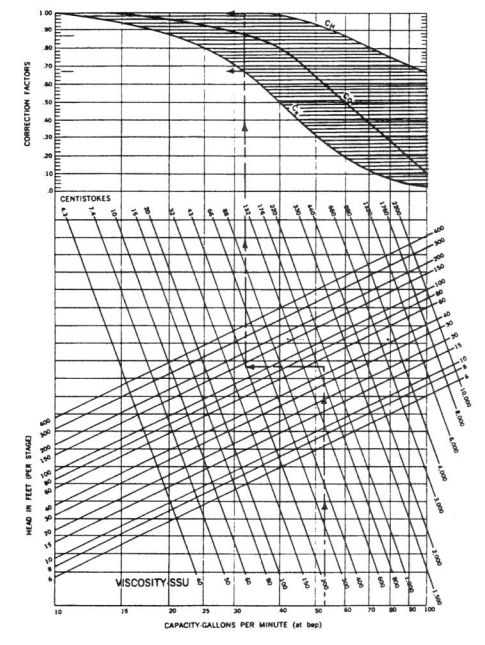

Figure 3.31 Viscosity corrections for small centrifugal pumps

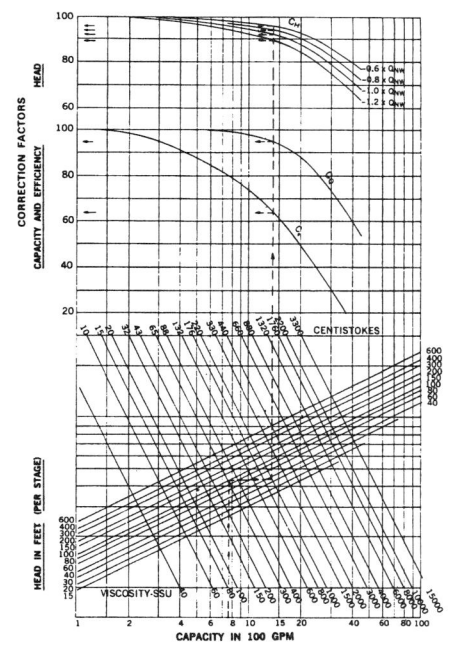

Figure 3.32 Viscosity corrections for large centrifugal pumps

3.2.3 Classification of rotodynamic pumps by design

Impeller

The impeller, together with the casing, constitute the most critical parts in a rotodynamic pump. The shape of the impeller is determined by a combination of factors; flow, differential head and speed. These three parameters are used to define the specific speed in accordance with equation 3.15. In principle all pumps with the same specific speed can be geometrically similar. The variation, in practice, is due to different design techniques and manufacturing methods for small and large impellers and also the effect of surface roughness.

Experience teaches the hydraulic designer how the impeller should be shaped in order to obtain the greatest efficiency and suitable curve shape. The pump is named according to the flow path through the impeller, such as; centrifugal, mixed flow and axial, see figure 3.33. There are unlimited numbers of intermediate impeller shapes to those shown in order to meet every performance requirement.

Two impellers can be manufactured as one for parallel operation, two single stage impellers back to back, a double suction impeller as shown in figure 3.33. A double suction pump should have a lower $NPSH_r$ than an equivalent single suction pump. For liquids contaminated with solids or sewage and where production must be simplified due to manufacturing problems of the tough materials used, the design principle of "greatest possible efficiency" is discarded and the impeller is given a much simpler shape. A distinction is made in theory between closed, semi-open and open impellers as well as special shapes for contaminated liquids. See also "non-clogging" pumps.

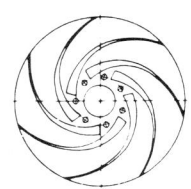

Closed centrifugal impeller. The illustration of the right shows the impeller with the front shroud removed

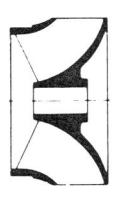

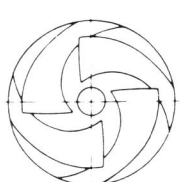

Closed mixed flow impeller. The illustration on the right shows the impeller with the front shroud removed

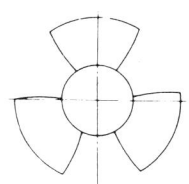

Axial impeller (propeller)

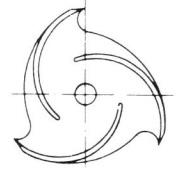

Open centrifugal impeller

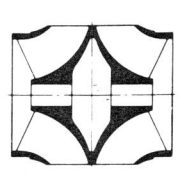

Double suction mixed flow impeller

Figure 3.33 Examples of different impeller types

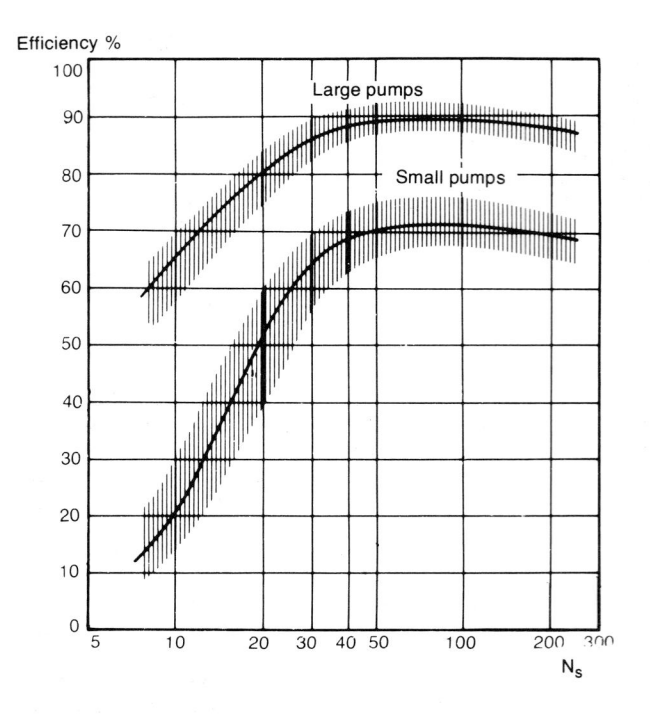

Figure 3.34 Pump efficiency as a function of specific speed

will have low specific speeds when calculated by equation 3.15. High differential heads can be produced by using several low head impellers in series. The specific speed will now be higher because the flow is the same as before but the differential head per impeller is smaller. A multi-stage pump consists of several similar impellers working on the same liquid consecutively. Doubling the number of impellers increases the specific speed by 20.75 = 1.68. The following efficiencies, for example, can be read off from figure 3.34 for a small pump:

No. of stages	N_s	Effy %
1	10.0	21
2	16.8	42
4	28.2	63

The range of application of multi-stage pumps for delivery heads over 50 and up to 100 m, is, as a general rule, dependent upon the flow and speed.

3.2.4 Types of rotodynamic pumps

Classification of rotodynamic pumps by application

In addition to a classification in accordance with different hydraulic designs, it is practical to apply a classification for different areas of application. Within each such area, there are special design and construction features dependent upon the:

- liquid properties,
- allowable leakage,
- driver type,
- installation arrangement,
- allowable noise level,
- operational safety.

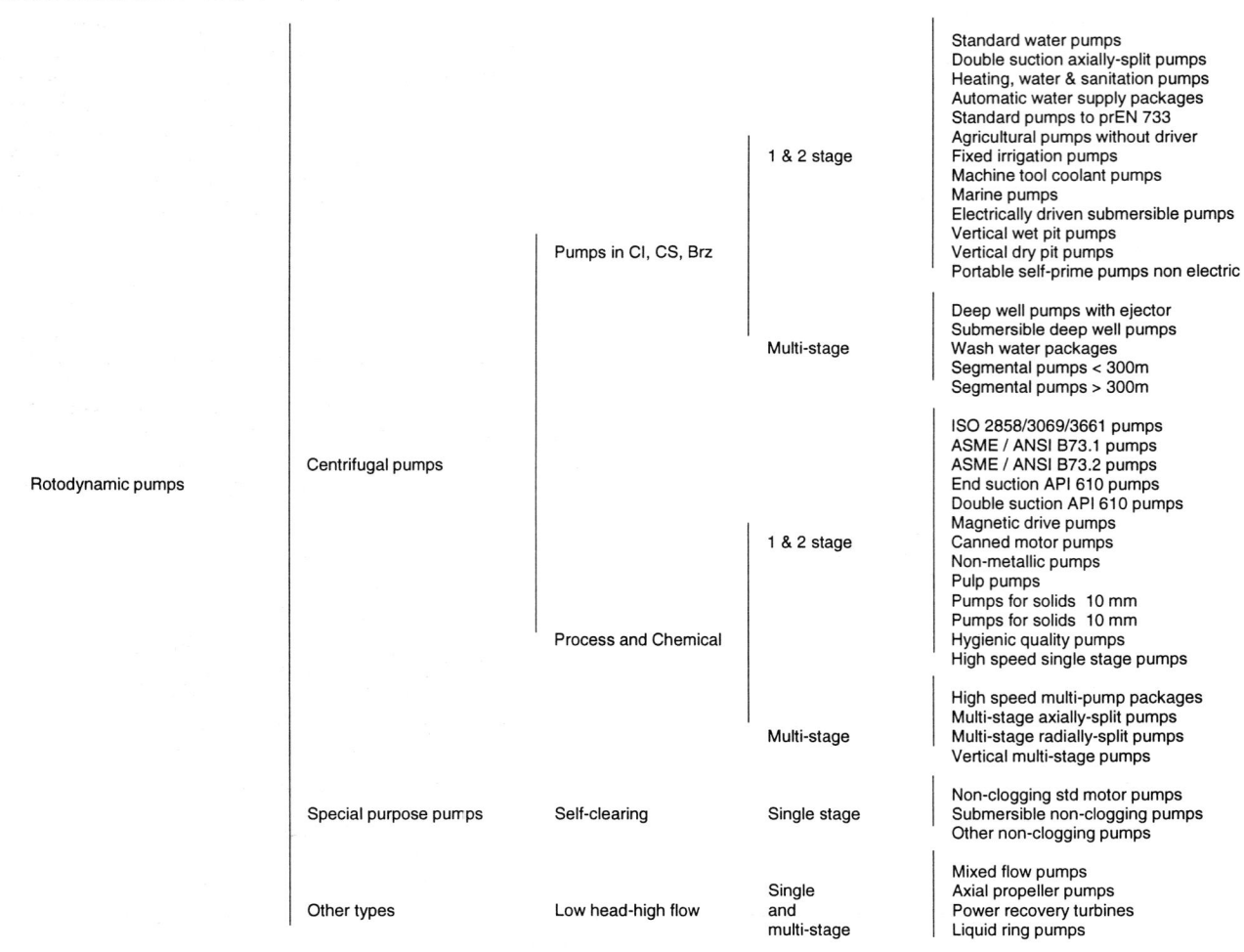

Multi-stage pumps

The efficiency of rotodynamic pumps is dependent upon the specific speed, as shown in figure 3.34. Single stage pumps, required to produce high differential heads or handle low flows

Figure 3.35 Classification of rotodynamic pumps

Rotodynamic pumps
- Centrifugal pumps
  - Pumps in CI, CS, Brz
    - 1 & 2 stage
      - Standard water pumps
      - Double suction axially-split pumps
      - Heating, water & sanitation pumps
      - Automatic water supply packages
      - Standard pumps to prEN 733
      - Agricultural pumps without driver
      - Fixed irrigation pumps
      - Machine tool coolant pumps
      - Marine pumps
      - Electrically driven submersible pumps
      - Vertical wet pit pumps
      - Vertical dry pit pumps
      - Portable self-prime pumps non electric
    - Multi-stage
      - Deep well pumps with ejector
      - Submersible deep well pumps
      - Wash water packages
      - Segmental pumps < 300m
      - Segmental pumps > 300m
  - Process and Chemical
    - 1 & 2 stage
      - ISO 2858/3069/3661 pumps
      - ASME / ANSI B73.1 pumps
      - ASME / ANSI B73.2 pumps
      - End suction API 610 pumps
      - Double suction API 610 pumps
      - Magnetic drive pumps
      - Canned motor pumps
      - Non-metallic pumps
      - Pulp pumps
      - Pumps for solids 10 mm
      - Pumps for solids 10 mm
      - Hygienic quality pumps
      - High speed single stage pumps
    - Multi-stage
      - High speed multi-pump packages
      - Multi-stage axially-split pumps
      - Multi-stage radially-split pumps
      - Vertical multi-stage pumps
- Special purpose pumps
  - Self-clearing
    - Single stage
      - Non-clogging std motor pumps
      - Submersible non-clogging pumps
      - Other non-clogging pumps
- Other types
  - Low head-high flow
    - Single and multi-stage
      - Mixed flow pumps
      - Axial propeller pumps
      - Power recovery turbines
      - Liquid ring pumps

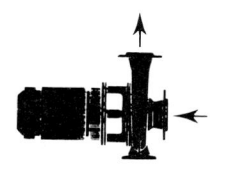

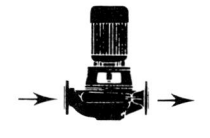

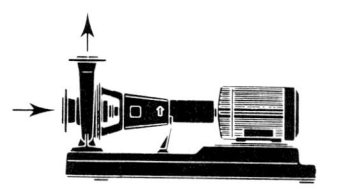

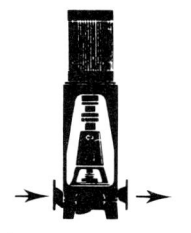

Horizontal compact pump with impeller mounted on the motor shaft (close coupled). End suction, top discharge; central or tangential.

Vertical compact pump with impeller mounted on the end of the motor shaft (close-coupled). In-line connections.

Horizontal pump with bearing bracket mounted with motor on baseplate. End suction, top discharge; central or tangential.

Vertical pump with support feet, bearing bracket and extended motor stool for spacer coupling. In-line connections.

Figure 3.36 Various designs of standard pumps

The following descriptions of the various types are made in accordance with the classifications as shown in figure 3.35. This classification is based upon both technical and marketing considerations and is subsequently applied in the chapters following. For this reason different types of pumps have been allocated numbers in key tables and product guide. In the following sections, therefore, the section number is marked for example, with **005** for the group "Standard water pumps".

005 Standard water pumps

Standard water pumps, i.e. pumps designed to operate with clean water, at temperatures up to 80/120 °C are used for clean liquids compatible with cast iron and in some special cases bronze. Impellers may be cast iron, bronze or plastic. Small pumps will generally not have wear rings. There is a wide range of styles and designs available, nevertheless, a classification into main categories can be made in accordance with figure 3.36.

Compact pumps with the impeller mounted directly on the motor shaft end, close-coupled, are suitable for flows up to a maximum of approximately 300 m^3/h and for differential heads of up to 100m. Casing pressure ratings can be up to 10 barg. Motor output can be over 100 kW however, motors exceeding 20 kW will require lifting facilities. Close couple pumps rely on the motor bearings to absorb axial thrust and radial loads. Suction pressure may be restricted. Most of the hydraulics specified in section **025** for prEN 733 pumps are also available in close coupled units which do not comply with the dimensional part of the standard. Smaller vertical compact pumps can often be mounted directly in pipework without supports, this of course assumes that the pipework has sufficient rigidity. Larger vertical compact pumps have a foot to rest on a support. Normally the pump can be mounted at any angle, with the exception that the motor must not be located under the pump because of the risk of motor damage in the event of seal leakage. Mounting, other than vertical, can cause maintenance problems. Small pumps may have female screwed connections as an alternative to flanges. Space around the stuffing box is restricted and the choice of packing and seal arrangements will be limited.

Pumps with bearing brackets, housing their own radial and axial bearings, are available to cover most of the operational range of close coupled pumps. Pumps with bearing brackets will not have the same restrictions on suction pressure and sealing arrangements. Somewhat larger pumps are produced in this configuration. A coupling is required between pump and motor and a baseplate for horizontal units. These units are slightly longer than the equivalent compact close coupled pump but more versatile. Some pumps are designed to be "back-pull-out"; the complete rotating assembly plus the bearing bracket can be removed leaving the casing connected to the pipework. If a spacer coupling is used to the motor, the pump can be maintained without disturbing the pipework or the motor. These pumps are capable of 2000 m^3/h at differential heads of 160 m and pressure ratings up to 16 barg.

010 Double suction axially-split pumps

Twenty-five years ago the double suction pump was the predominant water pump for applications where the pipe size exceeded 200 mm. The classic design with a large impeller between two bearings, two stuffing boxes and axially-split pump casing with the pipe connections in the lower half, figure 3.38, is still used for large flows at low heads. Normally used for large and very large water services pumps, for example, clean water pumps for large water works and distribution pumps for industrial water.

The specific speed per impeller half for double suction impeller type pumps is less than that of similar single suction pumps and this results in a corresponding flattening of the H-Q curve. In borderline cases the lower NPSH$_r$ value means that higher speeds can be used. In special instances where it is desirable to maintain as near pulsation-free flow as possible, as for example, fan pumps for paper machines, double suction pump impellers with staggered blades are used. Sometimes the location of the pump connection flanges can make the use of a double-suction pump advantageous.

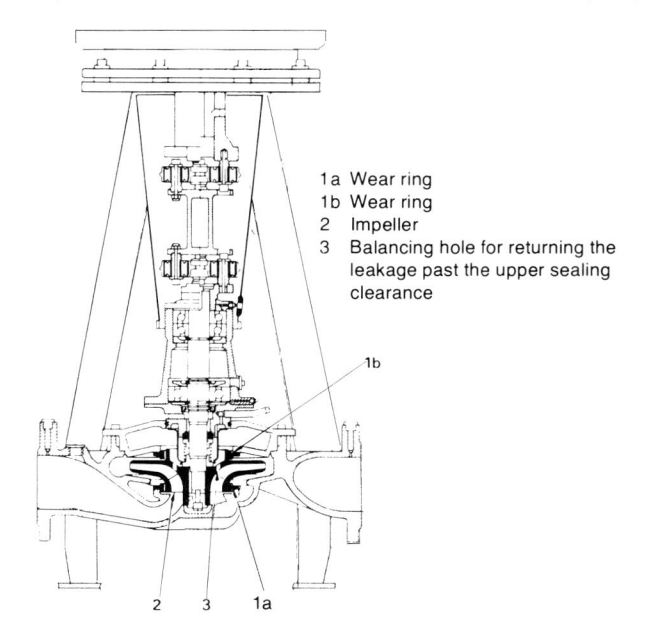

1a Wear ring
1b Wear ring
2 Impeller
3 Balancing hole for returning the leakage past the upper sealing clearance

Figure 3.37 Vertical inline pump with support feet, bearing bracket and extended motor stool for spacer coupling.

Figure 3.38 Double suction axially-split pump with sleeve bearings and tilting pad thrust

Water temperature can be up to 120 °C with standard cast iron casings. Special casings of bronze are sometimes supplied for seawater applications. Impellers can be cast iron or bronze. Pumps with 250 mm connections can handle 1200 m³/h at 100m. Larger pumps are available for up to 40000 m³/h at 40m. Small pumps have pressure ratings of up to 17 barg reducing to 5 barg for the largest sizes. Power requirements are up to 5.5 MW. Packed stuffing boxes or mechanical seals can be fitted as options.

This style of pump has been available in the past in two slightly different two stage versions. A two stage pump with single entry second stage was used as a condensate extraction pump for steam turbo-generators. The pump grew to unmanageable proportions in the 1960's as generators became much larger. The design concept changed and the condensate extraction pump became a vertical multi-stage pump. A two stage pump with two double entry impellers has also been available, particularly in America. This style of pump was designed when axial thrust was not fully understood and resulted in a hydraulically balanced rotor. Neither of the two stage designs described here are popular in the current realm of pump applications.

015 Heating, water and sanitation pumps

There are specially developed pumps, so-called heating, water and sanitation pumps for the different pumping requirements in buildings. One of their general characteristics is a low noise and vibration level, the requirements varying in accordance with the size of the building from a noise level of approximately 25 dBA for a private house, to approximately 65 dBA for pumps placed in a well insulated machine room in a larger building. These different pump requirements are covered by:

- Heating circulation pumps for circulating water in a central heating system. Smaller sizes up to a power requirement of normally 100W are of the "wet" type, i.e. "wet rotor" motors, figure 3.39. All the rotating parts, including the motor rotor are sealed inside a stainless steel "can". The stator windings are placed around the outside of the can and the rotating magnetic field passes through the can wall. No seals are necessary; the bearings are usually ceramic lubricated by the hot water. Pump casings are usually cast iron or aluminium alloy with bronze or stainless steel impellers. The pumps are mass produced with fixed performance, but in order to match the circulator to the system and avoid noise in the piping system, they are supplied with a two or three speed motor.
- There are so-called twin pump packages, i.e. a pump casing with two pumps built as a unit ready for installation in one suction and discharge pipe. The pumps can be operated together for parallel operation, see figure 3.40, or series operation. The unit includes a non-return valve system, controlled by the flow of liquid so that either pump can be operated independently or together with the other pump.

Figure 3.40 Heating circulation pump of the twin design

- Hot water pumps for circulating domestic hot water in larger buildings so that the hot water is almost immediately available when the tap is turned on. In contrast to "central heating pumps" the parts in contact with the fluid are made of bronze or stainless steel instead of cast iron. Wet motors are available, as well as dry motors using special designs for motor and shaft seals in order to avoid blockages due to "furring", deposition of the natural salts.
- Other pumps in buildings are pressure boosting pumps, see multi-stage pumps, ground-water pumps and to some extent also standard water pumps for fire systems.
- Complete pump packages are supplied, in buildings with water central heating systems, to maintain the static pressure in the heating circuit and make up any leaks.

Most of these pumps are of the inline design. The pump and motor form an integral package and the suction and discharge connections are inline so that the package can be fitted into a straight pipe run. Most pumps do not require external support but rely entirely on the rigidity of the pipework. Pumps can be mounted with the motors vertical or horizontal. In most designs the pump cannot be mounted vertically above the motor.

020 Automatic water supply packages

Automatic water packages are used for supplying drinking water to households, summer cottages and smaller farms which are not connected to the national water distribution network. The water is normally taken from a well with a suction lift for the pump of 3 to 7 m. The pump's flow and head are suitable for 1 to 3 outlets connected to the pump by relatively short pipes or hoses.

The package consists; a self-priming pump, (centrifugal or liquid ring type), an electric motor, an accumulator (pressurised holding tank) and usually a pressure switch. The pressure switch ensures that the pump starts when the pressure falls due to water being drawn off and stops when the accu-

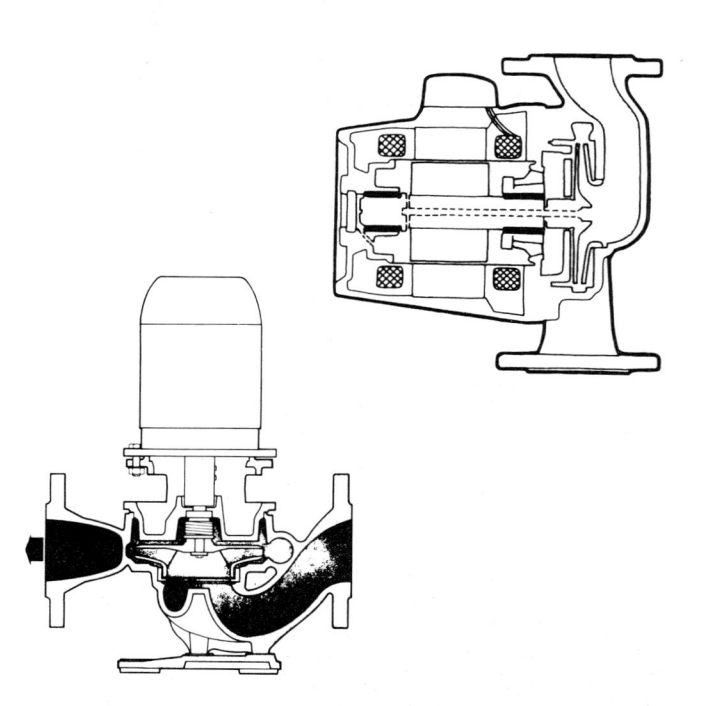

Figure 3.39 Central heating circulator pumps with in-line connections. The upper illustration is the "wet" type. The lower illustration is the conventional dry motor

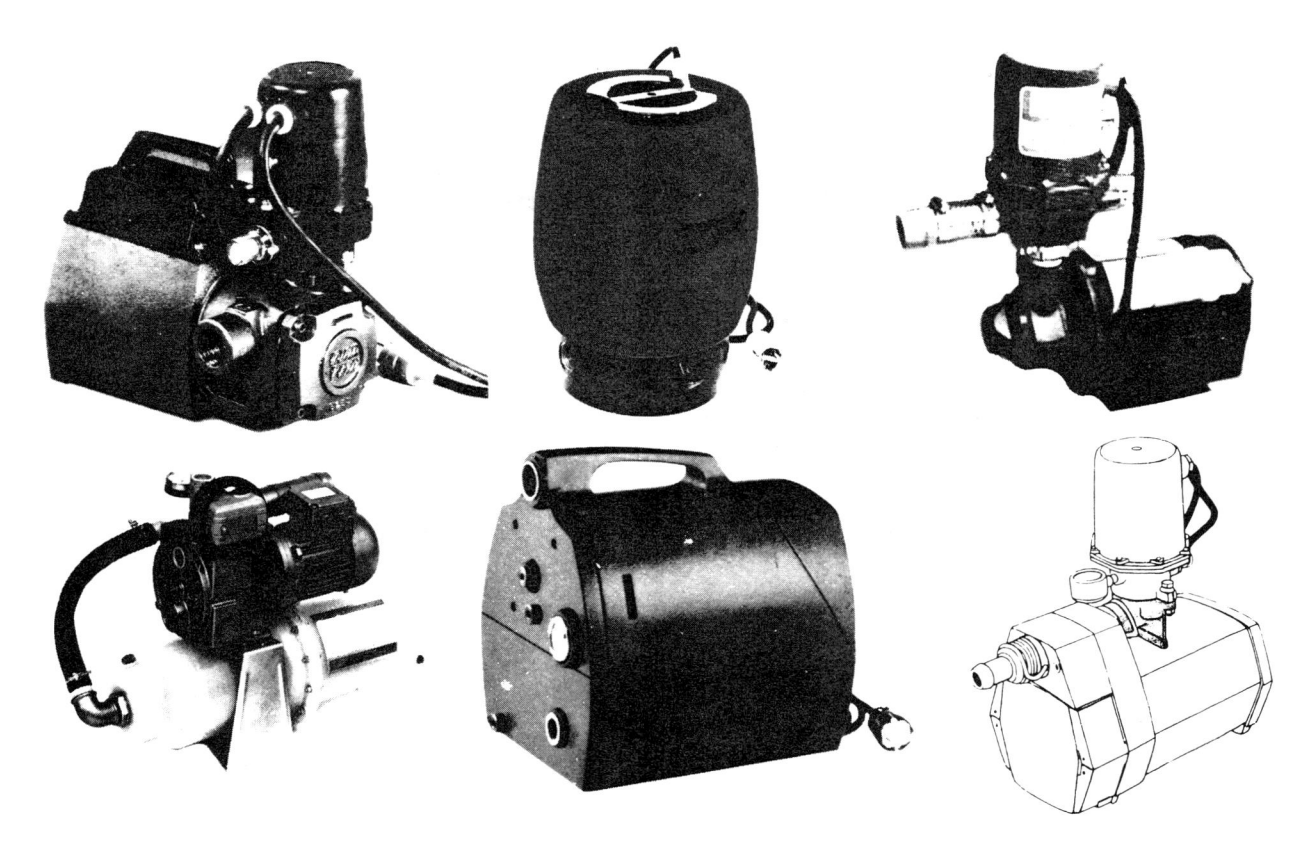

Figure 3.41 Automatic water supply packages - a selection of models

mulator has filled up due to the corresponding increase in pressure.

There are many different makes and models available as shown in figure 3.41. When purchasing, account should be taken of the noise level as well as the fact that the water package should fulfil current electrical and hygienic legal requirements.

025 Standard pumps to prEN 733

In West Germany and a few other European countries, there were standardised pumps for non-hazardous liquids. The German standard DIN 24255 was probably the most popular. The German standard is in the process of being replaced by a very similar European standard, prEN 733. The standard relates to foot mounted horizontal pumps with a bearing bracket, see figure 3.42, and a "back-pull-out" facility when driven via a spacer coupling.

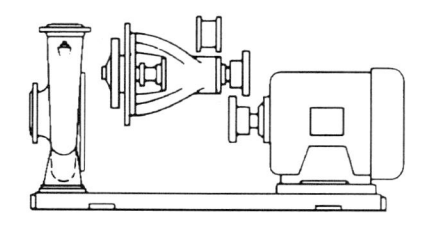

Figure 3.42 A pump in accordance with prEN 733

The standard does not define materials of construction, only performance and physical dimensions. The standard comprises a total of 29 sizes with discharge connections from 32 up to 150 mm. The nominal performance at BEP specifies flow of 6 to 315 m^3/h and the differential head of 5 to 80 m. The pump casing must have a pressure rating of 10 barg. The pump's design, figure 3.43, consists of a system of units based on three basic shaft sizes.

The pumps are generally built with cast iron casings, cast iron or bronze impellers and stainless steel shafts. Some manufacturers have bronze and steel casings available, as well as plastic impellers. The space allocated for soft packing and mechanical seals is small limiting the use of the pumps to

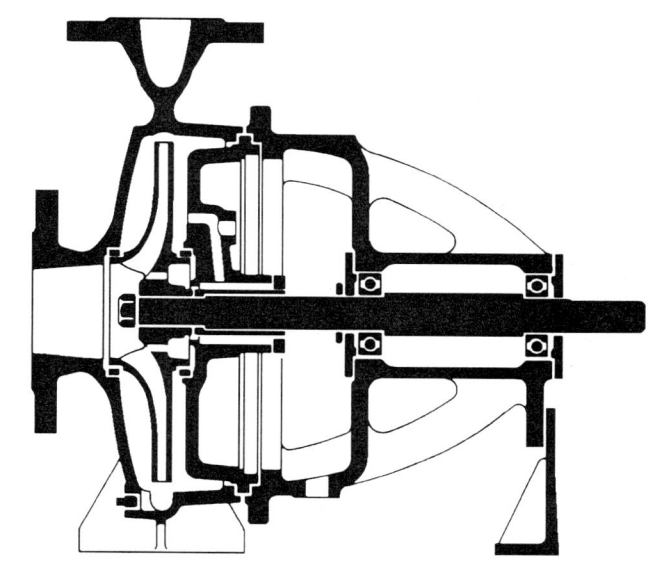

Figure 3.43 Design of a pump in accordance with prEN 733

Figure 3.44 A centrifugal pump driven from the power take-off of a tractor

non-hazardous liquids. The intent of the standard was to allow interchangeability between manufacturers equipment and provide the user with greater choice.

030 Agricultural pumps without driver

Pumps specially adapted to be driven from the power take-off of a tractor, see figure 3.44, are used for temporary installations, during the preparation period for a permanent installation and for mobile applications. The speed of the tractor power take-off is too slow, approximately 500 rpm, for a small centrifugal pump. The specially adapted pumps have built-in gears to increase the speed to 3000 to 4000 rpm, see figure 3.45.

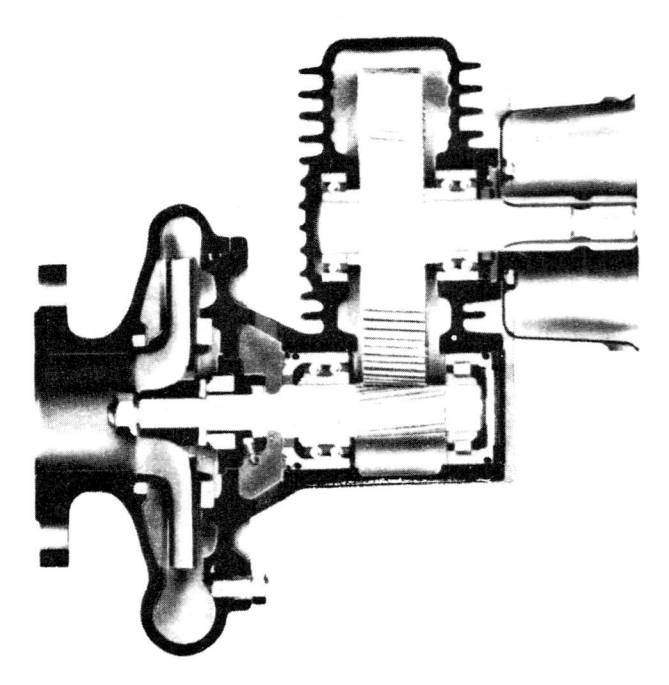

Figure 3.45 Typical design of a tractor driven centrifugal pump with built-in gears to increase speed

035 Fixed irrigation pumps

Irrigation and other applications within agriculture make use of different pumps depending upon the location and what power sources are available. Permanent surface installations normally have single-stage pumps in accordance with groups **005** and **025** or multi-stage pumps in accordance with group **085** and **090**. When wells have been sunk, submersible deep-well pumps in accordance with groups **055, 070** and **075** are used. The drive is generally an electric motor although anomalies in the electric tariff system can also make diesel or LPG operation a favourable alternative. Some large installations in very remote locations use crude oil burning engines or natural gas. The required pump performance depends upon the type and size of the installation. Normal duties lie within the range 10 to 100 m^3/h with heads of the order of 50 to 120m.

040 Machine tool coolant pumps

Special pumps are used for pumping coolant/cutting oil, White Water, for machine tools, lathes, grinding machines, etc. The liquid can be a specially blended oil or an emulsion of in water, the oil content being from 2% to 15%. The flow usually varies between 0.3 and 20 m^3/h with heads varying between 2 and 20m. This type of pump is standardised in Germany in accordance with DIN 5440. This standard indicates both performance and those dimensions which have an influence on interchangeability. The pump unit is constructed so that the pump casing is submerged in the liquid with the motor placed outside the tank, figure 3.46. The immersed depth is up to 500 mm. Due to this construction, no shaft seal is required. BS 3766:1990 specifies ten sizes in the form of dimensions and minimum performance characteristics for vertical top and side mounted units. DIN 5440 specifies six pump sizes in sub-

merged and external forms. The French standard, NF E44-301 incorporates DIN 5440 requirements. Normally coolant pumps are manufactured in cast iron or aluminium alloy but there are special types available in plastic.

045 Marine pumps

Marine pumps are included as important components in, not only ships and oil rigs, but also underwater equipment. Pumps for ships can be classified as follows:

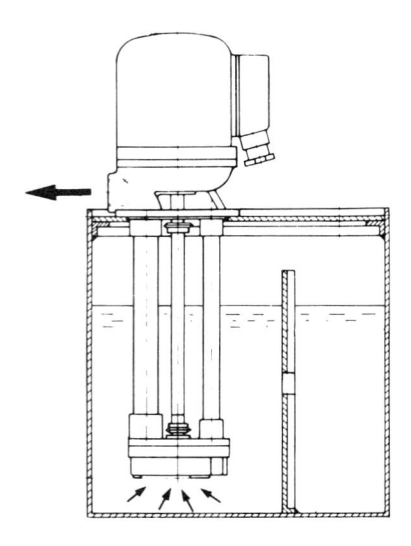

Figure 3.46 Coolant pump for machine tools

- Systems for propulsion, e.g. pumping fuel, lubricating oil, cooling water, feed water and condensate.
- Systems dependent on the cargo.
- Safety systems. e.g. for bilge pumping, ballast pumping, for fire fighting.
- Domestic housekeeping type systems for making and pumping drinking water, hot and cold fresh water, sewage etc.

The type of pumps used for fuel is dependant upon the type of fuel. A lot of ships use heavy fuel which is about 3500 Redwood No 1. Rotodynamic pumps would be inappropriate. Double acting piston pumps act as transfer and low pressure feed pumps. Plungers pumps, usually built by the engine manufacturer, inject the fuel into the cylinder. Special lubricators, driven by eccentrics or geared to an engine shaft, inject lubricating oil into the cylinders. Screw or gear pumps provide pressurised lubricating oil. Fresh water for engine cooling, and seawater for cooling are provided by centrifugal pumps. Packages are available using double ended motors which supply fresh and seawater simultaneously from two pumps on a common baseplate. Vertical units have been built with fresh and seawater pumps driven by a normal motor. Steam turbine ships use multi-stage centrifugal pumps for boiler feed and condensate extraction duties.

Again, the type of cargo pump is dependant upon the cargo. Crude is a common load but is extremely variable, low to very high viscosity. Steam driven, direct acting, vertical piston pumps are available which can operate on petrol to the most viscous crude oil. Electrically driven piston pumps are also available but are not so versatile as the steam versions. Liquified petroleum gas and anhydrous ammonia have become popular as a sea cargo. Special centrifugal pumps are available for loading and unloading.

Safety systems for marine applications are usually designed around sea-water because it is readily available. Bilge pumps largely handle sea-water. Some modern pumps have monitoring systems fitted which shut down the bilge pumps if the oil content is higher than 15 parts per million (ppm). The amount of liquid able to be pumped from the bilges is dependant upon the age of the ship. Modern ships tend to have mechanical seals on the stern tube so that very little leakage enters via that route. Soft packed stern tubes require some leakage for lubrication. In some ships the level in the bilges must be watched very closely as it has a great effect on the stability of the ship. Depending on the age of the ship the bilges may be pumped every two hours, at the end of each watch or automatically by float controls. The pumps may be reciprocating or centrifugal or ejectors powered by sea-water. Recips and centrifs would have a priming system to evacuate the suction line. Ballast and fire fighting pumps operate on sea-water and would be centrifugal pumps.

Turbine and engine driven ships can make fresh water from sea-water by evaporation or reverse osmosis. Evaporation is a low pressure process requiring single-stage centrifugal pumps. Reverse osmosis for sea-water operates at 70 barg and requires a recip pump or a multi-stage segmental pump. Other domestic water circuits would have single-stage centrifugal pumps similar to those used onshore.

Due to the fact that ships at sea should be able to function safely and entirely independently, special requirements are made on equipment which is involved with the safety and reliability of ships. These requirements are determined by marine insurance companies (classification societies) and each country's respective maritime authority. In certain cases additional requirements are imposed by, for example, port and canal authorities. Various types of pumps are used for the pumping systems mentioned earlier in the same manner as in the case of land based systems.

A typical feature of rotodynamic pumps on board, is that they are often of the vertical type in order to reduce the space requirements, mechanical design is adapted, generally reinforced, to take account of the ship's movement at sea and automatic, separate or built-in air ejection systems are present in order to avoid operational problems in the case of long suction lines or if the suction intake is momentarily above the surface of the liquid. As a general rule, bronze or gunmetal is used when pumping sea-water and the same material as in the case of land based systems is chosen for other liquids.

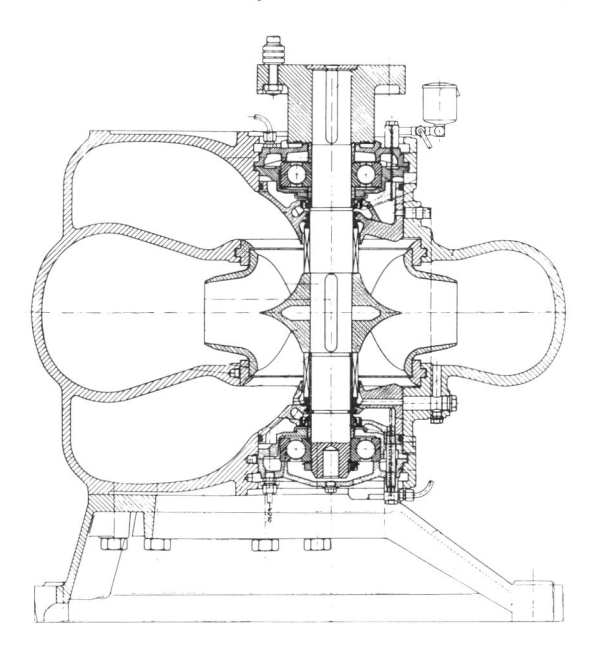

Figure 3.47 Vertical ship's pump with double suction impeller

Figure 3.47 shows a typical vertical pump for general use on board ship. This is a double suction pump with the pump casing divided axially (in a vertical plane) in order to facilitate servicing. Single suction impellers with axially or radially split casings are used for smaller flows. In the case of a radially split casing, the pump's rotor should be able to be disassembled in a simple manner, like for example, figure 3.37. Chapter 15 shows the installation of cargo oil pumps as a practical example.

For marine use, there are many interesting applications of pumps, for example, for steering/propulsion jet operation of non-anchored oil rigs with automatic positioning. A pump system as shown in figure 3.48 and constructed of axial pumps designed to operate with any flow direction, is used to limit heeling when loading, for example, or for the intentional generation of heel angle in the case of icebreakers.

050 Electrically driven submersible pumps

Submersible pumps placed in ponds, ditches and sumps are used to drain, mostly, water from locations which are deeper than the sewage system. Three types are available:

- simple, small, self contained for clean water,
- heavy duty for contaminated water,
- special purpose for sludge.

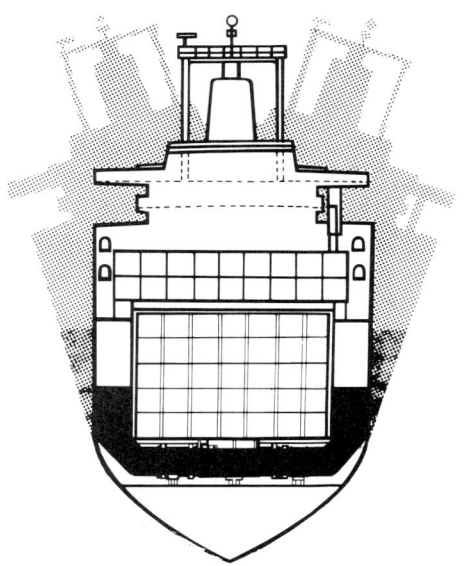

Figure 3.48 Heeling pump for ships

Small submersible pumps tend to be made of plastic where ever possible. This style of construction removes many of the insulation problems associated with electrical equipment. The pumps are close-coupled and completely self contained. The motor is totally enclosed and cooled by the water surrounding the pump. Level switches, either internally or mounted externally on the casing, detect the water level and switch the pump on and off. Low voltage versions, 110V and 240V, single phase as well as 380V three phase are available. The smallest units will pump 10m^3/h against a head of 7m. Pipe connections are usually screwed and a non-return valve is required on the smallest units.

Heavy duty electrically driven submersible pumps are close coupled pumps and comprise, together with external level controls, a complete unit. Their weight is the lowest possible due to the use of light metal alloys although their resistance to corrosion is low. The parts which are exposed to wear from the liquid are, on the other hand, of high quality. Typical features are hard metal impeller, rubberised wear parts around the impeller and hard-faced seals. A non-return valve is built into the discharge connection. The electric motor is completely encapsulated and cooled by the pumped fluid through a double jacket. In order to protect the motor in the case of blockage or operation without fluid, the motor is provided with a specially built-in motor protection unit. Normally the motor windings have thermistors which stop the motor when the temperature is too high by means of an inbuilt contactor. The shaft seal is nearly always of a double type with an intermediate oil chamber, see figure 3.49. Normally construction pumps have suction strainers which limit the particle size to 5-10 mm, but there are special models with a through flow of up to approximately 100 mm for pumping larger solids. Figure 3.50 shows a heavy duty pump located in a sump with external level switches. Heavy duty pumps can handle up to 7000 m³/h at heads up to 100 m.

Heavy duty submersible pumps are used for draining all types of building sites from the smallest ditch or hole to the large projects such as tunnelling in mountains or harbour building. This type of pump is practical in the event of flooding and for all other temporary pumping requirements. Of particular convenience is the fact that pump companies in this field offer both small and large pumps for hire for periods ranging from one day to several months.

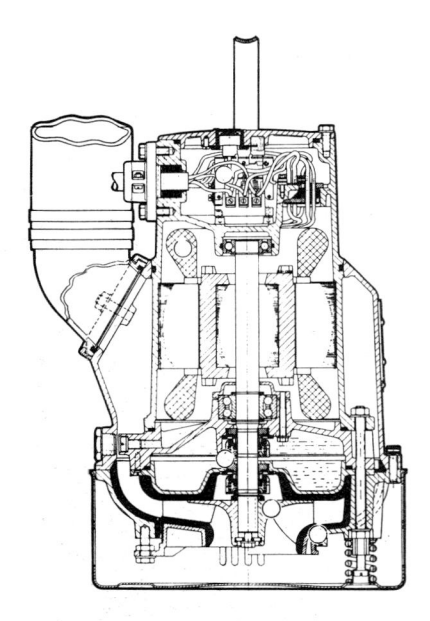

Figure 3.49 Cross sectional arrangement of a heavy duty submersible pump

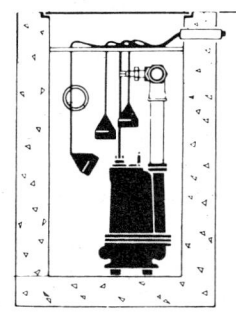

Figure 3.50 Installation of submersible pump in sump

A special adaptation of the heavy duty submersible pump has been developed for sludge and heavy muds. The impeller and casing are made from hard wear resistant iron alloys, not rubber coated, with an agitator mounted on the shaft end where an inducer would be fitted. The pump is intended for use on liquid solid mixtures with very high solids content where the pump can sit, initially, on the surface. The agitator beats the mixture and induces moisture to travel towards the impeller. The extra moisture locally helps the mixture to flow and the agitator action allows the pump to settle in the mixture. The novel design of the agitator allows these pumps to handle liquid solid mixtures which would otherwise be moved by shovels.

055 Vertical wet pit pumps

The pump type "vertical wet pit" comprises all those pumps designed to be suspended into the liquid with the drive motor on top. Figure 3.51 shows a typical single-stage design where the pump discharge is piped separately through the mounting plate. The pump seal is mounted in the pump casing below liquid level. The thrust bearing is included in the mounting plate assembly.

The style of construction shown in figure 3.51 is limited to fairly low power pumps. The column surrounding the shaft is relatively small in diameter and the discharge pipe helps to increase rigidity of the assembly. As the shaft length increases intermediate bearings must be added to reduce shaft deflection and vibration. Special rubber bearings are available which can operate with water as the sole lubricant.

Larger pumps; centrifugal, mixed flow and axial propeller; are constructed with a column which encloses the shaft and also acts as the discharge pipe. The column diameter is increased to accommodate a reasonable liquid velocity. The column is extended through the mounting plate to allow a branch to be fitted for the discharge connection. A seal, rated for discharge

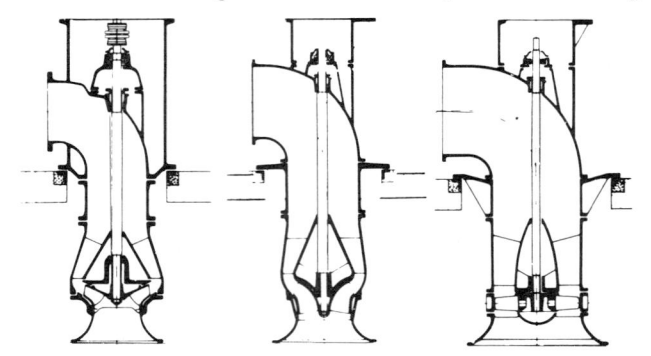

Figure 3.52 Single-stage vertical wet pit pumps in centrifugal, mixed flow and axial propeller designs

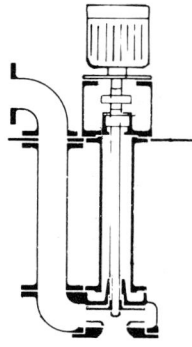

Figure 3.51 Single-stage vertical wet pit pump

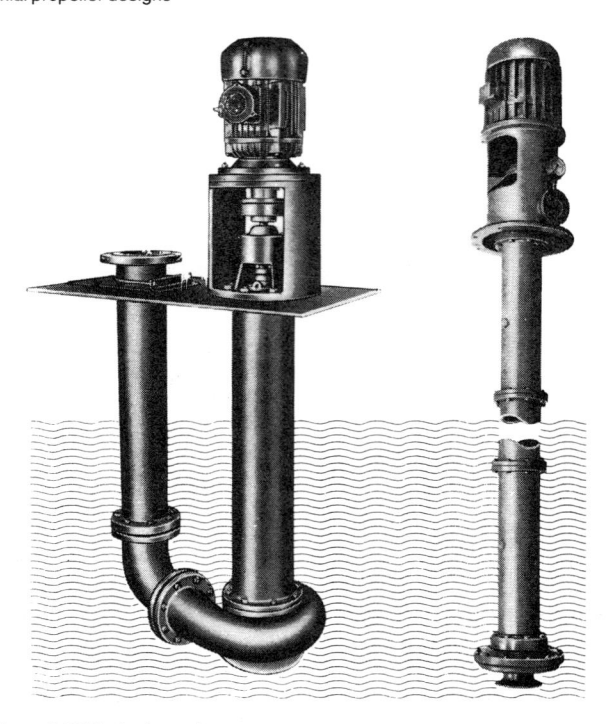

Figure 3.53 Vertical wet pit pumps

pressure, must be fitted above the discharge connection. Again, intermediate bearings must be fitted as the column length extends. Pumps are available for flows in excess of 6000 m³/h at heads up to 150m.

Vertical pumps, especially in the larger sizes, can be prone to vibration problems. The problems are caused by the pump or motor shaft unbalance occurring at a frequency which coincides with a natural frequency of the column. Because of this problem it is essential that large pumps are tested in the manufacturer's works at conditions which duplicate, as close as possible, the site operating conditions. Extensive testing and modification is much easier in the works than at site.

A variation of the vertical wet pit pump is the vertical canned pump. A vertical wet pit pump is built inside a "can" or pressure vessel which is also suspended from the mounting plate. The can has a flanged pipe connection, pump suction, which may be attached above or below the mounting plate. This type of design is used for volatile liquids with low $NPSH_a$. The liquid is taken from the suction line elevation down the can to the first stage impeller eye. The $NPSH_a$ is increased by the difference in levels between the suction line and the impeller eye. This variation is used extensively for liquified gas applications.

060 Vertical dry pit pumps

Vertical dry pit pumps are quite similar to wet pit pumps but the pump casing in not immersed in the liquid. The liquid is carried from the pit to the pump suction by concrete ducts or metal pipework. Figure 3.54 shows two examples of dry pit pump installations. The left hand installation shows a pump with metal suction pipe and a separate discharge pipe. The right hand installation shows a concrete suction duct and the discharge within the column. Pump sizes are similar to wet pit pumps.

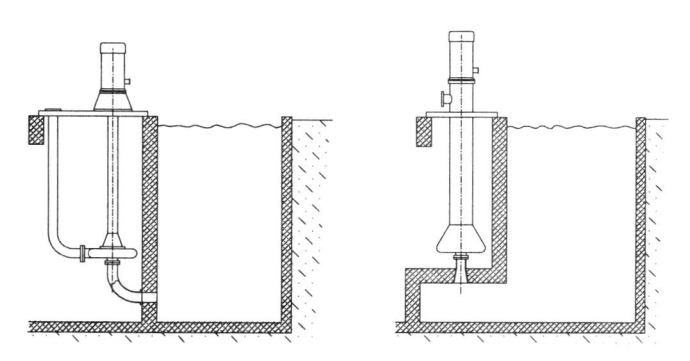

Figure 3.54 Vertical dry pit pumps

065 Portable self-priming pumps and submersible pumps not electrically driven

Most pumps must be vented and primed prior to start-up. For pumps which have flooded suctions this usually involves opening small valves, on top of the discharge pipework, so that all the air trapped in the suction pipework and in the pump will be forced out and replaced by liquid. Some pump designs require vent valves on the casing to fill the pump completely. When the pump suction is not flooded the vent valve in the discharge pipework is coupled to an air ejector. The air is withdrawn by a partial vacuum and the liquid fills the void. Submersible pumps do not suffer these inconveniences as the pump is always full of liquid. Self-priming pumps are also designed to overcome these problems. The casing of a self-priming pump is designed to retain enough liquid so that the impeller is always flooded. The casing needs priming only once provided the liquid does not evaporate or leak away through poor seals. The pump casing is shaped so that any gas or air bubbles become completely surrounded by liquid and are "pumped" from the suction to the discharge. Since there is only a limited amount of liquid during evacuation the gas and liquid must be separated, whereupon the liquid is returned to the suction side to recirculate. The internal design and shape of the pump casing to achieve this effect varies considerably depending on the manufacturer. When choosing a pump it is important to take into account the wearing effect which sand, for example, may have upon the pump's evacuating capabilities.

When electric power is not available, on building sites for example, other sources of power must be employed, such as engines, compressed air from mobile compressors or hydraulically from a separate power pack or one which is built into a vehicle. This type of pump is also suitable for intermittent use in construction work such as road making, rock blasting and excavating or as a fire precaution when carrying out stubble burning or site clearing.

Engine driven self-priming pumps are usually single-stage centrifugal pumps as shown in figure 3.55. In order to be self-priming the pump must be filled with liquid prior to starting. The evacuation times quoted usually refer to the time taken to obtain the full pressure at the pump discharge for a specific suction lift, using a hose having the same diameter as the pump's suction connection and whose length exceeds the suction lift by two meters. The pump performance can be adapted to actual requirements by varying the engine speed.

Figure 3.55 Engine driven self-priming pump

Engine driven pumps can handle solid particles of up to 10 to 15 mm. The suction hose is reinforced because of the vacuum caused by pumping. The normal suction strainer can usually be replaced with a much shallower design in order to facilitate pumping down to within a few millimetres of a floor. The pump should be equipped with a protective safety frame and runners or skids or a wheeled trailer to enable it to stand on soft ground. Priming of engine driven units can be assisted by using the engine inlet manifold vacuum or a separate air pump.

Engine driven single cylinder reciprocating pumps are also used as self-priming pumps especially for construction sites and road work. Engine driven pumps are available with suction sizes up to 150 mm which can handle flows of 400 m³/h at heads up to 30m. Engines sizes up to 30kW are not unusual.

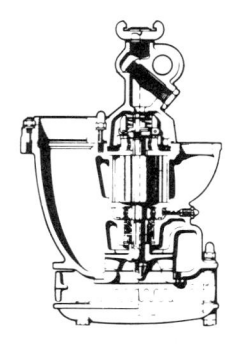

Figure 3.56 Compressed-air driven submersible pump

Engine driven units can be very noisy. Current designs pay special attention to noise reduction, levels of below 80 dBA at 1m are desirable. To achieve this level of noise it is usually necessary to fit silencers to both the inlet and exhaust sides. The engines are normally splash-lubricated which means that the pump must be placed on a relatively flat surface the inclination may seldom exceed 20°.

Compressed-air driven rotodynamic submersible pumps are constructed similarly to those which are electrically driven. They are equipped both with or without special heavy-duty components for pumping large solids. The high speed compressed-air motor, lubricated by oil mist, can compete on performance with electrically driven units, see figure 3.56.

070 Deep well pumps with ejector

Deep-well pumps with ejectors complement the previously described automatic water supply packages, **020**, when the level of the water in a well or, primarily in a borehole lies more than 5 to 7 metres below the pump. An ejector, or jet pump, see **290**, is placed below the surface of the water in the borehole, which is supplied with motive water from a pump located at ground level. By means of the ejector, the water from the borehole together with the motive water is transported to the pump. It follows that it is necessary to have two hoses or pipes, as shown in figure 3.57, between the pump and the ejector. Discharge water is taken from a separate outlet on the pump. For multi-stage pumps this outlet is located approximately in the middle of the pump at a suitable pressure.

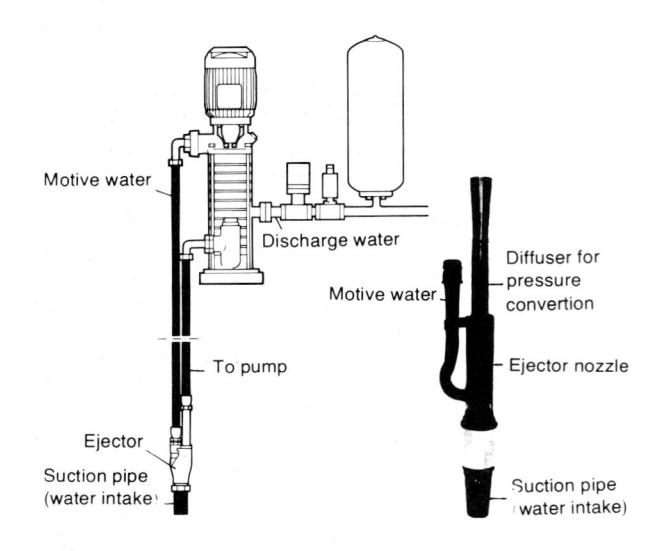

Figure 3.57 Deep-well pump with ejector

By using an ejector the level of the water can be more than 100 metres below the pump. In as much as the water supply to the borehole is small, dry running of the pump system can be avoided if the ejector is equipped with a suction pipe which is approximately 10 m long. A self-regulating effect is thus created due to cavitation in the ejector. The pump system is primed prior to initial start by filling a small reservoir on the pump. During normal operation, start and stop is performed automatically by the pressure in an accumulator.

075 Submersible deep-well pumps

Submersible deep-well pumps in smaller sizes have the same range of applications as deep-well pumps with ejectors. The difference here being that the pump is placed directly in the borehole connected by means of a delivery hose and electric cable. Since the smallest borehole diameter is 4"; the diameter of the pump must be somewhat less, see figure 3.58. Impellers are between centrifugal and mixed flow designs with as many as 100 stages. The pump is a multi-stage centrifugal pump driven by a special type of long thin electric motor located beneath the pump and cooled by the water in the borehole. For small pumps, suitable for domestic applications,

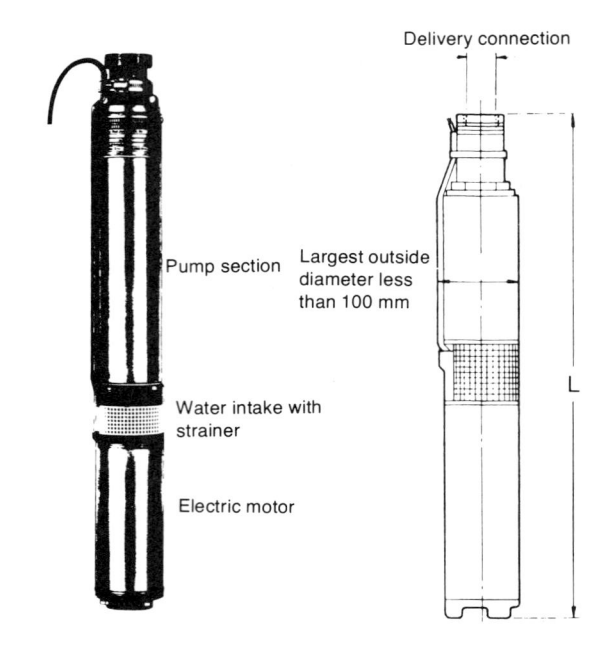

Figure 3.58 Small submersible deep-well pump

the pump length 'L' varies according to the depth of the borehole between 500 and 2500 mm.

Submersible deep-well pumps of larger sizes have outside diameters of up to approximately 500 mm and lengths of up to 5 metres. Their hydraulic data varies considerably, the maximum values being; flow ≈ 5000 m^3/h and differential head ≈ 1000 m. Motor outputs of several MegaWatts are available when using high voltage motors.

They are used for municipal drinking water supply, reduction of ground water levels and mine drainage. Deep-well pumps find favour in oil field applications, on and off-shore. Units are used for secondary oil recovery prior to water injection. Typical applications are to be found in restricted wells or shafts, figure 3.59. They can also be used for pressure boosting in drinking water supply networks, whereupon the pump may be integrated in a section of pipe.

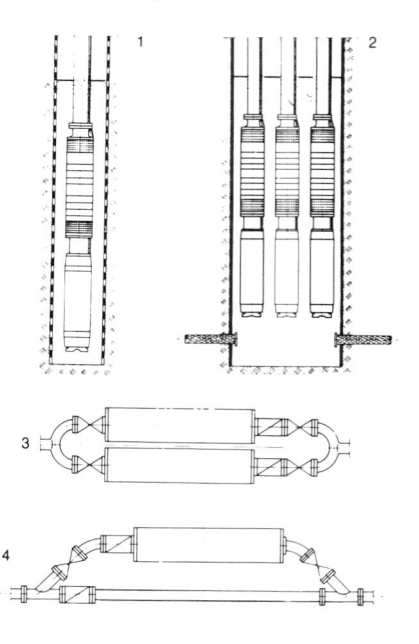

Figure 3.59 Deep-well pumps placed in wells - illustrations 1 and 2 - and placed in-line in pipe section for pressure boosting - illustrations 3 and 4

The electric motors are always squirrel-cage liquid-filled induction motors, which naturally makes very great demands upon the motor's electrical insulation. The liquid filling is either clean water, water-oil emulsion or oil depending on the requirements

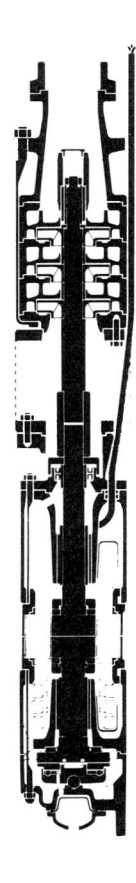

Figure 3.60 Principle construction of larger deep-well pump

of each specific case. The liquid filling is separated from the pumped liquid by a mechanical seal and a diaphragm to compensate for variations in volume, due to temperature changes.

The pump and motor rotors run in liquid-lubricated plain bearings. The axial thrust bearing is the most heavily loaded, which is why some manufacturers employ opposed impeller construction to reduce unbalanced axial forces. Special models are available for sand contaminated or corrosive water.

080 Wash water pump packages

For washing purposes, car washing for example, there are off-the-shelf packages available. Typical operating data; flow ≈ 40 litres/min with discharge pressure ≈ 1.5 MPa or 15 barg, equivalent to a differential head of 150 m. The pump normally supplied is an electrically driven multi-stage centrifugal pump. To prevent zero flow operation while the fluid flow is shut-off, which would cause overheating of the pump, a minimum flow sensor is included in the package. This monitors the fluid flow through the system and prevents the pump from operating in the event of low water flow. A typical arrangement of washing package is shown in figure 3.61.

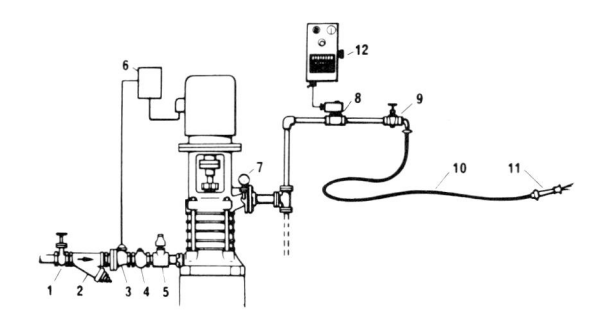

Figure 3.61 Wash water pump package

1. Shut-off valve. Required for servicing of pump and equipment.
2. In-line suction filter. Prevents contaminants from entering the minimum flow sensor and the pump.
3. Minimum flow sensor. Starts and stops the pump automatically. Connects with single-phase cable to the motor safety switch. The pump must operate with a positive supply pressure where the sensor is fitted.
4. Non-return valve. Prevents water from flowing back into the supply pipe. Essential when connected to municipal drinking water system.
5. Vacuum valve. Is activated in the event of the water supply pipe failing to supply the quantity of water required by the pump.
6. Motor safety switch. Switches off electric supply to the motor and triggers alarm signal in the event of a fault arising. Must always be used when a minimum flow sensor is fitted.
7. Pressure gauge. Fitted to pump's delivery flange.
8. Solenoid valve. Water-proof. Operated by coin slotmeter.
9. Cock. To isolate hose.
10. High pressure hose. If the delivery pipeline is long, a larger diameter should be fitted in order to reduce the pressure drop.
11. Wash-gun. Completely rubber covered with shut-off and regulating trigger. Can be adjusted for high velocity jet and fine spray.
12. Coin slotmeter. For self-service vending. Can be set to 10, 15, 20, 25, 30 and 35 minutes. Coin setting as required.

Self contained wash packages are also available utilising reciprocating pumps, see **265** and **270**.

085 Multi-stage segmental pump with heads <300m

The differential head produced by a centrifugal pump is dependant upon the impeller diameter and the speed. In very simple theory, it would be possible to increase the diameter or the speed to produce any desired head. In practice life is more complicated. If a pump for a small flow and high differential was designed with a single impeller running at a 2-pole motor speed, a large diameter and narrow tip width would result. At some stage in the range progression the impeller would become impossible to cast or machine. Also the pump would be large in diameter requiring lots of material and friction losses between the sides of the impeller and pump casing increase proportionally to the fifth power of the impeller diameter resulting in low efficiency. The pump speed cannot be increased indefinitely because impellers can disintegrate under the influence of centrifugal forces which increase with the square of the speed. The solution to the problem is to divide the total differential head across several impellers working in series. The impellers, usually identical, each do a proportion of the work. The segmental pump, or ring section pump, is a popular mass produced design to fulfil these requirements.

Very small vertical segmental pumps are produced for low flow high head applications. These pumps are similar, in some ways, to the vertical in-line pumps shown in figure 3.37. A base, with in-line suction and discharge connections and a mounting foot, is used to support the stack of segments and the electric motor. Each segment consist of a diffuser, an impeller and a spacer. The diffuser forms the outer pressure casing and collects the flow from the tip of one impeller and redirects into the eye of the next impeller. The spacer is used to separate the impellers. On small pumps the seal between diffusers is metal-to-metal; good surface finish and accurate machining being essential. Larger pumps use 'O' rings. The bearings and seals are housed in the base and a motor mounting stool at the top. The complete pump assembly being held together by long tie-rods around the outside of the diffusers. Very small units of this style are limited to 5 to 10 kW. Pumps in cast iron and bronze with pressures up to 25 barg are standard.

Slightly larger versions of segmental pumps have the suction and discharge connections at opposite ends of the pump; horizontal and vertical variations are available in these sizes with from one to thirteen stages, see figure 3.62. Power requirements would be up to 300 kW. Pumps are available in cast iron, steel and bronze with pressure ratings up to 40 barg.

Various differential heads are achieved by varying the number of stages. In order to maintain good efficiency it is necessary, for a given head, to increase the number of stages as the volume flow decreases, figure 3.63.

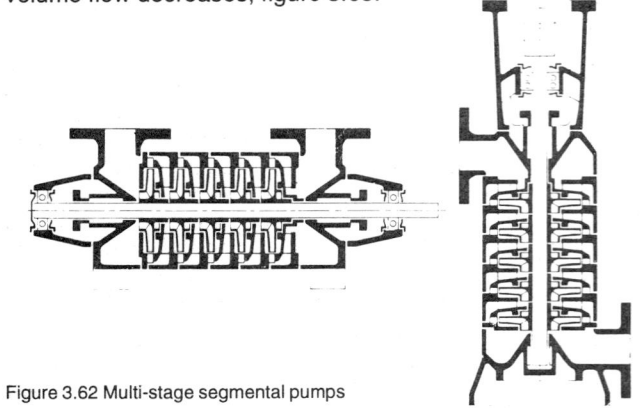

Figure 3.62 Multi-stage segmental pumps

Multi-stage segmental pumps are used in commercial and domestic installations in multi-storey buildings for; water distribution, central heating, steam condensate, fixed fire fighting pumps.

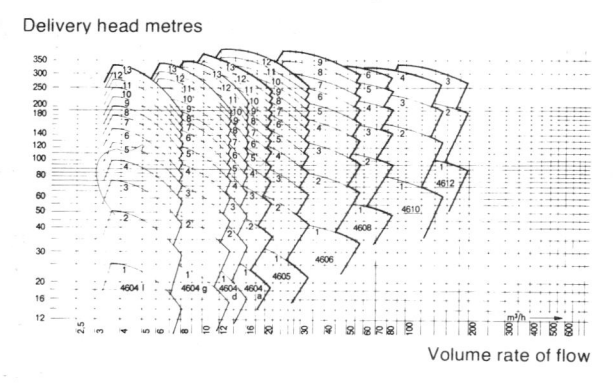

Figure 3.63 Range chart at 2900 rpm for segmental pumps

One disadvantage of the segmental pump design is the number of seals, and potential leak paths, in the pressure casing due to the assembly of individual diffusers. For this reason maximum operating temperature is limited to around 170 °C and applications on hazardous liquids extremely restricted.

090 Multi-stage segmental pump with heads >300m

These pumps are a larger version again of **085**; horizontal pumps only with flows up to 3000 m³/h at heads up to 4500 m are available requiring drivers up to 10 MW. Special designs for boiler feed pumps allow operation over 200 °C at 500 barg. Larger pumps are available in a wider range of materials; cast iron, steel, bronze, 18-10-2 austenitic stainless steel and duplex stainless steel. The previous comments regarding hazards liquids still apply. Higher head segmental pumps are popular as boiler feed pumps and sea water pumps for desalination and reverse osmosis. Also useful in mining, irrigation and gas scrubbing processes.

095 Standard pumps to ISO 2858 / ISO 3069 / ISO 3661

The International Standards Organization, ISO, has prepared a standard recommendation, ISO 2858 - for end-suction centrifugal pumps - pressure rating PN 16 (maximum pressure 1.6 MPa = 16 barg). This is included in British Standard - BS EN 22858 and equivalent to the German standard - DIN 24256. The standard specifies the principle dimensions and nominal duty point - figure 3.65. The standard covers differential heads of 5 to 125 m and flow rates of 6 to 400 m³/h. Some manufacturers can exceed the standard performance requirements without changing the physical dimensions. Pumps are available with flows up to 1700 m³/h and heads to 150m. Operating temperature requirements are not specified in the standard pump are available for up to 350 °C.

Pumps to this standard are intended for chemical and corrosive applications, although pumps are available in high silicon iron and Ni-Hard for solids handling applications. Most pumps are built in stainless steel although the standard does not specify materials. Stuffing box and seal cavity dimensions are to ISO 3069, which is based on parts of DIN 24960. See also **100** for ANSI inch chemical pumps and **110** for higher pressure and heavier duty applications.

The design and construction of the ISO pumps is of the "back pull-out" type, figure 3.66, i.e. removal of the pump rotor is carried out from the motor side, after the spacer element in the shaft coupling has been removed. The construction is based on a unit system with 3 bearing assemblies comprising shaft and bearing housings. Flanges are to ISO 2084. Overall package dimensions, including motors and baseplates are given in ISO 3661.

Most manufacturers, at least 30 in Europe, build a comprehensive range with regard to material, quality and shaft seals. There are also glandless pumps which conform to the ISO dimensions and performance characteristics. It should be noted that the ISO standard is only applicable to dimensions and performance, not materials, flange loadings, applications or quality. When purchasing therefore, it is necessary to refer to quality standards or to at least establish that minimum requirements, regarding shaft deflection and bearing life are suitable.

General note on ASME/ANSI standards

The ANSI B73 standards were produced in response to requests by users to standardise manufacturers products for the chemical industries. The end suction pump standard was

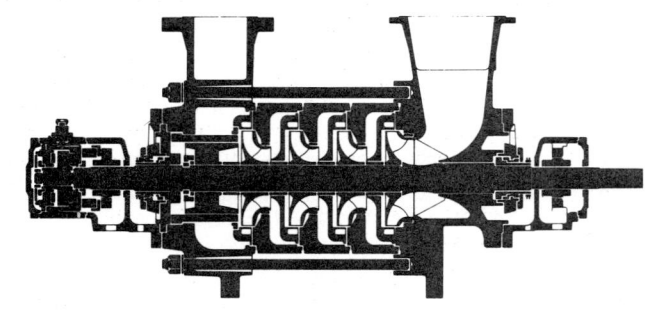

Figure 3.64 Large 4 stage segmental pump with balance drum, sleeve bearings and tilting pad thrust bearing.

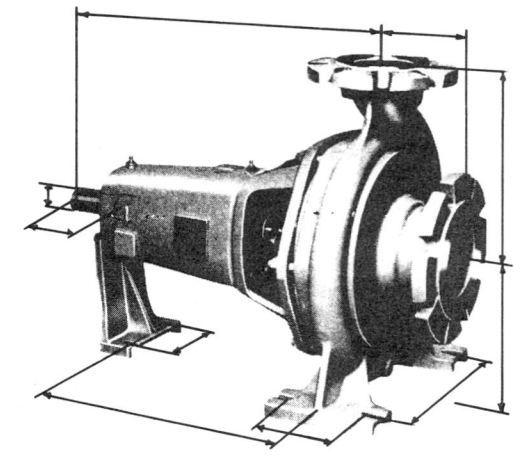

Figure 3.65 Standard pump according to ISO 2858

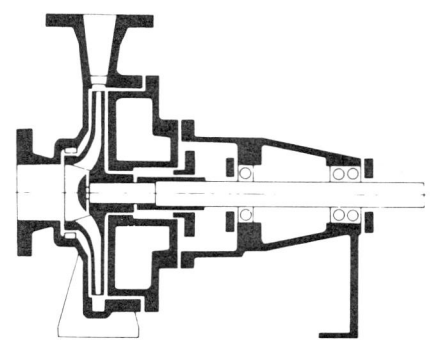

Figure 3.66 "Back pull-out" construction

introduced in 1974 and the vertical inline in 1975. The standards were based on recommendations by users but compiled by a committee of mostly manufacturers. There are many American chemical installations in Europe and European manufacturers and users should be aware of these standards.

100 End suction pumps to ASME/ANSI B73.1

ANSI, the American National Standards Institute, has a philosophy more similar to DIN than ISO in that it specifies materials and mechanical seal options. B73.1 is a standard for single stage centrifugal pumps, with back pull-out, intended for chemical and corrosive duties. Construction is similar to **095**. The standard specifies bareshaft pump dimensions, also installation dimensions for complete packages with motor and baseplate. The standard does not specify hydraulic duties. Mechanical seal arrangements, with various piping plans, are detailed to cover most applications.

Depending upon materials and cooling arrangements pumps must be suitable for at least 260 °C. The standard specifies material columns with casings in:
- cast iron,
- ductile iron (spheroidal graphite iron),
- carbon steel,
- alloy steel,
- 18-8 stainless steel,
- 18-10-2 stainless steel.

Many manufacturers can supply pumps in more exotic materials such as; Carpenter 20, Hastelloy B & C, duplex stainless steels; as well as non-metallic materials like glass reinforced polymer. The standard requires a corrosion allowance of 0.125", 3 mm.

Pump sizes vary from 1½" to 10" suction connections with flanges to ANSI B16.1, B16.5 or B16.42. Impeller diameters range from 6" to 15". At 3600 rpm, pumps are capable of 900 m^3/h at heads up to 120 m. At 3000 rpm the performance falls off slightly to 750 m^3/h at 83 m. ANSI does specify pressure ratings for the pumps; 125LB/150LB flange rating, 12 barg and 20 barg. An option for 250LB/300LB flanges is given.

The standard specifies a minimum bearing life of 17500 hours at maximum load conditions. On test, with a speed of ± 10% of operating speed, the peak velocity of vibration shall not exceed 0.25 in/s, 6.35 mm/s.

End suction pumps can suffer high axial thrust due to high suction pressure. Shortened bearing life may result. Manufacturers should confirm the bearing life if it is less than the standard requirement.

105 Inline pumps to ASME/ANSI B73.2

The basic philosophy of vertical inline pumps is identical to pumps in category **015**. The pump is mounted in the pipework and is not bolted down to foundations. The ANSI B73.2 pumps are larger than those in **015**, all pumps must be mounted with the motor vertically above the pump and the pump can be optionally supported by a 'stool'. The weight of the pump package is supported by the 'stool' but the pump is free to move to accommodate pipe movement. The standard is dimensional, to allow interchangeability between manufacturers, but does not specify hydraulic performance. Pump sizes with suction/discharge from 2"/1½" to 6"/4" are covered; length across flange faces is from 15" to 30".

Three pump designs are included within a common envelope. A close coupled design, VM, using a NEMA JM or JP flange mounted motor, with motor shaft and mounting dimensions and tolerances to NEMA MG 1-18.614. Obviously the motor must be moved to inspect the seal or remove the rotating assembly. The second design, VC, is a short pump shaft rigidly coupled and spaced to a NEMA P-base flange mounted motor. Motor flanges, dimensions and tolerances to NEMA MG 1-18.620. Removal of pump rotating assembly or seal does not need the motor to be moved. The third design option, VB, is for pumps with bearing housings driven through flexible spacer couplings, see figure 3.37. These motors shall be NEMA C-face to NEMA MG 1-11.35 with solid shafts. Again removal of pump rotating assembly or seal does not need the motor to be moved. Designs VC and VM have specified bearing L10 lives of 26000 hours at rated duty or 17500 hours at maximum load. Pumps to VB shall have bearings suitable for 17500 hours at maximum load. On test, running at 100/105% speed and ± 5% rated flow, the unfiltered vibration on the top pump or motor bearing housing shall not exceed 0.25 in/sec or 0.0025 in peak-to-peak displacement.

The standard specifies pressure ratings of 125 LB for cast iron pumps and 150 LB for steel and stainless steel; 12 and 20 barg. An option is given for 250 LB and 300 LB flanges. Pressure/temperature ratings in accordance with ANSI B16.1 and B16.5. Material requirements are similar to category 100 with a 0.125" corrosion allowance. Allowable nozzle loads must be specified by the manufacturer for the material offered at the specified operating temperature.

The standard specifies a seal cavity which must be capable of accepting double seals with a pumping ring for forced circulation of buffer liquid. A packed stuffing box is optional.

At 60Hz pumps are capable of 320 m^3/h at heads of 120 m; 50Hz performance is reduced to 265 m^3/h at heads of 83 m. The standard specifies the maximum pump absorbed power to be 172.5 hp, 128 kW.

Some manufacturers build a heavier version of vertical inline pump to the VC design which complies with the requirements of API 610. Removal of the coupling allows seal inspection and replacement without disturbing the motor. In some designs the rotating assembly cannot be removed without detaching the motor stool. Capacities up to 1150 m^3/h at heads up to 230 m are possible at 60 Hz. Suction/discharge sizes go up to 14"/8" and pressure ratings of 100 barg are available on some sizes.

Vertical inline pumps are really end suction pumps with an extension on the suction nozzle. High suction pressure creates axial thrust which will reduce bearing life. The manufacturer should be asked to confirm the bearing life if any doubt exists about the magnitude of the suction pressure.

General note on API 610

API is the American Petroleum Institute which is predominantly a trade association of pump users concerned with oil and gas production and refining. Pump manufacturers are represented on the API committees but are outnumbered by the users. For example, API 610 has a complete section for "Vendor's Data"; it does not have a section for Purchaser's Data. API is a long established organisation and has a large catalogue of standards, publications and recommended practices. American technology is used world-wide in oil and gas related sectors and it is logical to use proven methodology where appropriate.

API 610 is a mature standard currently up to the seventh edition published in February 1989. API is unlike the majority of DIN, ISO and EN pump standards in that it is more a philosophy to achieve results rather than instructions. It is not a dimensional standard which would allow interchangeability between manufacturers. It does contain requirements but also a lot of 'by agreements' which are frowned upon by European standards authorities because of the lack of control and uniformity. However API 610 does specify materials of construction. The API 610 and the ISO/DIS 9905 committees are discussing the possibilities of an international pump standard combining the best of both existing standards.

API does not specify the types of pumps which are covered by 610. However, they do specify the types of pumps which can only be used when approved by the purchaser; close coupled, two stage overhung, double suction overhung, multi-stage segmental single casing pumps.

The first two requirements are a 20 year service life and the capability of 3 years uninterrupted operation. API does follow these requirements with a statement that these are design criteria. From a European viewpoint both are not verifiable and are application specific. Bearing this in mind, API 610 pumps are heavier and more robust than most other pump designs, no pressure or temperature limitations are specified, and are suitable for the most arduous duties. Drain rim baseplates are standard and most pumps are centre-line mounted to reduce alignment problems caused by thermal growth. Allowable forces and moments on process nozzles are large necessitating substantial baseplates. Pump selection is guided by a requirement to have BEP between the normal operating point and the rated duty point.

API 610 specifies the size of seal cavities and the distance to the closest obstruction and specifies mechanical seals, single balanced, as a minimum. The space available allows for complex double or tandem seal arrangements. Auxiliary systems for mechanical seals are covered in detail showing 18 arrangements. Mechanical seals for pumps, and associated systems, are soon to have a separate API standard, 682, see Chapter 7. The draft currently circulating is 180 pages. Arrangements for other auxiliary piping, cooling water for seals, bearing housings, jackets and pedestals, are also shown in 610.

Spacer couplings are standard with a minimum spacer length of 127mm. Bearings, seal(s) and rotor must be removable without disturbing the driver or the casing.

API 610 specifies 11 material columns covering the following range of casing materials:
  cast iron,
  carbon steel,
  12% chrome steel,
  18-8 stainless steel,
  18-10-2 stainless steel.

110 End suction pumps to API 610

The most popular process pump. Basic pump design is similar to prEN 733, 025, and ISO 2858, 095, with a back pull-out feature. In practice, the minimum material readily available is carbon steel. Some manufacturers will make specials in bronze when commercially attractive. Standard flanges are inch to ANSI standards, B16.1, B16.5 or B16.42. Both suction and discharge flanges must be the same pressure rating. Rolling bearings, typically ring oiled, are supported in a cast housing bolted to the stuffing box or casing. Bearing housings are steel for hazardous liquids.

Pumps are available with flows over 2000 m^3/h at differential heads of 300 m. Operating temperatures from -30 to 425 °C are common, down to -180 °C is possible. Popular pumps range from suction branch sizes of 2" to 12". Pressure ratings of smaller pumps are up to 100 barg reducing to 40 barg for the larger sizes. Pressure/temperature rating variations with materials are important; check flange specifications.

115 Double suction pumps to API 610

Not as popular as the end suction pump, probably 10% of the market size. Similar in some respects to category **010** pumps

Figure 3.67 End suction process pump to API 610

but centre-line mounted radially split casings. Generally with top-top process connections, but side-side and other variations are available. The double entry impeller mounted between two bearings. Rolling bearings are standard, sliding bearings and various lube oil systems to order. The mechanical seals can be serviced by removing the coupling and bearing housings. A spacer coupling is required. The rotor assembly is withdrawn from the casing, away from the driver.

These pumps tend to be slightly larger than the end suction pumps. Flows over 5000 m^3/h are possible at heads up to 500 m. Suction sizes from 8" to 16" are popular with pressure ratings up to 60 barg. Materials and flanges as for **110**.

120 Magnetic drive pumps

Magnetic drive pumps are a type of glandless pump; glandless pumps are characterised by the fact that their design and construction physically prevents leakage to the surrounding environment. Packing or mechanical seal type of shaft seal cannot in this context in any way be considered as leakage free, even if it is barrier liquid or very dilute process liquid/vapour in minute quantities which escapes. The description "glandless pumps" is therefore associated with the alternative description "hermetically sealed pumps" or "canned pumps".

Glandless pumps must be used for the most dangerous liquids from the point of view of toxicity, environmental hazards and radioactivity. In some cases for high or very low temperatures. Leakage may occur however when handling certain types of highly penetrative liquids due to penetration through minute faults in static seals. Magnetically driven pumps can avoid this problem by having the containment 'can' welded to the casing. When ordering pumps for these types of liquids the level of acceptable leakage should be specified.

Two types of magnetic drive are available; synchronous magnetic couplings and magnetic induction couplings. In both cases the pump is driven by a standard motor. However the motor does not drive the pump shaft carrying the impeller(s) directly. The motor drives a rotating magnet assembly. In synchronous magnetic couplings the rotating magnetic field drives another set of magnets coupled to the impeller(s). In magnetic induction couplings the rotating magnets create a magnetic field in a rotor of copper bars, very similar to a standard squirrel cage motor rotor. The magnetic properties of materials decay as temperature increases so that synchronous magnetic couplings suffer from a loss in torque. Magnetic induction couplings are better at elevated temperatures because the magnets are not in contact with the liquid.

The pump shaft 'proper' is totally immersed in the process liquid and runs in bearings which are process lubricated. Silicon carbide ceramics and some carbon graphites are popular as bearing materials. Solids in the liquid can cause high

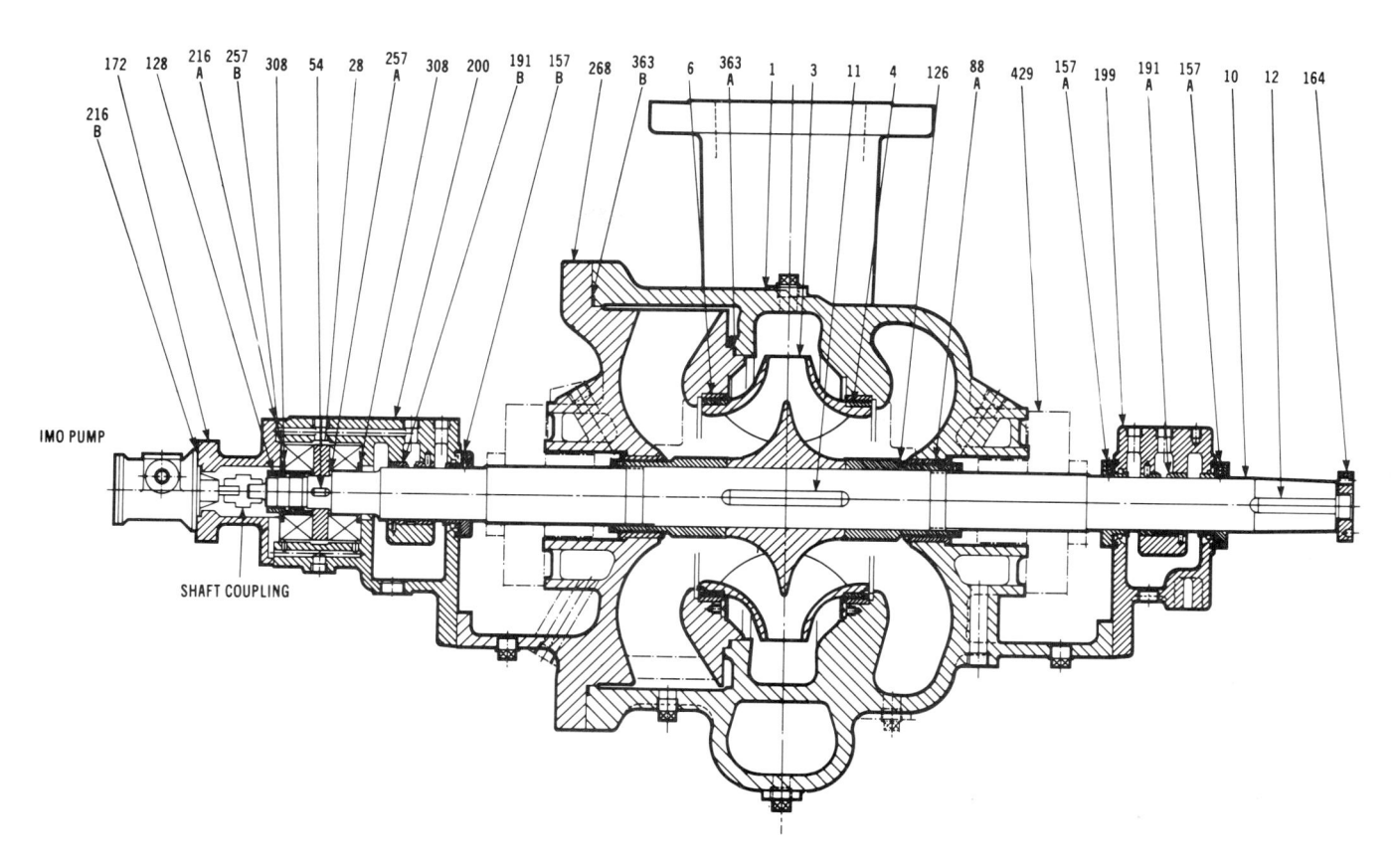

PART NO.	NAME OF PART	PART NO.	NAME OF PART	PART NO.	NAME OF PART
172	SUPPORT HEAD	12	COUPLING KEY	308	SEALING RINGS-THRUST BAG
164	COUPLING NUT	11	IMPELLER KEY	268	SUCTION END BRACKET
157B	FLINGER-THRUST BRG.	10	SHAFT	257B	SHIMS-THRUST BRG. TO SUPPORT HEAD
157A	FLINGER-PLAIN BRG.	6	CASING RING	257A	SHIMS-THRUST COLLAR
128	LOCKNUT	4	IMPELLER RING	216B	GASKET-IMO PUMP TO SUPPORT HEAD
126	IMPELLER LOCKNUT	3	IMPELLER	216A	GASKET-SUPPORT HEAD TO THRUST BRG.
88A	STUFFING BOX BUSHING	1	CASING	200	BEARING BODY & CAP-THRUST
54	THRUST COLLAR KEY	429	MECHANICAL SEAL-COMPLETE	199	BEARING BODY & CAP-PLAIN
28	THRUST COLLAR	363B	GASKET-SUCTION END BRACKET (OUTER)	191B	BEARING LINING-THRUST
		363A	GASKET-SUCTION END BRACKET (INNER)	191A	BEARING LINING-PLAN

Figure 3.68 Double suction process pump to API 610

bearing wear and in some designs can accumulate in the 'can' and cause obstructions.

A variety of magnetic materials are used of which aluminium-nickel-cobalt are probably the most widely and successfully employed. Some designs use rare earth elements, such as samarium cobalt and neodymium iron boron, for the upper end of the torque range, the higher costs involved being more than offset by improved performance.

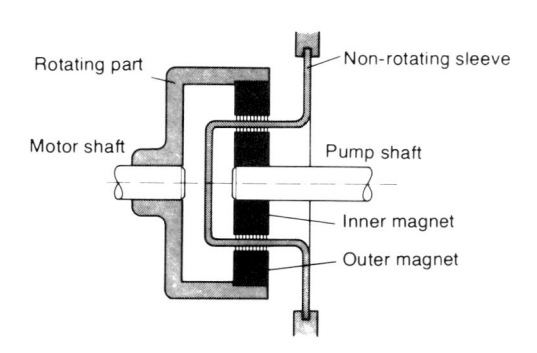

Figure 3.69 Synchronous magnetic coupling for "seal-less" torque transfer

Magnetically driven pumps have become increasingly popular recently for critical applications where unscheduled process stoppages create additional plant problems as well as loss of revenue. Mechanical seals are seen as a maintenance intensive item, difficult to install and set-up, with costly spares and attendant inventory costs. Removing the mechanical seal and operating pumps of a modified design can save overall costs. Paul Barnard, of Exxon Chemicals, writing in World Pumps January 1992 cites a case where a magnetic drive pump costs 30% of a conventional pump over a ten year operating life.

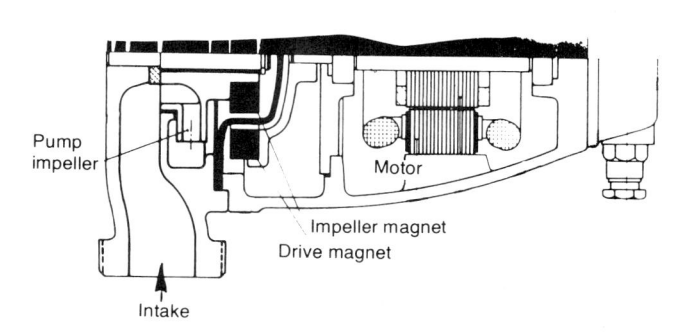

Figure 3.70 Close coupled magnetically driven pump

Magnetically driven pumps are available from several manufacturers, pumps up to 350 kW are currently built. Hydraulic capacities of 550 m^3/h at 500m are possible, temperatures of -80 to 450 °C are advertised. At least one manufacturer can supply to API 610 7th edition.

125 Canned motor pumps

Canned-motor, 'wet-motor', driven pumps generally have the stator windings sealed from the pumped liquid by means of an annular tube - hence the name 'canned' - pump - figure 3.71. There are many types available with power outputs of up to

Figure 3.71 Canned-motor pump

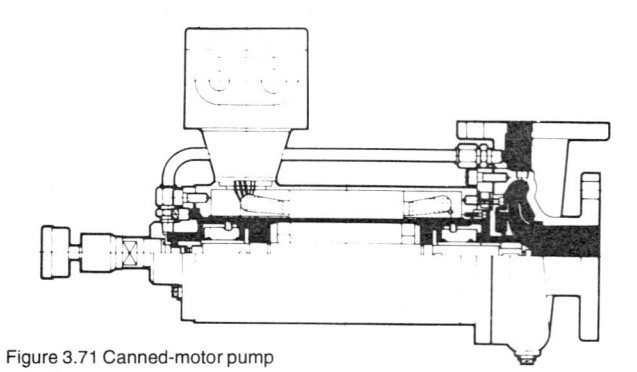

several thousand kilowatts. Smaller sizes, up to about 500 W are generally used as domestic hot water pumps and central heating circulating pumps, see group **015**. Canned-motor pumps are produced as both single and multi-stage types with various types of drive motor. Liquids with temperatures ranging from -200 °C to +500 °C and pressures up to 1000 bar can also be handled. Heating jackets are available for handling liquids with high melting points, whilst built-in filters are primarily used to cope with contaminated liquids although trickle feeding to the rotor chamber may also be necessary for liquids containing high concentrations of solid particles.

Larger canned pumps were primarily developed for boiler circulation pumps and liquid metal circulating pumps for nuclear reactors. Circulating pumps are required on modern large multi-tube boilers to ensure a reasonable water velocity inside the tubes to improve heat transfer and eliminate hot spots and problems with localised vaporisation. The suction pressure of the pumps is boiler pressure, 150 barg or more, and the differential only 20 or 30 metres. The high suction pressure produces a large axial thrust requiring enormous, by comparison to the standard, thrust bearings and also causes mechanical seal problems. Liquid metal pumps for nuclear reactors must be positively leak-free because of the radioactive nature of the liquid. The canned motor pump solves these problems.

Now the canned motor pump is available as a chemical and process pump, very similar to magnet drive pumps and in special variants for sewage, effluent and paper stock. Flows of 300 m^3/h at 110m with pressure ratings of 25 bar are fairly common. Some canned motor pumps are available to DIN 24256 standard pump performance and dimensions.

Postscript for Magnetic drive and Canned motor pumps

The American standards body, the Hydraulic Institute, has just released a standard covering sealless pumps. The new standard is designed to eliminate misunderstandings between purchasers and manufacturers and to help the purchaser specify a pump correctly. The standard covers; pump types, definitions, applications, installation, commissioning, operation, maintenance and testing. Details of monitoring devices unique to sealless pumps are given. Testing specific to sealless pumps; can integrity and winding integrity for canned motor pumps. The application section deals with; gaskets and joints, venting and draining and internal process liquid lubricated bearings. The standard includes descriptions of piping arrangements for process liquid circulation and recommendations on how to prevent problems caused by solids, magnetic material and gas/vapour evolution.

130 Non-metallic pumps

Non-metallic pumps are most suitable when used for pumping acid, alkalis and corrosive salt solutions. A general review of physical properties and corrosive resistance of various non-metallic materials is given in section 6.2.12 and 6.3. Larger sizes of non-metallic pumps must, because of the physical properties of the material, be equipped with an outer reinforcement casing to carry the nozzle loads and to absorb forces due to the fluid pressure, figure 3.72. This outer casing is not

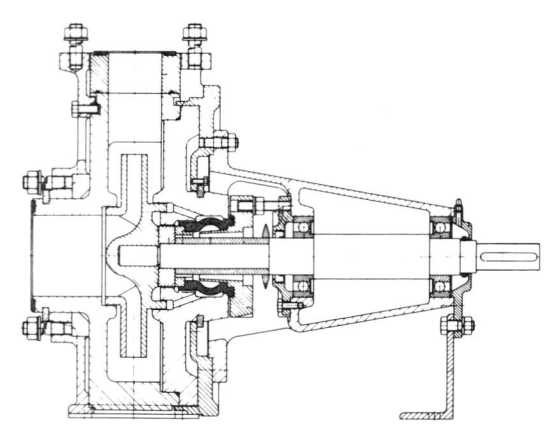

Figure 3.72 Non-metallic pump with surrounding reinforcement casing

necessary for smaller pumps or if the particular non-metallic material has suitably good physical properties or is fibre reinforced internally. Since the liquids handled are of a dangerous nature, great care must be given to the type of shaft seals. The construction generally being similar to that of the ISO 2858, ANSI B73 and API 610 standard pumps. The pump shaft is usually 18-10-2 stainless steel minimum and can be completely isolated from the liquid by a non-metallic shaft sleeve under the mechanical seal.

Pumps are available for 400 m^3/h at differential heads up to 80m; pressure ratings of 7 bar and temperatures up to 100 °C are possible. Pumps complying with the dimensions of ANSI B73.2 are available.

135 Pulp pumps

For handling paper pulp, depending on the concentration of fibre in suspension, it is necessary to use specially designed pumps as follows:

- Up to 0.5%   standard pumps
- 0.5 to approx. 2%   pulp pumps or standard pumps fitted with special semi-open impellers
- Approx 2 to approx 5%   pulp pumps
- Approx 5 to 6-7%   pulp pumps with specially shaped inlet blading
- Above 6-7%   positive displacement "dense pulp" pumps.

For unscreened pulp of concentrations above 3% the pump flow passage should be at least 40 mm. The impellers for pulp pumps are semi-open with back vanes for simultaneous balancing of axial thrust and "cleaning" behind the impeller - figure 3.73. The choice of material is normally stainless steel and for raw product sometimes grey cast iron. Due to the presence of entrained air in pulp suspensions it is necessary to design the blades of the impeller intake so as to prevent air locks. The

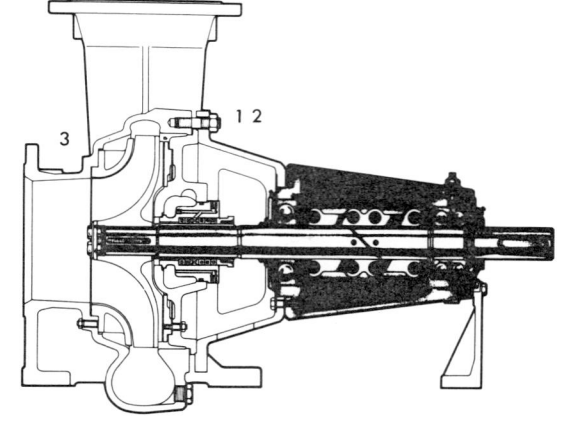

Figure 3.73 Pulp pump with semi-open back vanes. Replaceable wear plate in the pump casing and axial adjustment of the rotor

concentration of air also causes "rough" running with increased levels of vibration and mechanical stress. Compared to a water pump, the mechanical construction of a pulp pump requires an extra safety factor of approximately 2 in order to achieve satisfactory reliability. For pumping twigs and waste from the macerater, non-clogging free-flow pumps are best. Horizontal end suction pumps with mixed flow impellers are capable of 6000 m³/h at 35 m up to 110 °C with pressure ratings of 7.5 bar.

140 Pumps for solids >10mm

Solids handling pumps are used for pumping suspensions and liquid/solid mixtures of various solid particle sizes up 70% solids by weight. Two types of construction for rotodynamic pumps are available; all metal and rubber coated or rubber lined. The particle sizes and the abrasiveness of the solids determine the choice of construction. Abrasiveness can be measured by the Miller Number according to ASTM G75-82, Test method for slurry abrasivity by Miller Number.

All metallic materials are used in the case of larger particles and higher liquid temperatures, see also Section 6.5. Some manufacturers restrict the size of small particles which can be mixed with the large. With particles over 50 mm it is unlikely that 70% concentrations can be achieved. Pumps are available in various materials from plain cast iron to chrome irons with 2.5% to 25% Cr. Despite the use of extremely wear-resistant materials the operational life is relatively short - of the order of a few months and depending greatly upon the speed of flow, rpm, particle size and nature.

For delivery heads above 50 to 60 m it is normal to employ several pumps in series. Pump impellers are of the closed or semi-open type with suitably thick shrouds and vanes. The number of vanes are relatively few to reduce the chances of impact. Some models have adjustable casings to correct for wear. Shafts, bearings and bearing housings are dimensioned for heavy-duty loading due to the out-of-balance forces and high SG. Some manufacturers limit pump speed as SG increases. Solids pumps are usually V-belt driven, speeds from 2600 to 400 rpm, so as to be able to adjust the flow without machining the impeller. Throttling, to correct pump performance, is particularly difficult and prone to constant adjustment; variable speed is preferred.

Due to the abrasive nature of the liquid mixture particular attention has to be given to the design of the shaft seals. Design solutions such as stuffing box flush with clean liquid; hydrodynamic seals as in section 7.3; centrifugal seals with back vanes and as glandless pumps with overhung impellers for vertical models, as in figure 3.74, are available. The pump in this figure is particularly suitable for frothing fluids, where air can rise and not block the impeller intake. A certain degree of self-regulation is obtained by the air mixing at low level.

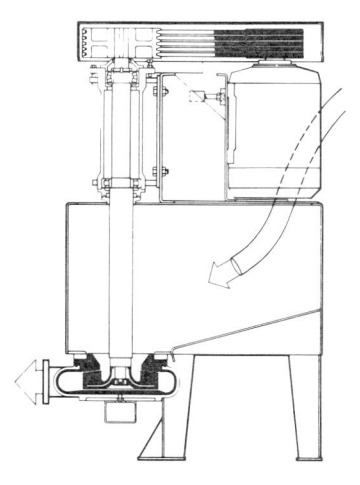

Figure 3.74 Vertical glandless solids pump combined with sump

The maximum solid size handled continuously is a function of the pump size and impeller design; open or closed. The following table can be used as a guide for sizing pipework.

Suction nozzle	Discharge nozzle	Maximum particle size mm	
		Closed impeller	Open impeller
2"	1½"	11	24
3"	2½"	16	32
4"	3"	20	41
6"	5"	38	62
10"	8"	50	82
14"	10"	70	90
16"	12"	80	98
20"	16"	100	120

Figure 3.75 Solid sizes for various pump sizes

A 2" suction pump can handle up to 30 m³/h at heads of 60 m; a 20" suction pump can handle 2250 m³/h at heads up to 38 m. Pumps are available which can handle solids larger than 120mm. Obviously to pump large solids in quantity, not an occasional piece as with a contractor's self priming site pump, a lot of liquid is required and a large pump. Pumps up to 36" or 900 mm suction are available. Efficiency tends to be low, although one pump is advertised at 82% on water between 500 and 900 rpm, because the kinetic energy given to the solids cannot be converted into increased pressure. Consult manufacturers very early in the system design and feasibility studies and be prepared to supply plenty of NPSHa, up to 10 m.

145 Pumps for solids <10 mm

For very abrasive mixtures with particle sizes of up to approx 10 mm the components in contact with the pumped fluid are rubberised or rubber coated. Typical solids handled would be some sands, iron and copper ore and mine tailings. Pumps which are rubber coated can have the coating repaired or replaced if the damage is not severe. Rubber lined pumps are built so that the lining of the casing can be removed and completely replaced.

Smaller pumps, 2" to 4" suctions, may limit solid sizes to 6 mm or less. Stuffing box requirements and drive arrangements are similar to category **140** pumps. Small pumps are similar in hydraulic performance to all metal pumps. Larger pumps can handle more flow; a 16" suction will pass 5000 m³/h at up to 48 m.

Figure 3.76 Solids pump with replaceable rubber lining

150 Hygienic quality pumps

Within the food processing industry and chemicals and pharmaceuticals, special centrifugal pumps are used, figure 3.77, their shape and construction being distinctive and ordained by the hygienic and sterile requirements. The materials used are usually stainless steels with O-ring gaskets etc. of approved hygienic quality. Both exterior and interior surfaces must be smooth and polished to approved hygienic standards, it should also be possible to disassemble the pump quickly for cleaning and washing. When cleaning without disassembling, 'cleaning in place', CIP, is specified, it must be possible to carry out this process quickly and effectively and requires that all components and clearances are designed for this purpose, especially with respect to temperature. To simplify external cleaning, the electric motor is often encased in a polished stainless steel casing. For reasons of hygiene many foodstuffs pumps may not have oil lubricated shaft bearings. Food processing centrifugal pumps are normally available up to approximately 30 kW. Centrifugal pumps cannot be used for viscous and shear sensitive products such as yoghurt, cheese, liver paste etc.. In such cases lobe rotor pumps are generally used, see group **240**.

CEN has prepared a draft European standard covering the requirements for hygiene for machinery used in the preparation of food for human or animal consumption. The standard describes methods of construction and the types of acceptable fittings, couplings and flanged joints.

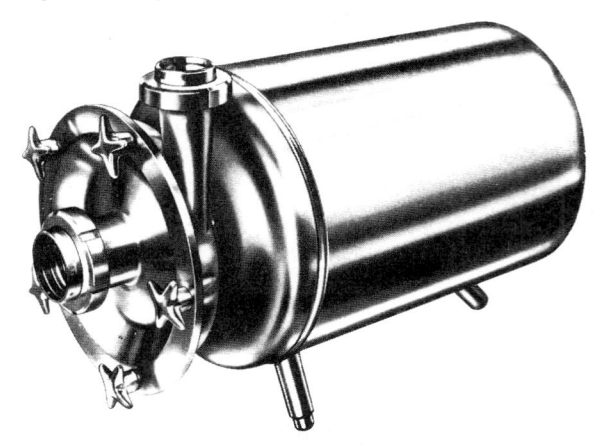

Figure 3.77 Hygienic quality centrifugal pump

155 High speed single-stage pumps

The process industries have shown considerable interest in high speed pumps. Lower speed multi-stage pumps can be replaced with smaller, more compact packages. Pumps consist of specially designed radial vane centrifugal impeller and casing which incorporates one or two step-up gears driven by a standard 2 pole motor. The maximum, practical, gear ratio for a pair of gears is 6:1 so that 60 Hz pumps could run up to 120000 rpm and 50 Hz to 105000 rpm. A very wide speed range is available at both frequencies. Both horizontal end suction and vertical inline styles are available. When seal flushing is fitted pumps can run dry for short periods.

Some operational problems have occurred with some installations. The high pump speed can cause mechanical seal, no soft packing, and bearing problems. Due to the speed, rolling contact bearings may be replaced by sliding bearings with attendant lube systems. High speed pumps can be relatively intolerant of rapid process transients and particularly distressed by $NPSH_a$ inadequacies and solids. Liquid viscosity can be a problem. The variation between manufacturers in the pump's resilience to cope with operating problems has lead users to consolidate with a specific manufacturer for specific applications.

Although the possible speed range is great, most pumps operate in the range from 15000 to 30000 rpm. Differential heads of 1500 m are common and powers up to 500 kW are possible.

Radial vane impellers can have a smaller stable operating range than the standard backward swept vane impellers. A variable flow by-pass line may be necessary for some applications. Speed-up gear boxes tend to give more problems than speed reducing gear boxes. Contemporary developments in electronics may make this mechanical technology redundant. Variable frequency invertors are used routinely to vary the speed of standard squirrel cage induction motors. Many existing invertors produce frequencies up to 120 Hz. A

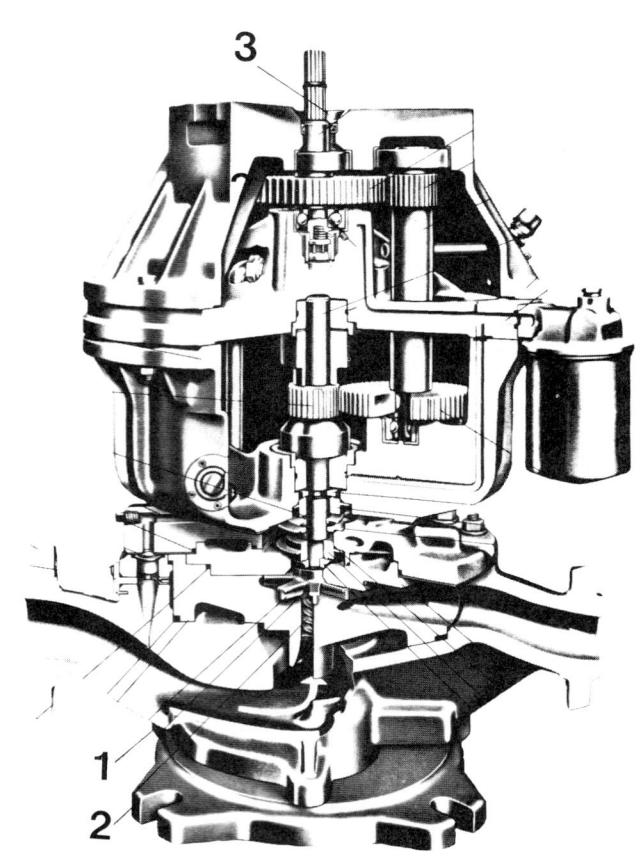

Figure 3.78 Vertical inline high speed single-stage pump. The pump impeller, 1, is equipped with $NPSH_a$ reducing inducer, 2 and is driven with a two-stage gear. The motor drive is connected to the shaft 3

high speed pump could become a pump driven directly by a high speed motor.

160 High speed multi-pump packages

High speed multi-pump packages are an extension of the design philosophy of the high speed single-stage pumps. A speed-up gear box is driven by an electric motor. The gear box has two or three output shafts, each of which drives a single stage centrifugal pump. The pumps can be connected in series or parallel to produce higher flow or higher head. An option is available to have a slower speed first stage pump which is specially designed for low $NPSH_r$. Inducers are available for some impellers. The multi-pump packages are restricted to horizontal pumps.

This type of packaging results in a compact powerful unit. Mechanical seals must be used as there is no access to the stuffing box for adjustment. Seal cavities are large enough for double and tandem seal configurations with flush and buffer liquids. The individual pump casings must be removed in order to inspect or maintain the seals; the process pipework and the inter-stage pipework must be removed to accomplish this. A special liquid injection feature allows the use of standard mechanical seals.

The individual pumps run at speeds from 12000 to 25000 rpm. Bearings may be a combination of rolling and sliding or all sliding. Lube systems can be tailored to customers requirements including API 614 compliance. Flows up to 225 m^3/h are possible at differentials up to 4500 m. High suction pressure capabilities are available up to 150 barg with casing pressure ratings up to 310 barg. Absorbed power can be to 1150 kW.

Most pump users tend to standardize on regularly used wear parts such as mechanical seals and coupling diaphragms. The choice of manufacturers for high speed seals may be very restricted. The comments at the end of category **155** regarding by-passes and gear boxes also apply here.

High speed pump packages have replaced multi-stage centrifugal pumps and reciprocating pumps in some applications. High speed multi-pump packages are smaller but not as

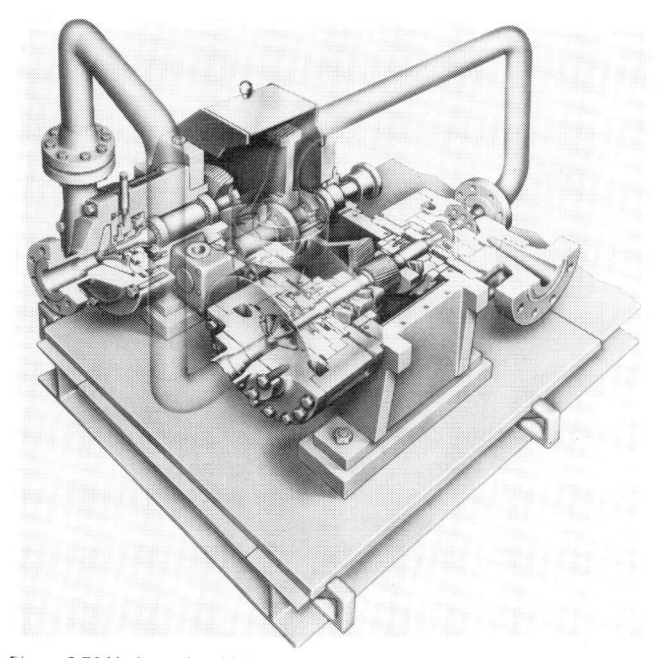

Figure 3.79 Horizontal multiple high speed pump

efficient. High speed multi-pump packages cannot reproduce the reciprocating pump characteristic but only cover the duty point.

165 Multi-stage axially split pumps

Large multi-stage pumps have a very wide range of applications, notably mine drainage, boiler feed and process industries. They are also used extensively in the refining, petrochemical industries and in oil field production. The design and construction being determined by the specific functions and components, see figure 3.80:

- diffuser type,
- first stage impeller,
- thrust balancing,
- bearing types.

As we have seen from the theory of rotodynamic pumps the kinetic energy of the liquid produced in the impeller must be converted to pressure energy by efficiently reducing the velocity of the liquid. Two methods are used; vaned diffusers and vaneless diffusers called volutes. The two main groupings of axially-split pumps are diffuser pumps, meaning vaned diffusers, and volute pumps. Modern trends are leaning towards volute pumps as these are cheaper, in theory, to produce.

There are also two choices of first stage impeller; single entry and double entry. Double entry impellers are used in low $NPSH_a$ applications to try to obviate the need for separate booster pumps.

For moderate pressure increases it is possible to absorb the axial thrust directly in the thrust bearings. However, it is generally necessary to balance out the thrust. The most common method of balancing thrust is by opposing the direction of the impellers, balance discs or by means of balance drums. Opposing impellers leads to some complications with the casing castings due to the transfer passages and can lead to poor casting integrity and weld repairs. The experts argue endlessly about the merits of balance disc and drums; personal experience with the specific application may be the best guide.

Depending upon the pump size, ball bearings are probably the standard with angular contacts, back-to-back, for the thrust. Sliding bearings with tilting pad thrust bearings may be an option or become the standard as pump speed increases. At higher speeds the life of rolling contact bearings is reduced whereas sliding bearings can have an almost unlimited life. Lube system options will cover all requirements.

Small multi-stage pumps, like borehole pumps, can have many stages, 50 or 100. Larger multi-stage pumps are limited by shaft deflection and critical speeds and the number of stages rarely exceeds 15. It is often required to run at higher speeds than those produced directly by electric motors, this necessitates the use of steam or gas turbines or geared-up electric motors. The speed is generally 5000 to 8000 rpm. Some recent development trends are towards high speed single stage pumps but this is not universal.

Pumps are available in many material combinations with casings in; carbon steel, 13% Cr steel, 18-10-2 stainless steel and duplex stainless steel. Pumps at 2900 rpm can handle 2400 m^3/h at heads over 1400 m. Running at higher speeds will produce more flow at higher heads. Standard casing designs are available for pressure ratings up to 280 barg. Connections are usually in the bottom half so that all maintenance can be accomplished without disturbing the process pipework. Other connection positions are sometimes optional. Stuffing boxes and seal cavities tend to be big enough to cope with most requirements. Some manufacturers can incorporate intermediate connections on the pump so that liquid can be added or extracted at an intermediate stage.

Pumps are available to API 610 requirements. API excludes the use of axially-split pumps when:

- liquid temperature of 205 °C or higher,
- flammable or toxic liquid with SG 0.7,
- flammable or toxic liquid over 69 barg.

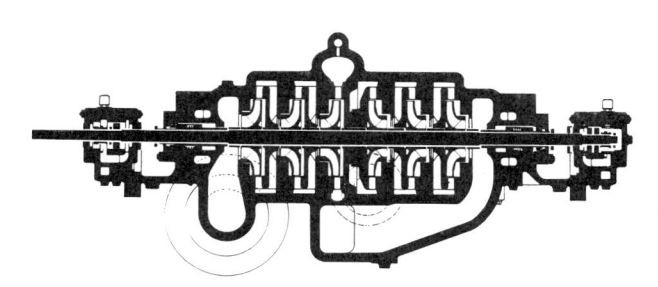

Figure 3.80 Multi-stage axially-split pump with thrust balancing by means of opposed impellers. Pump fitted with sleeve bearings and tilting pad thrust bearing

The main weakness in the axially-split design is the joint between the casing halves. Some designs employ gaskets and others metal-to-metal joints. As the pump pressure increases the joint bolting becomes more heavily loaded and tends to stretch reducing the interface pressure at the joint faces. The number and size of bolts or studs which can be fitted around the joint is finite. Capnuts can be used to eliminate the space required for spanners but the highly loaded stud/bolt cross section still sustains the tensile stress. This weakness in design is cured by the pump design in the next category.

170 Multi-stage radially split pumps

As pressures within the process industries increased so the range of hazardous liquids increased. The limitation of the axially-split multi-stage pumps was addressed to remove fears of potential leaks of hazardous liquids. The radially-split pump design was evolved.

Consider the multi-stage segmental pump shown in figure 3.64; it has already been stated that this pump design has many potential leak paths. However if the stack of segments is housed with a pressure vessel any leakage would be contained; this is the basic design principle of radially-split pumps. The small clearance around the outside of the segments is usually at discharge pressure so that the segments do not have to be designed to cope with high tensile stresses caused by internal pressure. Unlike axially-split pumps, the radially-split pumps all have vaned diffusers and usually balance compensation by disc or drum. Some very special pumps have opposed impellers. The segmental stack, pumps have opposed impellers. The segmental stack, together with seals and bearing assemblies, is build as a complete assembly, called the cartridge, which is inserted into the pressure vessel from the non-drive end. This allows all work on the pump to be carried out without disturbing the motor or the pressure vessel and process pipework.

Three types of construction of radially-split pumps are built as standard, the variation being in the pressure vessel which is normally called a 'barrel'. Hence the popular name for this type of pump, the barrel pump; not to be confused with very small barrel emptying pumps described in category **300**. Small and low pressure pumps can have the barrel fabricated from standard pipe or rolled plate; this type of construction has been called 'segmental-pump-in-a-pipe'. Larger pumps have cast or forged barrels. Forged barrels may have the process connections welded to the barrel requiring additional quality control. The barrel may be foot or centre-line mounted depending upon the application and inter-stage connections may be possible. Double suction first stage impellers are an option as is compliance with API 610.

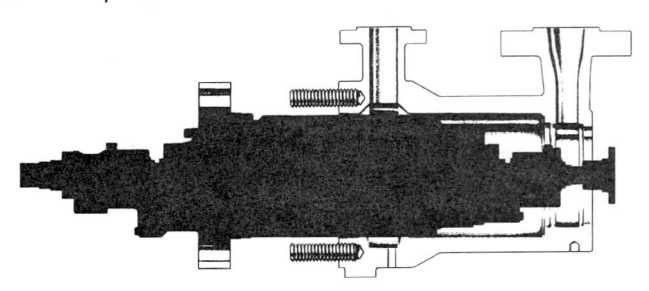

Figure 3.81 Radially-split pump showing cartridge removal

Figure 3.82 Cross section of a multi-stage radially-split pump

Radially-split multi-stage pumps are used for the most arduous high pressure applications. Temperatures over 400 °C have been accommodated. Many material combinations are available to cope with temperature, pressure and corrosion. Flows over 400 m³/h at 3000 rpm are possible; operating speeds up to 6500 rpm are not uncommon. Special pumps are available for direct drive by gas turbines. Total head rise of over 3000 m is not unusual. Pressure ratings up ANSI 2500LB, 414 barg, are available. The largest barrel pump to date is a boiler feed pump in a power station and is rated at 52 MW. Process connections can be to any standard or special. Boiler feed pumps tend not to be flanged but butt welded directly to the pipework; as the casing is not disturbed during maintenance this is not a problem. Stuffing boxes and seal cavities can be made to suit any sealing requirements.

175 Vertical multi-stage pumps

Vertical multi-stage pumps tend to be 'wet pit' pumps. Some versions are available where the pump is suspended within its own tank and effectively becomes a 'dry pit' pump. These vertical pumps are used for high flows in applications such as cooling water where one or two pumps feed a ring main which supplies a whole site. Construction details vary considerably including centrifugal and mixed flow impeller types. Pumps are produced individually to order based on a range of standard designs. Countless material combinations are available at every conceivable level of quality.

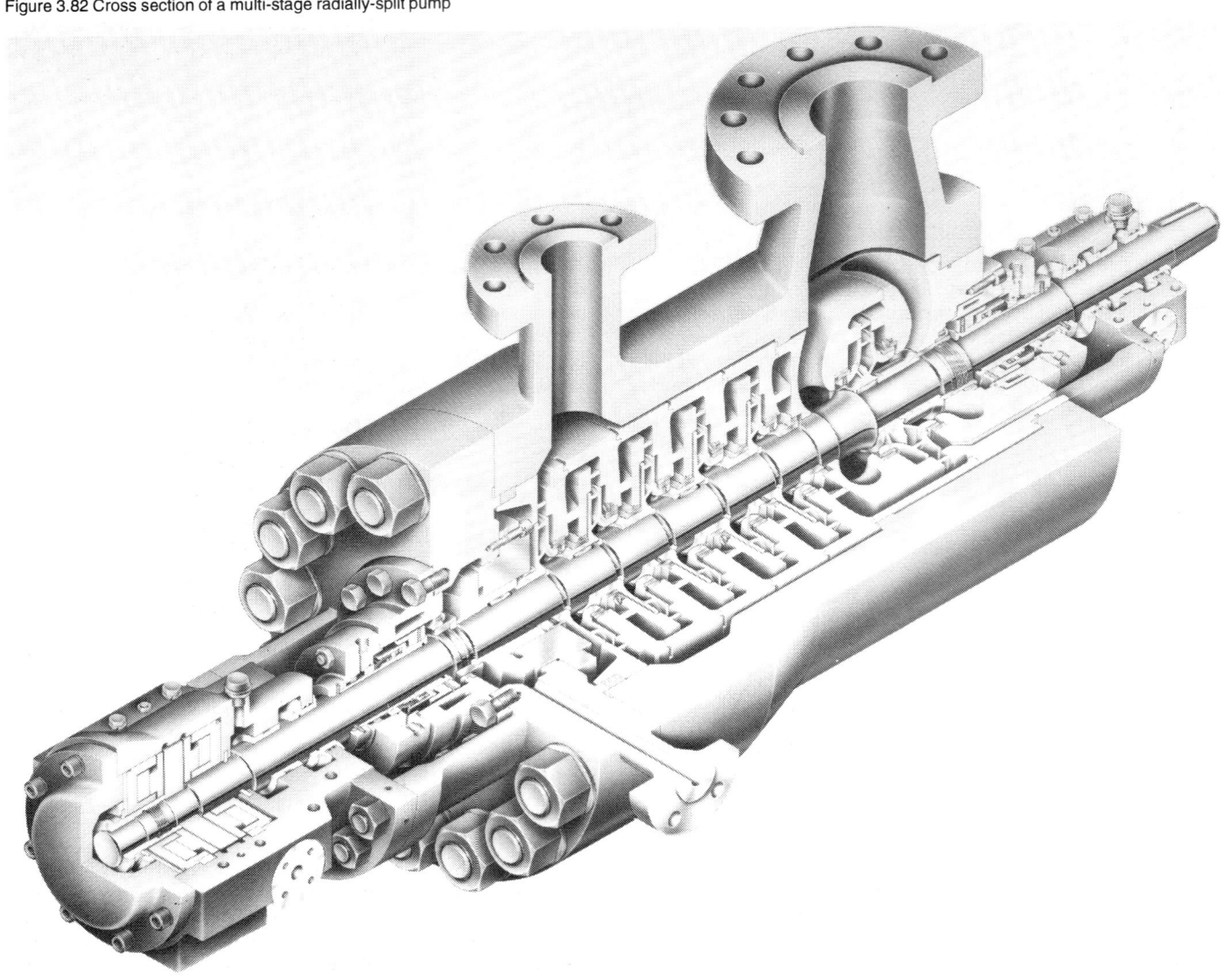

Pumps can be built with flanged vertical motors mounted directly on top of the pump. Small pumps can be coupled with rigid couplings utilising thrust bearings built into the motor. Larger pumps have self contained bearing assemblies, rolling or sliding, and are driven through flexible couplings. Some pumps are built with a right-angle gear box mounted on top; these pumps must have self contained bearings. The pumps are driven via flexible couplings by motors or engines. This style of pump is popular off-shore.

Flows over 70000 m³/h are possible, differential heads vary from 50m to over 400m.

Vibration can be a problem on vertical pumps. It would be reasonable to purchase large pumps from manufacturers who have facilities to test pumps in the fully assembled condition. Balancing requirements on motors and rotating elements may need tightening to reduce problems.

180 Non-clogging pumps with standard motor

Non-clogging pumps are mainly used as sewage pumps for pumping untreated sewage in sewage treatment plants. Group **180** deals with normal models for "dry" installation and with only the pump section immersed in the liquid. Group **185** covers submersible models where both pump and drive motor can be operated below the surface of the liquid.

Sewage handling pumps are completely dominated by rotodynamic pump designs, the following types are used:

- Centrifugal pumps with through-flow impellers, special enclosed impellers with large almost straight passages,
- Free-flow pumps, recessed open impeller,
- Pumps having semi-axial impellers and axial pumps (propeller pumps) for larger flow rates and lower differential heads.

The free area, the through-flow for non-clogging pumps, is normally 60 to 100mm for the smaller pump sizes and 125 to 150 mm for the larger pumps. The obviously desirable feature that particles as large as the connection diameter should be able to pass through the pump is not usually possible because of the hydraulic problems and pump efficiency. Any solids should, in theory, be surrounded by liquid. See the table in figure 3.75 in Group **140** for guidance. For small flow requirements, in order to obtain sufficient liquid flow, oversize pumps must be used.

All sewage handling pumps comprise single suction impellers. This is to avoid the necessity of locating the pump shaft in the intake. Pump materials chosen are usually grey cast iron or SG iron for municipal sewage and grey cast iron or stainless steel for industrial effluent.

Pumps with channel impellers have long since been the most well developed and constitute a special class of their own with regard to the number of different types and the number of pumps in use. In practice most pumps have impellers with single or double channels, although larger pumps sometimes have three channels. As implied by the description "through flow impeller" the pumped liquid after leaving the pump intake, passes through the pump impeller and out through the delivery connection, figure 3.83.

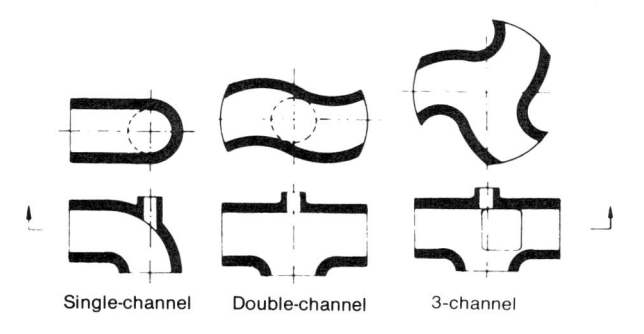

Single-channel   Double-channel   3-channel

Figure 3.83 Various types of channel impeller

Pumps with through flow impellers are characterised by their high degree of efficiency over a large part of the H-Q curve. The overall efficiency at the nominal duty point for medium size pumps is about 60%. The wear resistance of the impeller is moderately good when used in combined systems and good in separate systems.

The shape and design of free-flow, recessed impeller, pumps differs from through-flow pumps in that the location of the impeller and the utilisation of the pump casing is different, figure 3.84. The pump impeller is symmetrical with open vanes. The profile is low and its pulled backed location, recessed in the stuffing box wall, leaves the pump casing completely or partially free. The free flow pump can be described as a centrifugal pump having very large sealing clearance. The liquid and contaminants flow freely under the pump impeller and out through the delivery connection. The absence of sealing clearance results in a reduction in efficiency compared to through flow pumps. The overall efficiency at the nominal duty point for small and medium size pumps is 40 to 42%.

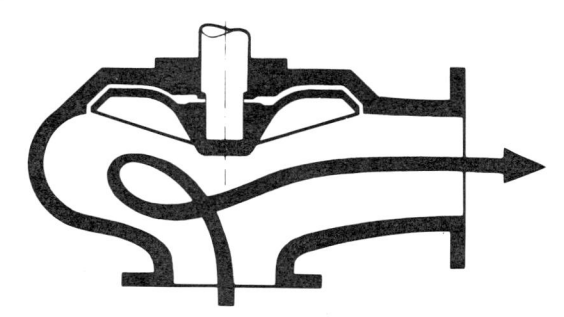

Figure 3.84 Free-flow pump

185 Submersible non-clogging pump

The submersible pump is today of predominant importance compared to the conventional dry pump type. The submersible pumps are usually capable of functioning in dry locations and thus can be used in low locations where there is risk of flooding, as in the case of an electricity power cut for example.

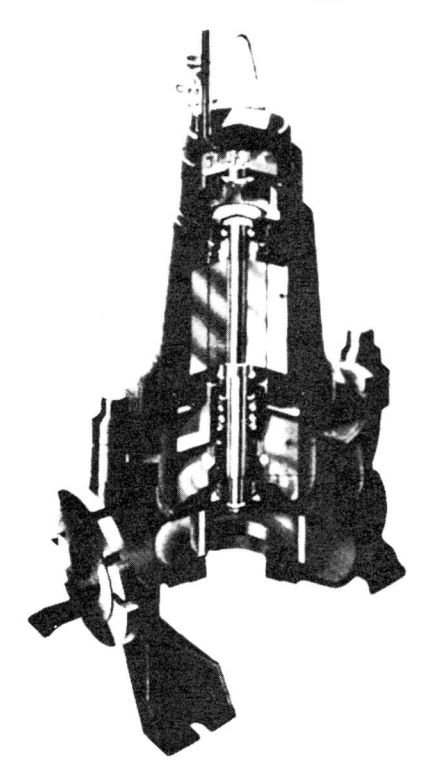

Figure 3.85 Submersible sewage handling pump suitable for both dry and submersed installation

Submersible pump construction is recognised by the fact that the impeller is located directly onto the combined motor and pump shaft, figure 3.85. The drive motor, a squirrel cage induction motor, is completely sealed by axial and radial O-rings and the pumped liquid is sealed from the drive motor by double mechanical seals which run in an oil-bath. The type of bearings, size of shaft and method of sealing are deciding factors in reliability and operational safety of the whole system. Deflection of the shaft due to radial loads should not exceed 0.05 mm at the shaft seal. The shaft seal components should be assembled in a seal cartridge, which enables preassembly and pressure testing of the shaft seal prior to fitting a replacement, figure 3.86. From the point of view of pump service and repair, it is advantageous to be able to fit separate motor units of varying sizes to a number of different pump casings. Such a range system makes it possible to maintain a complete community with a small stock of motor parts. Some pump systems also require facilities for converting on site from through flow to free flow impeller.

Figure 3.86 Seal cartridge with double mechanical seals for submersible sewage handling pump

190 Other non-clogging pumps

Pumps, similar in design to sewage handling pumps, are used for transporting larger solids by means of water, objects such as large, whole fish, fruit and root vegetables. The channel areas and shapes being designed to cause the least amount of damage to the product being pumped.

In direct contrast there are other types of non clogging or chokeless pumps where deliberate attempts are made to finely disintegrate the material being transported. Apart from sewage handling pumps with a integral fine disintegration device or macerater there are pumps specially developed for the handling of sludge and waste within the food processing industry, figure 3.87, and figure 3.88.

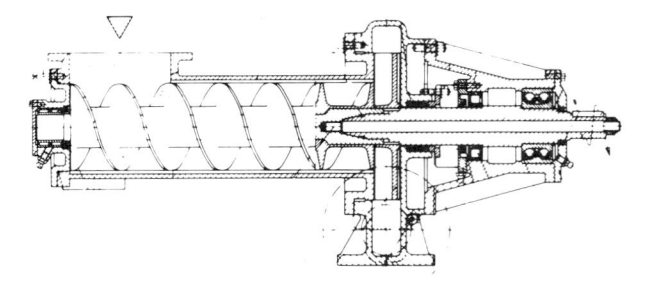

Figure 3.87 Waste pump with "compression screw" for feeding the centrifugal pump unit

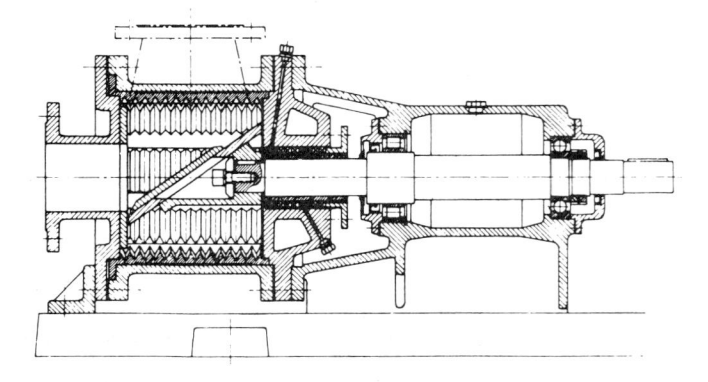

Figure 3.88 Oblique disc pump (Gorator pump) for simultaneous disintegration and pumping

195 Mixed flow pumps

Mixed flow pumps are used for large flows and low differential heads. The differential head can be increased by using multiple impellers - a multi-stage mixed flow pump. These pumps are nearly always installed with the impeller immersed in the liquid, flooded, and the motor located above, so called extended shaft pumps for immersed installation. For relatively clean fluids the central support tube also functions as a delivery pipe while for contaminated liquids the delivery pipe is separate, similar to figure 3.53.

Typical of immersed installation pumps, at least for the larger sizes, is that the delivery connections and mounting plates are located according to each particular installation. If the total height of the pump exceeds the available height required for assembling and disassembling then the support and the delivery pipe as well as the shaft may be divided into segments. For long shafts it is necessary to fit intermediate bearings to prevent vibration and to increase critical speeds. There is a clear trend towards replacing extended shaft pumps with submersible pumps in the smaller sizes.

The hydraulic design and number of stages is varied according to the duty requirements. Typical applications include wells and boreholes, condenser circulating water, condenser extraction and drainage pumps. Pump sizes start at 20 m^3/h and approximately 30 m differential per stage. Pumps are available in all popular 'pumping' materials.

200 Axial propeller pumps

Many of the comments in category **195** can be applied to axial flow pumps. Most pumps are vertical, the vast majority are single stage. Some horizontal versions are available which look like pipe elbows. Axial pumps can be equipped with adjustable blades which can be set, either when the pump is at rest, pre-set, or during operation, variable, by means of mechanical or hydraulic actuators. The differential head produced by axial flow pumps is much less than mixed flow or centrifugal pumps; typically 15 m per stage. Because of the simplicity of the construction they are easy to build in large sizes. Axial flow pumps are generally used for flows greater than 400 m^3/h; flows of 30000 m^3/h are not uncommon.

205 Power recovery turbines

Rotodynamic pumps can run in reverse as turbines. This fact is long established and is one of the reasons why non-return valves are fitted in pump discharge lines. Pumps and their drivers have suffered damage as a result of "turbining" due to process upsets or non-return valve failures. The power available from pumps running in reverse can be captured and utilised for useful work.

Recent serious interest in power recovery turbines stems from the oil crisis of the 70's and the escalating cost of energy. Some widely used processes are ideal for considering power recovery. In oil refining, the hydrocracker process operates at over 100 barg and gas scrubbing at various pressures over 50 barg, in drinking water manufacture by reverse osmosis from sea-water the process operates at 70 barg. In these cases the process only degrades the pressure slightly leaving a large amount of pressure energy to be throttled or recovered.

When considering power recovery from high pressure process streams there are three basic choices for installation:

- use the power as the sole power supply to drive a pump or other machine,
- use the power as the sole power supply to drive a generator,
- use the power to assist in driving a pump or other machine.

Standard mixed flow and centrifugal pumps are popular for these applications; both types are in effect fixed geometry Francis turbines. The Francis turbine is an inward flow radial turbine and, depending upon the head to be converted, can be mixed flow or truely radial. Power generation turbine manufacturers, in general, do not build small machines thus making standard pumps an ideal option. The slight loss of efficiency, 2 or 3%, is more than offset by the economic benefits of purchasing a standard piece of equipment. If the power recovered is used as the sole power supply it must be remembered that the turbine will not produce power until the flow reaches 40% of design. Also, speed control may be necessary. If the recovered power is used to assist in driving then the speed control of the prime driver should suffice.

Assisted driving is probably the most common application because of the simplicity. Consider a pump driven by a squirrel cage motor which has two drive shafts, a double extended motor. The extra drive shaft is coupled, via an automatic clutch, to the turbine. The pump is started in the normal manner. If there is no flow through the turbine it remains stationary. The pump process conditions can be adjusted and allowed to stabilise. If during this period flow, and some pressure, appear at the turbine inlet the turbine will start to spin. As the turbine design conditions are approached the turbine speed will increase until it tries to run faster than the motor. As the turbine tries to overspeed, the clutch locks and power is transmitted to the motor. As the turbine transmits power into the motor, the motor speed increases slightly, unloads by the amount of power supplied by the turbine and the motor power consumption is reduced. If the turbine liquid supply fails the turbine will reduce in speed and when the motor attempts to drive the turbine the clutch will disengage leaving the motor to supply all the power.

In typical applications, such as hydrocracking and reverse osmosis, the pump and turbine are the same pump type. Hydrocrackers use radially-split, barrel, machines. Depending upon the size, reverse osmosis systems use multi-stage segmental or axially-split machines. It is possible to purchase multi-stage segmental machines with the pump and turbine in the same casing; precautions must be taken to prevent the turbine running dry.

When considering power recovery, also refer to category **600**, Pelton Wheels, as these are an option. Review Chapter 15 on typical case studies.

3.3. Liquid ring and peripheral pumps

Liquid ring and peripheral pumps occupy a position between rotodynamic pumps and rotary positive displacement pumps. Their theory of operation having some of the characteristics of both types. In principle, liquid within the casing is separated into "elements" by means of rotating vanes, as in the case of displacement pumps with rotating vanes, vane pumps, although radial clearance permits a large percentage of recirculation. Energy is successively imparted to the liquid by contact with each vane in a manner which is reminiscent of rotodynamic pumps. When pumping, energy is transferred to each "enclosed" element of liquid by the action of the passing vanes. The increase in head for a given outer diameter and speed is 5 to 10 times that of an equivalent rotodynamic pump. The positive displacement feature makes it possible to pump air or gas provided that a certain quantity of liquid is available to act as a seal. Liquid ring pumps and peripheral pumps have therefore good self-priming characteristics.

3.3.1 Liquid ring pumps

210 Liquid ring pumps

The working principle of a liquid ring pump is illustrated in figure 3.89. The pump impeller is fitted with straight radial vanes 'A', which cause the liquid in the pump casing to rotate. The duct 'B' in the wall of the pump casing has its greatest depth at point 'C', thus causing the volume of the pocket between two adjacent vanes to vary as the pump impeller rotates. The volume between two adjacent vanes is increased as the duct depth reaches a maximum at 'C'; this causes a partial vacuum which draws more liquid in through the suction port 'D'. The depth of the duct decreases at 'E' thus reducing the volume of the liquid pocket between two adjacent vanes. This causes a build up of pressure which forces the liquid out through the exhaust port 'F'. The shape and depth of the duct relative to the intake and exhaust ports determines the evacuation capacity and, to a certain extent, the performance characteristics. The ducts or side channels can be replaced by equivalent channels located outside the periphery of the pump impeller. The side channels give the pump its alternative name, Side Channel Pump, which is the one used by CEN European standards. As we now know, the side channels can be around the periphery without changing the principle of operation so that Liquid Ring Pumps is a better description.

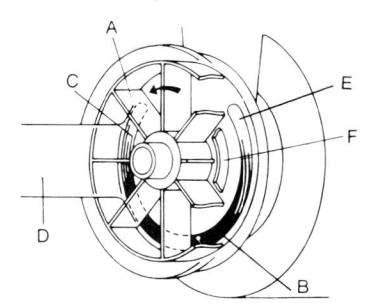

Figure 3.89 Working principle of liquid ring pump

The H-Q curve of liquid ring pumps is characteristically steep, figure 3.90, which is ideally suitable for automatic operation in conjuction with an accumulator. This type of pump is not usually permitted to operate at very low flow rates.

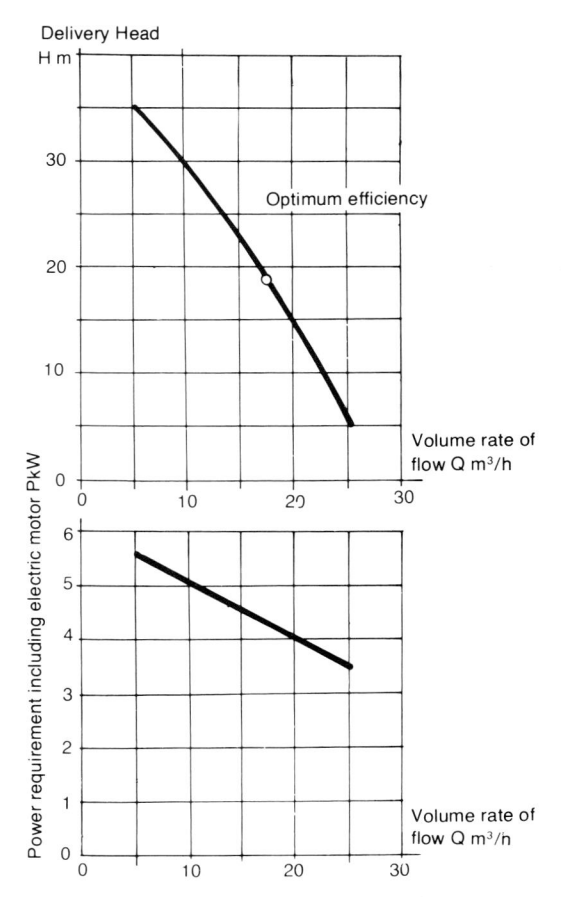

Figure 3.90 Example of performance curves for a large single-stage liquid ring pump

Because of the liquid ring pump's steep H-Q curve the power demand increases with reduced flow rate and increased differential head. This can cause overloading of electric drive motors designed to meet the demand at the design operating point. Reduction of flow rates to low values should therefore be carried out by means of by-pass control i.e. by returning liquid from the delivery side to the suction side using a control valve connected in the return pipe. Special designs with integral by-pass valves are available, figure 3.91. However, operation under integral by-pass control should be carefully monitored to avoid excessive temperature rises. If prolonged operation at low flows is necessary, a by-pass which recycles to the suction source, not the pump suction, is preferred.

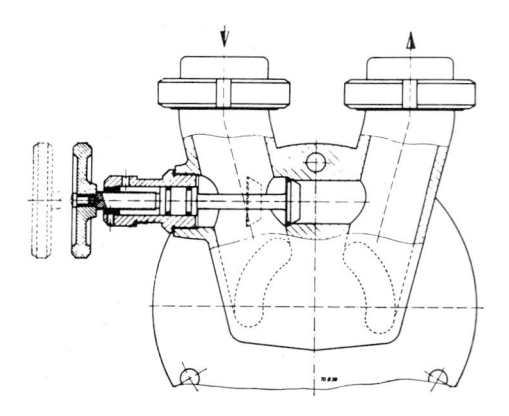

Figure 3.91 Liquid ring pump with integral by-pass valve

Liquid ring pumps produced by different manufacturers are so similarly constructed that it is even possible to interchange some of their spare parts, not during the warranty. The most usual type is illustrated in figure 3.92 and consists of an intermediate section and one impeller per stage together with suction and discharge end covers with connections. This style of construction is very similar to multi-stage segmental pumps, **085** and **090**. The shaft runs in a liquid lubricated plain bearing at the discharge side and a greased ball-bearing at the suction. The most common material is grey cast iron with bronze impellers. Special stainless steel types are available, primarily for use in food processing.

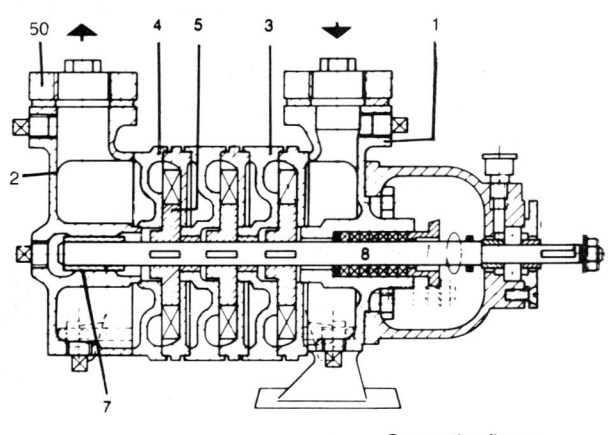

1. Inlet cover
2. Outlet cover
3,4. Intermediate section
5. Impeller
6. Connecting flange
7. Plain bearing
8. Stuffing box

Figure 3.92 Three stage liquid ring pump

Liquid ring pumps are used in great numbers for domestic and agricultural water supply from wells and also as small high pressure pumps. Maximum differential head for multi-stage pumps is about 350 m and flows up to 350 m^3/h are possible. Disadvantages are wear, in the case of solids contaminated water, and the relatively high noise level despite the low speeds. Efficiency is lower than for a centrifugal pump of the same performance. The big advantage of the liquid ring pump is that it is self-priming if it contains liquid prior to starting, and also that it can pump liquids containing quantities of entrained gas.

3.3.2 Peripheral pumps

215 Peripheral pumps

Peripheral pumps are also called regenerative pumps. The principle of the pump is similar to that of a liquid ring pump. The difference is that the vanes are greater in number and are mounted on the periphery of the rotating disc. In practice the vanes are formed by making cut-outs in the periphery of the disc and supply energy during almost a complete revolution. The head increase is greater for a given outer diameter than for liquid ring pumps. Other characteristics and applications are however similar. A longitudinal section through a single stage peripheral pump is shown in figure 3.93. A transverse section through the stage is shown in figure 3.94.

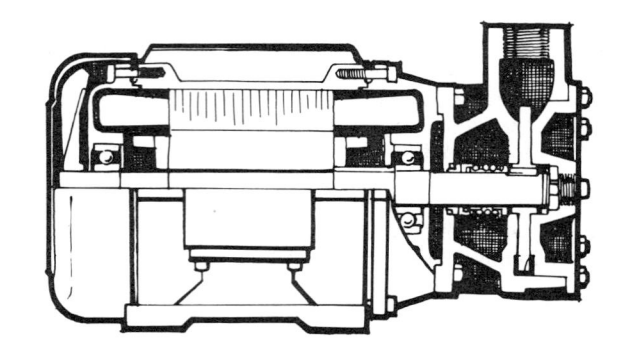

Figure 3.93 Section through a single stage close coupled peripheral pump

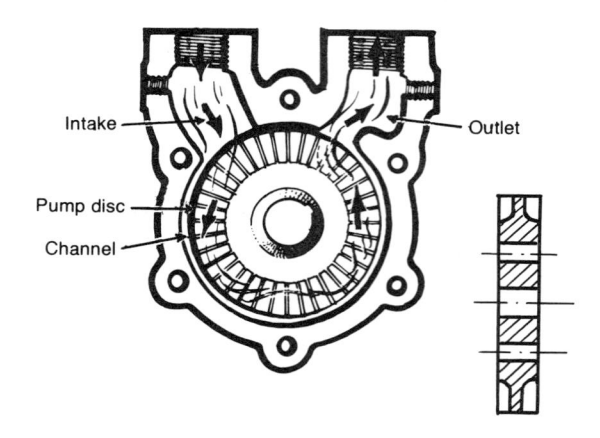

Figure 3.94 Peripheral pump stage

3.4 Positive displacement pumps

Positive displacement pumps rely on completely different principles of operation than rotodynamic pumps. The most important difference is speed. Positive displacement pumps do not rely on speed to develop pressure. And now we talk of pressure, not head, because the small velocity head is not converted to static head. Velocities tend to be very low compared to rotodynamic pumps. Also differential head is replaced by discharge pressure. Positive displacement pumps do not try to increase the suction head by a fixed differential head. Positive displacement pumps develop sufficient discharge pressure to force the liquid into the discharge system. In broad terms, positive displacement pumps compress the liquid from suction conditions to discharge conditions and can achieve this at any speed. As already stated, velocities tend to be low so there are no problems in ignoring the kinetic energies in the suction and discharge pipes. The pressures developed tend to be very large in comparison to the physical dimensions of the pumps; so there are no problems ignoring potential energy due to elevation in the suction and discharge systems.

Positive displacement pumps can produce very high pressures, over 10000 barg is possible but uncommon. Standard pumps used for high pressure cleaning and descaling operate at 1000 barg. Liquids are compressible. The liquid volume at discharge conditions may be considerably smaller than at suction conditions; less than 85%. If the discharge flow rate is important, rather than the mass flow, compressibility cannot be ignored and must be included in pump and system design calculations.

3.4.1 Rotary positive displacement pump theory

Basic principles

One of the simplest rotary pumps, and probably the easiest to understand, is the external gear pump, see figure 3.95. One gear is driven by an external power source the other meshes with the driven gear. The gears are enclosed in a close fitting casing. The radial clearance around the gear teeth is controlled by machining; the axial clearance down the sides of the gears is usually controlled by shimming during assembly. Clearance is necessary to avoid seizure during operation. The axial clearance is also necessary to allow lubrication.

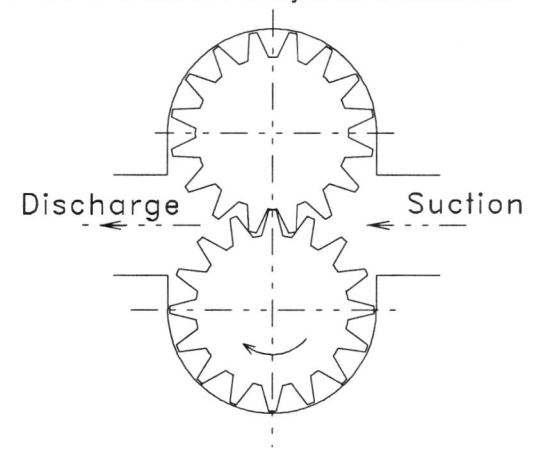

Figure 3.95 Diagrammatic external gear pump

Liquid is carried around the periphery of the casing in the voids between the teeth. The theoretical flow is given by:

$$Q_{th} = 0.001 \times a \times w \times (2 \times N) \times rpm/60$$

Equ 3.24

where
- Q_{th} = theoretical flow (cc/s)
- a = area between teeth (mm^2)
- w = tooth width (mm)
- N = number of teeth in each gear
- rpm = speed (1/minute)

Equation 3.24 applies to gear pumps where identical gears are used; this is not essential for gear pump designs but it is the most usual. Theoretical flow cannot be realised in practice. Leakage occurs directly from discharge to suction where the gears mesh; this leakage can be very small and is dependant upon the tooth form and the clearance. Most leakage occurs around the periphery between the tips of the teeth and the casing. This leakage is complex to analyse but approximates as follows:

- $\propto$ differential pressure
- $\propto$ gear width
- $\sim\propto$ tip clearance^3
- $\propto$ 1/μ
- $\propto$ 1/N
- $\propto$ 1/gear tup length
- $\sim\propto$ 1/function of tip speed

Gear pumps rely on viscosity to reduce leakage. As viscosity reduces the leakage becomes a greater proportion of the theoretical flow. The gear pump characteristic shown in figure 3.106 in 3.4.3 indicates the effect of speed, viscosity and differential pressure. Leakage is reduced at higher speeds for constant viscosity and differential pressure. A popular light oil grade for centrifugal pumps with rolling bearings is ISO 32. At 20 °C these oils would have a viscosity of 85 mm^2/s. Water has a viscosity about 1 mm^2/s. It can easily be seen in figure 3.106 how the leakage losses escalate as less viscous liquids, more like water, are pumped.

The mechanical efficiency of gear pumps is controlled by several mechanisms and assumptions should not be made when calculating power required. At low viscosities the efficiency tends to improve with increasing speed. At high viscosities the efficiency may be constant or reduce with increasing speed. The distinction between high and low viscosity is a function of pump size. Mechanical efficiency usually increases as the differential pressure increases. Small pumps can have efficiencies of 25% and large pumps 70%.

Because gear pumps have multiple pumping elements, the teeth, the flow pattern may have a slight ripple which can cause pressure pulsations. Pumps with few teeth are worst. Tests by Professor Edge, at Bath University, have indicated peak to peak pressure pulsation levels of 3.25 bar with a mean suction pressure of 3 bara.

Wear in rotary displacement pumps

The internal leakage varies as the cube of the clearance between the rotating components and the casing. Displacement pumps are thus susceptible to increased clearances due to wear. Displacement pumps used for "abrasive" liquids are constructed from wear resistant materials and operated at lower speeds. Exceptions to these guidelines can be made by utilising principles of construction which eliminate leakage by the use of an elastic element, peristaltic pumps, for example.

Considering elementary principles, wear is a function of:

- Surface pressure between components
- Relative rubbing speeds between components
- A dependance upon the coefficient of friction

or with more usual wear theory, the combination of surface pressure and rubbing speed, PV.

The heat produced per unit of surface area is dependant upon PV and the coefficient of friction and can be a significant factor.

The technique of creating wear-resistant rotary pumps is concerned with reducing the product PV. Low values of velocity can result in large pumps, the only solution then is to reduce the surface pressure.

The driving force for rotary pumps can be transmitted to the various components in different ways. In the case of gear pumps the external drive is normally applied to only one of the gears, while the other is driven by engagement with the gear teeth of the first. However, if both the gears are driven externally, the load relationship within the process liquid is completely different. It is necessary to synchronise the shafts by means of a separate synchronising gear in order that the teeth shall pass within each other when rotating without contact. The surface pressure between the interactive gear components is thus reduced to nil. The shape of the interactive gear surfaces can now be chosen with regard to the pumped fluid. Instead of gear teeth, the pump elements can now be given smoother shapes or radically different shapes, as in the case of lobe pumps and rotary piston pumps.

Within the rotary pump group there is a special "family" of pumps whose interactive components are guided by external devices as in the case of the lobe pump mentioned above.

Even pumps where the rubbing pressure is eliminated in other ways also belong to this group. Examples of such pumps are rotary piston pumps and twin screw pumps.

With the rubbing pressure eliminated or greatly reduced the duty range can be increased. Depending upon other design features, this range of application can be extended to contaminated liquids, low viscosity, some dry-running capability or high speed.

Rotary pumps suitable for abrasive liquids should be constructed according to the following:

- The active components should be positively located by external means or have greatly reduced sliding pressure.

- The forces generated on the pumping elements by the liquid pressure should be absorbed by bearings which are not in contact with the liquid.

3.4.2 Reciprocating positive displacement pump theory

The theoretical principles of positive displacement pumps are basically very simple. For each working cycle, each stroke, a certain volume of liquid is drawn into and expelled from the cylinder virtually irrespective of the differential pressure. The volume of liquid depends upon on the displacement (swept volume) and the volume of the cylinder. The maximum attainable pressure is primarily dependent only upon the mechanical construction and the driving power available. The maximum pressure can be limited by means of a pressure relief valve installed in the system or integrated in the pump. Because reciprocating pumps can develop such high pressures they must always be operated with over pressure protection. A relief valve or bursting disc must be fitted. In hydraulic power applications it has become standard practice to incorporate the over pressure protection with other functions, such as unloading, by using a common pilot operated valve. In process applications this approach would not be tolerated; a dedicated device must be fitted solely for protection.

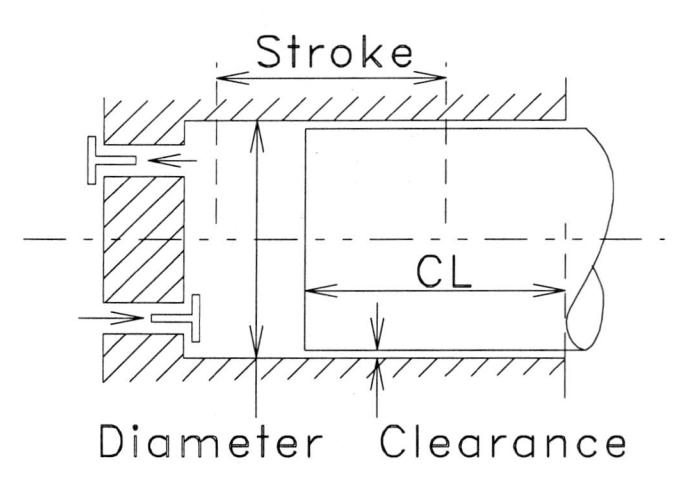

Figure 3.96 Basic reciprocating pump

Consider the single cylinder pump shown in figure 3.96. Ignoring leakage and compressibility, the ideal flow delivered is calculated from:

$$Q_{th} = 0.001 \times \text{displacement} \times n/60$$

and

$$\text{displacement} = s \times \pi d^2/4 \qquad \text{Equ 3.25}$$

where

Q_{th}	=	theoretical flow	(cc/s)
n	=	strokes per minute	(1/minute)
d	=	piston diameter	(mm)
s	=	stroke	(mm)

The displacement of the piston establishes the name of the pump type, positive displacement. Displacement is also called swept volume as it is the volume swept out by the face of the piston.

The ideal power required for the pump is given by:

$$P_{th} = Q_{th} \times (p_d - p_s) \qquad \text{Equ 3.26}$$

where

P_{th}	=	ideal power	(W)
Q_{th}	=	ideal flow	(m³/s)
p_d	=	discharge pressure (gauge)	(Pa)
p_s	=	suction pressure (gauge)	(Pa)

Equation 3.26 is not very useful for engineering applications. The Pascal is an extremely small unit of pressure, the kPa is very small. In order to avoid confusion at site, where suction pressure gauges may be calibrated in kPa and discharge gauges in MPa, the pump industry prefers to use the "bar". 1bar = 100kPa. The following equations are more practical.

$$kW = \frac{m^3/h \times \Delta bar}{36} \qquad \text{Equ 3.27}$$

$$kW = \frac{L/s \times \Delta bar}{10} \qquad \text{Equ 3.28}$$

The ideal power is based on ideal flow. This section on theory commenced by excluding leakage and compressibility. Obviously real pumps leak and real liquids are compressible. Compressibility is the easiest to tackle.

It can be seen from fig 3.96 that the piston stroke does not extend right to the end of the cylinder. The volume of liquid trapped in the cylinder, together with liquid compressibility, effect the volumetric efficiency. The volumetric efficiency is the ratio of the liquid volume pumped to the displacement of the piston. We have already seen that the discharge volume may be smaller than the suction volume, this leads us to two definitions of volumetric efficiency, suction and discharge.

$$ve_s = 100 \times \left[\frac{\left(1 - \Delta p \times \chi \times \left(1 + \frac{V_{int}}{V_{sw}}\right)\right)}{(1 - \Delta p \times \chi)}\right] \qquad \text{Equ 3.29}$$

$$ve_D = 100 \times \left(1 - \Delta p \times \chi \times \left(1 - \frac{V_{int}}{V_{sw}}\right)\right) \qquad \text{Equ 3.30}$$

$$ve_s = 100 \times \left(1 - \frac{V_{int}}{V_{sw}} \times \left(1 - \frac{\rho_D}{\rho_S}\right)\right) \qquad \text{Equ 3.31}$$

$$ve_D = 100 \times \left(1 - \left(\left(\frac{V_{int} + V_{sw}}{V_{sw}}\right) \times \left(1 - \frac{\rho_S}{\rho_D}\right)\right)\right) \qquad \text{Equ 3.32}$$

where

ve_S	=	suction volumetric efficiency	(percentage)
ve_D	=	discharge volumetric efficiency	(percentage)
Δp	=	differential pressure	(bar)
χ	=	compressibility	(*)
V_{int}	=	clearance volume	(cc)
V_{sw}	=	swept volume	(cc)
ρ_D	=	discharge density of liquid	(kg/m³)
ρ_S	=	suction density of liquid	(kg/m³)

(*) the units for compressibility are "contraction per unit volume/bar differential pressure".

Compressibility is the inverse of the bulk modulus which has units of "N/m²". The bulk modulus of water at 20 °C from atmospheric pressure to 100 barg is approximately 2.1×10^4 bar. The presence of entrained and dissolved gas or air in the liquid increases the compressibility and reduces the volumetric efficiency.

The formulae for volumetric efficiency indicate the importance of keeping the clearance volume as small as possible. Also the volumetric efficiency reduces as compressibility increases. At low differential pressures the compressibility of water can be used for most other liquids without creating significant errors. Some liquids are much more compressible than water, liquid gases for example can be six times as compressible, and using water compressibility will create problems of undersizing and low flow.

Leakage further reduces the actual liquid volume delivered by the pump. The term "slip" is used to collect together all the liquid losses which can be evaluated and those which must be compensated by an experimental allowance. Slip is evaluated as a percentage and subtracted from the volumetric efficiency.

Because there is a clearance around the periphery of the piston, leakage can occur due to the pressure rise within the cylinder; this leakage can be approximated from:

$$Q_L = 0.00156 \times \left(\frac{d}{CL}\right) \times \frac{H^3 \times \Delta p}{\mu} \qquad \text{Equ 3.33}$$

where
- Q_L = leakage flow (cc/min)
- d = cylinder bore (mm)
- CL = contact length (mm)
- h = clearance (mm)
- Δp = differential pressure (bar)
- μ = viscosity (cP)

As the piston and cylinder wear the leakage rate rises dramatically. The liquid viscosity may be lower in the clearance due to frictional heating. This leakage is fairly constant and independant of pump speed, strokes per minute.

The valves shown diagrammatically are spring loaded automatic valves. The differential pressure across the valve, in the direction of flow, opens the valve and liquid flow forces hold the valve open. As the flow reduces the valve starts to close due to the spring force. The pressure losses and leakage through valves of this generic type are impossible to generalise due to the infinite range of proportions and shapes. However, the following relationships can be used for guidance:

- for fixed valve lift

    $\Delta p \propto Q$

    $\propto \mu$

- for fixed flow

    $\Delta p \propto \mu$

    $\propto l/\text{lift}^3$

One valve loss, particularly troublesome at high speed, is valve backflow. Automatic valves are closed by the spring; the spring applies a force to the mass of the valve and the valve accelerates. At higher speeds, there is insufficient time for the valve to close, unless a stronger spring is fitted. The valve remains slightly open when the piston travel has reversed allowing some liquid to escape. After inception, this loss varies with speed and is almost linear.

Some pumps are fitted with actuated valves or valve timing is incorporated in the rotating cylinder block of hydraulic pumps. These valves are usually rotary and driven directly by the main shaft. The valve timing is optimised for the design conditions. Operating at other conditions causes backflow or over-compression.

Valve flow losses are complicated because the liquid flow is approximately sinusoidal. The reciprocating pumps described here are usually used for hydraulic power applications. The liquid is oil or water-oil emulsion providing good lubrication for the close fitting piston. The pumps are axial or radial piston. The pistons are driven by a mechanism which produces simple harmonic motion. The piston dynamics, sine and cosine functions, are shown in figure 3.97.

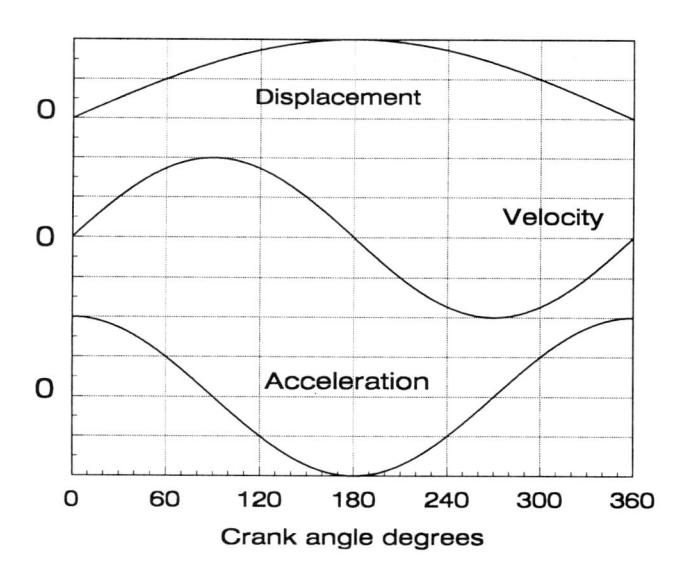

Figure 3.97 Simple harmonic motion piston dynamics

The liquid is unable to follow the piston dynamics exactly due to inertia. The liquid inertia causes the liquid to accelerate and decelerate faster than the piston causing pressure pulsations. Liquid flow variations and pressure pulsations caused by reciprocating pumps can not be approximated easily as they are a function of the system as well as the pump. In critical applications, and with low viscosity liquids, an acoustic analysis should be performed which will evaluate pressure pulsations, flow variations and acoustic resonance.

Two variations of reciprocating pump design used for process applications where the liquid may not have good lubricating properties, are shown diagrammatically in figure 3.98. The plunger pump is single acting whereas the piston pump is double acting.

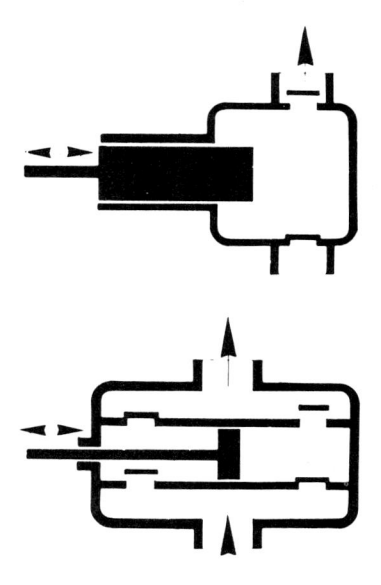

Figure 3.98 Process plunger and piston pumps

The reciprocating motion is produced by a crankshaft with connecting rods and crossheads, the slider-crank mechanism. There is no side thrust applied to the piston or plunger from the driving mechanism, this is applied by the crosshead to the crankcase. The length of the connecting rod compared to the throw of the crankshaft is important because it affects the piston/plunger dynamics and the crosshead side thrust. The sine and cosine functions of simple harmonic motion are modified by connecting rod proportions as shown.

$$x = r(1 - \cos\theta) + l(1 - \sqrt{1 - \lambda^2 \sin^2\theta}) \qquad \text{Equ 3.34}$$

$$v = \omega r\left(\sin\theta + \frac{\lambda \sin\theta \cos\theta}{\sqrt{1 - \lambda^2 \sin^2\theta}}\right) \quad \text{Equ 3.35}$$

$$a = \omega^2 r\left(\cos\theta + \frac{\lambda(1 - \sin^2\theta(2 - \lambda^2 \sin^2\theta))}{(1 - \lambda^2 \sin^2\theta)^{3/2}}\right) \quad \text{Equ 3.35}$$

where

x	=	travel	(m)
v	=	velocity	(m/s)
a	=	acceleration	(m/s/s)
r	=	crank throw	(m)
l	=	connecting rod centres	(m)
ω	=	angular velocity	(rad/s)
λ	=	r/l	(non-dimesional)
θ	=	crank angle	(degree)

The three parameters are plotted in fig 3.99, for λ=0.25, alongside the simple harmonic motion curves. Notice that the:

- curve shapes are modified,
- maximum velocity is not at 90°,
- forward and return acceleration is different,
- maximum acceleration is greater.

Having an infinitely long connecting rod produces simple harmonic motion. In practice, connecting rods are 4 to 7 times the crankshaft throw.

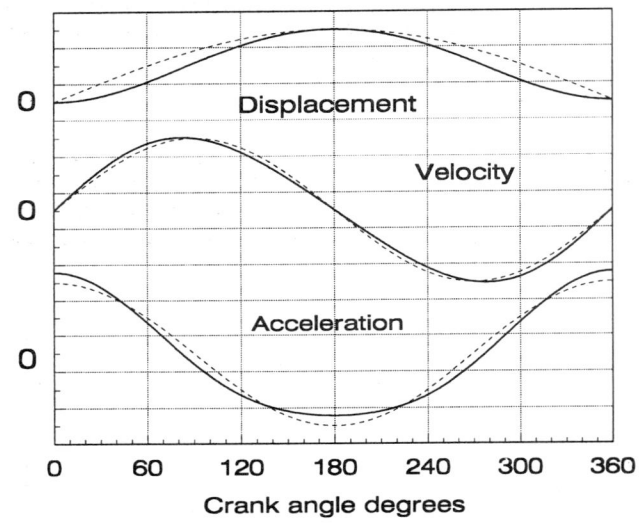

Figure 3.99 Travel, velocity and acceleration of slider-crank mechanism

A plunger pump is very similar to the piston pumps described previously for hydraulic fluid power except a stationary sealing stuffing box prevents most leakage. The piston pump shown is double acting and has moving seals on the piston as well as a stationary sealing stuffing box. The prime distinction between piston and plunger pumps is the absence of moving seals in the plunger pump. Plunger pumps are not a type of piston pump.

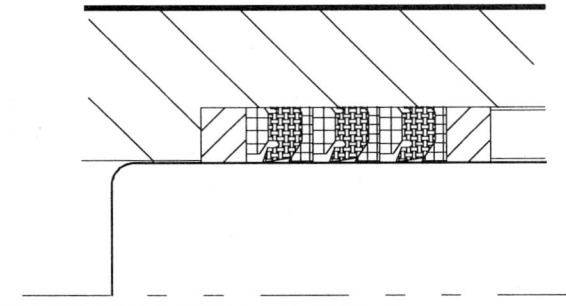

Figure 3.100 Non-adjustable stuffing box for water

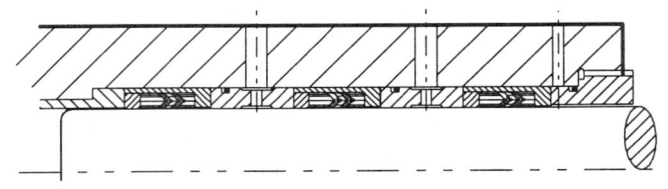

Figure 3.101 Self-adjusting stuffing box for hazardous liquids

Piston pumps are used for pressures up to about 150 barg and plunger pumps for higher pressures. There is no absolute limit on plunger pumps, 10000 barg is possible, but popular pumps operate up to 1200 barg. Fig 3.128 shows a plunger pump for about 350 barg and fig 3.132 shows a piston pump for about 25 barg. Plunger pumps are also built as vertical pumps, see fig 3.129.

The stuffing box prevents leakage of process liquid to the atmosphere. Simple seal designs are used for non-hazardous liquids such as water and ethylene glycol. Hazardous liquids; petrol, methanol, lpg; or liquids contaminated with hazardous gases like hydrogen suphide, can use very complicated stuffing boxes depending upon the location.

Auxiliary systems can be fitted to stuffing boxes, see Section 7.5.

Reciprocating pumps for process applications can be fitted with various valve designs to suit the liquid handled, fig 3.130.

Mechanical losses

Reciprocating pump crankcases and actuating mechanisms are designed to exert a specific force on the piston or plunger, the rod load. Pumps are selected so that they operate as close to the design rod load to minimise mechanical losses. Different piston/plunger diameters are fitted to match the operating pressure. Figure 3.102 indicates the effect of discharge pressure on a 3" stroke plunger pump, with a fixed plunger diameter and a simple single packed stuffing box as shown in figure 3.100.

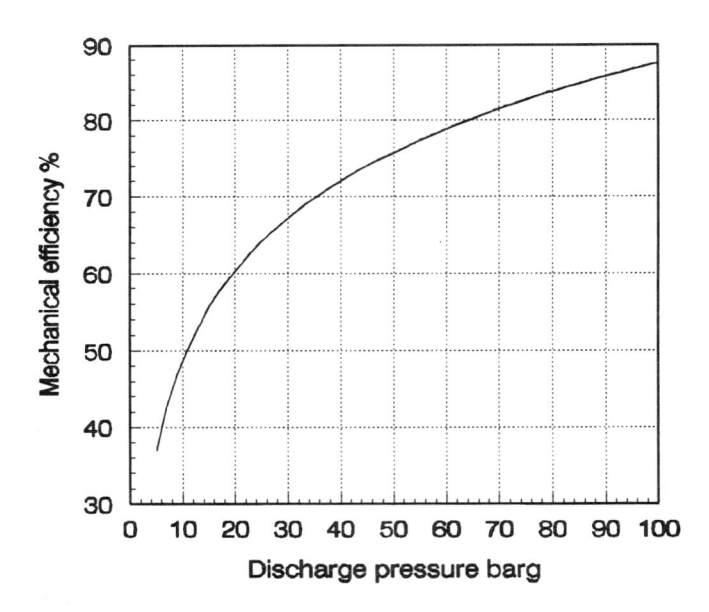

Figure 3.102 Discharge pressure and mechanical efficiency

At pressures below the design pressure, friction becomes a more significant influence and reduces the mechanical efficiency. In general, the mechanical efficiency of reciprocating pumps is not effected by pump speed or viscosity. Larger pumps may be more efficieny; pumps of 1000 kW can have a mechanical efficiency of 94% at design pressure. Complicated, multi-seal stuffing boxes increase friction and reduce the mechanical efficiency.

A style of reciprocating pump, without any rotating parts, is available for special applications, the direct acting pump. The process piston or plunger is connected to another piston, in a separate cylinder, which is driven by steam, air or liquid. Small plunger pumps driven by 7 barg compressed air are capable of producing pressures over 10000 barg. Process pumps, driven by steam, are available for difficult applications; very low NPSHa and high viscosity.

Wear in reciprocating pumps

The main wearing parts in reciprocating pumps are piston seals, stuffing box packing and valves. Wear in stuffing box packing produces an external leak which operators or maintenance staff can recognise and act upon. Wear in piston packing and valves produces internal leaks, increased slip, and can only be detected as a reduction in pump flow.

Reciprocating pumps can handle contaminated liquids with solids. The Miller Number should be evaluated. Piston pumps in standard materials are adequate for Miller Numbers up to 50. Above 50, severe wear problems may occur. Plunger pumps can handle the most abrasive liquid/solid combinations by running at reduced speeds, using ball valves and flushing the stuffing box to protect the packing.

Both piston and plunger pumps can be modified for solids handling by introducing a diaphragm to completely isolate the reciprocating driving element from the process liquid. Diaphragm pumps without pistons or plungers are also possible. Diaphragms can be driven directly by another fluid, liquid or gas, which has timed control valves to regulate the motion.

Flow and pressure pulsation

For many displacement pumps the flow varies during the period of the working cycle. Extreme examples of flow variation occur in single and twin cylinder reciprocating pumps, see figure 3.103.

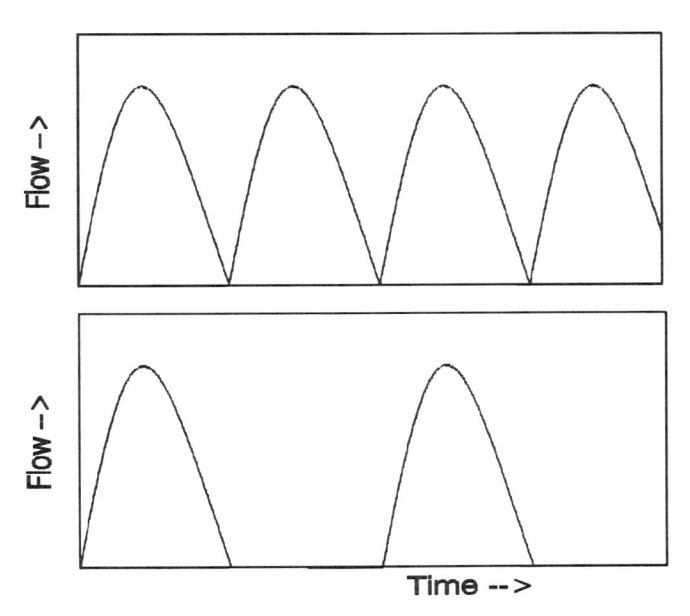

Figure 3.103 Flow variation of single and twin cylinder single acting reciprocating pumps

During a working cycle the instantaneous flow varies, to a greater or lesser extent, for most positive displacement pump types. In order to reduce variations in flow it is possible to increase the number of cylinders or elements or to provide pulsation dampeners at the suction and discharge connections. These can be filled with air/gas or equipped with air cushions or springs separated from the liquid by an elastic membrane. The size of the flow variation is an important feature, which unfortunately cannot be easily specified by the manufacturer. The flow variation, and the resulting pressure pulsations, are a function of the pump design, the liquid properties and the installation design. The same pump will have different characteristics in a different installation. Flow variations and pressure pulsations measured during factory testing will not be representative of values experienced at site, unless the pipework and liquid conditions are identical. Flow resistance in a test rig consists largely of throttling losses in control valves i.e. with a resistance varying as the square of the flow. Dampening in, for example, rubber hoses also has considerable influence. If flow variations and pressure pulsations are critically important for an installation, then a computer simulation, analogue or digital, must be performed of the complete system to verify instantaneous flow and pressure values together with sources of acoustic resonance.

IMPORTANT NOTE

The flow variations shown in figure 3.103 are based on an incompressible liquid, an ideal pump and no suction or discharge systems. These conditions cannot be reproduced in practice. Flow diagrams of this type are purely indicative and very approximate. No practical data can be derived from such diagrams and no useful calculations can be performed for flow variations, pressure pulsations or pulsation dampener sizing.

The minimum flow for single and twin cylinder pumps is zero so the total flow variation is 100%. Screw pumps have flow variations of a few percent. Pressure pulsations caused by flow variations are the primary cause of noise and vibration in positive displacement pumps. In the case of gear pumps for example, it is very easy for a small volume of liquid to become trapped in a pocket at the moment of gear contact, the volume of which changes during the course of rotation. This gives rise to harsh running and high noise levels. The problem can be considerably reduced by introducing a slight helix angle on the gear; by providing relief grooves in the side plates or by increasing the side clearance.

3.4.3 Displacement pump curves

Figure 3.104 shows a positive displacement pump curve drawn in the same manner as a rotodynamic pump curve. Little detailed information can be gained from the curve and two curves must be drawn separately to show the power relationship.

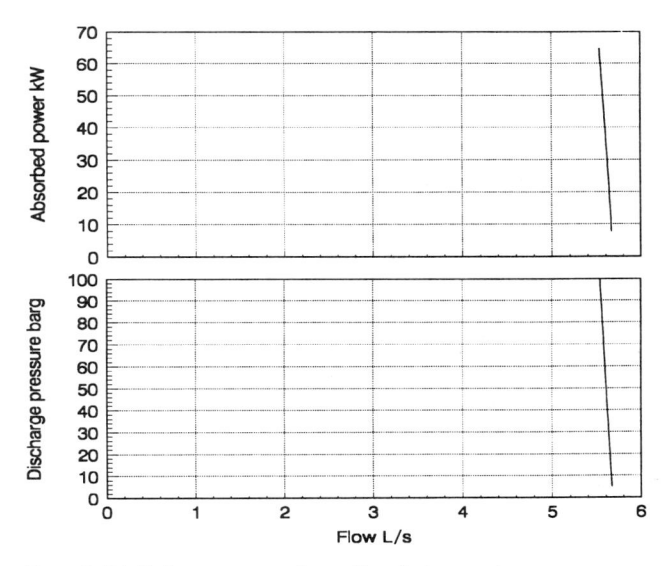

Figure 3.104 Performance curve for positive displacement pump

If curves are drawn it is better to plot the discharge pressure along the abscissa. This then conforms to mathematical standards; the abscissa being the independant variable. Flow, power and mechanical efficiency can be plotted on the same curve, see figure 3.105. For sensible flow information to be retrieved an expanded scale must be used and zero flow is not necessarily shown. The characteristic curve of flow v pressure may not be linear. Compressibility and slip effects may be more pronounced at higher pressures. Curves for different speeds may not be the same shape. When various liquids are pumped, or operating conditions vary, additional curves may be required. For fixed speed pumps with a single duty point, tabulated data is preferred.

To make pump selection easier performance curves can be drawn differently.

Figures 3.107 and 3.108 show the curves for a lobe pump for two different viscosities.

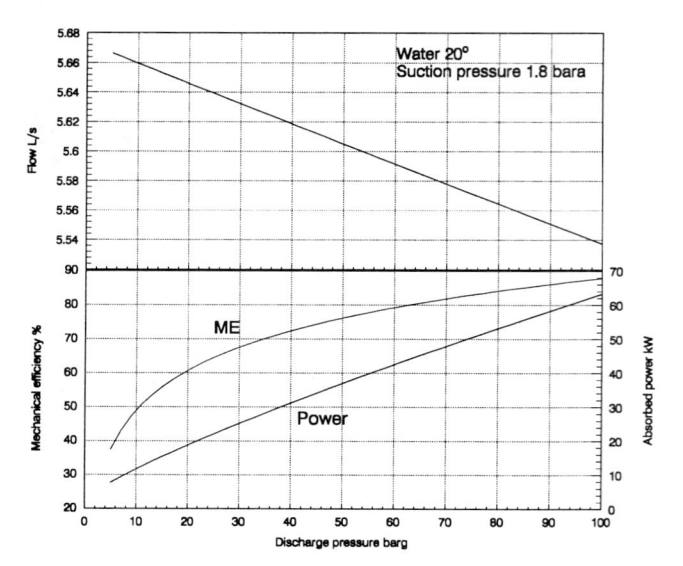

Figure 3.105 Performance curve for positive displacement pump

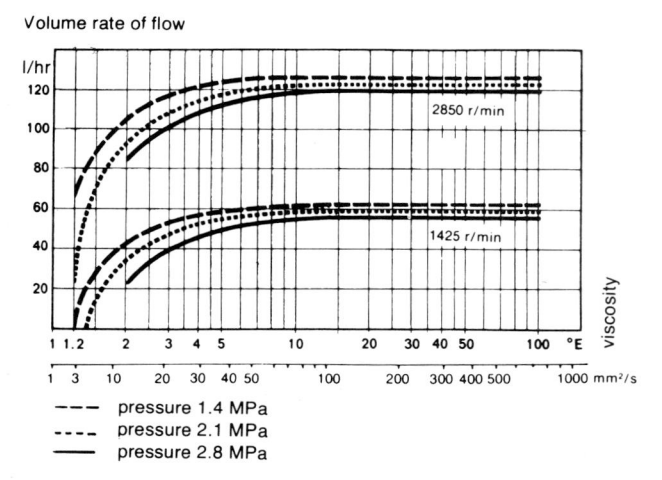

Figure 3.106 The effect of viscosity, speed and dp on a small gear pump

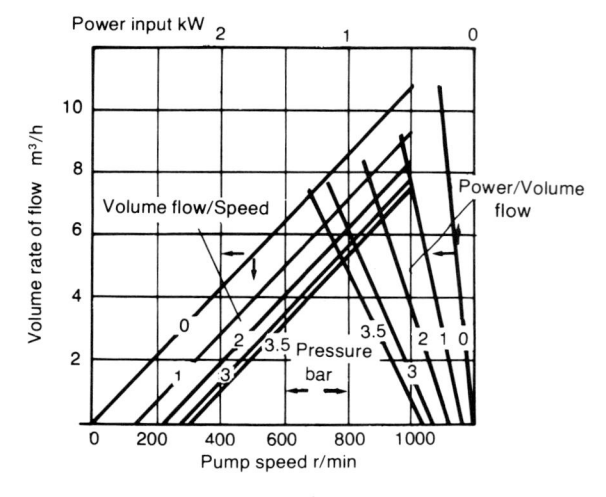

--- pressure 1.4 MPa
---- pressure 2.1 MPa
—— pressure 2.8 MPa

Figure 3.107 Performance curve for a lobe pump in water

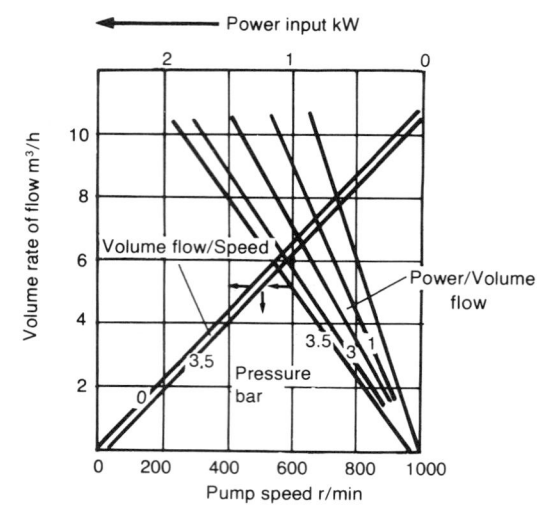

Figure 3.108 Performance curve for the lobe pump in fig 3.111 on liquid at 100 cSt

3.4.4 Classification of positive displacement pumps

There are many different types of positive displacement pump. The following section comprises a summary of various types and begins with a review of positive displacement pumps arranged basically in accordance with their mode of operation. Each type of pump has been designated a section number in the tables of contents, product review tables etc. Reference is therefore made to this section number in the following; for example **220** for "External gear pumps".

NOTE: Reciprocating pumps for hydraulic power applications usually have no seals and rely on the pumped liquid for all lubrication. Reciprocating pumps for process applications do not rely on the pumped liquid for any lubrication. Process pumps can be used for hydraulic power applications.

3.4.5 Types of positive displacement pumps

Rotary positive displacement pumps are generally self priming i.e. they can begin pumping with a "dry" suction pipeline and pump casing. Reciprocating positive displacement pumps are self priming, in theory, although they tend to be poor compressors. In general, reciprocating pumps should be vented and primed prior to operation unless the manufacturer specifically states that self priming is allowable.

All positive displacement pumps can pump a wide variety of liquids of varying characteristics, from very low viscosity liquids to very viscous pastes. The construction, application and pumping characteristics of different types of displacement pumps varies considerably however. It is therefore necessary to thoroughly understand the specific characteristics of each type of pump in order to be able to determine the suitability of a particular pump for a particular application. Gear pumps are not suitable for abrasive liquids for example. A progressing cavity pump with its flexible stator is more suitable. If the discharge pressure is relatively high a vane pump with flexible vanes is unsuitable, a gear pump or piston pump being a better choice.

The specified volume flow per working cycle is relatively constant and reduces insignificantly for moderate pressure increases. Liquids with high compressibilities, liquid gases, should always be evaluated for density changes. Control of the flow should be carried out by increasing or decreasing the pump speed. The flow can also be controlled by arranging a by-pass between the discharge and suction source. Do not, however, attempt to throttle the discharge in order to reduce the flow, as in the case of centrifugal pumps. This results in

As mentioned earlier, viscosity and speed can play an important role in the performance of rotary pumps such as gear pumps. Figure 3.106 indicates the speed, viscosity and differential pressure effects on a gear pump.

increased pressure, increased power consumption and increased loads on the working components, without significantly reducing the flow. Serious damage and personnel injury can occur if the pump discharge should become completely blocked, by the closing of a valve for example. The pressure will increase rapidly and can become so great that it causes failure of pipelines and other vital components. A dedicated pressure relief device should always be fitted as a safety precaution. The opening pressure of which should be selected so as to provide a good margin of safety for the protection of the system and its components.

A certain amount of internal leakage (slip) is always present in displacement pumps, this is due to the clearance between the working elements and inefficiencies of internal valves. The clearance can be of the order of 0.01 to 0.9 mm and is necessary in order to counteract galling and to reduce friction within the pump. Leakage occurs from the discharge side back to the suction side, i.e. against the direction of flow. The amount of slip, which reduces volumetric efficiency, varies greatly according to the type of pump construction. A progressing cavity pump with its long rotor and effective sealing rubber stator has less slip than a conventional gear pump having only line contact between the gears. The pumped liquid, if it is sufficiently viscous, can help to "seal" the clearances and reduce internal leakage.

Positive displacement pumps should not be installed so as to utilise their maximum suction lift capacity. Pump wear is unavoidable and this means that in time, the maximum suction performance will be reduced. This means that the pump loses its priming ability with subsequent loss of its pumping function. Unnecessary service costs are thus incurred in order to constantly maintain the pump in an "as new" condition.

220 External gear pumps

Gear pumps generally have two gears, but designs are available with three. One gear being driven by the motor, the others

Rotary	Multi-rotor	220 External gear pump	straight gears / helical gears / double hilical gears
		225 Internal gear pump	with crescent / without crescent
		230 Screw pump	2 screw / 3 screw
		240 Lobe pump Rotary piston pump	
	Single-rotor	235 Progressing cavity pump	
		245 Vane pumps	vanes in casing / rigid vanes / flexible vanes
		250 Peristaltic hose pump	
		255 Rotary peristaltic pump	
Reciprocating	Single-acting	260 Piston pumps (hydraulic power)	radial / axial
		265 Portable plunger pumps	complete packages
		270 Plunger pumps	horizontal / vertical
		275 Piston pumps (process)	slider-crank
		280 Diaphragm pumps	fluid actuated / mechanically actuated
		285 Metering pumps	dosing pumps
	Double-acting	275 Piston pumps (process)	slider-crank

Figure 3.109 Positive displacement pumps - classifications and groupings

by meshing with the driven gear. The driven gear usually runs in plain bearings. The bearings and shaft journals are located in the pump casing and surrounded by the pumped liquid. These bearings are thus dependent upon the lubricating qualities of the pumped liquid. Gear pumps are not therefore suitable for handling non-lubricating liquids. For reasonably acceptable pump life, gear pumps should not be used for such "dry" liquids as water or petrol. Paraffin and diesel oil are examples of liquids whose lubricating properties, whilst not being good, are perfectly acceptable for gear pumps.

The simplest gear pump has two external gears, figure 3.110. This type is used for relatively free-flowing liquids and can produce quite high pressures. Pumps for process applications start at 7 bar, hydraulic power applications have a standard rating of 170 bar but pumps are available for 270 bar. Gear pumps are self-priming with a suction capacity of 4 to 8 metres. Pumps are built in cast iron, steel, stainless steel, titanium, Hastelloy and non-metallic materials.

Figure 3.110 Gear pump with external gears

Simple gear pumps are used for hydraulic power application, where the motor can be a gear pump running in reverse. Figure 3.111 shows a low power application for driving a radiator fan on an engine. The thermostatic control valve provides speed control for energy saving and faster engine warm up. This type of drive is useful in hazardous areas where electrical equipment is bulky and costly.

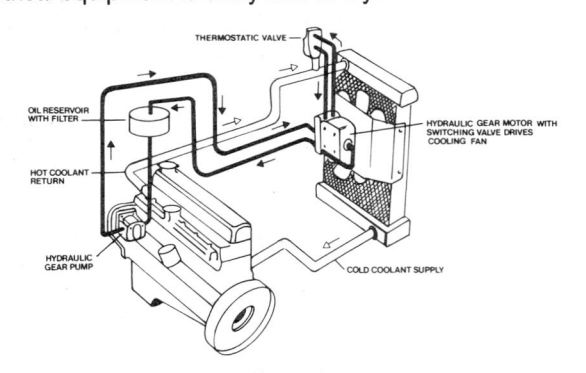

Figure 3.111 External gear pump - hydrostatic fan cooling drive

(Photograph and diagram supplied by Sauer-Sundstrand Ltd)

225 Internal gear pumps

A far greater range of application is possible with gear pumps having a small rotor mounted eccentrically within a larger external gear. A crescent shaped partition separates the two gears, figure 3.112. When the rotor is driven, the gear also rotates. The difference between the diameters of the rotor and gear and the eccentric location means that the gear teeth engage only at one point. The tooth of the gear successively rotates out of the gear pocket in the rotor during the first half of rotation. This induces a vacuum and the gear pocket is filled with liquid from the suction line. The liquid is forced out into the discharge line by the gear teeth during the second half of rotation. The number of teeth is kept to a minimum in order to make maximum use of material to obtain deep pockets between the teeth and achieve large displacement. By using sophisticated tooth profiles the number of teeth for external gear pumps can be as low as 10 to 15. Internal gear pumps can utilise the advantages of more favourable gear tooth engagement and therefore have even less teeth.

The pressure of fluid causes side forces on the gear which are taken up by the bearings. The axial force is usually relatively small, although the variations in side clearance and resulting pressure variations can cause wear. Wear which causes increased gear clearance does not affect the internal leakage as much as wear at the teeth tips and sides.

It is important to know the temperature range within which a gear pump is going to operate. Working components must take up greater clearances than normal if the pump is to operate with liquids at high temperatures. Some manufacturers supply pumps suitable for operating at temperatures of up to 300 °C.

Gear pumps should be used with care for liquids containing solid particles and for abrasive liquids. Pumps are available for hard solids up to 15 μm and soft solids up to 100 μm. Slow speeds are recommended. Wear can be reduced, when handling abrasive liquids, by selecting a pump having a somewhat larger capacity enabling it to operate at lower speeds. Gear pumps are not suitable for use within the foodstuffs industry where compliance with hygiene regulations is mandatory. They should not be used for handling shear-sensitive liquids.

The respective uses and applications for "external" and "internal" gear pumps varies somewhat. Greater pressure increases and increased flow is obtained with external gears, whereas internal gears are more suitable for high viscosity liquids and have better suction capabilities. To obtain a good volumetric efficiency it is always necessary to adjust pump speed or suction pressure to suit the viscosity of the liquid. A high viscosity requires a lower speed or higher suction pressure.

Shaft seals for pumps handling lube oil can be as simple as a 'U' ring. Stuffing boxes or mechanical seals are available to cope with many liquids. To increase pump life when handling non-lubricating liquids, special precautions such as hardfaced shaft journals, self-lubricating plain bearings and special gear tooth surface treatment may be necessary.

There are a great number of models suitable for small flows designed for laboratory use, whilst larger industrial pumps have maximum flows over 3000 m^3/h.

External gear pumps offer an excellent middle of the road choice with regard to performance and cost. Gear pumps are extremely compact for the power they develop. Although unable to match the high pressure capability of piston pumps, gear pumps develop higher pressures than vane or lobe types.

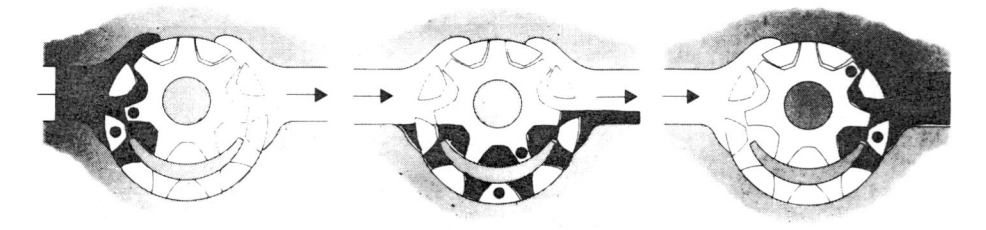

Figure 3.112 Internal gear pump showing the 3 working phases - filling, transfer and delivery

By the use of precision cut gears and close tolerance assembly, particularly between the tips of the teeth and the casing, volumetric efficiencies in excess of 90% are normal.

Gear pumps must be one of the most popular pump types, every car engine has at least one for lube oil. Some gearboxes and differentials have separate pumps for their lube systems. A major area of application for external gear pumps is on mobile plant and machinery. They provide a power source for a variety of lifting services, as well as acting as power assisted steering pumps. When assembled as a tandem unit the pump becomes a compact, economic solution serving a number of circuits. In the agricultural industry, tractors often depend upon engine driven external gear pumps for powering on-board services. These pumps are also popular for use with PTO shafts, with or without assistance from a gearbox.

A number of magnetically driven sealed gear pump designs are available. These are especially useful for hazardous viscous liquids or liquids which must not come into contact with air. A disadvantage with some designs is that bearing wear can be quickly followed by damage to the magnetic coupling with the result that the complete unit is unfit for further service. With other models, should wear reach the point at which the gears come into contact with the pump casing, the magnetic coupling, by far the most expensive component, will immediately spin free without damage.

230 Screw pumps with two or more screws

All screw pumps are grouped together in this section even though the operating principles of two screw pumps is different to three and five screw pumps. Two screw pumps consist of two identical profiles meshing together and sometimes with external timing gears. Without external timing, the idler screw is driven by liquid pressure between the profiles. The two screws do not touch and a minute flow path from discharge to suction always exists. The most common type of screw pump has three rotating screws of two different profiles, figure 3.113. The central screw is the driver whilst the basic function of the two idler screws is to provide sealing by metal to metal contact. The screw helixes form a number of sealing elements with the pump casing which transports the liquid axially when the screw rotates. The liquid flow is smooth without disturbing flow variations. The screw pump is quiet running and can be run at speeds up to 4500 rpm. By constructing pumps with longer screws more sealed elements can be obtained resulting in higher discharge pressures.

Figure 3.113 Screw pump with three screws

Screw pumps rely on viscosity for sealing, and in triple screw pumps for driving the idle screws. Figure 3.114 shows the effect of viscosity on flow and power at constant speed and differential pressure.

Axial thrust is taken up partly by hydraulic balancing and partly by thrust bearings. Large pumps are constructed with two opposed screws on each shaft. Liquid enters at the centre of the shaft and flows outward to the discharge. Equal and opposite flows creating a high degree of hydraulic balance. With regard to the construction of thrust bearings, screw pumps are normally designed for one direction of rotation and flow.

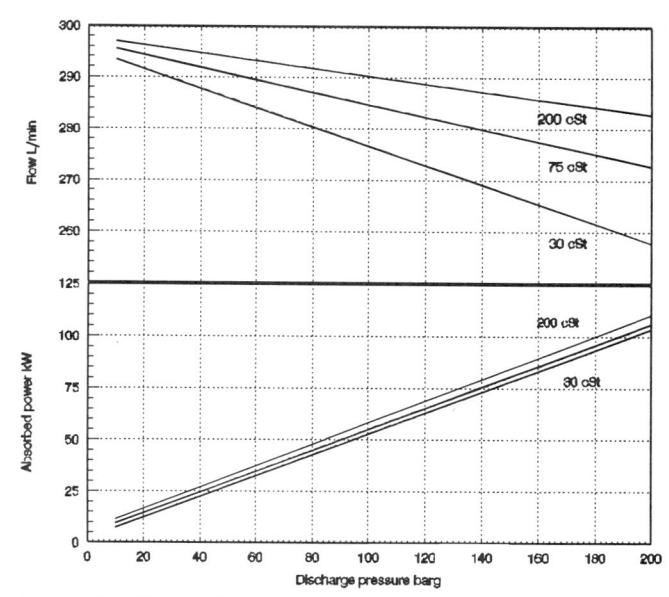

Figure 3.114 Viscous effects on triple screw pumps

Screw pumps with two screws and an external synchronising gear can handle liquids containing small quantities of solid particles, 2 mm may be possible. Pump users have been looking for pumps which can handle mixtures of liquids and gases, two phase flow. Crude oil production, from some wells, varies considerably with water and CO_2/H_2S content. Centrifugal pumps can only accept approximately 15% gas percentage by volume before losing prime. Twin screw pumps have been developed to handle continuously variable mixtures from 100% liquid to 97% gas.

Triple screw pumps are used primarily for oil, and other lubricating liquids like ethylene glycol, which should be free from contaminants. They are self-priming but must not be run dry as this can cause damage to the bearings.

Screw pumps are made from grey cast iron, nodular cast iron and carbon steel. Special surface treatment of components can improve wear resistance. Shaft sealing is by means of stuffing boxes or mechanical seals.

Triple screw pumps are available for flows ranging from 300 l/h to over 3000 m³/h. Low pressure ranges cover 7 to 10 bar discharge pressure. High pressure pumps are available for 250 bar. Viscosity should not be too low and should be in excess of 1.5 cSt. The maximum permissible viscosity is about 5000 cSt; factory approval should be sort for viscous applications over 400 cSt on higher pressure applications. Screw pumps can be used for liquid temperatures ranging from -20 °C to +155 °C. Mechanical efficiency can be over 85%.

Twin synchronised screw pumps are available up to approximately 2200 m³/h at pressures of 240 bar. These pumps can handle water and there is no maximum limit on viscosity. Gas entrained in viscous liquid tends to reduce the effective viscosity. Slugs of 100% gas and entrained abrasive solids, such as sand, can be handled.

(The assistance of IMO AB and Multiphase Pumping Systems Ltd is gratefully acknowledged.)

235 Progressing cavity pumps

Progressing cavity pumps may seem to be a new pump type. This is the modern name for an old pump. In the past, these pumps have been called, 'eccentric screw pumps', 'eccentric gear pumps' and 'helical gear pumps'. The new name is better as it descibes the action of the pump. The principle of operation was first established in the 1930's by Rene Moineau.

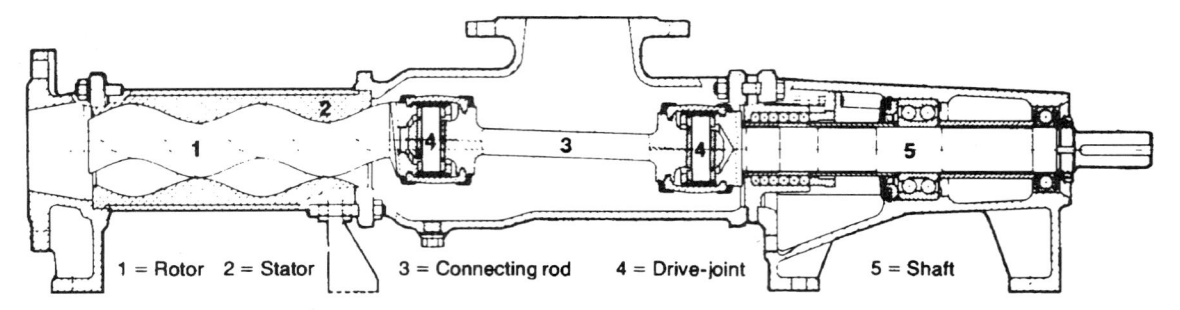

Figure 3.115 A progressing cavity pump

Progressing cavity pumps have only one rotor working within a rubber stator. The rotor, which looks something like an elongated corkscrew, rotates in a flexible stator having double internal helixes. The pitch of the stator helix is twice that of the rotor's. The difference in pitch forms sealed cavities between the rotor and the stator which, with rotation of the rotor, are caused to travel axially along the stator resulting in a smooth axial flow. Since the centre of the rotor describes a circle when rotating, drive is usually by means of a cardan shaft with standard universal joints (Hooke's joints) at both ends which lie in the fluid flow, figure 3.115.

Alternative methods of flexible power transmission are used by some manufacturers, see figure 3.116. A modified version of Hooke's joint with only one pin driven by a slot is available; this design may be prone to rapid wear. A better alternative is the crowned gear driving a toothed annulus; the load is distributed over many tooth flanks and angular motion is guided by a spherical bearing. All joints require good seals to prevent the process liquid attacking the bearings or allowing solids to be ingested between the bearing surfaces.

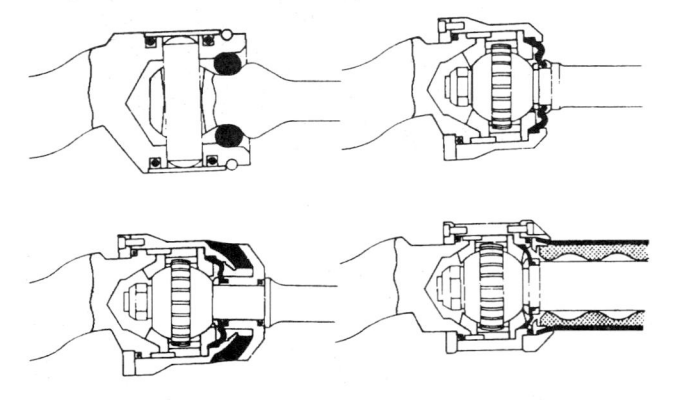

Figure 3.116 Variations of flexible couplings for progressing cavity pumps

Increasing the length of the rotor and stator makes possible greater differential pressures. Pressure differential for a low pressure pump is 5 bar. Differential pressures up to 25 bar can be obtained by increasing the length. Pumps are available for flows over 300 m³/h.

Progressing cavity pumps are used for practically all types of liquid from very fluid to very viscous. They can handle liquids containing abrasive contaminants and are relatively insensitive to solid particles. Larger pumps can allow the passage of random hard particles of 30 mm and soft particles over 100 mm. Special adaptations can be made to feed very viscous product in to the pump suction. Progressing cavity pumps have good suction capacity up to 8 m but their extreme sensitivity to "dry" running requires venting and priming prior to starting.

Many different materials are used in the manufacture of the pump casings and rotors ranging from cast iron to titanium. The stator can be made from a wide variety of elastomers, for example natural rubber, nitrile rubber, Neoprene and Viton®. In order to select the correct elastomer it is necessary to know the chemical properties of the liquid to be pumped, the nature of any solids and the temperature at which the pump is to operate. The choice of material is often a very difficult and complex matter due to the wide range of materials available. When in doubt, consult a manufacturer who has experience of the duty.

240 Lobe pumps and rotary piston pump

Lobe pumps have two rotors, which, unlike gear pumps and similar to synchronised twin screw pumps, operate without metallic contact with each other. Both rotors are driven by synchronising gears which are completely separated from the pump chamber, see figure 3.117. The pump shaft bearings are also situated in the gear case. The pumped liquid does not therefore come into contact with the bearings. Synchronising drive gears cause the rotors to rotate in opposite directions. The inlet liquid flow is divided into two halves, trapped in the space formed between the rotor and the pump casing and transferred, without change in volume, towards the outlet where the rotors meet, thereby reducing the cavity and forcing the liquid out, figure 3.118. Lobe pumps are not usually applied to pressures over 40 bar so there are no problems with volume changes due to compressibility.

The absence of metallic contact between the surfaces of the rotors themselves or between the pump casing and the rotors means that wear of the rotating parts is insignificant. The only wear which occurs is due to erosion of the pumped liquid or

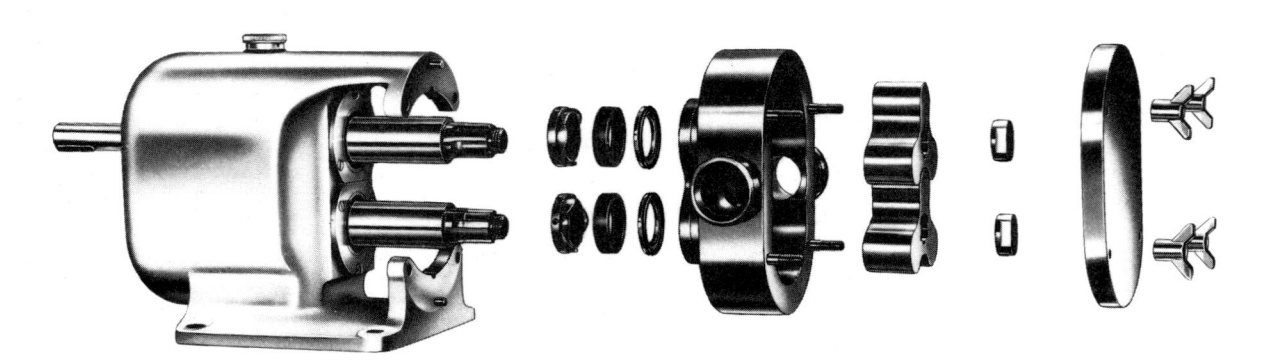

Figure 3.117 Lobe pump

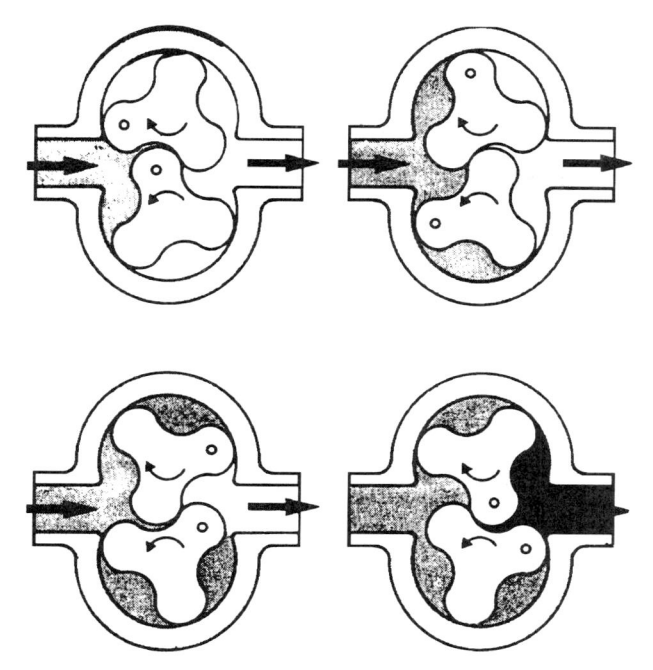

Figure 3.118 Lobe pump working principles

entrained solids. Since the shaft bearings are normally mounted outside the pump casing, the shaft ends are relatively long and unsupported which imposes limits in terms of working pressure. For pressures in excess of 12 bar some manufacturers fit plain bearings inside the pump casing. Others mount an extra bearing bracket on the outboard side of the pump casing. The latter case giving rise to four shaft seals.

The profile of the rotors varies from manufacturer to manufacturer. The most usual is shown in figure 3.119. The profile does not alter the principle of operation. It can be said however that rotors with one or two lobes give rise to greater pressure pulsations than rotors having three lobes. For gentle handling of fluids, rotors having one or two lobes should be chosen.

Rotary piston pumps are lobe pumps with greatly modified lobe profiles. The rotors are designed so that much more tip surface area is presented to the bore of the casing. This modification changes the pump viscosity characteristic and also the pumping action. The liquid is no longer squeezed out of the pockets between the lobes but is pushed directly out of the discharge. Figure 3.120 shows two typical rotor profiles.

Lobe and rotary piston pumps are suitable for both fluid and viscous products. Special feeding arrangements being necessary for extremely viscous liquids which cannot otherwise flow. Liquid temperatures of up to 200 °C can be handled if the clearances between the rotors and the covers are increased. By increasing the rotor side clearances it is possible to handle liquid at temperatures of down to minus 40 °C. To maintain constant temperature, pumps may be fitted with heating or cooling jackets.

Lobe pumps and rotary piston pumps handle the pumped liquid very gently. Examples of this phenomenon can be found within the foodstuffs industry where pumps are used for pumping cooked pea soup, preserves containing whole berries and other similar applications including handling whole fish. In these cases the pumps are specially constructed to fulfil the hygiene requirements; being easily dismantled and suitable for washing by hand or in accordance with the CIP method. Pumps can also be equipped for completely aseptic pumping for use within the pharmeceutical industry. They are also extensively employed in the chemical industry for handling both corrosive and non-corrosive products. Special dense-pulp pumps have been developed for the cellulose industry, these being suitable for pulp concentrations ranging from 6 to 30%, figure 3.121.

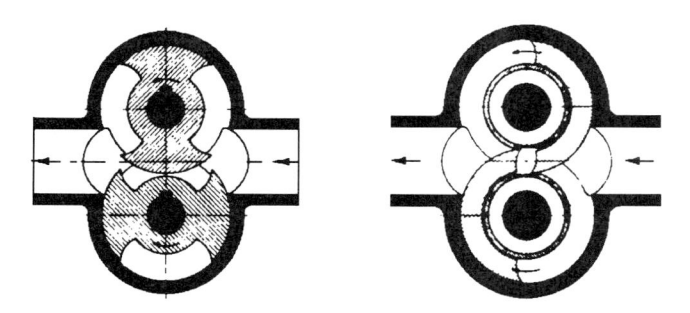

Figure 3.119 Lobe pumps - popular rotor profiles

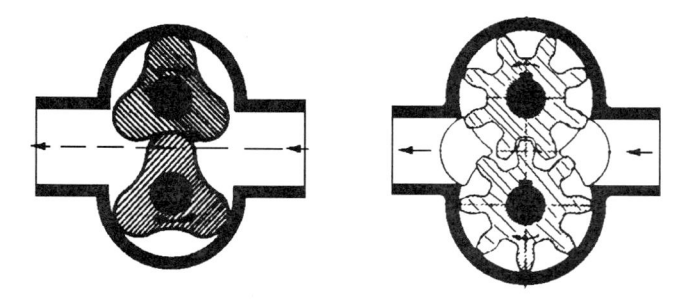

Figure 3.120 Rotary piston pump rotor profiles

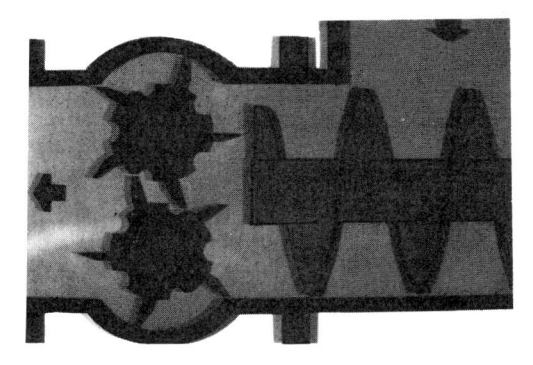

Figure 3.121 Dense-pulp pump with feed screw

Pumps usually operate at relatively low speeds and they are often used for high viscosity fluids. The pump is quiet running and delivers a largely pulsation-free flow. For most pumps the components in contact with the liquid are manufactured in stainless steel. Pumps for non-corrosive applications can have a cast iron casing with rotors and shafts made of steel. Sometimes the rotor material can be varied in order to increase the pump's suction capacity, by the use of nitrite rubber rotors, for example. Shaft seals consist of various forms of mechanical seals and stuffing boxes. Since the pump has two shafts which pass through the pump casing, two sets of seals are needed for each pump.

The rotary piston pump is a successful variant of the lobe design. Also called the circumferential piston pump, it uses arc shaped pistons or rotor wings, as shown in figure 3.120. The rotor wings are manufactured in a corrosion resistant, non-galling copper free nickel alloy and are designed to operate with very close clearances. This feature, combined with the geometry of the rotor wings, produces a long sealing path between inlet and outlet resulting in minimal slip when pumping low viscosity fluids. On viscous liquids and when handling products with solids in suspension, the large fluid cavities in the rotor wings combined with low operating speeds and carefully profiled anti-cavitation ports produces a smooth, pulsation free, low shear pumping action. Twin wing rotors are fitted for the majority of duties; single wing rotors are preferred for easily damaged products.

Lobe pumps and rotary piston pumps are available in a large variety of sizes from 0.1 to over 300 m³/h. Discharge pressures are normally up to 40 bar. Suction capacities vary between 1 and 5 m, depending upon internal clearances, pump size and speed. Solids handling capabilities are good, solids up to 100 mm on the largest pumps.

245 Rigid vane pumps and flexible vane pumps

In a rigid vane pump the rotor is eccentrically located in the pump casing and the vanes slide in and out to maintain contact with the casing wall. In a flexible vane pump the rotor is mounted eccentrically in the pump casing and the vanes flex to maintain contact with the casing wall. The pumping action is created by the variation in volume between the vanes, figure 3.122.

In both types of pump the liquid is drawn into the pump by an increasing volume between the vanes on the suction side, transported to the discharge side whilst trapped between the vanes and a proportion forced out by a decreasing volume.

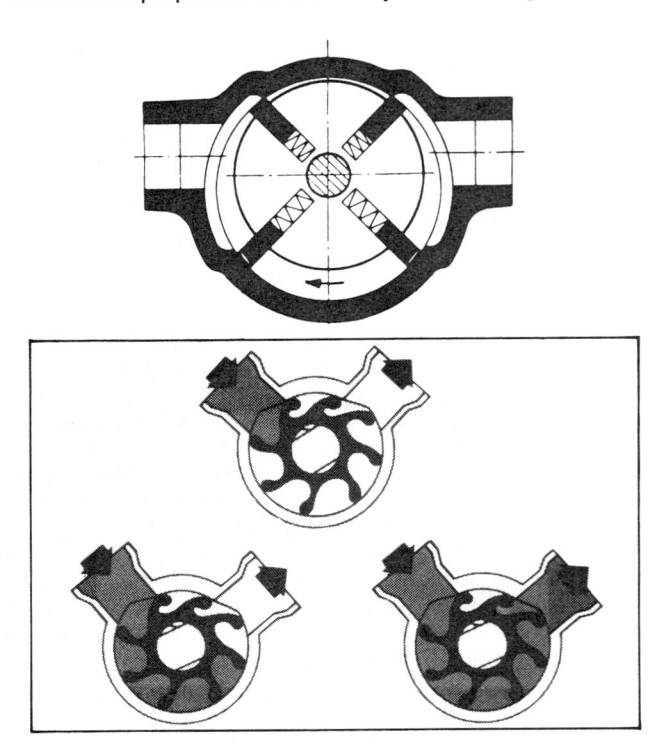

Figure 3.122 Operating principles for vane pumps

The vanes in sliding vane pumps may be controlled by springs, hydraulic pressure or rely entirely upon centrifugal forces induced by rotation. Some types utilise a rotating cam to guide the movement of the vanes.

Figure 3.123 Vane pump with sliding vanes in the rotor

There are many types of vane pump where the vanes are located in the pump casing (stator), figure 3.124.

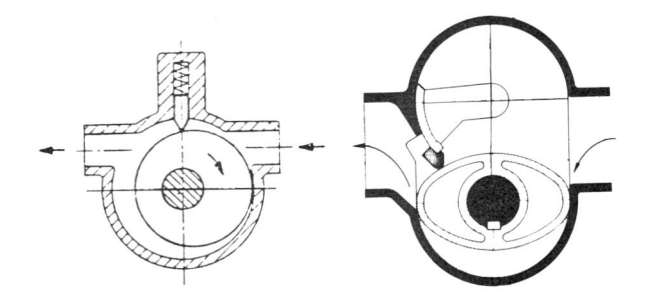

Figure 3.124 Vane pump with vanes in the stator

Another variant of vane pump is the sliding shoe pump where the "vanes" are U-shaped and operate against two separate surfaces, one against the rotating cam, whilst another surface slides against a valve plate, figure 3.125.

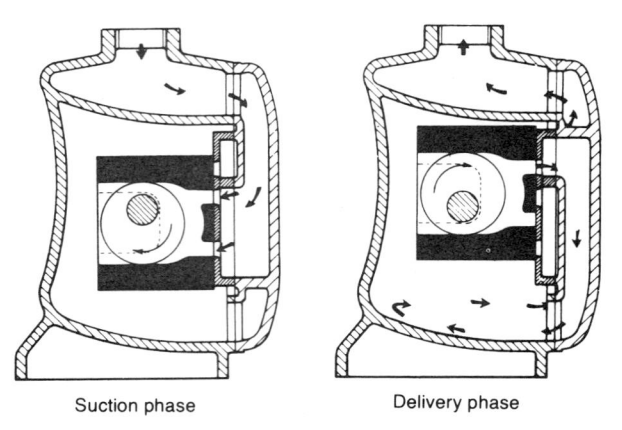

Figure 3.125 Sliding shoe pump

The pump casing and rotors for vane pumps with sliding vanes are usually made of cast iron, gunmetal or stainless steel, whilst bronze or glass fibre reinforced PTFE is used for the vanes.

Vane pumps with sliding vanes are suitable for most clean liquids and especially suited for those with entrained gases or those having a low latent heat, petrol or ammonia for example. Maintenance costs are relatively low due to the ease of replacement of the vanes, even when being used to pump liquids which are somewhat contaminated, waste oil for example. Vane pumps can operate within a large range of viscosities, although it is necessary to compensate the speed, i.e. the higher the viscosity the lower the speed. This phenomenon being common to most displacement pumps. Vane pumps are capable of suction lifts of 2 to 5 m. Flexible vane pumps are suitable for discharge pressures up to 5 bar, sliding vane process pumps of over 10 bar. Sliding vane pumps for hydraulic power applications can run at over 3000 rpm at pressures to 200 bar.

In a flexible vane pump, the rotor (flexible impeller), made of a synthetic elastomer, creates a seal between suction and discharge side. The pumps are therefore dry-self priming with a suction lift of 4 to 5 m, i.e. do not require to be vented and primed. With liquid in the pump it is capable of a vacuum of up to 500 to 600 mm Hg.

As the rotor material depends on the liquid pumped for lubrication, it is generally recommended that the pump should not operate without liquid for more than 30 seconds. This is more than sufficient time for self priming with properly designed pipework.

Flow capacities for flexible vane pumps range from 0.1 to over 30 m^3/h, sliding vane pumps are avaialable for over 300 m^3/h.

The single rotor, multi-vane principle, provides largely pulsation-free flow and enables the pump to handle products susceptible to damage.

Chemical resistance information, compiled by manufacturers of flexible vane pumps, provides a guide to the compatibility of various elastomers with liquids commonly used in industry.

Broadly speaking the temperature range of some of these rotor materials is as follows:

Neoprene	4 to 80 °C
Nitrile rubber	10 to 90 °C
Viton®	4 to 90 °C

However, as with any type of pump when pumping liquids which are not cold, great care must be taken to observe suction limitations and provide sufficient NPIP.

250 Peristaitic hose pumps

A peristaltic pump transports liquid by mechanically squeezing a space enclosed by a flexible element. The most usual peristaltic pump, flexible hose pump, operates by means of rollers or cams acting directly upon the hose containing the liquid, figure 3.126. The number of rollers or cams vary depending upon the manufacturer, usually 2 or 3. A pump may be fitted with one or more hoses, sometimes as many as thirty.

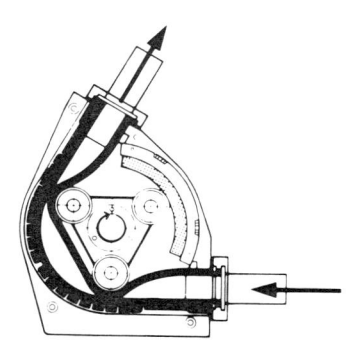

Figure 3.126 Peristaltic hose pump

At one time the use of peristaltic pumps was confined largely to laboratory and similar specialised applications but developments in the field of rubber technology have resulted in a new generation of heavy duty pumps and peristaltic units will now be found on a wide range of industrial duties involving viscous and/or abrasive sludges and slurries as well as products with high solids content.

An important factor in favour of the peristaltic concept is the pump's ability to handle solids up to the full bore of the hose tube. With most pump designs, solids handling capacity is restricted to approximately 15% of the pump branch sizes.

In one modern heavy duty design the roll/squeeze peristaltic movement takes place in a sealed housing partially filled with a mixture of glycerine and glycol. This not only lubricates the cams or sliding shoes from which the roll/squeeze motion is derived; it also acts as both coolant and lubricant to the exterior of the pump hose tube. This prolongs hose life and reduces incidence of mechanical failure. Hose life in excess of 3,000 hours is quite common.

The same design uses a hose tube with an inner core of soft natural rubber or NBR rubber with an outer covering of hard natural rubber reinforced with braided nylon. During the roll/squeeze phase of the peristaltic cycle, solids present in the pumped liquid are cushioned in the soft inner core of the tube with minimal damage or erosive effect and are released gently back into the liquid stream when the squeeze cycle is concluded. This enables delicate, shear sensitive solids to be pumped without damage and abrasive material to be handled with little or no effect on the hose interior.

The peristaltic design is inherently sealless with the pumped liquid completely enclosed with no possibility of leakage except in the case of a hose failure. Failure from a cause other than mechanical stress is rare and, under stable operating

Viton® is the registered trade name of E.I. DuPont de Nemours & Company.

conditions, can often be predicted with considerable accuracy. Regular hose inspection is of course advisable.

Peristaltic pumps can be used for all types of liquids and to some extent even gases. The suction capacity varies greatly for different constructions. Discharge pressures over 9 bar are normal. The capacities vary from very small metering quantities of 0.001 l/h up to flows of 50 m³/h. Special pumps are available for over 3000 m³/h.

255 Rotary peristaltic pump

Another type of peristaltic pump, sometimes called an orbital lobe pump, works by means of an eccentric rotor operating within a flexible elastomeric element with the liquid trapped between the elastomeric element and the pump casing. The rotation of the eccentric rotor transfers the liquid from the inlet to the outlet, figure 3.127.

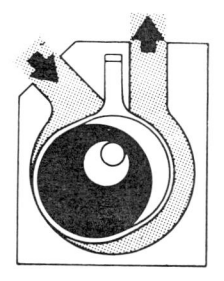

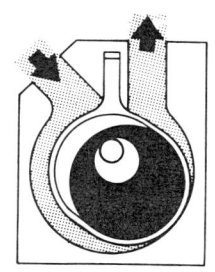

Figure 3.127 Rotary peristaltic pump

This pump is also sealless and leak-free, on condition that the flexible elastomeric element does not fail as a result of the mechanical stresses. The elastomeric element should therefore be replaced regularly to avoid unexpected failures. In order to select the correct elastomer and the correct pump casing material it is important to know the chemical properties and temperature of the liquid to be handied. Capacities range from 0.06 to over 50 m³/h. Suction capacity 1 to 3 m. The pump creates a strongly pulsating flow making it advisable to use flexible hoses in the suction and delivery lines. If fixed pipes are used some type of pulsation dampener should be fitted between the pump and the pipe. These pumps are suitable for pressures over 10 bar.

260 Piston pumps for hydraulic power applications

These pumps are characterized by their lack of external lubrication and not having any seals on the piston. All moving parts are lubricated by the product. Early systems all used mineral oil as the motive liquid and this may have been a standard industrial lubricating oil. Installations in hazardous areas, where the problems caused by any fire were great, look for replacement liquids to reduce fire risks. Mixtures such as water/glycerine and water/glycol were tried but equipment needed derating to cope with the poor lubrication properties. Soluble oils were developed which led eventually to 95/5 and 97/3 water/oil emulsions. Modern oils are biodegradable and can be dumped in normal sewers. However, they are also costly and leaks are kept to a minimum. Piston leakage is controlled by the clearance between the piston and the cylinder bore. Any leakage returns to suction. Piston pumps for hydraulic applications fall into two broad categories; axial piston and radial piston. Inline piston pumps, the forerunner of modern axial and radial pumps, are still used for a few applications.

Axial piston pumps consist of a number of cylinder bores in a rotating block. The pistons can be anchored in various ways to produce the stroke as the cylinder block rotates. Most pumps have pistons which are driven sinusoidally. The stroke can be variable. Variable flow can be produced from fixed speed pumps. Some pumps can adjust the stroke so that the pump runs in reverse; the suction becomes the discharge. Suction and discharge valve functions are performed by ports in the cylinder block support plate which connects cylinders to the suction or discharge at the correct part of the stroke.

Radial piston pumps have multiple cylinders spaced radially around a central cam shaft. Most pumps are fixed stroke. Valves can be poppet, like engine valves, or ball; both spring loaded and automatic.

Inline pumps are similar in construction to inline car engines. Several cylinders are mounted above a cam shaft. Pumps tend to be fixed capacity.

Pumps are available for pressures up to 670 bar for flows up to 8 l/s and operate at speeds up to 5000 rpm. Applications for these pumps are the hydraulic systems found in all parts of industry. The list below is not exhaustive:
- agricultural machinery,
- coal mining,
- crude oil production,
- earth moving equipment,
- machine tools,
- power transmission,
- variable speed drives,
- raw steel products,
- steel making.

The latest developments in these pumps are axial piston designs in ceramics for use on sea-water. Low cost off-shore developments tend to be seabed mount wellheads with pipelines linking to centralised facilities. Wellhead hydraulic requirements, serviced by sea-water would have no leakage or liquid shortage problems. Ongoing research is optimising proportions and material combinations to reduce wear and extend operational life.

265 Portable plunger pump packages

High pressure water is used for descaling and cleaning. The high pressure produced by the plunger pump is converted to kinetic energy in a nozzle. The high velocity water jet impacts on any surface converting the kinetic energy to a destructive force. Pump packages are produced in several configurations.

- Electric motor driven
- Engine driven
- Skid mounted
- Two or four wheel site trailer
- Two or four wheel road-going trailer with brakes and lights
- Vehicle mounted

Packages may be complete with a water tank or have suction hoses for connection to a water supply. Facilities may be incorporated for hose reels and lance storage. Units may be capable of supplying more than one lance simultaneously.

These packages are produced to standard designs by pump manufacturers and by specialist firms who purchase all the components, including the pump. Horizontal and vertical pumps are used for these applications. If a standard package is not available to fulfil a particular requirement it is possible to buy a purpose built unit.

The following equipment should be fitted, as a minimum, to all systems:

- suction strainer with Δp indicator,
- pressure control valve for each operator,
- relief valve,
- low suction pressure cut-out and indicator,
- driver high temperature cut-out and indicator,
- pump high temperature cut-out and indicator.

Pumps for high pressure descaling and cleaning usually run at 690 bar to 1050 bar. Portable electric motor driven packages are available from 7.5 kW. Skid mounted engine units are available to over 1000 kW.

These packages are used for many applications:
- concrete cleaning,
- drain cleaning,
- graffiti removal,
- heat exchanger tube cleaning,
- insulation removal,
- paint stripping,
- ship hull cleaning,
- vessel cleaning.

270 Plunger pumps - horizontal and vertical

Plunger pumps are available from single cylinder hand pumps, capable of pressures over 300 bar, to nine cylinder vertical pumps absorbing over 3000 kW. Pumps are built with 1, 3, 4, 5, 6, 7 or 9 cylinders. Horizontal pumps have 1, 3 or 4 cylinders. Vertical pumps have 3 or more cylinders. Increasing the number of cylinders increases the pump capacity if speed and stroke remain constant. Also increasing the number of cylinders tends to smooth the flow variations, reduces pressure pulsations and torque variations at the crankshaft. Pulsation dampeners can be fitted to both suction and discharge pipework to further attenuate pressure pulsations.

It must be stressed that flow variations and pressure pulsations are not just a function of the number of cylinders in the pump. Pump speed and the pipework system design have significant effects which cannot be ignored. Changes in pipework can eliminate the most serious problems. Acceleration head losses in the suction pipework are always considered when assessing NPIP available.

The horizontal pump shown in figure 3.128 has suction valve unloaders fitted and would be driven through an external gear box or vee-belt drive. Liquid ends can be built in different configurations for various applications. The monoblock liquid end shown is typical for standard pumps on non-corrosive applications. Other designs are available with separate manifolds and externally clamped valves.

Vertical pumps are used for the most ardous applications. More space is available for special liquid end designs, bigger valves and longer stuffing boxes. The vertical pump shown is a 'modern' pump designed in the late 1930's. The original vertical pump designs had the liquid end at ground level with the crankshaft above, see figure 3.131.

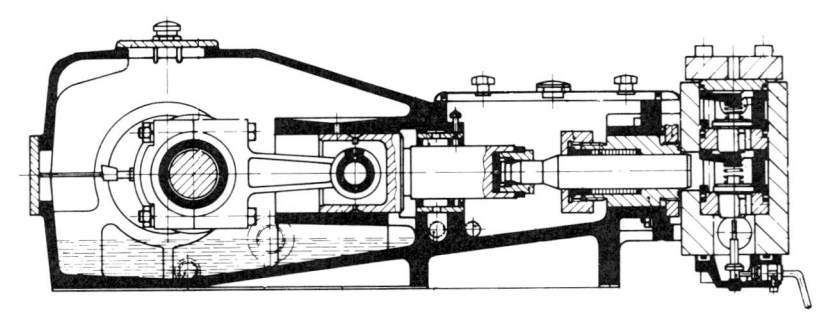

Figure 3.128 Horizontal plunger pump with mono-block liquid end and wing guided valves

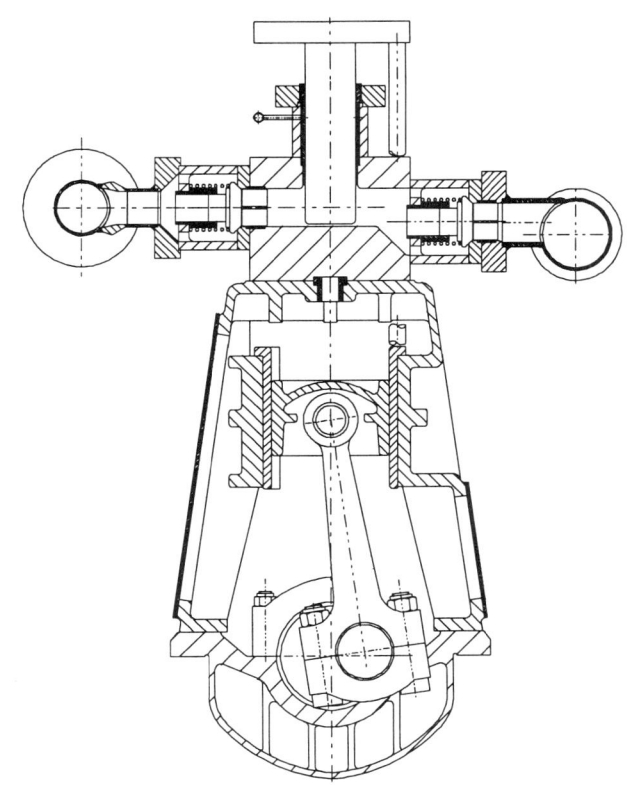

Figure 3.129 Vertical plunger pump

Plunger pumps can be fitted with a variety of valves to suit the application. Standard horizontal pumps, with monoblock liquid ends usually have plate valves, see figure 3.130.

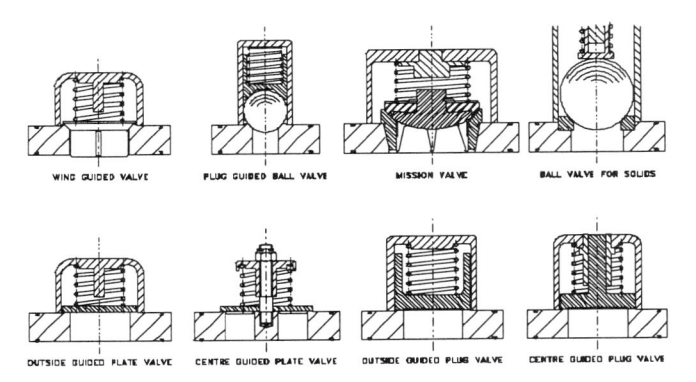

Figure 3.130 Valves for plunger pumps

- **Plate valves**
  General purpose valve for clean liquids with excellent dynamics, can have low NPSHr, suitable for speeds over 300 rpm. Mass produced by proprietary manufacturers as well as individual pump makers. Generally limited to 400 barg.

- **Plug valves**
  Heavy duty valve for clean liquids, medium NPSHr characteristics, usually operated below 350 rpm. Specially made by pump manufacturers for applications up to 550 barg.

- **Wing guided valves**
  Can be designed as light or heavy duty, has good dynamics, low NPSHr and suitable for all speeds. Available as a proprietary design and also as specials by pump makers. Suitable for pressures up to 550 barg, some designs can have an elastomeric seal for solids handling, 5% by volume approximately up to 500 μm.

- **Plug guided ball valve**
  Suitable for clean liquids at the highest pressures. Dynamics are poor and should operate below 300 rpm.

- **Mission valve**
  Proprietary valve manufactured by TRW Mission specifically for oil well mud pumps. Poor dynamics, high NPSHr and should operate below 200 rpm. Can handle solids, up to 1 mm in the larger valves, at concentrations up to 65% by weight. Designed for low abrasion solid mixtures, less than Miller Number 50. Available world-wide.

- **Ball valves for solids**
  Special valves made to order. Almost the ideal solids handling valve. Suitable for the most abrasive solids at concentrations of 65% by weight. Solids up to 6 mm in the larger valves. Dynamics can be poor, NPSHr can be very low, largest valves suitable for speeds below 150 rpm. Long parts life, but requires lots of space and is costly. Also very good for high viscosity liquids.

Figure 3.131 Low pressure vertical plunger pump for sewage applications

The choice of material for the valve and the valve seat is critical. The valves are opened by a liquid differential pressure and held open by drag forces. Closure is spring assisted. Allowances must be made in the valve design for shock loading due to rapid closure at higher speeds and pressure pulsation loading. Sometimes valves can be actuated hydraulically for special applications.

Modern pumps are designed to operate at mean plunger speeds of 1.5 to 2 m/s. Pumps may operate a lower speeds to achieve better NPIPr or to suit operating conditions. The pressure developed by any plunger is related to the plunger load; the force applied by the crankshaft/connecting rod/crosshead to the plunger. Current designs include plunger loads over 400 kN.

The liquid is prevented from escaping the liquid end by the stuffing box. Various seal designs, from simple to complex, can be fitted to suit the liquid and the environment. Pumps handling abrasive solids in suspension can have the front of the stuffing box flushed with clean liquid to protect the packing. Pumps with very low suction pressures can have a quench across the back of the packing to prevent air being induced. A quench can also be used to prevent hot liquid vapourising. Liquid can be circulated between seals to cool, lubricate and/or remove vapours. Single shot lubrication, where the lube oil mixes with the product and is lost, is the most popular stuffing box system.

Pumps are selected by adjusting all the variables; speed, stroke, plunger diameter; to suit the application. Knowledge of the liquid compressibility is essential for proper selection when the differential pressure exceeds 40 bar.

The plunger diameter is calculated from the plunger load. This diameter is rounded down to a standard seal diameter. The volumetric efficiency is calculated based on the pump proportions, the differential pressure and the liquid compressibility. The flow per revolution is then calculated based on the plunger diameter, stroke, volumetric efficiency and number of plungers. The pump speed can then be calculated for the rated flow. If the pump speed is too fast, more cylinders can be considered or a larger pump with a higher plunger load and consequently larger plunger diameter. For pumps to run at full speed adequate NPIP must be available.

Plunger pumps can operate with high suction pressure; over 135 bar in some applications. High suction pressure is defined as a suction pressure which produces a plunger load higher than a prescribed limit. Some pump designs use 10% as the benchmark, others 20%. High suction pressure must be considered in the selection process because it effects volumetric and mechanical efficiency. Mechanical efficiency is reduce because of increased friction losses during the suction sroke.

Plunger pumps should not be used in self priming applications unless the manufacturer has fully approved the operating conditions. Pumping air or vapour corresponds to 100% cavitation and will do no damage. The damage will occur during the transition from 100% cavitation to zero cavitation. Very high internal pressures may be created, far higher than the normal discharge pressure. This will lead to increased wear and component failure due to fatigue. The commonest failures are listed below.

- High bearing wear and failure of bearing shells or ball/rollers
- Crankshaft failure on pumps with two or three throws between two bearings
- Crosshead failure
- Very short packing life
- Liquid end stud/bolt fracture
- Suction valve failure, rapid seat wear
- Cylinder cracking

Plunger pumps are capable of working at the highest pressures, with the most difficult liquids and with abrasive solids if necessary. Pressures to 500 bar are common in process applications. Discharge pressures of 1000 bar are common in descaling and high pressure water jetting. Pumps for 3500 bar are available to order. Flows up to 500 m^3/h are possible at pressures up to 100 bar.

In addition, plunger pumps are used for low pressure, difficult applications where good suction capabilities are required and low speed is essential to reduce wear. Tank bottom residues, sediment, industrial and municipal wastes and sewage are typical mixtures handled by the original vertical pump design. See figure 3.131.

The crankcase assemblies are standard using mostly cast iron. Liquid ends, stuffing boxes and valves can be made in almost any material. Standard materials include cast iron, bronze, steel, stainless steel, duplex stainless steel. High pressure pumps require materials with high fatigue endurance stress levels, such as 1% CrMo, 15-5PH and 22Cr 13Ni 5Mo.

Plunger pumps applications are very diverse. The following list highlights some of the important areas;

boiler feed pumps,

carbon dioxide injection for crude oil recovery,

chemical processing,

crude oil/water emulsion recovery,

domestic high pressure cleaners,

ethylene glycol injection at wellheads,

garage car washes,

gas drying,

gas sweetening,

high pressure water jetting,

industrial waste,

LNG reinjection,

methanol injection at wellheads,

municipal waste,

oil/water emulsion hydraulics,

reverse osmosis,

sewage,

soap powder manufacture,

tank residue disposal,

viscous crude oil pipelines,

water injection for crude oil recovery.

275 Piston pumps

Piston pumps are similar to plunger pumps but have a moving seal on the piston. Piston pumps can be single acting or double acting. Double acting pumps have a stuffing box similar to a plunger pump stuffing box. The piston is driven by means of a rotating crank or eccentric, connecting rod and crosshead, see figure 3.132.

Figure 3.132 shows ball and centre guided plate valves as alternatives and has an internal single reduction gear set built into the crankcase. The pump has a motor mounting bracket, with tension adjustment, on top of the crankcase for belt or chain drives. The most popular valve for piston pumps is the Mission valve shown in figure 3.130.

Applications for piston pumps are wide ranging but two applications far exceed all the others. Duties on drilling rigs handling drill mud and crude oil production. Piston pumps are ideal for viscous liquids and handling solids which are not abrasive. Special pumps have been developed for pumping coal. Coal can be transported as a suspension in water or can be blended with oil to form a fuel for direct injection into boilers. Most applications are below 140 bar.

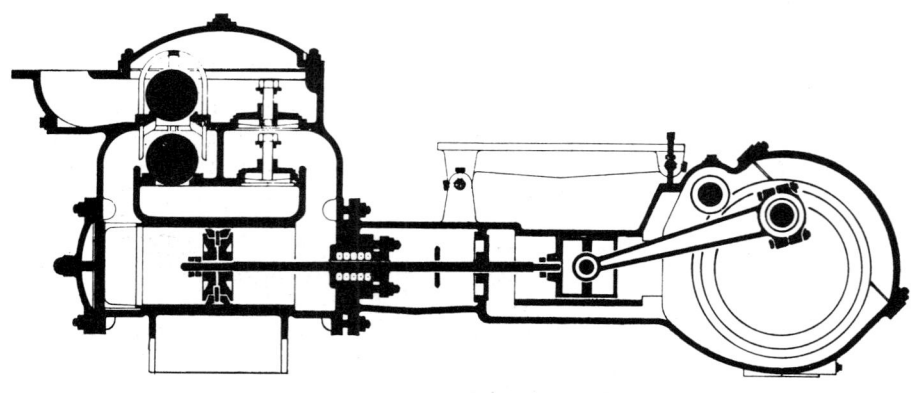

Figure 3.132 Crank driven piston pump

The comments in section **270** regarding self priming and cavitation apply equally to piston pumps. Piston pumps are manufactured mainly in cast iron, cast steel and abrasion resistant chrome iron.

280 Diaphragm pumps

Diaphragm pumps are reciprocating pumps where the working element is a flexible diaphragm. The drive can be by means of a motor driven crank arrangement or by means of compressed air acting directly upon the diaphragm, as in figure 3.133, or hydraulically instead of compressed air. Two valves, one on the discharge and one on the suction side, are necessary. Several types of valves are used, see figure 3.130. The most popular valves are flap, plate and ball. Hydraulically actuated diaphragm pumps for concentrated solids handling applications use Mission valves.

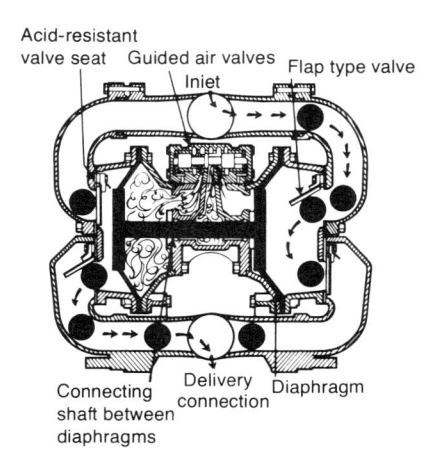

Figure 3.133 Compressed air driven diaphragm pump

Diaphragm pumps are leak-free since they have no plungers or piston rods which pass through the pressure boundary. Leakage can however occur if the diaphragm breaks. It is therefore necessary to replace the diaphragms in the course of preventative maintenance to avoid breakdowns. Some manufacturers can fit double diaphragms with rupture detection systems on the primary diaphragm.

A variation on the standard diaphragm principle is the tube diaphragm pump. The flat diaphragm is replaced by a flexible tube. The tube is compressed and distended by the action of surrounding compressed air or hydraulic liquid.

The material used for the diaphragm is usually an elastomer such as nitrile rubber, Neoprene or fluorocarbon. Stainless steel is also possible. The diaphragm material can limit the temperature of the liquid.

Diaphragm pumps have good suction capabilities varying from between 1 to 8 m depending upon the size of the pump, speed of stroke and type of valves. Only use as self priming pumps with the manufacturer's agreement. With the exception of the smaller pumps, diaphragm pumps can be used to handle practically any kind of liquid; fluid, viscous, clean or concentrated solids. The type of valve, however, often imposes limitations upon the maximum size of solid particle which can pass through the pump and the concentration. Flap valves allow the largest particles and a pump having 50 mm connections can, if fitted with flap valves, handle particles of 40 - 45 mm in diameter.

Diaphragm pumps always produce a pulsating flow. The pulsations can be reduced by fitting pulsation dampeners before and after the pump. Flow rates can vary between 0.4 and 300 m^3/h for compressed air powered pumps. Hydraulically actuated pumps are capable of 850 m^3/h at pressures up to 350 bar. Pump casings can be made of plastic, aluminium, cast iron or stainless steel.

285 Metering pumps

Metering pumps are also called 'Dosing Pumps' and 'Proportioning Pumps'. Metering pumps are used to add small quantities of liquid to other liquid streams or vessels. Figure 3.134 shows examples of this type of process requirement.

Pressure bar	Flow L/h	Temp °C	Liquid	Process	Problem
5	300	room	hydrochloric acid	water preparation	corrosion
5	1	room	mercury	chemical	high density
5	500	500	lead	metallurgical	high temperature
10	10	120	sodium	chemical	explosive
10	100	room	plutonium salt	nuclear fuel recovery	radioactive
10	20000	room	sugar solution	soft drink manufacture	hygenic
100	200	room	tungsten carbide suspensions	spray drying	abrasion
200	1500	-30	chlorine	chlorination	toxic
300	500	-200	oxygen	chemical	low temperature explosion cleanliness
500	1000	room	vinyl acetate		toxic high pressure
3000	30	room	catalyst	polythene manufacture	high pressure

Figure 3.134 Examples of processes in which metering pumps are used

Other typical applications include the injection of biocide into water and sea-water pipelines to prevent the growth of organisms. Injection of methanol or ethylene glycol into wellheads to prevent the formation of hydrates. The injection of chemicals into sea-water prior to reverse osmosis treatment.

The basic requirements of a metering pump are to deliver measured volumes of liquids accurately without significant changes in volume as discharge pressure varies. Typical accuracies would be ±1.5% of the rated flow. Positive displacement pumps are ideally qualified and specific variants of reciprocating pumps have evolved to fulfil the functions. Metering pumps are usually small plunger pumps, diaphragm pumps or peristaltic pumps for low pressure applications. Plunger pumps used as metering pumps are modified versions of the pumps described in **270**. The main modification involves providing a variable stroke mechanism so that the flow can be adjusted while the pump runs at constant speed. In theory, pumps can operate from zero to maximum flow. In practice, accuracy can only be maintained from 10% to 100% flow.

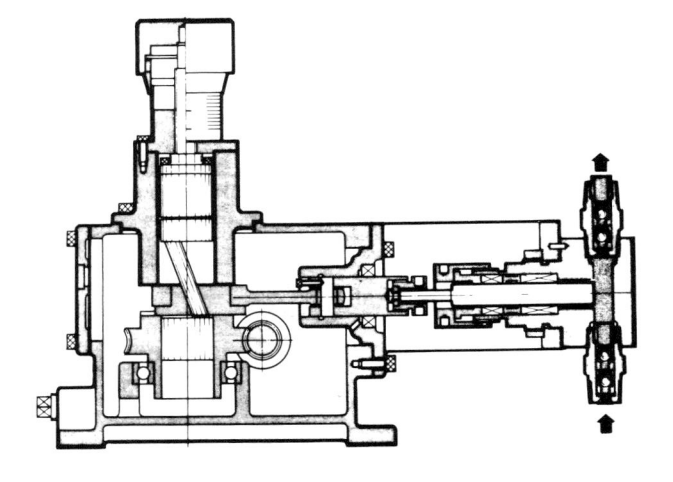

Figure 3.135 Plunger type metering pump with variable stroke

Figure 3.135 shows a plunger pump type metering pump fitted with double ball valves on suction and discharge.

Raising or lowering the "Z"-shaped crankshaft changes the length of the stroke. Metering pumps are single cylinder pumps. However, pumps can be ganged together on one drive shaft to form multiple units. This type of construction is useful when more than one liquid is metered and when several metering rates must be adjusted, in the same proportions, simultaneously. Stroke lengths and plunger diameters of each cylinder can be different. The stroke of each cylinder is adjusted individually to deliver the correct volume; the speed of all cylinders is varied simultaneously for collective flow adjustment.

Diaphragm metering pumps are built in two different designs. Small pumps can have the diaphragm driven directly by an electric solenoid. Stroke length is adjusted mechanically and stroke speed electrically. Larger diaphragm metering pumps have the diaphragm actuated hydraulically, the hydraulic power being provided by a plunger pump. Figure 3.136 indicates the construction of a plunger pump liquid end. Figure 3.137 shows the modifications necessary to add a diaphragm.

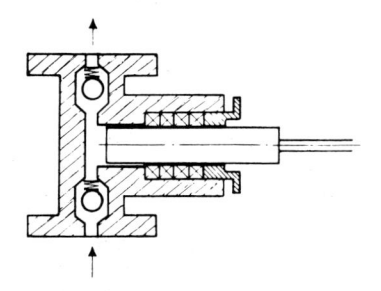

Figure 3.136 Plunger type metering pump

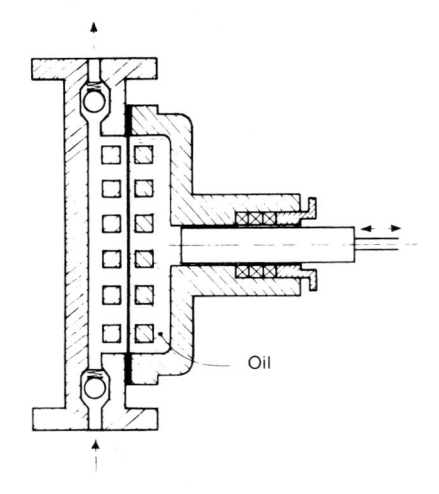

Figure 3.137 Hydraulic diaphragm type metering pump

The movement of the diaphragm being limited by supporting walls with slots on each side of the diaphragm or by profiled cylinder walls. Hydraulically actuated diaphragm pumps can become complicated. Provision must be made to maintain the oil volume behind the diaphragm. Leakage through the stuffing box must be replenished to preserve accuracy. Also overpressure of the hydraulic oil may pose problems.

Ball valves are the most popular choice for metering pumps. Ball valves have the longest possible life while preserving good sealing capabilities. Metering pumps always run slowly and ball dynamics are not a problem. Some designs incorporate soft seals within the valve seat to improve accuracy. Pumps are often fitted with double spring-loaded ball valves in both the suction and discharge in order to reduce the risk of backflow through the valve after each respective suction and delivery stroke.

Plunger style pumps are used for all pressure ranges when the liquid is not toxic and there is no risk of corrosion of the pump external components. Diaphragm pumps are used when any leakage from the pump is unacceptable. Because metering pumps run slowly there are no real problems with viscous liquids. The ability to handle solids in suspension is dependant upon the pump size and should be confirmed with the manufacturer. Metering pumps can cause pressure pulsations. Pulsation dampeners may be required on the suction to preserve the NPIPa. Dampeners may be required on the discharge to eliminate pipework vibration and smooth out the dosing rate.

Metering pumps are available in a wide range of materials including complete construction in PTFE. Flow rates can be as low as 0.1 l/h and over 10 m^3/h per head.

290 Jet pumps

Jet pumps or injectors operate by converting the potential energy of a fluid to kinetic energy of driving jet thus reducing the static pressure and moving directly the pumped liquid without the assistance of moving mechanical components. The motive medium can be gas, air for example, steam or a liquid, water for example. There are different combinations of motive-pumped media. The most usual are water steam - air, water steam - water, air - water, water - water. Jet pumps working with gases as the pumped medium are called ejectors.

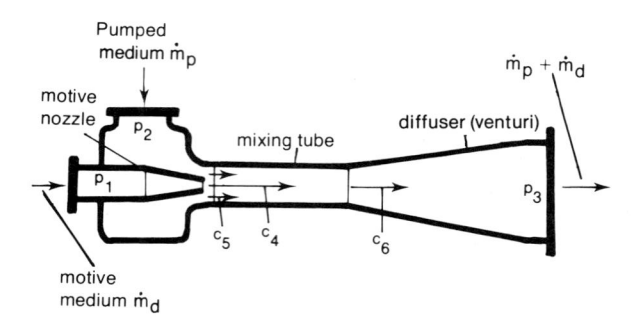

Figure 3.138 Diagrammatic principles of a jet pump

By applying the momentum equation, equation 2.7, to a control volume, which surrounds the medium in the mixing tube:

$$p_4 \cdot A - p_6 \cdot A - \pi \cdot D \cdot l \cdot \tau_0 = (\dot{m}_p + \dot{m}_d) \cdot c_6 -$$
$$- \dot{m}_d \cdot c_4 - \dot{m}_p \cdot c_5 \qquad \text{Equ. 3.37}$$

or

$$(p_6 - p_4) \cdot A = \dot{m}_d (c_4 - c_6) - \dot{m}_p (c_6 - c_5) - \pi \cdot D \cdot l \cdot \tau_0$$
$$\text{Equ. 3.38}$$

Certain characteristics of jet pumps can be learned by studying equation 3.38. In order that the pressure p_6, exiting the mixing tube, should be greater than p_4 then the velocity leaving the jet nozzle c_4 must be considerably higher than c_6 and c_5. Considerable mixing losses are therefore also unavoidable. Pressure increase is greatest when $\dot{m}_p = 0$. The shear stresses at the wall of the mixing tube have a tendency to reduce the pressure increase. A diffuser is mounted after the mixing tube where velocity is reduced and static pressure increases. The relationship between the pumped mass flow $\dot{m}_p$ and the motive fluid mass flow $\dot{m}_p$ is called the flow coefficient and is designated:

$$q = \frac{\dot{m}_p}{\dot{m}_d} = \frac{\rho_p \cdot Q_p}{\rho_d \cdot Q_d} \qquad \text{Equ. 3.39}$$

The pressure relationship for a jet pump can further be defined as:

$$z = \frac{\text{total pressure increase of the pumped media}}{\text{total pressure increase of the motive media}} =$$
$$= \frac{p_{03} - p_{02}}{p_{01} - p_{03}} = \frac{H_p}{H_d} \qquad \text{Equ. 3.40}$$

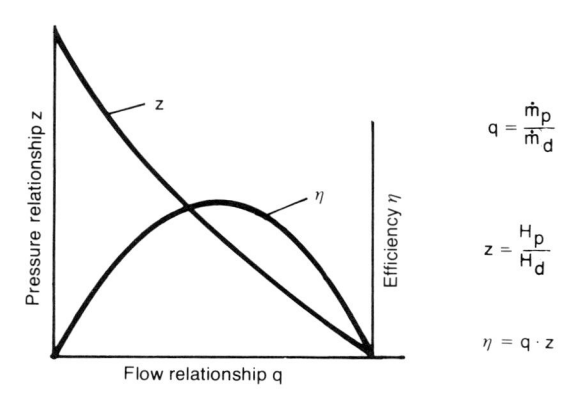

Figure 3.139 Performance curve for a jet pump

Using these designations, jet pump efficiency can be expressed as:

$$\eta = q \cdot z \qquad \text{Equ. 3.41}$$

Jet pump losses consist of flow losses in the nozzle, intake chamber, mixing tube and diffuser. The mixing losses constituting the greatest loss. Mixing losses are primarily dependent upon the area relationship:

$$a = \frac{\text{motive fluid nozzle outlet area}}{\text{mixing tube cross sectional area}} = \left(\frac{d}{D}\right)^2 \qquad \text{Equ. 3.42}$$

For each combination of pressure relationship "z" and flow relationship "q" there is an optimum area relationship "a", see figure 3.140. Other important design parameters are the distance between the mouth of the nozzle and the start of the mixing tube; the length of the mixing tube and the angle of the diffuser. The lowest pressure in a liquid jet pump occurs in the upstream section of the mixing tube. If the lowest pressure reaches the liquid's vapour pressure then cavitation will occur. The cavitation number is defined as:

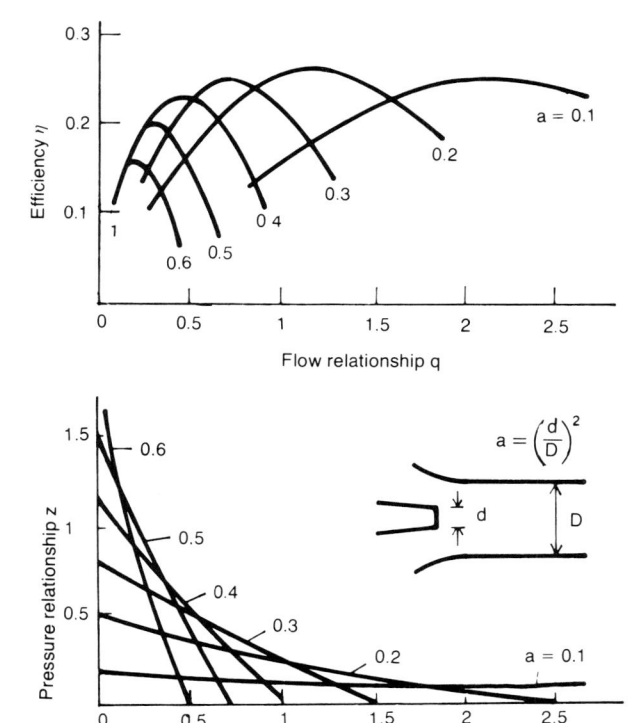

Figure 3.140 Examples of performance curves for an oil jet pump for various area relationships "a"

$$\sigma = \frac{p_{02} - p_v}{p_{01} - p_{02}} \qquad \begin{array}{l} p_0 = \text{total pressure} \\ p_v = \text{vapour pressure} \end{array} \qquad \text{Equ. 3.43}$$

If, for a given jet pump, p_{02} is reduced at the same time as p_{0I} and p_{03} are adjusted so that the pressure relationship "z" is

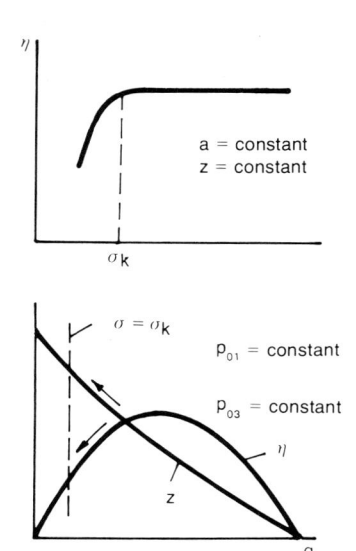

Figure 3.141 Cavitation in liquid jet pumps

maintained constant, the cavitation number σ will reduce without initially changing the value of q and η. Further reduction of p_{02} causes successively more cavitation in the mixing tube and a rapid reduction in efficiency. This value of cavitation number is designated σ_k.

Another way of illustrating the onset of cavitation in a liquid jet pump is that at constant motive pressure and back-pressure, p_{0l} and p_{03} respectively, p_{02} reduces. The pump's optimum operating point will then follow down the performance curve, "z" increases, "q" and σ reduce. For $\sigma > \sigma_k$ the pump operates without any effects of cavitation. At $\sigma = \sigma_k$ the performance is reduced in relation to the cavitation free operation. Prolonged operation where cavitation is present can cause damage to the mixing tube and the diffuser. Jet pumps have certain fundamental advantages:

- No moving parts
- No lubrication requirements
- No sealing problems
- Self-priming, can evacuate the suction line
- Non electrical, no temperature or sparking problems

Deep-well pumping

Sludge pumping

Evacuating suction piping

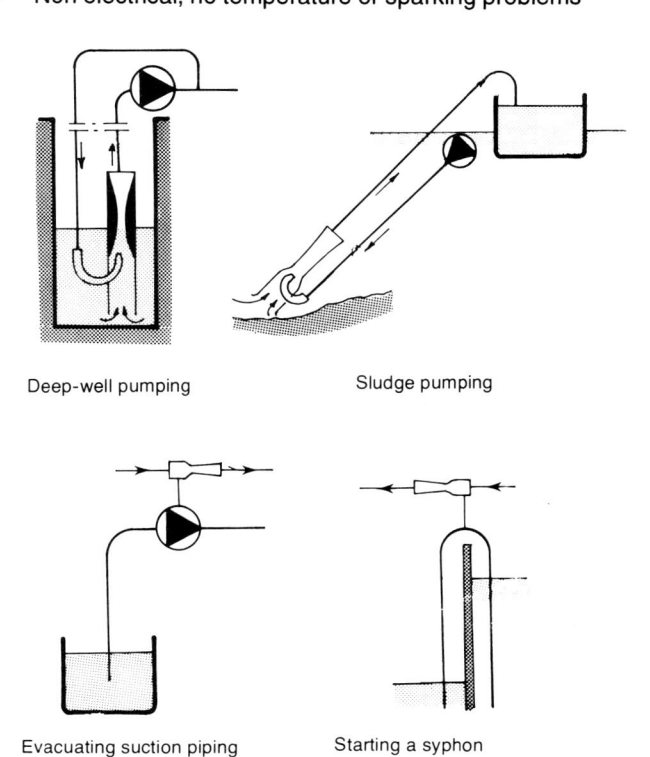

Starting a syphon

Figure 3.142 Liquid jet pumps - practical examples

The most obvious disadvantage is the low efficiency. Maximum efficiency being 25 to 30%. With small flows the power absorbed will be small therefore low efficiency may be of no consequence.

Steam jet pumps are normally used in order to achieve low pressure at the inlet side. For inlet pressures of down to 103 Pa absolute water steam is generally used as the motive medium. For even lower pressures, down to hundredths Pa, oil steam is used as the motive medium.

Practical examples: De-airing condensers, evacuating flammable gases, liquid transportation coupled with simultaneous heating requirements. Compressed air is often available and is the most usual motive medium for gas jet pumps. Some common practical examples of liquid jet pumps, usually using water as the motive medium, are illustrated in figure 3.142.

Especially advantageous use can be made of jet pumps if a requirement to mix the various combinations of motive and pumped fluid can simultaneously be fulfilled. Examples of this are steam jet pumps used in ventilation systems where simultaneous humidifying is required and liquid jet pumps for transporting liquids which require simultaneous dilution. A very interesting application is possible where the use of normal refrigerants is prohibited or a "green" policy is imposed. Water as a refrigerant with water as the motive fluid. Water temperatures down to 5 °C could be produced without difficulty. Leakage would be easily observed, without any hazards, and topping up would be simple.

295 Air-lift pumps

The air-lift pump consists of a riser tube immersed in the pumped liquid, a compressed-air line and a pressure chamber where air forces the liquid, via small holes into the riser tube, figure 3.143. The air/liquid mixture is lighter than the surrounding liquid and therefore rises up the tube. The flow of liquid up the riser tube increases as the flow of air is increased up to a certain maximum value, after which it begins to decrease again.

The air supply must be compressed to a pressure which is equivalent to the immersed depth plus the losses in the piping and inlet holes. The product of the work done during compression and the air mass flow, neglecting the efficiency of the compressor, is equal to the power input to the pump P_{input}. The pump efficiency

$$\eta = \frac{\dot{m}_{Liquid} \cdot H}{P_{input}} \qquad \text{Equ. 3.44}$$

is greatly dependent upon the immersed depth in relation to the delivery head, figure 3.144. The quantity of air required in kg per cubic metre of pumped liquid, designated L, is greatly affected by the immersed depth.

The advantages of the air-lift pump are:

- Simple construction, no moving parts
- No sealing problems
- Small risk of blockage
- Not sensitive to temperature.

The disadvantages of the air-lift pump are:

- Low efficiency
- Immersed depth requirements, S/H > I
- Compressed air can be expensive.

Air-lift pumps are used for pumping sludge, contaminated liquid, large particles, sugar beets, and also hot or corrosive fluids.

300 Barrel emptying pumps

Barrel emptying pumps are intended for emptying small containers and tanks. Consisting either of an externally mounted, hand operated semi-rotary hand pump or of a motor section, immersion tube and a pump section. The pump section is located in the lower end of the immersion tube and is driven by the motor via an extended shaft. The shaft is protected by a sealed column. The pumped liquid flows between the column and the extended shaft to the pump's outlet at the motor end. For obvious reasons, a barrel emptying pump should be light weight and easy to transport to the next barrel which is to be emptied.

Barrel pumps are manufactured with a variety of immersion tube lengths. They can be made from many materials having

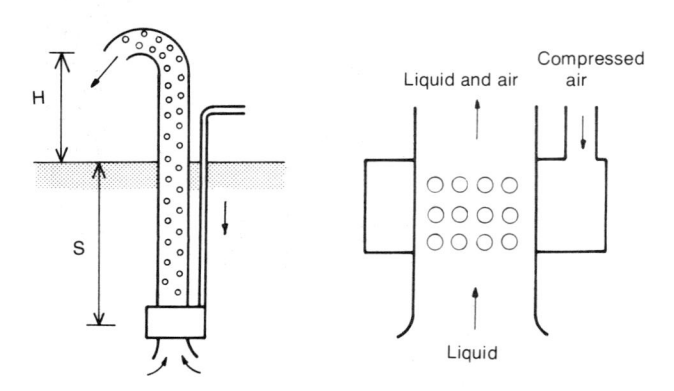

Figure 3.143 Air-lift pump

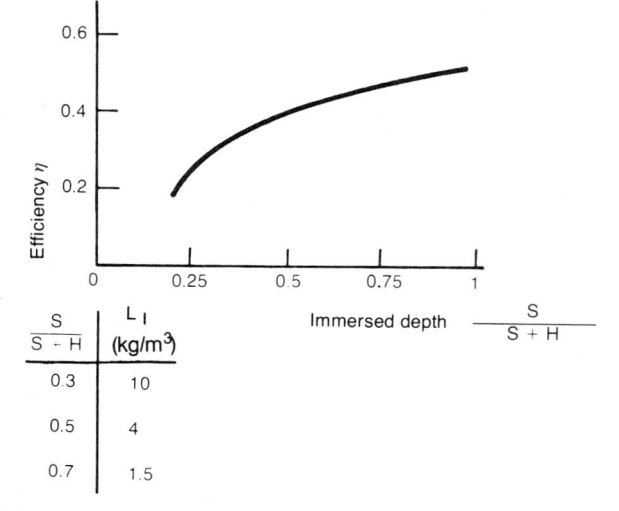

$\frac{S}{S+H}$	L (kg/m³)
0.3	10
0.5	4
0.7	1.5

Figure 3.144 Example of maximum efficiency and air requirement for air-lift pump - Pump efficiency neglects compressor efficiency

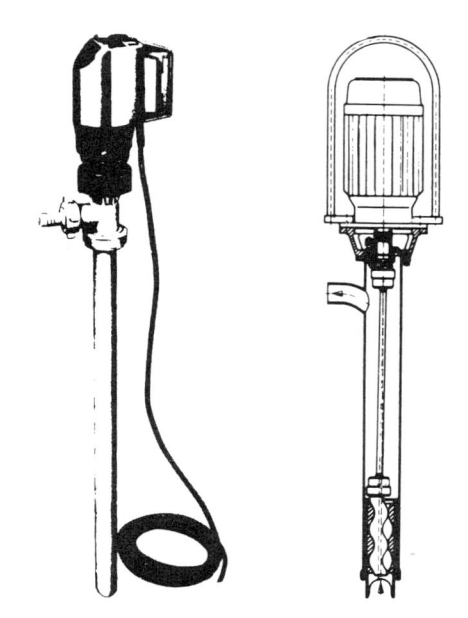

Figure 3.145 Barrel emptying pumps for fluid and viscous liquids

good resistance to chemical corrosion. Pumps can be fitted with low voltage, totally enclosed electric motors, motors for use in potentially hazardous atmospheres and air motors. The pump design for fluid liquids is of the centrifugal or mixed flow type whilst screw or progressing cavity pump types are used for viscous liquids.

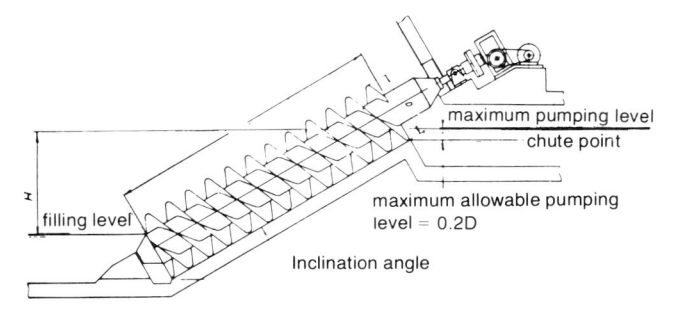

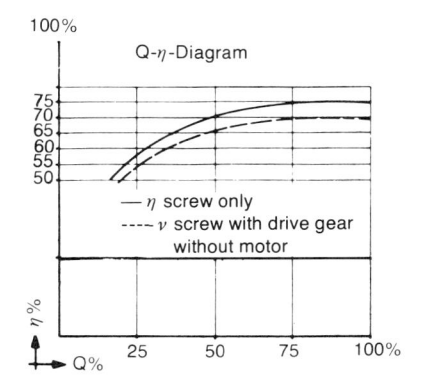

Figure 3.146 Rigid screw pump - construction and efficiency

305 Rigid screw pumps

The rigid screw pump is a development of the Archimedean screw, which is the oldest known type of rotating pump. The rotor is in the form of a thin helical screw. The thread form being made by plate wound around the shaft. The number of threads varying from one to three. The rotor usually rotates in an open channel having a circular cross-section, enclosing about three-quarters of the rotor periphery. The rigid screw shaft is usually inclined at an angle of 30° to the horizontal, the length of the screw being approximately twice the lift obtained. The lift heads are moderate, rarely in excess of 10 m. Increased nominal flow is obtained by increasing the rotor diameter and reducing speed. Speeds range between 20 and 80 rpm.

Delivery heads are virtually independent of flow up to a maximum flow, maximum capacity loss factor or filling coefficient. The efficiency curve is flat. As shown in figure 3.146. The efficiency is only reduced by about 5% from maximum flow to 50% flow.

There are special types of rigid screw pumps, like that shown in figure 3.147, where the "channel" is totally enclosed and rotates with the screw. Provided that the drive and shaft bearings are completely isolated from the liquid it is thus possible to transport a large variety of liquids and mixtures.

Rigid screw pumps are used for transporting liquids containing large solid contaminants, untreated sewage for example. Rigid screw pumps are ideal for low delivery heads and large flows. As a sewage pump the rigid screw pump is self-regulating according to the supply flow. When the level in the pump sinks then the capacity loss factor, filling coefficient, automatically reduces causing the pump flow to decrease.

The advantages of the rigid screw pump are as follows:

- High efficiency
- Simple robust construction
- No seals
- Can run dry
- Flat efficiency curve
- Self-regulating for varying output
- Not sensitive to contaminants

310 Pitot tube pumps

Within the rotating casing of a Pitot tube pump the pumped fluid is accelerated to a high tangential velocity. The velocity depends upon the shape of the blades in the walls of the casing and the effects of frictional forces. This velocity is captured by a stationary tangential Pitot tube which delivers a flow at increased pressure, figure 3.148.

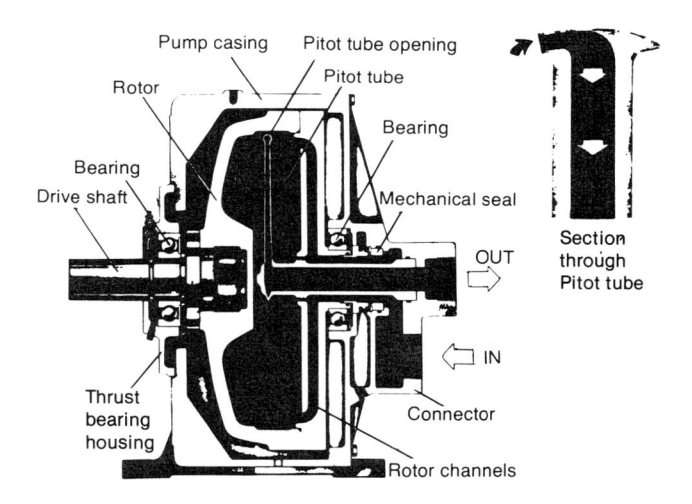

Figure 3.148 Pitot tube pump

Assuming that the tangential velocity of the liquid at the level of the Pitot tube is equal to the peripheral velocity of the casing, u, then the theoretical pressure rise is:

$$\Delta p_{theor} = \rho \frac{u^2}{2} + \rho \frac{u^2}{2} + \rho u^2 = (\frac{\pi}{60})^2 \cdot \rho \cdot (D \cdot n)^2$$

Equ. 3.45

The first part of the equation represents the pressure rise imparted to the liquid due to rotation, the second part of the equation represents the deceleration at the Pitot tube opening. Normally pressure is reduced as the result of flow losses and due to rotational slip between the liquid and casing. The design of the Pitot tube and blades being of critical importance with regard to losses.

For small flows and high pressures, low specific speed centrifugal pumps exhibit low efficiency due to the large impeller friction losses. The Pitot tube pump is more efficient in this respect. Maximum efficiencies of 50 to 60% are possible. The

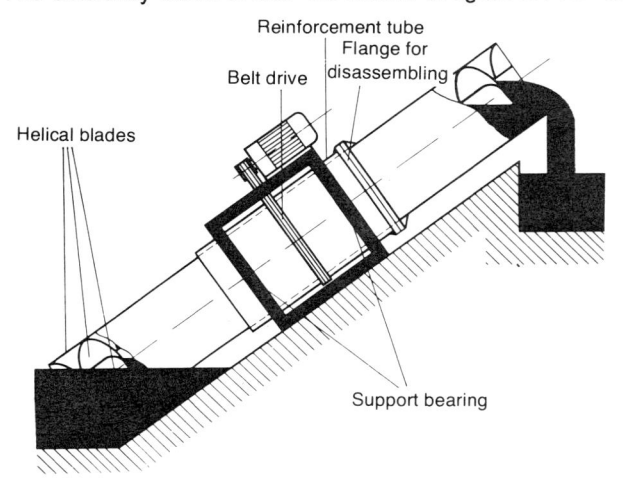

Figure 3.147 Rigid screw pump with rotating "channel"

performance curves for a Pitot tube pump are similar to those for a centrifugal pump. Safety valves, as a protection against over loading or over pressure, are not necessary and the pump can be regulated by throttling.

Advantages of the Pitot tube pump are:

- High pressure in a single stage
- Good efficiency for small flows
- Non-lubricating liquids can be pumped

The Pitot tube pump cannot compete with the centrifugal pump for "normal" flows. Pitot tube pumps are not suitable for contaminated, abrasive or viscous liquids. Capacities of up to several hundred litres per minute and pressures of up to approximately 70 bar are usual. Practical applications are high-pressure washing, water spraying and feed water etc.

3.5 Other pump types

The word pump appears in other contexts which are beyond the scope of this book. Air pumps and vacuum pumps are compressors. There are pumps which are used for energy transfer from lower to higher temperatures, heat pumps which are compressors in a refrigeration cycle. This section deals with liquid pumps which cannot easily be categorised with other pump types.

3.5.1 Theory

The pumps described in this section rely on liquid inertia, compressibility and the elasticity of non-metallic materials

3.5.2 Classification

315 Miscellaneous types

Semi-rotary hand pumps

Powered barrel emptying pumps are covered in category **300**. When power of any type is not available "muscle-power" must be used. Semi-rotary hand pumps are similar to vane pumps but with valves. One set of valves are in the vanes, the others in the casing. On the filling stroke, the valve in the vane opens to allow liquid to change sides. When the vane is moved in the opposite direction, the liquid is squeezed in the casing opening the casing valve. Pumps normally have two chambers providing a suction and discharge stroke for each action. Semi-rotary hand pumps have good suction lift capabilities and work well with fairly viscous liquids. As well as barrel emptying applications, these pumps are often used for priming lube and fuel oil systems on large engines, pumps, compressors and other equipment.

Hydraulic ram pump

The principle of construction of a hydraulic ram is illustrated in figure 3.149. The supply line must be of a certain length, usually 5 to 20 m. The air vessel is located 0.5 to 3 m below the water surface and pumps water up to a higher level, usually a tank for fresh water supply.

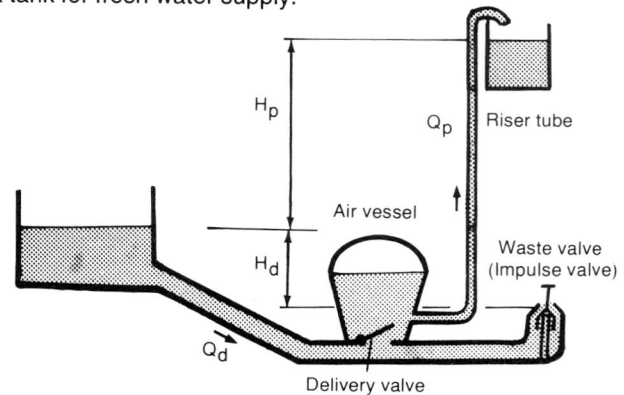

Figure 3.149 Diagrammatic arrangement of the hydraulic ram pump

If the waste valve suddenly closes, the mass of water in the supply pipe is subjected to rapid deceleration which causes a rise in pressure at the waste valve and the delivery valve. The delivery valve opens and water is forced up into the air vessel. When the water in the supply line momentarily changes direction, a partial vacuum is created at the waste valve causing it to open, whereupon water begins to flow in the supply line again. The action of water flowing through the waste valve causes it to close again and the cycle is repeated. The following applies for the hydraulic ram:

$$Q_p \cdot H_p = \eta \cdot Q_d \cdot H_d \qquad \text{Equ. 3.46}$$

or

$$Q_d = \frac{Q_p}{\eta} \cdot \frac{H_p}{H_d} \qquad \text{Equ. 3.47}$$

where

Q_d	=	Motive water velocity in supply line	(m/s)
Q_p	=	Delivered velocity in delivery line	(m/s)
H_d	=	Supply head	(m)
H_p	=	Delivery head	(m)
η	=	Efficiency	non-dimensional

The efficiency depends upon the quotient z, H_p/H_d according to figure 3.150. By means of comparison, it may be noted that for large scale water delivery via water turbine, electric grid system and water pump, efficiencies of up to about 50% can be obtained, whereas for domestic use via electric grid and water pump the efficiency is usually less than 10%.

The hydraulic ram has long been used for transporting fresh water for domestic and agricultural purposes. There are installations which have been in use for more than 50 years. Hydraulic rams are sold in many sizes for supply water flows Q_d from 60 l/h to 9 m³/h with total delivery heads of up to about 40 m, see figure 3.151.

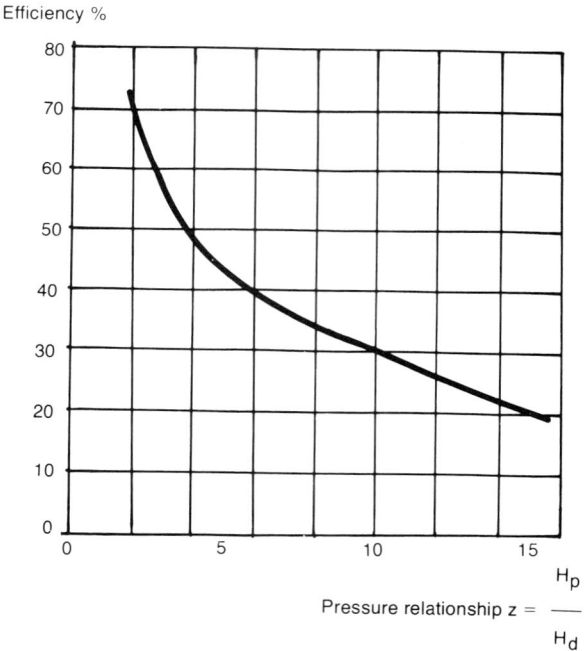

Figure 3.150 Efficiency of hydraulic ram pump

Contraction pump

A contraction pump consists of a special rubber hose which is reinforced in such a way that when it is stretched by means of applying an axial force the diametral contraction is such that the volume is reduced. Pumping effect is achieved by fitting non-return valves to each end of the hose, suction and discharge valves. Suction is achieved as a result of the hose returning to its original length and volume as a result of releasing the axial force.

Efficiency is very good since the only losses, valve losses and hysteresis losses in the rubber, can be reduced to a minimum

Figure 3.151 Hydraulic ram pump

within the pump's most suitable operating range. Typical values are 95% to 98%.

Typical applications are hand operated or motor driven deep-well pumps for well depths of 30 to 100 m and borehole diameters of 50 to 150 mm. Pump performance depends upon the type of hose. Normally hose stretch is 10% of its length and the transported volume is 5 to 20% of the internal hose volume. For hand operated pumps the flow volume is within the region of 5 to 20 l/min depending upon the depth of the well. A particular advantage is the simplicity of installation and removal and the single tube, which functions both as a pull-rod and a delivery pipe, resulting in considerable weight and cost savings.

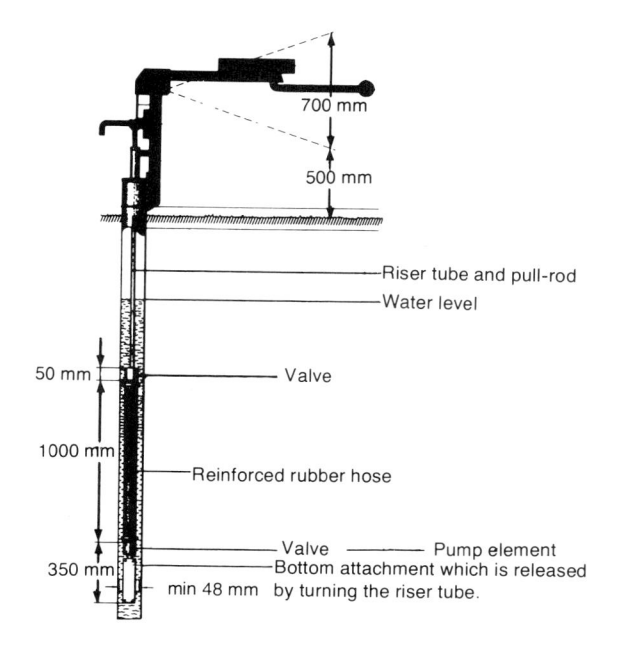

Figure 3.152 Deep well contraction pump

Macerators

Some solids handling pumps include a macerator in the pump inlet. A macerator is a type of solids grinding or cutting machine which ensures solids entering the pump are below a certain size. Macerators cannot normally be used for hard strong solids such as gravel, copper ore or tailings. They are most often used for long stringy solids as found in pulp, sewage and rubbish handling. The macerator ensures that solids entering the pump will pass right through the pump without fouling. Some macerators may be capable of handling soft coal. Extra power must be available to cover the macerator's requirements.

3.6 Suction performance
3.6.1 Cavitation

The word cavitation originates from the Latin meaning "hollow". Cavitation occurs when the static pressure some where within the pump falls to or below the vapour pressure of the liquid. Some of the liquid then vaporizes giving rise to the formation of vapour bubbles. These vapour bubbles are entrained in the liquid and transported until they reach an area where the static pressure is greater than the vapour pressure. At this point the vapour bubbles implode since they can no longer exist as vapour. Each implosion causes a severe pressure wave. The continuous repetition of this sequence of events occurring many times and at high frequency can causes mechanical damage to the pump materials. Furthermore, the hydraulic performance of the pump is reduced by the onset of cavitation. Cavitation is therefore an undesirable phenomenon in connection with pump operation and should be avoided.

The risk of cavitation is greatest where the pressure within the pump is lowest. The lowest pressure in a centrifugal pump occurs on the suction side of the impeller blades, slightly downstream of the front edges where the liquid has started to accelerate, see figure 3.153. At the pump's inlet connection at the same elevation as the shaft the pressure is somewhat greater, by Δp. At this point, which lies at the static suction head h_s above the liquid free surface, the flow velocity is c_s. By applying Bernoulli's equation, see equations 2.4 and 2.5, the flow through the suction pipe gives:

$$p_s = p_a - \rho h_s - \rho g h_{fs} - \rho \frac{c_s^2}{2} \qquad \text{Equ 3.48}$$

$$p_{min} = p_s - \Delta p \qquad \text{Equ 3.49}$$

where

p_{min}	= Lowest pressure within the pump	(N/m²)
p_s	= Suction pressure	(N/m²)
Δp	= Local pressure reduction	(N/m²)
c_s	= Flow velocity in inlet pipe	(m/s)
h_s	= Static suction head	(m)
h_{fs}	= Friction losses in suction pipe	(m)
p_a	= Pressure on liquid surface	(N/m²)

Δp covers any losses within the pump, after the suction connection, such as velocity increases or friction losses.

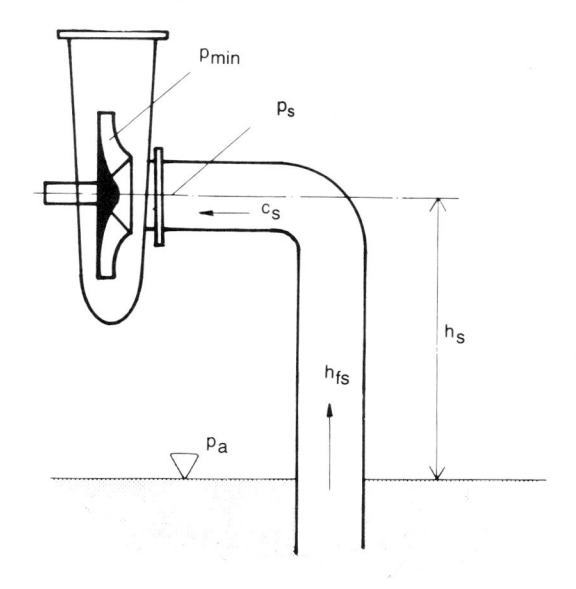

Figure 3.153 Cavitation model

Positive displacement pumps require a corresponding pressure differential, Δp above vapour pressure, in order to avoid cavitation within the pump. The lowest pressure in a pump is determined partly by external factors such as suction pressure and partly by factors concerned with construction and design. The latter consists of the local pressure drop Δp and the flow velocity at the inlet connection c_s. In order to avoid cavitation it is necessary to ensure that p_{min} is greater than the vapour pressure of the liquid p_v.

3.6.2 Net positive suction head

In connection with cavitation in rotodynamic pumps, the concept of NPSH, Net Positive Suction Head, is encountered. The definition of NPSH is as follows:

NPSH = suction pressure - vapour pressure.

$$NSPH = \frac{p_s - p_v}{\rho g} \qquad \text{Equ 3.50}$$

In this context the suction pressure is the pressure at the NPSH datum. For rotodynamic pumps the NPSH datum is the shaft centre-line, except for vertical pumps when it is the front face of the impeller, see figure 3.154. In some cases when purchasing a new pump, the pump user will not know where the NPSH datum is going to be. In these cases, the NPSHa and the suction pressure should be referred to the top of the foundations. NPSH is a differential head, therefore the pressures are converted to metres of liquid when required.

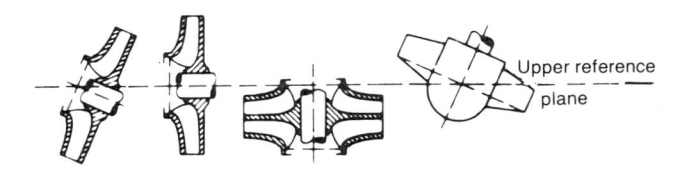

Figure 3.154 NPSH datum for rotodynamic pumps

Some text books state that NPSH refers to the suction total pressure, the suction stagnation pressure. This makes life very complicated. The pump suction pressure is the static pressure; head has been deducted to produce the velocity head of the flowing liquid. This is the value to which the pump differential head is added. It does not make sense to calculate a different suction pressure, the stagnation pressure, for NPSH purposes.

NPSH has two forms; one is a system form and the other a pump form. The system form is designated NPSHa, Net Positive Suction Head Available. NPSHa is calculated from the suction system configuration and is the suction pressure minus the vapour pressure at the NPSH datum. NPSHa informs the pump manufacturer how much Δp is available for losses in the pump suction. The pump form is designated NPSHr, Net Positive Suction Head Required. NPSHr is a characteristic of the pump and specifies how much Δp is necessary for losses in the pump suction. NPSHa and NPSHr are difference between two pressures or heads; there is no requirement to specify gauge or absolute.

Both forms of NPSH are variables, not constants. Both are a function of flow. In a fixed suction system configuration, NPSHa will increase as flow is reduced because friction losses and the velocity head requirement are reduced. Conversely, NPSHa will reduce as flow increases because the same losses increase. NPSHr for a pump is predicted during the design phase and confirmed by testing. NPSHr increases with flow and, at some point, increases as flow is reduced. Rotodynamic pumps are optimised for a specific operating condition; flow and differential head; the Best Efficiency Point. When operating at other conditions parasitic losses increase and blade/liquid angles are mismatched. As flow is reduced below BEP angular mismatch increases. At a certain point, the losses created by mismatch become larger than the gains produced by reduced friction losses and NPSHr increases, see figures 3.10 and 3.12.

In an ideal world, the NPSHr value for a pump would signify the initiation of cavitation. However it is extremely difficult, if not impossible, to detect the first indications of cavitation. Because of this problem various criteria have evolved as bench-marks for defining cavitation. The most popular definition is for 3% head loss. The pump is tested at constant flow and the suction pressure is reduced until the differential head has decayed by 3%, see figure 3.155. The NPSHr is calculated from the suction pressure and the vapour pressure.

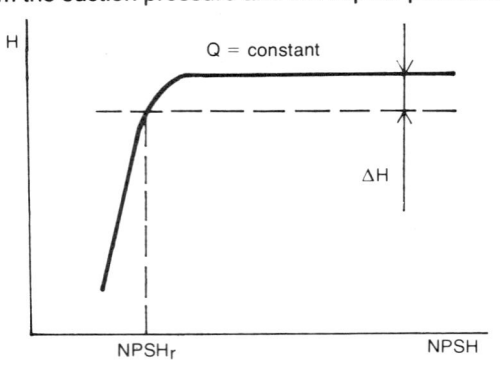

Figure 3.155 Test determination of required NPSH, NPSHr

A complication to the 3% head loss criterion arises with multi-stage pumps; 3% head loss of what? Some manufacturers use 3% of the pump differential head. In fact it should be 3% of the first stage differential head as this is the only impeller which is cavitating. How strictly this requirement needs to be enforced depends upon the size of the pump and how much margin exists between NPSHa and NPSHr.

As mentioned earlier, other NPSHr definitions exist. Cavitation causes damage due to the implosion of the vapour bubbles; this damage results in a loss of performance and increased power consumption. The concept of damage limitation was studied and a 40000 hour NPSHr criterion has evolved. If effect, the pump manufacturer warrants that the pump will suffer no loss of performance or require parts replacement due to cavitation damage for 40000 hours of operation. This value of NPSHr is somewhat greater than the 3% criterion and tends to be applied to large pumps and pumps built of costly materials. Obviously, the 40000 hour criterion can not be applied to pumps which will suffer extensive corrosion damage or be eroded by abrasive solids.

Acoustic measuring techniques have advanced during the past few years and the general opinion is that cavitation produces a distinctive noise at about 400 kHz to 500 kHz. This is obviously outside the range of human hearing and detectors are required. Devices are available which include detectors and suitable circuits to initiate alarms or trips. It is possible in the future that an NPSHR definition will be based on the strength of the cavitation signal.

NPSHr is a function of pump speed. The affinity laws, equations 3.16, can be used for small speed changes. The effect of impeller diameter changes should be referred to the manufacturer.

3.6.3 Permissible suction lift

The permissible suction lift of a pump is controlled by:
- the liquid vapour pressure
- the liquid density
- the pressure on the liquid surface
- the velocity in the pipe
- the friction losses in the pipe
- the NPSHr of the pump.

Equations 3.44 and 3.45 indicate the mathematical relationship of the variables. p_{min} must never fall below p_v if cavitation is to be avoided. We can rewrite equation 3.44 in terms of p_v and NPSHr and rearrange to evaluate the limiting value of h_s:

$$h_s = \frac{p_a}{\rho} - gh_{fs} - \frac{c_s^2}{2} - \frac{p_v}{\rho} - \text{NPSHr} \qquad \text{Equ 3.51}$$

where p_a and p_v are pressures and NPSHr is in metres.

If the pump is drawing liquid from a vessel which is open to the atmosphere then the maximum suction lift will vary with atmospheric pressure. Atmospheric pressure varies, for example, due to the height above sea-level. This is illustrated in figure 3.156. The height of the barometer at sea-level is assumed to be 760 mm Hg. Atmospheric pressure also varies according to climatic conditions. At sea-level the atmospheric pressure does not normally fall below 720 mm Hg, 960 mbar, 9.8 m H$_2$O.

Height above sea-level m	Atmospheric pressure Pascal Pa	Millibar mbar	Torr mm Hg	Water column m H$_2$O
0	$1.013 \cdot 10^5$	1013	760	10.3
1000	$0.899 \cdot 10^5$	899	674	9.2
2000	$0.795 \cdot 10^5$	795	596	8.1
3000	$0.695 \cdot 10^5$	695	521	7.1

Figure 3.156 The variations of atmospheric pressure due to height above sea-level (760 mm Hg at sea-level)

The vapour pressure of a liquid is dependent upon temperature. Figure 3.157, for example, shows the vapour pressure for water at various temperatures. When pumping warm water or hydrocarbons for example the term pv can be large. When pumping boiling liquids the pump must be placed below the liquid free surface so that static head will be greater than the losses in the pipework and the NPSHr. Vapour pressures for a large selection of liquids appear in Section 1.6.2.

It should be noted that neither ISO or CEN have approved steam tables for the properties of water. Care must be taken to ensure all interested parties work to the same figures.

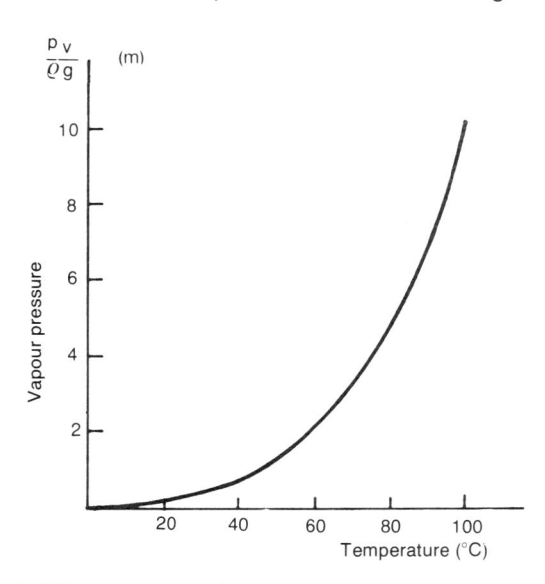

Figure 3.157 Vapour pressure for water

The losses in the suction pipe should be kept as low as is practically possible, since they have the effect of reducing the permissible suction lift for cavitation-free operation. The suction line should be of large diameter, short and without unnecessary bends, valves etc. Suction pipe losses and NPSHr vary with flow. It is therefore important to know the minimum and maximum flow at which the pump is to operate and check for adequate NPSHa.

The preceding regarding cavitation and NPSH have been explained in the context of rotodynamic pumps. Positive displacement pumps have similar phenomena and criteria although not identical.

Cavitation in positive displacement pumps causes a reduction in flow rather than head or pressure. Any vapour bubbles which form fill part of the displaced volume and reduce the liquid capacity. The bubbles collapse when the pressure increases and flow is reduced by the vapour bubble volume. The imploding bubbles cause material damage as in rotodynamic pumps. Additionally, in pump types such as reciprocating, severe shock waves are produced which can cause fatigue failure in components and bearing failure.

The concept of NPSH becomes NPIP, Net Positive Inlet Pressure, again in two forms, NPIPa and NPIPr. Pressure is used rather than head because all the other pump parameters are expressed as pressure. Positive displacement pump types which are used for process applications handling a wide range of liquids consider NPIP to be an equivalent to NPSH; NPIP is quoted in a completely analogous manner. However, some positive displacement pump types are used mostly on one type of liquid and a different approach has become accepted. Gear pumps and screw pumps are used mostly on lubricating oil. Lube oil has a very low vapour pressure and NPIP testing is difficult due to the high vacuums which must be achieved and attendant air leaks. Also, air and gases evolving from solution tended to mask the true NPIPr value. It has become the practice for these pump manufacturers to specify a minimum suction pressure. Unfortunately some still call this NPIPr and quote values in bara. The picture is further complicated by some positive displacement pump manufacturers who do not fully understand NPSH or NPIP and quote values of NPIPr in bara. It is worthwhile double checking with manufacturers to be sure exactly what figure is being quoted.

Plunger pumps are used extensively for process applications and NPIP data is readily available. Pumps are tested at constant discharge pressure while the suction pressure is gradually reduced. When cavitation is initiated, the flow per revolution and the volumetric efficiency are reduced. It is possible that the pump flow may increase slightly due to a very slight increase in speed caused by reduced power consumption. Pumps driven by fixed frequency squirrel cage motors will run at slightly different speeds depending upon the absorbed power. The flow per rev or volumetric efficiency can be plotted against NPIP, see figure 3.158.

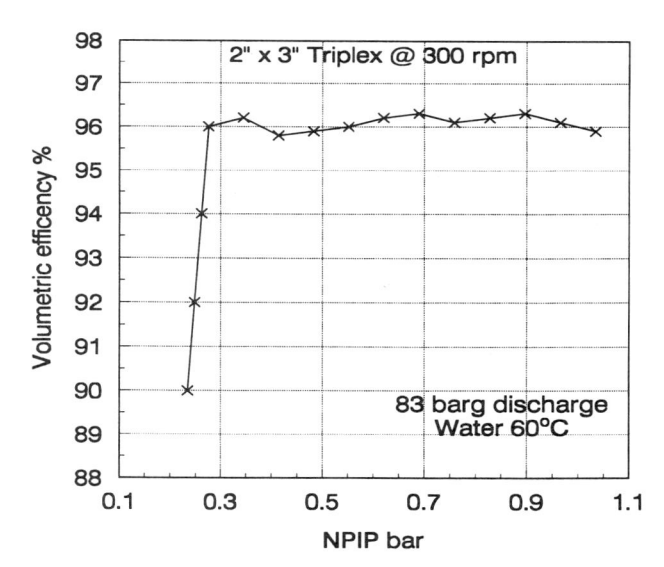

Figure 3.158 Plunger pump NPIP test results

Reciprocating pumps with plate or plug valves have very distinctive NPIPr characteristics as can be seen from figure 3.158. API 674 and 676, reciprocating and rotary positive displacement pumps, advise that NPIPr should be specified at 3% flow loss. This is obviously a direct translation of the rotodynamic pump criterion of 3% head loss. 3% flow loss can not be applied universally to reciprocating pumps. The Hydraulic Institute is considering a 5% flow loss criterion for rotary pumps. Most reciprocating pumps are built with plate valves. The NPIPr characteristic is very clear. NPIPr values should be quoted for 0% flow loss.

Cavitation in reciprocating pumps can produce catastrophic failures due to fatigue. Pumps should not be operated routinely while cavitating. The effects of cavitation on rotary positive displacement pumps should be discussed in detail with the manufacturer, on a case by case basis, when deciding on the operating conditions and the purchasing specification.

Reciprocating pump NPIPr characteristics are plotted for constant differential pressure against speed. Some curves may show more than one viscosity or more than one plunger diameter. Increasing viscosity increases NPIPr although the relationship is not easily defined and varies considerably from pump to pump. Increasing viscosity has a more pronounced affect on small pumps than large pumps. Viscosity can also affect the valve dynamics resulting in reduced speed operation. The NPIPr curve does not approach zero at zero speed. A minimum value, depending upon the pump and valve design, is required to avoid cavitation. It may be possible for the NPIPr value to increase at very low speeds, similar to the effect in centrifugal pumps. No test results have been reviewed which show this effect but the possibility should not be ignored if pumps are required to operate below 10% speed.

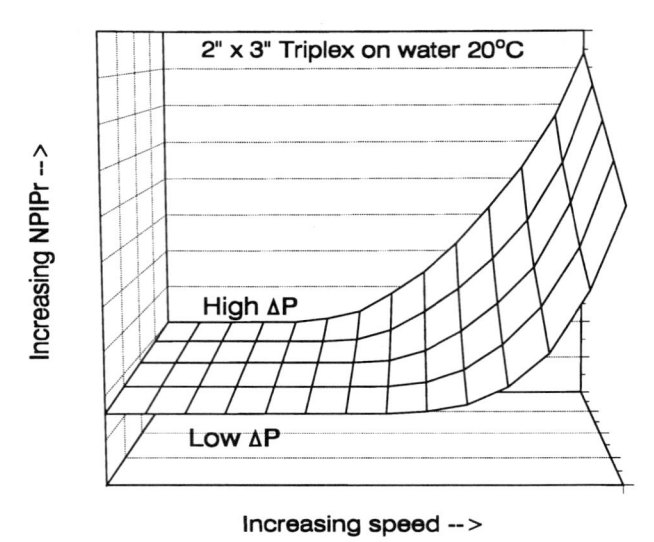

Figure 3.160 NPIPr surface

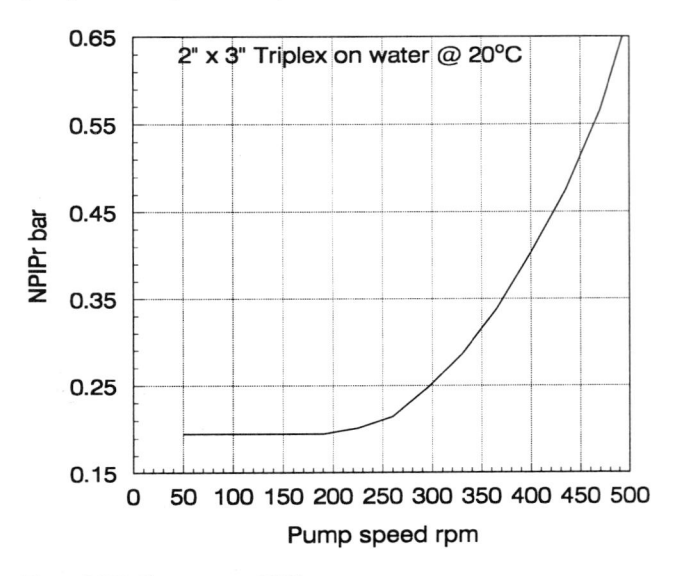

Figure 3.159 Plunger pump NPIPr curve

As explained in 3.2.1.5, and shown in figure 3.12, NPSH can be considered as a 3D surface. The same concept can be applied to reciprocating pumps showing the relationship between speed, differential pressure and NPIPr, see figure 3.160. The surface shown can be considered as the relationship for a specific viscosity. A higher viscosity would produce another surface, not necessarily the same shape, above the surface shown. The effect indicated where NPIPr reduces as differential pressure reduces can be completely negated by poor suction valve design.

Rotary positive displacement pumps tend to have a linear NPIPr relationship with speed. The slope of the curve increases as viscosity increases. Usually two or three viscosity curves are shown on one graph to allow interpolation for the required value.

3.6.4 Cavitation effects on pump operation

Cavitation in rotodynamic pumps, if allowed to continue for prolonged periods of time, can cause mechanical damage to the pump. Surface material is removed by the implosion of the vapour bubbles and characteristic damage is visible when inspected. Clearances can be increased causing permanent reduction in performance. Mechanical seals are prone to physical damage. The hydraulic performance is also reduced by the onset of cavitation. The performance is affected in different ways depending upon the type of pump. In general, there is an accompanying increase in noise and vibration.

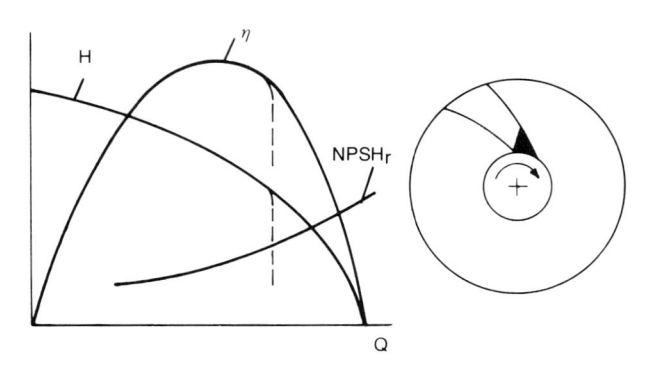

Figure 3.161 Cavitation in a centrifugal pump

The full lines shown in figure 3.161 represent the pump's differential head and efficiency for cavitation-free operation. The pump's cavitation sensitivity in the form of NPSHr is also shown. The pump has a low specific speed, $N_s = 20$. The pump is intended for a system having a certain static suction head. When the flow increases, the NPSHr. When the positive suction static head is reduced eventually the limits of cavitation-free operation are reached and the pump performance declines rapidly (dashed lines). The rapid decline in performance is associated with the narrow vane passages which are characteristic of pumps having low specific speed. The complete length of the front edges of the vanes are at approximately the same diameter and are subjected to the same velocity. With the onset of cavitation it requires only a slight increase in flow to completely fill the vane passages with vapour. The pump's delivery head can thus be reduced to zero.

The relationship for axial flow pumps is different. Here cavitation occurs on the suction side of the profiles at the tips of the

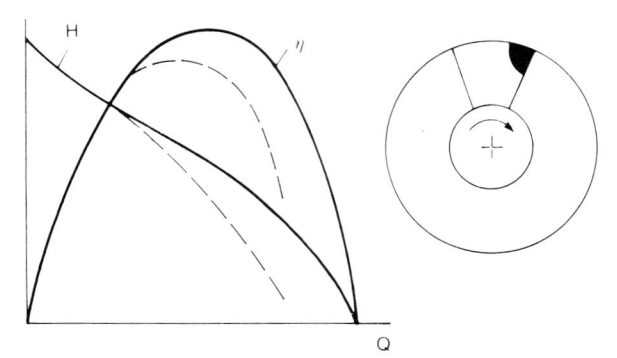

Figure 3.162 Cavitation in an axial pump of $N_s = 200$

blades where the relative velocity, Δp is greatest. Despite cavitation at the tips of the blades there still remains a large usable amount of blade passage, figure 3.162. The decline in performance is therefore a more gradual process in the case of axial pumps.

The onset of cavitation in positive displacement pumps causes a reduction in flow rate since part of the displacement becomes filled with vapour. When the vapour bubbles collapse, severe pressure waves can be created which result in fatigue failures. In reciprocating pumps, classic cavitation damage includes:

- cylinder cracks
- broken bolts and studs
- broken valve plates and springs
- high bearing wear.

Externally signs of cavitation include a characteristic noise "pinking", increased pump vibration and increased pipework vibration due to increased levels of pressure pulsations in both the suction and discharge.

For reciprocating pumps, and some rotary pumps with only a few lobes or teeth, an additional loss, to those already explained, is experienced in the suction pipework. The flow variation caused by the small number of pumping elements requires some additional head to be available to accelerate the flow to keep up with the changing velocity of the pumping elements. This phenomenon has been studied by the Hydraulic Institute in the USA and the VDMA in Germany. The VDMA work is only available in German although an English translation may be available soon. The extra head required is a function of the:

pump type
number of pumping elements
pump speed
velocity in the suction pipe
length of the suction pipe
nature of the liquid.

This extra loss is called Acceleration Head Loss and can be approximated for water, amine, glycol, from the following equation.

$H_{acc} = 4LUnC$ Equ 3.52

Where:

H_{acc} =	acceleration head loss	(m)
L =	suction pipe length	(m)
U =	average velocity in pipe	(m/s)
n =	pump speed	(s^{-1})
C =	pump characteristic	non-dimensional

The pump characteristic values for single acting reciprocating pumps are:

1 cyl	C = 0.4
2 cyl	C = 0.20
3 cyl	C = 0.07
4 cyl	C = 0.11
5 cyl	C = 0.04
6 cyl	C = 0.06
7 cyl	C = 0.03
8 cyl	C = 0.045
9 cyl	C = 0.025

For other liquids multiply the water value by:

urea, carbamate, liquids with traces of entrained gases	1.5
de-aerated water	1.1
hydrocarbons	0.75
hot oil	0.6

The approximation given in 3.52 must be used with caution. It is, of necessity, an extremely simplistic evaluation of a very complex phenomenon. It was proposed for use on relatively short pipe runs, 15 to 20 diameters. On long runs it tends to be pessimistic, on short runs optimistic. The formula tends to zero as the variables reduce. In practice, acceleration head loss tends to a finite minimum depending upon the pump detail design. If acceleration pressure loss will cause pipe vibration or reduce NPIPa below NPIPr then pulsation dampeners may solve the problem. A correctly selected, installed and maintained pulsation dampener can attenuate pressure pulsations by 90%.

Positive displacement pumps should always operate with a safety margin between NPIPa and NPIPr. A fixed value of safety margin would be too simplistic and a calculated value is very difficult. The following guide-lines, for reciprocating pumps, were proposed by the British Pump Manufacturers Association. (See Figures 3.163 - 3.165.)

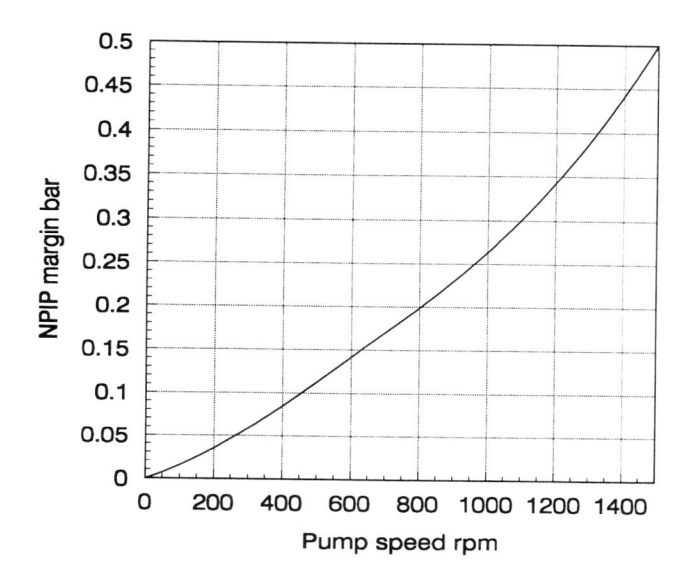

Figure 3.163 Recommended margin of NPIPa over NPIPr for pump speed

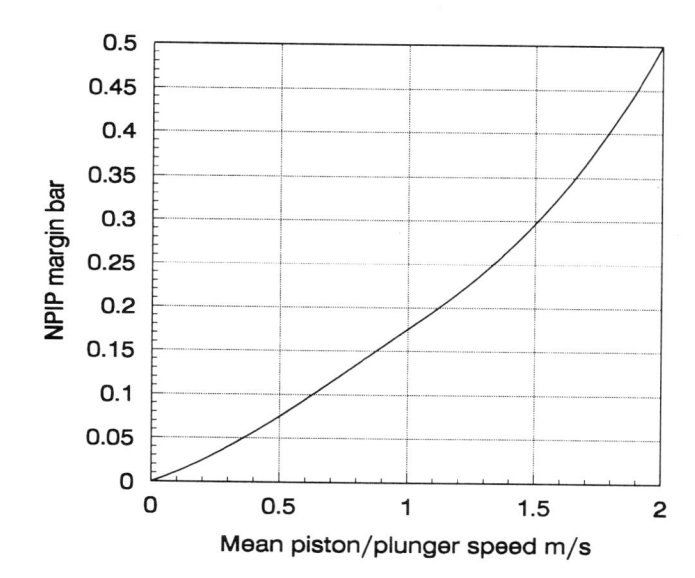

Figure 3.164 Recommended margin of NPIPa over NPIPr for mean piston/plunger speed

Values of margins are obtained from both figures 3.163 and 3.164. A multiplier is obtained from figure 3.165 depending upon how highly loaded the pump is at the rated conditions and applied to the highest value obtained from 3.163 or 3.164.

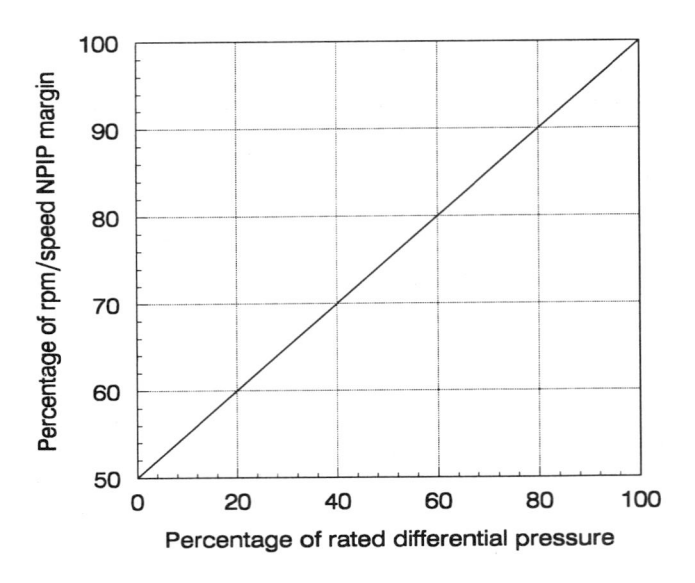

Figure 3.165 Percentage of recommended margin for percentage of rated differential pressure

3.6.5 Self priming

A rotodynamic pump operating with air rather than water produces a head rise which is about 1/1000. If the pump is placed above the level of the upper liquid surface and the suction line is filled with air the pump is thus incapable, when starting, of removing this air by itself. One refers to a centrifugal pump as being unable to evacuate it's own suction line. Before pumping can start it is therefore necessary to fill the pump with liquid, venting and priming. This can be arranged in a number of ways, see figure 3.163. Placing the pump below the level of the upper liquid surface, flooded suction, eliminates the problem of evacuation and reduces the risk of cavitation. The NPSH can be achieved for example, by immersing the pump or both the pump and the drive motor, so-called submersible pumps.

As we have seen earlier in chapter 3, rotodynamic pumps are available in a tremendously wide range of designs and configurations. **Not all pumps are self venting.** Pumps installed as indicated in figure 3.166 as "dry" may still require venting and priming. Venting and priming will only be necessary once, after installation or maintenance, unless the liquid level falls below the top of the pump. The manufacturer's operating instructions should be checked to ensure proper pump operation. Submerged pumps are designed to be self venting and priming.

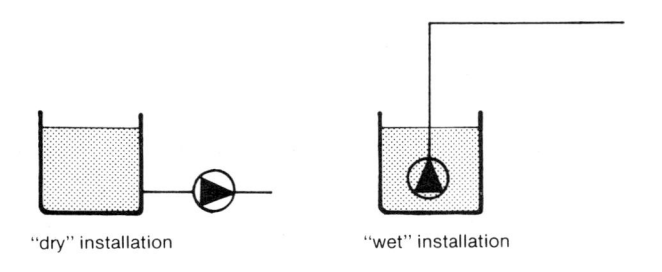

Figure 3.166 Pump installations without suction line evacuation

An alternative method of keeping the pump filled with liquid is shown in figure 3.167. The non-return (foot) valve prevents liquid draining from the suction line when the pump is stopped. The foot valve is however, always subject to a small amount of leakage and also has the disadvantage that it causes large pressure losses within the suction line when pumping. Provision must be made to fill the suction line initially. The volume of the evacuating tank must be several times greater than the volume of the suction line. It is also necessary to check that cavitation does not occur in the suction line.

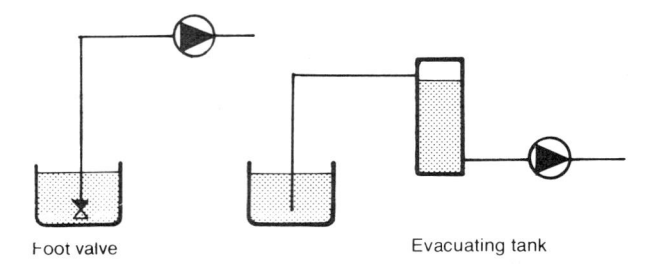

Figure 3.167 Foot valve and evacuating tank

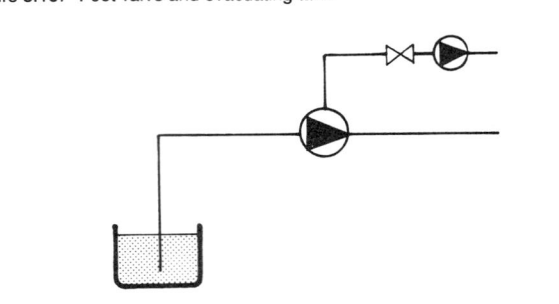

Figure 3.168 Evacuating pump

The air in the suction line and pump casing can be evacuated with the help of a smaller self-priming pump, figure 3.168. The priming pump must be connected to the highest point on the pump casing. A large variety of pump types can be used as evacuating pumps. The most usual are liquid-ring pumps, side-channel pumps and jet pumps.

All positive displacement pumps are in principle self-priming providing that the inner seals are sufficiently effective and that "dry" running is acceptable during starting. It is therefore advisable to check with the manufacturer with regard to "dry" running in the case of applications having long suction lines. Reciprocating pumps should always be vented and primed prior to starting.

In the case of specially designed self-priming centrifugal pumps the pump is built into a tank containing liquid. The evacuating process is then carried out automatically if the pump is first filled with liquid, see section **065**. The suction line must be dimensioned with regard to both flow losses and evacuation time and must rise continuously towards the pump. Too large a suction line makes evacuation more difficult. A useful rule is to dimension for a suction head loss of 1 to 2 m. If the volume of the suction line above the surface level of the liquid is very large, for example, for suction pipe lengths of several hundred metres, a foot valve must, without exception, be placed below the liquid surface.

3.6.6 The effects of dissolved and entrained gases

The pumped liquid may contain certain quantities of dissolved gas or entrained gas. Lubricating oil, for example, at atmospheric pressure and room temperature can contain up to 10% air by volume, petrol can contain up to 20% air by volume. The ability of a liquid to dissolve gas reduces with reduced pressure and with increased temperature. Gas can therefore evolve in areas of localised low pressure and re-dissolve when the pressure increases again. This process is similar in some ways to the formation and implosion of vapour bubbles, cavitation, but is considerably less violent and causes no mechanical damage unless the pressure changes are large, as in positive displacement pumps. When gas separation and cavitation occur simultaneously in rotodynamic pumps the gas has a dampening effect on the implosion process and in this way helps to limit the extent of damage caused by cavitation. Small amounts of air have been introduced into the pump suction on purpose to alleviate certain site problems.

The pumped liquid can also contain entrained gas, usually air. This entrained gas has considerable influence on pump performance. In rotodynamic pumps both the delivery head and the efficiency are noticeably reduced by increased quantities of entrained gas. In positive displacement pumps entrained gas reduces the liquid flow rate and creates pressure pulsations which can cause premature component failure and severe pipe vibration.

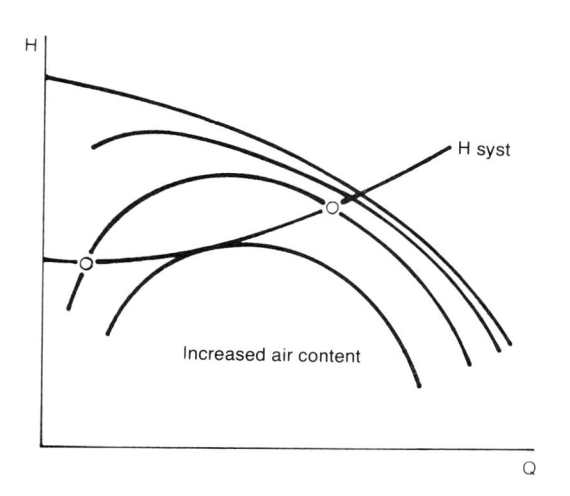

Figure 3.169 The influence of entrained air in water on a centrifugal pump's H-Q curve

As shown in figure 3.169, the originally stable H-Q curve becomes unstable due to the influence of entrained air. For the system curve shown there are, for a given concentration of entrained air, two alternative intersection points and there is risk of unstable operating conditions.

Certain conditions can cause air to collect in the pump impeller, which can lead to loss of prime. The risk of air collecting is greatest for flows which are less than the design flow. It is normally possible to pump without other difficulties arising, other than reduced efficiency, for air concentrations of up to 2 to 4% by volume measured at the pump intake connection.

Positive displacement pumps usually cope with larger concentrations of air better than centrifugal pumps. For displacement pumps operating with a suction lift, the theoretical volume flow is reduced considerably, up to 50%, if the liquid contains entrained or dissolved gases. Entrained gas will expand and dissolved gas will evolve and partially fill the displacement volume. Severe pressure pulsations, causing vibration and fatigue damage, may result. Consult pump manufacturers with full operating conditions, including details of gas contamination, before deciding on a pump type and selection.

Pumps handling cold fresh water from rivers, reservoirs and tanks can suffer damage and short component lives due to entrained and dissolved air. Experience has shown that an allowance must be made for air evolution. The vapour pressure of the water should be increased by 0.24 bar.

Development of rotodynamic pumps in recent years has tried to increase the gas handling capabilities. To date, rotodynamic pumps are limited to 15% by volume. Special versions of positive displacement pumps have been developed to cope with entrained/dissolved gas continuously. Twin screw pumps, with synchronised screws, can handle 97% gas at pressures up to 40 barg.

SPECIAL NOTE

It cannot be stressed too often or too strongly that successful pump installations are based on good communications between the user and the pump manufacturer. The pump manufacturer issues guarantees on the basis of operational data provided. If important data is withheld, the guarantee will be invalidated and the manufacturer will not supply replacement parts or service engineers free-of-charge. The user should not judge which information is important and which is trivial. Forward all information to the pump manufacturer and let the manufacturer decide. This course of action will provide the user with a "water-tight" guarantee. If certain aspects of operation are impossible to quantify, discussions with manufacturers will indicate whether the problem is important or insignificant and if there could be an effect on performance or reliability. It is in the best interests of all concerned to discuss points which need clarification. In general, data sheets for pumps are poor. Full operational conditions are not described in any detail. Filling in an API 610 or API 674 data sheet will not assure a "water-tight" guarantee.

3.6.7 Examples

Example 1

A pump is required to lift water from a river. The minimum barometric pressure is 735 mm Hg and the maximum water temperature is 10 °C. At the rated flow the suction pipe losses, including velocity head and entrance losses, are 1.5 m and the pump NPSHr is 2 m. What is the maximum suction lift possible?

Water vapour pressure = 0.01227 bara
Water density = 999.8 kg/m^3
Mercury density = 13600 kg/m^3
Atmospheric pressure ≡
  735 x 13600 / 999.8 / 1000 = 9.997 m
Water vapour pressure ≡
  0.01227 x 100000 / 999.8 / 9.80665 = 0.125 m
Allowance for air evolution ≡
  0.24 x 100000 / 999.8 / 9.80665 = 2.448 m
NPSHa = 9.997 - 1.5 - 2 - 0.125 - 2.448 = 3.924 m

The maximum suction lift is 3.924 m without any safety margin between NPSHa and NPSHr. If a self-priming pump was used, then the margin for air evolution would not be required and the maximum suction lift would increase to 6.372 m.

The density used for water is compressed water data, not saturated liquid data. The water is not at saturated pressure therefore it is compressed.

Example 2

Water at a maximum temperature of 60 °C is to be pumped from an open tank. The minimum barometric pressure is 735 mm Hg and the pump suction will be 3 m below the minimum liquid surface level. If the suction pipe losses, including velocity head and entrance losses, are 4 m at the rated flow what is the NPSHa for the pump ?

Water vapour pressure = 0.1992 bara
Water density = 983.2 kg/m^3
Mercury density = 13600 kg/m^3
Atmospheric pressure ≡
  735 x 13600 / 983.2 / 1000 = 10.167 m
Water vapour pressure ≡
  0.1992 x 100000 / 983.2 / 9.80665 = 2.065 m
NPSHa = 10.167 + 3 - 4 - 2.065 = 7.102 m

Notice that the head equivalent of atmospheric pressure has changed even though the barometric pressure is the same. The water density is reduced because of the increased temperature. Allowance for gas evolution is not necessary because the temperature is high enough to drive off any gas.

Example 3

Water at a temperature of 100 °C is to be pumped from an open tank. The minimum barometric pressure is 735 mm Hg and the suction pipe losses, including velocity head and entrance losses, are 4 m at the rated flow. If the pump has an NPSHr of 5 m at the rated flow, how much static suction head is required to prevent cavitation ?

Water vapour pressure = 1.01325 bara
Water density = 958.1 kg/m^3
Mercury density = 13600 kg/m^3
Atmospheric pressure ≡
  735 x 13600 / 958.1 / 1000 = 10.433 m

Water vapour pressure ≡
 1.01325 × 100000 / 958.1 / 9.80665 = 10.784 m
NPSHa = 10.433 - 4 - 5 - 10.784 = -9.351 m

The minimum static head required is 9.351 m without any safety margin. The head equivalent of atmospheric pressure is increased due to the reduced water density.

Example 4

Water is to be pumped from a closed tank which is pressurised to 4 barg 0.1 barg. The minimum barometric pressure is 735 mm Hg. The maximum water temperature is 150 °C. At rated flow the suction pipe losses are 3 m and the NPSHr of the pump is 3 m, how much positive static head is required to prevent the pump cavitating ?

Water vapour pressure = 4.76 bara

Water density = 916.7 kg/m^3

Mercury density = 13600 kg/m^3

Atmospheric pressure ≡
 735 × 13600 / 916.7 / 1000 = 10.904 m

Pressure in tank = (3.99 × 100000 / 916.7 / 9.80665) + 10.904 = 55.287 m

Water vapour pressure ≡
 4.76 × 100000 / 916.7 / 9.80665 = 52.949 m

NPSHa = 55.287 - 52.949 - 3 - 3 = -3.662 m

The minimum static head required is 3.662 m without any margin.

Example 5

A three cylinder single acting pump, running at 600 rpm, is to be supplied with water at 60 °C from an open tank. The minimum barometric pressure is 735 mm Hg. The minimum water level above the pump suction is 5 m. The suction line velocity is 0.75 m/s and the line length is 10 m. The suction line losses total 2 m and the pump NPIPr is 0.7 bar. Will the system operate successfully ?

Water vapour pressure = 0.1992 bara

Water density = 983.2 kg/m^3

Mercury density = 13600 kg/m^3

Atmospheric pressure ≡
 735 × 13600 × 9.80665 / 1000 / 100000 = 0.980 bara

Static pressure =
 5 × 983.2 × 9.80665 / 100000 = 0.482 bar

Friction pressure drop =
 2 × 983.2 × 9.80665 / 100000 = 0.193 bar

H_{acc} = 4 × 10 × 0.75 × 10 × 0.07 = 21 m

H_{acc} ≡ 21 × 983.2 × 9.80665 / 100000 = 2.025 bar

NPIPa = 0.98 + 0.482 - 0.1992 - 0.193 - 2.025 = -0.955 bar

As originally described, the system will not work. The acceleration head loss is too great. A larger suction pipe could be used; this would reduce the friction loss and the acceleration head loss. This would be the best solution for really critical installations. Or a pulsation dampener could be fitted; the acceleration head loss could be reduced to 0.2025 bar which would provide an NPIPa/NPIPr margin of 0.867 bar. Figure 3.163 recommends a margin of about 0.15 bar for pumps at 600 rpm. So 90% attenuation of H_{acc} is not necessary for NPIPr purposes. Check effect of pressure pulsations on pipework vibration before deciding on attenuation required.

3.7 Standards

3.7.1 General

Within any civilised community there is legislation which controls relationships between individuals. Successful technology must also be controlled and guided by standards and regulations to ensure its safe application and allow some degree of interchangeability.

Standards were initially applied to the finished dimensions of machine components to ensure easy replacement of worn or broken parts. The scope and activities of standards have been successively extended to cover the complete machine not only with regard to dimensioning but also to such parameters as performance, reliability and safety. Within the pump industry standardisation has developed in a similar fashion and today there are a considerable number of standards and regulations. There are also a large number of new standards which are in the process of preparation and old standards in the process of revision.

Some standards have legal authority and can be enforced by the process of law; these are mostly restricted to safety requirements. Most standards are issued for guidance. If, for example, one issued an enquiry to ten pump manufacturers which simply stated "5 pumps required for 100 m^3/h at 100m ΔH" one would receive ten quotations for ten totally different pumps. Comparison between the quotations would be difficult or impossible. Selection of the best pump would be dubious without very detailed design and construction information on each pump. The solution to the problem is to issue a specification with the enquiry detailing the important characteristics required. To avoid the necessity of every purchaser writing their own specifications, groups were formed to write a single common specification for certain areas of application.

Because of the diversity of issuing authorities, the style and concept of standards is not uniform. CEN have decided upon rigourous control of all aspects of standards; presentation, structure and content. CEN have produced a manual for all concerned with standard writing and it is worthwhile reviewing for the structure and content requirements.

CEN standards consist of requirements; these are basically statements controlling design, construction and use. Requirements must be verifiable. A competent person, inspecting the equipment/installation/service, must be able to see that a requirement has been met or not. For example, if a standard specifies that pumps must painted with a minimum thickness of 200 μm, an inspector can measure the paint thickness and say categorically that it complies or not. If an inspector visits a manufacturer's works to witness all stages of manufacture and testing and the standard specifies "Equipment shall be suitable for three years uninterrupted operation", what can the inspector do ? We will have to wait until inspectors are issued with crystal balls. Standards cannot address concepts; rugged, durable, heavy-duty; are all subjective and cannot be verified.

3.7.2 Issuing authorities

In principle, a standard can be issued by anyone. In order that the standard shall carry some influence, however, the issuing organisation should be one that is recognised by both the manufacturer and the user of the object to be standardised. It is therefore normal practice for new standards to be developed by these three parties. The organisations which issue standards consist basically of the following categories:

- International standard organisations
- Geographic country group standard organisations
- National standard organisations
- Trade societies and associations
- Corporations, large companies, government departments
- Classification societies (insurance companies)
- Certification authorities (insurance companies)

International standard organisations require the cooperation and support of many countries. As the world becomes, effectively, smaller more and more international trade transactions will be regulated by international standards.

Geographic country group organisations are made up of countries, usually close to each other, which form an economic trading block. This type of cooperation is becoming popular and could result in an expansion of issuing authorities.

Corporations issue standards to control important functions. In most cases these standards form part of a purchaser order and are binding on all parties. Chemical companies, who develop proprietary manufacturing processes, often issue standards covering size, speed and rating of important equipment.

Classification societies are generally one, or more, insurance companies who issue rules for design and construction. This type of standard is used for marine installations and the most notable authorities are: Lloyds Register of Shipping, Det Norsk Veritas and the American Bureau of Shipping. If a ship owner tries to insure a ship, the first question is "Can I see your Classification Certificate".

Certification authorities are very similar to classification societies but work in conjunction with the purchaser on large projects which have some degree of risk or hazard to the purchaser's capital investment or personnel safety. Certification authorities are regularly used for large petrochemical installations and off-shore platforms. Specific standards are not always issued. Experienced inspectors scrutinise and witness all relevant aspects of the equipment; design, raw material, construction, testing, safety; and issue a certificate, if completely satisfied, when the equipment is ready to be dispatched.

The number of issuing authorities can therefore be very large and varying which often presents difficulties when developing new standards. Since there are many conflicting interests, especially when applied to international standards, the result is often a compromise with which none of the involved parties is completely satisfied. The advantages to be gained by the specification and general acceptance of standards and regulations however usually outweigh the disadvantages.

At present, there are many new standards being issued in two important areas. CEN is issuing standards to control the trade within the EEC and EFTA. From 1995, it will not be possible to use national standards for trade between member countries; EN's or ISO's will be used. ISO is issuing new standards to control world trade. ISO is in discussion with all the standards organisations and trade associations to rationalise world-wide standardisation. ISO and CEN have an understanding and try not to duplicate standards.

Standards usually contain a immunity clause. Strict compliance with any standard does not ensure success or guarantee success. Standards are written with the best of intentions with the best information available. It is the responsibility of the user of the standard to ensure the standard's requirements are compatible with the equipment and operation thereof in each individual case.

The most important bodies for issuing standards impacting on the manufacture and use of pumps are, in alphabetical order:

	Body	Standards
ANSI	American National Standards Institute	ANSI
API	American Petroleum Institute	API
ASME	The American Society of Mechanical Engineers	ASME
ASTM	The American Society for Testing Materials	ASTM
BSi	British Standards Institution	BS
CEN	Comité Européen de Normalisation	EN
DIN	Deutsches Institut for Normung	DIN
EUROPUMP	The European Committee of Pump Manufacturers	
HI	The Hydraulic Institute (USA)	HI
ISO	International Organisation for Standardisation	ISO
SAE	The Engineering Society for Advancing Mobility Land Sea Air and Space	SAE
VDI	Verein Deutscher Ingenieure	VDI
VDMA	Verein Deutscher Maschinenbau-Anstalten	VDMA

There are other standards which do not refer to pumps directly but contain essential information for both users and manufacturers.

	Body	Standards
ADR	See note	ADR
CENELEC	Comité Européen de Normalisation Electrotechnique	HD
EPA	The Environmental Protection Agency (USA)	
HSE	The Health and Safety Executive (UK)	EH
IEC	The International Electrotechnical Committee	IEC
NACE	The National Association of Corrosion Engineers (USA)	MR
NFPA	The National Fire Protection Agency (USA)	NFPA
OSHA	Occupational Safety and Health Administration (USA)	OSHA

NOTE: ADR is a very special standard. It is not published by a multinational standards authority but it is subject to agreement by many national authorities who publish their own copy.

The standards authorities which are multinational represent the following countries:

ADR   Austria, Belgium, Czechoslovakia, Denmark, Finland, France, Germany, Greece, Hungary, Italy, Luxembourg, the Netherlands, New Zealand, Norway, Poland, Portugal, Spain, Sweden, Switzerland, UK, Yugoslavia.

CEN
CENELEC   Austria, Belgium, Denmark, Finland, France, Germany, Greece, Iceland, Ireland, Italy, Luxembourg, the Netherlands, Norway, Portugal, Spain, Sweden, Switzerland, UK.

ISO   Argentina, Australia, Austria, Belgium, Brazil, Bulgaria, Canada, China, Czechoslovakia, Denmark, Egypt, Finland, France, Germany, Greece, Hungary, India, Indonesia, Ireland, Israel, Italy, Japan, Korea (DPR), Korea (REP), Malaysia, the Netherlands, New Zealand, Norway, Pakistan, Poland, Portugal, Romania, Russian Federation, Singapore, South Africa, Spain, Sweden, Switzerland, Turkey, UK, USA, Yugoslavia.

3.7.3 Safety

The standards and regulations applying to safety are primarily concerned with the way in which pumps and systems shall be designed and constructed so as to avoid accidental damage, personnel injury or environmental pollution.

The general principles to be used in the design of complete installations are given in many standards and guide-lines. The factors which should be considered include:

- access
- dust/liquid/gas hazards
- fire fighting
- noise
- proximity of adjacent property
- redundancy of equipment
- separation of safety equipment from process equipment.

BS 5908 covers fire precautions for chemical and similar installations. BS 8005 Part 2 gives guidance on the design and construction of pumping stations and pipelines for sewage.

The hazards posed by substances are given in:

    EH 40/93   Occupational exposure limits

    NFPA 325M   Fire hazard properties of flammable liquids, gases and volatile solids

    NFPA 704   Identification of the fire hazards of materials

    ADR   European agreement concerning the International carriage of dangerous goods by road.

The safety of all machinery in Europe is covered by the Machinery Directive. The Directive requires that all manufacturers shall make machinery, by design and construction, as safe as possible. If any hazards will occur during operation or maintenance the manufacturer must highlight these problems and state what precautions must be taken. General machinery safety is covered by EN 292.

Pump safety will be covered by EN 809, eventually. This standard has been rejected at the public comment stage. The original version of the standard shared the safety aspects between the manufacturer and the user. Once the pump is installed the manufacturer cannot be held responsible for the user's actions. This approach to life-cycle safety was rejected.

An important factor in applying safety considerations is the ability of the system to withstand pressure. Pumps, pipes, valves and flanges etc. are therefore constructed to withstand specific pressure ratings associated with temperature. The pressure/temperature ratings for flanges for pipes, valves and fittings etc. can be found in:

ISO 7005; ANSI B16.1; ANSI B16.24; ANSI B16.42; ANSI B16.47; ANSI B16.5; ANSI B93.75; BS 10; BS 1560; BS 3293; BS 4504

The design of pressure vessels and associated pressure/temperature ratings can be found in:

ASME VIII; BS 5500

The design of tanks are covered by:

API 620; API 650; BS 1564; BS 2594; BS 2654; BS 4741; BS 4994; BS 7777

The design of liquid pressurised systems are covered in:

ANSI B31; BS 806; BS 8010

The requirements of pressure relieving valves and systems are dealt with in:

ISO 4126; ISO 6718; ANSI N278.1; ASME PTC 25.3; API RP520; API RP521; API 526; API 527; API 2000; BS 2915; BS 6283; BS 6759

The construction and dimensions of pipe fittings are standardised in:

ISO 49; ISO /R508; ISO 2045; ISO 2531; ISO 3458; ISO 3459; ISO 3501; ISO 3503; ISO 4145; ISO 4179; ISO 8179; BS 143; BS 759; BS 1256; BS 1640; BS 1740; BS 1965; BS 2051; BS 3799; BS 4346; BS 4772; BS 5114

A variety of standards deal with material specifications for pressure retaining parts.

Pressure vessels: ISO 2604; ISO 9328; EN 10028; ASME VIII

Pipework: ISO 2604; ASME VIII

Pumps: API 610; API 674; API 675; API 676.

A draft document issued by BSi, BS Draft DD 38, now withdrawn still includes useful information on pump materials.

In order to ensure that a system is capable of withstanding the required nominal pressure it is necessary to pressure test all the equipment which is subjected to internal pressure. The standards dealing with pressure vessels contain regulations concerning pressure testing. For normal land use the pump must normally be tested to a pressure which is at least 30% greater than that of the specified nominal pressure. A new EN standard has been drafted for pressure testing pumps. For marine pump systems regulations are laid down by the classification societies which often demand a greater margin of safety.

The risk of explosion is an important consideration from the point of view of safety. There are no special regulations for pumps although stringent requirements are specified for electric motors and electrical equipment, see Chapter 9. There are many standards covering the safety of electrical equipment and these standards contain a wealth of useful information for the mechanical and process engineers designing, selecting and arranging equipment. The following should be reviewed; IEC 34-1; IEC 34-5; EN 60034 Part 5; BS 5345 and some sections of BS 4999.

The transmission of power between the motor and the pump presents another safety hazard. ANSI B15.1 and BS 5304 are examples of the standards which contain criteria for the design, construction and application of guards for shaft couplings and other mechanical hazards. The new EN standard for pump safety will address guards for Europe. See also API RP 11ER.

For marine installations the instructions laid down by the classification societies, Lloyds Register of Shipping for example, must be followed. The safety requirements being of particular importance, since a ship at sea cannot assume that assistance will be available in the event of an accident. It thus follows that the safety regulations on board ship are often more stringent than those of an equivalent installation for land use. In principle all equipment must be type-approved and also have passed special safety inspection controls before installation on board ship is permitted. Similar restrictions are imposed for equipment to operate on off-shore platforms.

3.7.4 Reliability and operational life

The concepts of reliability and operational life have come more to the forefront in recent years as the demands for operational reliability and economy increase. Emphasis has been placed on total life-cycle costs rather than initial capital costs. This trend has been followed up by the organisations responsible for issuing standards and resulted in the development of instructions relating to pump design in order to fulfil these demands.

The API 610 standard concerning centrifugal pump reliability gives instructions and recommendations regarding the pump components which are associated with safety. Also dealt with are the factors which must be considered in a pump installation together with comments concerning the components such as shaft couplings and electric motors. The API standard applies primarily to the petroleum industry but is also applicable in other areas.

ISO and API have formed a joint committee to prepare a technical specification of centrifugal, medium duty pumps; the next edition of ISO 9905. The new standard will try to combine the best features of API and current ISO pump standards. It is intended that this standard will become an accepted international guide-line in respect to pump design for, among other things, reliability in operation. As mentioned earlier, inspectors cannot look at pumps and say "this one will only last two years but that one will last three".

Specifications applying to the design, construction, performance, reliability and testing of small-bore domestic heating and hot water pumps are also to be found in EN 60335-2-42; EN 60335-2-51; BS 1394; BS 3456 Part 202 Section 202.41.

An ISO committee is reviewing the draft of the next edition of API 674 with a view to issuing a new ISO standard. It is hoped to have a standard ready for public comment in 1994.

A curious twist has occurred in the specification of reliability very recently. The major European oil companies have realised that if they tell the pump manufacturer how to build a pump for extended reliability and the pump is not reliable they have no claim on the manufacturer. A group of oil companies is discussing with certain manufacturers the possibility of purchasing pumps with specifications. The user completely specifies the duty and the time between overhauls. The pump manufacturer supplies a suitable pump. If the pump fails to perform the manufacturer has to cure the problems.

3.7.5 Dimensions and performance

EN 60335-2-42; EN 60335-2-51; BS 1394 and BS 3456 Part 202 Section 202.41 cover the requirements for circulators for domestic heating and hot water installations for glandless pumps up to 300 W.

The advantages of standardised dimensions for individual components has long been recognised. The scope of standardisation has been developed to cover all aspects of machine design and construction. Since size and performance are often related to each other, it has proved practical to develop standards which have regard to these two parameters.

An example of this is ISO 2858 which specifies the principal dimensions and nominal duty points of horizontal end-section back-pull-out centrifugal pumps for pressure rating PN 16 (16 bar), mainly intended for chemical applications. The British version of this standard is BS 5257 and the German is DIN 24256. The European standard for these pumps is EN 22858.

Water pumps of the same type but rated at pressure class PN 10 are covered by the German standard DIN 24255, prEN 733. This standard covers pumps of approximately the same size as ISO 2858 (BS 5257; EN 22858) but with reduced dimensions and higher port velocities.

BS 5257 also specifies dimensions for baseplate installations, ISO 3661, and seal cavities, ISO 3069.

The ISO 2858 standard pump has not been accepted in the USA because of its metric dimensions. There is however an equivalent inch standard, ANSI B73.1 which specifies dimensions and duty ratings in American units. ANSI B73.2 covers a range of inline vertical chemical pumps.

BS 4082 specifies external dimensions, pressure-temperature ratings and nominal duty points at various speeds for vertical inline centrifugal water pumps. These apply to maximum differential heads of 160 m and maximum flows of 384 m^3/h.

prEN 734 specifies the designations, dimensions and duty points of a range of multi-stage segmental pumps rated at PN 40.

BS 7117 deals with construction, installation and maintenance of metering pumps specifically for dispensing liquid fuel.

API 610 is a specification of pump construction and packaging. Dimensions of the seal cavity are specified together with materials and inspection requirements. Hydraulic duties are not specified but test tolerances for water are given. API 610 allows performance testing on water up to 66 °C which gives good results based on vapour pressure rather than gas evolution.

API 674, 675 and 676 cover positive displacement pumps; reciprocating, dosing and rotary. VDMA 24286 covers procurement, testing, supply and dispatch of the three positive displacement pump types. DIN 5437 covers semi-rotary hand pumps. DIN 24289 covers specification, installation and data sheets for reciprocating pumps.

There are no British Standards for specifying positive displacement pumps. A European Standard for rotary pumps will be issued for public comment soon. If this standard is accepted, a reciprocating pump standard will be produced using the same format. An ISO committee is reviewing API 674 for a possible new standard.

3.7.6 Vibration and noise

Vibration occurs in all rotating and reciprocating machinery when in operation. Vibrations are normally caused by imbalance of the rotating or reciprocating components, but can also be caused by hydraulic flow through a pump or valve. Since imbalance gives rise to vibration it is obviously desirable to maintain sufficiently low values so as to prevent mechanical damage. Noise is also created and is also undesirable from an environmental point of view. Standards and guide-lines are available which specify maximum levels of vibration to avoid mechanical damage, loss of reliability and also to prevent unacceptably high noise levels.

The standards dealing with noise and vibration are therefore basically concerned with the evaluation of maximum permissible vibration levels for various types of machinery and the requirements and specification of the necessary measuring equipment. Vibration levels are today measured as vibration velocity mm/s rms instead of amplitude and frequency. This simplifies measurement and the evaluation of vibration levels. The requirements for sensors and measuring equipment for vibration velocities in mm/s rms within the frequency 10 to 1000 Hz are dealt with in ISO 2954, BS 4675: Part 2: 1978. ISO 3945 specifies how to measure and evaluate vibration at site for machines with speeds between 10 rev/s and 200 rev/s. ISO 8821 and BS 7130 specify the convention to be adopted regarding shaft keys.

The evaluation and recommendation of preferred vibration severity ranges is complicated by the variations in design, construction and installation of different types of machinery. The VDI 2056; ISO 2372 and BS 4675: Part 1: 1976 have therefore divided up machines into groups. Rotodynamic pumps usually appear in group T (VDI) or Class IV (ISO and BSI), although some types may appear in other classification groups as a result of variations in design and construction. The pump manufacturer usually establishes the group to which a particular pump construction belongs.

The joint ISO/API standard currently being prepared, relating to the technical specifications of pumps, will also contain a section concerning vibration in horizontal pumps.

The vibration level measured on a pump during testing is greatly influenced by the pump mounting. Temporary mounts on test rigs can be quite flexible allowing the pump to register high levels. Positive displacement pumps, both rotary and reciprocating, can suffer from vibration due to imbalance which cannot be rectified due to the pump design. Vibration readings during tests can only be used as information if the pump is mounted on temporary stands. When installed at site, and grouted in, the vibration levels will be much reduced.

IEC 994 and EN 60994 give guidance on vibration and pulsation measurement at site for rotodynamic pumps and turbines. Other standards which may be useful include ISO 5343; ANSI S2.17; ANSI S2.40; API 678;

Noise emitted by machinery constitutes a considerable problem where it is necessary for personnel to be present in the vicinity of machinery for long periods. Consideration must be given to the complete pump package not just the pump. The driver, plus any power transmission equipment, and process pipework will contribute to the overall noise level. Specific areas can be designated "ear protection zones" where ear defenders must be worn. Typical noise level restrictions in current specifications call for 84 dBA.

The current standards deal basically with the measurement, evaluation and presentation of noise data. The following standards should be reviewed for suitability for any specific application: ISO 3740; ISO 3744; ISO 3745; ISO 3746; ISO 3747; ISO 4412; ISO 4871; ISO 6081/2; EN 27574; ANSI B93.71; ANSI B93.72; ANSI S1.4; ANSI S1.4a; ANSI S1.6; ANSI S1.13; API 615; BS 4196; BS 7025 and DIN 45635.

As most pumps are driven by electric motors the following standards on noise from electric motors should be given consideration; ISO /R 1680, IEC 34-9 and BS 4999 Part 109.

3.7.7 Forces and moments on connections

Pump installations should be laid out so that the loads transmitted from the piping system to the pump connections are as small as possible. Considerable forces are brought to bear on the pump connections as a result of liquid pressure and thermal expansion, for example. A knowledge of the extent to which the pump connections are capable of withstanding forces and moments without risk is therefore important, from the point of view of pump installation and construction, so as to avoid the unnecessary use of compensating devices etc.

However, the pump should not be looked upon as a pipe support or anchor. These are secondary functions for a pump and can be detrimental to the primary function.

Instructions regarding the permissible forces and moments are given in API 610. This standard determines the permitted forces and moments on the basis of the size of the flange connection and, to some extent, the weight of the pump. API 610 pumps are robust. Special reinforced baseplates are used because the nozzle load requirements are so high. Other standards use the maximum pump pressure and connection dimensions as the basis for determining the permitted forces and moments. The situation is further complicated by the fact that different manufacturers employ their own individual methods of determining the permitted forces and moments. It is therefore advisable to contact the manufacturer for advice in cases of doubt.

The joint ISO/API standard, which is being prepared, concerning the technical specifications for centrifugal pumps, will include a section devoted to forces and moments on flanged connections. It is hoped that this will result in the general adoption of more consistent regulations. The new EN for rotary positive displacement pumps includes forces and moments data.

3.7.8 Components

Many pump components and parts are not exclusively used on pumps, but are also used in many other types of machinery. The dimensions of such parts are usually the subject of separate component standards. Screws, plugs, keys and keyways for example, are basic pump components for which dimensional standardisation is a well known concept. The standards applying to certain other pump components, however, are worth studying more carefully.

The shape and dimensions of flanges is dealt with by many standards which are sub-divided according to material, type and pressure rating. The following standards are the most popular and cover working pressures up to 420 barg depending upon the material and the temperature.

Standard	Dimensions
ISO 7005	both
BS 4504	metric
DIN 2531 to 2533	metric
DIN 2543 to 2549 + 2551	metric
DIN 2628 + 2631 to 2638	metric
DIN 2565 to 2569	metric
ANSI B16.5 / BS 1560	inch
SAE	inch

Figure 3.170 Popular flange standards

Shaft end dimensions are given in ISO 775 and its equivalent BS 4506. These standards give specifications relating to various types of cylindrical and conical shaft ends with optional screw threads for retaining purposes.

Shaft centre heights and tolerances for driving and driven machines are standardised in ISO 496 and BS 5186. These standards should be read in conjunction with IEC 72 and IEC 72A which define shaft heights for electric motors. Mounting flanges, together with shaft sizes for hydraulic pumps and motors are covered by ISO 3091/1, ISO 3091/3, ANSI B93.81, BS 6276 and SAE J896.

Shaft seals are important machine elements in all pumps from a reliability point of view. Since there are many different types of seals manufactured by a great number of manufacturers, there has long been a necessity to establish some form of standardisation. DIN 3780 covers stuffing box dimensions for soft packing. DIN 24960 is a recent standard and is derived from an earlier VDMA standard for mechanical seal cavities and material codes. For further information see Section 7.3.2. The international standard ISO 3069 and sections of its BSi equivalent, BS 5257, include tolerance and surface finish requirements for shafts and cavities also refer to shaft seals. These standards only deal with the dimensioning of the actual sealing cavity. The manufacturer should be consulted about pressure/temperature/speed limitations. API 610 and ANSI B73.1 and B73.2 all specify seal cavity dimensions.

API 382 is a major standard covering all aspects of mechanical seals. This standard deals with problems associated with more stringent leakage and safety requirements including selection and testing.

There are no established dimensional standards for shaft couplings. Data concerning different types of shaft couplings are dealt with in Chapter 8. Some criteria concerning shaft couplings are covered by BS 3170 and API 671.

3.7.9 Testing

Mass produced pumps are subject to selective testing. Some manufacturers of mass produced pumps test all the rotating assemblies in fixed test casings; this is not a good policy and should be avoided. Pumps built specifically for contract will be tested when complete. The running test is the final check that the pump fulfils the agreed or specified requirements and is therefore of great importance to both the consumer and the supplier. Acceptance testing is usually carried out at the pump manufacturer's premises. It is obviously impracticable for the purchaser to fully specify the manner in which the acceptance test is to be carried out and evaluated for every order. Reference is normally made, therefore, to an acceptance testing standard which contains instructions regarding the testing procedure, permissible tolerances etc. Testing standards have been issued by many standard issuing organisations, as shown by the following list:

ANSI B93.95	hydraulic pumps & motors
API 610	centrifugal pumps
API 674	reciprocating pumps
API 675	metering & dosing pumps
API 676	rotary positive displacement pumps
API RP 11S2	submersible pumps
ASME PTC 7, 7.1, 8.2, 18.1	rotodynamic pumps
BS 4617 ≡ ISO 4409	hydraulic pumps and motors
BS 5316 Part 1 ≡ ISO 2548	rotodynamic pumps
BS 5316 Part 2 ≡ ISO 3555	rotodynamic pumps
BS 5316 Part 3 ≡ ISO 5198	rotodynamic pumps, laboratory precision
BS 5860 ≡ IEC 607	very large pump/turbines
BS 6335	pressure ripples in hydraulics
BS 7250 ≡ ISO 8426	capacity of hydraulic pumps/motors
DIN 1944	centrifugal pumps
DIN 14410	portable fire pumps
DIN 14420	fire pumps
DIN 19760	effluent pumps
Hydraulic Institute 1.6	centrifugal pumps
Hydraulic Institute 6.6	reciprocating pumps
ISO 8278	pressure compensate hydraulic pumps
VDMA 24284	positive displacement pumps

ISO has been working on a standard combining ISO 2548 and ISO 3555 since 1981; a version has been circulated as a committee draft since 1991, ISO/CD 9906. This is a poor standard and needs considerable work. ISO is also working on a standard for testing positive displacement pumps; this has been rewritten completely three times to date.

It is important to select a test standard which is compatible with the pump application to be tested. If a centrifugal pump is being used to pump oil at 300 cSt it would be a good idea to test with a liquid which approaches the viscosity. Most rotodynamic pump test specifications call for "clean cold water". API 610 allows performance testing on water up to 66 °C which gives good results based on vapour pressure rather than gas evolution. Some of the positive displacement pump test requirements do not specify anything about the liquid.

BS 5316: Part 1 is a direct equivalent of ISO 2548, which, since its introduction, has become the dominating international standard for rotodynamic pumps for industrial applications. Petrochemical pumps are tested to API 610 or the Hydraulic Institute. In all testing standards it is of course the maximum permissible tolerances of flow, differential head and efficiency which are of greatest interest. Some purchasers annex the testing standard to the purchaser order and add overriding clauses for the tolerances.

The correct choice of testing standard, when ordering and specifying, is obviously very important. The acceptance test standards for mass-produced pumps according to BS 5316 Part 1, can be about 10% higher than the specified duty point. The standards must be reviewed carefully to ensure the particular problems of a specific application are addressed.

Manufacturers should be assessed regarding their in-house process control systems. Great store is set by ISO 9000, EN 29000 and BS 5750. These standards specify rules regarding how departments should work, how departments should communicate with each other and most important, what to do when things go wrong. Pumps for critical applications; nuclear, pharmaceutical, off-shore; are normally purchased from manufacturers who operate one of the standard systems and have been vetted by an accredited third party assessor. All pump manufacturers should operate some type of formalised control system.

A traditional system can be as good as ISO, EN or BS when operated correctly and the Quality Control staff are empowered to scrap substandard components. Inquiries for important pumps should always request a Vendor Document Requirement form, VDR, and a Quality Control Plan, QC Plan, to be submitted as part of the quotation.

The VDR will show the purchaser what documentation; drawings, manuals, certificates; the manufacturer considers important and will supply with the pump. The QC Plan will show all the inspection operations, hold points and any conditions when Engineering must be contacted to exert overall control for important decisions. After carefully studying the VDR and the QC Plan, the true character of the manufacturer will be revealed.

VERTICAL AND HORIZONTAL PUMPS

▶ AS/NT

Centrifugal pump, vertical axis, single stage support plate, column pipe, overhung impeller, coupled to electric flanged motor.
Four construction versions, eight sizes, twenty types. Delivery heads up to 40 meters and capacity up to 90 m3/hr. Height of suction from 300 to 3000 mm, without intermediate bearings or positive seals. PFTE- filled bushings fluxed by the pumped liquid.

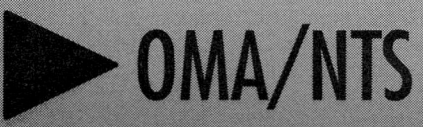

▶ OMA/NTS

Centrifugal pump, horizontal axis, close coupled, single stage, end suction, overhung impeller, coupled to electric flanged motor and food mounted. Four construction versions, 7 sizes, 20 types. Delivery head up to 40 meters and capacity up to 90 m3/hr. Its mechanical seal, of exclusive design, insures a higt safety degree and a remarkable chemical and mechanical strenght. The internal assembly optimizes the flushing action and the loss of heat and reduces the load on faces a to minimum.

MAIN CONSTRUCTION FEATURES

The centrifugal pumps Barbera, in plastic materials, heavy-duty designed, have beeen manufactured to operate in particular with corrosive and dangerous fluids, held in tanks, reservoirs or wells. As no metal part comes into contact with the pumping fluid, no corrosion or fluid pollution occurs.
To insure the higest consistency with the widest range of chemicals as possible and the maximun service reliability, the whole pump and not only the fluid-wetted parts, are made in Polypropylene, PVC or PVDF and gaskets, bushings and mechanical seal are in PTFE, CERAMIC, FPM, EPDM materials, interchangeable to each other, if required.
The electric driving motors are direct coupled to the pump shaft, sized according to international standards and protected from acid, alkaline or saline environments.
Even if exposed to vapors or drippings, the SAVINO BARBERA centrifugal pumps are designed to accomplish the most burdensome tasks under the maximum duration of life and reducing the servicing requirements to a minimum.

sb SAVINO BARBERA
di Barbera & Castiglioni
Via Torino, 12 - 10032 Brandizzo (TO) - ITALY
Tel. (011)913.90.63 - Tlx 22.22.35 SAV BAR I - Fax 39.11.913.73.13

RS ITALIA

MACKLEY PUMPS

CLARKE CHAPMAN MARINE, VICTORIA WORKS, GATESHEAD, TYNE & WEAR NE8 3HS
Telephone: 091 477 2271 Fax: 091 477 1009

Split Case and Multi Stage Ring Section Pumps: Specialising in a variety of liquid handling applications.

- Mine Dewatering
- Water Supply and Sewage Services
- Power Stations
- Chemical and Process Industries
- Irrigation and Fire Fighting
- Gas and Petroleum Industries

SCH 8/6
Diesel Driven Single
Stage Split Case Pump

ROLLS ROYCE

INDUSTRIAL POWER GROUP

Rolls-Royce Industrial Power Group, NEI House, Regent Centre, Newcastle upon Tyne, NE3 3SB

4 Pumps and piping systems

Chapter 4 - Pumps and piping systems - deals with the interaction of the pump and piping systems. The chapter is primarily directed to the pump sizing problem. Operational parameters at part-load are dealt with in Chapter 5 - Flow regulation. In order to determine the required pump performance it is first necessary to define the characteristics of the particular pipework system or installation. This applies to single pipe systems as well as branched systems. In this context, a single pipe system implies both suction and discharge pipework. An appreciable portion of the pressure losses in the system are caused by fittings and valves, especially control valves in throttle-regulated installations. Pumps working in a system operate in conjunction with the pipework and other components to deliver the required flow. Final choice of pump size must be made with due regard to the pump performance and efficiency curves and the hydraulic tolerances of the installation components. The chapter finishes with an explanation of the water hammer, pressure surge, phenomenon. Easily calculated estimations are given to show when the risks of water hammer are present and when the use of protective equipment is necessary. Various types of protection devices are reviewed.

Contents

4.1 System curves
   4.1.1 Pump nominal duty point
   4.1.2 Single pipe system
   4.1.3 Variable system curves
   4.1.4 Branched pipe systems
   4.1.5 Viscous and non-Newtonian liquids

4.2 Valve pressure drop
   4.2.1 General
   4.2.2 Isolating valves
   4.2.3 Non-return valves
   4.2.4 Control valves

4.3 Multiple pump systems
   4.3.1 Series pump operation
   4.3.2 Parallel pump operation
   4.3.3 Pressure boosting
   4.3.4 Pressure maintenance

4.4 Pump hydraulic specification data
   4.4.1 Hydraulic tolerances
   4.4.2 Pump Q and H at maximum efficiency
   4.4.3 Minimum pump flow

4.5 Water hammer in pump installations
   4.5.1 Hydraulic gradient
   4.5.2 Causes of water hammer
   4.5.3 Pump behaviour power loss
   4.5.4 Protection against water hammer

4.6 Pressure pulsations in piping systems
   4.6.1 Rotodynamic pumps
   4.6.2 Positive displacement pumps

4.7 Bibliography

4.1 System curves

The sections which follow deal with the relationship between fixed speed pumps and piping systems. Variable speed pumps produce an infinite number of pump characteristics, which can be derived using the affinity laws, and make it possible for a pump to operate virtually anywhere on any system curve.

4.1.1 Pump nominal duty point

When connected in a pipe system a pump will operate at a point of equilibrium between the pump and the pipe system. At the point of equilibrium the energy supplied by the pump is equal to the losses due to resistance in the system. The pump characteristics in this respect are usually represented by the H-Q curve. The equivalent curve for the pipe system is called the system curve and is designated H_{syst}. H_{syst} must include suction **and** discharge losses.

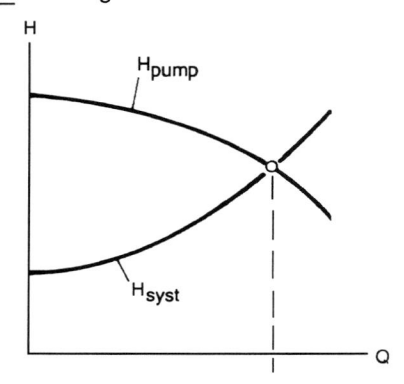

Figure 4.1 Pump nominal duty point

The flow at which the pump and system curves intersect each other, is the flow which will pass through the pipe system. In order to correctly determine the size and dimensions of the pump and pipe system a knowledge of the characteristics of their respective curves is necessary. This section deals with the characteristics of the pipework system.

Figure 4.1 shows typical curves for a centrifugal pump with a low specific speed, centrifugal pump, when pumping a water-like liquid. The curves vary in shape when pumping other liquids and when using other types of pumps. Generally, however, the flow is always determined by the balance between the pump and system.

4.1.2 Single pipe system

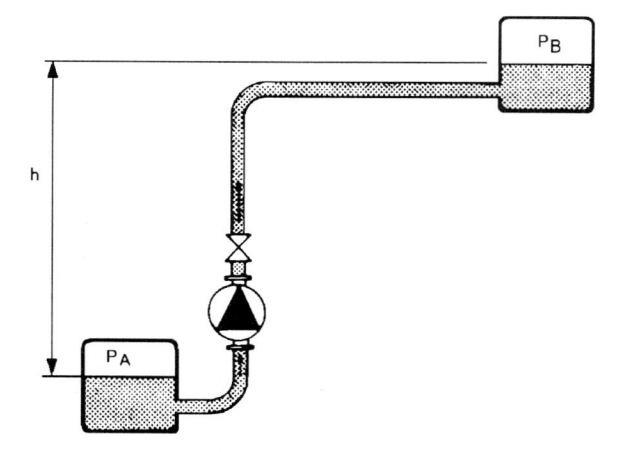

Figure 4.2 Example of single pipe system

The system differential head is usually divided into a static component H_{stat} and a loss component h_f.

$$H_{syst} = H_{stat} + h_f \qquad \text{Equ. 4.1}$$

The static component, which is generally independent of flow, comprises of the difference in static pressures and the level differences between the boundaries of the system. Using the designation as in figure 4.2:

$$H_{stat} = \frac{p_B - p_A}{\rho g} + h \qquad \text{Equ. 4.2}$$

where
- p = static pressure (N/m²)
- ρ = density of the liquid (kg/m³)
- g = acceleration due to gravity (m/s²)
- h = difference in elevation (m)

The losses include friction losses in straight pipe and losses in valves and fittings etc.

$$h_f = h_{(straight\ pipe)} + h_{(fittings\ and\ valves)} \qquad \text{Equ. 4.3}$$

Using the terms as defined in Chapter 2

$$h_f = \lambda \cdot \frac{l}{d} \cdot \frac{v^2}{2g} + \Sigma \zeta \cdot \frac{v^2}{2g} = (\lambda \cdot \frac{l}{d} + \Sigma \zeta) \cdot \frac{v^2}{2g}$$

$$= (\lambda \cdot \frac{l}{d} + \Sigma \zeta) \cdot \frac{Q^2}{\left(\frac{\pi \cdot d^2}{4}\right)^2} \cdot \frac{1}{2g} \qquad \text{Equ. 4.4}$$

where
- λ = loss coefficient for straight pipe
- ζ = loss coefficient for valves, fittings
- Σζ = sum of all loss coefficients
- l = pipe length (m)
- d = pipe diameter (m)
- Q = flow (m³/s)
- v = flow velocity (m/s)

For a given pipework system of l and d, with water-like liquids the loss coefficients λ and ζ are often independent of Q for large values of Reynolds Number. The loss of head then becomes:

$$h_f = \text{constant} \cdot Q^2 \qquad \text{Equ. 4.5}$$

If the pressures at the system boundaries are such that $P_A = P_B$ = atmospheric pressure, then H_{stat} equals the elevation difference h, the system differential head then becomes:

$$H_{syst} = H_{stat} + h_f = h + \text{const} \cdot Q^2 \qquad \text{Equ. 4.6}$$

or graphically

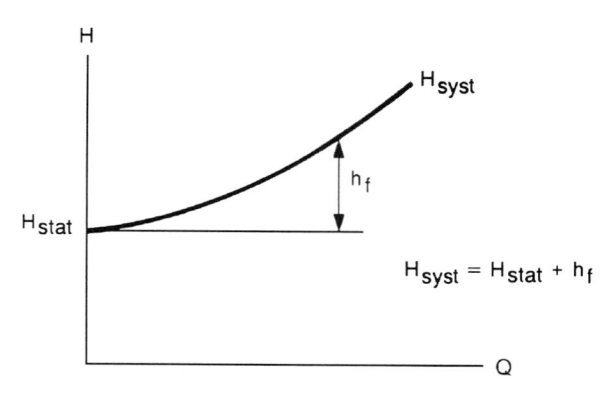

Figure 4.3 System curve

Note that the system curve represented in figure 4.3 is based on p_A, p_B and h being independent of the flow Q. It is also assumed that λ and ζ are independent of Reynolds Number. These conditions are often, but not always, fulfilled.

Figure 4.4 shows three different pipework systems having the same level differences and therefore the same differential static head.

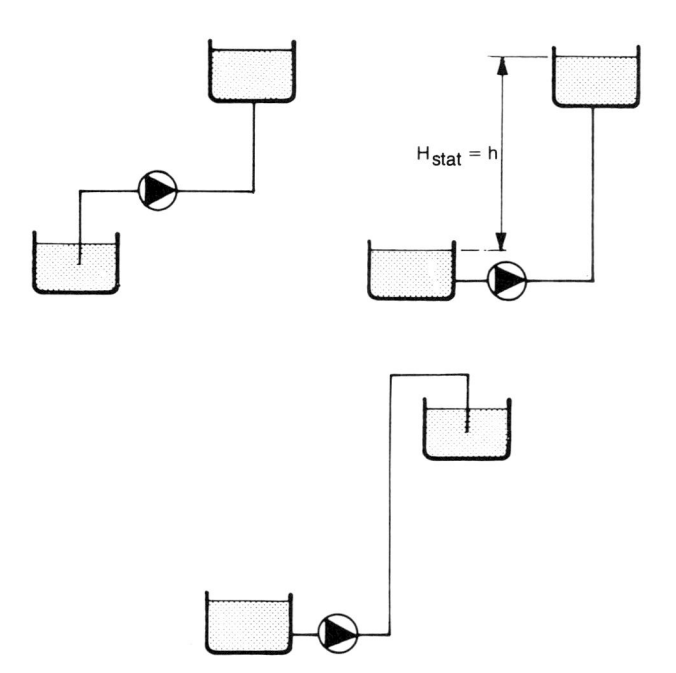

Figure 4.4 Pipework systems with the same static head

For certain piping systems, circulating systems for example, $H_{stat} = 0$ and the head of the system consists entirely of pipe flow losses.

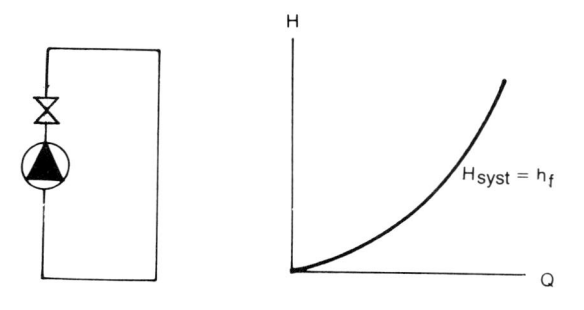

Figure 4.5 Piping system where $H_{stat} = 0$

For other piping systems with short pipes and considerable pressure or level differences the flow losses are negligible and $H_{syst} = H_{stat}$.

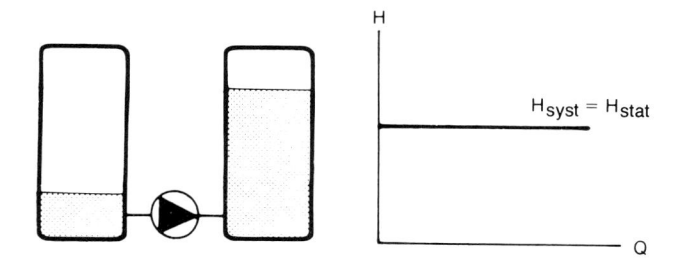

Figure 4.6 Piping system where $h_f = 0$

Systems similar to figure 4.5 are called 'pipeline' systems because all the losses are friction losses. Systems similar to figure 4.6 are called 'chemical' systems because it is typical of the arrangements found in chemical and petrochemical installations. When developing the system curve the flow losses are calculated according to Chapter 2 Liquid flow.

4.1.3 Variable system curves

Certain conditions can cause the system curve to vary according to the operating situation. Some examples of such situations are given below.

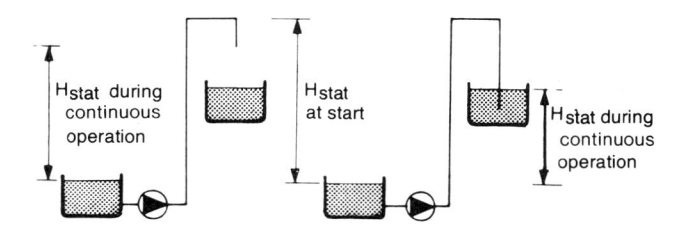

Figure 4.7 Static head at start and during continuous operation

When starting up, the discharge pipes are filled with air above the liquid level of the lower tank. The pump must lift the liquid to the highest point in the line. The static delivery head reduces when the line is completely filled with liquid. If the pump's differential head at the zero discharge point, shut-off head, does not exceed H_{stat} at the start then pumping cannot begin.

For installations similar to figure 4.8 the static head varies as the level difference between the liquid surfaces in the vessel changes. Similar system curve changes occur in enclosed vessels, in which the pressure is dependent upon the operational situation.

The system curve's loss component can also vary according

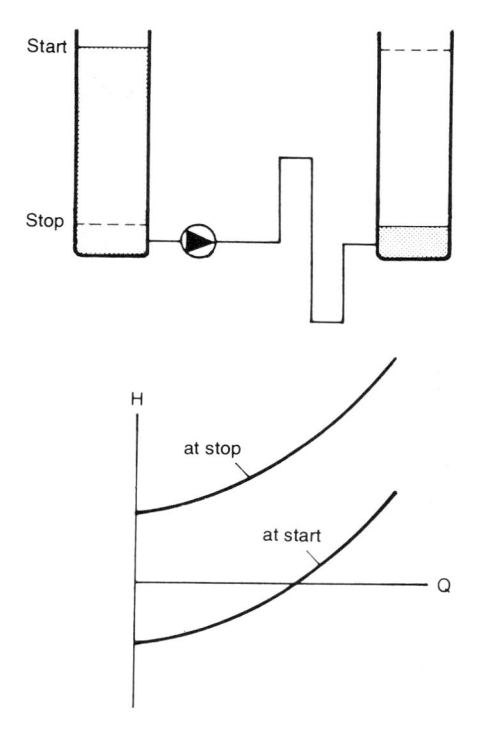

Figure 4.8 System curve with varying static head

to the operational situation. The most usual case being when rust or other deposits begin to build up in a pipeline after a period of operation. This causes an increase of flow resistance in the pipeline which results in a steeper system curve. Other examples of varying loss curves are temperature variations; viscosity, Reynolds Number; concentration variations in the case of liquid solid mixtures, dosage variations etc.

An intentional change to the system curve is brought about by throttle-regulation, see Chapter 5 Flow regulation.

The pump H-Q curve is also affected by viscosity and wear for example, see Chapter 3 Pumps.

The relevance of the system curve to operational situations can be illustrated by some examples:

The pump installation shown above has been designed to be on the "safe side", the pipe losses having been estimated too high. The actual system curve results in a larger flow than required, which means greater pump power requirement and reduced efficiency. This situation gives rise to the risk of pump overload and cavitation. The desired flow can be maintained by introducing a flow resistance into the piping system.

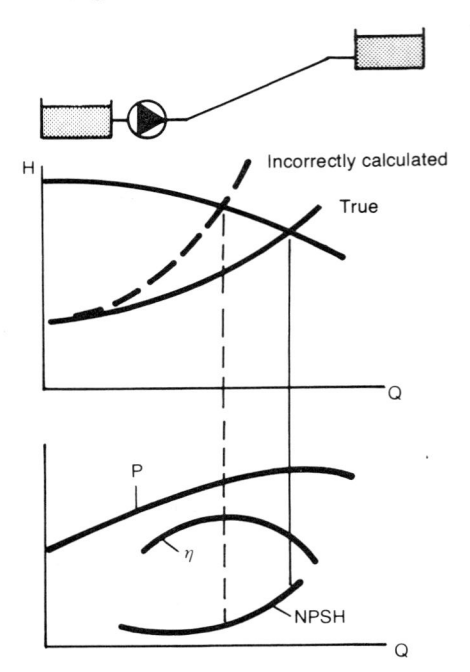

Figure 4.9 Consequences of incorrectly calculated pipe losses

Figure 4.10 illustrates the reduction of pump flow caused by the build up of deposits in the pipeline. Since the size of pump has been chosen in order to obtain optimum efficiency for operation with new pipes, the subsequent flow reduction results in reduced pump efficiency.

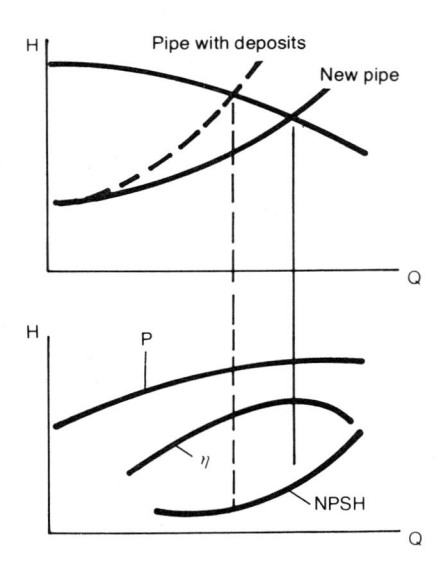

Figure 4.10 The effect of deposits (scale, rust etc.) on pipes

The pump in figure 4.11 has been chosen to suit a particular operational situation, which is equivalent to $H_{stat, max}$ i.e. when the liquid levels in the lower and upper vessels are lowest and highest respectively.

In the alternative case $H_{stat, min}$ a larger flow is obtained. When plotting pump and system curves, it is not sufficient to plot only the differential head. The NPSHa and the NPSHr should be plotted for the various flows considered. If allowances are added to the system losses, for eventual fouling for example,

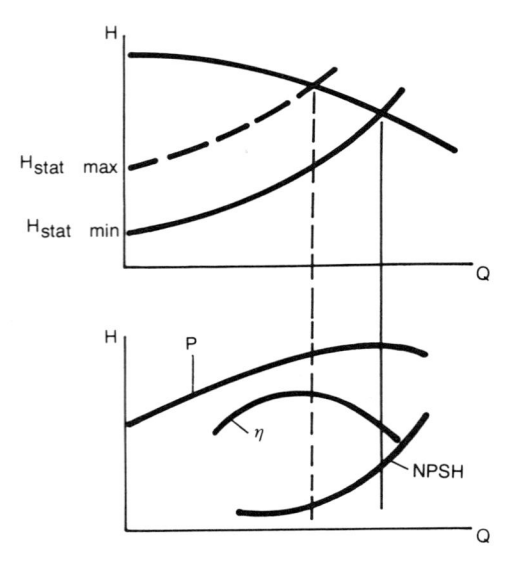

Figure 4.11 The effect of varying static head

the operating conditions for the new pipe must be checked if balancing valves are not fitted.

If the plotted characteristics indicate that there is no pump available to suit the system NPSHa the "system" must be modified. In this context, the system includes some operating conditions. In order to provide more NPSHa the suction pressure must be increased or the difference between the suction pressure and the vapour pressure must be increased. The following courses of action should be considered.

- Re-route the suction pipe to make it shorter.
- Move the pump closer to the suction source and reduce the length of the suction pipe.
- Lift the suction source up to increase H_{stat} on the suction side.
- Increase the suction pipe diameter.
- Reduce the liquid temperature to reduce the vapour pressure.

4.1.4 Branched pipe systems

The first example of a branched pipe system is a circulation system. In such a case $H_{stat} = 0$. The flow Q_p, which passes

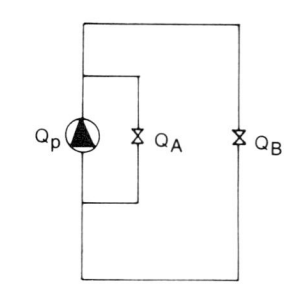

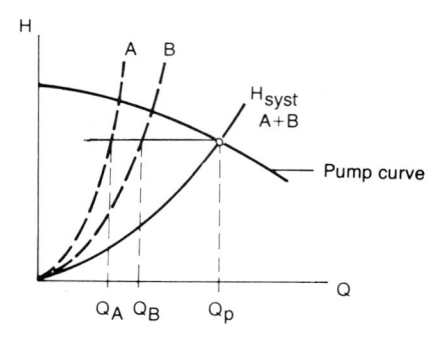

Figure 4.12 Branched circulation system

through the pump, is divided at the branch point. Continuity dictates that

$$Q_p = Q_A + Q_B \qquad \text{Equ. 4.7}$$

Both branches A and B have their own system curves, the sum of which determines the resultant system curve. The intersection point of the resultant system curve and the pump curve determines the pump's operating point.

Figure 4.12 also shows how large a portion of the pump flow flows through each respective branch. The division of flow depending upon the magnitude of the losses in each respective branch. In this example it is assumed that the branch point is near to the pump i.e. that flow losses between the pump and the branch point can be neglected.

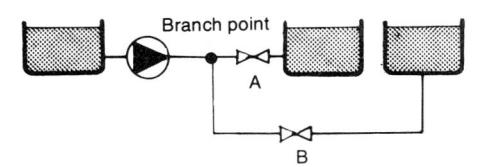

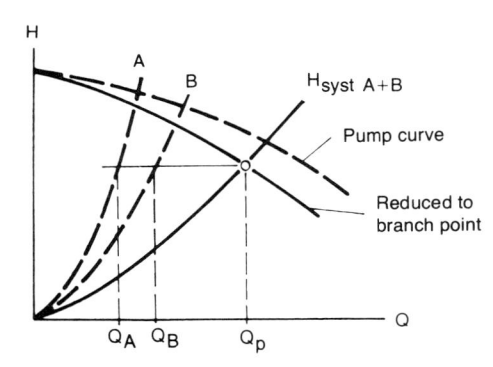

Figure 4.13 Branched pipe system, $H_{stat} = 0$

The next example, figure 4.13, also takes into consideration the losses in the main pipeline up to the branch point. This is done by reducing the pump curve by the magnitude of the losses up to the branch point. The "reduced" curve is then matched to the remaining system curve as before.

In the third example the pump curve is first reduced to the branch point. Thereafter the system curves A + B are added to the resultant system curve H_{syst} A + B. The intersection point between the reduced pump curve and the resultant system curve determines the pump's operational point, as shown in figure 4.14.

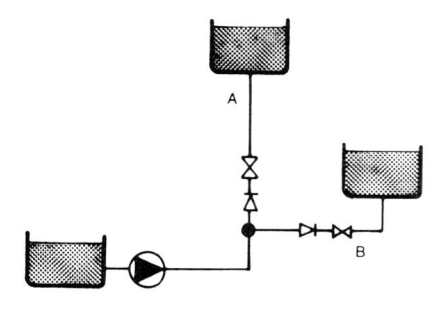

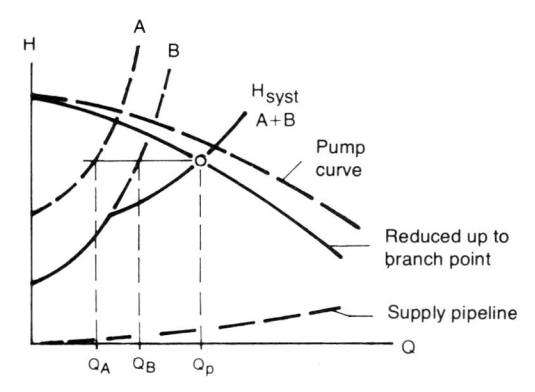

Figure 4.14 Branched pipe system with static differential head

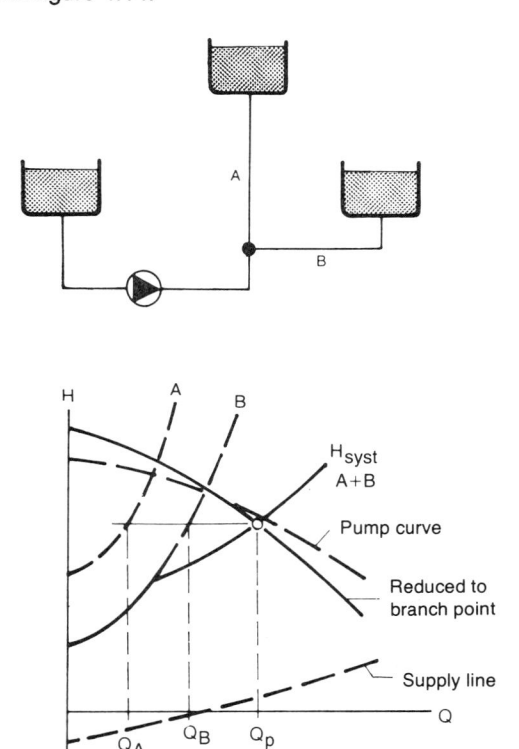

Figure 4.15 Branched pipe system with static supply line

In the fourth example there is a level difference, in this example a supply head, between the supply vessel on the suction side and the branch point. The system curve for the supply pipeline exhibits therefore, a static head, $H_{stat} < 0$. The method is the same as in the previous examples. First reduce the pump curve with the supply line curve up to the branch point. Then determine the system curves for the branches A + B from the branch point. The resultant branch curve is then matched to the pump's reduced curve as shown in figure 4.15.

This method can be applied in principle to calculate any pipework system, however complicated and containing any number of branch points. The first step is to determine the resultant system curve for the branch furthest from the pump in relation to the furthest branch point. The next step is to deal with the next branch point and so on until the branch point nearest to the pump is reached.

Remember the system curve is a combination of suction and discharge losses. The H for the vertical axis in figures 4.1, 4.3, 4.5, 4.6, 4.8 and 4.9 to 4.15 is really ΔH, the losses from the beginning of the suction pipe to the end of the discharge pipe. Rotodynamic pump curves indicate ΔH across the suction and discharge connections of the pumps.

A common problem is to size a piping network with many branches and many tapping points in such a way as to guarantee a predetermined flow and pressure at each tapping point. Sizing is carried out so that the pipeline requiring the most pressure, usually the pipeline to the furthest tapping point, is dimensioned first. The branch lines are then calculated on the basis of the excess pressure available at the branch points. The excess pressure at the branch points can be such that it is necessary to equip a branch line with extra throttling to prevent excessive flow and unbalancing the system. Throttle valves can be fitted to balance a complex system. As we have seen in figure 2.25, some applications have well defined velocity restrictions depending upon the quality of the pipework construction. Also, throttle valves can be used to correct imbalance caused by irregular fouling of different parts of the system.

4.1.5 Viscous and non-Newtonian liquids

The pump and system curves have, in the previously illustrated figures, been representative of the relationships for centrifugal pumps and water-like liquids. For viscous and non-Newtonian liquids the curves have different shapes.

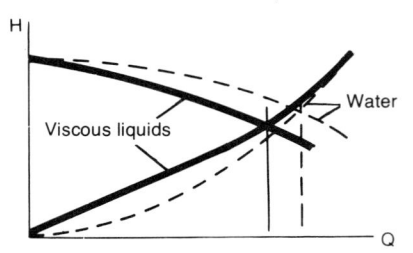

Figure 4.16 Rotodynamic pump and system curves for more viscous liquids compared to water

For viscous liquids the laminar flow regime is maintained at higher flow velocities. The system curve thus becomes modified. The centrifugal pump H-Q curve falls off more quickly with increased flow for higher viscosities.

High viscosity liquids are normally pumped by positive displacement pumps. Rotodynamic pumps can handle higher viscosity liquids by derating the pump, usually reducing the speed to reduce the torque required. A common bench-mark used to separate rotodynamic pumps, low viscosity, from positive displacement pumps, high viscosity, is 300 cSt. This is not an absolute number; it depends upon the size of the pump and can be exceeded when circumstances necessitate using a rotodynamic pump. The high viscosity tends to reduce the internal leakage, slip, of rotary positive displacement pumps making them more efficient. High viscosity causes higher pipe losses. The system curve becomes completely linear if laminar flow is maintained at the duty conditions. The increased flow and increased pressure necessitates greater power requirements.

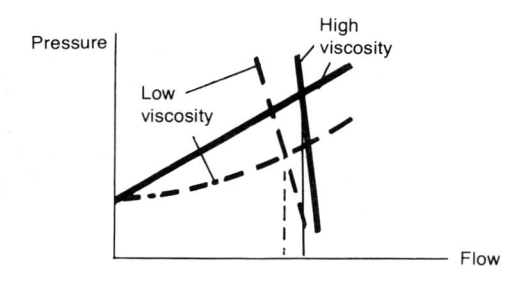

Figure 4.17 Rotary positive displacement pump and system curves for different viscosities

The pumping of chemical paper pulp is chosen here as an example of system curves for non-Newtonian liquids. The liquid characteristics of pulp suspensions is reflected in the shape of the system curve. The centrifugal H-Q curve also changes in relation to that for water.

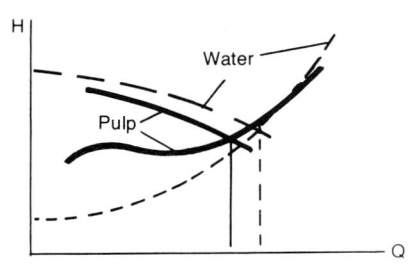

Figure 4.18 Rotodynamic pump and system curves for water and paper pulp

Since the pump's operating point is always determined by the equilibrium between the pump and system, the shape of these curves is significant from the point of view of pump installation performance. This applies also to the sensitivity to disturbances and part-load characteristics.

4.2 Valve pressure drop

4.2.1 General

For pump installations which use valves as a means of controlling the flow, the pressure drop across the valves may constitute a considerable proportion of the piping system head losses. The system curve which is the basis for the determination of the required differential head for the pump, cannot be defined with sufficient accuracy if the pressure drop across the valves, particularly control valves, is not considered. The following considerations apply to the choice of valves and the pressure drop across the valves for the purpose of determining the required pump performance. A more detailed treatment of flow control is to be found in Chapter 5.

4.2.2 Isolating valves

The term "isolating valves" applies to those valves which are intended to be either fully open or fully closed. Isolating or stop valves are installed in an installation in order to make components accessible for service; to direct flow in another direction and for other similar functions.

The crucial requirements of an isolating valve are low pressure drop in the open position and good leak-free sealing in the shut position. Information regarding the calculation of pressure drop in the fully open position is given in Chapter 2 Liquid flow. The most reliable method, however, is to obtain the value of the loss coefficient, Flow Coefficient, from the valve manufacturer.

4.2.3 Non-return valves

Non-return or check valves are installed in a pipe system in order to prevent the reverse flow of liquids. There are many different types of non-return valves available. Of the possible choices of valves for a particular application, it is normal to select the one which gives least pressure drop. In some cases it is necessary to have exceptionally fast closing check valves. These are biased, by means of spring loading for example, and give rise to large pressure drops. These valves can also give rise to horrendous pressure surges due to the water hammer effects they create. Very slow closing valves are available specially designed to eliminate water hammer. Since the effective flow area varies with the flow through the valves, the loss coefficient ζ also varies with flow in non-return valves. Pressure drop calculations are best made by using data supplied by the manufacturer of the valve in question.

4.2.4 Control valves

The purpose of control valves, in contrast to the other valve groups, is to create head losses and thereby regulate the flow in the pipework system. The choice of valve is therefore made on the basis of entirely different criteria than for isolating or non-return valves.

The principal requirements of a control valve is based upon its regulating characteristics. These characteristics are also dependent upon the interaction between the pump, control valves, pipe system and other control equipment. The valve can only regulate flow by increasing the liquid velocity locally to create a high pressure drop. The capability of the valve to regulate ceases when the pressure drop across the valve approaches zero. If it is required to sustain regulation capability even at Q_{max}, maximum flow in the proposed process, the valve must therefore have an available pressure drop at that flow which is somewhat greater than that for the fully open valve position. Control valves are sized so that they are never fully open or closed; a 10% margin of travel at each end of the stroke is a minimum safety margin.

The design of other control equipment is simplified if the flow through the pipe system has fairly linear characteristics in relation to the position of the valve. This requirement also makes a large pressure drop across the valve desirable at Q_{max}.

The flow through a valve is calculated using the formula:

$$Q = A_v \sqrt{\frac{\Delta p}{\rho}} \qquad \text{Equ 4.8}$$

where
- Q = flow (m³/s)
- A_v = valve flow coefficient (m²)
- Δ_p = pressure drop across the valve (Pa)
- ρ = density of the liquid (kg/m³)

A measure of the valve capacity and size is its flow coefficient A_{v100} at the fully open position, 100% stroke. In order to make a rapid assessment of a suitable valve size the following rule of thumb is used:

$$A_{v100} \geq \frac{Q_{max}}{\sqrt{g\,(0.1 \cdot H_{stat} + 0.3 \cdot h_{fpipe})}} \qquad \text{Equ 4.9}$$

- A_{v100} = flow coefficient at fully open position 100% stroke (m²)
- Q_{max} = maximum flow with regulation requirement (m³/s)
- g = acceleration due to gravity (m/s²)
- H_{stat} = system static head (m)
- h_{fpipe} = total head loss for pipe system at Q_{max} excluding valve losses (m)

If the two sides of the equation are equal then this indicates that the valve at full stroke would just precisely allow the flow Q_{max} to pass through and constitutes a theoretical lower limit of valve capacity. The given pressure drop at Q_{max} is a compromise between, achieving good regulating characteristics for the system without using too large a valve on the one hand and not necessitating too high an additional pump differential head with associated pump and energy costs on the other hand. Control valves are not sized by the pipe size. A control valve will frequently be two sizes smaller than the pipe.

The next step is to select the valve characteristic. The two most usual are linear and logarithmic, equal percentage. In principle a valve can be given any characteristic by specially shaping the internal components, the valve trim i.e. plug etc. The valve is described as having a linear characteristic if its flow coefficient increases in direct proportion to the stroke. If A_v increases logarithmically in relation to the stroke then the valve is referred to as having an equal percentage characteristic.

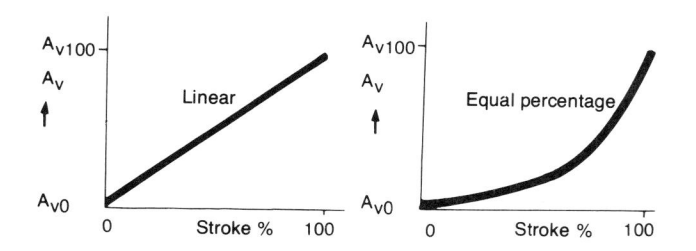

Figure 4.19 Linear and logarithmic control valve characteristics

Between these two types are valves with quadratic, V-port characteristics. Many valves display characteristics which differ to a greater or lesser extent from these mathematically defined forms.

The following rule of thumb is used as a guide when selecting a suitable valve characteristic:

If the relationship between the valve pressure drop at Q_{min}, minimum flow which is to be regulated, and at Q_{max} is less than 3, try a valve having a linear characteristic. If $\Delta pQ_{min}/\Delta pQ_{max} > 3$, choose a logarithmic characteristic.

With a lower limit of A_{v100} and with a preferred valve characteristic, a preliminary choice of control valve can be made. Alternatives to A_v are used in practice:

$$K_v = \frac{Q}{\sqrt{\Delta p \cdot \frac{\rho_{H_2O}}{\rho}}} = \left[\frac{m^3/h}{bar^{1/2}}\right] = \frac{A_v \cdot 10^6}{28} \qquad \text{Equ 4.10}$$

$$C_v = \frac{Q}{\sqrt{\Delta p \cdot \frac{\rho_{H_2O}}{\rho}}} = \left[\frac{US\ gpm}{psi^{1/2}}\right] = \frac{A_v \cdot 10^6}{24} \qquad \text{Equ 4.11}$$

In catalogues, the specified value for A_v, K_v and C_v are derived by testing with water at 5 to 30 °C, which must be considered when selecting a valve for high viscosity or non-Newtonian liquids.

Having made a preliminary selection of pump and control valve and after calculating the system curve excluding the valve, it is possible to check the pump installation regulating characteristics.

Example:

For a particular pump the system curve, excluding the valve, is determined, figure 4.20 upper illustration. The flow is to be valve regulated within the range 0.03 to 0.1 m³/s. Select valve and pump size and check the regulating characteristics of the pump installation.

At maximum flow with regulating requirements:

$Q_{max} = 0.1$ m³/s, $H_{stat} = 18$ m, $h_{fpipe} = 20$ m

According to the rule of thumb, Equ. 4.9, at Q_{max}

$h_v = (0.1 \cdot 18) + (0.3 \cdot 20) = 7.8$ m

Select a pump whose H-Q curve passes through the point, $H = 18 + 20 + 7.8 = 45.8$ m at $Q = 0.1$ m³/s.

According to the rule of thumb, Equ 4.9,

$$A_{v100} \geq \frac{0.1}{\sqrt{9.81 \cdot 7.8}} = 0.0114$$

From the pump and system curves the largest and smallest pressure drop across the valves within the regulating range can be read. The relationship between these is 35/7.8 = 4.5 which is > 3 and indicates a logarithmic valve characteristic. Choose a regulating valve with $A_{v100} = 0.013$ having a logarithmic characteristic. Calculate the valve setting, percentage stroke, at various flows by using the formula:

$$Q = A_v \cdot \sqrt{g \cdot h_v}$$

$$A_v = A_{v0} \cdot e^{\frac{s}{100} \cdot \ln\left(\frac{A_{v100}}{A_{v0}}\right)} \qquad \text{Equ 4.12}$$

where
- s = actual setting (stroke) (%)
- A_{v0} = flow coefficient at s=0 (m²)

The calculated result is shown by the lower illustration in figure 4.20. In the example $A_{v0} = 0.04 \cdot A_{v100}$, i.e. the valve regulating range is 1:25. For comparison the pump installation regulating characteristics have also been calculated for a valve having linear characteristics. For the linear valve:

$$A_v = A_{v0} + (A_{v100} - A_{v0}) \cdot \frac{s}{100} \qquad \text{Equ 4.13}$$

The largest flow, which can pass through the pump installation, is 0.102 m³/s and corresponds to the valve fully open condition. Within the regulating range the valve stroke varies between 35 and 96% of full stroke for the logarithmic valve. Corresponding figures for the linear valve are 87 and 9%. The logarithmic valve gives the installation an approximately linear characteristic.

The chain-dotted curves in the lower illustration in figure 4.20 show the installation's regulating characteristics at constant pressure drop across the valve, i.e. without considering the effects of the pump and the shape of the system curve.

Note that the pump data selected in the example neglects tolerances of the installation components, which is not to be recommended in practice.

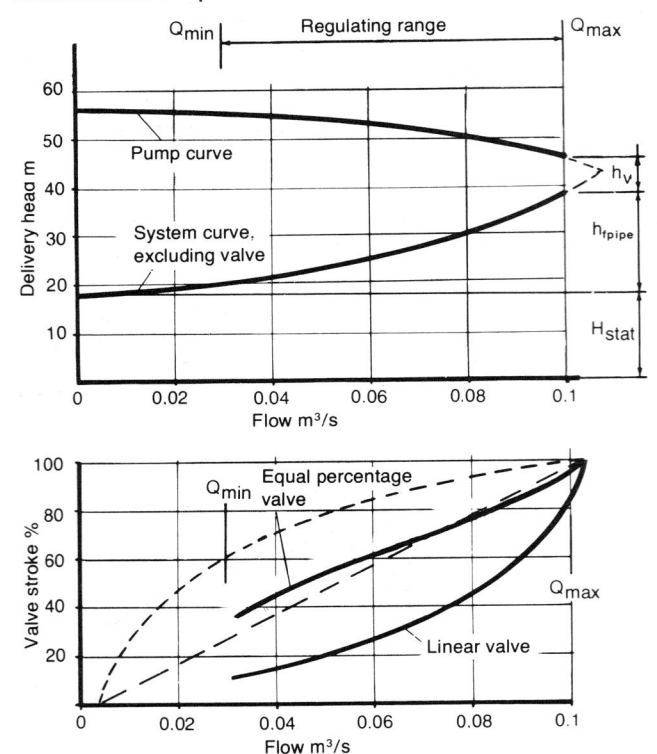

Figure 4.20 Evaluating the regulating characteristics of the pump installation

Summary

- The pressure drop across control valves can be considerable and cannot be neglected when determining pump operating characteristics.
- The regulating characteristics of the installation determines the pressure drop required across a control valve.
- The smaller the valve pressure drop in relation to the pipe system losses, the more distorted the valve characteristic.
- Valve pressure drop must be paid for with increased pump power and can generate considerable energy costs.
- Rules of thumb can never be applied universally.
- All components in the pump installation are subject to certain hydraulic tolerances. Deviations from the nominal performance for pumps, control valves and other fittings, together with inaccuracies when calculating pipe losses etc. are unavoidable and must be given careful consideration. In valve regulated systems the control valve is the component which is used to compensate for the component tolerances.

4.3 Multiple pump systems

4.3.1 Series pump operation

Series pump operation is utilised when high differential heads are required. Series operation is especially applicable for high differential heads in combination with comparatively small flows. By dividing up the differential head between more than one pump, a higher specific speed, N_s, is maintained for each individual pump, thereby increasing efficiency.

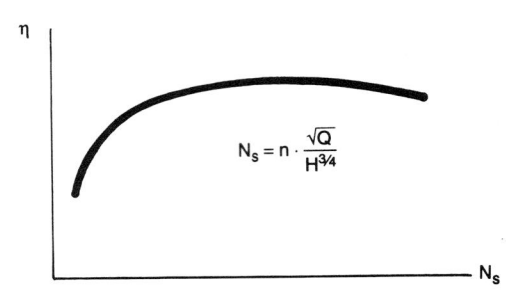

Figure 4.21 Maximum efficiency for centrifugal pumps at various specific speeds

$$N_s = n \cdot \frac{\sqrt{Q}}{H^{3/4}}$$

Multi-stage pumps are a compact, integral form of series connection.

Systems incorporating a number of pumps, where the pumps are situated in close proximity to one another are simply dealt with by calculating a common performance curve for the complete "pump package". The common pump curve is then compared with the system curve. For series operation the resultant H-Q curve is obtained by adding the actual differential heads for each value of flow, figure 4.22.

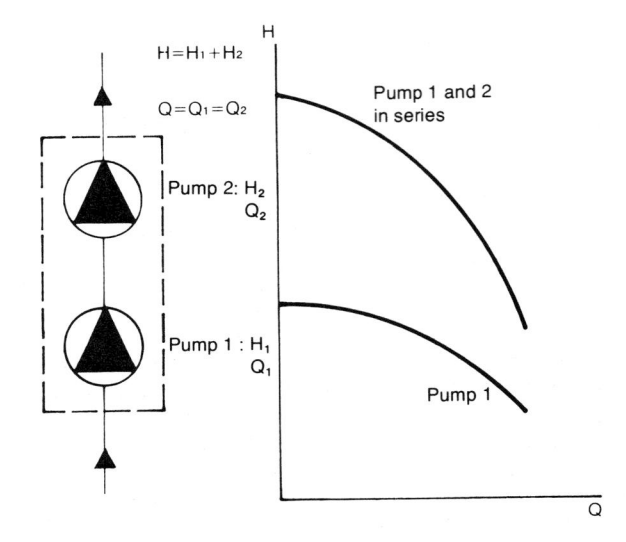

Figure 4.22 Resultant H-Q curve for series pump operation

The resultant efficiency at various flows is calculated from the relationship:

$$\eta = \frac{\rho \cdot g \cdot Q (H_1 + H_2)}{P_1 + P_2} \qquad \text{Equ 4.14}$$

When two similar pumps are connected in series the same efficiency curve applies for the "package" as for each separate pump unless the liquid is very compressible. For very compressible liquids, liquified gases are the worst, the flow throw the second pump may be considerably smaller than the first if the pump differential head is high.

The pumps shown in figure 4.23 cannot operate individually on the system because one pump does not develop sufficient differential to overcome H_{stat}. Depending upon the pump type and size, precautions may be needed to prevent one pump running without the other. The second pump may cavitate if the first pump is not running. Both pumps may tend to over heat. Instrumentation and control logic should be fitted to protect the pumps. It is possible to pump through a rotodynamic pump. The head loss through a pump is proportional to the specific speed. As the liquid flow through a pump increases it will begin to rotate in the normal direction. The liquid and impeller angles will match and the head loss will not rise as steeply. The pump should not overspeed.

For most liquids, the same flow passes through both pumps. The inlet pressure for pump 2 is the pump outlet pressure of pump 1. The pump pressure rating and shaft seals can thus

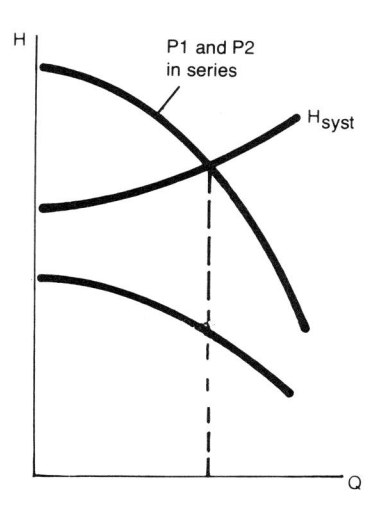

Figure 4.23 Pump nominal duty point for series operation

be affected. The axial thrusts, which are dependent on the differences between the internal and atmospheric pressure, will also be different for the two pumps. For low differential pressures, identical pumps may be used. As the differential pressure increases, the second pump may require significant modifications to accommodate the increased inlet pressure.

One very common occurrence of series operation is used to overcome the problem discussed in 4.1.3, low NPSHa. Some rotodynamic pumps are required to produce very large differential heads. Even with multi-stage pumps, 12 stages are common, the ΔH at normal motor speeds is not sufficient. The pumps are driven at higher speeds by step-up gearboxes or by steam/gas turbines. The NPSHr of the pump increases. To overcome the problem, and to avoid very costly solutions such as raising the suction source, a low power booster pump is operated in series with the main pump. The function of the booster is to supply NPSHa only. The differential head is therefore very small compared to the main pump and the power absorbed is low; efficiency is not really a consideration. The booster pump would generally be a single stage pump, perhaps double entry, running slowly to match the NPSHa. If the main pump is geared up from an electric motor, the booster pump may be geared down from the same motor. If the main pump is turbine driven, the booster will probably be motor driven. In this type of series pump installation the pumps will be completely different, reflecting their respective functions.

Booster pumps are also used on positive displacement pumps. Depending upon the nature of the liquid the booster pump may be another positive displacement pump or a rotodynamic pump. High pressure positive displacement pumps working with water-like liquids may use a centrifugal pump booster to provide adequate NPIPa. Pumps handling more viscous liquids would use a low pressure positive displacement pump. When two positive displacement pumps are used, the booster should be oversized slightly to allow for wear. A relief valve or back-pressure regulator should be fitted to the inter-connecting pipework to prevent excessive pressures. It is also a good idea to fit a pulsation dampener, with extra gas volume, to reduce any pulsation levels and function as an accumulator.

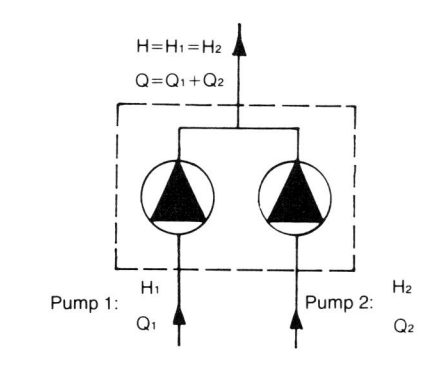

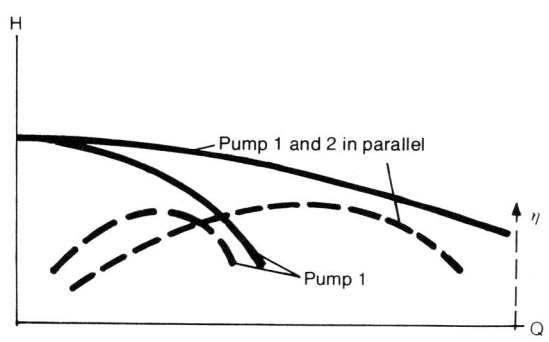

4.3.2 Parallel pump operation

For parallel pump operation, the resultant H-Q curve is obtained by adding the actual flows for each head value, figure 4.24.

Parallel pump operation is used for the purpose of achieving large flows and for dealing with greatly varying flows by operating a varying number of pumps. The effect of parallel operation is dependent largely upon the shape of the system curve.

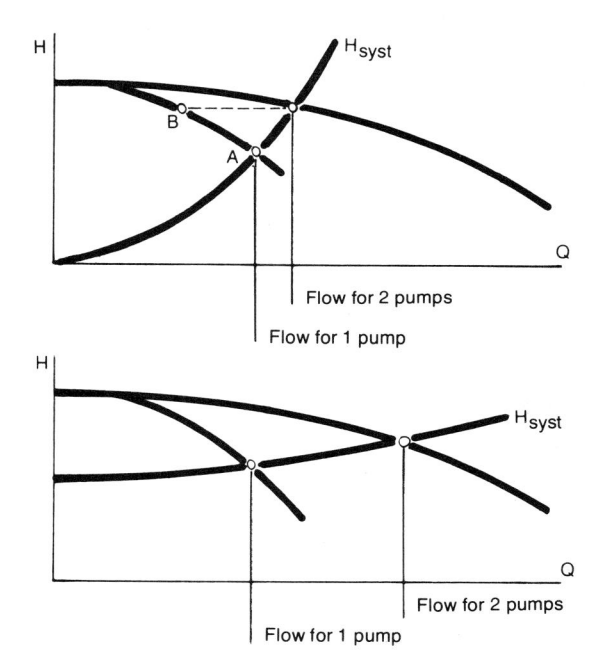

Figure 4.24 Parallel pump operation

Figure 4.25 Parallel operation of 2 similar pumps with different system curves

For steep, friction dominated, system curves only a small increase in flow is achieved by parallel operation. A much better effect is obtained if the system curve is dominated by static head.

The pump's specification, Q and H, is based upon the pump's nominal duty point, B in figure 4.25. If only one pump operates, the duty point will then be at point A. This can cause conditions where there is risk of cavitation, because of the lower NPSHa value at point A than at B, and risk overloading of the motor.

The resultant efficiency curve for the "package" is calculated from the relationship:

$$\eta = \frac{\rho \cdot g \cdot H (Q_1 + Q_2)}{P_1 + P_2} \qquad \text{Equ 4.15}$$

For two similar pumps the efficiency values are the same for each separate pump except that the Q values are displaced by a factor of two, figure 4.24.

Unstable pump curves should be avoided when pumping in parallel. In any case the intended operational range should not fall within the unstable region without taking special precautions. Special minimum flow valves or orifices can be fitted to prevent the pump reaching the peak of the curve.

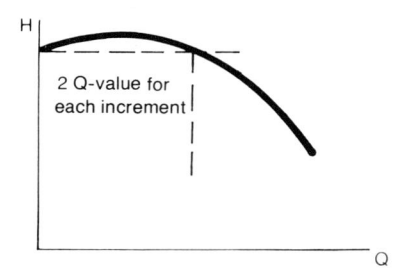

Figure 4.26 Unstable pump curve

The problems which can occur are, for example, irregular flow due to varying operational points and the inability of a pump to open it's non-return valve when starting. Non-return valves are fitted to prevent reverse flow through stationary or standby pumps.

4.3.3 Pressure boosting

Pressure boosting pumps are primarily employed in long pipelines and large circulation systems. A common case is that pressure in a distribution network is insufficient for certain higher locations, for example, the water supply to high-rise apartment blocks, or when an additional tapping point is connected to an existing network.

Pressure boosting is a form of series connection where the pumps are located at considerable distances from each other. The advantages of pressure boosting is a better distribution of the specified pressure in the pump installation is achieved and that the pump and energy costs can be reduced. The latter advantage applying to branched pipework systems.

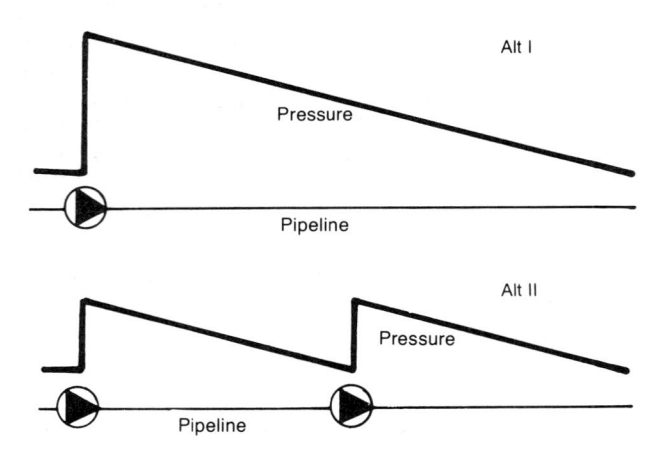

Figure 4.27 Schematic pressure distribution in a pipeline

A higher pressure rating may be required for the upstream section of the pipeline and pump for alternative I, or for the second pump in the event that the pressure boosting for alternative I is also divided into two pumps.

Branched pipework systems are usually sized by calculating for the pressure requirements of the furthest tapping point. The required pump specification is therefore based according to this pressure requirement and the total flow consumption. The application of pressure boosting principles can provide an alternative, as illustrated by the following example.

Pump data for alternative I becomes:

Q = 1, H = 0.5 + 0.5 = 1, P = 1

Pump data for alternative II becomes:

Q1 = 1, H1 = 0.5 + 0.1 = 0.6, P1 = 0.6

Q2 = 0.5, H2 = 0.5 - 0.1 = 0.4, P2 = 0.20

The total installed pump power P = 0.80

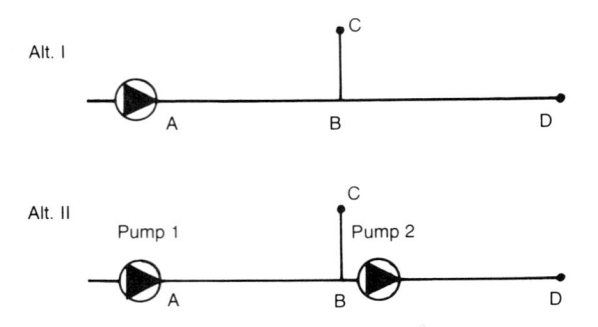

Data supplied		
Length	Flow	H_{syst}
AB	1	0.5
BC	0.5	0.1
BD	0.5	0.5

Figure 4.28 Example of conditions for pump sizing in a branched pipe system

The smaller power requirement for alternative II is explained by the fact that the surplus pressure in pipe BC is eliminated. In alternative I this excess pressure must be throttled to prevent tapping point C from overflowing and reducing the flow at D.

A better illustration of the pump's operating situation is obtained from drawing up the pump and system curves. In figure 4.29 the respective pump and system curves are shown both separately and in relation to each other. The curves are compiled in accordance to Section 4.1.3. The points marked indicate the situation at the specified flow condition.

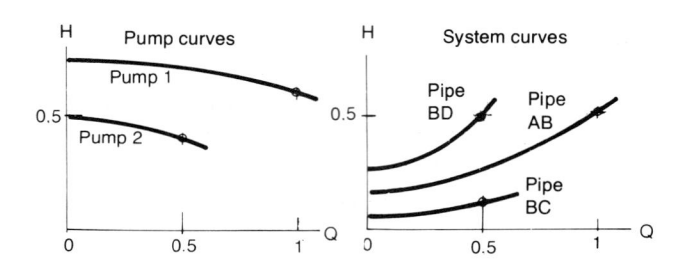

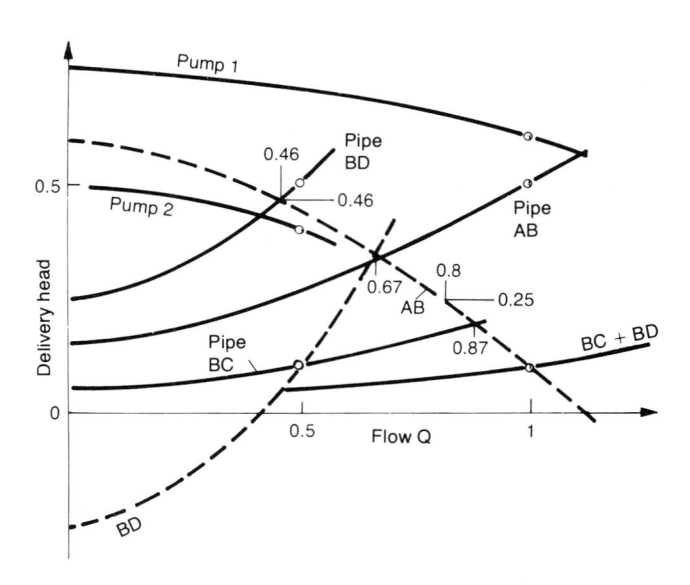

Figure 4.29 Operating conditions for pressure boosting example fig 4.28

The H-Q diagram shows, for example, that pressure boosting is not required at all for total flows Q << 0.46, even if the complete flow is tapped off at D. The largest flow which can pass through the pressure boosting pump is $Q \cong 0.67$. This occurs when C is completely closed and is sized for the pressure boosting pump motor. The largest flow at C, D completely closed, is $\cong 0.87$. For $Q > 0.8$, $H_B < 0.25$, then D is receives no flow unless the pressure boosting pump is started.

The pressure boosting point can be controlled in the following way. In order to guarantee full flow at D, $Q_D = 0.5$, independent of consumption at C, the pressure boosting pump is started by a signal from a pressure sensor at the branch point when H_B falls below 0.46 and is stopped when H_B exceeds 0.46 by a certain margin. Within the range $0.19 \leq H_B \leq 0.46$, the total flow $0.46 \leq Q \leq 0.87$, a situation where flow $Q_D = 0$ can occur. The input power is transmitted to the liquid, the temperature rises and the risk of seal leakage and damage due to cavitation increases. Excessive temperatures can be avoided by means of a temperature sensor which transmits a signal to a valve which opens a by-pass line. To protect the pressure boosting pump against cavitation in the event of low pressure on the inlet side, which could occur if pump 1 became damaged for example, a pressure sensor is installed which stops the pressure boosting pump when H_B falls below the required cavitation limit, H_{Bmin}.

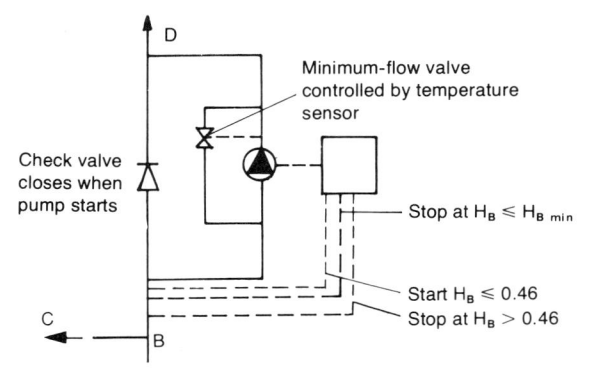

Figure 4.30 Installation of pressure boosting pump

The most energy conserving method of regulation would be to control the speed of the pressure boosting pump according to the constant pressure criteria in D.

4.3.4 Pressure maintenance

Pressure maintenance, as the name implies, is intended to sustain a certain pressure in a pump installation or circulation system. The liquid being pumped is often hot and, because of the high temperature, has high vapour pressure. The pressure maintenance prevents cavitation at high points in the pipework and pumps, in other words cavitation is avoided. Pressure maintenance also compensates for unintentional leakages and makes up variations in the volume of liquid and pipework due to changes in temperature and pressure. Normal cases, where pressure maintenance is necessary, are district heating networks and hot water circulation systems in buildings. Pressure maintenance is carried out at a point in the pipe network where it will be most effective, which will vary from case to case. If the pressure maintenance pump is placed on the suction side of the main pump then a constant inlet pressure is maintained, whilst the outlet pressure varies with change in flow and vice-versa.

The simplest method of providing pressure maintenance is by means of a high-level tank. A higher pressure can be provided if required by the introduction of a pressure maintenance pump.

The by-pass line ensures a minimum flow to the pressure maintenance pump, even though no liquid is supplied to the main pipeline, thus preventing unacceptable temperature rises in the pump. The by-pass line can also be used to maintain constant pressure for varying flows to the main pipeline.

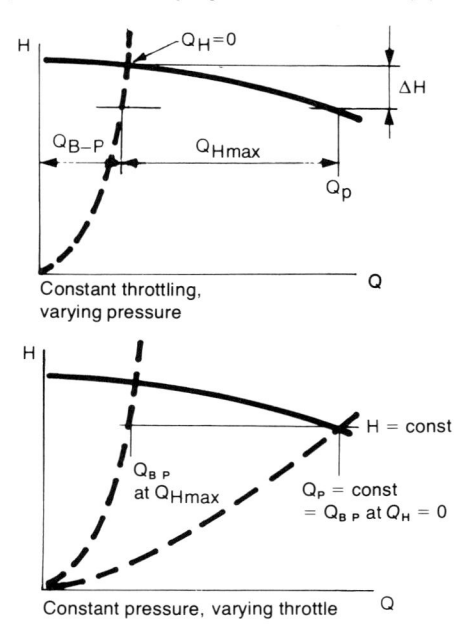

Fig. 4.32 Pressure maintenance pumps for different control methods

The main line pressure is equal to the pressure increase due to the pump plus the supply pressure. Constant throttling in the by-pass line causes the pressure level in the main line to vary by an amount ΔH dependent upon the flow in the main line. Variable throttling enables constant pressure to be maintained. Regulation of pressure maintenance can also be carried out in other ways, for example, by throttling the delivery pipe or by regulating pump speed.

4.4 Pump hydraulic specification data

4.4.1 Hydraulic tolerances

General

All components in the pump installation have tolerances in respect of hydraulic performance. To ensure achievement of a definite flow through the pump installation it is necessary to consider the hydraulic tolerances when determining pump data.

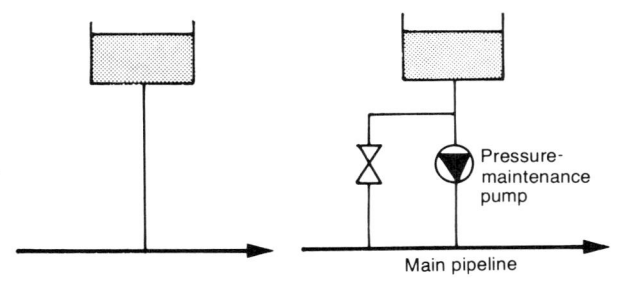

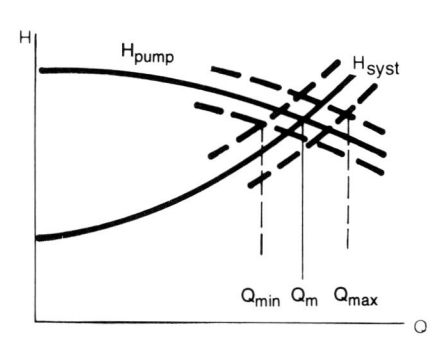

Figure 4.31 Arrangements for pressure maintenance

Figure 4.33 Illustration of hydraulic tolerances
——— calculated value − − − − − − − tolerance limits

Pump hydraulic tolerances are carefully set out in the pump testing standards. The maximum permissible deviations from the guarantee data are thus defined by specifying the standard according to which acceptance testing and approval is to be carried out. Special conditions must however be agreed upon in the case of high viscosity and non-Newtonian liquids.

It is however, considerably more difficult to determine the tolerance limits of the system curve, H_{syst}. Examples of the tolerances which accumulate in the complete system curve are:

- tolerances for pipe diameter,
- pipe bore roughness,
- valves, bends and fittings,
- flowmeters and instrumentation,
- variable filter Δp,

together with uncertainties in respect of the characteristics of the process liquid at various temperatures and pressures, and the methods of calculation.

Another type of uncertainty, when specifying the pump data, is the future flow and head requirement. This is however, a question which should be clarified before selecting a pump. Rotodynamic pumps can be supplied so that additional impellers can be fitted. Reciprocating pumps can be built to accommodate various sizes of pistons or plungers.

Because of the difficulty in determining the system curve tolerance limits a general allowance is often applied when calculating the necessary pump data. Such allowances are a safety margin and provisions must be made to accommodate them when they are not used usefully.

Rotodynamic pumps

The magnitude of hydraulic safety factors which are normally applied to the system requirements are:

- +5% of the calculated differential head due to losses other than H_{stat} for pumps, P > 30 kW
- +2% of the required flow to allow for leakage (may require a larger margin for large complex installations)
- +10% of the calculated differential head due to losses other than H_{stat} for pumps, P < 30 kW

The percentage values given apply to Newtonian liquids. The uncertainty increases in the case of non Newtonian liquids.

The consequences of the allowances are illustrated by means of an example. For a mass-produced centrifugal pump, class C, purchased according to a catalogue curve, the tolerances for differential head are ± 6%. For a tolerance of ± 8% for H_{syst} at the desired flow, the allowance will be:

$$\sqrt{0.06^2 + 0.08^2} = 0.10 \text{ i.e. given as 10\%}$$

The calculation means that there is a 95% chance that the actual flow will be equal to the desired flow. The formula also assumes that the production tolerances of the pump and system are independent of each other and equally distributed about their respective calculated values.

The smaller allowance, 5%, for larger pumps is influenced by the stricter delivery tolerance requirements and more extensive system curve calculation.

Another way of approaching the uncertainty of pump/system matching is to specify two duty points. A rated duty and a nominal duty. The rated duty is the flow plus a leakage allowance at the head required to cover all the highest losses. The pump characteristic must pass through this point. The nominal duty is the basic design flow without any leakage at the head required for a new clean system. This point must be on or below the pump characteristic.

Pumps built specifically for a contract can have their characteristics "fine tuned" to suit. Impellers can be machined to a specific diameter and the vanes can be modified slightly in the eye and at the tips to change the shape of the characteristic to some degree.

Positive displacement pumps

For positive displacement pumps, which have very steep H-Q curves, the system curve tolerances have little effect on the actual flow. The pump flow tolerances are completely definitive.

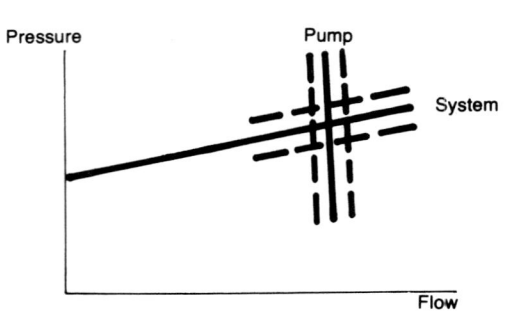

Figure 4.34 Illustration of hydraulic tolerances for positive displacement pumps

The required flow will most certainly be achieved if the following rules are applied:

- Specify the pump flow as -0% +3% of the design flow.
- Specify the pump suction pressure as the system minimum.
- Specify the pump discharge pressure as the system maximum.

Allowances must be made in the discharge system for overpressure when the relief device operates.

4.4.2 Pump H and Q at η_{max}

On the basis of the installation liquid requirements the mean flow Q_m and maximum flow Q_{max} are determined. Then:

$$H_{Qmax} = H_{stat} + h_{fpipe} + h_v + h_{tol} \qquad \text{Equ 4.16}$$

where

H_{Qmax}	= the pump head at Q_{max}	(m)
H_{stat}	= static head	(m)
h_{fpipe}	= head losses in the pipe system excluding the control valve at Q_{max}	(m)
h_v	= pressure drop across control valve at Q_{max}	(m)
h_{tol}	= allowance for hydraulic tolerances at Q_{max}	(m)

In this way a point Q_{max}/HQ_{max} is established on the intended pump H-Q curve. Except in the case of on-off control systems however the pump will very rarely, if ever, be required to operate at this point. The pump operating point, design point, η_{max}, should preferably occur at a lower flow, as near to Q_m as possible. If the difference between Q_{max} and Q_m is great, more than 15%, and if Q_m is chosen to coincide with the pump's η_{max} then there is risk of over loading the pump at Q_{max}. Normally, however, a flow which is 20% greater than the design flow is not considered to present any problem for rotodynamic pumps providing the manufacturer is consulted and power is available. With regard to the above, the following rules of thumb are given for the determining of pump hydraulic specification data for continuously regulated pump installations.

- Determine Q_m and Q_{max} according to the process requirements.
- Calculate H_{Qmax}.
- Look for a pump which can cope with Q_{max}, H_{Qmax}. The pump's η_{max} should lie as near to Q_m as possible.
- Check the risk of cavitation, $NPSHr_{Qmax}$, and the required shaft power at Q_{max}. Check with the pump manufacturer

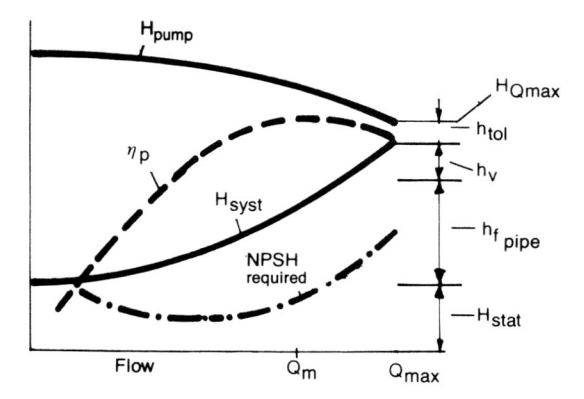

Figure 4.35 Illustration of recommended pump data

if there are any other reasons which may prevent operation at Q_{max}.

The matching of pump flow to Q_{max} when the production tolerances are known can be accomplished by:

- remachining the pump impeller(s)
- constant throttling or by-pass
- speed matching (belt or gear drive)
- normal regulating equipment.

4.4.3 Minimum pump flow

There are many reasons why it is necessary to ensure a minimum flow for a rotodynamic pump:

- Heating of the pumped liquid.
- Increased cavitation risk when Q approaches zero.
- Increased vibration levels when Q approaches zero.
- Axial thrust and radial force on the pump impeller are greatest at Q = 0.
- The power required for an axial pumps is greatest at Q = 0.

Pump efficiency is low in the vicinity of zero discharge, Q = 0. Most of the power is converted to heat, which must be carried away by the pumped liquid. The increase in temperature of the pumped liquid can be calculated from the relationship:

$$\Delta T = \frac{g \cdot H}{c_{sp}}\left(\frac{1}{\eta} - 1\right) = \frac{P(1-\eta)}{\rho \cdot c_{sp} \cdot Q}$$

$$\cong \frac{P_{Q=0}}{\rho \cdot c_{sp} \cdot Q} \qquad \text{Equ 4.17}$$

where
H	=	equals differential head	(m)
g	=	acceleration due to gravity	(m/s^2)
c_{sp}	=	specific heat	(J/kg °K)
η	=	pump efficiency	
Q	=	flow through pump	(m^3/s)
ρ	=	density of liquid	(kg/m^3)
P	=	supplied shaft power	(W)

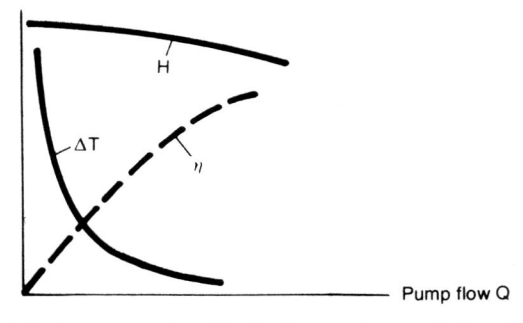

Figure 4.36 Temperature rise of liquid when pumping near to the point of zero discharge

ΔT increases rapidly as the flow through the pump approaches zero (Q →0, η → 0). Apart from increased material stresses due to temperature rise, the vapour pressure of the liquid also increases with corresponding cavitation risk and the shaft seal liquid film is also in danger, so that a breakdown of sealing function may be imminent.

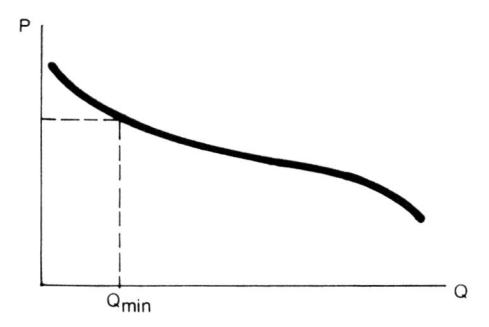

Figure 4.37 Power curve for axial pump

For axial pumps, propeller pumps, pumps with high specific speed, the required shaft power increases as flow decreases at constant speed.

By having a specified minimum flow through the pump a smaller motor can be used without risk of overloading.

Devices for obtaining a minimum flow

A minimum pump flow at Q = 0 in the system can be maintained by introducing a by-pass line into the system.

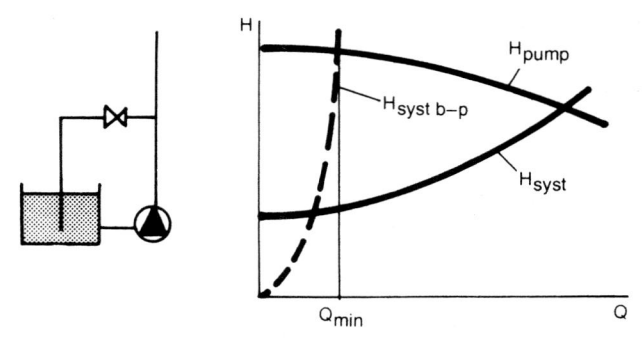

Figure 4.38 By-pass line to guarantee a minimum flow through the pump

The return line should return the liquid to the pump sump in order that a large liquid volume should be available if circulation pumping is protracted. The by-pass flow can be controlled in various ways:

- Constant throttling
- Loaded check valve
- Pressure sensor after the pump
- Flow sensor in the main line
- Temperature sensor after the pump

In cases employing constant throttling, an orifice plate or locked needle valve, a certain flow will pass through the by-pass pipe even at Q_{max} in the main line and should then be added to Q_{max} when defining the pump size. A loaded non-return valve or pressure sensor necessitates a stable pump curve whose differential head falls continuously as pump flow increases.

In the case of positive displacement pumps a relief valve or other overpressure protection device must always be provided as protection against blockage in the discharge line. Flow regulation cannot be carried out by throttling in the main line. Regulation of fixed speed pumps is usually by means of a by-pass line. Other alternatives are speed regulation and variable displacement.

NOTE: Speed regulation, using variable frequency invertors to supply standard squirrel cage motors, is becoming very popular. The reason for the popularity of this technique is; ease of use, wide range of choice, efficiency. Variable speed pumps can overcome all system mismatch problems, usually very efficiently.

4.5 Water hammer in pump installation

4.5.1 Hydraulic gradient

In order to obtain a complete picture of the distribution of pressure throughout a pump installation, use is made of the hydraulic gradient. By definition for pressure head H_t:

$$H_t = \frac{p^1}{\rho g} + z \qquad \text{Equ 4.18}$$

where
- H_t = pressure head above liquid surface (m)
- p^1 = static pressure above atmospheric (Pa)
- ρ = density of the liquid (kg/m³)
- g = acceleration due to gravity (m/s²)
- z = elevation above liquid surface (m)

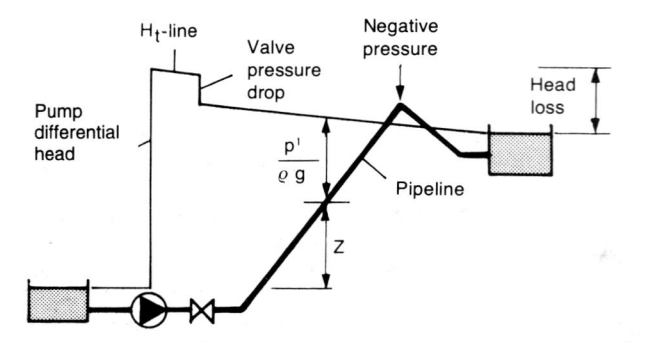

Figure 4.39 Illustration of hydraulic gradient for a pump installation

This method of representation presents the pipeline inner positive pressure as the difference between the H_t-line and the elevation of the pipeline. The figure also shows a high point where there is negative pressure, that is below atmospheric pressure. Parameters which should be observed are the prevailing inner positive and negative pressures as regards sizing for stress and margins of safety against the risk of cavitation. The cavitation risk is greatest in pumps, valves and other installation components having restricted sections together with high points in the system.

The hydraulic gradients for steady flow take on different appearances for different operating points. The extreme value for the H_t-line often occurs, however, in conjunction with load changes associated with a transient, a time dependent process.

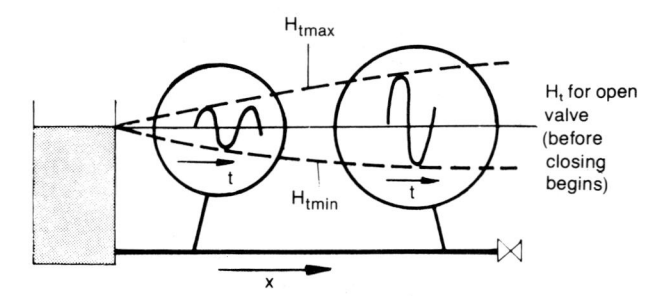

Figure 4.40 Pressure head lines for closing valve

When closing a valve the pressure increases in front of the valve. Increasing pressure gives rise to a pressure wave, which travels through the pipeline at the speed of wave propagation "a". The magnitude of the wave speed "a" is a function of the liquid and the pipe. Pressure waves would normally travel at the speed of acoustic waves in the liquid, which is a function of the liquid properties only.

$$c = \sqrt{\frac{K}{\rho}} \qquad \text{Equ 4.19}$$

where
- c = acoustic velocity (m/s)
- K = liquid bulk modulus (Pa)
- ρ = liquid density (kg/m³)

Water has an acoustic velocity of approximately 1450 m/s. Constraining the liquid within a pipe reduces the acoustic velocity because of the flexibility of the pipe wall, see equation 4.20.

$$a = \sqrt{\frac{g/w}{\frac{1}{K} + \frac{1}{8E}(d_o^4 - d_i^4)}} \qquad \text{Equ 4.20}$$

where
- a = wave speed (m/s)
- g = acceleration due to gravity (m/s²)
- w = specific weight (N/m³)
- E = Young's Modulus of Elasticity of the pipe (N/m²)
- d_o = pipe outside diameter (m)
- d_i = pipe inside diameter (m)

The elasticity of the pipe wall has a significant effect on the wave speed. Air or gas bubbles have a much more severe effect on the wave speed and can reduce it to 100 m/s. The pressure wave is reflected at the end of the pipe as a negative pressure wave, returns to the valve, is reflected again and so on. In this way a process of pressure oscillation is set up in the pipeline and continues, even after the valve is completely closed, until the oscillations are dampened out by friction. The hydraulic gradient lines H_{tmax} and H_{tmin} connect the highest and lowest pressures which occur at different points along the pipeline during the oscillation process. The greatest pressure variations are to be found in the valve and occur when the valve is closed rapidly. The greatest amplitude of pressure surge is:

$$\Delta p_{max} = \rho \cdot a \cdot v \qquad \text{Equ 4.21}$$

where
- Δp_{max} = max. pressure variation (Pa)
- ρ = density of the liquid (kg/m³)
- a = speed of wave propagation (m/s)
- v = velocity of flow before valve begins to close (m/s)

For water in steel pipes $\rho \cong 1000$ kg/m³, $a \cong 1100$ m/s, and a flow velocity of 1 m/s creates, for rapid valve closing, a water hammer pressure of $1.1 \cdot 10^6$ Pa = 11 bar $\cong$ 110 m. The pressure process at the valve cannot be affected before the pressure wave has been reflected from the end of the pipe and returned to the valve. This time is referred to as the pipeline reflection time and is designated t_r.

$$t_r = 2 \cdot \frac{L}{a} \qquad \text{Equ 4.22}$$

where
- t_r = pipeline reflection time (s)
- L = the pipe length (m)
- a = wave speed (m/s)

If for example the length of pipe is 1100 m, then $t_r \cong 2$ s. All valve closing times ≤ 2 s thus give maximum water hammer pressure.

For a plastic pipe $\Delta p_{max} = 3 \cdot 10^5$ Pa $\cong$ 30 m is obtained for valve closing times ≤ 7.36 s. Increasing the closing time reduces the magnitude of water hammer and very slow closing valves result in no line shock at all.

4.5.2 Causes of water hammer

Water hammer in pump installations is caused basically by changes in flow and especially by rapid flow changes. Rapid flow changes occur during:

- rapid valve movement
- starting and stopping of pumps
- special situations

Since the shock of water hammer can reach considerable magnitude there is risk of damage to the pump installation. Apart from the pressure oscillations which occur with normal variations, the prediction of unintentional operational conditions must also be considered. Accidental pump stoppage caused by power failure or power cut is one such situation which a pump installation must contend with sooner or later. An estimation of the magnitude of the line shock must therefore be carried out for every installation.

Water hammer caused by cavitation in the pipeline is difficult to estimate. If the absolute pressure somewhere in the pipeline is allowed to fall below the vapour pressure of the liquid, vapour is formed in the pipeline. The original column of liquid divides itself into two which, with different speeds and separated by a growing vapour bubble, flows on through the pipeline. Eventually the two columns of liquid begin to approach each other causing the vapour bubble to implode as the liquid columns reunite, resulting in a large pressure surge. The magnitude of the water hammer shock depends upon the instantaneous difference in velocity of the two columns of liquid at the moment when they reunite, which, in the worst case, can be greater than the steady flow velocity in the pipeline. Cavitation in pipelines must, therefore, be avoided which also substantiates the necessity for controlling H_{tmin}. The risk of unacceptable negative pressures are greatest at high points, down stream of valves and pumps and in long suction lines.

The following situations are especially susceptible to water hammer:

- Cavitation due to heating or insufficient cooling in thermal installations.
- Trapped air-pockets; pump starts against closed or nearly closed valves in an insufficiently vented system.
- Automatic re-start of pump, which rotates backwards after a power failure or after reverse flushing.
- Acoustic resonance effects due to periodical disturbances caused by valve oscillations, pump vane passing frequency, positive displacement pump flow variations.

4.5.3 Pump behaviour after power loss

Pump curves

The pump is sometimes compelled to operate at abnormal operating points especially in the case of pump installations without a non-return valve. Multiple rotodynamic pump installations with parallel operation always have non-return valves fitted to prevent reverse flow back to suction through any stationary pumps. Single rotodynamic pump installations must have a non-return valve fitted when there is positive static head in the discharge system. Reciprocating pumps do not usually need non-return valves unless a by-pass is used for starting. Other types of positive displacement pumps may need non-return valves, consult the pump manufacturer when system characteristics are known. Reverse flow can occur in the event of power failure for example. In as much as reversal of flow or direction of rotation are occurrences which must be considered, a knowledge of pump curve behaviour under these conditions is advisable.

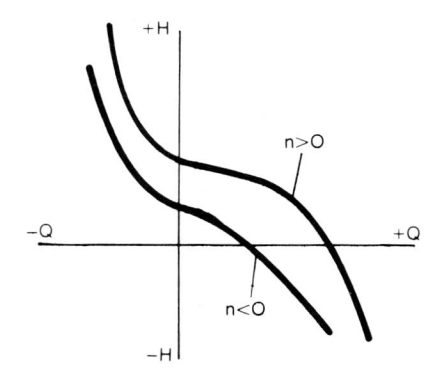

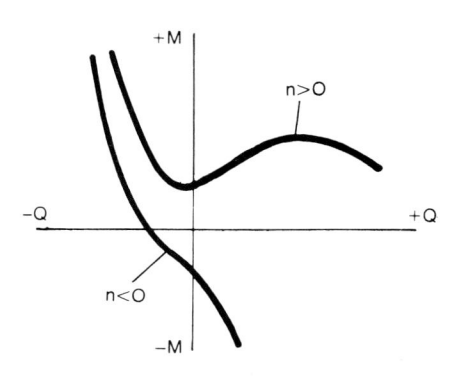

Figure 4.41 Complete performance curves for a centrifugal pump
(H = differential head, Q = flow, n = speed, M = shaft torque)

Figure 4.41 shows that a normal centrifugal pump possesses limited pumping capacity even when rotating in the reverse direction and that it functions exceptionally well as a turbine. Values of negative Q indicate reverse flow through the pump. Values of speed, n < 0, indicate reverse rotation of the pump. Reverse rotation can be incompatible with shaft driven oil pumps resulting in considerable bearing damage, damage to mechanical seals and overspeed may damage the pump. Pumps with low specific speed do not tend to overspeed; overspeed can be a serious problem with high specific speed pumps. An important consideration from the point of view of water hammer is that of the rapidly increasing pressure differential across the pump which occurs with reverse flow and pump speed. Another way of illustrating a possible pump operating condition is shown in figure 4.42.

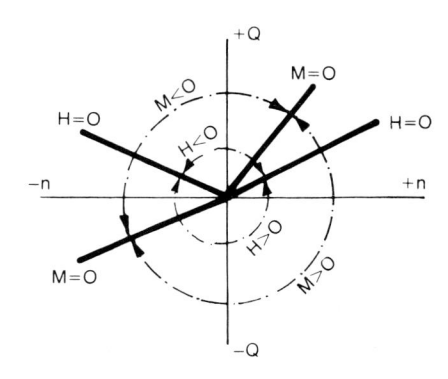

Figure 4.42 Possible centrifugal pump operating conditions

Pump stops without non-return valve

The transient process in the event of power failure is illustrated by an extreme example. The installation is equipped with a slow-closing valve which remains open during the process.

When the driving torque fails, the impeller is braked by the hydraulic torque. The decreasing speed also reduces Q and H for the pump. The flow through the pump changes direction whilst the speed is still positive. The pressure in the pump reaches its maximum value. The direction of rotation also changes a few seconds later. Very high pressures occur in the pump when retarding reversed flow at high reversed speed.

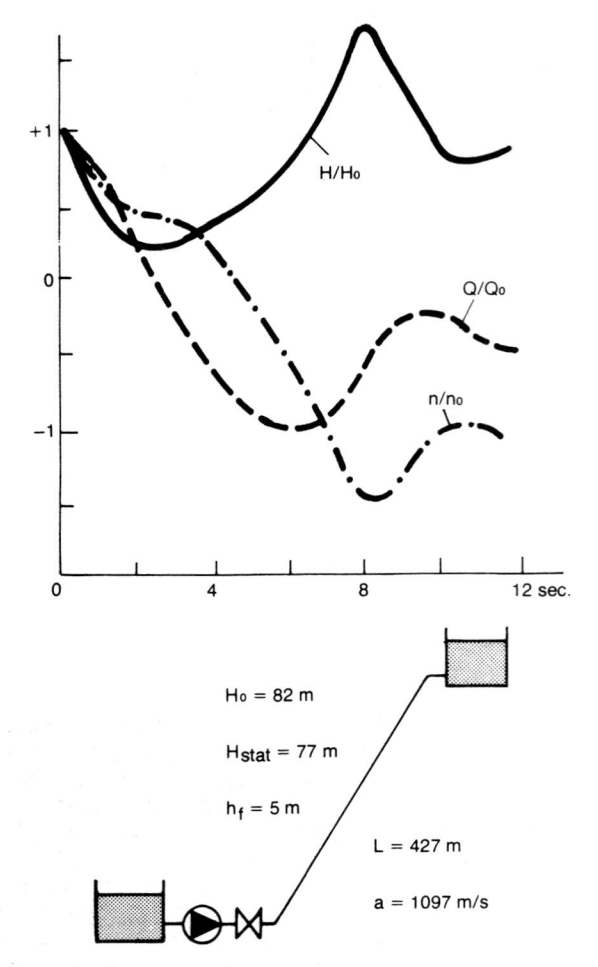

Figure 4.43 Transient process during power failure, conditions at pump

$H_0 = 82$ m
$H_{stat} = 77$ m
$h_f = 5$ m
$L = 427$ m
$a = 1097$ m/s

The pressure oscillations are eventually dampened out and a steady flow condition is obtained. The pump reaches its so-called runaway speed.

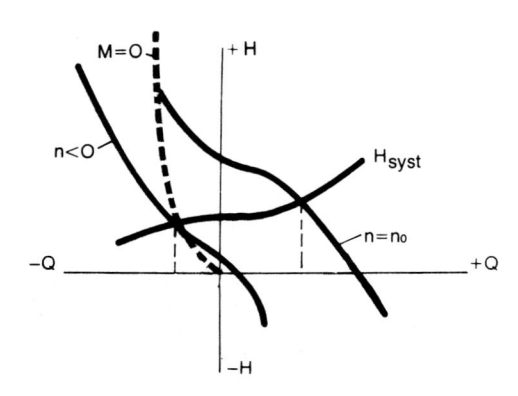

Figure 4.44 Steady, initial and final conditions during power failure

4.5.4 Protection against water hammer

When is protection equipment necessary?

Protection equipment is necessary when the pump installation would be damaged by the resultant:

- Excess pressure.
- Risk of cavitation.
- Negative pressure in the line.
- Reverse running and overspeed.

Rotodynamic pumps should nearly always be fitted with non-return valves; this is not considered protection in this context. Installations subjected to repeated water hammer must be designed to withstand fatigue.

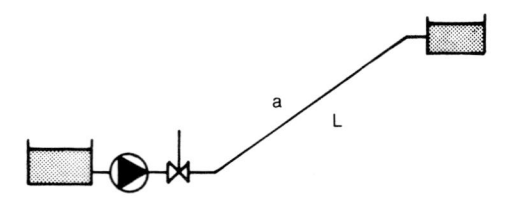

Figure 4.45 Pump installation with slow-closing valve

Detailed water hammer calculations are complicated and costly. Specialist consultants can be employed to investigate the effects and make recommendations for protection. This obviously raises the question as to when a detailed investigation is necessary. For certain installations simple estimations can give an indication as to the magnitude of water hammer.

Valve closing time T

The pump starts and stops against a closed valve. The pressure changes at the valve in connection with closing can be approximated by:

$$\pm \Delta H = \frac{2 \cdot L \cdot v}{g \cdot T} \quad \text{for } T > 2 \cdot \frac{L}{a} >$$

$$\pm \Delta H = \frac{a \cdot v}{g} \quad \text{for } T \leq 2 \frac{L}{a} \qquad \text{Equ 4.23}$$

where
- L = pipe length (m)
- v = initial flow velocity (m/s)
- g = acceleration due to gravity (m/s^2)
- T = valve closing time (s)
- a = wave speed (m/s)

The above formula assumes a linear valve closing characteristic, i.e. linear reduction in flow in relation to time.

Pump stops with non-return valve

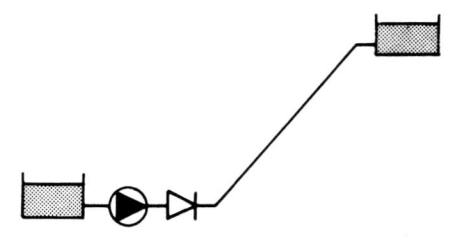

Figure 4.46 Pump installation with non-return valve

The pressure changes with non-return valves in conjunction with the pump stopping are approximated by:

$$\pm \Delta H = \frac{2}{\frac{g \cdot C}{v \cdot L} + \frac{K}{H}} \qquad \text{Equ 4.24}$$

Equation 4.24 does not apply to the special non-return valves which have delayed action and/or anti-slam features.

Example:

The following data applies to a pump installation with non-return valve; pipe length $L = 1700$ m, delivery head $H = 22$ m and

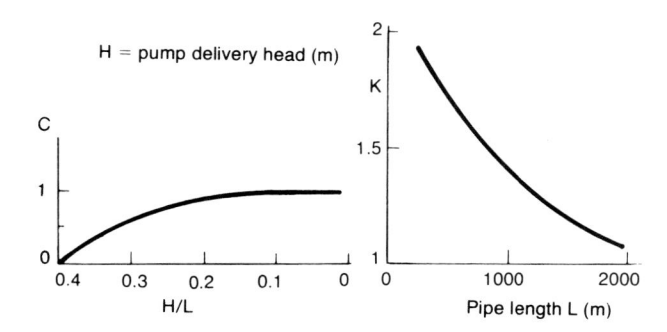

Figure 4.47 Values for C and K in Equ. 4.24

flow velocity v = 1.3 m/s in the pipeline. Estimate the magnitude of pressure surge when the pump stops.

L = 1700 m → K = 1.1, H/L = 22/1700 → C = 1

Then according to equation 4.24

$$\pm \Delta H = \frac{2}{\frac{g \cdot 1}{1.3 \cdot 1700} + \frac{1.1}{22}} = 37 \text{ m}$$

Pump stops without non-return valve

The installation is the same as in figure 4.45. The pump stops because of power failure. The slow-closing valve remains open.

If $h_f/H > 0.2$ Equ 4.25

then $H_{tmax} \leq H$

where
h_f = pipe system head loss   (m)
H = pump delivery head   (m)

The size of the pressure pulse, $-\Delta H$, is estimated from equ. 4.24.

Hydraulic gradients

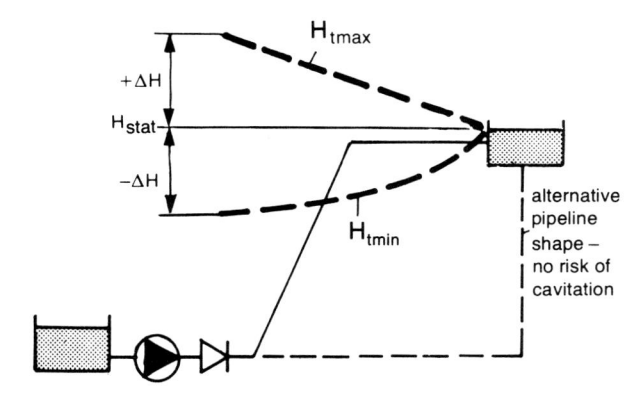

Figure 4.48 Reviewing hydraulic gradients

The rules of thumb take no account of the shape of the hydraulic gradient along the pipeline. The gradient shown in the figure are based on estimates derived from experience. It is, however, possible to conclude that there are definite risks of cavitation in the case of the pipeline shape as represented by the continuous line. The chain-dotted line routing eliminates this risk. It then remains to be shown whether or not H_{tmax} at the valve is acceptable.

By means of these simple estimations it is possible to judge many non-critical cases, usually having short pipe lengths and low flow velocities. In cases of doubt it is advisable, as the next stage of investigation, to refer to the sizing diagrams which are published in technical literature etc. These cover more thorough calculations with regard to pump stops without non-return valves for certain simple pump installations and also quote figures for H_{tmax} and H_{tmin} at mid-point in the pipeline, references 1 and 2 at the end of the chapter. If there is still cause for doubt then the only solution is to carry out a detailed digital computer simulation using the Method of Characteristics, see also reference 3, or an analogue computer simulation.

Protection equipment

Water hammer is avoided during normal operation by ensuring that load changes are carried out sufficiently slowly. Starting and stopping of pumps is carried out against closed valves and/or by using by-passes. Extra protection equipment is therefore usually installed in order to protect the pump installation against power failure.

One large group of protection equipment attempts to eliminate the basic causes of water hammer, the too rapid flow reduction resulting from a power failure.

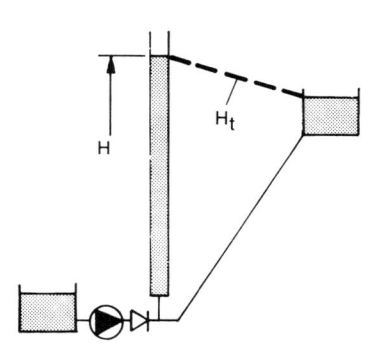

Figure 4.49 Surge column

A surge column of sufficient diameter and height provides good protection against water hammer. In the event of power failure the non-return valve closes and isolates the pump from the pipeline. Liquid from the surge column flows out into the pipeline and prolongs the retardation process, weakening the negative pressure wave. The column refills when the flow in the pipeline has changed direction. The mouth of the column can be throttled to increase dampening in the system. A disadvantage of the surge column is that it must be at least as high as the maximum pressure head at the non-return valve. This problem can be avoided by means of an air chamber.

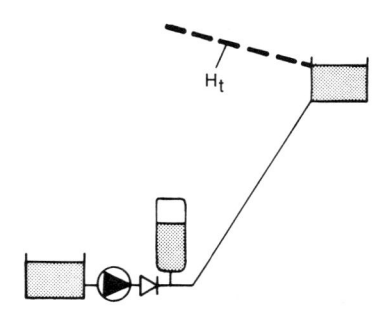

Figure 4.50 Air chamber

During normal operation the air pressure in the air chamber is equal to the pump delivery head. When pumping stops liquid flows out of the air chamber and the air pressure reduces. The greater the volume of air the smoother the process. The

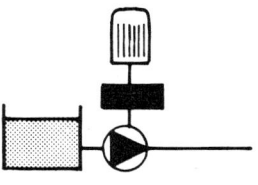

Figure 4.51 Pump with flywheel

system can be suitably dampened by throttling the chamber at the flow inlet. A disadvantage with the air chamber is the necessity of providing a compressor to replace the air which is dissolved in the liquid and carried away in the pipeline. This may still be a lot cheaper than building a much taller water column.

The retardation of the pump rotor is also affected by the moment of inertia of the rotating parts. By fitting the pump with a flywheel it is possible to prolong the run-down time. The method is most effective for high speed pumps and moderate pipe sizes. A disadvantage is increased starting difficulties. The starting problems can be alleviated by using modern soft starters. Variable speed pumps should not have starting problems.

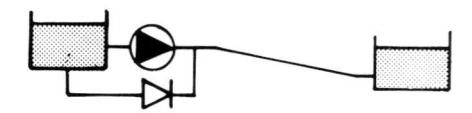

Figure 4.52 Pump with reverse by-pass

With low or negative static heads there is a risk of negative pressure on the pressure side of the pump. The pressure drop across the pump is reduced by means of a by-pass line to augment forward flow when the pump has stopped.

The other main group of protection equipment is intended to eliminate water hammer at a somewhat later stage of the transient process.

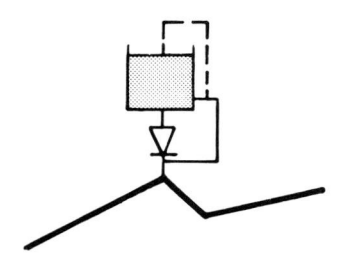

Figure 4.53 Surge tank

Negative pressure, with the associated cavitation risks can be avoided by placing a surge tank at critical points. In the event of negative pressure, liquid flows from the tank into the pipeline. The tank is refilled during normal operation via a separate level controlled filling line.

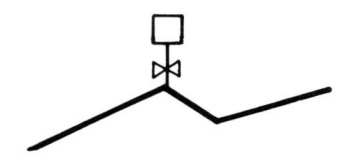

Figure 4.54 Automatic air bleed valve

An alternative method is to let air into the pipeline when the pressure falls below atmospheric pressure. The pipeline must be vented, however, prior to re-starting. In installations having

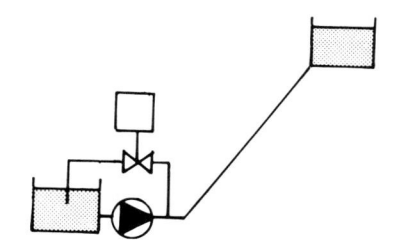

Figure 4.55 Controlled valve in by-pass line

large static delivery heads it is often the excess pressures which occur upon reversal of the liquid column and in bringing it to rest, which are most critical. These pressures can be reduced with the aid of control valves.

In order not to amplify the initial pressure wave, the negative pressure wave, the controlled by-pass valve remains closed, in principle, until the flow at the pump is reversed. It then opens rapidly, thereby relieving pressure in the pipeline and finally it closes slowly.

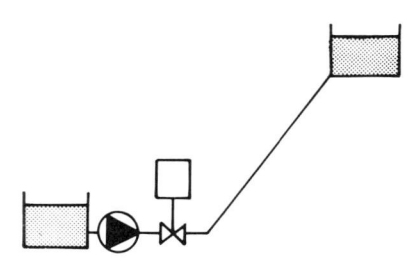

Figure 4.56 Controlled valve in main line

The controlled valve in the main line should, in principle, remain open until the flow at the pump reverses. It then closes rapidly to its optimum position, the ideal characteristics of which is to retard the reversed flow with optimum distribution of pressure drop between the pump and valve.

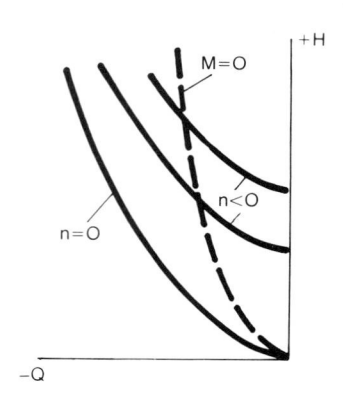

Figure 4.57 Effects of reverse running clutch

Pump reverse rotation can be prevented by means of a reverse running clutch. The reversed pump flow is then retarded more smoothly and the pressure surge is dampened. The risk of reverse overspeed is eliminated.

Apart from the pure pressure surge dampening characteristics, the choice of protection equipment is also affected by other factors. Liquid characteristics, construction, maintenance and procurement costs are examples of such factors. Each case must be judged on its own merits as to the most suitable choice of protection equipment.

4.6 Pressure pulsations in piping systems

We have seen in section 4.5 that pressure surges can occur due to isolated operating conditions such as start-up, shut-down and valve operations. These are not the only conditions under which rapid pressure changes occur. Some systems run continuously with pressure pulsations.

4.6.1 Rotodynamic pumps

It is generally thought, and often stated in literature, that rotodynamic pumps produce a continuous steady flow pattern without pressure pulsations. Axial flow pumps probably come closest of all rotodynamic pumps to fulfilling this requirement; mainly because of the low differential head produced.

Mixed flow pumps tend to be used for low differential head applications with high flow rates and consequently have wide impeller tip widths. Wide impeller passages, in the axial direction, produce uneven velocity streams with the liquid. The liquid leaving the pump discharge connection is not a steady stream but a mixture of several flow streams. As the liquid flow pattern stabilises in the discharge pipework pressure pulsations and flow variations can be measured. If flow measuring devices are to be used close to the pump discharge some form of flow straightening may be necessary.

Centrifugal pumps operate at a wide range of speeds. As speed is increased the circumferential flow pattern between adjacent vanes is modified. Liquid tends to be forced closer to the advancing vane and dragged away from the receding vane. It has been suggested that the flow between a pair of vanes will segregate into three distinct flow streams moving at different velocities. The action of one or two cutwaters produces additional pulsations. It has been proposed in new test standards that fluctuations in flow readings of $\pm 3\%$, smallest tolerances, are acceptable. Obviously flow variations of this magnitude will have associated pressure pulsations.

Centrifugal pumps can have additional problems. When operating at low flows some pumps surge in the suction pipework due to recirculation in the impeller eye. Pump manufacturers should be able to quantify the onset and magnitude of suction surge.

4.6.2 Positive displacement pumps

The flow variations of reciprocating pumps are well documented, even if incorrectly. These flow variations are of sufficient magnitude to produce pressure pulsations which create pipework vibration which does not require special instrumentation to detect. Some other positive displacement pumps produce pressure pulsations. Lobe pumps and gear pumps with few teeth deserve special attention.

Newer test standards recognize the problems caused by pressure pulsations and include test requirements. However, testing on any test rig will not reproduce site conditions including site pulsation levels. Pressure pulsations can be amplified by pipework resonance and cause severe problems. Problems with site pipework can only be detected by computer simulation or site operation.

Most pressure pulsation problems can be controlled by pulsation dampeners, see section 10.6, or by simple pipework modifications to eliminate resonance.

4.7 Bibliography

Kinno, H. — Water hammer.

Kennedy, J. F. — Charts for Centrifugal Pump Systems, J. Hydraulic Div., May 1965.

Parmakian, J. — Water hammer Analysis, Dover Publications, New York 1963.

Streeter, V.L. & Wylie, B.E. — Fluid transients, McGraw Hill. 1978.

Fox, J. A. — Hydraulic analysis of unsteady flow in pipe networks, Macmillan, London, 1984.

When the going gets tough...
Industrial Pumps from ITT Jabsco

PUREFLO 24 Series
Rotary Lobe Pumps. Rugged stainless steel process pumps able to cope with arduous duties. For positive displacement pumping of thin and viscous liquids.

Industrial Flexible Impeller Pumps
Produced in stainless steel or Epoxy, these pumps provide low cost solutions to many chemical pumping problems. Self priming with the ability to handle hard and soft solids easily, they are very versatile.

Drum Emptying Pumps
Ideal for both laboratory and factory use. "Seal-less" design ensures maximum durability for longer life.

ITT Jabsco maintains total quality management systems and is a Company Approved to BS5750 Pt 2: ISO 9002: EN29002

ITT Jabsco
ITT Fluid Technology Corporation
Bingley Road, Hoddesdon, Herts. EN11 0BU
Tel: +44-992-467191
Tlx: 263251 - G
Fax: +44-992-467132/451079

PERISTALTIC PUMPS

☆ Pumps with flow rate up to 20 litres/min
☆ Programmable, variable speed, or fixed speed
☆ Computer control options available ☆ Standard cased units or purpose-built to requirements
☆ 2 year guarantee ☆ Standard or long-life tubing.

AUTOCLUDE
Division of Victor Pyrate Limited

Arisdale Avenue · South Ockendon · Essex RM15 5DP · England
Tel: 0708 856125/6 · Telex: 897837 Pyrate G · Fax: 0708 857366

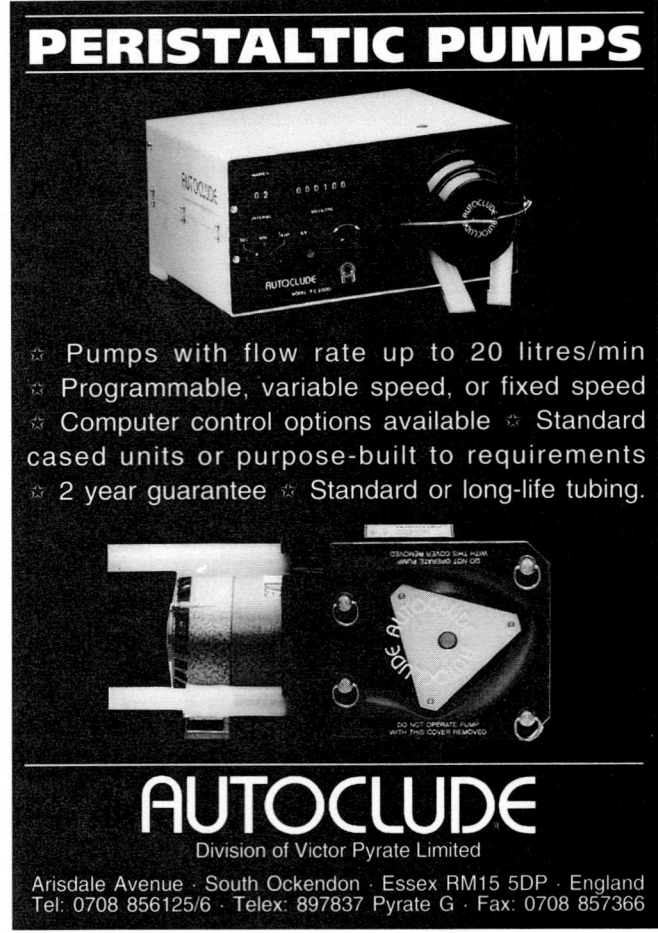

Practical reference books and buyers guides

for the worldwide oil, petrochemical, nuclear, power, process, water, paper and energy industries

●

Process Control Handbook

●

Energy Industries Council Catalogue

●

European Pumps & Pumping

●

For further details of these
and other publications
contact:

CHW Roles & Associates Ltd
PO Box 25, Sunbury-on-Thames
Middlesex TW16 5QB United Kingdom
Tel/Fax: 081- 783 0088

5 Flow regulation

Chapter 5, Flow regulation [1], deals mainly with the methods of controlling flow rate in pump installations. This chapter describes the various methods from the point of view of the principles involved, information needed for sizing, and the calculation of energy consumption. The chapter ends with "Choice of method of regulation". Flow regulation is considered in depth as there are many methods available. The method chosen for a particular installation will be based on the flexibility required, energy consumption and initial cost. A study of this chapter should not be made without a parallel study of Chapter 4 "Pumps and piping systems" and, above all, Chapter 13 "Pump economics".

(1) The term "flow regulation" in this context refers to the control of liquid flow rate through pumps and piping systems and not to "process control" involving complete industrial processes. The words "control" and "regulation" are otherwise considered to be synonymous in this and other chapters.

Contents

5.1 Introduction

5.2 Variable flow requirements

5.3 Flow regulation
   5.3.1 Methods
   5.3.2 Control signals

5.4 On-off control of constant speed pump
   5.4.1 Principle and application
   5.4.2 Costs
   5.4.3 Problems when starting/loading
   5.4.4 Problems when stopping/unloading
   5.4.5 Operational sequences
   5.4.6 Storage volumes
   5.4.7 Power consumption
   5.4.8 Examples

5.5 Pole changing induction motor
   5.5.1 Principle and application
   5.5.2 Costs
   5.5.3 Problems when starting
   5.5.4 Problems when stopping
   5.5.5 Storage volumes
   5.5.6 Power consumption
   5.5.7 Examples

5.6 Multi-speed gearbox
   5.6.1 Principle and application
   5.6.2 Costs
   5.6.3 Problems when starting
   5.6.4 Problems when stopping
   5.6.5 Storage volumes
   5.6.6 Power consumption
   5.6.7 Examples

5.7 Throttling by control valve
   5.7.1 Principle and application
   5.7.2 Costs
   5.7.3 Sizing control valve
   5.7.4 Power consumption
   5.7.5 Examples

5.8 By-pass return
   5.8.1 Principle and application
   5.8.2 Costs
   5.8.3 Sizing by-pass
   5.8.4 Overflow by-pass return
   5.8.5 Power consumption

       5.8.6 Example
5.9 Infinitely variable speed
       5.9.1 Principle and application
           5.9.1.1 Variable ratio transmissions
           5.9.1.2 Variable speed drivers
       5.9.2 Costs
       5.9.3 Conversion of pump curves to various speeds
       5.9.4 Efficiencies of various methods
       5.9.5 Power consumption
       5.9.6 Regulation system: schematics
       5.9.7 Schematics for various systems
5.10 Factors affecting choice of flow regulation method
       5.10.1 Direct flow regulation
       5.10.2 Speed regulation

5.1 Introduction

When dealing with liquid transportation, a pump installation is sized to cope with a maximum flow which, in practice, might never occur. This principle of sizing is nevertheless correct, since inadequate pumping capacity on isolated occasions can lead to the most damaging consequences. It is also obvious that sensible ways of varying the flow are important, not only for the functioning of the pump installation itself but also for the process for which the pump is included as a major or minor component.

Procurement costs for pumps generally amount to a small proportion of the total investment cost of a plant. But not always! Some pumps cost over $1M each and the remainder of the installation is just pipework. In most installations the functional quality of the pump may be the critical factor for the overall operation and it's associated running costs. Advanced methods of flow control, e.g. speed regulation of pumps, are today's best methods of varying the flow but with this disadvantage, the initial investment costs for the pump equipment are increased. The advantages are reduced running costs overall, smaller requirements for storage tanks and sumps, smaller electric cabling, and a reduction in starting current.

With regard to existing installations, the method of flow regulation affects, most of all, the energy costs. As a result of increasing energy costs, the optimum methods of pump control have changed greatly from those with low initial costs and high energy consumption to those with relatively high investment costs and good power economy. For pumps with a power requirement greater than 10 kW, it will even now pay to convert to more power-saving methods of regulation. See also Chapter 13, Pump Economics.

Pump flow can be controlled in steps, discrete flow regulation or smoothly, continuous flow regulation or infinitely variable flow regulation. Combinations of the two methods are also used. The choice of method is determined by the overall requirement of the pump installation, variation of flow, time scales, operating economics, initial cost and technical possibilities.

In this chapter, pump, may also mean pumps, as in multiple pumps of the same size operated in parallel. When the detailed description of the installation is explained it will become obvious whether a single pump or multiple pumps are involved.

The operating point of the pump at a particular speed is the intersection of the system characteristic and the pump curve as in figure 5.1. At this point, the energy dissipation of the system is equal to the energy supplied by the pump.

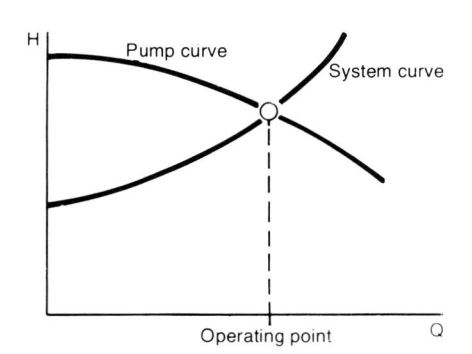

Figure 5.1 The operating point of pump and system

5.2 Variable flow requirements

Whether or not a pump needs flow regulation is dependant upon the pump type and the nature of the process involved. The pump flow may not match process demands because of:

- Safety margins and hydraulic tolerances.
- Failure to utilise full production capacity.
- Variations in requirement caused by climatic variations, e.g. summer, winter, rain, drought.
- Variations in production processes.

Flow variation with time can be described either by a running curve, figure 5.2 or by a constancy curve, figure 5.3.

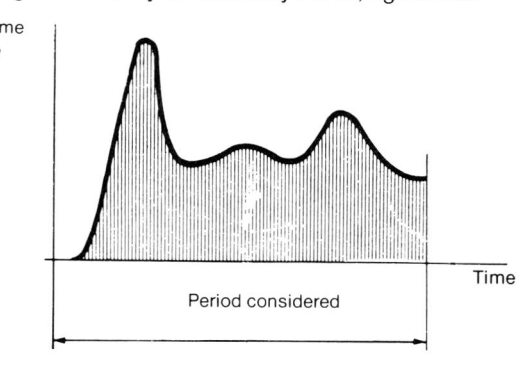

Figure 5.2 Flow demand as a function of time, running curve

In the constancy curve, time has been summated for all periods of flow having the same magnitude. A point on the constancy curve gives the time during which the flow at least amounts to a particular value.

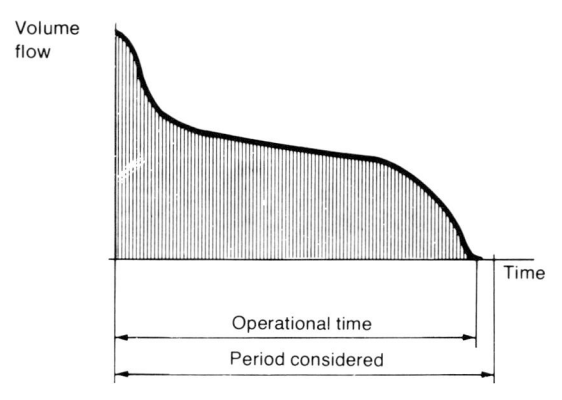

Figure 5.3 Constancy curve for flow

Constancy curves can be made up for various periods of time ranging from a full 24 hours to the lifetime of the installation. Constancy curves are usually made up for a period of 1 year.

Unfortunately, constancy curves are not that easy to compile, but a simple analysis of the actual liquid transportation often gives adequate guidance for the determination of maximum, minimum and average flow. The mean flow, for example can be determined from the annual production capacity and maximum flow from peak consumption.

The pump size is obviously determined by the maximum flow figure, whereas pump economy is determined by the duration of flow. Economic assessments are easily made if the volume flow is known at 100%, 50% and 0% constancy. It would be even better, of course, if the values at 25% and 75% are also known. The following simple equation is obtained by the use of Simpson's Rule for integration:

$$Q_{mean} = \frac{1}{6} Q_{100} + \frac{2}{3} \cdot Q_{50} + \frac{1}{6} \cdot Q_0 \qquad \text{Equ 5.1}$$

where

Q_{mean}	=	Mean flow	(m³/s)
Q_{100}	=	Flow at 100% constancy, i.e. the flow taking place 100% of operational time ($Q_{100} = Q_{min}$)	(m³/s)
Q_{50}	=	Flow at 50% constancy	(m³/s)
Q_0	=	Flow at 0% constancy ($Q_0 = Q_{max}$)	(m³/s)

or if more points on the curve are known:

$$Q_{mean} = \frac{1}{12} \cdot Q_{100} + \frac{1}{3} \cdot Q_{75} + \frac{1}{6} \cdot Q_{50} + \frac{1}{3} \cdot Q_{25}$$
Equ 5.2

where Q_{75} and Q_{25} are expressions for the flow at 75% and 25% constancy, see also figure 5.4.

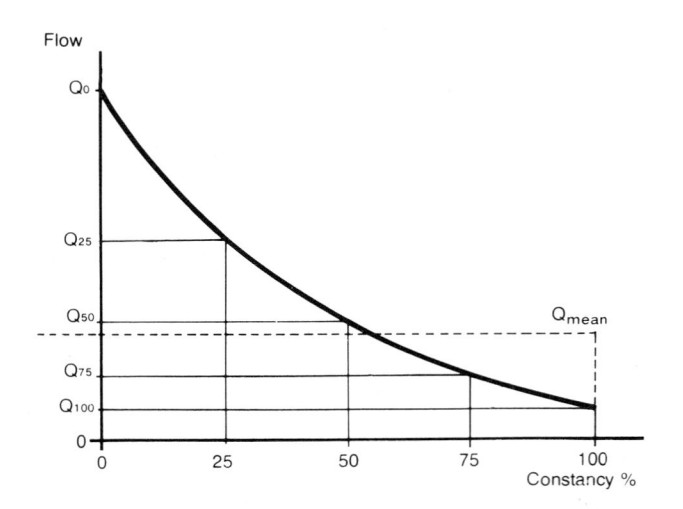

Figure 5.4 Graphical representation of Equations 5.1 and 5.2

5.3 Flow regulation

5.3.1 Methods

Generally speaking, flow can be regulated in many ways, see figure 5.5. Throttle regulation, by-pass regulation and overflow by-pass regulation all mean that the control takes place outside the actual pump itself. Some of the flow produced by the pump is thrown away, wasting energy. Other methods change the actual performance of the pump including power consumption, either within the pump unit or by means of variable filling, e.g. Archimedian pumps and displacement regulation. As can be seen from figure 5.5, the only universally applicable method is speed regulation, i.e. it can be used for all kinds of pumps, from the smallest dosing pump with power requirements of 10 to 100 W to the largest possible pump units in power stations with power requirements up to 60 MW.

There are no rules against combining various methods, e.g. the parallel operation of several units with continuous regulation by means of throttling or speed regulation. The choice of optimum methods requires great experience and often extensive economic and process technical analyses. Sometimes there are also special problems to be considered, such as noise levels from control valves when throttling.

As mentioned earlier, it is possible to combine control methods. If a modulating by-pass was fitted to most of the discrete variation methods the problems of surge and water hammer could be eliminated.

Series and parallel operation are dealt with in Section 4.3.1 and Section 4.3.2 respectively. Examples of pump performance for pumps with adjustable impeller blades, or with adjustable inlet guide vanes, pre rotation, are shown in figures 3.21 and 3.22 respectively. Archimedian, rigid screw, pumps and variable displacement regulation of metering pumps are dealt with further in Section 3.4.5. Where changes in flow requirements are needed over longer periods, more than about six months say, it may be economic to change the diameter of a centrifugal pump impeller or piston/plunger or modify the blade angles on a propeller. See also Section 3.2.2.

Methods	Advantages	Disadvantages
Discrete variations		
Start-stop control	Cheap, easy to operate	Flow surges, water hammer, wear and tear on switchgear
On-off by-pass return	Relatively cheap, easy to operate	Flow surge problems eliminated with slow acting valve
Pole changing motors	Relatively cheap, easy to operate	Must stop to change speed, limited number of speed options, flow surges, water hammer
Multi-speed gearbox	Easy to operate	Must stop to change speed, flow surges, water hammer, may need maintenance staff intervention (1)
Vee belts and pulleys	Cheap	Must stop to change speed, flow surges, water hammer, needs maintenance staff intervention
Continuously variable		
Throttling	Cheap, easy to operate	Wastes energy, not for solids handling applications
Modulating by-pass return	Easy to operate	Wastes energy, not for solids handling applications
Overflow by-pass return	Easy to operate	Wastes energy
Variable speed	Energy efficient	Can be costly
Adjustable blade angle	Energy efficient	Axial pumps only, low differential head
Adjustable inlet guide vanes	Energy efficient	Limited flow range, can be costly
Part loading	Energy efficient	Archimedean screw only
Adjustable stroke	Energy efficient	Reciprocating pumps only, very costly in larger pumps

(1) Portable engine driven pumps frequently use semi-automatic transmissions, no need to stop to change gear.

Figure 5.5 Pump control methods

5.3.2 Control signals

Control parameters may be flow, pressure, or some quantity derived from these such as level, temperature, concentration and so on. Some examples of the principles of control are shown in figure 5.6. The principles of regulation can also be shown by the H-Q diagram, see figure 5.7 for the simplest cases.

Control parameter	Sensor type	Example of application
Discrete level control	Level switch	Fresh water reservoirs Emptying separators
Proportional level control	Level transmitter	Process control on reactors and vessels
Constant flow	Flow meter	Process control for variable pressure systems
Discrete pressure control	Pressure switch	Charging vessels and accumulators. Pressure maintenance
Constant pressure	Pressure transmitter	Process control of vessels and accumulators
Proportion control for mixing	Flow meter in each circuit	Chemical processing Water treatment
Proportion control for mixing	Flow meter in mixer outlet	Chemical processing Water treatment

Figure 5.6 Some of the principles of regulation of pumping equipment

5.4 On-off control of constant speed pump

5.4.1 Principles and range of application

On-off control means that one or more pumps are started and stopped, or loaded and unloaded, systematically, according to the capacity requirement. Loading and unloading is accomplished by an isolating valve in a by-pass. When the flow

174

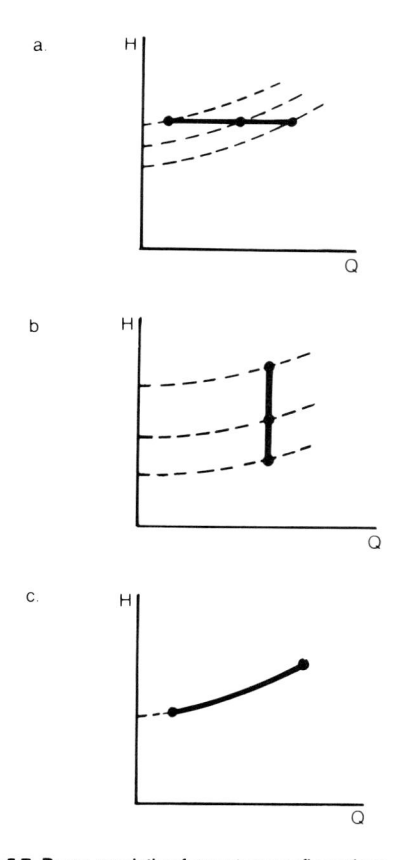

Figure 5.7 Pump regulation for system configurations
(a) constant pressure for variable system
(b) constant flow for variable system
(c) variable flow for a fixed system

requirement does not coincide with the pump capacities the pumps will work intermittently at a starting frequency of 1 to 15 starts per hour for small pumps, reducing to 1 to 4 starts per hour for larger pumps. In order to account for the difference between instantaneous flow demand and the supply flow, some form of liquid storage or reservoir must be provided.

On-off control and load/unload control is arranged automatically so that pumps are started at a high level in a storage vessel on the suction side of the pump or at low level in a storage vessel on the discharge side. When pumping between a suction tank and a discharge tank there is no need for other vessels or tanks to store the imbalance between supply and demand. The storage capacity will be included in the suction/discharge tanks. For systems using pressure control, rather than level, gas charged storage vessels, called accumulators, are used to store the imbalance. Pumping begins with a low pressure signal from the accumulator and stops at high pressure signal. The accumulator(s) must be sized so that motors do not start more frequently than their specification allows. A combination of on-off and load-unload can be very effective with larger pumps to eliminate starting problems and also increases the efficiency of load-unload used in isolation.

5.4.2 Costs

Of the discrete flow control methods, the on-off and load-unload methods are the most usual. The on-off control method is the cheapest of all methods to install. The additional costs for full automatic control may be limited to a single level switch, extra control contacts in the motor starters and some wiring. The starters for the pump motors are required for every installation, control contacts are standard with some models.

However, consideration must be given to the following when additions are necessary to the pump installation:

- cost of extra storage in which to hold the liquid during the regulation cycle,

- additional costs for heavy duty starting equipment since this has to be selected to cope with a very large number of regular starts,
- additional costs for equipment to assist in the alleviation of water hammer when starting/stopping and loading/ unloading,
- additional costs for pipework and fittings if these have to be designed to withstand the fatigue stresses and vibration induced by water hammer.

5.4.3 Problems when starting/loading

The number of permitted starts per hour is a critical factor in total initial cost for the pump installation. The type of starting used, direct-on-line, DOL, or star-delta, S-D, or a modern soft start system affects the frequency of starting. Small motors can be started 20 times per hour without problems depending upon the operating conditions. Larger motors, 125 kW and above, have strict restrictions on starting, 4 starts per hour maximum is very common. Experience shows that the number of starts chosen can be 5 to 10 starts per hour. Thus there will be 20,000 to 50,000 starts per year and up to the time of financial write-off, the number of starts will amount to several millions, thus some components have to be sized to withstand fatigue if precautions are not taken.

Even if the starting time of the pump is short, about 0.25 s with DOL and about 1 s to 1.5 s for S-D, this can put a heavy strain on the mains supply.

Loading and unloading, by automatic by-pass, removes the electrical problems associated with starting and stopping. This approach should be seriously considered as motor sizes increase. A slow acting by-pass valve can eliminate the surge and water hammer problems and remove the need for equipment to be oversized for fatigue protection.

5.4.4 Problems when stopping/unloading

A liquid flowing in a lengthy pipeline possesses considerable kinetic energy. When the liquid decelerates with the stopping of the pump, large pressure variations occur. This can be calculated and is, for abrupt stops:

$$\Delta H = \frac{a \cdot v_o}{g} \qquad \text{Equ. 5.3}$$

where
ΔH = change of pressure head (m)
a = wave speed (m/s)
v_o = initial liquid velocity (m/s)
g = gravitational acceleration (m/s^2)

In order to reduce the variation of head the flow has to be decelerated over a period of time, t_o (s), which can be roughly assessed in a pipeline of length L (m) as:

$$t_o = (15 \text{ to } 60) \cdot \frac{L}{a} \qquad \text{Equ. 5.4}$$

In the case of pipelines with high points or with low resistance to fatigue, the stopping time may have to be doubled. The normal arrangements used to increase the stopping time or to reduce the stresses are:

- slow-closing valves,
- air-chambers or pulsation dampeners,
- pumps with flywheels,
- surge columns,
- overpressure and underpressure valves.

See section 4.5.4 for full details.

In all cases the flow continues for a time, t_o. To avoid sucking air into the pump and pipeline, a sufficient supply of liquid must be available on the suction side of the pump, whilst that same volume of liquid must not cause trouble on the delivery side, through surge for example.

For air chambers, the volume of liquid is contained in the air chamber and in other cases, the storage tank or pump sump has to supply the corresponding increase in volume. When the flow requirement or inflow is negligible, which is the most unfavourable condition, the mean flow during the stop cycle can be estimated to be about 2/3 of the maximum flow. The partial storage volume, V_o, (m³), between stop signal and stop flow will, for the decelerating time t_o, s, be as before:

$$V_o = (10 \text{ to } 40) \cdot \frac{L}{a} \cdot Q_p \qquad \text{Equ. 5.5}$$

where Q_p = pump flow m³/s.

With longer pipelines, especially for installations with slow closing valves, a large part of the operational period may be taken up by the start and stop cycles, see figure 5.8. During these transients, throttle regulation or modulating by-pass takes place, see also Sections 5.7 and 5.8. Power consumption is determined by the relationship between the times for pure on-off operation and those for valve operation. Throttle regulation and modulating by-pass can increase energy consumption by anything up to 50 %.

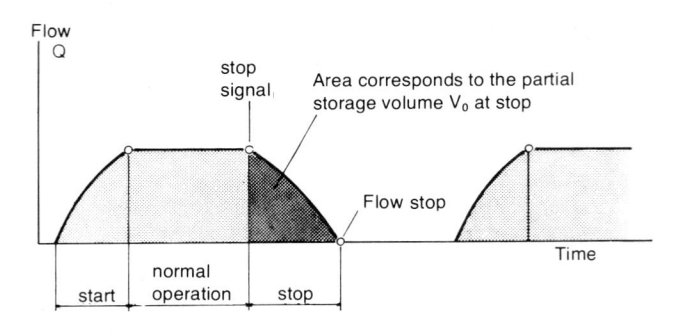

Figure 5.8 Flow variation for on-off and load/unload control with long pipeline

5.4.5 Operational sequences

There can be a choice of several different operational sequences for switching in and out parallel operation units. Critical factors to be considered are:

- the number of permitted starts per pump per hour,
- the number of permitted starts during the lifetime of the installation,
- permitted variations in flow,
- costs of storage tanks, sumps and accumulators.

The differences between various possible sequences, designated A, B and C, are shown in the following three schematic examples applied to an installation with two identical parallel operating pumps P1 and P2. It becomes clearest if the condition is examined when the supply flow is equal to 1.5 times the flow of one pump. The other operational possibilities; when the supply flow is less than the flow of one pump, and when the supply flow is greater than the flow of two pumps; must also be considered for a complete control strategy. In these examples, on and off can be synonymous with load and unload.

Operational Sequence A, Figure 5.9

P1 operates continuously and P2 operates on and off in response to a level switch signal. The cycle time is obtained when sizing P2's storage or partial sump volume. In this operational sequence there are moderate flow variations.

Operational Sequence B, Figure 5.10

P1 starts first, and then, at a higher level, P2 starts. Stop occurs simultaneously for both pumps at sump minimum level. The flow variation is very wide, but the storage or partial sump

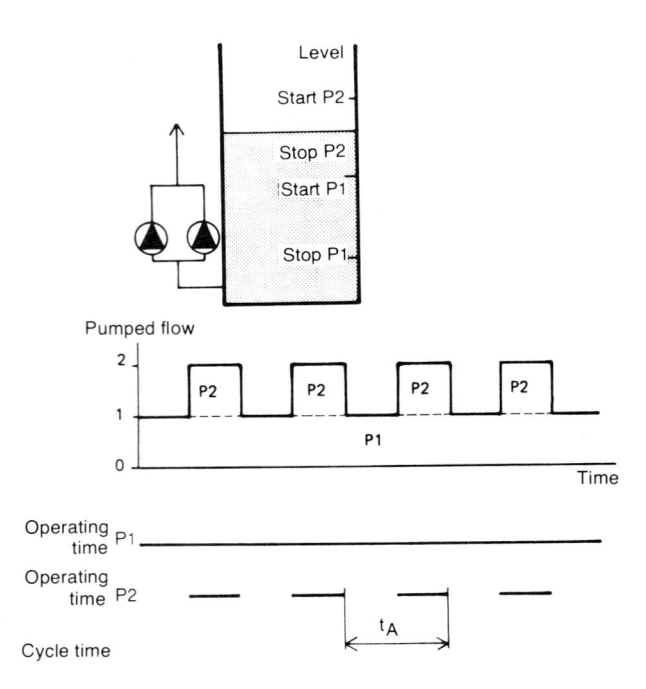

Figure 5.9 Flow variation and operating times for operational sequence A

volume for P2 is reduced to almost half of that for Sequence A. Because of the common stop, one sensor signal is saved. Operational Sequence B has been applied to about 90% of all sewage pump installations up to the present time.

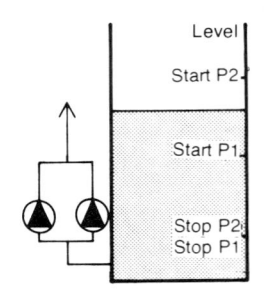

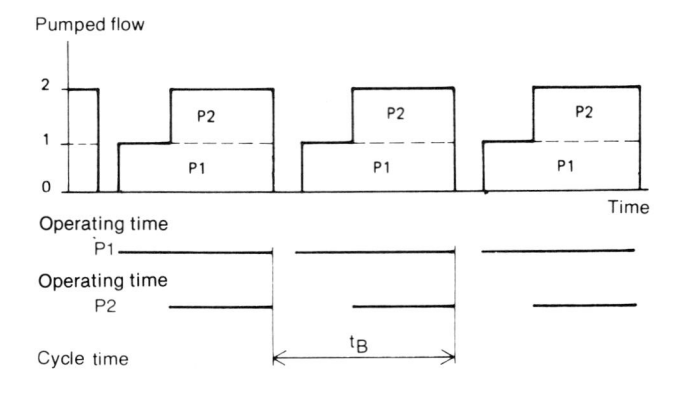

Figure 5.10 Flow variations and operating times for operational sequence B

Operational Sequence C, Figure 5.11

Here we get the same flow variation as in operational Sequence A, except that the stop sequence is the same as the start sequence, i.e. the stop signal always stops the pump which started first. Operational Sequence C means a doubling of the cycle time, i.e. $t_C = 2t_A$, meaning the halving of the storage or partial sump volume for P2.

Where there are more than two Pumps, the same principles as above are used, as also with the speed changing of pole-changing motors, for example. For a number of pumps working in parallel, the partial storage volume is then determined for each pump depending upon its respective incremental flow Ql, Q2, Q3, etc., as obtained from the pump and system curves, according to figure 5.12.

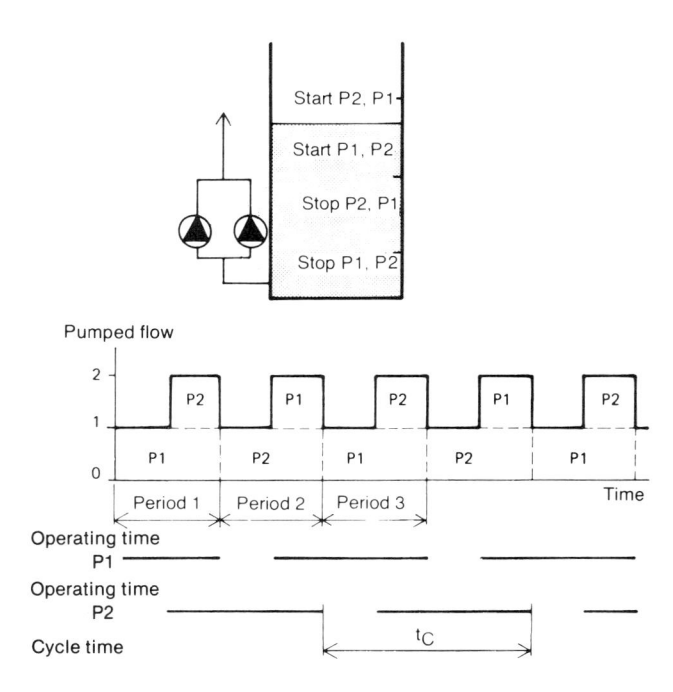

Figure 5.11 Flow variation and operating times for operating sequence C

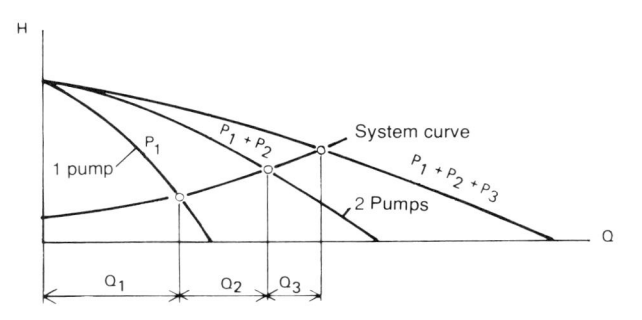

Figure 5.12 Flow increments Q1, Q2, and Q3 obtained by switching in various numbers of parallel operating pumps

In practice, to determine switching times and storage volumes, supply flows very close to pump capacities would be used, e.g. 99%, 101%, 199%, 201%, etc.

Recent advances in micro-electronics have permitted the development of small but powerful Programmable Logic Controllers, PLC's. Versions of PLC's are currently available with 8 digital inputs and outputs and cost approximately £150; about the same as a good pressure switch. These devices allow on-off control to be completely optimised. For example, the PLC could keep a running total of the "hours run" for each pump and start the pump with the least hours first and switch it off last. A simple programme would try to balance the operating time across all the pumps. Operating times could also be biased so that one pump had few hours and could be considered as a stand-by.

When considering how many pumps to install, the possibility of failure and allowances for unavailability due to routine maintenance must be considered. The need for stand-by capacity must be weighed against the consequences of unavoidable reduced capacity.

5.4.6 Storage volumes

Storage and pump tank/sump volumes, see figure 5.13, are most easily calculated by the use of a filling time, time constant, for the various partial volumes V, and V_2, and so on, defined as:

$$t_2 = \frac{V_2}{Q_2} \text{ etc.} \qquad \text{Equ. 5.6}$$

Time t_1 thus corresponds to the time taken to fill volume V_1 with flow Q_1.

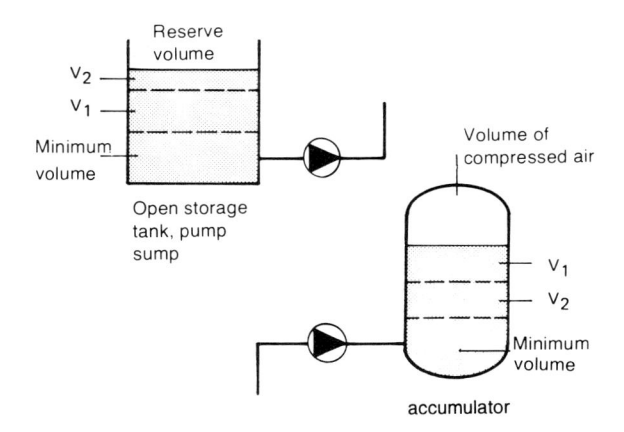

Figure 5.13 Storage with associated minimum volumes, partial volumes V1 and V2 and the volume of air for various arrangements

In the case of an accumulator, Q varies because of counter-pressure from the precompressed volume of air or nitrogen. The gas law proposed by Boyle at constant temperature, $p_1V_1 = p_2V_2$, should be used to calculate the change in gas volume with pressure. Time t_1 for the pump switched on first depends solely on the number of permitted starts per unit time, depending on whether starting is applied to several pumps alternately.

Time t_1 is determined by

$$t_1 = \frac{3600}{4 \cdot m} \quad (s) \qquad \text{Equ. 5.7}$$

where

m = number of starts per hour per pump if the pumps are not alternated or m = number of starts per hour for the whole installation, if the pumps are alternated.

Here the number of starts per hour is a maximum when the mean flow per hour is equal to half the maximum flow. The times t_2, t_3, t_4, etc., are heavily dependent upon the operational Sequences A, B and C. Figure 5.14 shows various values for up to 4 pumps operating in parallel. In the table it is assumed that no more pumps are included in each operational sequence range than are needed to satisfy the flow requirements, see also the example which follows.

Number of starts/hr per pump	5			10			15		
Sequence	A	B	C	A	B	C	A	B	C
Filling times for:									
pump 1: t_1	180	180	180	90	90	90	60	60	60
pump 2: t_2	180	72	90	90	36	45	60	24	30
pump 3: t_3	180	48	60	90	24	30	60	16	20
pump 4: t_4	180	40	45	90	20	22	60	13	15

Figure 5.14 Table for determination of filling time for partial volumes in pump tanks/sumps

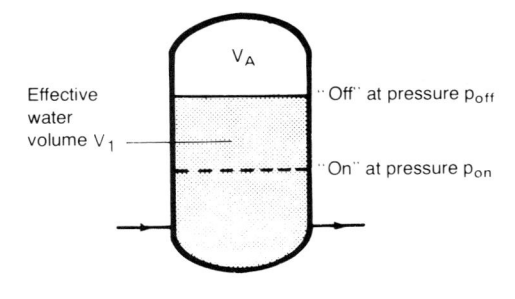

Figure 5.15 Accumulator with designations for working volumes

The air volume in an accumulator is calculated using Boyle's Law for isothermal compression, figure 5.15.

At the absolute pressures p_{off} and p_{on}, there are respectively the volumes of air V_{Aoff} and V_{Aon}:

$$p_{off} \cdot V_{Aoff} = P_{on} \cdot V_{Aon} = P_i \cdot V_{Ai} \qquad \text{Equ 5.8}$$

where
- p_i = initial absolute pressure (kPa)
- V_{Ai} = initial volume of air (m³)

Using the effective volume of water $V_1 = V_{Aon} - V_{Aoff}$ we get:

$$V_{Ai} = \frac{P_{on}}{P_i} \cdot \frac{P_{off}}{P_{off} - P_{on}} \cdot V_1 \qquad \text{Equ 5.9}$$

Combining equations 5.7 and 5.9, if only one pump of flow Q, delivers to the accumulator, we then get:

$$V_{Ai} = \frac{P_{on}}{P_i} \cdot \frac{P_{off}}{P_{off} - P_{on}} \cdot Q_1 \cdot \frac{3600}{4 \cdot m} \qquad \text{Equ 5.10}$$

where
- V_{Ai} = total volume of accumulator if not precharged (m³)
- V_{Ai} = volume of air in the accumulator when precharged (m³)
- p_{on} = absolute switch-on pressure (kPa)
- p_{off} = absolute switch-off pressure (kPa)
- p_i = absolute initial pressure (kPa)
- Q_1 = flow from pump, mean value of flow at p_{on} and p_{off} (m³/s)
- m = maximum number of starts per hour per pump (n/hr)

If several pumps are working on the same accumulator, the effective water volumes V_1, V_2, V_3, etc. are determined for each one as above. The various on and off switching pressures are then bound entirely by Boyle's Law, Equ. 5.8, which necessitates careful evaluation of the various quantities.

When liquid is supplied to processes which themselves are controlled by on-off valves the process control valves can create surges and water hammer effects. Large on-off process demands can have local accumulators to reduce peak loads on the supply system. The cost of local accumulators is offset by the reduction in supply pipe size necessary to feed the demand.

5.4.7 Power consumption

An on-off controlled pump operating at constant differential head has constant running power consumption, whilst the operational time varies with volume flow. The relative operational times, or factor of intermittency i_Q when transporting flow Q with a pump of flow Q_p will be:

$$i_Q = \frac{Q}{Q_p} \qquad \text{Equ 5.11}$$

The energy consumption E, i.e. the power requirement multiplied by time will be, for time t:

$$E = \frac{Q}{Q_p} \cdot P_{Qp} \cdot t \qquad \text{Equ 5.12}$$

In the case of varying flow the power consumption is summated for various intervals of flow. Simpson's Rule integration formula provides the total power consumption in simple terms thus:

$$E = P_{mean} \cdot t_o \qquad \text{Equ 5.13}$$

$$P_{mean} = \left(\frac{1}{6} \cdot \frac{Q_{100}}{Q_p} + \frac{2}{3} \cdot \frac{Q_{50}}{Q_p} + \frac{1}{6} \cdot \frac{Q_0}{Q_p}\right) \cdot P_{Qp} \qquad \text{Equ 5.14}$$

or at 5 points taken from the flow constancy curve:

$$P_{mean} = \left(\frac{1}{12} \cdot \frac{Q_{100}}{Q_p} + \frac{1}{3} \cdot \frac{Q_{75}}{Q_p} + \frac{1}{6} \cdot \frac{Q_{50}}{Q_p} \right.$$
$$\left. + \frac{1}{3} \cdot \frac{Q_{25}}{Q_p} + \frac{1}{12} \cdot \frac{Q_0}{Q_p}\right) \cdot P_{Qp} \qquad \text{Equ 5.15}$$

where
- E = Energy consumption (kWh/yr)
- P_{mean} = mean power throughout the year (kW)
- t_0 = operational time (hr/yr)
- Q_0, Q_{50} = flow at 0%, 50% etc. constancy (m³/s)
- Q_p = flow at pump operating point (m³/s)
- P_{Qp} = pump power @ flow Q_P (kW)

For pumps operating at variable discharge pressure, with accumulators for example, consider the power consumption to be the mean of the two values at low and high pressure. Power can also be "wasted" due to throttling or modulated by-pass during starting and stopping. Pumps which run continuously but are loaded and unloaded will consume some power when unloaded. This energy consumption should be added to the power calculated from equ. 5.15.

On-off control has good power economy when there is a high proportion of static differential head. The power economy may be expressed in terms of regulation efficiency η_r comprising the ratio between head requirement at each part of the system curve and the pump head at flow Q_P. The regulation efficiency is shown in figure 5.16. The total efficiency factor for liquid transportation is obtained by multiplying η_r by pump and motor efficiencies; see also Section 13.3.1.

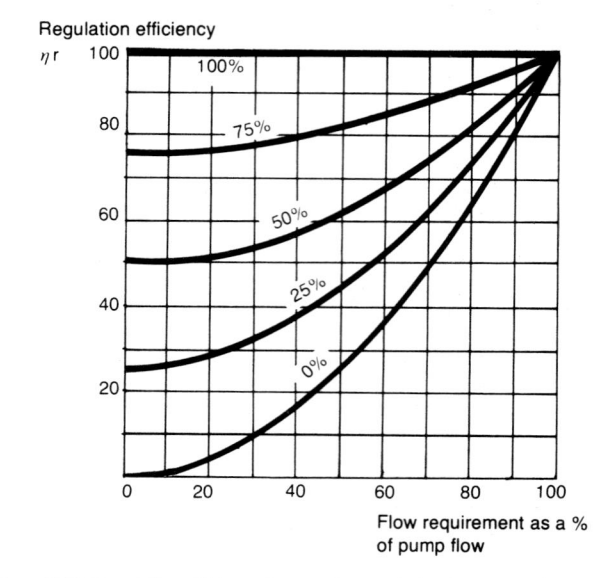

Figure 5.16 Regulation efficiency for on-off control of a single centrifugal pump working with water or liquid with characteristics like water for various ratios of H_{stat}/H_o. Ratio of the static delivery head as a percentage of total differential head H_o at the operational point of the pump

5.4.8 Examples

On-off control is used to a considerable extent in pumps for water and sewage installations and for domestic drinking water installations. Also in large hydraulic power packs used in steelworks and chemical processing. Some installations use 10 or 12 pumps to reduce the supply step size and also to limit the size of the liquid store. On-off control has a great variety of practical applications.

Certain special cases omit the liquid store, for example booster pumps, see Section 4.3.3, working in "dead" pipeline networks. In order to avoid too frequent starting, an electrical interlocking device with, say, a time relay must be fitted. In certain flow ranges, pressure variations due to starting and stopping of pumps is inevitable.

In the case of circulatory pumping in totally enclosed systems, there is no requirement for a flow storage reservoir, but there is often a requirement for storage of some other parameter, e.g. heat in a thermal store.

Calculating storage volumes

The following flows have been determined from the pump and system curves for a pump installation using three parallel connected pumps, see also figure 5.12.

Number of pumps operating	Total Flow	Flow increment
1	100 L/s	100 L/s
2	180 L/s	80 L/s
3	240 L/s	60 L/s

From figure 5.14, we get from 10 starts per hour using operational Sequence C;

Pump number	Filling time	Partial storage volume
1	$t_1 = 90$ s	$V_1 = 100 \cdot 90 = 9000$ L
2	$t_2 = 45$ s	$V_2 = 80 \cdot 45 = 3600$ L
3	$t_3 = 30$ s	$V_3 = 60 \cdot 30 = 1800$ L

The total storage volume will be
$9000 + 3600 + 1800 = 14,400$ litres $\equiv 14.4$ m^3.

Added to this is a minimum volume and, for closed storage, also a volume of air, see figure 5.13, as well as a storage volume dependent upon pressure surge, see Sections 5.4.3 and 5.4.4. The sensor signals for start-stop are arranged at levels or pressures corresponding to the various partial volumes. If operational Sequence A had been selected the storage volume would have increased to $(100 + 80 + 60) \cdot 90 = 21,600$ litres $\equiv 21.6$ m^3.

We obtain a storage volume for operational Sequence B = 100 - 90 + 80 - 36 + 60 - 24 = 13,320 litres $\equiv 13.3$ m^3, in the same way.

If only the number of starts for each pump and not the total number of starts for the installation has been considered, there can be reductions in storage volume. With complete rotation using 1, 2 and 3 pumps respectively in operation and applying the principle of operational Sequence C, $t_1 = t_2 = t_3$ can be reduced to 30 seconds, giving a total storage volume of 7.2 m^3.

Furthermore, in many cases the partial storage volumes can, to some extent, be absorbed within each other, in which case, the minimum total storage can be estimated to be approximately 4 m^3.

It should be noted that a part of the cost saving for storage will then contribute towards more complex automation, although these costs are dropping dramatically with modern developments.

5.5 Pole changing induction motor

5.5.1 Principle and application

Special versions of induction motors with arrangements for changing the number of poles can be used for pump control. Pump speed is inversely proportional to the number of poles, see Chapter 9. Two distinct versions of motor are available with similar, but not identical, properties. Pole changing motors achieve two or more speeds with one set of windings connected in different configurations. A dual wound motor has more than one set of windings; one for each speed. Small motors can have a switch built into the motor for local manual speed control. Usually all the winding terminations are brought out through the terminal box to the starter enabling remote automatic speed changes. Speed changing under load is normal. Star-delta starting on all windings of dual wound motors is possible. With pole changing motors, S-D starting is only possible on the lowest speed. Soft starts can be incorporated but the wiring becomes complicated. Pole changing motors, and dual wound, can be used in those situations when a limited range of speeds; 2:1 or 4:3, 4:2; can be useful.

In some cases motor power is not always proportional to speed. Rotodynamic pump performance changes in accordance with the affinity laws, see Section 3.2.1.6, equation 3.16. Positive displacement pumps react differently to speed changes. The pump flow is approximately proportional to speed, depending upon the pump type and the viscosity of the liquid; the ability to generate pressure is unaffected.

Pole-changing regulation with rotodynamic pumps is particularly favourable when the system curve has only pipe friction losses as in figure 5.17, in circulation installations for example. Systems with H_{stat} will limit the value of the lowest useful speed.

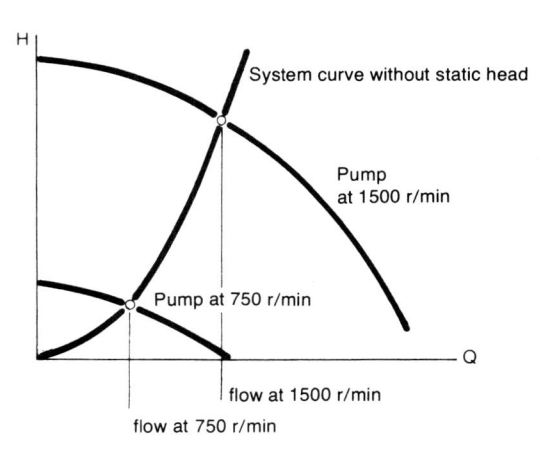

Figure 5.17 Example of change in duty point due to pole-changing 750/1500 r/min. In this case the flow is approximately halved by switching to 750 r/min

Positive displacement pumps can work with systems with low and high H_{stat}. This is why PD pumps are favoured in many chemical applications using constant pressure vessels and reactors. Gear, lobe, screw and progressing cavity pumps are used when viscosity is suitable; reciprocating pumps can be used for any viscosity. Figure 5.18 indicates the effect on the duty point of varying the speed of a PD pump when working with friction only systems and systems having a high H_{stat} value.

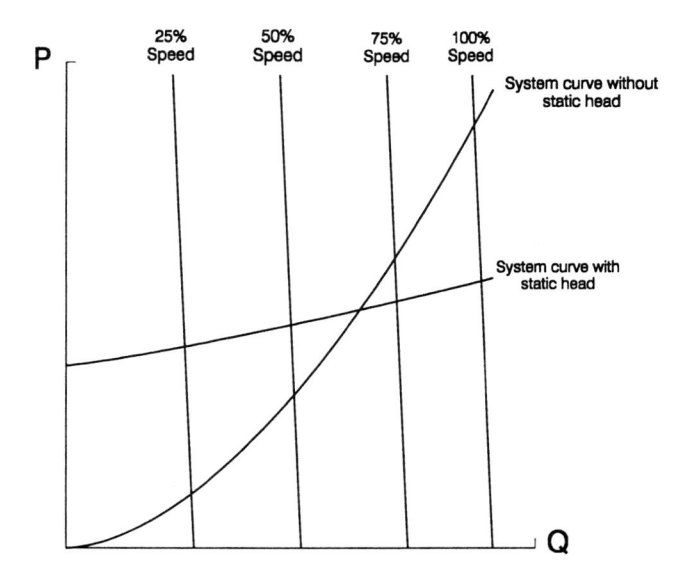

Figure 5.18 Positive displacement pumps at various speeds

The power required by the pump is dependant upon the flow, the pressure and the corresponding pump and motor efficiencies at the duty point. Irrespective of the pressure, the flow varies very much in proportion to the speed change. The change in pressure is controlled by the system characteristic.

5.5.2 Costs

From the point of view of cost, a variation factor of 2 is the most favourable since this gives pump speeds of, say, 750/1500 or 1500/3000 r/min, although other speed ranges are also possible. Some motors with four speeds are available, 1500/1000/750/500 r/min at 50 Hz. Pole changing motors are cheaper than a motor plus a frequency invertor. If continuously variable speed is not necessary, and all electric controls is an asset, the pole changing motor should be considered. At 20 kW a two speed motor is 25% more costly than a standard motor, at 135 kW it is 50% more. At 37 kW a three or four speed motor is double the cost of a standard motor.

5.5.3 Problems when starting

Starting problems are not as bad as fixed speed pumps with on-off or load-unload control because the pump can be started at low speed. Once the system has stabilised at low speed the motor can be switched to the next speed. The period of time required for stabilisation will depend upon the pump size and the system size.

5.5.4 Problems when stopping

As with starting, stopping problems can be reduced by changing speed down in steps, with stabilisation periods, before switching off at the lowest speed. Power failure, when operating at the fastest speed, will cause the same problems as other systems. The reliability of the power supply, together with the likely problems, must be assessed when deciding if precautions are warranted, see 5.4.4.

5.5.5 Storage volumes

Since pole-changing regulation is a discrete method of regulation, liquid storage is necessary. Sizing is carried out in the same way as for on-off control, Section 5.4.6, with extra steps to allow for the speed increments. The permitted number of pole changes is usually between 1 and 15 times per hour but should be confirmed with the motor manufacturer once specific running conditions are known.

5.5.6 Power consumption

Rotodynamic pump and motor efficiencies are for practical purposes the same at the various speeds, which leads to good operational economy, see figure 5.19. The efficiency of positive displacement pumps is dependant upon the discharge pressure, which is in turn, dependant upon the system characteristic. The cost of pole-changing motors is of the order of 20% to 40% higher than for normal, squirrel-cage induction motors. A limiting factor, to some degree, is the reduced availability of this type of motor from manufacturers.

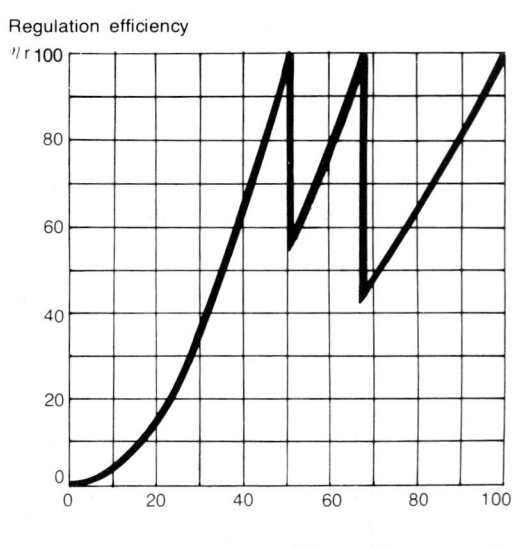

5.5.7 Examples

Consider the pumps and system curves shown in figure 5.12. We saw in 5.4.8 that the nominal storage volume was 14.4 m^3. If the pump curves shown referred to a 4 pole pump, 1480 r/min, then it may be possible to consider pole changing. Using the affinity laws, the pump curves were replotted at 6 pole speeds, 986 r/min. The closed valve head was greater than H_{stat} so the pumps would operate with the system curve successfully. The new values of Q_1, Q_2 and Q_3 became 60, 45 and 35 l/s. Using the same switching per hour and control sequence C, the volumes V_1, V_2 and V_3 become 5400, 2025 and 1050 litres giving a total volume of 8475 litres. Pole switching produces a reduction in basic volume of 41%. Further reductions in volume, considering combining partial volumes as in 5.4.8, will lead to further reductions. The flow range of the pump package has been increased from 240/100, 2.4:1, to 240/60, 4:1. Without considering mixed speed operation, the operating capacities available are 25%, 42%, 44%, 58%, 75% and 100%.

The example from figure 5.12 could not use pole changing if the pumps had been running a 2 pole speeds. The next lower speed would be 4 pole and the pumps would not produce sufficient differential to overcome H_{stat}.

Pole changing motors can offer increased flexibility and reduced storage volumes. It may be possible to run slow rotodynamic pumps in parallel with fast rotodynamic pumps; the relative shapes of the pump and system curves, together with the driver power installed, will determine whether this is feasible. Of course they are no problems running slow positive displacement pumps with fast positive displacement pumps.

5.6 Multi-speed gearbox

5.6.1 Principle and application

The principle and application of multi-speed gearboxes is very similar to that of pole changing motors except there are no strict limitations on the speeds available. Manual change multi-speed gearboxes are nearly always built specially to order so the gear ratios can be optimised for the application. If a two pump system is considered, speeds of 100% and 50% could be chosen. This would provide "pump package" capacities of 25%, 50%, 75% and 100%. Again considering two pumps, gear ratios of 33%, 66% and 100% would produce package capacities of 17%, 33%, 50%, 67%, 83% and 100%. Two and three speed gearboxes are relatively common; more ratios are rare. Manual change gearboxes usually have internal dog clutches, the gears are always in mesh, for selecting the ratio. One or two levers are moved, while the drive is stationary, to change a ratio. It may be necessary to jog or inch the motor to line up the dog clutches before engagement. On units up to about 250 kW, the high speed coupling can be turned by hand. On larger units facilities must be provided for barring, jogging or inching.

Semi-automatic gearboxes are frequently used on engine driven portable pumps. Versions of these gearboxes are available for motor drives. Just like the automotive counterparts, gears can be changed onload without stopping. The gearbox is fitted with a torque converter or fluid flywheel to smooth the transition from one speed to the next. Remote and automatic control of gear changes is simple. The drive can be disengaged to give on-off control as well as variable speed. Larger units are fitted with a flywheel lock to save the 2 to 4% speed, and energy, loss due to flywheel slip. Gearboxes can be fitted with auxiliary power take-off to drive associated equipment. Standard gearboxes are available with up to ten gears. Choice of ratios is limited but because of the multiple speed increments available this is not a problem.

Figure 5.19 Regulation efficiency of a centrifugal pump using a pole-changing motor with 4, 6 and 8 poles. There is no static delivery head

5.6.2 Costs

For small manual change gearboxes, about 40 kW for a reciprocating pump, a two speed box will cost approximately 4.5 times as much as a standard fixed ratio gearbox. A three speed box will cost 10 times as much. As the gearbox size increases the cost impact is reduced. For a 700 kW gearbox, two and three speed cost twice and 3.5 times as much.

Semi-automatic gearboxes start at approximately 8000 for an air operated four speed unit with fluid flywheel suitable for input speeds up to 3500 r/min.

The increased cost of the power transmission should be considered with the reduced costs of storage volumes. The overall scale of the pumping project will determine the relative cost proportions. The reduction in drive efficiency of semi-automatic gearboxes without fluid flywheel locks must be considered.

5.6.3 Problems when starting

Manual change gearboxes are similar to pole changing motors, see 5.5.3. Semi-automatic gearboxes, with more ratios, can have really low speed starts and smaller increments between the ratios effectively removing the problem.

5.6.4 Problems when stopping

Again, manual change gearboxes are similar to pole changing motors, see 5.5.4 and 5.4.4. Semi-automatics do not experience problems with normal stops but power failures can be a problem, see 5.5.4 and 5.4.4.

5.6.5 Storage volumes

Speed changing by fixed ratios is a discrete method of regulation and requires some form of liquid storage to compensate for the imbalance between supply and demand. Manual speed changing should only be considered when demand varies periodically rather than continuously. Manual speed changing can be viable if the demand only varies significantly say once or twice a day or less. At any given speed the pump should be considered as fixed speed and volumes calculated as shown in 5.4.6. Semi-automatic gearboxes have no restrictions on gear changes and remote automatic control is relatively simple which includes disengaging the drive. The flows possible are in discrete steps, however small, and even with the additional on-off control some small storage capacity may be necessary. However, the storage capacity will be based on the smallest pump flow thereby reducing the storage required dramatically compared to manual speed change, on-off or load-unload options.

5.6.6 Power consumption

Power consumption will depend upon the curve shapes and the intersection points. Pump efficiency depends upon flow and head or pressure, driver efficiency depends upon the load. Manual change gearboxes are relatively simple so the mechanical efficiency should be about 95 to 97%. The basic efficiency of semi-automatic gearboxes will be slightly lower because of the greater complexity. Also the fluid coupling/fluid flywheel may lose up to 4% when not locked up. Manual gear changing is very efficient at transmitting power but poor at changing gear. Semi-automatic gearboxes are good at changing gear at the expense of reduced transmission efficiency. The type and size of the pump package, limitations on storage volumes and cost, considered with the operational requirements will show which gearboxes are suitable.

5.6.7 Examples

Referring to the example shown in figure 5.12, the minimum pump speed possible, allowing a 10% margin for wear and increased losses, is 60% of the speed shown. This would give a slowest running speed of 888 rpm based on an original 4 pole motor speed. Unusually, in this case $Q_1 = Q_2 = Q_3 = 40$ l/s. The basic storage volume would be 6.6 m^3. The flow capacity increments would be 17%, 33%, 42%, 50%, 75% and 100%. Using this pump on this system would produce a minimum safe capacity of 17%. Positive displacement pumps do not have a closed valve head problem and any minimum speed could be selected to suit operations. Minimum speeds for reciprocating pumps can be as low as 5% of design.

5.7 Throttling by control valve

5.7.1 Principle and application

In the case of throttle regulation for rotodynamic pumps, figure 5.20, the operating point of the pump changes because the system curve is modified. As the flow area of the control valve is reduced the pressure drop increases, the flow through the pump reduces from Q_1 to Q_2.

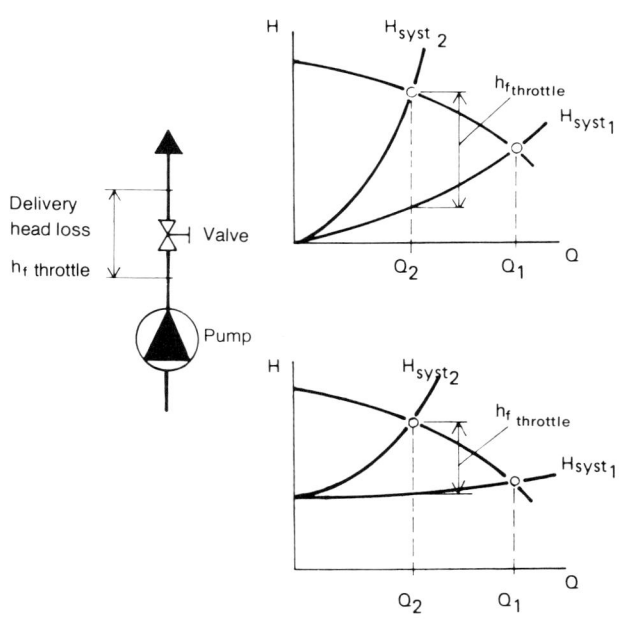

Figure 5.20 Throttling by control valve, schematic and curves

Throttle regulation creates an extra flow loss $h_{f\,throttle}$ in the system. To overcome this, power must be delivered to the pump, thus:

$$P_{throttle} = \frac{\rho \cdot Q_2 \cdot h_{f\,throttle}}{1000 \cdot \eta_2} \quad (kW) \qquad \text{Equ 5.16}$$

In a completely zero loss regulation system, the power supplied would be less by $P_{throttle}$.

As can be seen from figure 5.20, $h_{f\,throttle}$ and therefore $P_{throttle}$ are greater when the system curve is steeper for the same change of flow from Q_1 to Q_2.

Throttle regulation and by-pass regulation are the commonest of the continuously variable methods of regulation, but throttle regulation cannot be used for positive displacement pumps. In view of rising energy costs, however, the method is becoming uneconomic in spite of the relatively low procurement cost of a control valve.

Throttle regulation is a continuously variable flow method. Operation is possible at any point on the system curve where the pump curve is higher. If the pump can be operated at closed valve without problems there is no requirement for storage volumes. The problem is efficiency; all the throttling energy is wasted. The choice of throttling is dependant upon the size of the pump, the cost of energy, how often throttling will be used and for how long. If the pump is small and/or energy is cheap throttling may be attractive. If throttling is only necessary for start-up and shut-down it may be attractive. Most pumps are installed with isolating valves. The throttle valve can be fitted as a by-pass around the isolating valve.

After throttling for start-up the isolating valve can be opened fully for normal operation.

5.7.2 Costs

Installation costs are dependant upon the size of valve required and the mode of operation desired. Local manual control would be sufficient for start-up/shut-down that took one hour say every month. If continuous throttling was necessary a fully automatic control valve would be required. A 1" manual throttle valve, specifically designed for continuous operation would cost about £800. A 2" low pressure constant flow control valve in cast iron suitable for water would be about £600.

5.7.3 Sizing control valve

The control valve is sized so that the control function itself has the best possible regulation characteristics. This means that the parameters to be controlled, i.e. the flow or the head, should be as linear as possible in relation to the movement of the valve. It is a requirement that a loss of head should always exist across the valve even at maximum flow Q_0, figure 5.21.

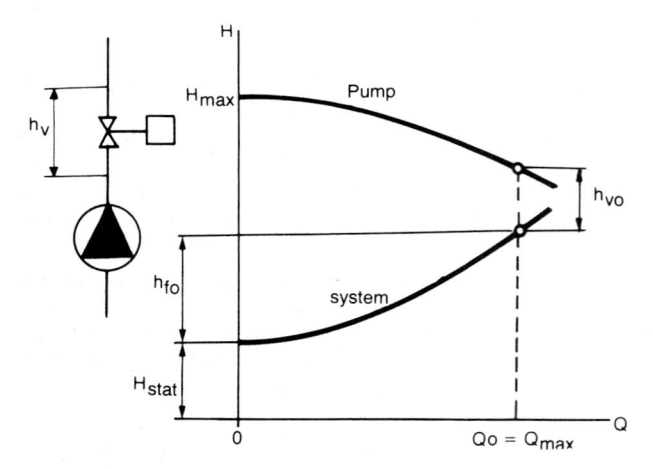

Figure 5.21 Relationship between pump and system characteristics for successful throttling

When regulation is necessary in a system which has a simple smooth characteristic, as shown in figure 5.21, the following approximation may be used for the minimum head loss at maximum flow:

$$h_{vo} = 0.1 \cdot h_{stat} + 0.3 \cdot h_{fo} \text{ (m)} \qquad \text{Equ 5.17}$$

The sizing of the control valve and its linearity for this case has already been dealt with in Section 4.2.4.

When the system curve is variable, throttling may be used to obtain a constant flow. The loss of head due to the valve must be selected at a considerably greater value than that given by the rule of thumb in equation 5.17. An analysis of regulation characteristics should be carried out in the manner shown in the example in Section 4.2.4.

In throttle regulation, the power requirements of the pump are determined entirely by its power curve, regardless of what the system curve looks like, see figure 5.22, i.e. the power requirement can be read directly from the pump power curve.

Energy consumption E is power multiplied by time, or

$$E = P_{mean} \cdot t_0 \qquad \text{Equ. 5.18}$$

where

E	=	annual energy consumption	(kWh/year)
t_0	=	operational time per year	(hr/year)
P_{mean}	=	mean power during operation	(kW)

The mean power is most simply determined by the use of Simpson's Rule integration formula in the same manner as flow in Section 5.4, see equ. 5.1 and 5.2.

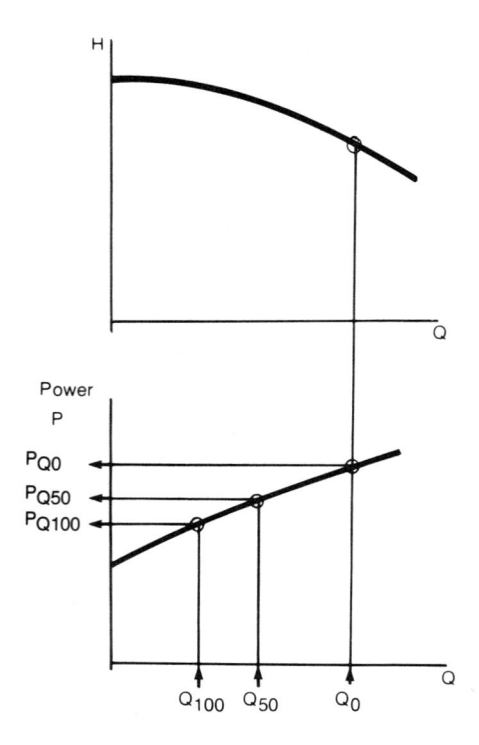

Figure 5.22 Pump power for throttle regulation

P_{Q0}, P_{Q50} and P_{Q100} are the power values corresponding to flow Q_0, Q_{50} and Q_{100} respectively, see figure 5.22. These flows apply to mean flows at 100%, 50% and 10% constancy.

For throttling, the regulation efficiency is the ratio between the differential head of the system curve compared to the pump differential head. Figure 5.23 shows the regulation efficiency for the special case of a centrifugal pump working with water or a liquid with water-like qualities. The loss of head due to the valve has been chosen in accordance with equation 5.17 and the pump curve is assumed to have its closed valve head some 20% higher than the operating point for maximum flow. In the case of throttle regulation, pump efficiency varies with flow and the motor efficiency varies with absorbed power.

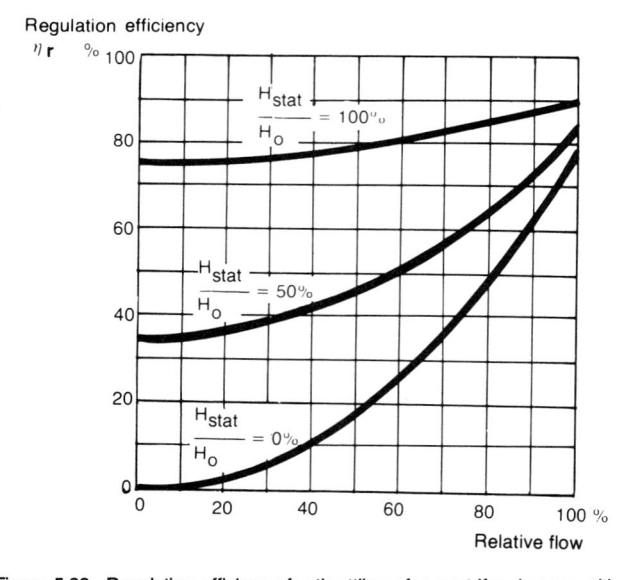

Figure 5.23 Regulation efficiency for throttling of a centrifugal pump with a shut-off head about 20% higher than the operating point at maximum flow. The head H_0 applies to the head loss of the valve, i.e. $H_0 = H_{stat} + h_{fo}$, see also figure 5.21

5.7.4 Power consumption

Throttling increases the power consumption. The head loss through the control valve is completely destroyed but must be paid for with energy from the pump. If throttling must be used continuously it will be a continuous waste of energy. If throttling is only required for start-up or shut-down it can be accomplished in a by-pass around the isolating valve and once the isolating valve is opened the throttling loss will be zero.

5.7.5 Examples

Consider the pumps and system shown in figure 5.12. The process has been redesigned and only one pump is required. The storage volume has been moved to another part of the process. During start-up and shut-down, which last 2 hours each every month the flow through the system must be reduced to 50 l/s. The head drop through the valve must be 60 m. What size of valve is required?

Using Equ 4.10 with Q = 180 m³/h and Δp = 60 · 1000 · 9.81/ 100000 = 5.89 bar we get a minimum K_V of 74.2. Reviewing manufacturers data reveals the smallest valve suitable to be 80 mm, which would be at 80% open. This is considerably smaller than the pump discharge which is 150 mm. With the valve wide open K_V = 110, trying to pass 100 l/s would result in a pressure drop of 14.3 bar, 146 m. This pressure drop is higher than the closed valve of the pump so the pump would be unable to run at the normal duty point. This is an ideal case for putting the throttle valve in a by-pass around the discharge isolating valve.

5.8 By-pass return

5.8.1 Principle and application

By-pass regulation is sometimes called shunt regulation in text books because of the similarity with shunts in electrotechnology. In practice, it is simply called by-pass. In by-pass regulation, as shown in figure 5.24, liquid is fed back from the delivery side of the pump into the suction side of the pump. In practice, as with relief valves, the by-pass should be piped back to the suction source not to the suction pipe adjacent to the pump.

Operating points are most easily determined by combining the curves for a rotodynamic pump and the by-pass line into a single curve.

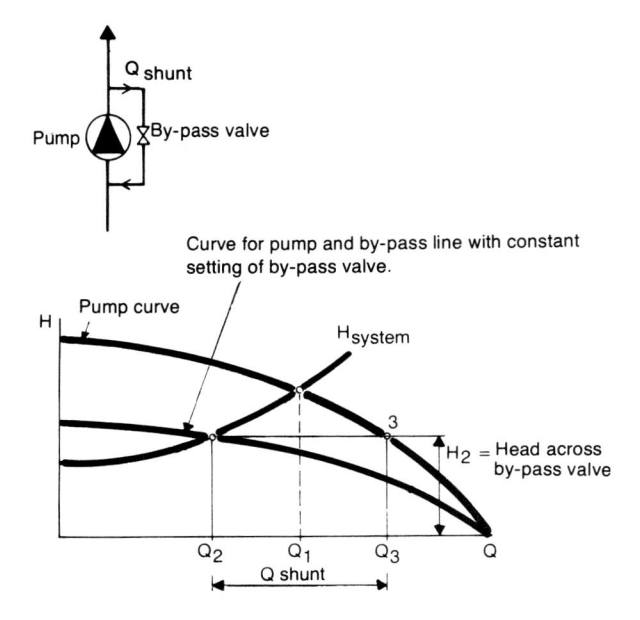

Figure 5.24 By-pass regulation, schematic arrangement and curves for centrifugal pump and by-pass line are reduced to a combined curve

By-pass regulation is used for positive displacement pumps and only rarely for rotodynamic pumps. The energy consumption for these is generally greater with by-pass regulation than it is with throttle regulation, see for example the power curves in figure 3.27.

By-pass control is used with positive displacement pumps in three main modes of operation:

- low pressure starting
- load/unload control
- small flow changes

Some positive displacement pumps operate in systems where the discharge pressure is almost constant, consequently starting takes place at high torque. This type of starting can generally only be accomplished with direct-on-line starting. Not all electric supplies can cope with the current demands of DOL and star-delta is preferred, especially for motors over 15 kW. Star-delta starting does not provide 100% motor torque so that starting the pump on load is not possible. A by-pass, returning to suction source, provides a low pressure, low torque starting option. The pump starts with the by-pass wide open, the discharge pressure generated is only sufficient to overcome friction losses of the by-pass line and valve. Once the pump is running at full speed the by-pass valve is gradually closed, the discharge pressure increases until the non-return valve opens and the pump is online. The by-pass valve is completely closed and incurs no energy losses. The pump may be stopped by gradually opening the by-pass valve to reduce water hammer and pressure surges.

Step capacity control can be achieved on positive displacement pumps by load and unloading the pump rather than starting and stopping by on-off. A by-pass valve is opened and closed and the pump flow varies from 100% to zero. The by-pass valve can be fitted with dampeners to ensure long opening/closing times to eliminate water hammer and pressure surges.

When only small flow reductions are necessary a by-pass valve can be economically attractive compared to other methods. As only a small flow passes through the by-pass the amount of energy wasted compared to the pump power is small and only a small valve is needed.

Positive displacement pumps working with non-Newtonian liquids rarely have sufficient information to accurately predict system curves to allow the control characteristics to be analysed. Large safety margins are advisable. It should be noted that the control valve flow area may not be adequate for liquids containing larger particles. The power requirement is determined by the flow, $Q_{max} + Q_{shunt}$, and the discharge pressure according to the system curve at the actual system flow. Because of this, the regulation efficiency will be relatively high, figure 5.25.

It is generally not necessary to have 20% extra capacity with positive displacement pumps. If the flow rate is critical then

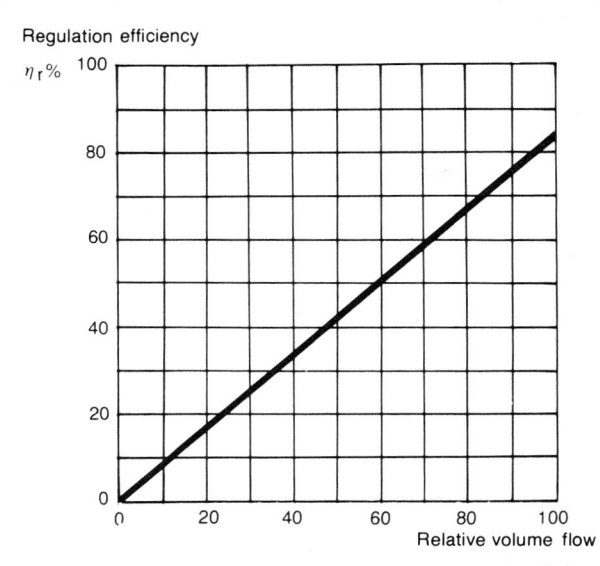

Figure 5.25 Regulation efficiency for by-pass regulated positive displacement pump with 20% by-pass at maximum required flow

an allowance should be made for wear in the pump between overhauls. Depending upon the pump type, and how corrosive/abrasive the liquid is, this should be between 2% and 5%.

5.8.2 Costs

Costs for by-pass valves are similar to throttle valves. A 1" manual valve specifically designed to cope with the throttling and high speed liquid would cost about £800. Valves are available up to 2.5"; a 1" valve has a C_v of 27, a 2.5" is 79.

5.8.3 Sizing the by-pass

The by-pass valve is sized in a similar manner to a throttle valve. Once the flow and pressure drop are known equations 4.10 or 4.11 can be used to evaluate K_V or C_V. A slightly bigger valve is no problem; it just requires closing a little. A slightly small valve will stop the pump from operating down the system where it was intended.

5.8.4 Overflow by-pass return

With overflow control the liquid is pumped into a reservoir at almost constant level. The required system flow is taken from the reservoir. The surplus liquid flows over a weir, which creates the constant head, and goes back to the suction side of the pump or escapes in to a drain pipe, figure 5.26. The pump in such cases operates at constant flow and constant power requirement.

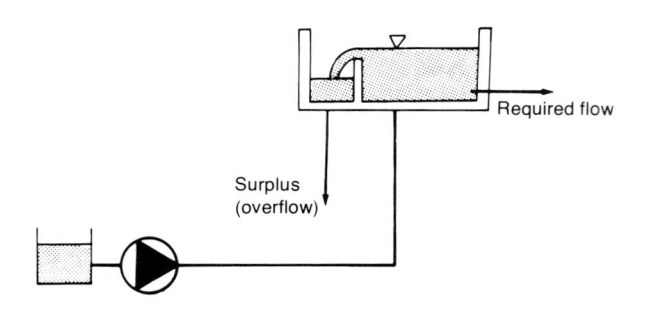

Figure 5.26 Principle of overflow control

Overflow control presents a simple and reliable method but uses a relatively large amount of energy. The regulation efficiency depends upon the proportion of static delivery head, figure 5.27. The figure makes the assumption that there are no unwanted level differences.

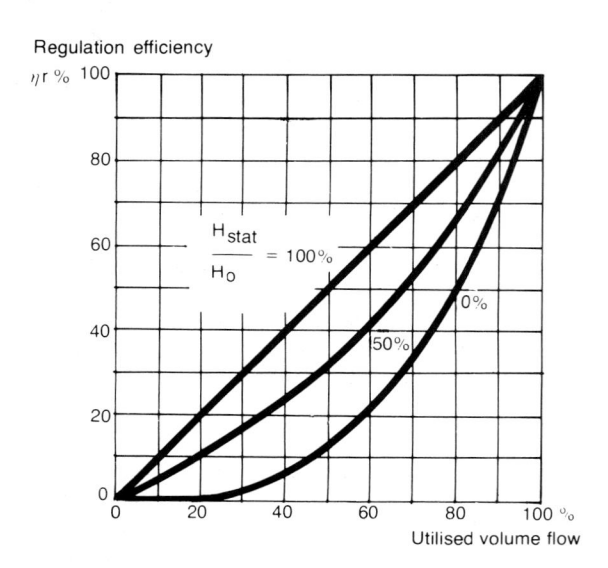

Figure 5.27 Regulation efficiency with overflow control, $H_0 = H_{stat} + h_{f0}$

5.8.5 Power consumption

For rotodynamic pumps, energy consumption is determined as for Section 5.7.4, i.e. operating points on the H-Q curve are established, thereafter the power requirement of the pump is derived from its power curve. The regulation efficiency for rotodynamic pumps is generally lower with by-pass regulation than it is with throttle regulation, figure 5.23. In spite of this, by-pass regulation of a rotodynamic pump may be justified because of the lower loss of head across the control valve with consequent lower noise levels and reduced risk of cavitation in the control valve. And by-pass valves can be closed completely so there is no energy wasted at all. The operating points of the rotodynamic pump should, however, always be checked with regard to cavitation and abnormal vibration levels.

When a by-pass is closed there are no energy losses. When the by-pass valve is open the pump must provide the system flow and the by-pass flow, both at system pressure. The wasted power loss with by-pass regulation is:

$$P_{shunt} = \frac{\rho \cdot g \cdot Q_{shunt} \cdot H_2}{100 \cdot \eta_3} \quad (kW) \qquad \text{Equ 5.19}$$

where η_3 is for flow Q_3.

For positive displacement pumps using by-pass regulation, the power consumption is determined by the back pressure, i.e. by the system curve. In the same way as for throttle regulation, the valve must be given priority in the system control loop. This is achieved by increasing the required flow, figure 5.28.

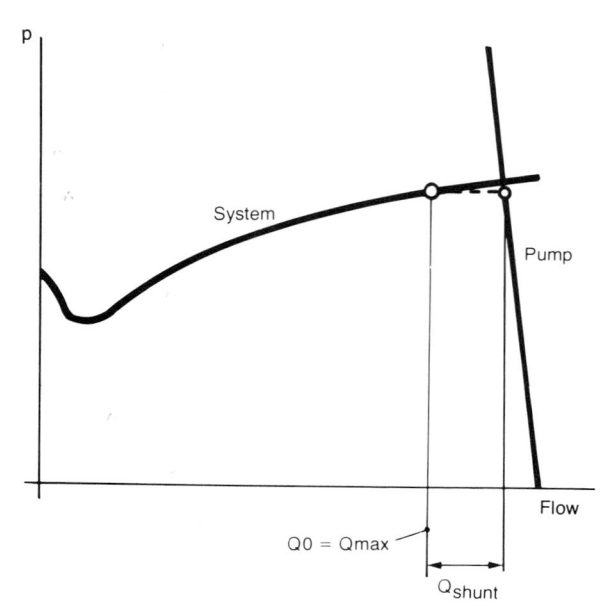

Figure 5.28 By-pass regulation of a positive displacement pump. The system curve shown applies to a non-Newtonian liquid

Example

A fixed speed reciprocating pump operates in a system with a high static pressure, see figure 5.29. The nominal duty point is 190 m³/h at 184 barg, the suction pressure is 1 barg. For starting purposes the system flow must be reduced to 100 m³/h, what size of by-pass valve is required?

Q_{normal} = 190 m³/h, p_{normal} = 184 barg, Q_{start} = 100 m³/h

From the curves, p_{start} = 132 barg. The flow to be by-passed is 90 m³/h and the differential pressure available is 131 barg. From equ. 4.11 the K_V of a suitable valve would be 7.86. Reviewing manufacturers data, a 25 mm valve has the required K_V at approximately 85% opening. The extra capacity of the valve will allow for other losses in the by-pass system.

5.9 Infinitely variable speed

This section deals with infinitely variable speed providing infinitely variable flow. There are some fixed speed pumps which can provide infinitely variable flow. Reciprocating metering and dosing pumps are usually variable stroke and flow can be varied from zero to 100%. For small flows category **285** should be consulted. Metering and dosing pumps can be surprisingly large. Plunger pumps, category **270**, can be built in variable stroke versions. These pumps were first developed as boiler feed pumps and pumps up to 75 kW were available. For hydraulic applications, liquids with very good lubricating qualities, variable displacement vane pumps and variable stroke piston pumps are available. Some variable stroke piston pumps can actually reverse the flow direction, -100% to 0 to +100% flow, while running at constant speed. For hydraulic applications consult categories **245** and **260**. Variable speed pumps will generally not offer zero flow. A minimum pump speed is required to assure proper lubrication of bearings. Variable speed pumps will tend to offer 5% to 100% or 10% to 100% maximum flow ranges.

The need to have infinitely variable speed to provide infinitely variable flow will ultimately depend upon:

- the liquid
- the size of the pump
- the type of pump
- the flow range required

5.9.1 Principle and application

Many pumping applications require the flow to be adjusted to several different values which sometimes cannot be accurately predicted. In these situations it is advisable to make the flow infinitely adjustable over a range. Changes in operating conditions due to fouling or corrosion in pipework, variations in process conditions or wear in the pump can be accommodated without compromising operational requirements.

With rotodynamic pumps, throttling and by-pass regulation could satisfy most of the needs. With positive displacement pumps, by-pass regulation could satisfy all of the needs. However, both throttling and by-passing are inefficient, energy supplied by the pump to the liquid is destroyed in a valve, and may not be economically viable in larger installations or when used continuously.

Pump speed variation allows the pump capacity to be matched to the process demand. We have seen in sections 5.5 and 5.6 that pump speed can be adjusted in discrete steps; pump capacity may or may not match the demand but it will be much closer than it would have been without speed adjustment. The ability to vary the pump speed infinitely allows pump capacity to match process demands exactly.

Pump speed can be varied in many different ways. Two main approaches are adopted; a fixed speed driver with a variable ratio transmission and a variable speed driver. The two approaches can also be combined.

5.9.1.1 Variable ratio transmissions

Some pumps, typically larger pumps, require some form of power transmission because the pump speed does not coincide with the driver speed. In these instances variable ratio can sometimes be incorporated in the speed reduction. The popular methods for varying speed ratios are:

- mechanical variators
- hydrostatic couplings
- fluid couplings
- torque converters
- eddy current couplings

Mechanical variators

Mechanical variators are a sub-group consisting:

- variable ratio belts
- variable ratio chains
- variable ratio friction drives
- variable ratio gearboxes

Belt and chain drives usually have fixed diameter pulleys and sprockets. By adjusting the diameter of the pulleys/sprockets the speed ratio between the two shafts can be varied. Variable speed belt drives use special cone pulleys to adjust the diameters. Adjustment can be manual or automatic. Power transmitted is up to 25 kW and speed ranges of 9:1 are possible. Variable speed belt drives are mounted like normal belt drives, inside a guard provided by the purchaser.

Variable ratio chain drives are constructed like gearboxes, housed within casing which acts as the oil reservoir. Fixed gear ratios can be incorporated to adjust the output speed. Some variable ratio chain drives are designed to be constant power transmissions rather than constant torque devices; this feature can be useful in some applications and must be considered in the cost evaluation. Small units are capable of 6:1 speed ranges, the larger units, 600 kW with a service factor of 1, only 3:1.

Variable ratio friction drives use discs, rollers or ball bearings clamped between two flat or specially shaped plates to transmit the power. One plate is mounted on the input shaft, the other on the output shaft. The variable ratio is created by adjusting the contact circles on the plates. Units up to 15 kW are popular; output speed ranges are about 6:1.

The variable ratio gearboxes described here should not be confused with multi-speed gearboxes in section 5.6. Variable ratio gearboxes are epicyclic gearboxes with small control drives attached to the outer gear annulus. The basic epicyclic gear is designed for the required full speed ratio. Then, by driving the outer gear annulus with a low power source, the gear ratio can be made infinitely variable. Usually built specially for order so any configuration is possible.

Hydrostatic couplings

Hydrostatic couplings consist of a variable displacement pump and a fixed or variable displacement hydraulic motor mounted in one housing which acts as the oil reservoir. The fixed speed driver provides the power to the variable displacement pump; variable stroke piston pumps are most usual. The high pressure oil drives the output hydraulic motor. Hydrostatic drives are constant torque devices, full load torque is available at all speeds. Up to 200% instantaneous torque can be used at start-up. Speeds down to 0 r/min are not usual, most units have a minimum continuous speed limitation dependant upon the input speed to the unit. With 1450 r/min motors, the minimum speeds would be about 50 r/min. Hydrostatic couplings are very stable and have a linear characteristic at constant torque. Over-running protection is inherent in the design as is controlled deceleration. Output speeds up to the input speed are standard, 1900 r/min maximum, speed range 27:1 with powers up to 22 kW. Speed control can be adapted to suit the process requirements. Cooling is by natural convection around the housing, high ambient temperatures would restrict coupling capacity.

Hydrostatic couplings can be built as separate units. A power pack, with reservoir, motor, pump and cooler, can supply high pressure oil to a remote hydraulic motor. This format can be useful in hazardous areas. All the electrical equipment can be in a safe area with only the hydraulic motor in the hazardous area. Standard packages of this style are available for powers up to 150 kW. Specials can be built to order to cover virtually any speed and power requirements.

Fluid couplings

Fluid couplings are one of the two types of coupling which use oil to transfer power from the drive to the pump. Because these are "torque" transmitters they are always fitted in the drive train at the highest speed. Fluid couplings would be fitted in the input side of reduction drives and the output side of speed increasing drives. The torque which is transmitted, and consequently the power, is dependant upon how much oil is circulating. A scoop in the oil reservoir adjusts the proportions of oil in the coupling to oil in the reservoir. When the coupling is full of oil 100% torque is transmitted. When used with centrifugal pumps on a pure friction system curve with no static head the fluid coupling is capable of speed ranges of about 8:1. For constant torque loads, due to oil cooling limitations, the speed range is only about 4:1. At speeds less than 100%, the excess power is given up to the circulating oil and must be extracted by suitable cooling. The fluid coupling must always operate with some slip, consequently the maximum output speed is about 98.5% of the input speed. Typical input speeds are up to 1800 r/min. Couplings over 650 kW are available as standard. Larger units, up to 1.5 MW, use a pump rather than a scoop to control the oil flow. Some of the larger units have built-in gearboxes. Controls can be supplied to suit any system.

Torque convertors

Torque convertors are the rotodynamic equivalent of hydrostatic couplings. A centrifugal pump on the inlet shaft transfers energy to the oil. Inlet guide vanes direct the oil through an inward flow radial turbine on the outlet shaft. Torque convertors transmit torque and should be in the highest speed shaft. Torque convertors are almost constant power devices, rather than constant torque. At low output speeds output torque is increased. The output speed can also be increased above the input speed. One problem with torque convertors on manual control is that they can run away it the output load is suddenly reduced. Oil cooling is by external heat exchanger. Couplings have been supplied for over 10000 r/min and powers over 10000 kW are no problem. Some units have integral gearboxes. Controls can be fitted to suit any system.

A special hybrid consisting of a fluid coupling and a torque convertor is available in certain sizes. The internal controls select the appropriate drive mechanism to optimise efficiency.

Liquid couplings have an important attribute which is not related to their variable speed capabilities. All liquid couplings effectively isolate the two sections of the drive train. If a torsional analysis is necessary, the total drive train can be treated as two separate systems divided at the coupling. A liquid coupling will not transmit torsional vibrations and it will act as a dampener for cyclic torque and speed variations.

Eddy current couplings

Eddy current couplings are the electrical version of fluid couplings but eddy current couplings can transmit more than 100% motor torque for starting. A small control current, typically less than 0.5 kW, adjusts the speed and torque relationship between the driven and the driving halves. Couplings are manufactured in standard sizes; the torque and speed range being a function of the driver/driven power to the coupling rating. Couplings are available up to 5 MW.

5.9.1.2 Variable speed drivers

Variable speed drivers are a very diverse group with very different characteristics. All of the following are currently used:

- petrol engines
- diesel engines
- gas engines
- crude oil burning engines
- air motors
- steam turbines
- gas turbines
- turbo-expanders
- hydraulic turbines
- AC squirrel cage induction motors
- AC slip ring motors
- DC motors

The choice of driver is dependant partially on the size required, the speed range and the energy source available.

Petrol engines

Petrol engines are used mainly for portable applications, both small light packages which can be man-handled and trailer mounted units. Speed adjustment is by throttle control of the engine. The useful range of speed available is dependant upon the torque characteristic of the pump. Rotodynamic pump torque, for pumps operating on purely friction systems, varies approximately as the square of the pump speed, so allowance must be made for changing efficiencies. As pump speed is reduced from the rated condition, torque reduces quite rapidly. In these situations it may be possible to achieve a 4:1 speed range. For rotodynamic pumps operating with high static heads compared to friction heads the engine may struggle to cover a 2:1 speed range depending upon the pump characteristic. Positive displacement pumps with frictional system losses may approach a 4:1 speed range; with constant discharge pressure systems the engine may have to be oversized to achieve any speed range if stalling is to be avoided. Small engines can operate up to 5000/6000 r/min, larger engines to 3000 r/min. A wide variety of sizes are available, up to 150 kW being the most popular.

Diesel engines

Diesel engines are used for fixed installations as well as portable machines. Diesel engines are used on sites where electricity is not available or is not capable of supporting the pump load. All the comments on petrol engines regarding speed ranges apply to diesel engines. The speed available from diesel engines is lower; small engines about 3000 r/min, large engines down to 600 r/min. Small diesel engines of 5 kW are available, large fixed installations use engines up to 3 MW.

Gas engines

Gas engines are popular in fixed installations where natural gas is readily available. The natural gas can be used locally, usefully, rather than incurring additional cost by transportation or wasting by flaring-off. Full details of the gas specification must be available to the engine manufacturer to allow engine outputs to be predicted. Gas engines are more similar to diesel engines than petrol engines. All the comments on diesel engines apply to gas engines.

Crude oil burning engines

Like natural gas engines, crude oil burning engines are popular in fixed installations where there is a good source of supply and other energy supplies are limited or non-existent. Crude oil burning engines are similar to diesel engines. Some types of crude oil may require considerable pre-treatment before being suitable for use in the engine. Some engines may require starting and warming-up on diesel or gas before the crude oil can be used. Decisions and costing for crude oil engines can only be made when all the details of the actual crude oil are available. Extra costs incurred for pre-treatment or starting on other fuels may indicate other energy sources to be more cost effective. The comments for speed and speed ranges given for diesel engines apply to crude oil engines except that small engines are not available.

Dual fuel engines

As indicated in crude oil burning engines it is possible to run engines on more than one fuel, not generally at the same time. The most popular combinations are diesel and natural gas, or crude oil and natural gas. Full details of both fuels must be given to the engine manufacturer.

Turbo-charged engines

The details given in previous engine categories apply to naturally aspirated engines; that is engines which draw the air supply through an air filter from the atmosphere. Turbo-charged engines are becoming very popular. Turbo-charging significantly improves the engine efficiency and increases the power output without much change in the space required. In some cases turbo-charging seriously modifies the shape of the engine torque curve so that torque reduces rapidly either side of the optimum operating point. This change in engine characteristic can considerable reduce the operational speed range of any pump with any system. This feature must be considered, with all the other points, when deciding upon the type and size of an engine.

NOTE: The use of the waste heat, from the engine exhaust, can significantly improve the overall operating efficiency of the pump installation. The heat can be utilised for water or process heating or steam raising. In general, the extra cost of heat exchangers will be quickly recovered due to the large amount of heat available.

Engines for hazardous areas

Internal combustion engines can be **modified** to be suitable for operation in some hazardous areas. This requirement must be specified at the initial inquiry stage. Not all engines can be modified; not all manufacturers are interested in modifications. The surface temperature requirements, together with any relevant approvals, must be specified. Hot areas of the engine are cooled or insulated, electrical auxiliary equipment is uprated. Special attention is paid to air filters and silencers for arresting sparks.

Problems with engines in hazardous areas can be alleviated, in some cases, by installing the engine in a safe area separated from the hazardous area by a fire wall. Power transmission is via a spacer coupling which has a special mechanical seal, set in the fire wall, to prevent direct leakage of the hazardous substances.

NOTE: All engines can be used with multi-speed gearboxes to increase the overall useful speed range. The engine provides the infinitely variable speed capability and the gearbox can be used to provide several operational speed ranges. This approach should be considered when wide speed ranges are required.

Air motors

Pumps can be driven by compressed air. Air motors are available as standard assemblies. Torque characteristics should be almost constant so wide speed ranges are possible with all pumps. Sizes are limited to about 5 kW, speeds up to 9600 r/min are possible.

Steam turbines

Steam turbines are popular for driving larger pumps, but also small pumps in situations, such as refining and chemical processing, where steam is plentiful. Continuous processes are best so that time waiting for warm-up is periodic rather than regular. Steam turbines can be condensing or pass-out, that is to say the turbine expands the steam to a lower pressure which is then used in another part of the process. The torque characteristic is fairly constant, wide speed ranges are possible. Steam turbines are built from about 25 kW; there is no practical limitation on the largest size. Small turbines tend to run quite fast, 6000 r/min and faster, but there are no fixed speeds, the governor can be adjusted for a range of speeds.

Gas turbines

Gas turbines are useful for producing lots of power from a small space; they also tend to produce a lot of noise. The turbine itself can be small, but has a tendency to be overshadowed by the air filters and exhaust ducts. Gas turbines are ideal for driving fast pumps, speeds of 10000 to 15000 r/min are common. Slower pumps will require reduction gearboxes. Warming up is generally not a problem; full load can be available in 1 or 2 minutes. Gas turbines can utilise a variety of liquid and gaseous fuels. Crude oil burning is an option from two manufacturers. The usefulness as a variable speed driver is dependant upon the torque characteristic which in turn is dependant upon the gas turbine design. A single shaft machine, one in which the compressor and the turbine are mounted on one shaft, or two separate shafts but geared together at a fixed ratio, does not have an ideal torque curve. Below the design speed, torque reduces at a power slightly higher than a square law. The point of zero drive torque occurs at about 20% speed. This type of gas turbine will be suitable for some rotodynamic pumps whose system characteristics have no static head. The other type of gas turbine is the free power turbine type; the compressor and its turbine are on one shaft, the turbine for external drives is mounted on a separate shaft and can rotate at virtually any speed required. The torque curve of this gas turbine is ideal. The compressor can be allowed to run up to full speed, if necessary, before the pump is started. Starting torque can be over double the normal running torque, reducing almost linearly to full load torque at full speed. Any type of pump or system can be operated with a free power turbine gas turbine. Speed ranges over 5:1 are possible. Turbine cooling may be a problem for low speed continuous operation with constant torque pumps. Gas turbines start at about 25 kW. Overall efficiency can be improved by utilising the waste heat in the exhaust gases. It is more cost effective to design the pump/system around a standard gas turbine than to ask the turbine manufacturer to modify the turbine to suit the pump/system. The world supply of gas turbines, in some sizes, is extremely limited. When discussing operational requirements for gas turbine drivers also ensure that machines will be available for delivery to suit your project timescale. Some projects have been postponed for two years waiting for turbine availability. Several manufacturers operate allocation systems and it is essential to reserve your place in the queue.

Turbo-expanders

Turbo-expanders are effectively the power turbines of gas turbines without the gas generator. Turbo-expanders can be utilised where a supply of gas is available which can be expanded. The gas utilised may be process gas which is expanded between process stages or waste gas which is expanded to atmospheric conditions. Speed and torque curves can be adjusted by control and by-pass valves; most pump requirements can be accommodated. Turbo-expanders are built in standard sizes which can be adapted to suit individual conditions. Turbo-expanders are also built especially for contract when enough machines are required.

Hydraulic turbines

Pumps can be driven by hydraulic turbines when a source of high pressure liquid is accessible. The turbine can be a Pelton Wheel, a Francis turbine or a pump running in reverse. The pump can be axial flow or radial depending upon the head and flow available. All hydraulic turbines can be controlled to produce variable speed outputs. The choice of turbine type will depend upon the characteristics of the driven pump system. Pumps running in reverse have become popular over the last 5 to 10 years. Many parts can be made common to the pump and only one manufacturer is involved. Size and speed is virtually unlimited.

AC squirrel cage induction motors

A standard, off-the-shelf motor, driven by a variable frequency invertor can produce speed ranges of 10:1, and over, without problems. This method of speed control is the most popular of all methods for motors up 125 kW, perhaps even 250 kW. Full load torque, plus a margin, is available for starting without transmitting large current demands into the supply system. Full load torque is available over the whole speed range.

Some motors may need derating by up to 10% to work with some types of invertors. There is a need to match the frequency invertor to the motor in the larger sizes. Care should be taken with continuous operation at low speeds. Some motors may need a separate motor driven cooling fan. Discussions should be held with the motor and invertor manufacturers regarding operating conditions. Special attention may be needed with starting torque or overload capabilities for operating at relief valve settings. If accurate speed control is required, for metering pumps for example, some feedback from the motor to the invertor may be necessary. Some small invertors can accept single phase power at 240 V and drive three phase motors at 380 V. Some invertors can produce output frequencies up to 120 Hz. The motor manufacturer must be consulted if motors are to be overspeeded on a regular basis. Virtually any control system can be interfaced to an invertor. Large invertors can transmit significant harmonic signals into the mains supply. Generating authorities should be consulted when contemplating drives over 500 kW.

AC slip ring motors

Slip ring motors are also called AC wound rotor motors. Slip ring motors were the standard before squirrel cage motors were invented. The lack of slip rings, plus simpler starting, made squirrel cage motors much more attractive for most fixed speed applications. However, slip ring motors continued to be used for limited variable speed applications and those applications which involved large inertias requiring long run up times. The external resistance, connected in series with the rotor windings, allows the speed/torque characteristic of the motor to be modified. At low resistance, when the slip rings are shorted, the torque characteristic is very steep giving a wide range of torque values for small variations in speed. As the rotor resistance is increased the speed/torque characteristic becomes shallower giving corresponding torque values at lower speeds. Full load torque can be made available at any speed. This method of speed variation can be applied to most practical pump installations. However, it is not efficient. At reduced speeds, energy is dissipated in the rotor resistances and external cooling, motor driven fans, may be necessary. The motor efficiency reduces approximately in proportion to the motor speed. The sliprings and brushes of wound rotor motors require maintenance and this is a disadvantage compared to squirrel cage motors. Problems can arise with hazardous area requirements. AC slip ring motors can be useful for large installations, with heavy inertias creating starting problems, when speed variation is only required periodically. If a pump was also required to run as a turbine, then the slip ring motor could become an alternator and produce power.

DC motors

DC motors were the principal electrical variable speed drivers before variable frequency invertors became economically viable. Most applications which used to have DC motors can now be driven by variable frequency AC motors. DC motors do have two advantages over AC motors; DC motors can be designed to run at any particular speed and DC motors have a faster response to transient conditions. DC motors do not have synchronous speeds like AC motors. A DC motor could be designed to run at 4500 r/min, for example, to eliminate the need for a gearbox. If rapid speed changes are required as part of the normal operating conditions then DC motors may be better than AC. However, consideration should be given to the flow variations and pressure pulsations caused by any rapid speed changes. The speed of a DC motor, up to its base speed, is controlled by the armature voltage. When the armature current equals the rated current, full load torque is available. DC motors can provide full load torque from zero speed up to base speed. Speeds above base speed, produced by field weakening, are at reduced torque. Above the base speed, DC motors are constant power, not constant torque. Motor efficiency is high over the whole speed range. Modern motors are powered via electronic controllers in a similar manner to variable frequency AC motors. DC motors have commutators and brushes which require maintenance attention similar to sliprings. Hazardous area requirements can cause problems. DC motors are not as popular as AC motors and are not mass produced in the same quantities.

5.9.2 Costs

Procurement costs for various speed regulation equipment depends upon the shaft power, maximum speed and the environment in which the drive is going to be located. To some degree, extra allowances must be made for ancillary cooling equipment and in the case of electrical methods, of the effects of feedback in the mains from thyristors, and protection against incoming mains transients. The method of speed regulation selected may affect the procurement costs by a factor of 5. Figure 5.30 indicates the cost of standard three phase squirrel cage AC motors. Figure 5.31 indicates values for additional cost for speed regulation in relation to the costs for standard constant speed squirrel cage induction motors. For choice of equipment see Section 5.10.

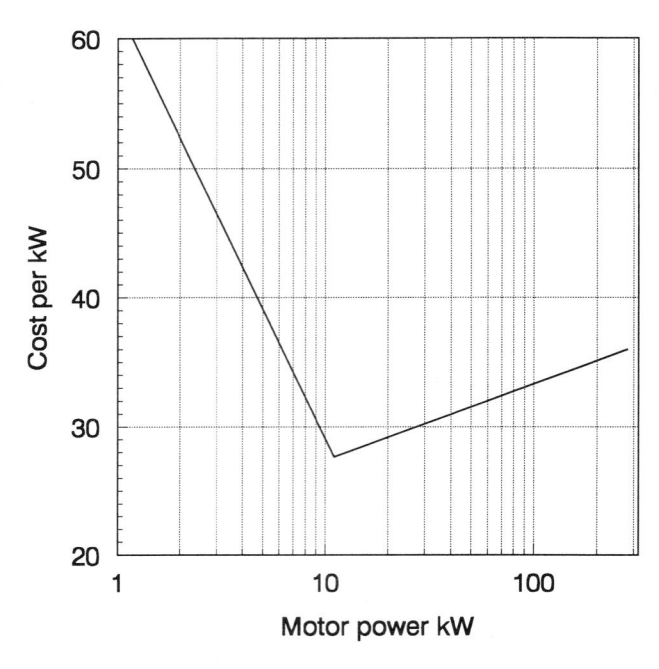

Figure 5.30 Costs of 2 and 4 pole standard squirrel cage AC motors

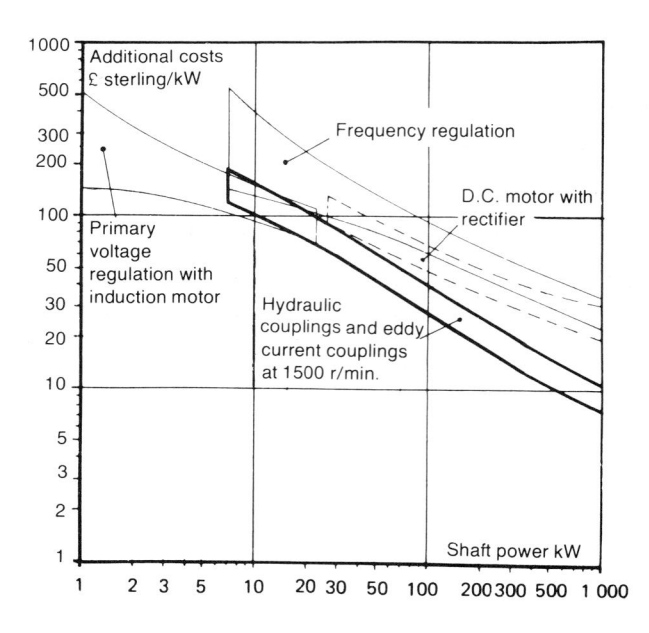

Figure 5.31 Additional costs for speed regulation over standard squirrel cage AC motors

5.9.3 Conversion of pump curves to various speeds

For rotodynamic pumps the affinity laws govern performance at speeds n_1 and n_2. If equation 3.16 is rewritten for constant diameter impellers then:

$$\frac{Q_2}{Q_1} = \frac{n_2}{n_1}$$

$$\frac{H_2}{H_1} = \left(\frac{n_2}{n_1}\right)^2$$

$$\frac{P_{p2}}{P_{p1}} = \left(\frac{n_2}{n_1}\right)^3 \qquad \text{Equ 5.20}$$

The third part of the equation for power applies only to the absorbed power of the pump P_p.

For any given pump in any given system, in accordance with figure 5.32, the minimum required speed, n_{min}, can be calculated from the following equation:

$$n_{min} = n_{max} \cdot \sqrt{\frac{H_{stat}}{H_{max}}} \qquad \text{Equ 5.21}$$

The lowest required speed has to be known when selecting speed regulation equipment.

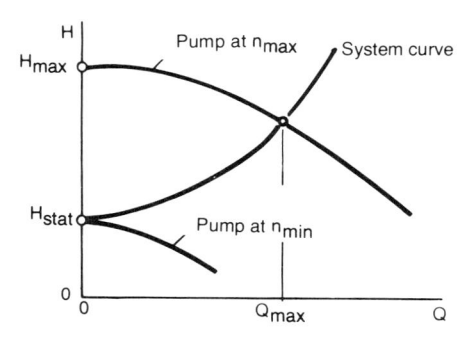

Figure 5.32 Relationship for lowest required speed, n_{min}

The performance of speed regulated pumps is usually described as a family of curves for H-Q and Q-P as shown in Chapter 3, figure 3.23. The curves there were determined by means of the affinity law, equation 5.20. The power requirement in accordance with equation 5.20 refers to the power requirement at the pump shaft. The power requirement for the electric motor is obtained by dividing by the efficiencies of both the motor and the transmission, see also figure 5.33. Frequency invertors and thyristor controllers for DC motors must be included. The overall efficiency is also called the "wire to water" efficiency.

For positive displacement pumps the flow is approximately proportional to speed, $Q \propto n$. The first part of equation 5.20 can be used but some allowances may be necessary for variations in slip. Slip will vary with speed and differential pressure. The differential pressure must be evaluated from the system characteristics, see figure 5.18. Pump power is the product of flow and differential pressure divided by the efficiency. Just as with rotodynamic pumps, pump efficiency varies with operating conditions.

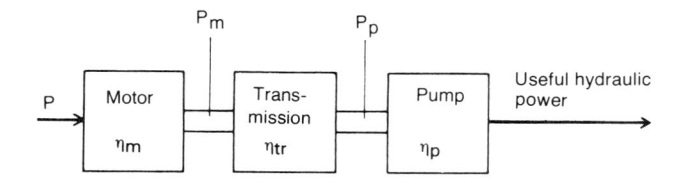

Figure 5.33 Pump unit with power transmission

Thus for rotodynamic pumps, drive power P:

$$P = \frac{P_m}{\eta_m} = \frac{P_p}{\eta_m \cdot \eta_{tr}} = \frac{\rho \cdot Q \cdot \Delta H \cdot g}{\eta_m \cdot \eta_{tr} \cdot \eta_p \cdot 1000} \quad \text{(kW)}$$

Equ 5.22

For positive displacement pumps with flow in m³/h and differential pressure in bar, drive power P:

$$P = \frac{P_m}{\eta_m} = \frac{P_p}{\eta_m \cdot \eta_{tr}} = \frac{Q \cdot \Delta P}{\eta_m \cdot \eta_{tr} \cdot \eta_p \cdot 36} \quad \text{(kW)}$$

Equ 5.23

5.9.4 Efficiencies of various methods

If the affinity laws are applied to a centrifugal pump then, when the pump speed is regulated down to half speed, the volume flow also reduces by a half, the differential head to a quarter and the power required at the pump shaft to an eighth based on a system without static head. Similarly, when regulating down to 20% of maximum speed the pump shaft power goes down to 0.8% of the power at maximum speed based on zero static head. The variable speed pump characteristics must be plotted against the system curve to evaluate the speed, torque, efficiency and power for each duty point. At low speed, losses within the pump, such as bearing friction and seal friction, may distort the power curve so that the affinity laws under estimate the power required.

For positive displacement pumps the procedure is exactly the same. The pump characteristic shapes will be different as will be the variation in pump efficiency. The pump torque and power for any duty point will be dependant upon the intersection with the system curve.

The characteristics of variable speed devices differ considerably. Not only is it essential to know the speed, torque and power of each duty point, but it is essential to know the timescale of operation at each duty point. It does not make sense to purchase a costly system which provides high efficiency at 10% flow if the only time low flow is necessary is during start-up and shut-down which occurs twice a year for a couple of hours.

Figures 5.34 to 5.38 indicate the general characteristics of the popular methods used for speed variation. The actual speed range possible is dependant upon the actual torque required at any speed. Efficiencies for drivers do not include the efficiency of any speed reduction necessary. Gearboxes should have transmission efficiencies of 95 to 98%. Efficiencies for variable ratio transmissions do not include the driver efficiency. Electric motors, the most popular driver, will have efficiencies between 80 and 96% depending upon the size.

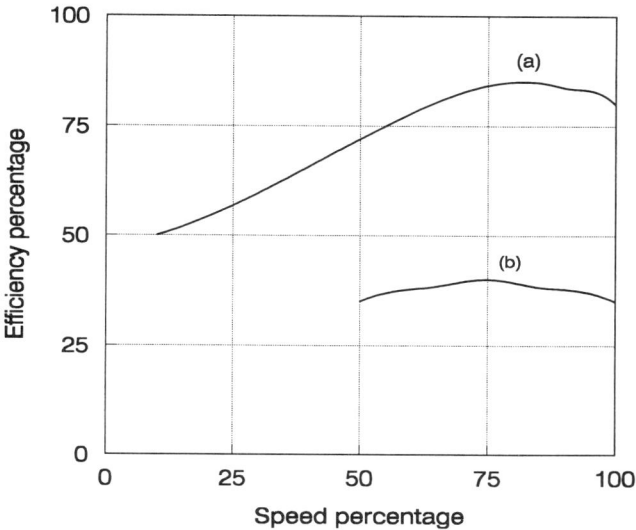

Figure 5.34 Speed range and efficiency —
(a) torque convertor (b) engine

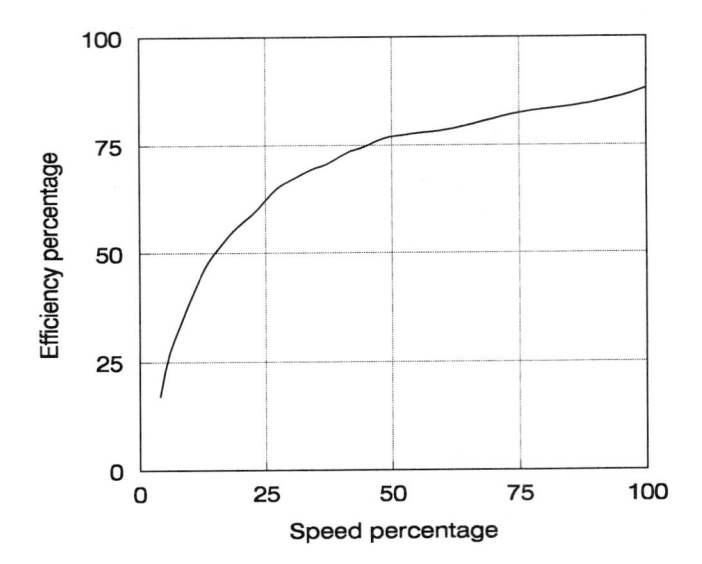

Figure 5.35 Speed range and efficiency of hydrostatic transmission

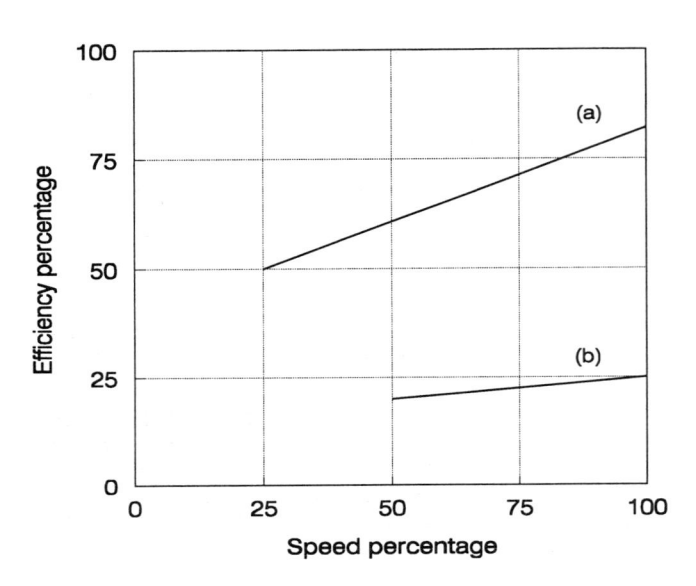

Figure 5.36 Speed range and efficiency —
(a) steam turbine (b) gas turbine with free power turbine

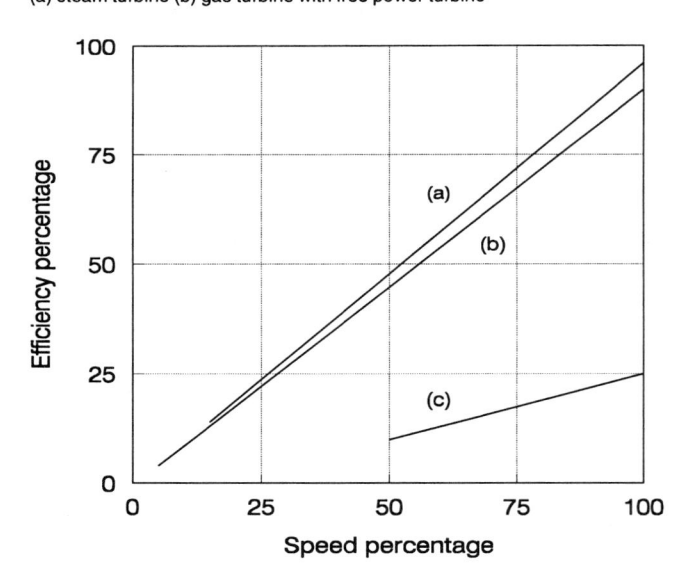

Figure 5.37 Speed range and efficiency —
(a) fluid coupling (b) eddy current coupling (c) single shaft gas turbine

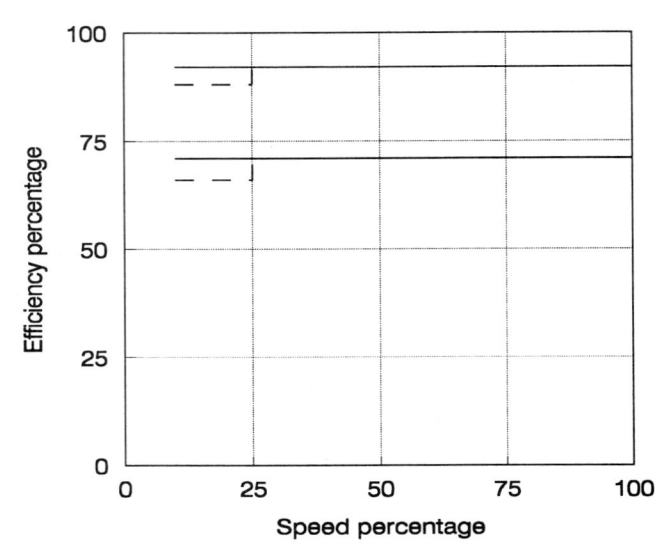

Figure 5.38 Speed range and efficiency of electric motors

The two curves in figure 5.38 represent the variation in efficiency due to size. The dotted lines indicate a possible reduction in efficiency at low speed due to the requirement to start a separate motor driven fan to cool the main motor. Any power required for cooling fans fitted to electronic control equipment must be considered in the overall efficiency assessment.

5.9.5 Power consumption

In recent years, users whose operating costs are dominated by pump energy consumption have pioneered the concept of 'wire to water' efficiency. The pump and drive train efficiencies are not considered in isolation but are combined to produce an overall efficiency. This process includes variable frequency invertors for AC motors and thyristor controllers for DC motors. Some users have expressed concern over the accuracy of power measurements, when using the pump manufacturer's

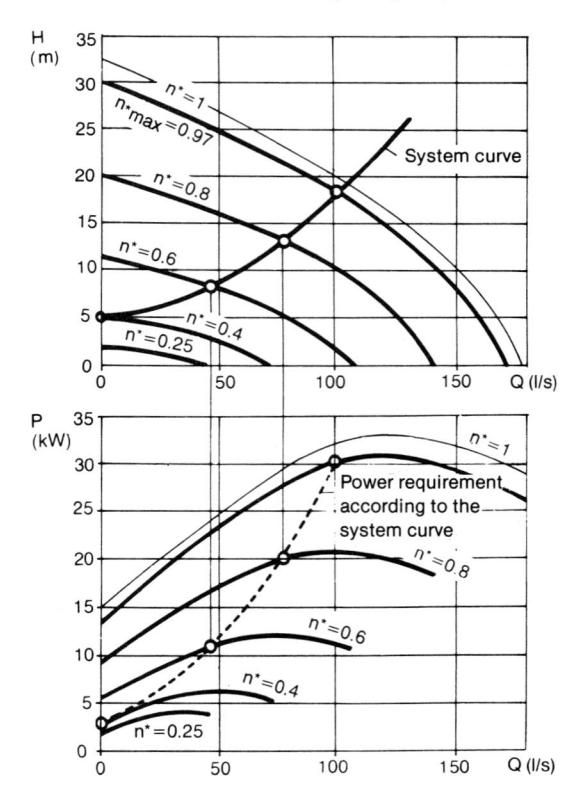

Figure 5.39 Example of performance curves for a speed regulated pump. The power requirement is for a pump with fluid coupling. The speed is expressed relative to maximum speed for a direct drive pump no = 1. Because of slip in the fluid coupling a slightly lower maximum speed pump curve is produced compared to an equivalent direct drive pump

standard instrumentation, for electronic motor speed controllers. The controllers distort the power supply by injecting harmonics. Some users have insisted that the pump manufacturer needs modern, costly, analysis equipment to accurately measure the power consumption. This may in fact be strictly correct and theoretically possible. However, the pump user will probably be charged for electricity based on measurements taken by an old fashioned kWh meter. This then, is the way to measure efficiency when the pump is being tested.

In operation, of course, the performance of a speed-regulated pump follows the requirement of the system curve in the H-Q diagram. By transferring the intersection points of the system curve with the various pump curves on the H-Q diagram to the power diagram the dotted line curve in figure 5.39 is obtained. This curve indicates the power requirement when operating along the system curve.

The energy consumption E in kWh/year will be the power multiplied by time, or

$$E = P_{mean} \cdot t_0 \qquad \text{Equ 5.24}$$

where t_0 = operational time per year (hrs) and where the mean power is determined by using Simpson's Rule:

$$P_{mean} = \frac{1}{6} \cdot P_{Q100} + \frac{2}{3} \cdot P_{Q50} + \frac{1}{6} \cdot P_{Q0} \quad (kW)$$

$$\text{Equ 5.25}$$

The values P_{Q100}, P_{Q50} and P_{Q0} are the values for power requirement taken from the system curve at volume flow Q_{100}, Q_{50} and Q_0 respectively, i.e. the flow at 100%, 50% and 0% constancy, see also figure 5.40.

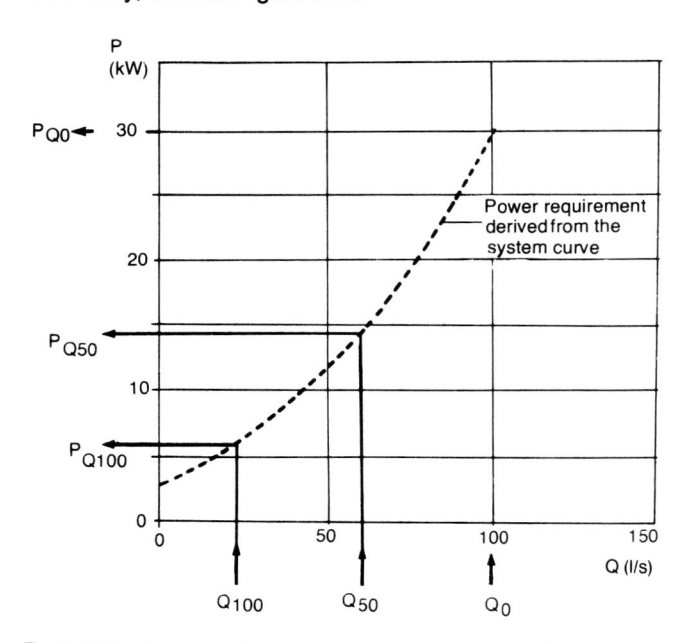

Figure 5.40 Power requirement for speed regulated centrifugal pump in accordance with figure 5.39. Taken at various values of Q provides the basis for determination of mean power in accordance with equation 5.25

In speed regulation, the regulation efficiency is always equal to one because supply always equals demand. The losses in speed regulation are instead described by a transmission efficiency η_{tr}, see also figures 5.33 to 5.38. The variations in pump speed and also transmission efficiency will depend upon the shape of the system curve. In some drive arrangements, AC motors with low speed pumps for example, transmission efficiency must include everything but the pump.

$$\eta_{tr} = \eta_i \cdot \eta_m \cdot \eta_g \qquad \text{Equ 5.26}$$

where
η_i = invertor efficiency (decimal)
η_m = motor efficiency (decimal)
η_g = gearbox efficiency (decimal)

The transmission efficiency for a large drive train would look like

$$\eta_{tr} = 0.97 \cdot 0.96 \cdot 0.97 = 0.903$$

Rotodynamic pump efficiency varies more favourably when speed regulated than with other methods of regulation, figure 5.41. With frictional losses only, ($H_{stat} = 0$), pump efficiency remains constant over the larger part of the flow regulating range. It should be noted, that if H_{stat} is not equal to zero, then better part-load efficiencies are obtained if the operational point of the pump at full load lies to the right of best efficiency point. Some process users will not relish operating passed BEP and detailed calculations will be required to confirm the financial benefits.

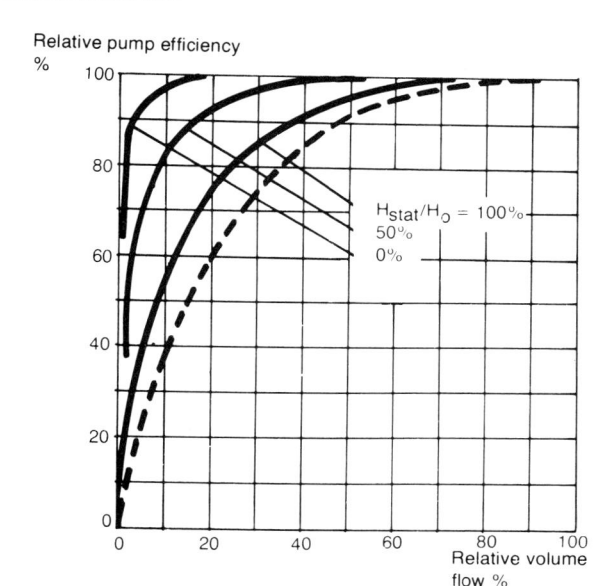

Figure 5.41 Examples of pump efficiency variation for a speed regulated centrifugal pump. The dotted curve represents the normal efficiency change for a constant speed pump when 100% flow is at BEP

Reciprocating pumps, not surprisingly, react differently to system variations. As static head increases in a friction system a reciprocating pump is able to operate closer to it's optimum rod load even as flow is reduced. The best system is one where the discharge pressure is constant irrespective of flow rate, see figure 5.42.

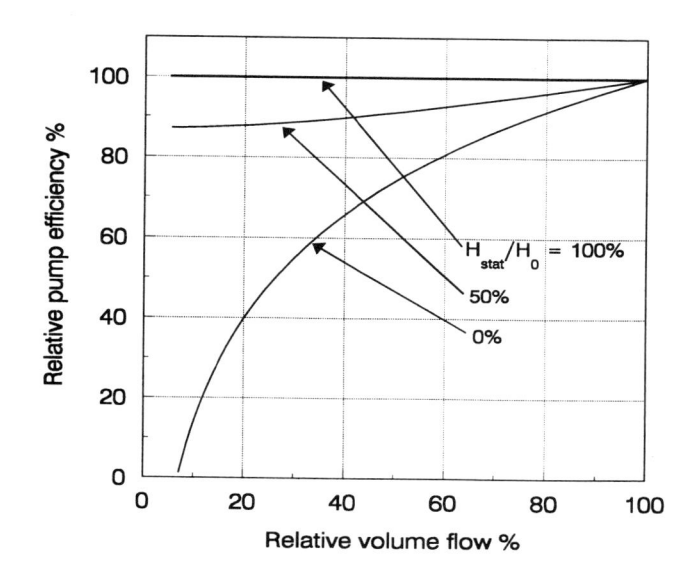

Figure 5.42 Examples of pump efficiency variation for a speed regulated reciprocating pump

5.9.6 Regulation system : schematics

A schematic showing the principle of operation for the complete regulation process of a speed regulated pump, by variable ratio transmission, is shown in figure 5.43. The control system can be local or more usually a sub-system in the plant control system. Linearity between signals and required pump function is of the utmost importance for functional stable regulation. This condition is emphasised particularly in public water supplies where very great variations of flow are common. In this application, unfortunately, a number of installations using speed regulation are unusable for the simple reason that the importance of the regulation function was not fully appreciated. In many cases it is necessary to measure additional parameters, e.g. pump speed, n, or flow, Q, and to use these as feedback into the regulation system, dotted signals in figure 5.43, if stability is to be achieved.

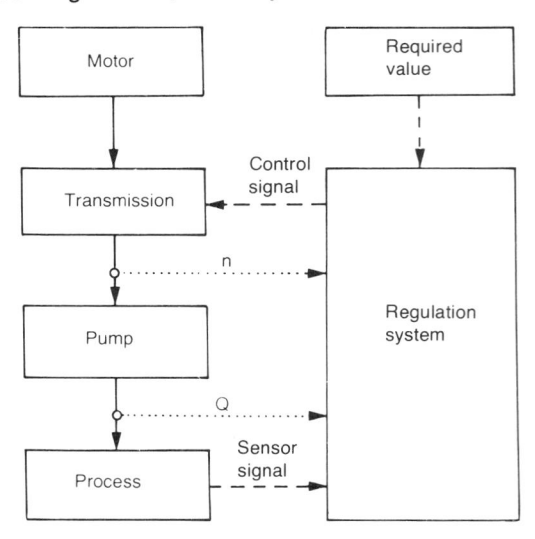

Figure 5.43 Block diagram for speed regulated pump

Linearity is exploited through comparison between the required control parameter, generally flow, Q, or speed, n. If both are measured, the control system can compensate for wear and report on pump condition. When H_{stat} is not equal to zero, Q and n are no longer linear, see figure 5.44. For rotodynamic pumps operating at high differential heads Q and n will never be linear. As can be seen from the table in the figure, it must be possible to regulate the speed at small values of flow with very great precision. Alternative methods are to linearise the speed by means of so-called indirect flow measurement or to eliminate speed completely from the control loop and to replace it by flow, for example. When precise flow control is required, the method adopted for flow measurement must be selected with great care. Rotodynamic pumps do produce some cyclic flow variations and pressure pulsations; centrifugal pumps at high speed can be quite objectionable. Flow measurement by orifice plates, for example, can be impaired by noise in the signal and may require averaging or smoothing.

Positive displacement pumps can suffer from complex relationships for speed and flow. Slip tends to be a variable rather than a constant. Changes in speed, differential pressure, and in some designs temperature, make it advisable to use flow measurement as the primary control signal. Metering pumps usually run slow enough for slip to be constant; direct speed control will produce proportional flow control.

The dynamic interaction between a speed regulated pump, reservoir or pump sump and pipeline can also be a cause of instability. In such case the problem has to be studied with regard to the response time of the speed regulation, time constant, delay time, filling time, of the reservoirs and transient changes in the flow requirement due to the opening or closing of valves, among other things. Relatively slow changes in pump speed are always preferred.

To build up an installation comprising of the various component parts, black boxes, etc., without carrying out a stability analysis from the point of view of regulation can involve considerable risks. When buying packaged solutions one

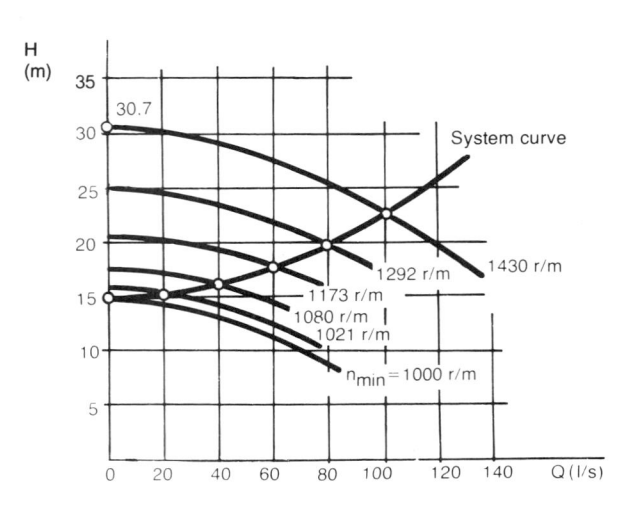

Figure 5.44 Examples of pump curves for determining the relationship between flow Q and speed n. With the help of additional pump curves and by interpolation, the following table can be compiled:

Q l/s:	0	5	10	15	20	40	100
n r/min:	1000.0	1001.3	1005.2	1011.7	1020.7	1080	1430

should for the same reason try to obtain functional guarantees based upon the actual case, the process parameters of which must then be clearly specified as regards pipeline characteristics, size of pumps, reservoirs, storage and operational conditions, including variations in the liquid properties.

Currently, local packaged systems are available commercially as an alternative to selecting component parts oneself in order to make up a complete system. These include pump, drive equipment with speed regulation and control system. In some cases several pumps are included, of which at least one is speed regulated. The control system then comprises facilities for the switching in and out as well as providing for interchange between the various pumps, so that the method of operation can be chosen to suit the total function of the process. An interface to plant control systems can be fitted by using one of the standard communication protocols; 4-20 mA, RS 232, RS 423 or IEEE 488; depending upon the nature and complexity of the data.

5.9.7 Schematics for various systems

Schematics are an important tool in system design and control logic. A large amount of information can be communicated regarding the various piped and electric/electronic systems using simple symbols and boxes. The symbol used for a pump, in the following examples, complies with several national standards. This symbol should not be used in actual schematics because it does not convey anything about the pump type. If engineers of different disciplines view and use a schematic, recommendations and improvements can be suggested based on the pump types shown. The symbols shown in figure 5.45 are proposed for consideration.

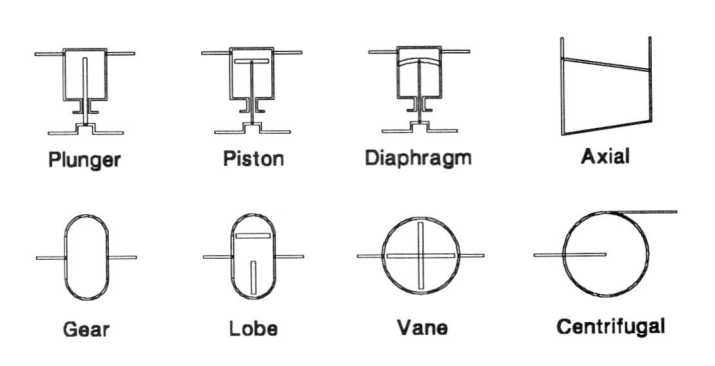

Figure 5.45 Symbols for different pump types

Variable ratio coupling

A variable ratio coupling can be hydraulic or electric. Fluid couplings and hydrostatic drives are torque devices, torque convertors are almost constant power devices and eddy current couplings are slightly better than constant torque. Fluid couplings, hydrostatic drives and torque convertors can be controlled electrically or pneumatically, eddy current couplings are usually controlled electrically.

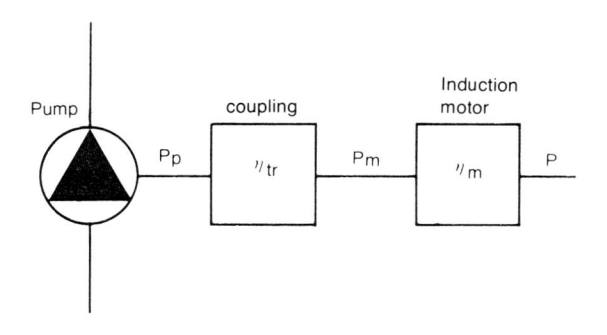

Figure 5.46 Schematic for speed regulation using variable ratio coupling

Semi-automatic multi-speed gearboxes can be controlled in a similar manner. The gearbox would be "pulsed" by a digital signal, which triggers the pneumatics, to change up or down in response to the demand. Mechanical variators can usually be controlled electrically or pneumatically.

Voltage regulated induction motor

For an AC induction motor the torque is proportional to the square of the applied voltage. Speed regulation can be obtained by varying this. A limiting condition, however, is that the motor should be equipped with a rotor having increased resistance, so-called high resistance rotor, in order for the losses to be reasonable. The variable voltage is obtained by means of a transformer or a thyristor connected in the motor supply, figure 5.47. The minimum speed depends on the motor characteristics, mainly on the magnitude of the rotor resistance. Control would normally be electric.

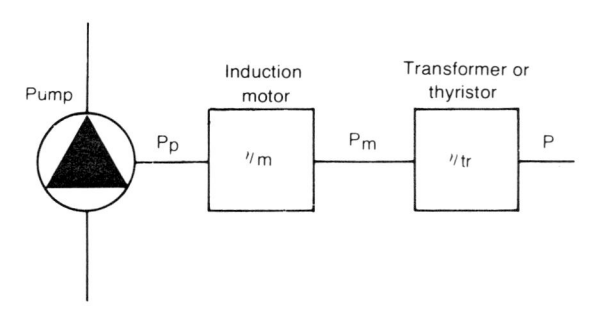

Figure 5.47 Schematic for voltage regulated AC motor

Frequency regulated induction motor

The speed of an AC induction motor is directly proportional to the frequency. In frequency regulation the frequency and also the voltage to the motor is varied simultaneously. The frequency converter, or invertor, is connected in the motor supply, figure 5.48. In the frequency converter, the mains power is first rectified and then the DC current is chopped up to form AC at the required frequency and voltage. Because the frequency converter is electronic, electric control is the ideal; all the protocols listed in 5.9.6 are used depending upon control source.

DC motor drive

For a DC motor with constant field, the speed is approximately proportional to the armature voltage. The armature voltage can be controlled quite simply for any required value by using a thyristor rectifier to convert AC to DC, figure 5.49. Electric control signals similar to frequency converters.

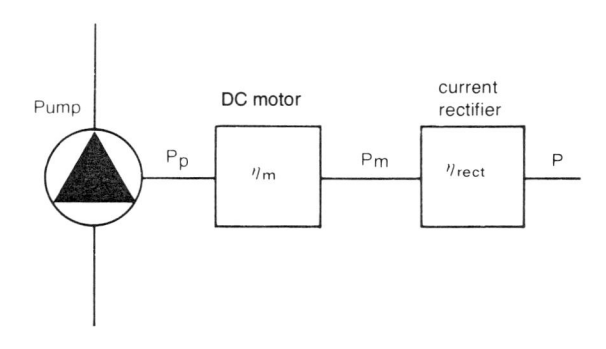

Figure 5.49 Schematic for DC motor drive

Slipring induction motor

The speed of an AC slipring or wound rotor induction motor can be controlled by taking power from the rotor via the sliprings, figure 5.50. The power can be dissipated in an external regulating resistance or fed back into the mains via a thyristor rectifier. In the latter case the rotor voltage is rectified and then chopped back into AC and is fed back into the mains via a transformer or sub-synchronous cascade rectifier. Electric control similar to frequency converters is standard.

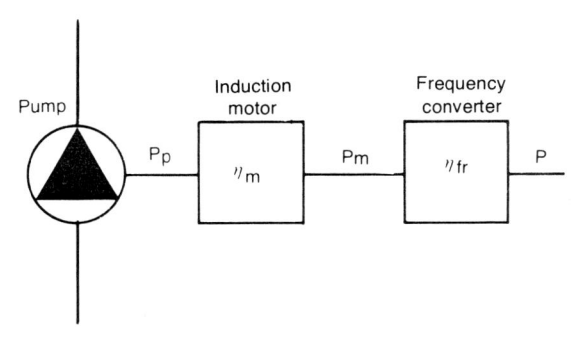

Figure 5.48 Schematic for frequency regulation

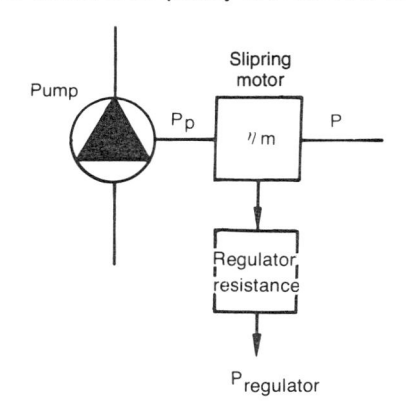

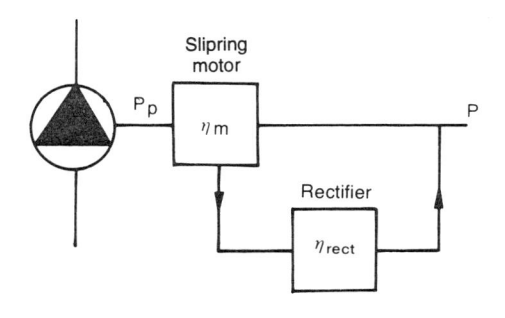

Figure 5.50 Schematic for speed regulation using a slipring motor Upper figure with external regulator resistance. Lower figure with sub-synchronous cascade rectifier.

Engines

Industrialised engines are fitted with governors which control the speed. The governor is set to a specific speed and the governor adjusts the fuel system to maintain the speed as the load varies. External speed control systems send signals to the engine governor to adjust the set speed. Most governors have a "tick-over" or "idling" setting which allows the engine to run slowly during warming-up or when the pump torque is low, on by-pass or unloaded. Governors are available to match any control signal. Figure 5.51 shows a typical engine arrangement.

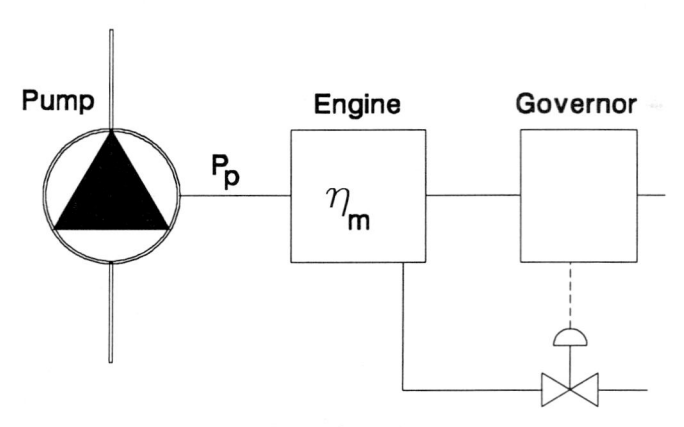

Figure 5.51 Schematic for engine speed regulation

Air motors

The speed of air motors is regulated by a flow control valve in the air supply. Control valves may be actuated by air, liquid or directly by an electric positioner. The control signal can be pilot air or liquid or digital or analogue electric. The pump flow should be measured and the speed adjusted to produce the correct flow, see figure 5.52.

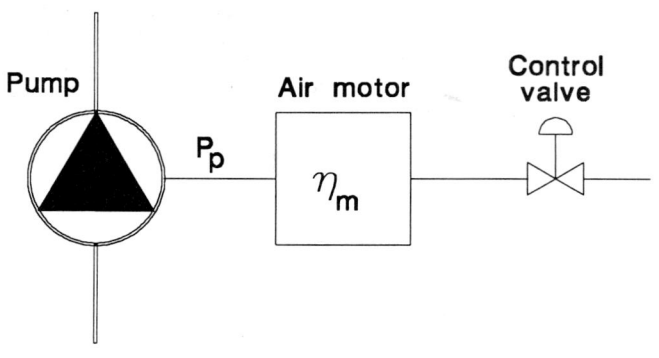

Figure 5.52 Schematic for air motor control

Turbines

Steam and gas turbines are fitted with governors in a similar manner to engines. The governor is set to the desired speed. In steam turbines, the governor actuates steam valves which control how much steam and how many nozzles are in operation to provide the power to operate at the set speed. Gas turbine control is identical to engines, the governor controls the fuel supply, see figure 5.51.

Turbo-expanders and hydraulic turbines

Both turbo-expanders and hydraulic turbines derive their power supply from flowing pressurised fluid streams. Neither device has any control over the fluid source so a flow control valve must be used to regulate the proportion used. Additional controls may be required in the primary supply to divert or by-pass fluid which is not used because of load conditions. The pump user and the manufacturer must agree who supplies the valves. In principle, control is exactly the same as for air motors, see figure 5.52.

5.10 Factors affecting choice of flow regulation method

In the past, the choice of flow regulation in pumping has been made largely on the basis of the initial investment required for the equipment. This method of assessment can be blamed on design/procure/build contractors who are not responsible for the ultimate running costs of the plant. As a result of increased energy costs, however, it has become essential to take into account the total costs during the whole life of the equipment, "life cycle cost", see also Chapter 13. This approach brings with it a drastic change of established practice, meaning among other things, that throttle regulation is often being replaced by speed regulation.

In order to make a correct choice of regulation method, it is necessary to carry out an extensive technical/economic analysis where consideration is given to:

- system requirements for cyclic flow variations and pressure pulsations and flow range,
- constancy of flow throughout the life of the equipment,
- shape of the system curve and the variation in the system curve due to start-up/shut-down, corrosion, fouling, seasonal production,
- criticality of service, how much of the process depends upon the operation of this unit,
- reliability, i.e. the general requirement for back-up and stand-by equipment,
- the ability to modify the equipment slightly or significantly to cope with process changes due to future refinements or modifications,
- primary control system,
- sensitivity to environmental conditions, particularly humidity, dust, temperature range, gases such as methane, ammonia and hydrogen sulphide,
- serviceability, particularly with regard to the various skills required of local personnel involved, e.g. electrical, electronic and mechanical,
- energy costs of all available sources, including variations during the life of the equipment and/or seasonal variations,
- initial cost of equipment,
- life, cost and availability of spare parts,
- service life of the equipment and estimated profits.

The flow/head required and constancy of flow decide whether the flow shall be distributed among several pump units, of which in its turn only one may need to be flow regulated.

The shape of the system curve and constancy of flow will guide the selection of pump type and flow regulation method.

The need for a stand-by unit, and if the stand-by units requires regulation, is dependant upon the criticality of the service. Reliability of the unit may be improved by imposing limitations on the normal operation; reduced speed, reduced power, reduced operating pressure. The requirement to have a stand-by unit affects the investment calculation significantly; additional space, piping, valves, cables, control equipment; since the cost savings from maintaining production have to cover the investment and operational costs of the stand-by unit.

As can be seen in Chapter 13, the cost of energy is the dominating operational cost, for which reason a partial optimization of investment for regulation equipment and energy savings will give satisfactory results. The energy costs can be determined by use of the total efficiency factor, see Section 13.3.1, or by use of total power consumption at various flows.

5.10.1 Direct flow regulation

Figure 5.53 compares the power requirements for flow control of a centrifugal pump using different speed methods and with flow regulation using throttle and on-off regulation to the hydraulic power. The comparison applies to one specific case in which the system's static differential head is zero, i.e. the only losses are pipe frictional ones. For other operational cases, particularly with increasing proportion of static differential head, the power savings are less for speed regulation. When the differential head consists only of static head, there is in certain cases, a 10% to 20% greater power consumption for speed regulation than for on-off control. In these circumstances a positive displacement pump may give better results, see figure 5.42.

	Flow %				
	20	40	60	80	100
Hydraulic power kW	0.8	6	22	51	100
Variable frequency AC (large) motor or DC motor (small)	1 / 1.5	6.5 / 8.5	24 / 31	55 / 72	109 / 140
Mechanical variators	8	16	33	61	110
Fluid coupling	10	23	44	74	115
Torque convertor	8	18	37	61	125
Hydrostatic coupling	10	19	38	72	129
Throttling	84	90	97	106	110
On-off control (see fig 5.8)	24	51	76	97	110

Figure 5.53 Comparison of power requirement with various methods of flow regulation of a centrifugal pump with zero static differential head

The loss of head in the control valve in accordance with Equation 5.17 is not included. This corresponds in this case to a power of about 35 kW.

From figure 5.53 it can easily be seen that throttling is inefficient. Throttling, or by-passing, should only be undertaken when the flow reduction necessary is small and/or the operating time at reduced capacity is short. On-off control is inefficient for flow regulation on purely friction systems. If a cheap reliable source of energy can be obtained the cost of inefficiency can be negated and the choice of method will be based on initial installation costs.

5.10.2 Speed regulation

In order to determine energy consumption it is necessary to summate power consumption over a period of time using a constancy diagram. If there is no constancy diagram, a rough estimate can be made using mean flow, or better still that flow which has 50% constancy. When studying the constancy of flow together with the performance of the pump already chosen, it is usually difficult to differentiate between the extreme flow requirement and the approximate safety margin. The normal maximum flow is generally less than 80% of the pump flow. The flow with 50% constancy is therefore usually quite low and can be judged to be:

- about 60% of maximum pump flow in process industries,
- about 30% of maximum pump flow in public water supplies,
- about 40% of maximum pump flow in heat transport (central heating and district heating)

Reducing safety margins and operating pumps closer to their BEP will lower operating costs.

A comparison of the power requirements in figure 5.53 at 60% of mean flow shows a difference between throttle regulation and speed regulation by fluid coupling of 97- 44= 53 kW. For example, for a capitalised cost for 1 kW of £300, the difference in the capitalised energy costs will be about £16,000. Since the additional costs for speed regulation by fluid couplings is about £4,000, see figure 5.31, the economic superiority of speed regulation is indisputable.

The example shown here illustrates that in many cases speed regulation is well worth considering. The cost of energy is critical. If energy is virtually free, waste steam or high pressure gas/liquid, there is no need to do any calculations. The energy source must be available for the lifetime of the pump though. As a result of general studies, it can be said that speed regulation of rotodynamic pumps for a continuous flow requirement is the best method of regulation at virtually all powers, if the system curve comprises pipeline losses only. A review of other cases for rotodynamic pumps is shown in figure 5.54. It can be seen that the capitalised value of energy costs per kW and the shape of the system curve, H_{stat}/H_{max} as in figure 5.55, has considerable effect.

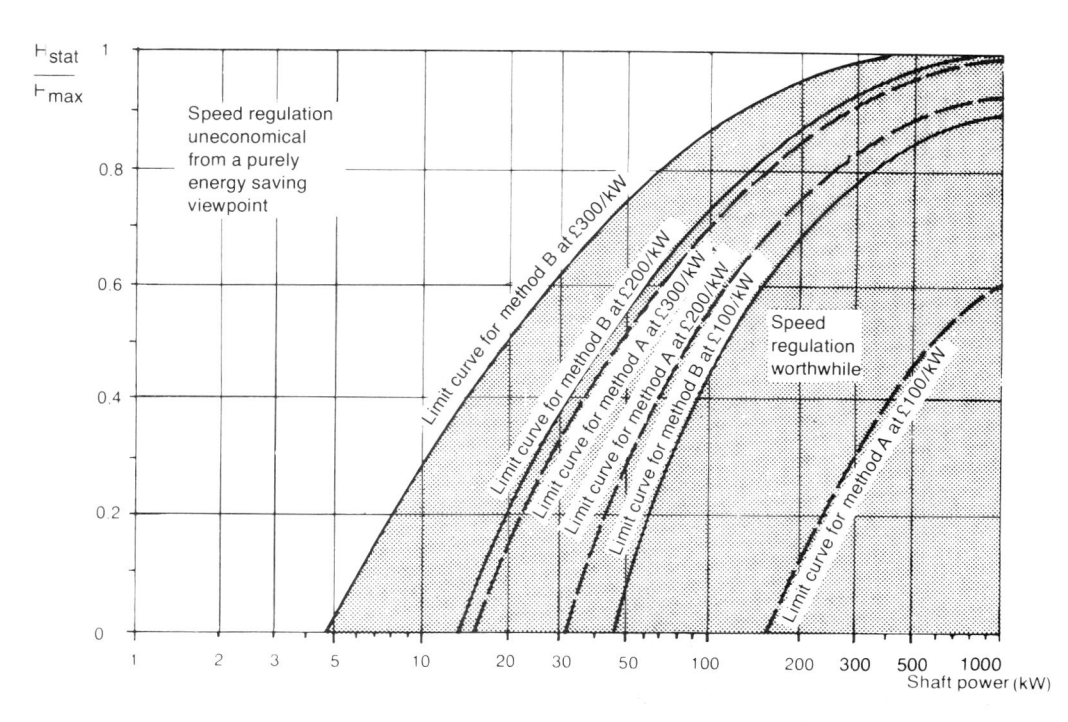

Figure 5.54 Approximate regions of viability for speed regulation. Method A applies to AC frequency regulation and B to fluid couplings. Figures in £/kW comprise capitalised energy costs based on maximum shaft power, see figure 13.16.

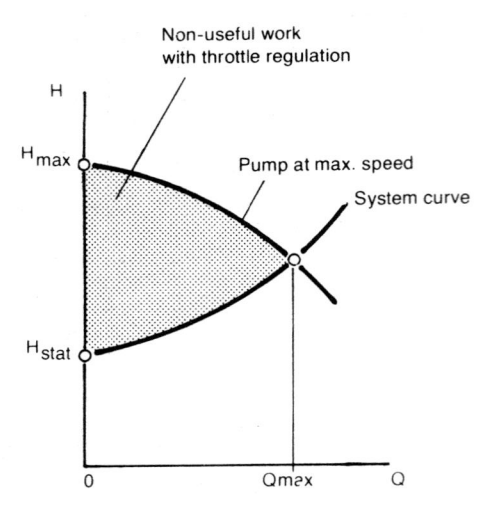

Figure 5.55 The system curve and centrifugal pump curve and their effect on method of flow regulation determined by the parameter H_{stat}/H_{max}

The limit of economic viability for speed regulation according to figure 5.54 gives a first guide regarding the choice of method of regulation for rotodynamic pumps. As H_{stat}/H_{max} increases, positive displacement pumps should be considered. As a second step in the assessment, the investment costs for the actual installation can be compared with the possible energy savings using speed regulation. Intensive studies of normal pump installations actually show that possible energy savings using speed regulation compared to throttle regulation amount to 20% to 60% and depend heavily upon the shapes of the system and pump curves. The values shown in figure 5.56 can serve as a guide at the initial project stage and apply most closely to the process industries. For water mains, central heating and domestic purposes, the energy savings are greater because of low mean flow. When the flow constancy is known, it is then of course possible to determine energy consumption for various regulation methods for the actual pump installation. The calculations can be simplified to a great extent by using Simpson's Rule as shown in previous sections.

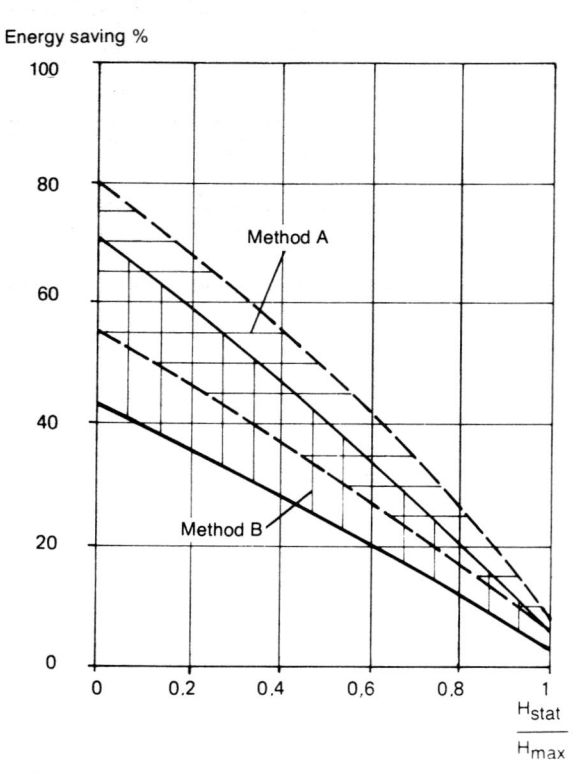

Figure 5.56 Guide-line values of normal energy savings in the process industries with speed regulation of centrifugal pumps relative to throttle regulation.

Method A applies to mechanical variators, frequency controlled AC motors and thyristor controlled DC motors.

Method B applies to fluid couplings and eddy current couplings.

Safety factors applied to flow and differential heads carry with them dramatic increases in costs of pump installations. A flow safety margin of just 25% can double both the investment and the operational costs for a pump installation. The consequence of the increased power consumption due to the safety margin is eased to some extent with speed regulation as can be seen from the following summary of costs per kW of shaft power.

	Throttle regulation £/kW	Speed regulation £/kW
Increase in pump costs	15-60	15-60
Increase in motor costs	20	20
Increase in costs for regulation equipment, approx.	10	50
Capitalised energy costs, approx.	300	50
Total of additional costs approx.	345/390	135/180

The necessity of determining real pump data as early as the initial project stage, is clearly illustrated by the above costing example which also underlines the importance of modifying pump data in accordance with possible changes in operational conditions.

The initial cost of the actual pump equipment increases with speed regulation by 80 to 150 £/kW shaft power in small pumps, about 10 kW, and by 20 to 80 £/kW for larger pumps, about 100 kW, depending upon the type of speed converter and ancillary equipment, see figure 5.31. All of the previously mentioned types of drive equipment are suited to various types of application. Figure 5.57, which is by no means exhaustive, shows just how widely varied applications and sizes are.

When comparing various methods of speed regulation it should be noted that it is the speed regulation itself which is responsible for the greater part of the energy savings. The method of speed regulation and its transmission efficiency only have marginal effects, as can be seen from the following examples of typical investment calculations.

	DC motor drive AC vari freq	Fluid cplg Eddy current
Capital investment costs	£15,000	£8,000
Saving in energy costs £/annum	£5,000	£4,000
Pay back time	3 years	2 years

On the basis of these calculations both groups seem to be entirely acceptable. If the marginal investment is now studied, we get:

Marginal investment	£15,000 - £8,000 = £7,000
Marginal savings	£5,000 - £4,000 = £1,000 p.a.
Pay back time	7 years

The pay back time of 7 years is matched roughly by the write-off time of 30 years at the interest factor 15% and means that variable speed motors in this case are uneconomical compared with variable speed couplings.

In the same way as in the marginal calculation above one may find that variable speed motors offer the same economic viability as variable speed couplings only when the procurement costs for motors do not exceed couplings by more than 20% to 30% i.e. the same value as corresponds to the greater energy savings of motors.

Method	Pumps	Power kW	Application
Multi-speed gearbox	Positive displacement pumps	10 - 2000	Pipeline pumps Transfer pumps
Semi-automatic gearbox	Positive displacement pumps	100 - 400	Engine driven portable test pumps Oil-well fracturing
Mechanical variator	Metering pumps Positive displacement pumps	1 - 10 10 - 200	Chemical dosing General purpose Variator can include speed reduction
Hydrostatic transmission	Centrifugal	5 - 100	Ship bilge pumps One power pack drives several remote motors
Hydrostatic coupling	Centrifugal Positive displacement	5 - 30 5 - 30	Drive prevents over-running and turbining Constant torque drive for high H_{stat}
Fluid coupling	Centrifugal Centrifugal	5 - 100 10 - 200 50 - 5000	Sewage pumps Liquor & pulp pumps Hostile environment Circulating pumps for district heating Very reliable
Torque converter	Centrifugal Centrifugal	50 - 20000 50 - 5000	Boiler feed pumps Circulating pumps
Eddy current coupling (*)	Centrifugal Mixed flow	5 - 300	Water works & sewage
Multi-pole AC motor	Centrifugal Mixed flow Axial	0.05 - 25	Short pipeline pumps Transfer pumps Circulating pumps
Frequency controlled AC motor (*)	All pump types	0.1 & up	Most applications
Primary voltage controlled AC motor	Centrifugal Mixed flow Axial	0.01 - 20	Short pipeline pumps Circulating pumps Booster pumps
Slipring AC motor	All pump types	20 - 500	Water works Sewage
DC motor (*)	All pump types	1 - 1000	Applications requiring precise speed control and rapid response
Petrol engine Diesel engine	Centrifugal Mixed flow Positive displacement	1 - 300	Portable pumps for site work, drainage Portable hydrotest pumps
Gas engine Crude oil engine	Centrifugal Positive displacement	100 - 3000	Pipeline pumps
Air motor	Centrifugal Mixed flow Axial	1 - 25	High speed pumps up to 9600 r/min from std compressed air
Steam turbine	Centrifugal Mixed flow Axial	25 & up	Boiler feed pumps Used in refineries as main pump, electric stand-by
Gas turbine	Centrifugal Mixed flow Axial	25 & up	Used where steam & electricity not available, popular off-shore
Turbo-expander	Centrifugal Mixed flow Axial	10 & up	Used in continuous processing, refineries with hp gas streams
Hydraulic turbine	Centrifugal Mixed flow Axial Positive displacement	10 & up	Used in continuous processing, refineries, RO, with hp liquid stream

Figure 5.57 Typical applications for speed regulation

RO ≡ reverse osmosis fresh water production

(*) Control equipment requires cooling air free from excess humidity

VOGEL Pumps
Tradition and quality by future technology

Pumpenfabrik ERNST VOGEL, A-2000 Stockerau, Ernst Vogel-Str. 2, Tel. 02266/604, Telefax 02266/65311

Once upon a time, a long, long time ago, a little Dutch boy put his finger in the dyke of the Zyder Zee to stop it leaking…

More recently 1.2 billion m³ of water from the Zyder Zee has been drained in 150 periods of 24 hours, thanks to Stork Pumps.

A good case of getting your finger out and getting on with the job.

Stork know a great deal about pumping water and everything else for that matter.

Call Stork Pumps now on 0293 553495 or 0274 742247 to find out more. Pumping is a question of Stork.

STORK®
Stork Pumps Ltd.

Meadow Brook Industrial Centre, Maxwell Way, Crawley, West Sussex RH10 2SA.

6 Pump materials

This chapter reviews the very wide range of materials used in pump technology. Strength of materials, pressure containment and creep are dealt with briefly in the special context of pumps. Corrosion is a big problem in this field, because of the substantially greater velocities within some pump designs compared to pipework. The form of corrosion due to this, erosion - corrosion, has therefore been given a relatively large amount of space, with regard both to the mechanism of corrosion itself, with build-up of protective film, and to testing at high liquid velocities. Judgement regarding the choice of materials suitable for abrasive liquids and cavitation resistance, have had to be made based on pump design and manufacturers' experience.

There are check-lists and cost relationships for various combinations of materials used in pumps, and also typical applications. Above all, pH value is given good coverage with regard to the effect of corrosion on choice of materials. The choice of materials for concentrated acids, hot liquids and liquids containing halogens, or contaminated liquids, usually require a detailed study of corrosion tables in special literature intended for that purpose, see Section 6.8. Please note, however, that this literature deals almost exclusively with corrosion in stationary liquids, to which the additional effects of velocity; erosion, abrasion; must be included.

Contents

6.1 Introduction

6.2 Typical materials
    6.2.1 Grey cast iron
    6.2.2 Spheroidal and nodular cast iron
    6.2.3 Low alloy steel
    6.2.4 Alloyed cast iron
    6.2.5 11-13% Cr steel
    6.2.6 Stainless steel
    6.2.7 Super stainless steel
    6.2.8 Copper alloys
    6.2.9 Aluminium alloys
    6.2.10 Nickel alloys
    6.2.11 Other metallic materials
    6.2.12 Non-metallic materials
    6.2.13 Coatings

6.3 Material strength and integrity

6.4 Corrosion and erosion
    6.4.1 Liquid corrosion of metals
    6.4.2 Corrosion rate
    6.4.3 Types of corrosion
        6.4.3.1 Galvanic corrosion
        6.4.3.2 Erosion corrosion
    6.4.4 Corrosion testing

6.5 Abrasion resistant materials
    6.5.1 Pump selection
    6.5.2 Hard metallic materials
    6.5.3 Rubber cladding

6.6 Materials resistant to cavitation damage

6.7 Material selection
    6.7.1 Basic information
    6.7.2 Material combinations
    6.7.3 Corrosion resistance
    6.7.4 Liquid effect on materials unknown

6.8 Bibliography

6.1 Introduction

Material losses from corrosion in industry amounts to a very large sum, hundreds of million pounds per annum in Britain alone for example; in layman's terms, this is the same as saying that the output from every fifth steelworks emerges in the form of rust. The nature of the problem of materials in the field of pumps is generally quite different from that of simple common rusting, because of the complex effects brought about by the pumped liquids themselves.

Since pumps are utilised in most processes, on-going process development creates new problems for which solutions based on experience do not yet exist. Examples of these are:

- In catalytic manufacture of sulphuric acid, the production rate has been increased by removing various cooling operations. Through this, the temperature rises by 50 °C, which increases the material corrosion problems of the pump by a factor of three.
- Plating workshops are adding fluoric acid to chromic acid baths in order to improve the shine on chromium surfaces. The new acid mixture is much more corrosive than before.

Individual materials and combination of materials for pumps are determined by a number of interactive factors, see figure 6.1. The end result is usually an inspired compromise between all of these, whilst at the same time, unfortunately, some of the factors may be more or less unknown quantities.

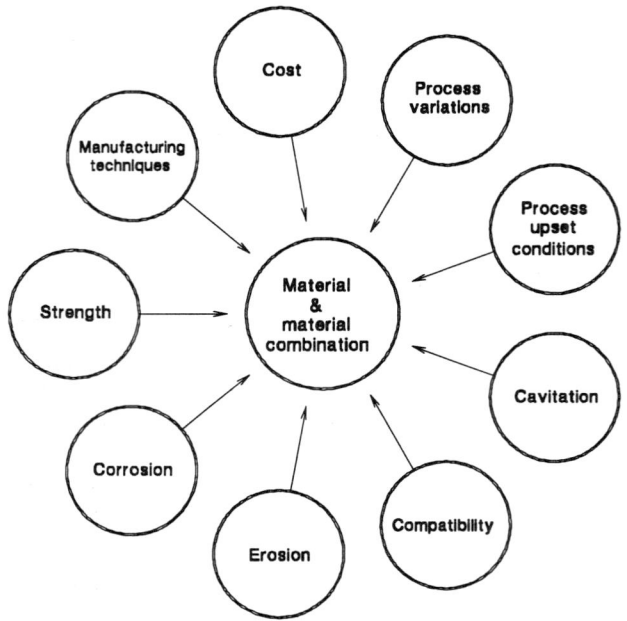

Figure 6.1 Factors effecting material suitability for pumps

Manufacturing techniques is quite a broad heading and includes some aspects which affect the useful life of a pump and the ability of the user to repair a pump. It is preferable to use materials which are easy to machine. If a material is so difficult to machine that "as cast" parts must be used then pump performance will normally be reduced. It is preferable to use materials which can be welded. Intricate components tend to be cast. Casting is not 100% successful. Provisions must be made to allow for approved repair methods. Welding is the most useful repair technique. Components made in materials which cannot be welded may have to be scrapped, increasing the overall cost. Pumps manufactured from materials which are difficult or impossible to weld will prove difficult or impossible to repair when faults occur in service. Some materials which can be welded should not be repaired by welding. This is controlled by the Quality Control Plan, QC plan, see chapter 11. Stationary materials in close proximity to moving materials must be compatible. Attention must be paid to the "bearing" qualities of some materials, which may be affected by the liquid properties. In some cases, non-metallic materials in particular, process upset conditions can change the physical properties of the material. In these cases, pump design conditions should protect the integrity.

The strength of materials can be judged by the allowable stress values given in pressure vessel codes. Local laws may specify a particular code to follow. ASME VIII and BS 5500 are popular and are specified as the limitation in many rotodynamic pump specifications. Pressure vessel stress levels are useless if components must be designed for fatigue resistance, reciprocating pump cylinders for example. The fatigue endurance limit of the particular material, in contact with the particular liquid, will dictate the allowable stress levels.

When reviewing different specifications for materials it is important to consider materials which are "equivalent". General purpose materials may not be equivalent to materials specifically designated for pressure containment. Manufacturing methods can impart integrity which is not easily indicated by physical properties or chemical composition.

6.2 Typical materials

The vast number of materials used currently in pumps can be classified within the following main groups:

- Material with iron as its main constituent
- Material with significant proportions of chrome and nickel
- Material with copper or aluminium as its main component
- Other metallic materials
- Non-metallic materials.

The mechanical properties are defined in the usual way by the tensile strength and elongation. Sometimes there are other physical properties which are of interest such as impact resistance and hardness. The pump pressure rating, PN, see also Section 6.3, is largely determined by the strength of materials and the temperature. For plastics at room temperatures as for metals at higher temperatures, the resistance to deformation must be considered. Creep resistance describes the effect of time under stress on permanent deformation.

Temperature plays an important part in the properties of solid materials. In general, as temperature increases materials become more elastic and deform more easily. This effect is reflected in the allowable stress which should be used for design purposes. Well respected pressure vessel standards are a source of good information. Pumps are not pressure vessels; the pump casing or cylinder must do much more than just retain internal pressure. However the data in pressure

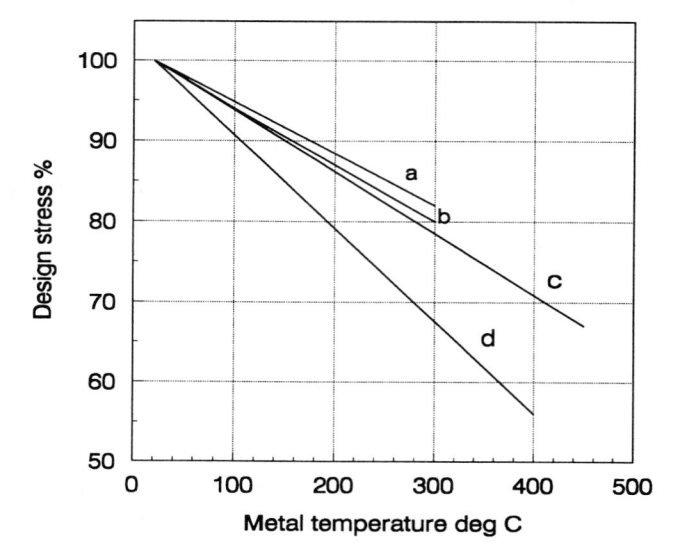

Figure 6.2 Variation in material design stress with temperature
a = spheroidal graphite iron
b = grey cast iron
c = austenitic chrome iron
d = cast steel

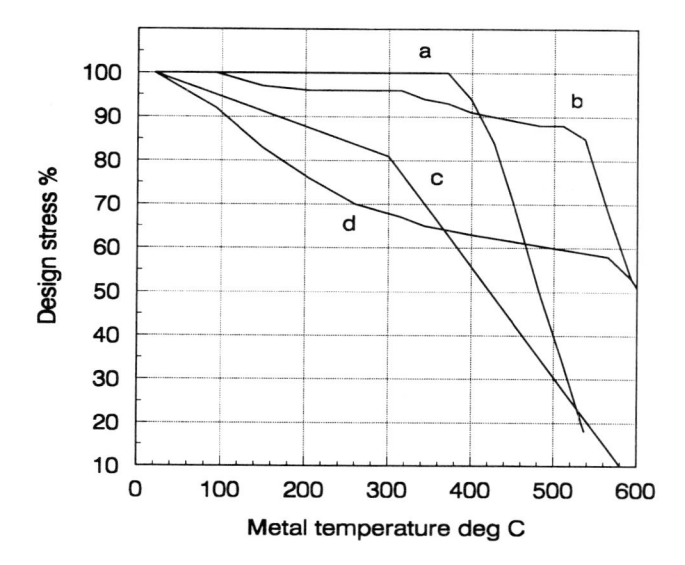

Figure 6.3 Variation in material design stress with temperature
a = high tensile steel bolting (B7)
b = cast 18/8/2 austenitic stainless steel, large deformation
c = cast 11-13 Cr steel
d = cast 18/8/2 austenitic stainless steel, small deformation

vessel standards can be used for guidance on the temperature effects. Figures 6.2 and 6.3 are based on data published in ASME VIII and BS 5500.

Low temperature can be a problem. The definition of low temperature varies with material types. Most materials become brittle and become susceptible to shocks. Both extremes of the temperature range need effective consideration. Creep can be a problem with metals at high temperature. Creep can be a problem with non-metallic materials at relatively low temperatures, see figure 6.4.

The tables presented in the subsequent sections, in some cases, show several material designations in one category. National and International standards are generally not identical for requirements of production and quality control. The designations quoted are extracted from other publications, such as ISO 7005, and must be treated with some caution. The decision as to the suitability of any material for a specific application rests with the design engineer responsible for product liability.

6.2.1 Grey cast iron

General purpose cast iron is manufactured in various strengths, BS 1452, Grades 180-300, the grades referring to the tensile strength. Cast iron specifically for pressure containing parts, ASTM A278, has similar grades. See figure 6.5. In general, cast iron is not defined by a chemical composition only by physical properties. Some specifications limit proportions of certain elements. The tensile strength depends primarily upon the content of carbon and reduces as the carbon content increases. The carbon occurs as graphite flakes enclosed in a steel-like matrix. The flake formation means that the value of elastic limit is low. The impact strength of cast iron is affected by the flakes but cast iron can be used down to -50 °C. Maximum temperature about 200/250 °C for general purpose cast irons. Because of the relatively low tensile strength, grey cast iron impellers are restricted to 'low' peripheral speed pumps. Note also that the strength of cast iron is related to the thickness; specifications must be reviewed during the design process.

Grey cast iron is the most commonly used material in centrifugal pumps. From the corrosion aspect, it has a very wide field of application. Grey cast iron can be used without any great risk in water, or aqueous solutions, with pH values between 6 and 10. Grey cast iron can be used in more acidic media, but it is dependent to a high degree upon what has caused the lower pH value. Temperature dependence is also more accentuated in this region. Because if its high carbon content,

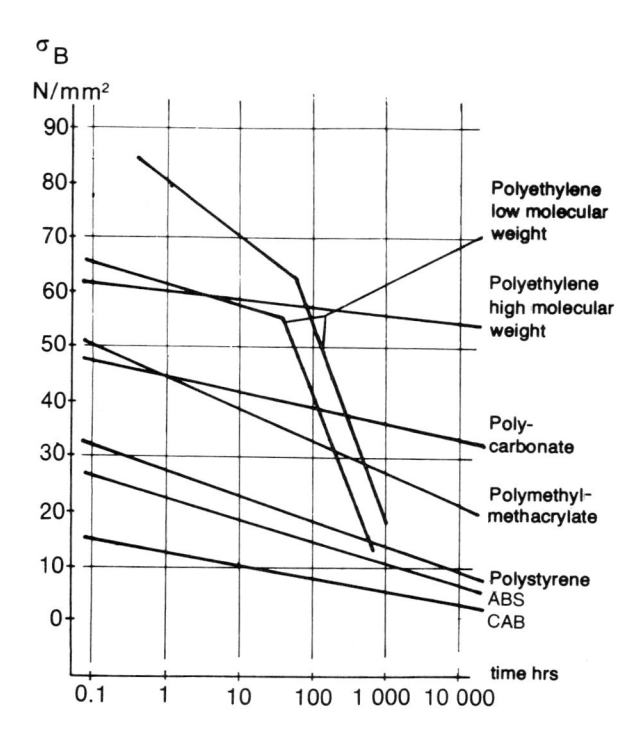

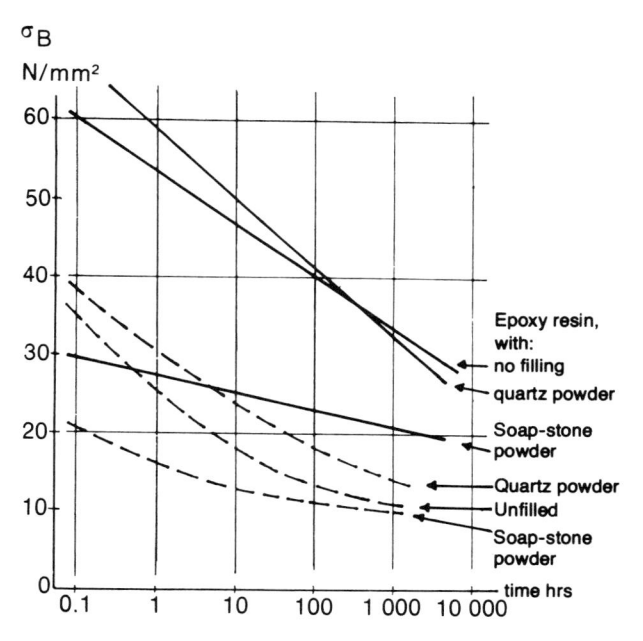

Figure 6.4 Creep of non-metallic materials
——— 23 °C   - - - - - - - 60 °C

grey cast iron quickly acquires a protective graphite layer, which means, from the strength point of view, that poorer quality cast iron with higher carbon content has better protection against corrosion. Grey cast iron is also superior to normal steel from the corrosion point of view for the same reason.

Grey cast iron is used for bearing housings because of its compressive strength and damping qualities. A standard material for crankcases for all types reciprocating pumps.

Meehanite® is a proprietary process for producing cast iron under more rigourously controlled manufacturing cycle. Meehanite tends to be stronger and can be cast in thinner sections. Meehanite should be considered when ever cast iron is appropriate.

Cast iron can be welded or repaired by welding. However, the weld material will have a different chemical structure, and perhaps physical properties, to the parent material. Extreme caution should be exercised if cast iron is to be welded.

NOTE: When designing pressure vessels, or pump casings, the maximum stress due to internal pressure occurs at the inner wall. For vessels, casings, without holes through the wall, the inner tensile stress will be approximately equal to the internal pressure. The internal pressure can never be greater than the design stress. Holes through the wall will create stress concentrations, values of 1.5 to 3.5 are typical, further reducing the maximum internal pressure. Design tensile stress values given are for room temperature conditions.

Standard	Grade	Tensile strength min MPa	Max temp °C	Approx max tensile design stress MPa	Remarks
BS1452 See DIN 1691 ASTM A48	150	150	≈ 300	39	As cast design values
	180	180	≈ 300	46	
	220	220	≈ 300	54	
	260	260	≈ 300	66	
	300	300	≈ 300	77	
	350	350	≈ 300	90	
	400	400	≈ 300	100	
ISO 185	200	200	300	50	As cast
	250	250	300	62	
ASTM A126	A	145	230	36	As cast
	B	214	230	53	
ASTM A278	C120	138	230	35	
	C125	172		43	
	C130	207		52	
	C135	241		60	
ASTM A278	C140	276	345	70	Restricted carbon content $P \leq 0.25\%$ $S \leq 0.12\%$
	C145	310		77	
	C150	345		86	
	C155	379		94	
	C160	414		103	
Meehanite	GF150	150	≈ 350	40	As cast design value
	GE200	200	≈ 350	50	
	GD250	250	≈ 350	62	
	GB300	300	≈ 350	75	
	GM400	400	≈ 350	100	

Figure 6.5 Specifications for grey cast iron

Standard	Grade	Tensile strength min MPa	Max temp °C	Approx max tensile design stress MPa	Remarks
BS 2789 See DIN 1693	370/17	370	≈ 350	129	Ferrite 17% elongation
	420/12	420	≈ 350	138	Ferrite 12% elongation
	500/7	500	≈ 350	145	Pearlite & ferrite
	600/3	600	≈ 350	156	Pearlite & ferrite
	700/2	700	≈ 350	173	Pearlite
	800/2	800	≈ 350	198	Pearlite or tempered structure
ISO 1083	350-22	350	350	87	Elongation 22%
	400-15	400	350	100	Elongation 15%
	500-7	500	350	125	Elongation 7%
	600-3	600	120	150	Elongation 3%
ISO 2531	400-5	400	350	100	Elongation 5%
ASTM A536	60-40-18	413	≈ 350	102	Elongation 18%
	65-45-12	448	≈ 350	112	Elongation 12%
	80-55-06	551	≈ 350	137	Elongation 6%
	100-70-03	689	≈ 350	172	Elongation 3%
	120-90-02	827	≈ 350	206	Elongation 2%
Meehanite	SFF350	350	350	87	Elongation 24%
	SFF400	400	350	100	Elongation 20%
	SF400	400	350	100	Elongation 17%
	SFP500	500	350	125	Elongation 8%

Figure 6.6 Specifications for spheroidal and nodular cast iron

6.2.2 Spheroidal and nodular cast iron

Spheroidal or nodular cast iron, also called SG iron and Ductile Iron, is a relatively new material. By analysis, this grade of cast iron is very like ordinary cast iron with regard to carbon and silicon content. In contrast to ordinary cast iron, however, the graphite is not in flakes but occurs in the form of small spheroidal nodules in a mainly pearlite matrix. This is brought about by adding small quantities of magnesium to the melt before pouring. Because of the spheroidal form of the graphite, spheroidal or nodular cast iron has a substantially greater strength than ordinary cast iron and is much more ductile. Furthermore, the elastic limit is greater and the resistance to impact is better. Further improvement in the elastic limit and impact resistance can be obtained by heat treatment, whereby the basic matrix is changed towards a ferritic structure. Spheroidal or nodular cast iron is ideally suited for pressure vessels subjected to internal pressures, where these were formerly always assigned to cast steel. Spheroidal or nodular cast iron also has some welding properties. The resistance to corrosion is very similar to that of grey cast iron at low flow velocities. At higher flow velocities this material often has a lower resistance than ordinary cast iron.

6.2.3 Low alloy steel

Low alloy steel castings and wrought material only differ slightly in chemical analysis because of certain additives intended to improve castability. Some low alloy steels have small additions of nickel and chromium in order to improve their strength at elevated temperatures. Steel castings are used mainly for high pressure and hot liquid pumps but are being replaced more and more by spheroidal or nodular cast iron and alloyed qualities of steel. From the corrosion aspect, steel castings are not as good as grey cast iron and spheroidal or nodular cast iron, because, by preference, the carbon content should not exceed 1.5%, which is insufficient to provide a protective graphite layer. Carbon content of steels affects the heat treatment of the steel and the weldability. Trace quantities of chrome and nickel improve through hardening qualities, ductility and impact resistance. High strength bolting materials have about 0.4% carbon and 1% chrome. Low alloy steel plate and forgings are used for various components.

6.2.4 Alloyed cast iron

The properties of flaked and spheroidal cast iron can be altered to a large extent by the addition of alloying elements affecting strength, workability, and resistance to corrosion and abrasion. The following elements can be used:

Copper, Cu (for flaked graphite cast iron only)

Copper additives give the basic matrix of cast iron a more uniform pearlite structure with less ferrite and a better form and distribution of the graphite. The tensile strength increases by 10% to 20% with additions of 1% and 2%, with corresponding increases in hardness. Corrosion resistance to very concentrated sulphuric acid and solutions of sulphuric acid, for example, is markedly improved.

Nickel and chromium, Ni and Cr

Small additions of nickel and/or chromium cause the structure to be more even and finely distributed giving advantages similar to those for copper. Furthermore, casting material integrity is improved in this way.

The additions of 20% Ni (Ni-Resist) causes the matrix to become austenitic and non-magnetic. This quality has shown itself to be very useful for sea-water and other chloride containing liquids, even at high temperatures. The strength is similar to the corresponding grades of grey cast iron and spheroidal or nodular cast iron.

Ni-Resist, with alloying proportions of about 6% Ni and 9% Cr produces a martensitic matrix with chrome carbides, s resulting in hardness figures between 500 and 600 BHN. The hardness can be further increased by heat treatment or by increasing the chrome content up to 30%. Thus a material can be produced with very good resistance to abrasion and erosion. Because of the considerable hardness, the material can only be worked by means of grinding so that special designs are required for pumps.

Standard	Grade	Major constituents	Tensile strength min MPa	Max temp °C	Remarks
BS 1504 DIN 1681 Werkstoff ISO 4991 ASTM A216	161-430 GS-45 1.0435 C23-45B WCA	0.25C 0.25Cr 0.15Mo 0.4Ni	430	400	Cast RT min (1)
BS 1504 DIN 1681 ISO 4991 ASTM A216	161-480 GS-52 C26-52H WCB	0.30C 0.25Cr 0.15Mo 0.4Ni	480	390/480	Cast RT min
BS 1503	164-490	0.25C 0.25Cr 0.1Mo 0.4Ni 1Mn	490	480	Forged
BS 1504	161-540	0.35C 0.25Cr 0.15Mo 0.4Ni	540	400	Cast
BS 1503	221-550	0.35C 0.25Cr 0.1Mo 1.1Mn 0.4Ni	550	350	Forged
BS 3100 ISO 4991 ASTM A352	AL1 C23-46BL LCB	0.2C 1.1Mn		≈ 425	Cast -46 °C min
BS 970 Werkstoff	070M20 'N' 1.0402	0.2C 0.7Mn	400	≈ 200	Wrought
BS 970	080M30 'N'	0.3C 0.75Mn	460	≈ 200	Wrought
BS 970 Werkstoff	080M40 'N' 1.0503	0.4C 0.8Mn	510	≈ 200	Wrought
BS 970 Werkstoff	708M40 'S' 1.7225	0.4C 0.85Mn 1Cr 0.2Mo	770	≈ 500	Wrought (2)
AISI Werkstoff	1040 1.1186	0.4C 0.75Mn	551	≈ 200	Wrought
AISI Werkstoff	4140 1.7225	0.43C 0.8Mn 1Cr 0.2Mo	860	≈ 500	Wrought (2)
BS 4360 Werkstoff	43C 1.0483	0.18C 1.5Mn	430		Structural plate
BS 4360 Werkstoff	50D 1.0570	0.18C 1.5Mn + Nb + V	500		Low temp structural plate -30 °C
BS 4882 ASTM A193 DIN 17200	B7 B7 42Cr.MO.4	0.4C 0.85Mn 1Cr 0.2Mo	860	400/500	Bolt/stud -45 °C min
BS 4882 ASTM A194 DIN 17200	2H 2H C45	0.4C	PLT (3)	450	Nut 0 °C min

Figure 6.7 Properties of low alloy steel — (1) RT ≡ room temperature — (2) reduced allowable stress over 350 °C — (3) PLT ≡ proof load tested

Standard	Grade	Major constituents	Tensile strength min MPa	Max temp °C	Remarks
ASTM A436	Type 1	2.1Cr 15.5Ni 1Mn 6.5Cu 1.9Si	172		Wear resistant, best anti-galling properties
BS 3468 DIN 1694	L-NiCr 20 2 GGG-NiCr20 2	1.8Cr 20Ni 1Mn 0.3Cu 1.9Si	170	650	Good for alkalis -50 °C min
ASTM A436	Type 2	2.1Cr 20Ni 1Mn 0.3Cu 1.9Si	172	705	Good for corrosive environments
BS 3468 DIN 1694	S-NiCr 20 2 GGG-NiCr20 3	1.8Cr 20Ni 1Mn 0.3Cu 2.2Si	170	650	
ASTM A439	Type D-2	2.25Cr 20Ni 1Mn 2.25Si	400	815	8% elongation, wear/galling resistant
BS 3468	S-NiMn 23 4	0.2Cr 23Ni 4.2Mn 2Si	440		-196 °C min
ASTM A439	Type D-2M	0.2Cr 22.5Ni 4Mn 2.2Si	417		-196 °C min wear/galling resistant
BS 3468 ASTM A436	L-NiCr 30 3 Type 3	3Cr 30Ni 1Mn 0.3Cu 1.5Si	172	815	Erosion resistant, good corrosion resistance
ASTM A439	Type D-3	3Cr 30Ni 1Mn 1.9Si	379	815	6% elongation, superior erosion resistance
ASTM A439	Type D-4	5Cr 30Ni 1Mn 5.5Si	414	815	Wear/galling resistant
ASTM A436	Type 4	5Cr 30.5Ni 1Mn 0.3Cu 5.5Si	172		Best grade for erosion, corrosion, oxidation resistance
ASTM A436	Type 5	0.05Cr 35Ni 1Mn 0.3Cu 1.5Si	138	425	Dimensionally stable
ASTM A439	Type D-5	0.05Cr 35Ni 1Mn 1.9Si	379		Elongation 20%

Figure 6.8 Typical alloy cast irons

Silicon, Si

Silicon alloyed cast iron, silicon iron, is highly resistive to acids, if the silicon content is greater than 13%. This material is resistant to sulphuric acid at all concentrations and also to many other non-organic and organic acids, though not to hydrofluoric acid, hot concentrated hydrochloric acid, sulphurous acids, sulphites and hot alkalis. An addition of 3% molybdenum makes the material resistive also to hydrochloric acid and solutions containing chlorine. Silicon iron is very difficult to cast and displays poor integrity under pressure. It should never, for this reason, be used in pumps with an internal pressure greater than 4 barg. As the material is hard and very brittle, special design principles are normally used for pumps, e.g. liners of silicon iron in a steel casing to take the pressure stresses.

Standard	Grade	Major constituents	Tensile strength min MPa	Max temp °C	Remarks
BS1504 ASTM A296 Werkstoff DIN 17445	420C29A CA-15 1.4027 G-X 20 Cr 14	0.16C 1Mn 12.5Cr 1Ni	620	600	Castings
BS 970 Pt4 Werkstoff DIN 17440	420S45 1.4028 1.4034 X 30 Cr 13 X 46 Cr 13	0.32C 1Mn 13Cr 1Ni	690	400	Bar
BS 1503 ASTM A182 Werkstoff	410S21 F6b 1.4001	0.12C 1Mn 12.5Cr 1Ni	590	400	Forgings
BS 4882 ASTM A193 Werkstoff DIN 17440	B6 B6 1.4006 X 10 Cr 13	0.15C 12.5Cr 1MN	758	425	Fasteners, -28 °C min

Figure 6.9 11-13 Cr materials

6.2.5 11-13% Cr steel

Rotodynamic pump manufacturers use 11-13 Cr steel as an improvement on low alloy steel when full stainless steels are not required. Material integrity is vastly improved due to carbide formation due to the addition of chrome. The structure is martensitic. Strength and hardness are improved, as are heat resistance and scaling properties. Can be used up to 600 °C but strength falls of rapidly over 300 °C. Weldability suffers slightly as the chrome content is increased. Traces of nickel improve hardenability and also impact properties. Hardness values between 180 and 450 BHN can be obtained. Used for pump casings, impellers, shafts and wear rings. Can be a good material choice for boiler feed water. Alloys with 4% Ni have improved corrosion resistance and better low temperature properties. 11-13 Cr is not used by reciprocating pump manufacturers.

6.2.6 Stainless steel

It is a general rule for steels that the content of alloying elements shall not exceed 50% of the total. A characteristic of all stainless steels is that they contain a minimum of 15% chromium. In the presence of oxygen this forms a thin invisible film of chromic oxide which is chemically resistant to a high degree and which inhibits any direct contact between the surrounding medium and the steel. It is, however, necessary that oxygen is present in order to maintain the film. The steel is said to be in a passive state. In a reducing atmosphere, on the other hand, there can be no build-up of the chromic oxide layer and the steel is then said to be in an active state.

The designation "stainless" steel is not altogether correct, as these steels can easily be subject to corrosive attack depending on external circumstances relative to alloying content, degree of heat treatment, welding and so on. In general, stainless steels exposed to normal atmospheric conditions will not corrode or discolour.

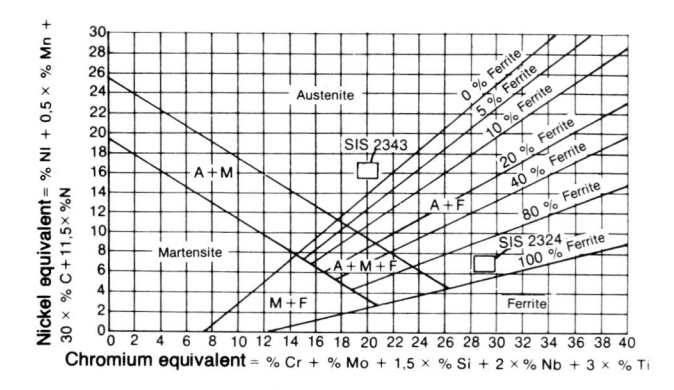

Figure 6.10 Modified Schaeffler diagram for determining structure of a stainless steel

The corrosion resistance of stainless steels is dependent upon the rate at which the passive oxide layer is dissolved into solution. This rate is slowed down mainly by alloying with metals which themselves have high resistance to corrosive media. Such alloying elements are, in the main, nickel, molybdenum and copper. The effects of chromium are then less important from the point of view of chemical resistance.

The presence of molybdenum in significant quantities greatly enhances the corrosion resistance. The structure of the iron matrix is changed by the various alloying additives. The type of structure is used to divide stainless steel and acid resistant steels into the following groups, see also figure 6.10:

- ferritic, which can be hardened by heat treatment
- martensitic, which can be hardened by heat treatment
- austenitic, which are totally non-magnetic and cannot be hardened by heat treatment
- ferritic-austenitic, which are paramagnetic

The austenitic stainless steels have lower tensile strength and greater corrosion resistance and toughness than the other groups. This makes them easier to manufacture into the various forms, such as sheet metal and piping. Because pumps are largely produced from castings, the chemical analysis can be chosen with greater freedom.

Using the same main analysis there are today a number of standardised variants of materials. The main reason for this is the modification of chemical structure around a weld, usually carbide precipitation. Carbide precipitation is eliminated in the various standard variants in the following ways:

- by low or extra-low carbon content
- by the addition of titanium or niobium in proportions of 5 to 10 times the carbon content to stabilise the carbides

Carbide precipitation can be corrected by a post weld heat treatment. In general, heat treatment is not preferred. Components which have welded attachments, or may require weld repairs, should be made from low carbon varieties.

Cast and wrought versions of the same variety may have slightly different chemical compositions to accommodate the different manufacturing processes

Austenitic stainless steels have low strength. The strength can be improved by inoculating the steel with nitrogen or by cold working.

Most stainless steels are recognized by their designations proposed by the American Iron and Steel Institute, AISI, and the Alloy Casting Institute, ACI, of America. The most well known are the AISI designations for wrought products. The majority of the popular stainless steels belong to the 300 series. The inclusion of 'L' means low carbon version; 'N' means nitrogen strengthened. The AISI designations are generic, a full product specification is required for purchasing. Figure 6.11 indicates the relationship between the popular AISI and ACI grades.

AISI	ACI	Nominal composition Cr/Ni/Mo
303	CF-16F	18/9/0.6
304	CF-8	19/9
304L	CF-3	19/10
316	CF-8M	17/12/2
316L	CF-3M	17/12/2
317	CG-12	19/13/3
347	CF-8C	18/11

Figure 6.11 Popular austenitic stainless steels

6.2.7 Super stainless steel

Materials are always required with better corrosion resistance. The expansion of off-shore oil production, together with secondary recovery by sea-water, led to a demand for materials with superior pitting resistance. Molybdenum and nitrogen content were found to be of critical importance. The super stainless steels are the result of much research by oil companies, pump manufacturers and steel makers. Super stainless steels can be welded by most popular techniques; resistance to stress corrosion cracking is not impaired, some loss of toughness and resistance to intercrystalline corrosion may be apparent.

Standard	Grade	Major constituents	Tensile strength min MPa	Max temp °C	Remarks
BS 1504 Werkstoff DIN 17445 ISO 4991	304C15 1.4308 G-X 6 CrNi 18 9 C47	0.12C 17Cr 7Ni 2Mn	460	700	Castings
BS 1503 Werkstoff DIN 17440	304S11 1.4306 X 2 CrNi 18 9	0.03C 18Cr 10.5Ni 2Mn 0.3Mo	480	450	Forgings
ASTM A351 ISO 4991	CF-3M C61LC	0.03C 18Cr 9Ni 2Mo	480	454	Castings
BS 1504 Werkstoff DIN 17445	347C17 1.4552 G-X7 CrNiNNb 18 9	0.08C 19Cr 8.5Ni 2Mn	480	680	Castings
BS 1503 Werkstoff DIN 17440	347S31 1.4550 X 10 CrNiNb 18 9	0.08C 18Cr 10.5Ni 2Mn 0.3Mo	510	450	Forgings
BS 1504 Werkstoff DIN 17445 ISO 4991	316C16 1.4408 G-X6 CrNiMo 18 10 C60 C61	0.08C 19Cr 10Ni 2.5Mo 2Mn	480	700	Castings
BS 1503 Werkstoff DIN 17440 ISO 2604-1	316S31 1.4401/36 X 5 CrNiMo 17 12 2 F62	0.07C 17.5Cr 12Ni 2Mn 2.2Mo	610	700	Forgings
ASTM A351 DIN 17445 ISO 4991	CF-8M G-X3 CrNiMoN 17 13 5 C57	0.08C 19.5Cr 10.5Ni 1.5Mn 2.5Mo	483	815	Castings
SIS Werkstoff DIN 17440	2343 1.4436 X 5 CrNiMo 18 12	0.05C 17.5Cr 12Ni 2.8Mo	450	400	Castings
ASTM A351	CG-6MMN	22Cr 13Ni 5Mo + N_2	517	565	Castings
ASTM A182 ISO 2604-1	F316L F59	0.03C 16Cr 12Ni 2Mo	448	454	Forgings
BS 1503	318S13	0.03C 22Cr 5Ni 2Mn 3Mo	640	300	Forgings
ASTM A182	XM-19	22Cr 13Ni 5Mn 2Mo + N_2	689	565	Forgings
SIS Werkstoff	2324 1.4460	0.05C 25Cr 5.7Ni 1.5Mo 2Mn	600	315	Cast
BS 1506	304S71	0.07C 18Cr 9.5Ni 2Mn			Fasteners B8N
BS 1506	316S51	0.07C 17.5Cr 12Ni 2.2Mo 2Mn			Fasteners B8MH
BS 4882 ASTM A193	B8 B8	0.08C 19Cr 9Ni 2Mn	517	425/575	Fasteners, -254 °C
BS 4882 ASTM A193	B8M B8M	0.08C 17Cr 12Ni 2.5Mo 2Mn	517	425/575	Fasteners, -196 °C
ASTM A193	B8R	0.06C 22Cr 12.5Ni 5Mn 2Mo + N_2	689		Fasteners

Figure 6.12 Austenitic stainless steels

Standard	Grade	Major constituents	Tensile strength min MPa	Max temp °C	Remarks
	SF22/5	0.03C 22Cr 5.5Ni 3Mo 2Mn N_2	750	≈ 300	Wrought
	FMN	0.05C 25Cr 5Ni 2Mo 0.8Mn N_2	695	≈ 300	Castings
	Ferralium 255	0.08C 25.5Cr 5.5Ni 3Mo 2Mn N_2	830	≈ 300	Castings
	Amazon 256-Cu	0.04C 25.5Cr 5.5Ni 3Mo 1.5Mn N_2	760	≈ 300	Wrought
	Zeron 25	0.03C 25Cr 6.5Ni 2.5Mo 1.5Mn N_2	650	≈ 300	Castings
	Zeron 100	0.03C 25Cr 7Ni 3.5Mo 1Mn N_2	756	≈ 300	Castings
	Ferralium 288	0.08C 27.5Cr 7.5Ni 2.5Mo 2Mn N_2	800	≈ 300	Castings
	254SMO	0.02C 20Cr 18Ni 6.1Mo 0.5Mn N_2	651		Castings

Figure 6.13 Super stainless steels

6.2.8 Copper alloys

Copper alloy components are used largely for salt water and weak cold chloride solutions; for corrosive properties of sea-water and choice of materials for use with sea-water see Section 1.2.4. The alloys have varying casting characteristics associated with solidifying temperatures, which in turn are determined by the alloying elements. For tin-bronze with about 10% tin, the temperature range is 200 to 250 °C. The prime result of this is that a certain amount of microscopic porosity cannot be avoided in the solidifying alloy. This porosity need not mean that the casting is not pressure-tight. The wide solidifying temperature range makes the alloy relatively easy to cast.

Brass with 30% to 35% zinc and aluminium-bronzes with about 10% aluminium have solidifying temperature ranges between 2 and 10 °C. The structure is fairly non-porous although large cavities due to coring may occur which can be eliminated by careful metal feeding. These alloys with narrow solidifying temperature ranges are more dependent upon the shape of the casting and involve higher costs.

Gun metal made up of about 5% each of tin, lead and zinc, is easily cast. The pressure-tightness quality increases with lead content up to about 7%.

From the point of view of corrosion, copper and its alloys give widely varying results. Here, as with other materials, it is a solid oxide film which provides the protection against further corrosion. The rate of build-up of this protective film varies and is affected by flow velocities. Figure 6.14 lists some commonly used alloys.

Standard	Grade	Major constituents	Tensile strength min MPa	Max temp °C	Remarks
BS 1400 ISO 1338 DIN 1714 ASTM B148	AB1 CuAl10Fe3 CuAl10Fe C95200	9.5Al 2.5Fe 1Ni	500		Castings
BS 1400 ISO 1338 DIN 1714 ASTM B148	AB2 CuAl10Fe5Ni5 CuAl10Ni C95500	9.5Al 4.5Fe 4.5Ni	640	≈ 300	Castings ≈ -150 °C
BS 1400 ISO 1338 DIN 1705 ASTM B145	LG2 CuPb5Sn5Zn5 CuSn5ZnPb Gr836	5Sn 5Zn 5Pb	200	≈ 300	Castings -196 °C min
BS 1400	PB4	9.7Sn	190		Castings
BS 2872 ISO	CA 107 CuAl7Si2	6Al 2.2Si	520		Forgings
BS 2872 ISO DIN 17665 ASTM B150	CA 104 CuAl10Ni5Fe4 CuAl10Ni C63200	9.5Al 4.5Fe 4.5Ni	700	≈ 500	Forgings
BS 2874	CA104	9.5Al 4.5Fe 4.5Ni	700	400	Fasteners, suitable for cryogenics

Figure 6.14 Common copper alloys

6.2.9 Aluminium alloys

Aluminium alloys, figure 6.15, are not used to any great extent in pumps, because of their low resistance to corrosion and low hardness. The exceptions are those areas where low weight is a prime factor, for example in building site ground drainage, and also where manufacturing costs are important as in domestic pumps. The range of application can be increased by inhibiting electrolytic corrosion or actually making use of it, for example by:

- insulating from more noble materials such as steel
- connecting to less noble materials, so-called sacrificial anodes of zinc or magnesium

Standard	Grade	Major constituents	Tensile strength min MPa	Max temp °C	Remarks
BS 1490 ISO 3522	LM 4 Al-Si5Cu3	5Si 3Cu	230		Castings
BS 1490 ISO 3522 DIN 1725 ASTM B85	LM 9 Al-Si10Mg AlSi10Mg SG100A	11.5Si 0.5Mn 0.4Mg	240	150	Castings
BS 4300/11 ISO	5454 AlMg3Mn	2.7Mg 0.7Mn	275		Forgings
BS 1472 ISO	2014A AlCu4SiMg	4.5Cu 0.8Mn 0.5Mg 0.7Si	370		Forgings
	AGS 6060		300		Fasteners
	AZ5GU-7075	5.5Zn 2.3Mg 1.5Cu	550		Fasteners

Figure 6.15 Common aluminium alloys

6.2.10 Nickel alloys

Nickel base alloys, as also pure nickel itself, come as a natural extension at a point where stainless steels no longer hold up from the corrosion point of view. They are not easy to cast as a rule, and require high annealing temperatures, about 1200 °C, in order to maintain the correct structure and grain size which are necessary for good corrosion resistance. Nickel alloys are dominated by a group usually called by the trade name of Hastelloy. Similar alloys are manufactured under different trade names. See figure 6.16 for a list of the chemical composition of the commonest nickel alloys.

Hastelloy B has high corrosion resistance in strong hydrochloric and sulphuric acids. Unfortunately, the alloy is sensitive to high flow velocities and is rarely used in centrifugal pumps.

Hastelloy C is the most common of the nickel alloys because of its resistance in both oxidising and reducing atmospheres. It is used mainly for chloride solutions, liquids containing free active chlorine, bleaching fluids for example, of the type hypochlorite and chlorine dioxide solutions, where the chlorine content should not exceed 15 g/l at room temperature. It also has a good resistance to acids at temperatures below 70 °C.

Hastelloy D possesses good resistance to erosion, due to its great hardness, approximately 400 BHN. It has its best application in sulphuric acid, against which it has high resistance at all concentrations and high temperatures, and within certain limits for boiling acids also. Maximum corrosion resistance is achieved only after about six weeks in operation, which is the time it takes for a protective sulphate film to build up on the surface.

Monel is more resistant than nickel under reducing conditions and more resistant than copper under oxidising conditions. In summary, it can be said that monel is more resistant to corrosion than are its main components. Re-precipitation of dissolved copper cannot occur, since nickel, copper and monel lie very close to each other in the electrolytic series, galvanic series. Monel however, is somewhat difficult to cast and often results in porous castings. It is therefore used mainly for non pressurised components, such as impellers, or in wrought form for shafts. Monel has very good resistance to salt water especially at higher flow velocities. In stationary sea-water, biological growth may cause local concentrations of oxygen which lead to localised attacks.

6.2.11 Other metallic materials

A variety of metallic materials are available for special purpose use. Amongst these are:

Lead and lead alloys

Because of its poor mechanical properties, pure lead is not used as a material for pumps. A 10% to 25% addition of

Standard	Grade	Major constituents	Tensile strength min MPa	Max temp °C	Remarks
	Hastelloy B	28Mo 5Fe 1.2Co 0.5Cr	≈ 950	760	Cast, Very ductile
	Hastelloy C	17Mo 16.5Cr 5.7Fe 4.3W	≈ 490	980	Cast, Very ductile
	Hastelloy C	17Mo 16.5Cr 5.7Fe 4.3W	≈ 690	980	Wrought, Very ductile
	Alloy 20	36Fe 20Cr 3.5Cu 2.5Mo 2Mn 1Si	670		Cast, Very ductile
	Langalloy 7R	23Cr 6Mo 6Cu 5Fe 2W	420		Cast, acid resistant
BS 3076	NA 18	30Cu 2.7Al 1Fe 0.8Mn	1000		Fasteners
BS 3076	NA 14	15Cr 8Fe	830	600	Wrought
	Inconel 625	22Cr 9Mo 5Fe + No + Ta	725/900	≈ 800	Fasteners
	Inconel X750	15.5Cr 7Fe 2.5Ti	1130	≈ 650	Wrought
	Monel K-500	30Cu 2.7Al 1Fe 0.8Mn	900		Wrought
	Incoloy 800	32Ni 21Cr 1.5Mn	517/690	815	Wrought

Figure 6.16 Common nickel based materials

antimony, Sb, improves its strength characteristics considerably. The alloy is called hard lead. For higher pressure applications, a structural pressure retaining casing should be lined with lead. Lead has a very good resistance to acids, with the one exception of nitric acid. In sulphuric acid, lead sulphate is built up on the surface of the lead which is difficult to dissolve and acts as a protective film. Lead can be used for sulphuric acid at a concentration of 30% and temperature 100 °C, but temperatures have to be reduced at higher concentrations. In hydrochloric acid a film of lead chloride builds up, which protects the underlying metal to some extent. If the acid is in motion however, the surface layer wears away easily, so that lead pumps are not recommended for hydrochloric acid.

Titanium

Titanium has low density, 4500 kg/m^3, and very good tensile strength. Titanium alloy castings can have strengths up to 890 MPa, forgings up to 1200 MPa. Titanium is relatively difficult to work however. It is cast centrifugally in vacuum chambers in the presence of inert gas with graphite moulds. Titanium can be wrought and is used because of its low density for dynamic components such as valve plates.

The metal titanium is being used more and more in the chemical industry mostly because if its good corrosion resistance to solutions containing free active chlorine. Corrosion is completely excluded in such solutions as chlorine dioxide in water, hypochlorites, chlorates and metal chlorides.

Titanium can be a more economic solution, in some cases, than the proprietary super stainless steels listed in figure 6.13. Because of its very good strength to weight ratio titanium can be used to advantage in corrosive applications where weight is important.

Tantalum

Such exotic metals as tantalum, hafnium, niobium and germanium have lately come more into their own for modern industrial products, where very high corrosion resistance is required from the particular aspect of reliability. The metal tantalum has a corrosion resistance very nearly equal to that of glass. It is resistive to boiling sulphuric and hydrochloric acids. The density of tantalum is 16,600 kg/m^3. The ability to cast and weld tantalum is extremely limited, so that parts with complex shapes do not lend themselves to economic production. For rotodynamic pumps tantalum is generally used to provide an internal lining. Some reciprocating pumps can be easily manufactured solely from wrought products without any welding.

Tungsten carbide

Tungsten carbide is a sintered material originally produced as a high quality cutting tool material. Its hardness, and abrasion resistance, and improved manufacturing techniques have led to its adoption in several key areas of pump construction. Initially produced only with a cobalt binder but now available with nickel as an alternative. The main area of use is in seal faces for mechanical seals. It is also found as valve seats in reciprocating pumps operating on abrasive liquid-solid mixtures. Tungsten carbide can be sprayed as a surface coating, see section 6.2.13.

6.2.12 Non-metallic materials

Non-metallic materials are extremely important in pump construction. Most pumps utilise some non-metallic materials even though most components are metallic. Most pump designs embody gaskets and seals which rely on the very different properties of non-metallic materials. Non-metallic materials can be divided into five main categories:

- Thermoplastics
- Thermosetting plastics
- Elastomers
- Ceramics
- Minerals

Thermoplastics, thermosetting plastics and elastomers are often confused; in some cases the physical properties are very similar and all three groups of materials are used for the same function. Gaskets, seals and 'O' rings are examples of components which are produced in all three groups of material. These materials also tend to absorb liquids when in contact; this property must be considered during the design and application stages. Ageing, and the effects of ultraviolet radiation must also be studied. Some materials are only suitable for relatively low temperatures, up to 70 °C for example. Care must be exercised to ensure that no operating conditions occur outside the temperature range.

Plastics can be filled with various materials, either for bulk or to improve the strength. Glass, carbon and Kevlar fibres are used for strengthening.

Thermal expansion, especially in the axial direction of the plastic pump, has to be taken into account when routing the pipework. Similarly, the pump must be relieved of pipework forces and moments to the fullest possible extent.

Many companies have registered trade names for materials. The same material may be known by many different names depending upon the company which produced it. Just as with metals, some materials may be slightly different and possess remarkably different physical and chemical properties. Great care should be exercised when specifying specific materials.

Thermoplastics

The following popular materials are thermoplastics:

Acetal

Acrylic

acrylonitrile butadiene styrene (ABS)

aromatic nylons (ARAMID)

fluorocarbons

fluorinated ethylene propylene (FEP)

Nylon

polyester ether ether ketone (PEEK)

polyacetal

polycarbonate

polychlorotrifluoroethylene (PCTFE)

polyether

polyethylene

polyethylene chlorotrifluoroethylene (PECTFE)

polyethylene terephthalate (PETP)

polyethylene tetrafluorethylene (PETFE)

polyfluoroalkoxytetrafluoroethylene (PFA)

polyimides

polymethyl pentene (PMP)

polymide

polypropylene

polysulphone

polytretrafluoroethylene (PTFE)

polytrichloroethylene (PTFCE)

polyvinylfluoride (PVF)

polyvinylidene fluoride (PVDF)

polyphenylene oxide PPO

polyvinyl chloride (PVC)

Thermoplastics are characterized by their ability to be repeatedly softened by heating and then hardened by cooling; their shape can be changed many times. Notice that PTFE is a thermoplastic and not an elastomer, even though it is used in many applications where elastomeric properties are required.

A sub group within thermoplastics are the thermoplastic elastomers. These are man made materials which have rubber-like properties with enhanced chemical resistance. The thermoplastic elastomers can have their chemical compositions adjusted to bestow particular properties to the material.

Some thermoplastics have excellent impact, abrasion and tear resistance. Compounds of this type would be used for internal liners on some solids handling applications.

The fluorocarbons are a sub group containing FEP, PCTFE, PECTFE, PETFE, PFA, PTFE and PVDF.

Acetal is a high melting point plastic with high strength and rigidity. Acetal can be extruded, cast and moulded. It is used for a wide variety of purposes including; impellers, casings, bearing housings and bearings.

Aromatic nylons are usually formed as fibres, ARAMID®, KEVLAR® or NOMEX®. These fibres are used to reinforce and strengthen moulded or cast materials. The fibres can also be used for braiding in the case of soft packing.

Popular thermoplastic materials are filled or alloyed with other materials to improve specific properties. Nylon is a favoured bearing material; it can be reinforced with glass to increase strength and stiffness; it can be alloyed with molybdenum disulphide also to improve mechanical properties, increase high temperature capabilities and improve bearing performance.

PTFE is probably the most well known thermoplastic after Nylon. PTFE has many trade names, TEFLON® for example. Renown for it's corrosion resistance and chemical inertness, PTFE has many uses in pumps. Pump casings, impellers, seal faces, gaskets and 'O' rings. It's wide temperature range, -250 to 250 °C, make it suitable for a wide range of applications. Alloying, or sintering, with other compounds can extend the properties and applications of PTFE. Typical additions include; glass fibre, bronze, carbon, graphite, molybdenum disulphide and stainless steel. PTFE is also machined for the construction of pistons, bellows and other components which need to be corrosion resistant. PTFE has the advantage of having a very low coefficient of friction. PTFE is generally not recommended for pumps designed for handling liquids containing abrasive solid particles.

PEEK is attracting attention because of its chemical resistance and physical properties. Some manufacturers are using it for wear rings, stuffing box bushings and valve plates.

Thermoplastics such as polyvinyl chloride, polypropylene, polyethylene and PTFE are used as blow-moulded parts and also in sheet or tubular form for coating.

Thermosetting plastics

Thermosetting plastics are formed to the final shape during manufacture and the shape remains fixed until the material is destroyed. The following materials are typical thermosetting plastics:

Alkyd

Amino

epoxides, epoxy resins

furanes

Melamine

phenolic resins

polyester

polyurethane

silicones

urea

Just like thermoplastics, thermosetting plastics can be filled or reinforced with other materials. Asbestos, and other mineral fibres, together with cotton are used. Selected parts of components can be treated. Some surfaces can be inoculated with PTFE or graphite to reduce friction and wear.

Pump bodies and impellers are produced in GRP, glass reinforced plastic. A matrix of glass fibres is coated with epoxy resin. A chemical reaction between the two components of the resin, filler and hardener, causes the mixture to solidify. Many resin formulations are available including casting versions.

Elastomers

Some elastomers occur naturally, natural rubber is the most obvious, others are manufactured for specific physical and chemical properties. The most popular elastomers are listed below:

ethylene propylene (EP)(EPDM)(EPT)

fluoroelastomer (FPM)

natural rubber (NR)

NEOPRENE® (CR)

nitrile rubber (NBR)

perfluoroelastomer (FFKM)

silicone (Si)

styrene butadiene rubber (SBR)

urethane (AU)(EU)

Natural rubber is compounded with fillers and other materials and then vulcanised, heated with sulphur compounds, to produce a solid elastic material for processing. The hardness and chemical resistance being defined by the vulcanisation process. Modern materials have been developed for increased resistance to oils, sunlight and high temperatures.

Modern compounds tend to be known by the registered trade name of the manufacturing company. Attempts have been made to standardise the nomenclature of generic groupings, ISO and ASTM abbreviations are used. The list below indicates the general progression to higher temperature operation and increased chemical resistance. Each individual case must be evaluated on its own merits.

NR, NBR, SBR Perbunan®, Buna N®, Buna S®

EPDM Dutral®, Vistalon®

FPM Viton®, Fluorel®

PTFE (thermoplastic) Teflon®, Fluon®

FFKM Kalrez®, Chemraz®

Some of the latest man-made compounds are very costly. Elastomer seals and gaskets can sometimes be replaced by metal parts. Silver plated stainless steel 'O' rings are much more cost effective than exotic elastomer compounds.

Ceramics

In pumps, ceramics are used in three forms, cast sintered and sprayed, see section 6.2.13 for sprayed applications. Cast ceramic is manufactured just like pottery. A wet clay mix is poured into moulds and then fired. The resulting component is very hard and almost completely chemically inert. Industrial ceramics are not glazed like pottery and some grades exhibit slight porosity; this can be remedied by suitable sealants. Some ceramic compounds are sintered from granules to produce the desired shape. Machining to finished sizes is by grinding and lapping.

Silicon carbide, silicon nitride and alumina, aluminium oxide, are the commonest ceramics. In rotodynamic pumps, silicon carbide is alloyed with silicon and carbon to produce mechanical seal faces and bearings in magnetic drive and canned pumps. In reciprocating pumps, alumina is used for plungers, bushings and ball valves.

Minerals

Asbestos used to be one of the most widely used minerals; an excellent gasket material for high temperature applications. Due its unacceptable health hazards, asbestos is rarely used in its free state. Asbestos is used as a filler or reinforcing in some grades of thermosetting plastics and elastomers. Typical applications include dry running and product lubricated plain bearings.

Carbon and graphite, because of their resistance to chemicals and because of their low coefficients of friction are used for mechanical seal components and product lubricated bearings.

6.2.13 Coatings

Various pump components are coated to achieve different objectives. Coatings are applied to achieve corrosion/abrasion/wear resistance and running compatibility. Coatings can be applied by various techniques; brushed, dipped, physical vapour deposition, sprayed, weld deposit, vacuum fused and electroplating.

Like elastomers, coatings are best known by their trade names rather than by chemical composition or standard designations. Metallic, non-metallic and ceramic compounds are used for different applications. Figure 6.17 lists sprayed coatings and indicates the major constituent.

Trade name	Major constituent
Stellite 6	cobalt
Stellite 12	cobalt
Colmonoy 5	nickel
Colmonoy 6	nickel
Metco 130SF	aluminium oxide/titanium dioxide
Colmonox	chrome oxide (dipped)
UCAR LW-15	tungsten carbide

Figure 6.17 Common coating materials used in pumps

Colmonox is a dipped coating which is built up in several layers. Chemical treatment between dip cycles eliminates any porosity.

Pure elements, such as nickel and chromium, are deposited electrolessly for corrosion and wear resistance. Titanium nitride is applied by physical vapour deposition in a vacuum chamber. A very thin hard coating is applied to finished components without dimensional change. Thermoplastic materials can be applied to metal components to provide corrosion resistance and compatible bearing surfaces.

Coatings, such as epoxy resins, can be applied by brush. An extension of this idea is to incorporate solids within the resin. Glass beads have been used with great success.

Coatings are used extensively on new components such as:

shafts,

sleeves,

wear rings,

balance drums,

bushings,

plungers,

piston rods,

valve seats.

Coatings can also be used to reclaim damaged or worn components. Coatings with glass beads have been used to repair pump casings. Worn shaft bearing lands can be reclaimed with hard chrome plate. Special nickel alloys are available to coat cast iron where wear has occurred.

6.3 Material strength and integrity

There is no European or International standard for hazards caused by liquids. National standards exist, see section 3.7, but there is little agreement or correlation of test methods and permissible doses. In Germany, for example, it is not permissible to pump petrol using cast iron pressure containing components. In France all road petrol tankers have cast iron pumps. In general, most proprietary specifications do not allow the use of cast iron for flammable liquids. The strength of material used related to the working pressure and the level of integrity required for the material are both related to the perceived hazards and the cost of failure.

Ratings of pumps follow pipe flange ratings. ISO 7005 has flanges which are suitable for metric and inch sizes. The following nominal pressure ratings, barg, are covered:

2.5

6

10

16

20

25

40

50

110

150

260

420

We have seen from figures 6.2 and 6.3 that the strength of materials is dependant upon temperature. In ISO 7005, PN 50 refers to a set of dimensions, not a guaranteed pressure rating. With some specified materials, PN 50 flanges are only suitable for 42.5 barg at room temperature. PN 420 flanges are suitable for 431 barg in certain materials at room temperature. Some users and manufacturers will think 431 barg a very high pressure of only academic interest. This is not so. Oilfield applications regularly operate at pressures over 400 barg and standard reciprocating pumps go to 1000 barg.

High pressure pumps require high strength materials. If a pump casing was infinitely thick, the highest tensile stress would be in the bore and equal to the liquid pressure. If cast iron was used, the theoretical maximum allowable pressure, not considering fatigue, would be about 200 barg. As the wall thickness was reduced to economical proportions the maximum stress would increase. Figure 6.18 shows the stress values for a pump casing of 250 mm bore subjected to an internal pressure of 200 barg.

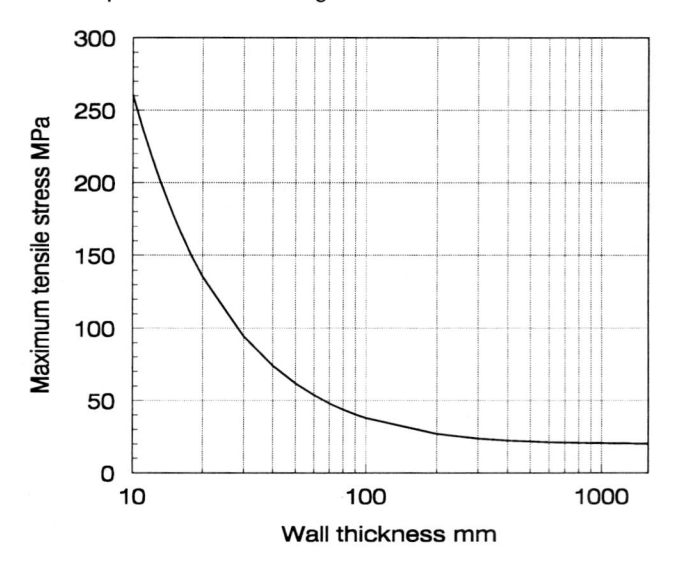

Figure 6.18 Casing tensile stress for various wall thicknesses

If it was necessary to only have a wall thickness of 10 mm then a much stronger material would be required, 13Cr alloy steel for example. If the casing was to be cast in 18/8/2 austenitic stainless steel, then the minimum wall thickness would be about 17 mm.

The design and manufacturing processes are complicated by the vast range of materials available. Manufacture at constant thickness and vary the pressure rating; manufacture at various thicknesses for constant pressure rating. Obviously the cost of the material plays an important role as does the batch size.

For any given application the pump designer must ensure that the pressure retaining materials are thick enough and strong enough to withstand the loads and stresses imposed on the pump. The design process is further complicated by the imposition of corrosion allowances. The pump user may require material to be added in the bore to allow for corrosion and erosion, 3 mm is a typical value. When the pump is new, it is stronger than necessary. In the corroded condition, the casing has reduced to the design thickness.

Integrity of pressure retaining parts is checked during production by hydrotesting; pressure testing with water. Castings are only machined sufficiently to provide good seals and then hydrotested. This first test is used to find casting weaknesses and porosity. The length of time a component must be held under pressure is a function of the material thickness. Thin casting can be checked in 10 minutes, heavier castings are usually tested for 30 minutes. Castings which fail this first test are evaluated to see if reclamation is cost effective. Depending upon the material, thickness, position and extent of fault, the casting will be reclaimed or scrapped. For mass produced standard pumps, the pump user has no control over this aspect of production. The pump manufacturer will have standard procedures to be adopted for various failure modes. All repairs implemented will be compatible with the published range of operations for that product. Repair techniques may include welding, impregnating, caulking, plating or plugging. Purchasers of standard products accept the manufacturer's warranty of suitability for service.

The position with pumps built specifically to order is different. The purchaser may chose to limit the type of repair techniques used and also limit the extent of repairs. Pumps for hazardous liquids are generally not repaired by impregnating, caulking or plating. Plugging may be allowable if the thread in the bore can be seal welded. The purchaser may define repairs as 'minor' and 'major', depending upon the position and nature of the fault and the extent. Minor faults may be repaired by agreed techniques, the purchaser is informed of the location but production is not impeded. The agreed techniques include:

- The inspection methods used to determine the extent of the fault and the qualifications of the operators.
- The repair techniques to be used, approved procedures, including post repair heat treatment, and qualifications of staff.
- Post repair inspection to prove the integrity of the repair.

When a major fault is detected production stops, the purchaser is informed and nothing happens until the fault has been inspected by a representative of the purchaser. Then discussions between the two parties decide if, and how, the fault can be repaired. One important part of any such discussion is the effect on delivery.

Castings will be hydrotested again in the fully machined condition, usually when the pump is fully assembled. This final pressure test assures that all material is sound and that bolted and screwed joints do not leak. Mechanical seals may or may not be fitted as agreed. Leakage through mechanical seals may or may not be permissible. Leakage through soft packed stuffing boxes must be permissible.

We have discussed castings here but the same principles apply to fabrications and forgings. Forgings, however, should only be repaired with great reluctance, a 'major' repair, and never in highly stressed regions. Cosmetic repairs can be made in any low stress zone which is not in contact with the process liquid.

The pressure at which hydrotests are conducted is subject to agreement and depends upon the design and construction and operating temperature. Small pumps tend to have suction and discharge flanges with the same pressure rating; the complete casing may be tested at one pressure. Larger pumps may have a low pressure suction section and a high pressure discharge section; the casing would be divided by a suitable diaphragm and tested at two pressures. It is not unusual for pumps to be fitted with high pressure suction connections when the suction pressure is low. The user and the manufacturer must agree whether the hydrotest pressure will be based on the suction pressure or the flange rating.

Some flange specifications, ISO 7005 and ANSI B16.5 for example, specify hydrotest pressures as part of the flange specification. In general the hydrotest pressure is 1.5 times the rated pressure. Components with standard flanges attached cannot be hydrotested at higher pressures. Other components, mechanical seals say, may limit the hydrotest pressure to a much lower value. Hydrotest pressures and pressure ratings must be clarified during the quotation stage. Individual components may be hydrotested at higher pressures than the assembled pump. There is no theoretical basis for having lower test pressures as the rated pressure increases. If the perceived hazard is as great, or greater because of the higher pressure producing increased leakage through similar faults, the safety margins should be maintained across the pressure range.

Mass produced pumps may not be hydrotested in the final assembled condition; clarification from the manufacturer should be sort if problems with leaks occur at site.

Integrity can be interpreted in another manner as well as material integrity, design integrity. Design integrity, or proof testing, is when the manufacturer shows the basic design and construction to be sound. This problem arises particularly when corrosion allowances are added to pump casings. The hydrotest proves the new casting is good, it does not shown how the casting will react when it is 3 mm thinner. Discussions are taking place regarding a new European standard to specify methods of design proving. It is probable that calculation and physical testing will be allowed. The manufacturer would show by calculation, computer model or finite element analysis, that stresses and strains in the 'thin' component would not impair its ability for pressure containment and other essential functions. Alternatively, the new 'thick' component could be subjected to higher test pressures to create the stresses and/or strains which would exist in the 'thin' component. For prototype pumps, and extensively modified standard pumps, discussions during the quotation stage should clarify the need, or otherwise, of design proving. Consideration must also be given to standard and prototype pumps which are designed and constructed to long established rules. There would be no point in forcing a manufacturer to produce a finite element analysis of a pump which had been operating successfully for twenty years.

6.4 Corrosion and erosion

6.4.1 Liquid corrosion of metals

The corrosion of metals in liquids is largely an electrochemical process, i.e. there is a positive electrode, the anode, and a negative cathode, between which an electric current will flow. If a ferrous component is immersed in acidic water, the anode-cathode process will occur over its entire surface. The cathode in this case comprises local elements of more noble metals in the galvanic series segregated in the alloying process, or non-metallic inclusions or by local reaction with oxides, sulphides and so on. Metal is dissolved by the generation of metal ions at the anode. Oxygen dissolved in the liquid is reduced by the generation of hydroxyl ions, OH$^-$, at the cathode, or in the absence of oxygen, hydrogen ions, H$^+$, are used up in the generation of hydrogen gas. Where metallic and hydroxyl ions come into contact, corrosion products are generated, usually in several stages of chemical reaction. The final corrosion product usually comprises metal oxides, which are difficult to dissolve, or metal hydroxides. Corrosion products generate a more or less dense film on the surface.

The incidence of protective films resulting from corrosion, or those generated in other ways, are necessary as a rule for most metals in use in order that further corrosion should be inhibited, or contained within reasonable limits. Only the most noble metals, like gold and platinum, are immune within the greater part of the stable region of water.

Protective films can be removed by fast-flowing liquid. If a metal surface has a protective layer of corrosion products, this may be one of two types.

Porous, relatively thick films

Examples are the film on copper alloys and graphite on cast iron. Thick protective films often have complex constructions, because contaminants in the liquid, such as lime, may be included as a significant component. It is for this reason, difficult to predict whether any effective film will be generated at all. At speeds normally found in centrifugal pumps, these films often have a limit, which is a step function, above which they get carried away by the flow of liquid. Thick films are relatively sensitive to solid particles, cavitation and even air bubbles in the flowing liquid. Even if such disturbance is sporadic, the damage may be great because the films take a relatively long time to regenerate, days or weeks even.

Dense, thin films

Examples are those passive films normally found on stainless steel, titanium and so on. The very dense and thin passive films usually consist of one single metal oxide only, for example chromic oxide on stainless steel. These films have very good adhesive qualities and are not removed even by very high liquid velocities. On the contrary, a flowing liquid tends to keep the film intact by preventing crevice corrosion and pitting. It is only when the flowing liquid contains solid particles or when flow conditions are such that cavitation occurs that a passive film will be damaged. Passive films usually regenerate quickly after damage and occasional penetrations by solid particles or cavitation do not usually lead to further corrosion.

If, for any reason, a passive film does not exist on stainless steel, titanium and so on, then heavy corrosion attack can be expected.

Chemical reactions increase exponentially with temperature. As temperature increases, however, secondary effects can come into play, resulting in reduced corrosion. The oxygen content of water at atmospheric pressure decreases as temperature increases, thus reducing the corrosive effect on, for example, iron. If there is positive pressure, dissolved gases such as O_2 and H_2 will be present in the water even at high temperatures and will still influence corrosion.

When a centrifugal pump operates against closed valve, the absorbed motor power will be converted to heat in the pump, which rapidly causes high temperatures. Especially in the case of corrosive liquids, this can lead to heavy attacks because a passive film can be destroyed. Heating can also mean mechanical damage resulting in deteriorating material strength and properties. Plastic pumps are especially prone to softening with heat and should therefore be fitted with temperature sensors. Metallic pumps often use elastomer gaskets; overheating can result in leaks. When considering chemical attack and temperature ratings all pump components must be evaluated.

6.4.2 Corrosion rate

For a material to be considered in practice as being completely resistive, the material lost from the surface should not exceed about 0.1 mm per year. If the material losses are of the order of magnitude of 1 mm per year, the material can still be used but must obviously be checked at regular intervals.

Other chemical reactions occur at the cathode with certain other liquid-metal combinations. The chemical rule of thumb that a reaction speed doubles with a temperature increase of 10 °C also applies to the corrosion process. Corrosion rates are expressed either as penetration depths or loss of weight per unit area per unit time. The conversion factors for some commonly used units are tabulated in figure 6.19.

Unit	Conversion factors		
	g/m^2/h	g/m^2/day	mm/year
1 g/m^2/h	1	24.0	1.12
1 mg/dm^2/day	0.004	0.10	0.0046
1 mg/cm^2/day	0.417	10.00	0.465
1 mm/year	0.890	21.4	1
1 mm/month	10.800	258.0	12
1 inch/year	23.0	554.0	25.4
1 mil/year	0.023	0.554	0.0254
1 oz/ft^2/year	0.035	0.84	0.0394

Figure 6.19 Conversion of various units for corrosion rate — 1 mil = 0.001 inch. For depth of penetration, the conversion applies only to steel or substances of the same density

6.4.3 Types of corrosion

Units for rate of corrosion assume an even attrition over the whole surface, i.e. uniform corrosion. This often does not happen and other forms of corrosion can occur:

- Electrolytic (galvanic) corrosion, which occurs when different metals come into contact in the presence of a conducting medium.

- Crevice corrosion, which occurs in crevices and small confined spaces, where quantities of liquid may become more or less isolated from the main body of the liquid.

- Pitting, meaning that the corrosion is concentrated at certain points, pits.

- Intergranular or intercrystalline corrosion usually associated with redistribution of alloying elements during heat treatment or welding.

- Selective corrosion, where one single alloying element is dissolved out, e.g. dezincification in brass.

- Erosion-corrosion, which occurs in flowing media and is the commonest complication in pumps, see also section 6.4.3.2.
- Stress-corrosion, where a combination of mechanical stress and attack by the liquid may lead to fracture.

Crevice corrosion and pitting are related phenomena, and are most troublesome for stainless steel in a chloride containing environment. These attacks occur in crevices or holes where the conditions for setting up a protective film are absent because of screening from the surrounding solution. Special care must be taken in pumps to prevent crevice corrosion at rotor fixings or on shafts rotating within seals. The risk of crevice corrosion is reduced in flowing liquids. Crevice corrosion will not usually occur so long as the pump is working. During extended stationary intervals in chloride-containing liquids, however, the pump should be emptied and flushed out with clean water.

Corrosion fatigue is also of interest in pumps and especially for shafts. Even seemingly insignificant corrosion may reduce the fatigue strength of a material by a factor of 3 to 10.

6.4.3.1 Galvanic corrosion

Where there is a combination of different metals, galvanic corrosion may occur. This means that the less noble metal, the anode, has a higher corrosion rate than the more noble metal, the cathode, which is protected. An indication of the risks of galvanic corrosion can be obtained by studying the galvanic or electrolytic series, see figure 6.20. In practice, metals with a potential difference of 0.2V can be connected without any trouble. Galvanic corrosion is of great significance in pumps be cause of the principle of using different materials for various parts, see section 6.7.2. However, it is apparent that the effects of erosion-corrosion motivate relatively wide departures from good practice from the galvanic corrosion viewpoint.

CORRODED END - ANODE (Least noble)	Electrode potential V
Magnesium alloys	- 1.51
Zinc	- 1.09
Cadmium	- 0.73
Aluminium alloys	- 0.75 to - 0.64
Aluminium alloys, cast	- 0.70
Iron and steel	- 0.55
Stainless steel, all types, active	- 0.53
Lead-tin solder	- 0.52
lead	- 0.51
Tin	- 0.47
Nickel	- 0.25
Inconel	- 0.25
Titanium active	- 0.25
Brass	- 0.22
Copper	- 0.22
Bronze	- 0.22
Nickel-copper alloys	- 0.20
Stainless steel, all types, passive	- 0.15
Monel	- 0.10
Silver solder	- 0.07
Silver	- 0.05
Gold	+ 0.18
Titanium, passive	+ 0.19
Platinum	+ 0.33
PROTECTED END - CATHODE (Most noble)	

Figure 6.20 Galvanic series for metals with potential values measured in sea-water (source, S. Bartha)

Where the anode surfaces are large and cathode surface is small, in the case of pump casings and impellers, for example, the potential differences for certain favourable combinations can be twice as much. Figure 6.21 shows a drastic case of galvanic corrosion.

The shaft had been lying in stock fitted with graphite containing soft packing, the corrosion being caused by condensation from humid atmosphere. The attack has been accelerated by crevice corrosion.

Figure 6.21 Galvanic corrosion on 13Cr steel pump shaft

6.4.3.2 Erosion corrosion

Erosion-corrosion is particularly common in rotodynamic pumps where the flow velocity nearly always exceeds 20 m/s. Characteristic of erosion-corrosion are the horseshoe-shaped patterns scavenged in the direction of flow and bright surfaces free of corrosion products, see figure 6.22.

Figure 6.22 Erosion-corrosion on a 13Cr + Mo steel impeller used for sulphate liquor. Note that the rivets are completely intact. These were made of SIS 2343. For this application the impeller should have been made of SIS 2343 or SIS 2324

When metal comes under attack, oxidisation media O_2 or H^+ for example, are consumed at the surface of the metal and metal ions are generated. Thus the concentration of metal ions will be greater and the concentration of oxidisation media will be lower at the metal surface than in the main body of the liquid, see figure 6.23. The difference in concentration at the metal surfaces is dependent upon the difficulties of transportation of the participating compounds which is necessary for the continuation of the corrosion process.

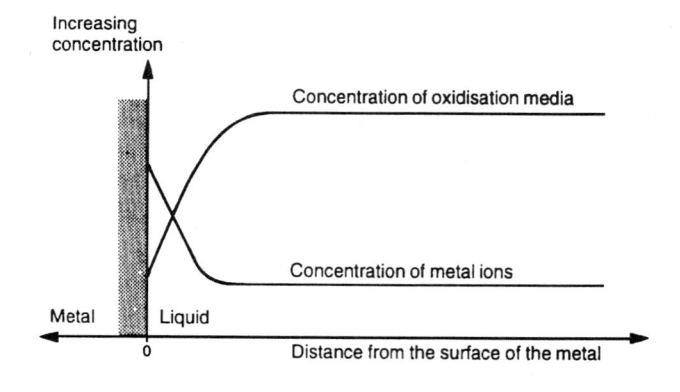

Figure 6.23 Concentration differences of general corrosion at a metal surface

Obviously the transport will be aided and, at the same time, the rate of corrosion will be accelerated if the liquid is in motion, which is generally the case in pumps, see figure 6.27. Care should thus be taken when using corrosion data from tests in stationary liquids for the selection of pump materials, since corrosive attack inside a pump may be ten times worse than it is in a still liquid, see also figure 6.24.

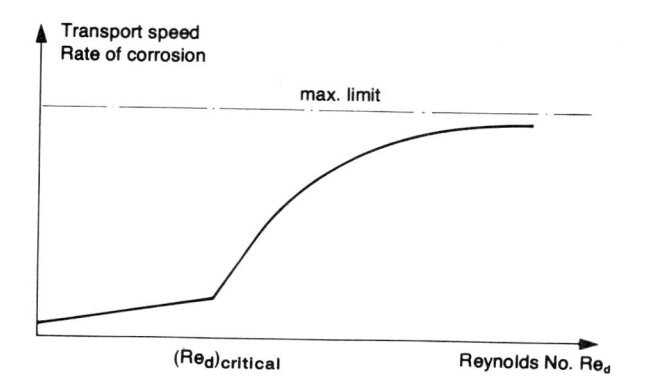

Figure 6.24 The principle of increased rate of general corrosion with liquid velocity. At (Re_d)critical, the change over from laminar to turbulent flow occurs

6.4.4 Corrosion testing

There is a choice of a number of methods of corrosion testing in flowing liquids:

- pipe flow
- rotating cylinder, freely rotating or enclosed
- rotating disc
- spraying a sample piece with a liquid spray.

If the purpose of the test is to determine the probable corrosion rate then the test rig should as far as possible resemble the condition intended to be tested. This is so that the test shall take place at the same Reynolds Number, i.e. similar flow. Otherwise the result will only be a comparison between different materials.

When the corrosion test is intended for pumps, it would be advantageous to use an enclosed rotary disc at a speed such that the Reynolds Number $Re_d = u_2 d_2 / \mu$ is about the same as that for the pump impeller. The units in the Reynolds Number formula are:

u_2 = impeller peripheral velocity (m/s)
d_2 = impeller tip diameter (m)
μ = liquid viscosity (m²/s)

Instead of rotating a complete disc of test material, small electrically connected samples can be set into the fixed wall opposite the disc, see figure 6.25, which are then exposed to the same velocity as the rotating disc. This positioning of the sample pieces has the advantage that electrical measurements can be carried out simply, see figure 6.18. Beside which, surfaces subjected to various liquid velocities can at least be of the same size. In a solid rotating disc, the larger part of the surface area is subjected to the highest liquid velocities, whereas in the pump the opposite is the case, because of the large surface area of the pump casing.

The test arrangement shown in figure 6.25 and figure 6.26 will resemble the corrosion process in a pump, since all the relevant liquid velocities are present simultaneously, and because redistribution of metal losses due to galvanic currents will occur between parts of the same metal subjected to various liquid velocities.

The galvanic currents are among the items measured in corrosion testing, apart from the usual weighings, potential measurements and so on. If metal corrodes, there are actual electrolytic currents flowing between parts of the same metal surface subjected to differing liquid velocities. Thus, by recording the currents, it is possible instantaneously to check the occurrence of corrosion, and at the same time qualitatively to follow changes in the rate of corrosion, build-up of protective film of corrosion products and so on. Unintentional changes in the rate of corrosion can also easily be recorded.

Materials generating thick protective films require relatively long test times, and worthwhile laboratory tests can be difficult to carry out, since it is not easy to maintain control over the composition of the liquid over long periods of time. In these cases it is better to move the laboratory to the liquid rather than the other way round. When there are decidedly passive films generated normally on the test samples by contact with the air, it is best to remove this film, i.e. activate, in order to establish

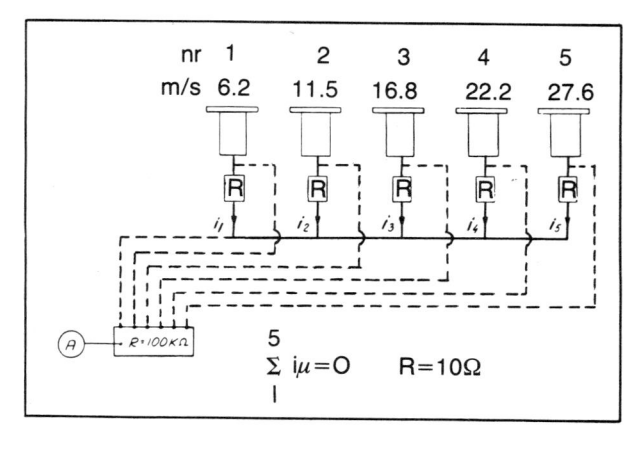

Figure 6.26 Electrical connection of test pieces subjected to various liquid velocities   A = amplifier and recorder

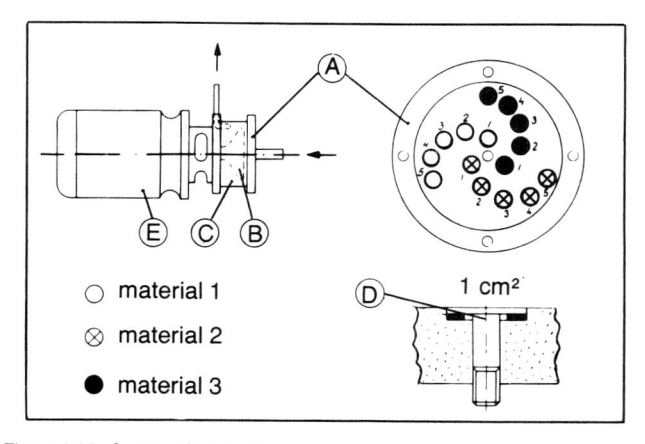

Figure 6.25 Corromatic test set
A - test plate   B - rotor   C - test chamber
D - test piece with fixing   E - speed regulated motor

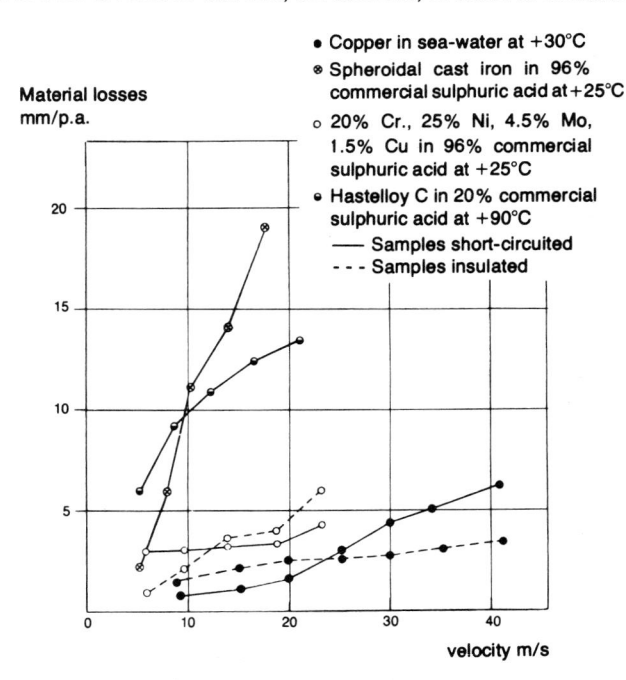

Figure 6.27 Onset of corrosion as a function of flow velocity

whether the passivity is restored in the actual liquid. After activation, passivity will be restored after only a short time, within hours, if it is going to be restored at all. If the general corrosion is uniform with time after a certain stabilising period, this is confirmed by the fact that the electrolytic currents also are uniform. The test time is then regulated so that stable conditions apply during as great a part of the time as possible.

Figure 6.27 shows some examples of test results using the equipment in figure 6.25.

6.5 Abrasion resistant materials

6.5.1 Pump selection

Liquids containing solid particles can present considerable problems as regards materials. To some extent the problems can be eased by choosing a pump design suitable for the abrasive conditions or reducing the velocities involved. Surfaces subject to wear can also be fitted with replaceable liners or designed with a good margin for wear. Examples of this are those pumps which are used for transporting abrasive solid materials such as ash, sand, asphalt, etc. In these pumps, not only are the wear margins large but even the pump shaft systems are designed to cope with the imbalance resulting from such wear.

Two main groups of material are used for abrasive liquids, hard metals and various types of rubber cladding. Wherever rubber cladding can be used, with regard to particle size and liquid temperature, it is superior to the hard metal materials. Hard metals can be used as coatings over normal metals. In this respect, hard coatings can be replaced to extend the lives of worn components. Ceramics are also very good but, so far, plastics have not given very good results.

Wear increases greatly with flow velocity for all qualities of material. Wear increases to a power between 2.5 and 5 with velocity. This relationship infers that wear for a centrifugal pump with differential head H increases in the proportion $H^{2.5/2}$ to $H^{5/2}$. Delivery head per pump stage is limited therefore and is a maximum of about 50 m for rubber-clad pumps and about 100 m for pumps made of wear-resistant metals. Positive displacement pumps can produce pressure without recourse to high velocities. Positive displacement pumps are ideally suited for liquids with high concentrations of solids because these liquids react with an apparent high viscosity. High viscosity reduces efficiency of rotodynamic pumps.

6.5.2 Hard metallic materials

The following applies to hard materials, mainly steel with a high carbon and chromium content:

- Material losses increase greatly when the hardness of the particles is greater than the material in the pump or liner.
- Material losses increase with the size and mass of the particles.
- Material losses increase if the particles have sharp edges and the particle material has sufficient strength to maintain the edges.
- Material losses increase as the particle concentration increases.
- Wear-resistance is improved by increasing the hardness of the pump material, but this effect can only be significant if the pump material hardness figure exceeds about 300 BHN.
- Combination of general hardness in the pump material with wear-resistance in individual elements in the pump material is a good thing. Examples of this are Ni-Hard, a martensitic chromium steel, where the wear-resistance is obtained from chromium carbide granules embedded in a hard matrix.

6.5.3 Rubber cladding

The wear-resistance of rubber cladding is due to the fact that solid particles more or less bounce on the rubber surface without actually damaging it. The following comments apply to rubber claddings:

- Soft rubber is better than hard rubber.
- Rubber gives favourable results if the particles strike the surface at right-angles, angle of incidence 90°, or if they move along the surface of the rubber. If the angle of incidence lies between 5° and 30° the results will be much less favourable.
- The thickness of rubber should be 2 or 3 times the size of the particle in order to exploit the bouncing effect.
- Sharp particles can cut the surface of the rubber to pieces.

Natural rubber is suitable for temperatures up to 70 °C, whilst nitrile rubber can be more suitable when mineral oil contaminants are present, although it is not so wear-resistant. Urethane rubber has become popular for parts of pumps used for abrasive pumping for temperatures up to about 50 °C, see figure 6.28. Urethane rubber has excellent wear properties and good resistance to oils, but polyester-based compounds are sensitive to hydrolysis. In this respect, polyester based urethane rubber, Adiprene® is better. Typically, urethane rubbers are considerably higher in price than natural rubber; however, this is compensated for with a simpler manufacturing process by forming in open moulds.

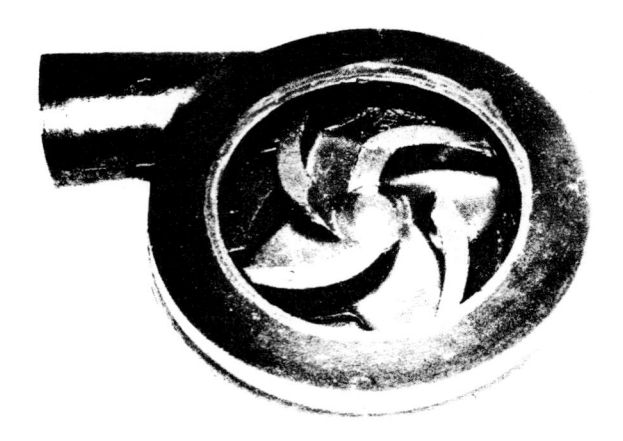

Figure 6.28 Pump impeller made of urethane coated steel and pump casing of solid urethane rubber

Glass beads set in resin have been applied to various metal casing parts to protect against corrosion and erosion.

6.6 Materials resistant to cavitation damage

When cavitation occurs inside a pump, the vapour bubbles implode resulting in local pressure waves up to about 10000 MPa. Implosions close to a metal surface in a pump impose repeated loads on the surface with the result that fragments of material are loosened. Cavitation damage thus has a typical appearance, see figure 6.29. The surface becomes rough and pitted in appearance, and is limited to those zones in a pump where low pressure can occur.

In reciprocating pumps the continuous action of the repeated implosion loads cause fatigue failures in important components. The classic failures include:

suction valve plates,

liquid end studs,

cylinders.

If cavitation damage has been caused in a pump, the first thing to do is to improve the NPSHa of the installation by:

Figure 6.29 Pump impeller exhibiting cavitation damage

increasing the available suction head or pressure or reducing the suction lift;

lowering the temperature of the liquid in the pump intake either by direct or indirect cooling.

If neither of these actions are feasible the pump speed should be reduced and consequently production. If this is not feasible or if the cavitation process is purposely used for regulation, as is the case for certain condensate pumps, then cavitation-resistant material must be used. Choice can be made by studying the results of comprehensive experiments, see figure 6.30. As shown in this figure, it takes a certain amount of time before cavitation damage appears. The process is similar to the incubation time for some infectious diseases.

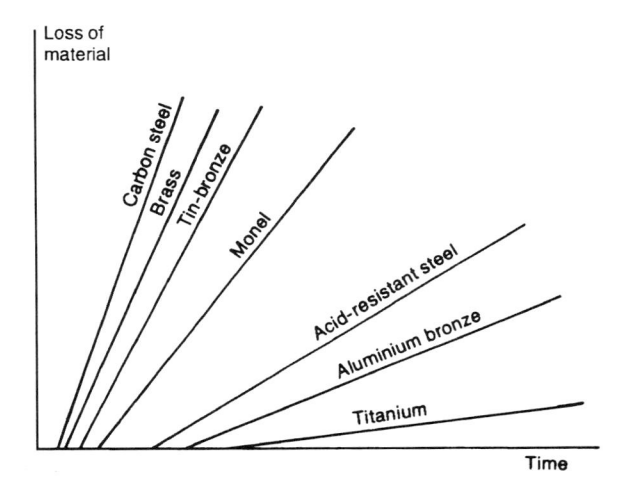

Figure 6.30 The resistance of various materials to cavitation damage

6.7 Material selection

6.7.1 Basic information

A review of preliminary material choice can be found partly in Section 1.2 and partly in Section 6.7.3. This information enables the selection of materials suitable for a wide range of liquids. In more specific cases, it is recommended that a study should be made of special literature such as that referred to in Section 6.8. It is, however, inevitable that even this background will not be adequate in many special cases, so that judgement has to be made based on practical experience of similar cases. The information required for choice of material has been collected in figure 6.31 in the form of a check list. It is the users/purchasers responsibility to provide all relevant data to allow pump type and material selection to be accomplished. Successful pump operation depends upon the initial decisions made. Decisions based on incorrect or incomplete data will tend to be suspect.

Liquid
Full description

trade name, chemical formula
      concentration
    hazardous/flammable/toxic
    controlled by legislation
    abrasive/corrosive
    all constituents including any trace elements
    vapour pressure
    viscosity
    viscosity characteristic with shear
    compressibility
    solidify temperature
    crystallisation temperature
    SG/Cp/pH

Liquid contaminants

dissolved/entrained gas
      release pressure
    solids
      size/concentration distribution
      hardness
      rigid/deformable/friable
      abrasiveness (Miller Number)

Operating conditions

max/min temperature
    max rate of temperature change
    max/min pressures
    max/min flow
    respective NPSHa's
    continuous or intermittent
    time between routine maintenance
    stand-by cycles
      full of product/flushed
      hazards during stand-by
    dry running

Special requirements

hygienic
    sterile
    allowable contamination of liquid
    cleaning of pump
      chemical/steam

Environment
Site conditions

indoor/outdoor/onshore/offshore
    attended/unattended operation
    min/max ambient temperature
    max black bulb temperature
    min/max relative humidity
    altitude
    hazardous area classification
    atmospheric pollution
    allowable product leakage rate
    conditions during transport if different to site

Material
User/purchaser material recommendations
recommended materials   prohibited materials
Physical properties

strength
    ductility
    impact resistance
    creep resistance
    erosion resistance
    thermal shock resistance

Chemical properties

resistance to corrosion
    corrosion allowance
    predicted life

Figure 6.31 Check list for material choice

6.7.2 Material combinations

Since some components more than others in a pump are subjected to stress, it is natural that these should be made of better materials. Some components are subject to compressive stress and others tensile. Some components are pressure retaining while others are completely surrounded by liquid or are external to the pressure boundary. Some components are subject to dynamic stressing with obvious considerations to fatigue. It may also be necessary for parts which are in permanent or occasional contact to have good bearing properties. This is the case in displacement pumps, plungers and bushings or pistons and liners, or with wear rings and balance drums in centrifugal pumps. Generally speaking, the materials should differ completely, but if they are the same, their hardness should differ by at least 50 BHN.

There is a certain established practice for combinations of materials in centrifugal pumps. This requires that pump impeller and sundry small parts shall be made of better material than the casing. In the case of shafts, attention must be paid to corrosion fatigue, which occurs with relatively small corrosive attacks and, in time, causes the shaft to fail. An alternative is to seal the shaft off completely from the liquid.

There are several different impeller materials for use with the simplest pump casing material grey cast iron, as shown in figure 6.32. Wear ring material combinations are listed in figure 6.33. When the impeller is made of grey cast iron, there is a risk of rusting after a test run, or whenever the pump is stationary. A better choice of material for the impeller would be gunmetal, plastic or stainless steel. If the pH value of the liquid is more than about 8 to 9 then gunmetal is not sufficiently resistant. Figure 6.32 shows the relative costs of the various materials. This intends only to show an approximate picture of price levels, since the cost ratios depend on size of pump, basic prices for the various metals and so on.

Casing	Impeller	Shaft	Approximate relative cost
Grey cast iron	Grey cast iron	Stainless steel	0.97
Grey cast iron	Gunmetal	Stainless steel	1.0
Grey cast iron	Plastic	Stainless steel	0.95
Grey cast iron	Stainless steel	Stainless steel	1.08
SG iron	Stainless steel	Stainless steel	1.25
Cast steel	Grey cast iron	Steel	1.1
Cast steel	Cast steel	Steel	1.2
Cast steel	Bronze	Steel	1.25
Cast steel	13Cr steel	Steel	1.3
Cast steel	Stainless steel	Stainless steel	1.5
Cast steel	Super stainless steel	Super stainless steel	2.0
13Cr steel	13Cr steel	13Cr steel	1.4
13Cr steel	13Cr steel	Stainless steel	1.4
Stainless steel	Stainless steel	Stainless steel	1.8
Super stainless steel	Super stainless steel	Super stainless steel	3.0
Hastelloy C	Hastelloy C	Hastelloy C	4.0
Titanium	Titanium	Titanium	10

Figure 6.32 Some commonly used material combinations for centrifugal pumps

If the shaft is protected against the liquid, it can be made of lower grade metal.

Casing wear ring	Impeller wear ring
Cast iron	Cast iron
Ni-resist	Ni-resist
Bronze	Bronze
13Cr steel	13Cr steel
Stainless steel/Colmonoy 6	Stainless steel/Colmonoy 5

Figure 6.33 Standard wear ring material combinations

Rotary positive displacement pump materials are so varied it is not wise to try to generalise. Reciprocating pumps have evolved to a point where manufacturers offer very similar material groups. Figure 6.34 lists cylinder materials in ascending order of corrosion resistance.

Low pressure	High pressure
Carbon steel	High tensile steel
Cast iron	
28Cr iron	28Cr iron
Aluminium bronze	Nickel aluminium bronze
AISI 304/CF-8 stainless steel	N2 strengthened austenitic stainless steel
AISI 316/CF-8M stainless steel	N2 strengthened austenitic stainless steel
Super stainless steel	Super stainless steel

Figure 6.34 Reciprocating pump cylinder materials

6.7.3 Corrosion resistance

From the point of view of corrosion, the choice of materials is made from two entirely different viewpoints:

- The material selected must last the intended life of the pump.

- The material should also be such that the liquid handled by the pump is not contaminated. Examples of effect on processes can be any thing from rusty water in a drinking water installation, to "poisoning" of catalysts in a chemical process because of corrosion products from the pump. The process effect can thus rule out some materials in certain cases, even though they, in themselves, have an adequate resistance to corrosion.

When it comes to corrosion from chemicals in aqueous solutions, it is the pH value and the liquid temperature which provide a preliminary guide-line for choice of materials, see figure 6.35.

pH	Material
0 - 4	Stainless steel Nickel alloys Silicon iron Plastic
4 - 6	Bronze Stainless steel
6 - 9	Grey cast iron Non-alloyed steel Bronze
9 - 14	Stainless steel Nickel alloys Plastic Grey cast iron

Figure 6.35 pH value and choice of material for water temperature 0 °C to 50 °C

At very low and very high pH values, i.e. with strong acids or strong alkalis, the resistance of the various materials must be studied more closely, e.g. by the use of corrosion tables and diagrams, see also the literature listed in section 6.8. In figure 6.36 there is a review of the typical fields of application for various materials. Within the pH range 6 to 9, grey cast iron is the most commonly used material. Exceptions are certain salt solutions, mainly chlorides, sea-water for example, where bronze is used with acceptable results. For the use of grey cast iron for ordinary cold water see also section 1.2. When a material is not completely resistant to a liquid, the effect of dissolved or solid contaminants, from erosion and cavitation and occasional temperature increases, can be important. This effect can mean that practical performance may vary from installation to installation.

Materials	Applications
Steel	De-aerated hot water, boiler feed pumps
SG iron Nodular iron	As above. Better resistance to corrosion at high liquid velocities than cast steel but not as good as cast iron
Grey cast iron	Has good resistance to many liquids assuming the graphite film is not damaged by high velocities, aeration, cavitation or solids. 120 °C max, pH >5.5

Materials	Applications
13Cr steel	De-aerated hot water, boiler feed pumps 350 °C depending upon O_2 treatment. A good replacement for steel for temperatures over 200 °C, better thermal stability, higher pressure capabilities
Silicon iron ≈ 14Si	Resistant to acid corrosion, HCl, H_2SO_4, and strong salt solutions. Not suitable for thermal shock
Ni-resist 20Ni 3Cr	Hot NaOH, sea-water
Gunmetal, bronze	Salt water, sea-water, brine and other moderately corrosive water solutions. Al bronze and NiAl bronze better than tin bronze in sea-water. NiAl bronze for higher pressures
Monel	Sea-water, brines
Austenitic stainless steels	Boiler feed water, general corrosive applications, poor in chloride solutions
SIS 2324, SIS 2343	Good for high velocity applications. Not recommended for some acids and chloride solutions
Alloy 20	Better resistance to acids and chlorides
Hastelloy C	Very resistant to corrosion. For H_2SO_4 at all concentrations and moderate temperatures. Dilute HCl, strong chloride solutions at higher temperatures
Titanium	Chloride solutions, chlorine dioxide and other liquids for bleach manufacture
Non-metallic materials	Acids, alkalis, corrosive reagents and solvents at moderate temperatures. Some also resistant to solid erosion. Temperature and pressure limitations

Figure 6.36 Materials for centrifugal pumps and their main areas of application. See also the respective descriptions of materials in section 6.2

6.7.4 Liquid effect on materials unknown

In some cases it is practically impossible to specify the liquid to be pumped. A typical example of this is the various residues and effluent from the plating industry. In this case, the only known conditions may be that the liquid contains troublesome constituents of chlorides, acids and perhaps even hydrofluoric acid, and that the corrosion resistance of materials available at reasonable cost levels will be inadequate. The problem is further complicated if the liquid also contains solid abrasive particles. Problems of this type have to be resolved by the use of common sense and a certain amount of experimentation during the early operational period. When the corrosive effects of liquids are uncertain, it is not only the pumps which will give problems. Pumps do not work without pipework. Pipework may have flanged or screwed connections. An early decision on the connection type and other hazards may help solve this dilemma and will prove very educational. Many installations require tanks; tanks are not easy to change once production has started. The pumps may prove to be a minor problem compared to the other parts of the installation. Here are some factors worth noting:

- Any particles should be separated if possible before the pump. If this is feasible, then various plastics for example would make a good choice in this instance. But if solid contaminants are present, then a metallic material should be chosen, such as silicon iron, chrome iron or the highest grade of high alloy stainless steel that is considered to be economically viable. Perhaps a rubber lined pump may be suitable.

- If possible, corrosion tests should be made on small material samples. The samples should be suspended where they can be retrieved and in such a way that galvanic corrosion does not take place.

- Choice of supplier of the pump should be made so that several alternative materials can be obtained for the component parts. In this way, individual parts or even the whole pump, can be replaced with better materials in the event of unexpected corrosion damage. Pump internals can usually be replaced easily. A long lasting casing can have many sets of internals. Internal pump corrosion does not lead to external leaks. The best casing and seal possible should be used.

- Since some experimental error may occur, the pump installation should be set up with both pump servicing and the consequences of breakdown firmly in mind.

- When practical operational results are available later, it will then be easier to optimise the economic choice of materials, balancing useful life and procurement costs of various materials.

6.8 Bibliography

La Que, Copson — Corrosion Resistance of Metals and Alloys, Reinhold, N.Y. 1963

Uligh — Corrosion and Corrosion Control, John Wiley and Sons, N.Y. 1963

Wranglen — Corrosion of metals and surface protection, Almquist and Wiksell, Uppsala, 1967

National Association of Corrosion Engineers, Corrosion Data Survey, Houston, Texas

DD 38:1974, Draft for Development, Recommendations for cast materials for use in the pump industry, British Standards Institution, London WIA 2BS. — (Withdrawn, but worth reviewing.)

Ultimate Sealing Solutions

For all applications requiring superior seal life and unquestioned reliability there really is only one solution. KALREZ* ultimate-performance parts from DuPont.

KALREZ parts can operate continuously at temperatures up to 315 °C, and withstand exposure to more than 1,600 different chemicals. Manufactured to the stringent military US MIL-STD 413C standard, they are available in most standard sizes, while non-standard shapes can be made to order.

KALREZ parts are made from a range of perfluorinated compounds, including black, white and high-purity unfilled types.

Experience has shown that KALREZ seals can be 8 times less likely to fail due to chemical attack than other high-performance seals, so in any situation where a failure could mean a spell of unplanned down-time (or worse), don't take a chance. Specify KALREZ.

Because anything less just isn't worth the risk.

For more information on the DuPont family of high-performance sealing solutions, contact:

James Walker
(0483) 757575
AES Engineering
(0709) 369966

For details on distributors in other regions, contact:

DuPont de Nemours
(Belgium)
Tel.: ++32 15 44 15 29

**Specialty Elastomers
Performance Parts**

KALREZ*   **ZALAK™**

* Registered trademark of DuPont
™ Trademark of DuPont

7 Shaft seals

Chapter 7, Shaft seals, first gives a survey of existing types of seals. Then the most common seals, i.e. soft packings, rotary and reciprocating, and mechanical seals, are treated with regard to their principles of operation, design and the materials of construction. Leakage, trouble-shooting and several other factors, see list of contents, are reviewed. Technical guide-lines are given for the choice of seal type to enable the project engineer to make a preliminary selection. The final choice, however, comprises so many aspects that the whole chapter should be studied.

Contents

7.1 Introduction

7.2 Methods of sealing : Rotary shafts
- 7.2.1 Non-contacting seal
- 7.2.2 Lip seal
- 7.2.3 Hover seal
- 7.2.4 Auxiliary pump
- 7.2.5 Soft packing
- 7.2.6 Mechanical seal

7.3 Process liquid seals for rotary shafts
- 7.3.1 Soft packing
  - 7.3.1.1 Operating principles
  - 7.3.1.2 Design variations
  - 7.3.1.3 Packing material
  - 7.3.1.4 External systems
  - 7.3.1.5 Maintenance
  - 7.3.1.6 Trouble-shooting
- 7.3.2 Mechanical seal
  - 7.3.2.1 Operating principles
  - 7.3.2.2 Surface flatness
  - 7.3.2.3 Design principles
  - 7.3.2.4 Temperature considerations
  - 7.3.2.5 Design variations
  - 7.3.2.6 Materials
  - 7.3.2.7 External systems
  - 7.3.2.8 Maintenance
  - 7.3.2.9 Trouble-shooting

7.4 Methods of sealing : Reciprocating rods
- 7.4.1 Lip seal
- 7.4.2 Soft packing

7.5 Soft packing process liquid seals for reciprocating rods
- 7.5.1 Operating principles
- 7.5.2 Design variations
- 7.5.3 Packing material
- 7.5.4 External systems
- 7.5.5 Maintenance
- 7.5.6 Trouble-shooting

7.6 Selection of process liquid sealing
- 7.6.1 Process liquid
- 7.6.2 Size, speed and pressure
- 7.6.3 Local environment
- 7.6 4 Cost
- 7.6.5 Standardisation

7.7 Bibliography

7.1 Introduction

The seal is the most vulnerable element in a pump and it is well worth devoting a great deal of attention to the choice, style, installation and maintenance of this seal. By this means the cost for spare parts and repairs as well as leakage and down-time costs can be kept to a minimum. Apart from costs, consideration must be given to environmental damage due to leakage, to the toxicity of the liquid, radio-activity or flammability or to solely the unpleasantness caused by the effects of leakage.

The methods normally used for sealing shafts and reciprocating rods are:

- Lip seals
- Soft packings
- Mechanical seals, (rotary only).

In the special case when the liquid is so hazardous that leakage cannot be tolerated under any circumstances and the risk of seal failure is unacceptable, the following pump categories, in chapter 3, should be consulted:

   120   magnetic drive pumps
   125   canned motor pumps
   250   peristaltic hose pumps
   255   rotary peristaltic pumps
   280   diaphragm pumps
   285   metering pumps.

7.2 Methods of sealing : Rotary shafts

7.2.1 Non-contacting seal

A simple non-contacting seal can be constructed by passing the shaft through a tight clearance bore. The narrow passage acts as a long orifice and creates a pressure drop as liquid flows through. Some leakage must occur to create the pressure drop. Ignoring the rotation of the shaft, at constant differential pressure the leakage rate is proportional to the cube of the clearance and inversely proportional to the viscosity and the length. Eccentricity of the shaft tends to increase the leakage. If the flow in the seal is laminar, eccentricity can more than double the leakage rate. Figure 7.1 shows how a non-contacting seal could be applied to a vertical pump. If the leakage can be returned to the pump suction, this type of sealing can be cost effective and durable.

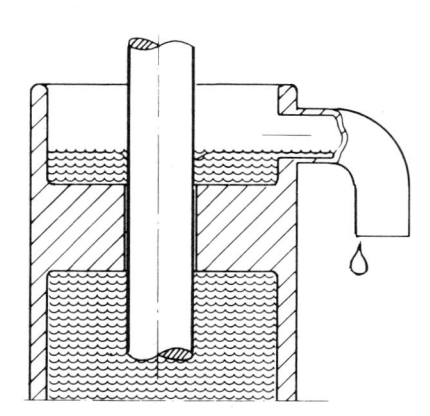

Figure 7.1 Shaft seal with radial clearance

The plain bore as shown is used as a back-up seal on some mechanical seal designs. In the event of mechanical seal failure, the throttle bush restricts the leakage rate by creating a high pressure drop.

The effectiveness of non-contacting seals can be greatly improved by using multiple seals connected by 'sudden enlargements'. This type of non-contacting seal is called a labyrinth seal and is used extensively in compressors and turbines. Having steps in the shaft, as well as the casing, further improves performance. When the shaft diameter is increased considerably, a flinger is created which prevents the axial motion of liquid by centrifugal force. Flingers are used in bearing housings to prevent oil leakage.

7.2.2 Lip seal

Figure 7.2 shows an elastomer lip seal which can be used at 0.2 to 0.5 bar differential at speeds up to 14 m/s. The lip is held in contact with the shaft by a garter spring. Designs are available which incorporate two lips on one seal. Seals can be manufactured without springs, which rely instead on the elasticity of the elastomer. Lip seals can be mounted in series for increased pressure capabilities; typically gear pumps for lubricating oil. Most applications are as a bearing seal and secondary seal for outer quenching of a mechanical seal. However, new designs are capable of sealing 10 bar differentials and surface speeds of 40 m/s are possible. Many compounds are available, including PTFE, which is not a true elastomer. Some designs can be fitted with back-up rings, for additional support, and the back-up ring can include a non-contact throttle bush.

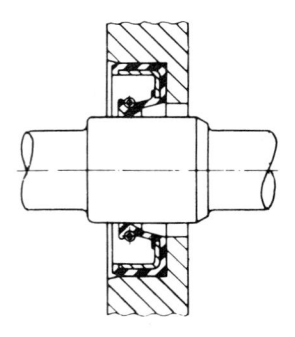

Figure 7.2 Lip seal

7.2.3 Hover seal

An elastomeric hover seal, see figure 7.3, can be used for moderate pressures and where dilution by the sealing liquid can be allowed in the process. It tolerates relatively large radial movements and can be used for solid-liquid mixtures.

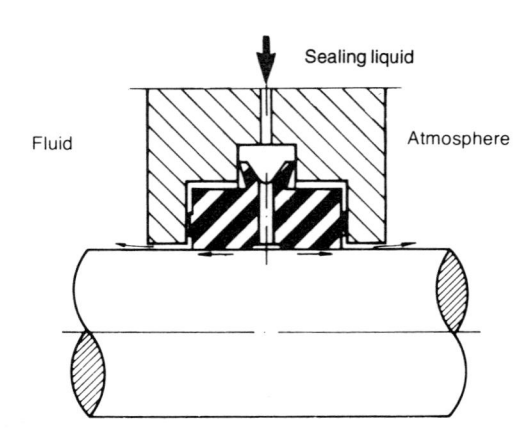

Figure 7.3 Hover seal

7.2.4 Auxiliary pump

An auxiliary pump can be constructed by fitting vanes on the back of centrifugal pump impellers, see figure 7.4; only the impeller next to the seal or stuffing box need be modified. These vanes are sometimes called 'pump out vanes'. The pumping action of the vanes reduces the pressure at the shaft and discourages flow down the back of the impeller. Some form of back-up seal, on the shaft itself, is necessary for when the pump is stationary. An auxiliary pump is useful when handling liquids difficult to seal and for solid-liquid mixtures.

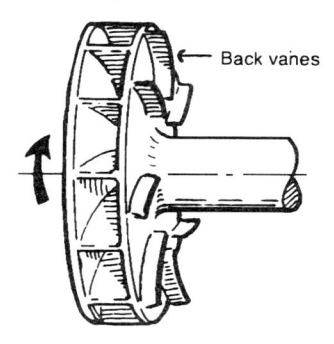

Figure 7.4 Auxiliary pump on centrifugal impeller

7.2.5 Soft packing

A stuffing box with a soft packing material is the traditional seal for pumps, the preload being applied by means of an axial compressive force, figure 7.5. A modern variant, figure 7.6, applies pressure radially and produces a better loading pattern.

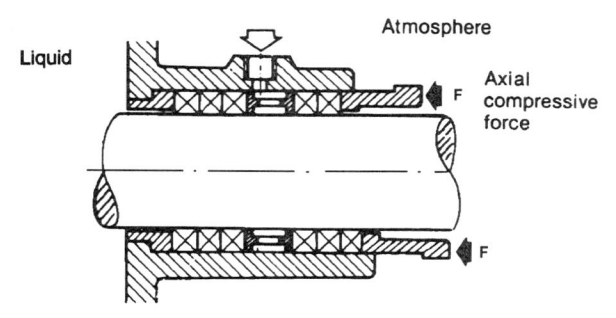

Figure 7.5 Traditional soft packed stuffing box

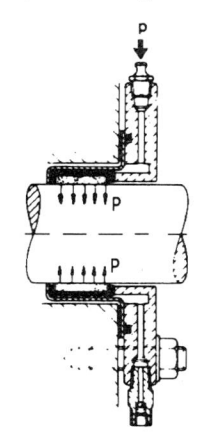

Figure 7.6 Modern soft packed stuffing box with hydraulic radial loading

7.2.6 Mechanical seal

Up till about 10 to 15 years ago, most rotodynamic pumps were fitted with soft packed stuffing boxes. Mechanical seals were available, but not mass produced and consequently costly. Manufacturing techniques and materials improved and simultaneously pump users were convinced of the operational advantages of reduced routine maintenance. Currently, probably 95% of all pumps built for process applications are fitted with mechanical seals.

In a mechanical seal, sealing takes place between a stationary and a rotating axial face, see figure 7.7. The basic concept is simple but fantastic accuracy is required for the seal to function. The important components are simple and relative small and can be used in multiple sets to build complicated seal units for hazardous liquids.

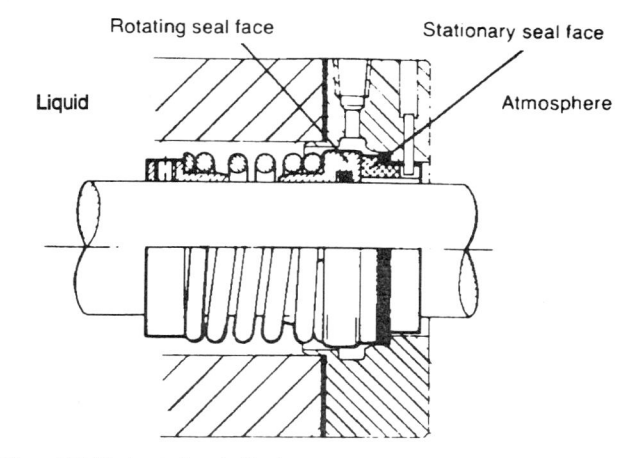

Figure 7.7 Mechanical seal with single driving spring

7.3 Process liquid seals for rotary shafts

A common feature of the different methods of sealing is that the condition in the seal area or stuffing box, see figure 7.8, plays a critical part in the design of the seal. The operating conditions are described by:

- the process liquid with possible solid contaminants and gas content,
- temperature, which can deviate from that of the process liquid by use of various cooling/heating or flushing arrangements,
- the shaft peripheral speed,
- fluid pressure at the shaft seal.

The pressure at the shaft seal p_s, for a single stage pump, can be calculated in the following manner:

$$p_s = p_i - p_d \qquad \text{Equ. 7.1}$$

where:

p_i = inlet pressure to the pump,

p_d = pressure differential within the pump between inlet and seal area, which varies according to the type of pump. In the case of single stage, single entry centrifugal pumps having normal axial thrust balancing by means of a back clearance with relief holes or back vanes p_d is approximately 30% of the pump's pressure rise.

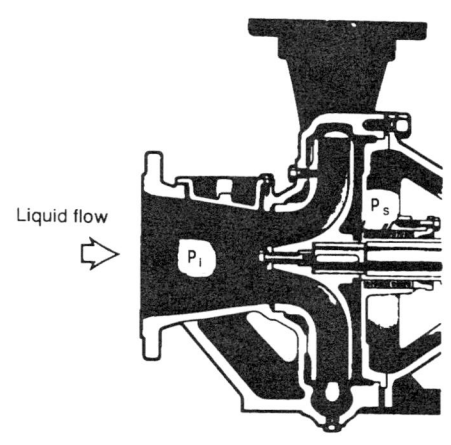

Figure 7.8 Pressure conditions in a centrifugal pump

For multi-stage pumps, the pressure at the seals is dependant upon the pump construction, i.e. the arrangement of the impellers and the type of axial thrust balancing employed. It is preferable not to have a high pressure seal and a low pressure seal. Identical seals reduce spares inventory and reduce possible errors due to fitting similar parts in the wrong seal.

The pressure in the seal area p_s and the peripheral speed v are usually combined into a pv value, in the same manner as in plane bearing calculations. At the same time, the pv value provides a measurement for the frictional heat developed in the seal. In addition to the frictional heat, the temperature at the surfaces of the seal itself are also dependent on the liquid temperature and the heat dissipation.

7.3 1 Soft packing

7.3.1.1 Operating principles

A soft packing set usually consists of two to six rings which are compressed by the axial tightening of a gland against the outside ring, see figure 7.9. A radial reaction pressure is exerted on the shaft and the bore of the stuffing box as a result of this compressive force. Due to the friction from, above all, the stuffing box, the radial reaction pressure is greatest at the gland and decreases towards the bottom ring.

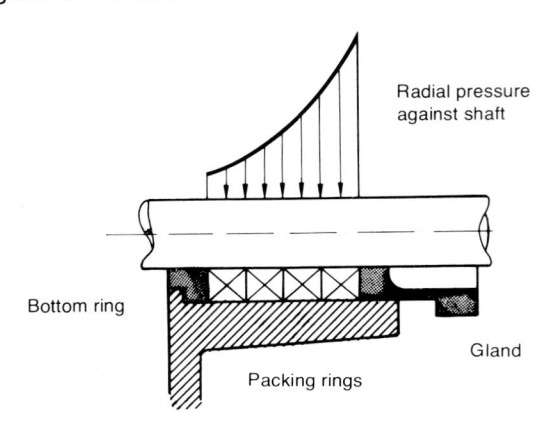

Figure 7.9 Soft packing with radial pressure distribution due to axial load

Figure 7.9 shows a typical simple soft packed stuffing box with a metal throat bush in the bottom of the stuffing box. In good stuffing box designs, a metal follower ring is fitted between the packing and the gland to promote even pressure distribution, circumferentially, around the shaft.

The liquid under pressure will penetrate between the packing rings and the shaft to form small fluid reservoirs. There are thus two different pressures to be considered, i.e. the liquid pressure in the reservoirs themselves, and the radial abutment pressure between the recesses. In the case of a soft packing charged with liquid pressure, the radial pressure varies as shown in figure 7.10. The liquid pressure reduces gradually through the soft packings to atmospheric pressure at the gland.

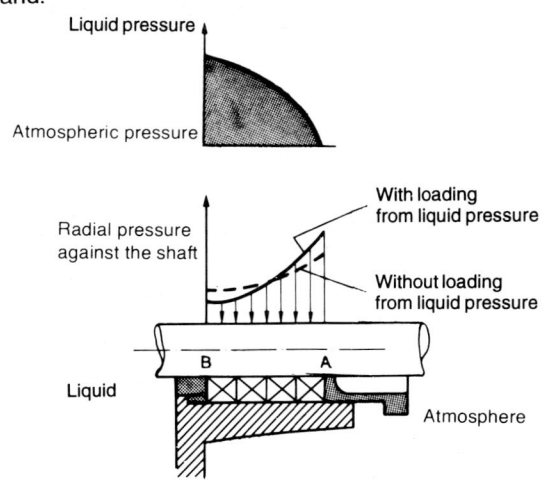

Figure 7.10 Pressure distribution in working soft packing

A direct consequence of the distribution of the radial pressure along the length of the soft packing is that wear on the shaft is greatest at the gland, at A in figure 7.10. If wear occurs at B then it is due to the action of abrasive particles in the liquid or shaft eccentricity.

The extent of leakage depends on the radial pressure which in its turn is regulated by means of the axial compressive force. Leakage must always be present in order to remove frictional heat and to make up the liquid lost due to evaporation. The leakage rate in the case of working soft packing which is properly adjusted is 10 to 60 drops per minute.

Soft packing sets with a lantern ring have a somewhat different pressure distribution, see figure 7.11. The barrier/buffer liquid or lubricant should be admitted in the middle of the box. In the case of a five ring set the lantern ring should be placed with two rings inboard and three outboard in order to avoid excessive axial movement of the lantern ring during adjustment, thereby blocking the supply of liquid.

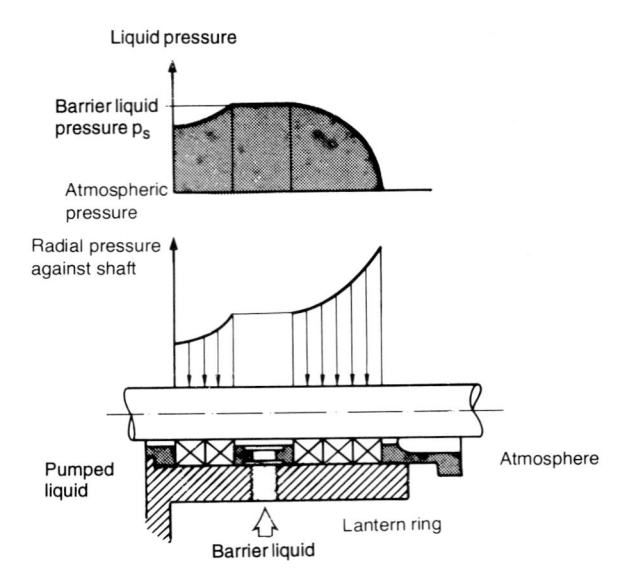

Figure 7.11 Pressure drop and radial pressure for soft packing with lantern ring

7.3.1.2 Design variations

Before looking at different stuffing box arrangements in detail, it is worthwhile clarifying some of the terminology which is used when stuffing box design becomes complex. It is essential that all parties concerned describe the same function or component by the same terms.

barrier liquid   A liquid, at a higher pressure than the process liquid, which is introduced behind the primary seal to prevent outward leakage of the process liquid. A barrier liquid can also cool and lubricate the seal. As no process liquid can leak into the seal, a barrier liquid effectively protects against solids ingress. The barrier liquid is constrained by some form of secondary shaft seal. The barrier liquid normally circulates through an external piping system driven by a pumping ring on the pump shaft.(*)

buffer liquid   A liquid, at a lower pressure than the process liquid, which is introduced behind the primary seal to lubricate and/or cool the seal, and/or dilute any leakage from the primary seal. The buffer liquid is constrained by some form of secondary shaft seal. The buffer liquid normally circulates through an external piping system driven by a pumping ring on the pump shaft.(*)

flush liquid A flow of clean liquid, which is introduced in front of the primary seal, to cool and/or lubricate the seal or to prevent the ingress of solids lubrication	A small flow of suitable lubricating liquid introduced behind the primary seal. The lubricant may mix with the process liquid and contaminate it. Some lubricant will leak passed the secondary seal together with process liquid. All the lubricant is 'lost'.

quench fluid	A flow of cool liquid or steam, which is introduced on the atmospheric side of the last seal, to condense any escaping vapour or prevent the formation of crystals or solids which would interfere with the seal function. Alternatively, a flow of warm liquid or steam to prevent ice formation when the process liquid temperature is below 0 °C.

which may be contained in the process liquid. The flow rate of the flush liquid is dependant upon the differential pressure available.

(*) Pumping rings are only used with mechanical seals. Variable speed pumps may require an external motor driven pump for circulation to ensure adequate flow at low pump speeds. Soft packed stuffing boxes will require an external pump unless thermo-syphon circulation can be used.

Plain stuffing boxes

The process liquid forms a fluid film and must therefore be clean in order to avoid unacceptable wear. This design, see figure 7.12, can cope only with moderate temperatures where the dissipation of heat takes place by means of the leakage and conduction through the shaft and stuffing box. There is a risk of air being sucked in at low inlet pressures. Because process liquid leaks out of the packing, this type of packing arrangement is not suitable for any liquid which damages the environment. It may be suitable for some hazardous liquids depending upon the nature of the immediate surroundings.

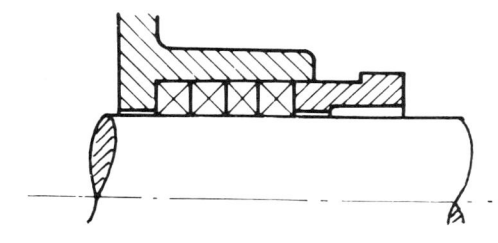

Figure 7.12 Plain stuffing box

The stuffing box arrangement as shown can be improved by the addition of a throat bush and a follower ring. If a liquid quench was fitted, the problem of air entrainment at low pressure would be eliminated

Stuffing box with lantern ring

The arrangement in figure 7.13 is also suitable for clean liquids. The lantern ring can perform several functions. Barrier or buffer liquid or lubrication. The barrier or buffer liquid could be the process liquid taken from a tapping on the pump. Both of these options would solve air entrainment at low operating pressures. If the process liquid was not oil, a separate oil supply would be required for lubrication.

If the barrier liquid was 'safe' this stuffing box arrangement could be used for hazardous liquids. The liquid supplied must be selected taking into account the process liquid and should maintain a pressure which exceeds the pressure at the seal by 1 to 1.5 bar. The quantity of barrier liquid which leaks into the process liquid is very small, a fraction of a litre/minute.

Flushed stuffing box

In the case of abrasive solid-liquid mixtures, the throat bush can be modified to allow the introduction of a clean flush liquid, see figure 7.14, in order to prevent abrasive particles from entering into the box. In this case the quantity of flush liquid which leaks into the process liquid can be considerable. The minimum flush required can be approximated by considering

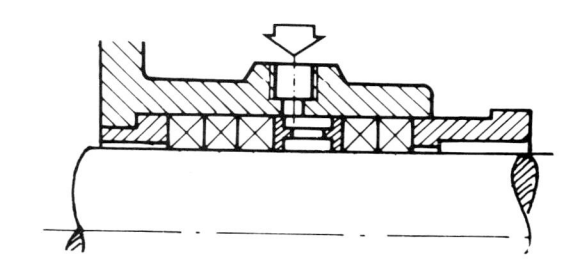

Figure 7.13 Stuffing box with lantern ring

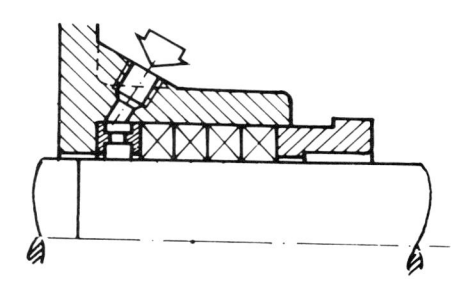

Figure 7.14 Stuffing box with flushed throat bush

a velocity of 0.1 m/s through the throat bush clearance when worn.

Stuffing boxes can be fitted with lantern rings as well as flushed throat bushes. Contaminated hazardous liquids can be pumped or contaminated liquids requiring lubrication.

Stuffing box with cooling

With liquid temperatures in excess of 80 to 120 °C, depending upon the packing material, cooling of the stuffing box should be considered. Cooling of the stuffing box and shaft can be accomplished by the barrier or buffer liquid pumped through the lantern ring. However this method does cool the bottom of the box or the front rings of packing. External cooling of the packing rings is relatively ineffective because of the poor thermal conductivity of the packing material. The cooling chamber should therefore extend in front of the packing rings in the case of highest temperatures, see figure 7.15. Cooling of the gland should also be provided if there is a possibility of vapour forming in the packing. Any vapour created in the packing will shorten packing life dramatically. Gland cooling prevents heat being conducted through the shaft to the front bearing. The gland shown is fitted with auxiliary packing, a smaller packing cross-section than the main packing.

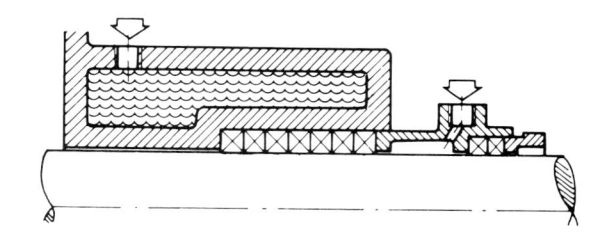

Figure 7.15 Cooled stuffing box with cooled gland

All the stuffing box arrangements shown, indicate the packing rubbing directly on the pump shaft. Wear of the shaft can be avoided by using a sleeve. The sleeve can be made of a different material to the shaft, a material with better wear and corrosion resistance properties. A pump with a carbon steel shaft could be fitted with a sleeve of hardened 12Cr steel or solid Stellite®, for example. When wear occurs, the sleeve is replaced, not the complete shaft.

7.3.1.3 Packing material

The most common packing material is plaited or braided fibre, but packing compounds and moulded rings are also used.

See figure 7.16 for a survey of packing materials. A braided packing consists of:

- a fibre, the basic material,
- an optional coating on each fibre,
- an optional impregnation or filler applied during braiding.

More than one type of fibre may be used. Modern designs sometimes incorporate a strong central core and/or reinforced corners. Fibres currently in use include:

Vegetable fibres: hemp, flax, linen, cotton; used for temperatures up to 100 °C. Poor chemical resistance.

White asbestos and **blue asbestos** have a wide operating range and can be used for temperatures up to 500 °C. The use of asbestos is now forbidden by legislation in certain countries and asbestos has largely been replaced by other materials.

Synthetic fibres: Aramid®, Kevlar®, PTFE; have excellent sliding properties and are resistant to a very wide range of liquids. Maximum temperature about 260 °C.

Graphite fibre and foil has a very low friction coefficient and good thermal conductivity properties as well as good chemical resistance, apart from oxidising acids such as, nitric acid. Maximum temperature from 500 to 3000 °C depending upon the liquid.

Glass fibre has largely replaced blue asbestos. Glass fibre is not resistant to hydrofluoric acid and strong alkalis. Maximum temperature about 480 °C but will reduce to 260 °C if PTFE lubricants are used.

In order to reduce friction the packing is impregnated with a suitable animal, mineral, vegetable or synthetic based lubricant. The natural and mineral lubricant impregnation withstands temperatures of up to 125 °C and a pH value between 5 and 10. PTFE impregnation withstands temperatures of up to 260 °C and pH 0-14. Synthetic silicone based lubricants can be used up to 480 °C.

The style of stuffing box required and the packing material necessary are both functions of all the operating conditions. The pump selection can be greatly influenced by operating conditions. No part of the pump application or design should be considered in isolation.

7.3.1.4 External systems

We have seen that stuffing boxes can have liquid or steam introduced from external sources; these obviously involve external systems. VDMA 24297 and API 610/682 have standard arrangements designated for most applications.

The most popular stuffing box system for soft packed boxes is actually an internal system, no external components. An internal drilling leads clean liquid from the pump discharge branch to the throat bush, a clean flush. This is classified as 'A' in VDMA and Plan 01 in API. If the process liquid was contaminated, or was a solid-liquid mixture, then feeding raw medium to the throat bush would be counter productive. In these instances, a completely internal system is not viable. An external piped system, including a cyclone separator to remove the solids, would be fitted; VDMA 'B' or API Plan 31. In its simplest form an external system can consist of a piece of pipe and an orifice, Plan 11. This would be used when an internal drilling was impossible. VDMA 24297 does not differentiate between internal and external systems, only functions.

Most modern stuffing box systems are designed for mechanical seals, see section 7.3.2.7 for a description of popular schemes. Pumping rings are usually not possible in soft packed boxes, external pumps must be fitted if required.

7.3.1.5 Maintenance

Mechanical seals have replaced soft packing, in many applications, because of reduced routine maintenance. Once fitted, mechanical seals cannot be adjusted. See section 7.3.2.7 about comments on strainers. Soft packing, on the other hand, must be inspected regularly and adjusted if required. Adjusting packing is a skilled job. Different applications and different packing types require different techniques.

Correct installation and bedding-in, after selecting the correct seal arrangement and materials, are a prerequisite of good packing life. It is difficult to assess the useful operating life of packing. In some hot applications, a good packing life will be three months. In some water pump applications, using large slow pumps, good packing life can be three years. In order to obtain the longest life possible the following procedure should be adopted.

- Read the pump manufacturer's instructions first.
  Some packing rings must be soaked in oil prior to installation. Failure to comply will inevitably shorten packing life.

- The stuffing box must be completely empty and perfectly clean before repacking.

- Inspect the old packing rings as removed, check for indications of overheating, extrusion and chemical attack. Note the order of component removal from the box and check that the box was assembled correctly the last time.

- Check the throat bush and gland bores for wear. Replace if the clearance is greater than the manufacturer's recommendations.

- Check the surface finish of the shaft and the stuffing box bore. The shaft should be better than 0.4 μm R_a, 16 micro inches CLA, but not too good. Not better than 0.2 μm R_a, 8 micro inches CLA. The stuffing box bore should be better than 1.6 μm R_a, 64 micro inches CLA. Both surfaces should be free of scratches

- Replace throat bush, whenever possible lubricate the bore. Special lubricants may be required for hot applications, and definitely for food applications.

- Whenever possible, lubricate the shaft and stuffing box bore prior to fitting the rings.

Packing	Fibre	Filler/coating	pH	Press bar	Speed m/s	Max °C	Min °C	Applications
Solid plaited	Flax	graphite	5 - 9	20	8	120		water, sea-water
Cross plaited	Ramiex	PTFE	4 - 11		15	120	-30	water, brine, water-oil emulsions
Cross plaited	PTFE	PTFE/PTFE	0 - 14	10	4	250	-100	weak and strong acids, weak and strong alkalis, solvents
Cross plaited	PTFE	PTFE/PTFE/lubricant	0 - 14	25	10	250	-100	weak and strong acids, weak and strong alkalis, solvents
Cross plaited	Aramid	PTFE/lubricants	2 - 13		20	250		water, oils, solvents, medium strength acids, food
Cross plaited	Aramid/-Graphite	PTFE/lubricants	2 - 13	20	20	260		Water, chemicals, mild acids, alkalis, sewage, abrasive slurries
Cross plaited	Aramid	PTFE/graphite/lubricants	1 - 14	20	25	290		weak acids, weak alkalis, oils, abrasive slurries
Cross plaited	BCX	PTFE/lubricants	3 - 13	100	10	300		acids, alkalis, oils, solvents, water
Compressed foil	Graphite		0 - 14	20	30	500+	-200	acids, alkalis, solvents, hot oil and water

Figure 7.16 Packing materials and operating parameters

- Fit each ring individually, tap into place with a piece of split pipe approximately the same proportions as the rings. Lubricate lantern rings when fitted. Stagger ring joints.

- Grease or lubricate the gland threads before the gland is tapped into place. Tighten gland nuts only hand tight.

- Vent and prime pump if necessary, start pump and run for about 15 minutes before tightening gland. Tighten evenly by 1/6th turn about every 10 to 30 minutes until leakage is reduced. If a tapping to the lantern ring is available, oil while running.

Hard packing will bed-in much slower than soft packing. Hot applications will probably need the packing adjusted as the pump warms up. Hot pumps will probably need the gland loosened as the pump cools down for shut-down. In some cases the packing bedding-in will take up to a week.

7.1.3.6 Trouble-shooting

If operating problems occur, there is one very important action which must be done immediately; check the operating conditions. If the pump is not running at the correct speed, suction head, discharge head and temperature, data sheet conditions, then it may not be surprising that the seal gives problems. Check the process liquid; is it the same, exactly the same, as the data sheet liquid? If **all** the operating conditions do not agree with the data sheet values then it is an upgrade problem not a trouble-shooting problem. If operating conditions are correct, figure 7.17 can be used for guidance.

Problem	Probable cause	Remedy
Packing begins to leak after a short period of operation.	Debris in process liquid or operating condition transients during start-up.	When process conditions have stabilised, disassemble and repack box.
	Packing materials not suitable for operating conditions.	Check operating conditions against specification. Consult pump and packing manufacturer.
	Packing materials not suitable for cleaning or stationary conditions.	Consult pump and packing manufacturer.
Packing extrudes into throat bush and/or gland.	Large clearance.	Check clearance, if too big replace throat bush and/or gland. Check life of component, is it short or adequate.
	Packing too soft.	Fit hard end rings to packing set. Glass reinforced PTFE is common. Metal rings with tight clearance can be used but must be a very good bearing material; phosphor bronze or cast iron.
Leakage from outside of gland.	Packing not fitted correctly.	Strip stuffing box and repack in accordance with manufacturer's instructions.
	Packing wrong cross-section.	Fit correct packing size.
	Stuffing box or gland worn.	Strip stuffing box and check dimensions and surface finish. Contact manufacturer for replacement parts if badly worn.
Packing scored on outside surface.	Packing wrong cross-section, rotating with shaft.	Fit correct packing size.
	Stuffing box worn.	Check stuffing box dimensions. If badly worn, oversize packing may be possible. Fit 1/4" instead of 6 mm.

Problem	Probable cause	Remedy
Packing rings extrude or flow into adjacent rings.	Packing rings cut too short.	Fit die formed rings prepared by packing manufacturer.
	Packing too soft, or overheating slightly.	Fit hard spacers between rings.
	Slight attack by liquid.	Upgrade packing material.
Packing burnt or charred in the bore.	Dry running.	Prevent pump operating without process liquid.
	Lubrication failure.	Change packing type to grade with more or different lubrication impregnated. Fit lantern ring and lubricate or supply buffer liquid.
Packing seriously attacked, reduced in volume.	Packing incompatible with process liquid, or barrier, buffer, lubricant, when used.	Check process liquid quality and barrier, buffer, lubricants. Change packing grade.
Excessive leakage.	Packing attacked by liquid(s).	Check liquid(s). Change packing grade.
	Leakage through ring joints, rings too short.	Fit die formed rings prepared by packing manufacturer.
	Stuffing box incorrectly packed or reassembled.	Strip stuffing box and repack in accordance with manufacturer's instructions.
	Shaft eccentricity.	Check shaft run-out. Check pump bearings.

Figure 7.17 Trouble-shooting guide

NOTE: Only use spare parts supplied by the original equipment manufacturer. It is not possible to completely 'reverse engineer' a component from the finished part. It is easy to check the chemical composition and physical properties. It is impossible to detect if the material was vacuum remelted, or X-ray or ultrasonic quality, or the full heat treatment history. Dimensional inspection will not reveal whether the part is on top or bottom tolerance. Fitting parts, made by others, will invalidate any manufacturer's warranty.

The correct functioning of any external system must be checked regularly and liquid supplies for barrier, buffer and lubrication must be maintained.

7.3.2 Mechanical seal

7.3.2.1 Operating principles

The sealing of a rotating shaft passing through a pump casing is accomplished by means of a rotating, optically flat seat,

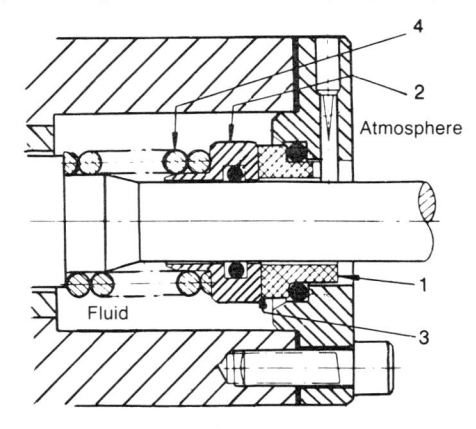

1. Stationary seat sealed with an 'O' ring.
2. Rotating seat sealed with an 'O' ring which allows axial movement to compensate for wear.
3. Sealing takes place here, axially, between rotating and stationary seats.
4. The rotating spring, maintains contact between the seat faces and to drive the rotating seat.

Figure 7.18 Single mechanical seal

rubbing against a stationary, optically flat, seat with a fluid film in between them, see figure 7.18. Sealing between the rotating seat and shaft or sleeve, and the stationary seat and gland, is usually performed by means of an elastomer such as an 'O' ring, 'V' ring, etc. The primary purpose of the spring is to provide an initial pressure between the sealing seats. One of the great advantages with this type of sealing is that it does not cause wear to the shaft or the sleeve.

The theory as to what happens between the seat faces is explained schematically in figure 7.19. The seal shall be considered as drip-free but not leak-free. The very low leakage rate, diffusion, will escape in the form of vapour on the atmospheric side of the seal.

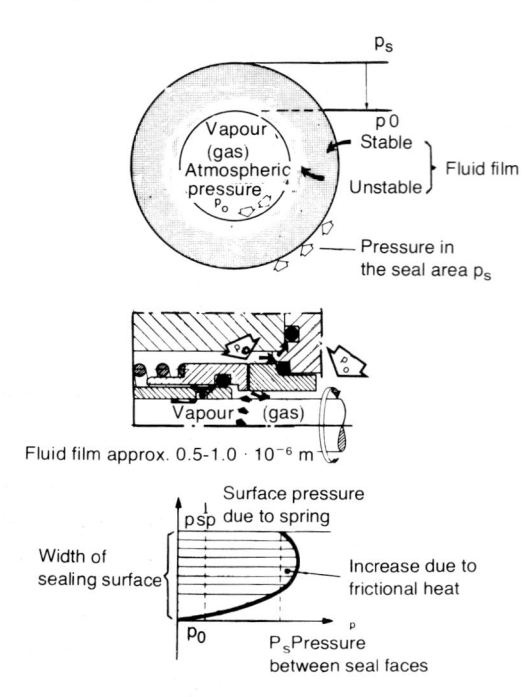

Figure 7.19 Fluid distribution across seat faces and operating pressures

Leakage through the seal faces in the case of mechanical seals amounts to 0.05 to 400 cm^3/hour depending on operating conditions, dimensions and choice of seal face materials. Manufacturers have formulae or graphs to evaluate leakage. No dripping, however, should be observed, only the formation of vapour which can lead to a certain amount of crystal formation. A general and very important rule is that a mechanical seal should never be run dry, there must always be a fluid film between the faces. Dry running can occur even when the pump is full of liquid if the temperature is so high that the liquid vaporises in the stuffing box. Only specially designed seals can be run on the vapour phase of a liquid. These dry running seals are available as back-up and standby seals for specific applications.

At high pressures and high speeds the sliding surfaces must be relieved so that the fluid film is kept stable. The frictional heat must be limited at temperatures approaching the vapour temperature. In such cases, the face pressure is reduced hydraulically. This is done by balancing the ratio between the sealing diameter "S" and the inner diameter of the stationary seat. The pressure between the seal faces is thus varied according to the liquid pressure. Dependent upon the relationship between the two surfaces "A" and "B" in figure 7.20, the seal is termed, balanced or unbalanced.

The limit between balanced and unbalanced seals is determined by the pv factor and the operating temperature. Balanced standard seals can be used for pressures up to about 16 bar at speeds of 20 m/s. In the case of higher pv values, hydrodynamic seals having a controlled leakage must be used. See also the list of references in Section 7.7.

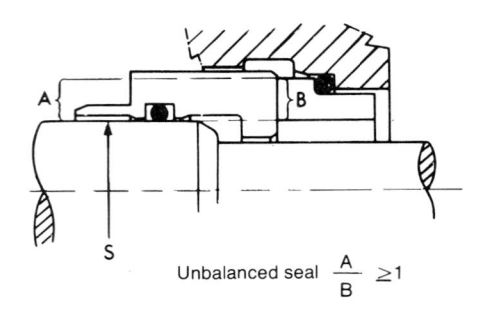

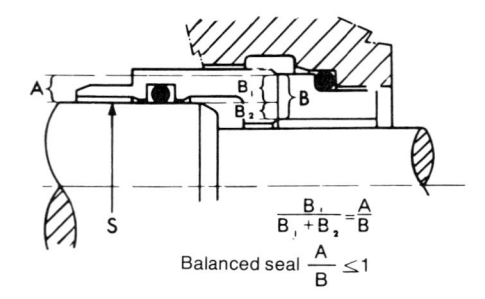

Figure 7.20 Hydraulic balancing of mechanical seals

7.3.2.2 Surface flatness

The flatness of the seat faces must be of a very high accuracy. The leakage is approximately proportional to the third power of the thickness of the fluid film which normally is in the order of 0.001 mm. Flatness is normally expressed in light bands, since checking normally takes place using an optical flat and monochromatic light, usually sodium light. A deviation of one light band corresponds to an out of flatness of 0.25 μm.

Normal standards of flatness are better than 1 to 2 light bands with variations up to approximately 5 light bands in the case of shaft diameters over 100 mm or in the case of special face materials. In addition to the number of light bands the light pattern must also be considered, see figure 7.21. Large differences in seal tightness occur, for example, between a spherical surface, light pattern 3, and a wavy surface, light pattern 5. A check of seat faces is recommended as a standard measure prior to the installation of a mechanical seal.

Because of the accuracy of mechanical seal components, any work done on seals must be carried out in clean conditions. All components must be thoroughly cleaned prior to assembly.

7.3.2.3 Design principles

The generation of the initial sealing pressure, i.e. the initial face load is provided by using different types of springs. Single springs, multiple springs, Bellville washers and wave springs are the most common, see figure 7.22. The drive, i.e. transmission of the frictional torque to the rotating seat can, when single springs used, be carried out by the spring itself, separate driving elements are used with other spring types. At least two driving points are necessary in order to guarantee a satisfactory loading of the seal faces. Single springs should generally be chosen in preference to multiple spring types. The reason for this is that driving and face loading can be combined in one element, that the single spring is robust and withstands corrosion, as well as the fact that the risk for blocking is less than in the case of small springs. In addition, the single spring is easier to handle. The single helical spring has one major disadvantage; the spring handing must be changed if the direction of rotation is changed. For example, a double suction between bearings pump, with two stuffing boxes, would have a different seal at each end, different springs. To overcome this problem, many standard mechanical seals are fitted with Bellville washers. The springs are usually inside the stuffing box in the process liquid. Special mechanical seals can have the springs outside the liquid on the atmospheric side.

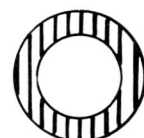

Light pattern 1
The surface is flat within 1 light band - 0.25 μm. The distance between the bands does not tell us anything about flatness only about the size of the air gap.

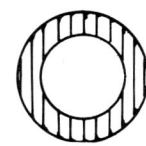

Light pattern 2
The surface is almost flat but the edges have been rounded, probably in connection with hand polishing.

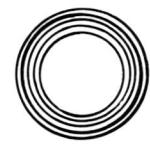

Light pattern 3
The surface is concave or convex to 3 light bands. This can be decided by pressing on the centre of the glass. If the rings then move outwards the surface is convex.

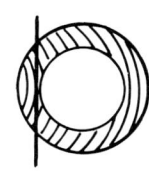

Light pattern 4
The surface is not completely flat. The tangent is drawn to the wavy bands. Half the number of light bands which are bisected by the tangent is a measurement of the flatness.

Light pattern 5
The surface goes in waves. It is difficult to achieve tightness. The cause can be residual tension after tightening in a chuck. The flatness fault is evaluated by comparison with light pattern 1.

Light pattern 6
The surface goes in waves. The same phenomena as light pattern 5 but 4 high points.

Figure 7.21 Seat flatness checks using light and an optical flat

The shape of the rotating seat varies with the method of driving, engagement and the loading, as well as the materials of which the seal faces are made. In order to take up any vibration in the shaft, parallel deviations in the 'O' ring grooves as well as any effects of the driving arrangement, the stationary seat should be mounted as flexible as possible. Figure 7.23 shows various methods of attachment.

'a' is DIN standard construction and provides the greatest flexibility. (Source Mayer)

(Static seals are produced in a great many varieties by different manufacturers. 'O' rings dominate, see figure 7.24. 'O' rings, 7.24 'a' are standard fittings, 7.24 'm' for temperatures up to 250 °C. Wedge shaped seal rings are often used for high temperatures, see figure 7.24 'c'.

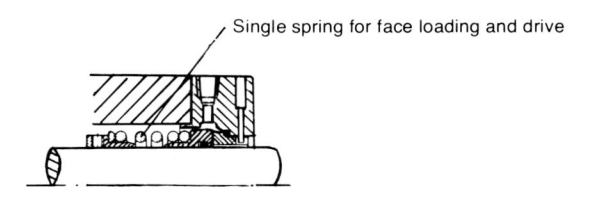

Figure 7.22 Examples of face loading and drive

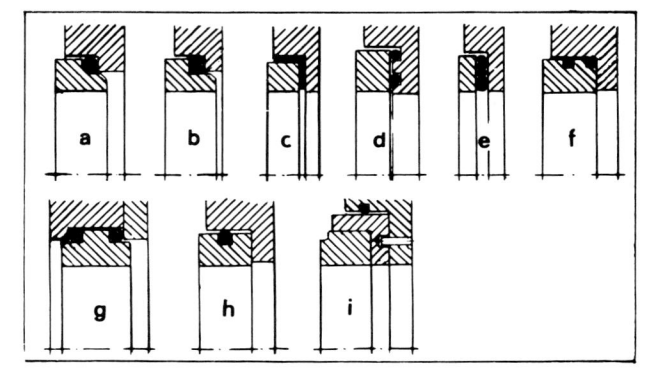

Figure 7.23 Some examples of how a stationary seat can be fitted and sealed.

Figure 7.24 Some examples of sections of static seal rings (Source Mayer)

7.3.2.4 Temperature considerations

In order to achieve acceptable temperature conditions at the seal faces, frictional heat must be dissipated in order to avoid dry running due to loss of liquid between the faces. Heat dissipation takes place by means of heat transfer through the seats and by convection to the surrounding medium. See figure 7.25.

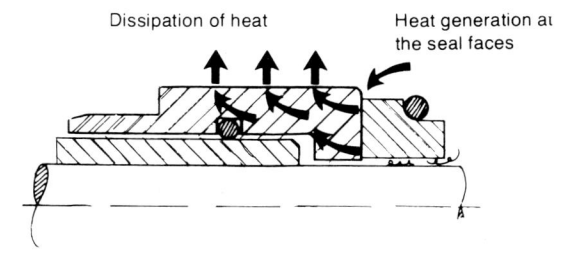

Figure 7.25 Heat dissipation in a mechanical seal

Each seal and combination of materials has what is called a minimum ΔT in order to be able to function, see figure 7.26. ΔT is the difference in temperature between the boiling point of the liquid at the seal cavity pressure and the actual temperature of the liquid. Usually, ΔT should be at least 20 °C. The seal manufacturer can provide information for each type of duty and seal arrangement. By means of various installation options, internal and external circulation or cooling, the available ΔT can be increased, see the definitions at the beginning of section 7.3.1.2. The required ΔT of a seal can also be reduced by using materials with good thermal conductivity or with low friction, etc. In such special cases the ΔT required can be reduced to approximately 5 °C.

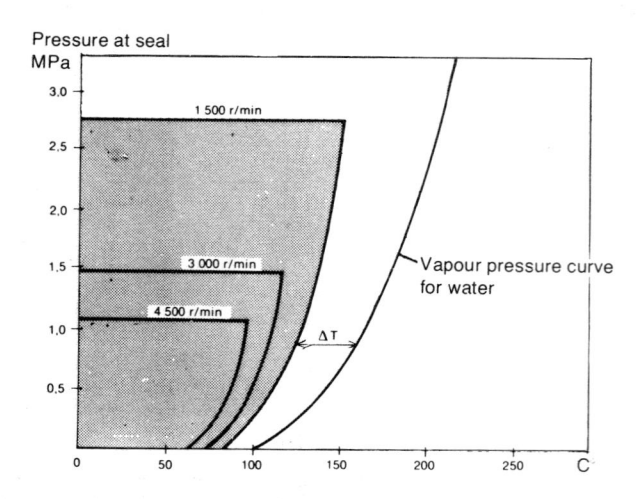

Figure 7.26 Pressure and temperature relationships at three different speeds for a balanced seal. The operation range of the seal is to the left of each respective speed boundary. The liquid is water. The distance between the vapour curve and the seal's limit is the required ΔT for each speed (Source Lymer).

7.3.2.5 Design variations

Single seal without circulation

A seal without circulation, see figure 7.7, can be used in favourable cases, clean liquid and moderate temperatures. Cooling or heating can take place in direct contact with the stationary seat, if channels are provided in the gland plate, see figure 7.27.

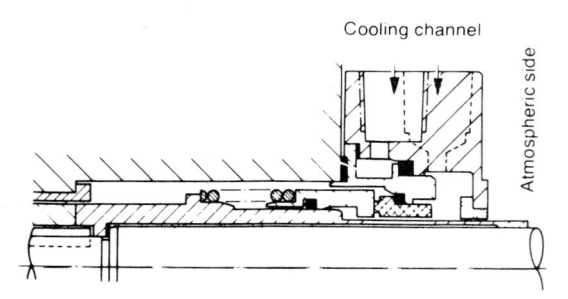

Figure 7.27 Seal without circulation but with external heating/cooling

Single seal with circulation of the pumped liquid

The purpose of the circulation is:

- to remove the frictional heat from the seats,
- to remove very fine particles and crystals from the area around the seal faces,
- to cool or to heat the seal cavity in relationship to the pump in general.

This is the most common seal arrangement and by varying the circulation system and the choice of material most liquids can be handled, both cold and hot, clean and slightly contaminated, see figure 7.28. See also section 7.3.2.7 for suitable external systems; Plan 11, Plan 12, Plan 21, Plan 22, Plan 31, Plan 52 and Plan 62. There are certain exceptions and they are described later.

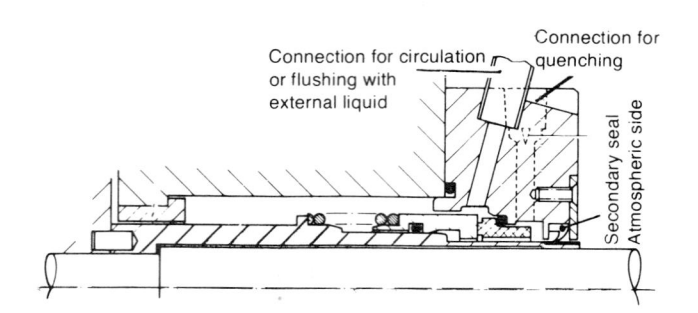

Figure 7.28 Seal with circulation or quenching

Single seal with flushing

Used for heavily contaminated liquids and suspensions, Plan 31, but the cleaned process liquid is piped to a modified throat bush, not the seal cavity. The flow of liquid through the throat bush discourages solids from entering the seal cavity. An external clean liquid may be used when dilution in the process is permissible. See figure 7.14 which shows a flushing throat bush in a soft packed stuffing box.

Single seal for slurries without flushing

If dilution cannot be allowed due to process, technical or economic reasons, a balanced reversed seal can feasibly be used with hard-wearing surfaces, e.g. tungsten carbide versus tungsten carbide, see figure 7.29. The spring is located outside the slurry and all the parts in contact with the slurry are so designed that clogging is prevented. A Plan 31 system would still be beneficial. Quench for cooling and keeping the atmospheric side of the seal clean is also recommended.

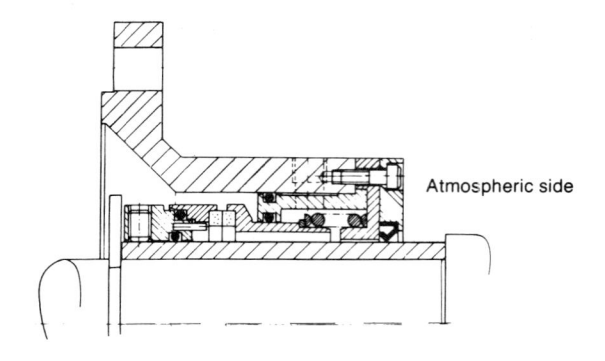

Figure 7.29 Balanced reversed seal for slurries

Single seal with quench

The quench, see figure 7.28, can be used to:

- remove dangerous liquids or gases which escape,
- cool or heat the seal from the atmospheric side,
- rinse away any wear powder and crystals,
- lubrication for dry-running protection.

See also section 7.3.1.2 and Plan 62 in section 7.3.2.7.

Double seals

A double seal, see figure 7.30, consists of two seals usually mounting back-to-back and having a selected barrier or buffer liquid in between. The pressure of the liquid placed between the seals is dependant upon its precise function, see 7.3.1.2. A double seal involves extra equipment in order to pressurise, circulate and possibly cool or heat the liquid. Due to these facts, double seals are only recommended where regarded as being necessary for protection of the pump, personnel or the environment.

Typical installations where double seals are required are as follows:

- suspensions and slurries in which the liquid is hazardous,
- liquids which are toxic, radio-active or liable to explode,
- liquids which have dissolved or entrained gases which are toxic or flammable,
- when there is a risk of severe crystal formation,
- when the liquid is very corrosive to pump external components and the surroundings.

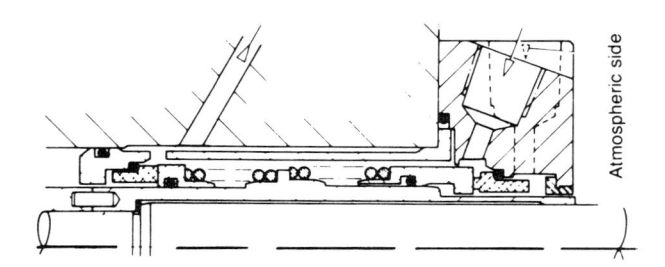

Figure 7.30 Back to back double seal

Double seals can be fitted facing the same way, tandem seals. In extreme cases, three seals can be fitted. For all hazardous liquid applications, the advice and experience of the seal and pump manufacturer should be obtained.

Single seal of bellows type

Used for very corrosive liquids and hot or cold liquids. The parts in contact with the liquid are made of ceramic and PTFE, or ceramic and CrNi steel, see figure 7.31. For PTFE bellows, pressure can be up to 6 bar with temperature between -20 °C and 120 °C. For metal bellows seals, pressures go up to 16 bar with a temperature range from -100 °C to 350 °C. Typical applications include asphalt, bitumen and proprietary heat transfer fluids.

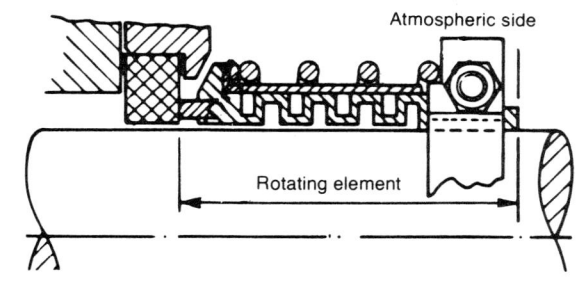

Figure 7.31 Chemical bellows seal

Some mechanical seal designs are now available as cartridge seals; the complete seal, including a shaft sleeve, are fitted as a complete assembly. This style of seal is intended to reduce failures due to incorrect fitting. A complete cartridge assembly also allows testing of the seal integrity; a gas test may be performed to assess the condition of the seal faces. Some seal designs are available as split seals. A cartridge seal with sleeve must be installed from an accessible bare shaft end, a split seal can be fitted at the seal site and does not require the removal of bearings.

7.3.2.6 Materials

The material of the two seat faces can be the same or different. The combination of materials is determined, amongst other factors, by wear, strength, friction conditions, lubrication, chemical resistance and thermal conductivity. Figure 7.32 shows some common combinations for the two seal faces. Tungsten carbide and ceramics can either be used in solid form or as a surface coating on, for example, stainless steel.

Elastomers of different types, see figure 7.33, are used as flexible elements and for static seals.

Seal face materials	Approximate price relation	Typical applications
Bronze - carbon Stainless steel - carbon	1	Clean liquids and moderate pressures at moderate periphiral speeds
Stellited Stainless steel - carbon	1.75	Clean and moderately contaminated liquids especially petroleum products
Ceramic - carbon	2	Clean and contaminated water-based liquids, chemicals. Sea water
Tungsten carbide - carbon	3 - 4	Clean and contaminated liquids. Moderate corrosion resistance but better thermal conductivity than ceramic - carbon
PTFE - ceramic	3 - 4	Strong chemicals, acids. Limited abrasion resistance
Silicon carbon - carbon	3	Clean and contaminated liquid chemicals
Silicon carbide - carbon	4	Clean and contaminated liquid chemicals. Better thermal conductivity and higher 'balancing' pressure (pv-factor) than Silicon carbon - carbon
Tungsten carbide - Tungsten carbide	4 - 6	Very abrasive media, suspensions. Moderate corrosion resistance
Tungsten carbide - Silicon carbide	4 - 6	Very abrasive media, suspensions. Moderate corrosion resistance. Better thermal conductivity and higher 'balancing' pressure (pv-factor) than tungsten carbide - tungsten carbide

Figure 7.32 Price relationships of seal material combinations

Material	Temperature of pumped liquid °C	Sealing element	Resistant to, e.g.
Nitrile rubber a Low nitrate b Medium nitrate c High nitrate	- 40 to + 80 - 35 to + 90 - 30 to + 100	O-rings, bellows, sleeves, etc	Water, mineral oils and other petroleum products
Ethylene-propylene rubber	- 40 to + 150	ditto	Animal and vegetable oil, chemicals. Not suitable for petroleum products
Silicone rubber	- 60 to + 200	O-rings	Weak and oxidising chemicals
Fluorocarbon rubber	- 5 to + 225	O-rings, bellows, sleeves, etc.	Chemically resistant wtih certain exceptions - e.g. ketones
PTFE Polytetrafluoro-ethylene	- 100 to + 250	ditto	Chemically resistant
Fibre reinforced PTFE	+ 150 to + 250	Sleeves, O-rings, etc.	Hot oils, asphalt, etc.
Perfluorelast (Kalrez)	- 30 to + 400	O-rings	Chemically resistant

Figure 7.33 Elastomers for mechanical seals

7.3.2.7 External systems

Mechanical seals can be fitted with various external systems to help the seal cope with the operating conditions. External systems are designed to protect the seal faces from adverse conditions and, in some cases, to assist in the sealing function. The popular systems, from API 610 and API 682, are listed below.

Plan 01 VDMA A	Internal recirculation from pump discharge to seal cavity. Liquid returns to pump through throat bush.
Plan 02	Seal cavity is fitted with two plugged connections for possible future use.
Plan 11	Recirculation from pump discharge through a flow control orifice to seal cavity. Liquid returns to pump through throat bush.

Plan 21 — Recirculation from pump discharge through flow control orifice and cooler to seal cavity. Liquid returns to pump through throat bush.

Plan 23 VDMA C — Recirculation from seal cavity by pumping ring through cooler to seal cavity.

Plan 31 VDMA B — Recirculation from pump discharge through a cyclone separator to the seal cavity. Unwanted solids/liquid returned to pump suction.

Plan 52 — Buffer liquid from external reservoir circulated through seal cavity. Usually used with double seals. Circulation can be by pumping ring or thermo-syphon. Reservoir is usually continuously vented to a vapour recovery system or flare stack. A cooler or heater can be fitted.

Plan 53 — Barrier liquid from external pressurised reservoir circulated through seal cavity, used with double seals. Circulation can be by pumping ring or thermo-syphon. A cooler or heater can be fitted.

Plan 61 — Gland plate fitted with two plugged connections, for possible future use, as quench on atmospheric side of seal.

Plan 62 — External fluid quench on atmospheric side of seal. Quench usually constrained by a throttle bush.

Plan 12 and 22 are similar to 11 and 21 but incorporate a strainer. Strainers, like filters, require emptying. Someone has to remember to check them. Cyclone separators are preferred because the solids removed from the seal auxiliary system are piped back into the pump. Plan 41 is similar to 31 but includes a cooler. API 610 has other systems which are used occasionally; 13, 32, 51, and 54. API 610/682 should be consulted for the optional equipment which can be fitted.

Plans 52 and 53 require an external liquid to be piped into the stuffing box cavity, a buffer or barrier liquid. The liquid must have some desirable properties, in areas where the process liquid is deficient. Some necessary qualities include:

- safe, i.e. not hazardous,
- clean,
- a better temperature, hotter or colder,
- good lubricant,
- good thermal conductivity.

Buffer liquids are generally a hydraulic oil or water. These two have successfully coped with most applications in the past. Both are obviously safe at reasonable temperatures.

The following table indicates popular barrier liquids which have been used extensively in the past.

Liquid	Temperature range
Ethylene chloride	-130 to -15 °C
Propanol	-120 to +70 °C
Ethanol	-105 to +50 °C
Methanol	-90 to +40 °C
Butanol	-80 to +90 °C
Paraffin	-30 to +160 °C
Hydraulic oils	-30 to +80 °C
Mineral oils	-10 to +200 °C
Water	+5 to +80 °C
Ethylene glycol	0 to +175 °C
Vegetable oils	+10 to +130 °C
Glycerine	+100 to +260 °C
Heat transfer oils	+100 to +350 °C

Figure 7.34 Traditional barrier liquids

One of the criterion we required was for the liquid to be safe. This requirement is necessary in case the liquid leaks out of the second seal. Let us review some of the liquids 'standardised' in past installations.

Ethylene chloride; ethyl chloride, chloroethane; highly flammable and toxic.

Propanol, flammable and slightly toxic.

Ethanol, ethyl alcohol, flammable, differing opinions about its health risk.

Methanol, methyl alcohol, flammable and toxic.

Butanol, butyl alcohol, flammable and toxic.

These liquids are covered by both UK and USA safety legislation, and by the ADR transportation regulations, see section 3.7.2. Obviously the use of these liquids as barriers or buffers needs very careful consideration of the immediate surroundings of the pump, especially if the pump is situated in an enclosed pump house.

Special attention should be paid to the external systems with regard to pressure and the consequences of unintentional shut-down or breakdown. The difficulties of achieving sufficient operating reliability in the external systems have recently led to a certain movement towards special single mechanical seals, which do not require barrier or buffer liquid.

7.3.2.8 Maintenance

The operating life of any seal is primarily dependent upon selecting the right seal and the right material, i.e. the optimum with regard to the type of duty. The service life as well as the leakage rate are further influenced by the condition of the pump and its design and by the manner in which it is operated, continuous or intermittent, etc. Depending upon the speed, a pair of seats should last for at least one year in a clean liquid environment.

Regular maintenance of a mechanical seal is limited to checking for leakage once a week. If the pump is handling a hazardous liquid controlled by safety legislation, the normal vapour leakage may have to be measured and logged. From the point in time when leakage appears as drops, the seal can usually be operated for a week or two before the seats need to be replaced.

It is necessary to check the function of any external barrier or buffer liquid system, etc.

7.3.2.9 Trouble-shooting

Leakage is the only indication of a faulty mechanical seal. In order to be able to evaluate if and when the seal must be dealt with, it is not sufficient to notice leakage. The occurrence of leakage does not provide sufficient information as to the course of action to be taken. It is also necessary to evaluate the circumstances, when and where, and the appearance of the leaking liquid. Figure 7.35 reviews some typical examples. In more difficult cases the pump and seal manufacturers should be contacted and should also be present when the seal is disassembled since the cause can easily be destroyed during disassembly. Always read the manual before carrying out any work on a pump or seal.

Problem	Probable cause	Remedy
Newly installed seal leaks badly when stationary.	Incorrect installation. Rotating seat hang-up.	Check manual, disassemble completely. Check cleanliness. Rebuild.
Newly installed seal leaks moderately when stationary and when running.	Damaged 'O' rings.	Strip, check cleanliness and rebuild.
	Seat faces not flat.	Depending upon the materials of the seat faces the seal may bed-in.

Problem	Probable cause	Remedy
Newly installed seal is pressure tight when tested but leaks when running	Single spring seal running in wrong direction.	Disassemble and check spring handing.
	Incorrectly fitted stationary seat.	Read manual, disassemble and rebuild.
	Shaft run-out too great.	Check run-out, check bearings. Replace parts as necessary.
	Eccentricity between shaft and sleeve.	Check run-out of sleeve to shaft. Remove sleeve and check sleeve bore to sleeve OD.
	Vibration.	Check for shaft or sleeve run-out.
Seal begins to leak after a short period of operation.	Debris in process liquid or operating condition transients during start-up.	When process conditions have stabilised, disassemble seal and rebuild.
	Seat or elastomer materials not suitable for operating conditions.	Check operating conditions against specification. Consult pump and seal manufacturer.
	Seat or elastomer materials not suitable for cleaning or stationary conditions.	Consult pump and seal manufacturer.
Seal was working correctly then suddenly starts to leak badly.	Cracked seat(s).	Thermal shock or dry running. Investigate cause and prevent repetition.
	Bearing failure.	Inspect bearings, replace as necessary.
	Loss of barrier/buffer liquid.	Add instrumentation to external system.

Figure 7.35 Trouble-shooting for mechanical seals

7.4 Methods of sealing: Reciprocating rods

There are two main differences between rotary shafts and reciprocating rods:

- a reciprocating rod tends to carry liquid through the seal,
- a reciprocating rod stops moving twice every cycle.

Because of these differences, the technology used for rotating shafts is not always applicable to reciprocating applications. Reciprocating seals can also be placed in one of two categories; stationary seals and moving seals. Moving seals are usually associated with piston seals.

In this section we are interested in seals for pumps. Modern plunger pumps operate at mean linear velocities of up to 2 m/s, piston pumps up to 1 m/s. The maximum instantaneous velocity is about 20% greater. A wide variety of seals are available, however some seals are for hydraulic power applications, piston/cylinders for hydraulic actuation. Hydraulic seals are generally suitable for 0.5 m/s and are intended to work with hydraulic oil, water-oil emulsions or water. Hydraulic seals may not be suitable for continuous operation in process pumps and seal materials may not be resistant to a wide range of liquids generally encountered.

The need for rod seals can sometimes be completely eliminated by the use of diaphragms or bellows. Short stroke pumps, typically metering and dosing pumps, have successfully used diaphragm and bellows to eliminate the leakage along rods. Precautions must be taken in the event of fatigue failures and extra instrumentation can be fitted for failure detection. Diaphragms and bellows cannot be applied to longer stroke machines because of the space required to accommodate the stroke.

7.4.1 Lip seal

Lip seals, similar in design to rotary seals, are used for preventing oil escaping along crosshead extensions where they pass through the crankcase wall. Similar seals are mounted in reverse to prevent the ingress of dirt and moisture. Seals for this type of application are limited to about 100 °C.

Special lip seals in the form of 'U' rings can be used for process sealing, see figure 7.36. Generally in FKM elastomers or PTFE and energised by pressure plus a stainless steel spring, these seals can be capable of sealing 300 bar. Like soft packing rings, 'U' rings can be stacked in series. The 'U' ring is not compressed axially and must be fitted with a gland which locks to provide a fixed installation gap. No adjustment is possible or necessary.

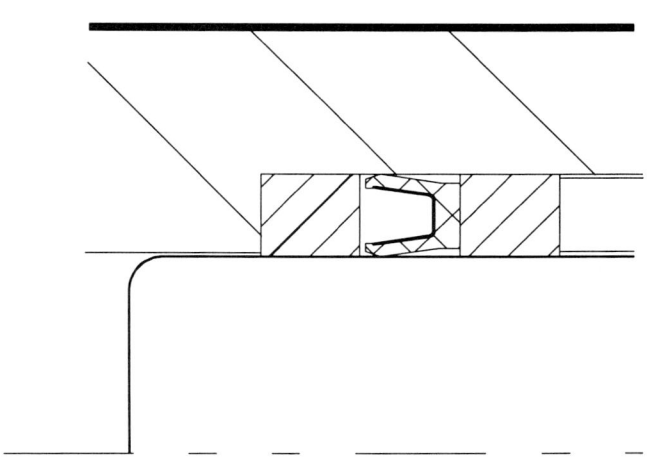

Figure 7.36 'U' ring for process seals

Another style of lip seal which has proved successful is the semi-chevron; the chevron style is discussed later. The semi-chevron is a combination of styles; soft packing and a chevron ring, see figure 7.37. The outside diameter of the packing is almost flush; the axial gland load compresses the OD against the bore of the stuffing box. The bore of the packing is recessed to form pockets for the sealing lip. The sealing force on the lip is initiated by interference and supplemented by liquid pressure. The packing is not adjustable and requires no maintenance. The gland is tightened to provide an initial compression on the packing but no further adjustment is necessary. Semi-chevrons are available for over 200 bar at temperatures up to 105 °C. A packing set normally consists of three units with three sealing lips. Packing sets can be arranged in series. A triple packed box has been operated successfully. Lantern rings between the sets can be used for buffer liquids or lubrication.

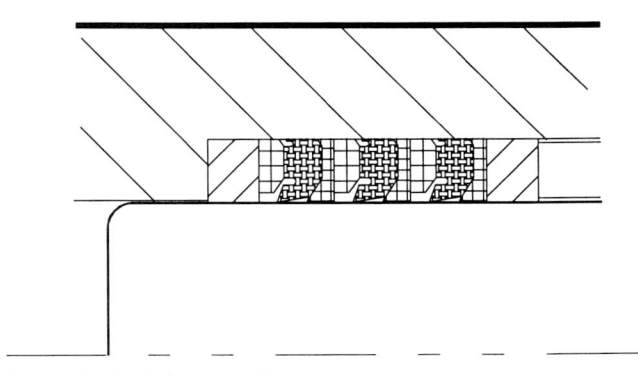

Figure 7.37 Semi-chevron packing set

Soft packing can be used for any pressure. At high pressures, over 300 bar say, the routine maintenance, adjustment, can become tiresome. Chevron packing was designed to remove the necessity for adjustment. Chevron packing has internal and external sealing lips. Packing sets can be compressed manually, but this requires skill and experience and is not recommended. Most sets are spring loaded and the gland is locked solid, see figure 7.38. Spring loading can be by a single large spring or multiple small springs. Chevron sets are available in various compounds and allow operation over 2000 bar at temperatures up to 290 °C. Pure PTFE is available but is not suitable for higher speed pumps. Chevron packing can be fitted as a cartridge assembly. Offshore pumps, handling hazardous liquids, have been equipped with three chevron sets in series with buffer and lubrication systems, see figure 7.39. For most high pressure applications, lubrication can greatly extend packing life.

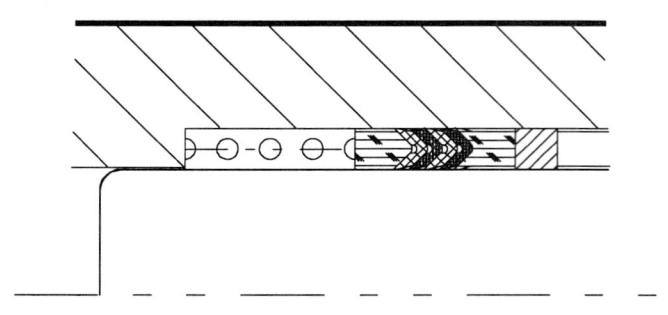

Figure 7.38 Spring loaded chevron packing

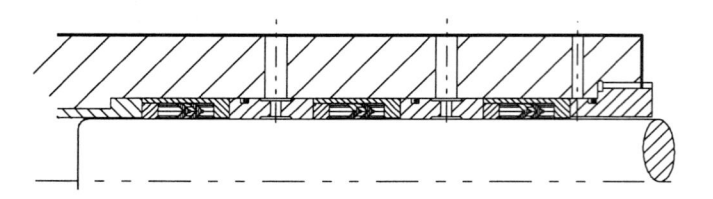

Figure 7.39 Triple packed cartridge chevron stuffing box

7.4.2 Soft packing

Soft packing for reciprocating rods is very similar to the soft packing used for rotary shafts. Because of the different running conditions the design limits are modified, figure 7.40 indicates the scale of the modifications.

Rotary		Reciprocating	
Speed m/s	Pressure bar	Speed m/s	Pressure bar
2	5	1	150
10	10	1	100
25	20	2	250

Figure 7.40 Comparison of soft packing capabilities

7.5 Soft packing process liquid seals for reciprocating rods

Seals on reciprocating rods, unlike rotodynamic pump shaft seals, operate alternatively at suction and discharge conditions. If multiple seals are fitted, the seals behind the primary seal may operate at reasonably constant conditions.

7.5.1 Operating principles

Soft packing seals for reciprocating rods operate on the same principles as the rotary seals. Allowances must be made for the rod attempting to drag the liquid through the packing. Most stuffing boxes have between three and seven rings of packing. Fitting too many rings is as bad as fitting too few. With too few rings it is difficult to create adequate pressure drop within the stuffing box and leakage is always great. With too many rings the outer rings receive too little lubrication and burn up. Proprietary pump specifications which specify a fixed number of rings in the stuffing box are misguided. Pressures higher than the pump operating pressure can be generated in the stuffing box and the packing rating must take this into account.

7.5.2 Design variations

The design of stuffing boxes for reciprocating rods generally follows the same pattern as rotary shafts. The major exception is that mechanical seals are not available as an option. Small pumps may have a diaphragm or bellows to eliminate the leakage problem. With packing, more sets are added as the liquid becomes more hazardous or problematic to seal. We have seen in section 7.4.1 that lip seals can be stacked in series, figure 7.39. Soft packing sets can also be stacked in series; the stuffing box design becomes complicated, correct adjustment is dubious and the rod sizes permissible within a fixed centre distance are greatly reduced. Figure 7.41 shows a stuffing box with two sets of independently adjustable soft packing, both rated at 100%. Lubrication or a buffer liquid can be supplied to the middle lantern ring. A quench could be applied across the atmospheric side of the second packing. Additional rings could be fitted after the second lantern ring but some liquid supply would definitely be required. Just as with rotary sealing systems, reciprocating rod seals can be multiple units; it is important to specify or know whether secondary/tertiary seals are rated for full pressure. When handling difficult liquids the use, if possible, of non-adjustable packing styles will achieve better packing life and pump availability.

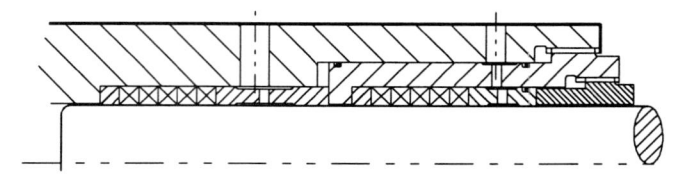

Figure 7.41 Stuffing box with adjustable primary and secondary packing

7.5.3 Packing material

Packing materials suitable for rotary shafts, see figure 7.16, in general, are suitable for reciprocating applications. Reciprocating rods do not have sleeves, the rod may be hardened or coated. Packing with Aramid® or Kevlar® fibres tends to be abrasive. Very hard coatings, such as chrome oxide or tungsten carbide, should be used.

Braided and plait packing is normally cut to length from a roll. Reciprocating rods also use moulded packing rings which are solid and must be assembled from the end of the rod. Rings can also be machined from solid material. Special profiles and diameters can easily be accommodated when the rings are machined.

7.5.4 External systems

External systems are similar to their rotating counterparts except barrier liquids are not used and the seal cavity cannot be flushed. Lubrication is the most common system. The lubricant is pumped by a special single cylinder reciprocating pump to each stuffing box. The lubricant mixes with the product and some is lost.

Low pressure quench is the second most popular system. The quench liquid is constrained by a close fitting bush in the gland or an auxiliary packing set.

Heating and cooling of the packing can be accomplished by circulating liquid through a lantern ring. Because the rod is reciprocating, liquid is carried to packing at both sides of the lantern ring. The rod is heated or cooled by direct contact with the liquid.

Flushing through the throat bush is popular on solids handling pumps and for liquids which may crystallise in the packing. Flushing can be low pressure on the suction stroke when supplied by a low pressure external source. High pressure flushing is like lubrication, a single cylinder from another pump is piped to each stuffing box. The flush injection is synchronised to flush on the back stroke when the rod would tend to draw solids into the box.

7.5.5 Maintenance

See the maintenance section for rotary shafts, 7.3.1.5. Maintenance of adjustable soft packing is a skilled job. The pump and packing manufacturer's instructions must be followed for good results. If site labour is poor or the site is inaccessible, non-adjustable packing should be specified.

7.5.6 Trouble-shooting

Read section 7.3.1.6. Except for the reference to bearing wear and packing rotating, all comments are valid. If spring loaded packing runs too hot and tries to stick to the rod it is possible to drag the packing axially in the stuffing box. This can cause longitudinal scratches on the packing OD. It is possible to have rods running eccentrically in the stuffing box. Periodically, the rod should be checked to ensure it is central. Some rod connections allow for radial misalignment, others do not. As the pump wears, crossheads may drop in their guides, effectively lowering the centre of the rod. If eccentric packing wear is a problem the cylinder alignment should be confirmed.

7.6 Selection of process liquid sealing

As we have seen from the preceding sections in this chapter, sealing technology is complicated and there are a wide range of choices open to the pump designer and the pump user. Environmental controls, together with increased personnel safety requirements, have imposed limitations on the allowable leakage rates of some liquids and vapours. A logical approach must be adopted for seal selection to achieve a sound, practical, operational solution. The pump and seal manufacturer must know everything the user knows about the liquid. If any approximations are needed, the user must do the approximations, ranges are better than single values, and the data must be distinguished as such.

7.6.1 Process liquid

The seal selection process must start with the process liquid. If the process liquid is legally registered as hazardous then limits on exposure and vapour concentrations will be published. From the design of the immediate pump surroundings, an allowable concentration can be converted to an allowable leakage rate. If a rotodynamic pump is being considered, and the liquid is so hazardous that leakage cannot be tolerated then read the introduction to this chapter again and select a seal-less pump. If a reciprocating pump is considered for very hazardous liquids, try to use small pumps with diaphragms or bellows. If the liquid is only flammable, then flammability limits will be published and an allowable leakage can be calculated.

Consideration must to given to what happens to the liquid when it does leak out of the seal; will it remain liquid or will it vaporise? If it remains liquid, will it cause damage to the pump or the immediate surroundings, can it be piped to normal drains or must it piped in closed drains for treatment? If the escaped liquid vaporises; is the vapour heavier or lighter than air, where will it accumulate, what damage will it cause?

Obviously the nature of the process liquid affects more than just the seal selection. However, the pump seal may be one of the few places where the liquid has a chance to escape.

7.6.2 Size, speed and pressure

The rate of leakage from different types of seal depends on the pressure at the seal, shaft diameter, speed and, in the case of soft packings, also to a very large extent upon maintenance. The following can be given as guide values:

- Mechanical seal, single, 0.2 to 20 cm³/hour
- Soft packing, single, functioning well and properly maintained, 5 to 50 cm³/hour

The values given refer to seals working correctly. When a seal begins to fail, the increase in leakage is dependant upon the whole stuffing box design. Leakage should also be considered as a proportion of the pump flow. Soft packing leakage rates can be as low as 0.01%.

Figure 7.42 shows guide-lines, presented by P. E. Bowden of Flexibox U.K., to the 13th BPMA International Pump Technical Conference in London in 1993, to assist with mechanical seal design choices. The graph applies to mechanical seals up to 152 mm diameter, pressures up to 40 bar and speeds up to 3600 rev/min.

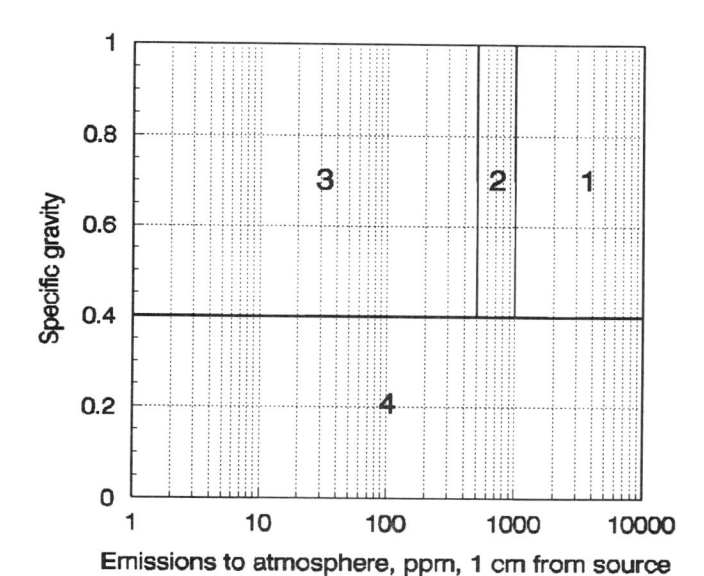

Chart area	Acceptable technology
1	Single, double or tandem seals acceptable
2	Single seals may be acceptable. Depending on actual operating conditions, seal size, pressure and temperature, tandem or double seals may be required to meet emission regulations.
3	Tandem or double seals acceptable.
4	Double seals required.

Figure 7.42 Mechanical seal selection guide-lines

Leakage should be a part of the design philosophy. Not 'What shall we do if the seal leaks' but 'What shall we do with the leakage when the seal leaks'. When treated correctly, leakage can be recycled and is not necessarily a waste disposal problem.

7.6.3 Local environment

The local environment surrounding the pump plays a important role in the consideration of leakage. If the pump is in an enclosed pump house, leakage will tend to accumulate within the pump house depending upon the arrangements made for ventilation. If the pump house is unmanned and personnel visits are once a week, the requirement for ventilation will be low. Vapour may accumulate to toxic concentrations for personnel or form an explosive mixture. Mechanical ventilation may be necessary, the outlet from the pump house may have to be ducted to a flare stack for disposal.

If the pump house is open, leakage will tend to disperse naturally. If the liquid has an obnoxious smell, this may not be acceptable. If the pump site is close to habitation, natural dispersion may cause problems in the community. Remote pump sites with open pump installations avoid local dispersion problems. The concentration is very dilute, perhaps undetectable, before reaching any community. One remote type of installation which has special problems is the off-shore installation. Offshore installations have limited escape possibilities if anything does go wrong. Although surrounded by water, the water is not always safe. Offshore in the Gulf of Mexico the water is relatively safe. Offshore in the North Sea, especially in the winter, the water is definitely not safe. Extra safety precautions must be taken to protect personnel.

If leakage is damaging to the environment then not just the local environment must be considered. The global environment may be at risk.

7.6.4 Cost

The initial cost for soft packing materials is relatively low compared with the initial cost for a mechanical seal. The initial cost is, however, not of prime importance and the life cycle cost is largely determined by a number of secondary costs associated with maintenance, leakage and external systems.

The level of routine maintenance required is dependant, not only on the seal, soft packing or mechanical seal, but also on the operating conditions. High pressure and high speed require more maintenance. The possibility of changing the working pressure is not usually within the options available. Changing the pump speed is a viable option. The increased cost of the pump must be weighed against the reduced maintenance costs and the extended operating time between routine shut-downs. Individual construction and operating costs are required to properly evaluate this option. The user should always have formed an operating strategy when pumps are purchased and the pump manufacturer should be aware of the target operating time between shut-downs. If the level of maintenance skills at the site are low, it may be best to fit cartridge seals where possible. Complete cartridges could be refurbished by the seal manufacturer ready for replacement.

The cost of liquid lost due to leakage may be considerable. Additional initial cost should be considered to reduce leakage or recycle leakage. The feasibility of either course will be dependant upon the difficulties of sealing the liquid. Suitable barrier or buffer liquids may not be available or cause as much trouble as the process liquid. Loss of any liquid in some closed loop processes may cause tremendous problems; a seal-less pump may be necessary even if the liquid is safe.

External systems may require extra instrumentation or maintenance. Some stuffing box lubricators may need the oil reservoir topping up every 24 hours. Extra storage space will be required for special lubricants and barrier/buffer liquids.

The true cost of sealing solutions is difficult to analyse. Very good data must be accessible and the data trends must be stable. Possible changes in economic conditions, over the life of the pump, may necessitate changes in sealing philosophy. Can the pump selected be adapted to other sealing systems?

7.6.5 Standardisation

The most widely used standard for seals is probably DIN 24960; parts of this were extracted for ISO 3069. DIN 24960 specifies dimensions for mechanical seals, seal cavities, seal designations and materials. DIN 24960 was referenced in the original DIN standards for standard water and chemical pumps, now ISO 2858 and ISO 5199. 24960 applies to shafts from 10 mm to 100 mm. Single seals, double and tandem seals are specified.

The Swedish forestry standardisation group has published a standard for soft packings, SSG 1300, 1320 and 1321 which comprises dimensions, installation and materials.

The most recent part of sealing standardisation is the introduction of API 682, Shaft sealing systems for centrifugal and rotary pumps. Even though the title states 'Shaft sealing systems', the standard only deals with mechanical seals; no mention of soft packed boxes. Eventually API 682 will replace the seal section of API 610. API 682 is referenced as a standard in the latest draft of the 8th edition of API 610; however it is not thought that it will be mandatory. It has already been stated that API is controlled by pump users. Pump manufacturers have made it known to API that they would like to stop redesigning their pumps every 4 or 5 years to comply with the users latest fashion. API 682 cannot be applied universally to all existing pumps because of the enlarged seal cavities required. While it is recognised by most people that larger seal cavities, not extended soft packed stuffing boxes, can extend mechanical seal life it is not possible to impose such drastic changes in a short time span.

API 682 is only applicable to seals between 30 mm and 120 mm, temperatures from -40 °C and 260 °C, and pressures from 0 to 34.5 barg. A list of applicable liquids includes; water, sour water, caustics, amines, some acids and most hydrocarbons. 'Some acids' obviously needs a little clarification. Only cartridge seals are acceptable to API 682. Hook sleeves are not acceptable. The complete seal, including sleeve, must be removable so that the complete assembly can be tested. The minimum requirement is a single balanced seal. API 682 has standardised 'qualifying' tests for seals based on four test liquids; water, propane, 20% NaOH and mineral oil. The results of the qualifying test must be submitted as part of any seal quotation. API 682 does allow pump testing with alternative face materials when the contract face materials are unsuitable for the test liquid, water.

API 682 does have a comprehensive seal selection procedure which runs to 10 pages. Seal type/style are covered, as are material recommendations, piping plans and guidance for selecting barrier/buffer liquids.

Pump users who have large numbers of pumps tend to standardise on one seal manufacturer. This type of standardisation is intended to reduce inventory, cost and space, and also reduce the chances of fitting incorrect parts to a seal. The pump user can also benefit by increasing his bargaining power when purchasing a large number of spares from one source. Seal manufacturers with a wide product range profit from this philosophy. In some applications, very special seals are required, and manufacturers must be selected for a specific product.

7.7 Bibliography

There is extensive literature for those who wish to go deeper into sealing technology and general sealing problems:

Fern, A.G., Nau, B.S. — Seals Oxford University Press.

Flitney, R.K., Nau, B.S., Reddy, D. — Seal Users Handbook, BHRA Fluid Engineering.

Mayer, E. — Axiale Gleitringdichtungen, VDI-Verlag.

Lymer, A., Greenshields, A.L. — Thermal Aspects of Mechanical Seals, Flexibox Ltd.

BPMA 13th International Technical Conference — Pumps for a safer future, London, April 1993.

BHRA Engineering is now the BHR Group. They have carried out a vast amount of research into sealing technology and have lots of papers available, at a cost. BHR Group can be contacted on 0234 750422, Fax 0234 750074.

8 Shaft couplings

Chapter 8 sets out to present the factors which influence the relationship between shaft couplings and the pump unit. Beginning with a short review of the different types of coupling the chapter continues with an explanation of the various types of misalignment and the forces and moments which are transmitted. Section 8.5 gives advice on "service factors" with special emphasis on the torque produced when starting electric motors. Several other factors are dealt with, see the contents, although Section 8.11, "Shaft alignment", is considered to be of importance, several different methods are explained with the intention of making alignment quicker, cheaper and better in comparison with current practice. Section 8.12, "Choice of coupling", contains a check-list of important factors related to couplings.

Contents

8.1 Introduction

8.2 Types of coupling

8.3 Misalignment

8.4 Forces and moments

8.5 Service factors

8.6 Speed

8.7 Size and weight

8.8 Environment

8.9 Installation and disassembly

8.10 Service life

8.11 Shaft alignment
    8.11.1 General
    8.11.2 Methods of alignment
    8.11.3 Determining shim thickness
    8.11.4 Graphical method of determining shim thickness

8.12 Choice of coupling
    8.12.1 Costs
    8.12.2 Factors influencing choice

8.13 Guards

8.14 Bibliography

8.1 Introduction

A shaft coupling is used to transfer torque between two in-line, or nearly in-line, rotating shafts. The magnitude of the torque in the shafts is equal, although slipping and disengagement can cause speed variations. In its simplest, and perhaps oldest form, the coupling acts as a means of joining shafts. That is one function; another function is to join two shafts which are not necessarily in perfect alignment with each other. The coupling in this case must be capable of absorbing such misalignment. Modern couplings, between pump and motor, must be capable of rapid disassembly.

Shaft couplings can perform many different functions and have varying characteristics. Shaft couplings are thus usually divided into three main groups with sub-divisions, namely:

Non-disengaging couplings

- solid
- torsionally rigid
- torsionally flexible

Disengaging couplings

- clutch with manual over-ride mechanism
- free-wheeling clutches

Limited torque couplings

- non-controlled
- controlled and variable

Some of the requirements for flexible couplings, including definitions, performance and operating conditions, dimensions of bores, reference to components together with an appendix on alignment is to be found in BS 3170. Friction clutches and power-take-off assemblies for engines, and their requirements are included in BS 3092. Process pumps to API 610 standard have spacer couplings in accordance with API 671.

For pump applications it is usual to utilise a coupling from the first group above, although special installations make use of disengaging clutches and limited torque couplings. Examples being centrifugal clutches to reduce starting loads when using a direct-on-line starting induction motor. Hydrodynamic clutches can be used for reducing starting loads and speed regulation. Combinations of brakes and reverse locks can be used in order to prevent reverse pump rotation for example.

Power recovery hydraulic turbines are often coupled to the non drive end of pump motors so that the turbine can unload the motor. The coupling used is a free-wheel type with manual over-ride so that the pump/motor can start-up before the turbine. Once the turbine runs up, as it tries to rotate faster than the motor the clutch locks automatically and power is transmitted.

8.2 Types of coupling

Non-disengaging couplings maintain, after assembly, a more or less flexible although continuous transmission of the rotational movement. The connection is only broken for disassembly, repair, etc. Flexible couplings of one form or another, which are capable of absorbing residual misalignment, are most common, although solid couplings do have their areas of use, see figure 8.1. One example is the split muff coupling, the main advantage being its ease of assembly. It is best used for low speed applications due to the difficulties in balancing. The sleeve coupling is mounted and removed by oil-injection; being almost symmetrical, balancing is easy.

Another example is in the case of long-shaft pumps and stirrers, figure 8.2, where, because of space requirements when disassembling, it is necessary to split the shaft. Production technicalities also necessitate the joining of long shafts. The most usual couplings in these cases are disc or flanged couplings.

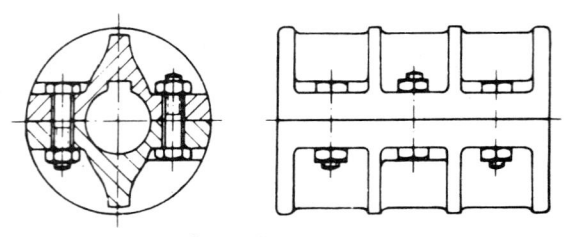

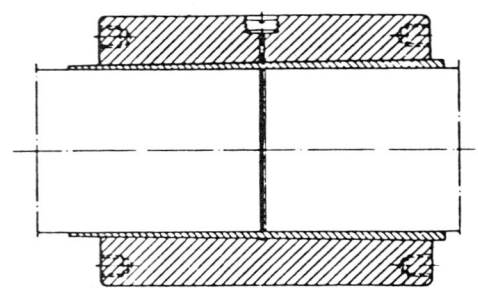

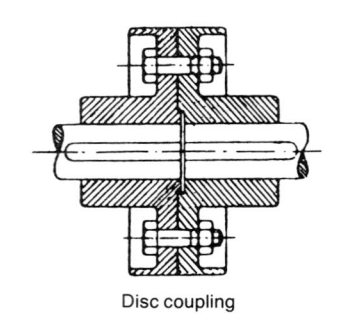

Figure 8.1 Examples of solid shaft couplings

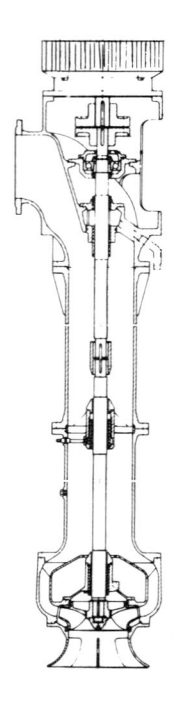

Figure 8.2 Long-shaft vertical pump

Torsionally-rigid flexible couplings consist of various types of diaphragm and gear couplings, see figure 8.3. Couplings with a single functional element have the ability to take up angular and axial misalignment. Couplings with two functioning elements separated by fixed 'spacer', are also able to cope with radial misalignment, whereby the magnitude of the radial misalignment is determined by the angular misalignment times the distance between the coupling elements.

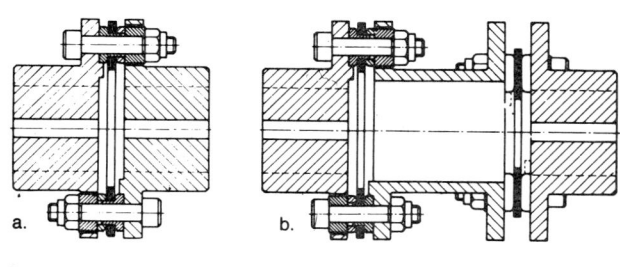

a. Single diaphragm coupling
b. Double diaphragm coupling

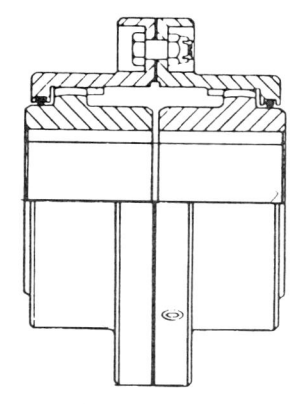

Double gear coupling

Figure 8.3 Examples of torsionally-rigid flexible couplings

Torsionally flexible shaft couplings usually consist of flexible rubber, plastic or even steel elements, figure 8.4. The first mentioned coupling elements require somewhat larger coupling diameters because of their lower load carrying capacity. Single element couplings can accommodate radial misalignment. The flexible spring coupling is interesting because it is designed to have a variable torque/deflection characteristic. Together with dampening provided by the grease lubricant, the variable torque/deflection characteristic furnishes a powerful torsional vibration dampener.

The torsionally-flexible couplings shown can be built with two working elements and a spacer to allow additional radial misalignment. In order to simplify disassembly and service of some machines, spacer couplings are preferred, figure 8.5. Removal of the coupling spacer enables service of the respective machines without necessitating their removal. API 610 specifies 5" spacer lengths as a minimum. EN 22858, ISO 2858, for 16 bar back pull-out pumps, specify spacer lengths from 100 mm to 180 mm depending upon pump size.

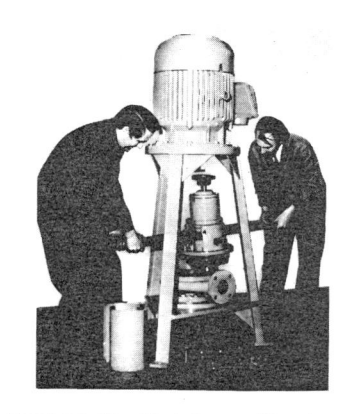

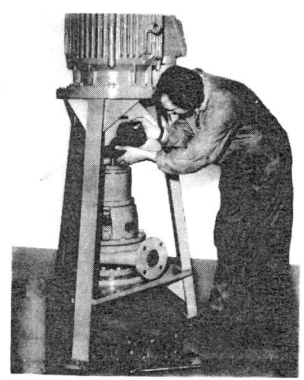

Figure 8.5 Simplified disassembly using a spacer coupling

8.3 Misalignment

Three types of movement or deviation can occur between two shafts, see figure 8.6, namely:

- Radial misalignment, where the shafts are parallel although not lying on a common centre line.

- Axial misalignment, end float, where the shaft centre lines are in alignment although the axial position is incorrect and axial movement may be possible.

- Angular misalignment, where the centre lines of the respective shafts are not parallel.

Rubber ring coupling

Flexible spring coupling

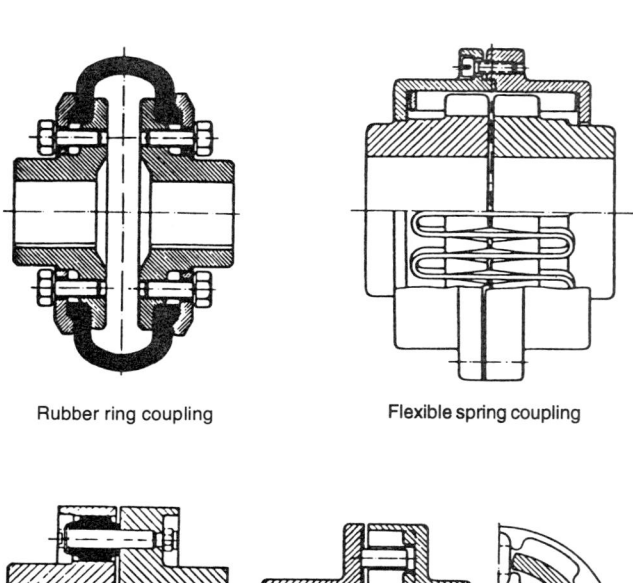

Rubber sleeve coupling

Rubber bush coupling

Figure 8.4 Examples of torsionally flexible couplings

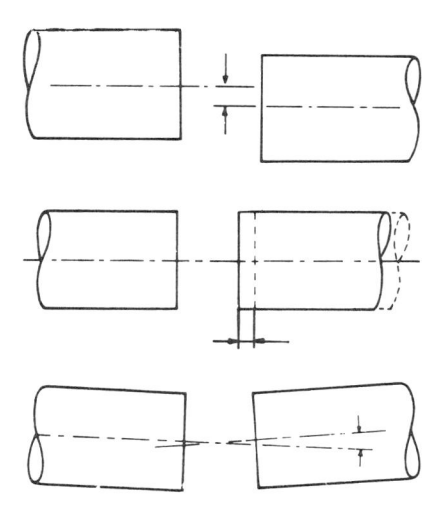

Figure 8.6 Types of misalignment

The deviations can occur singly or in combinations. Also the individual deviations can change with operating conditions. A typical changing condition is from cold to running temperature conditions. Thermal growth causes machine centre heights to increase slightly as they warm up. Process pumps, such as API 610 pumps, are centre-line mounted to avoid thermal growth of the pump casing effecting the pipework. However, the motor driving the centre-line mounted pump is foot mounted and does have thermal growth. In this situation, motors are mounted low so that the thermal growth will expand the motor to near perfect alignment. In large machines changes in ambient temperature or sunshine can affect the alignment. The thermal growth phenomenon can be further complicated when the drive and non-drive ends of a machine expand at differing rates. Not only the radial alignment changes, but also the angular alignment. Accurate on-line measurement is necessary to check for this condition.

Suppliers of couplings provide information relating to the maximum permissible deviations, usually stated for each individual type of deviation. It is important to know the maximum permissible values of combined misalignment, see figure 8.7, and how the maximum permitted deviations are influenced by speed and the torque transmitted.

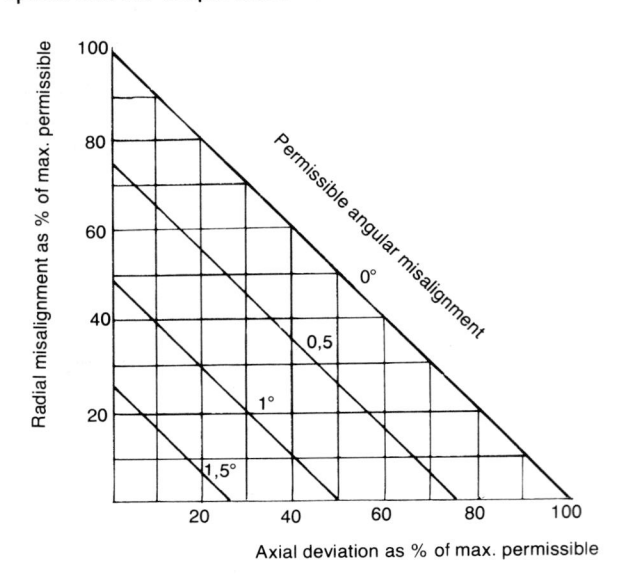

Figure 8.7 Permissible angular misalignment as function of axial deviation and radial misalignment for a particular size of double-diaphragm spacer coupling

The service life of both couplings and machines, normally machine bearings, are influenced by misalignment. Just how much the life of the machine is affected can only be judged when information regarding the precise magnitude of the torque and forces transmitted due to misalignment are known. It is usual to refer only to the amount of misalignment permitted for a specific coupling type, when it is the amount of misalignment tolerated by the machine, figure 8.8, which should really be investigated.

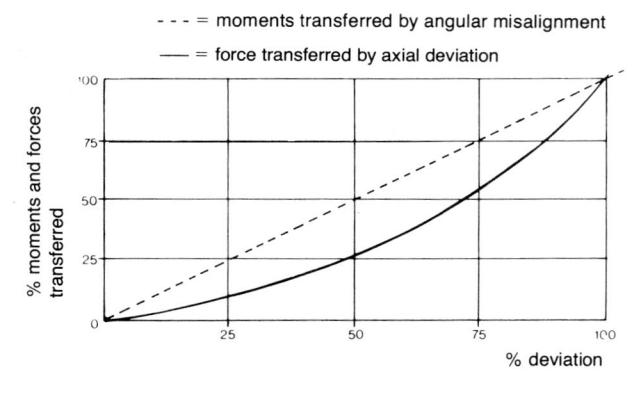

Figure 8.8 Relationship between misalignment and transmitted forces/moments

8.4 Forces and moments

A solid coupling is designed and constructed to only be subjected to torsional power transmission torques. Flexible couplings can be subjected to bending moments as well as axial and radial forces. The solid coupling does not allow the shafts to move independently of each other. Torque and axial movement are transmitted directly from one shaft to the other. Diaphragm and gear couplings transmit torque directly but react differently to axial and radial movement. A diaphragm coupling allows the shafts to move both axially and radially, the diaphragms are deformed, an axial force and a radial moment are generated. The double gear coupling also allows axial and radial movement. No axial force is produced, a radial load is produced rather than a moment. The torsionally flexible coupling produces radial loads rather than moments. The rubber ring coupling will produce an axial force in response to axial movement, the other couplings will slide to accommodate axial movement.

8.5 Service factors

When determining the size of flexible and solid couplings, it is usual to evaluate a service factor or safety factor, which primarily takes account of the type of driving and driven machines, the mode of starting and duty cycle. Most coupling manufacturers publish nominal ratings for each coupling type and size together with lists of service factors for various duty conditions. Drives with AC squirrel cage motors and rotodynamic pumps generally have a service factor of 1.0. Arduous duty cycles, like high pressure descaling, would increase the service factor to 1.25. If the pump driver was an engine, the service factor would increase between +0.5 and +1.0 depending upon the number of cylinders. Service factors for an AC motor driving a reciprocating pump can be as high as 3.0.

In order to compare different couplings objectively a new method has been developed which takes into consideration the frequency of starting, temperature, the moments of inertia of the driving and driven machine, normal torque and maximum torque.

This method has been presented in the German coupling standard DIN 740, which, apart from the method of calculation, also contains dimensional standards. There are, however, two additional service factors which should be considered.

The first is the effect that shaft misalignment can have on the coupling. A factor based on the extent of allowable misalignment expressed as a percentage of the maximum permissible deviation, should also be given.

The second factor should take into consideration the level of vibration of both machines, at least vibration velocities above 1.5 to 2 mm/s RMS. Note that for rotodynamic pumps, vibrations of up to 5 mm/sec RMS can be permissible, which is approximately equivalent to balancing degree Q16 according to VDI 2060 or quality class B for class IV machines according to BS 4675. The size of the various factors and their influence on coupling speed varies with different types, which is why the calculations and values given in DIN 740 must be used with a certain amount of caution and always with due regard to the suppliers instructions, which must apply.

A very important point in this context, to which too little consideration is given, is the magnitude of the starting torque, in the case of direct-on-line starting of a squirrel-cage induction motor. Measurements have shown that almost immediately after connection, approximately 0.04 s, a maximum torque is reached which is between 6-10 times the rated torque and even higher in some cases. This is a result of the electrical sequence in the actual motor and the fact that connection of the three phases does not occur absolutely simultaneously. The actual maximum torque thus being much greater than the starting torque quoted in motor catalogues.

An important factor for coupling calculations is the relationship between the moments of inertia of the driving and driven machine. This quotient determines the percentage of torsional moment which is to be used for the acceleration of the motor and pump rotors. When starting, the torque passing through the shaft coupling is:

$$M_k = M_i \left(1 - \frac{J_{mo}}{J_{ma} + J_{mo}}\right) = M_i \left(1 - \frac{t_o}{t}\right) \quad \text{Equ. 8.1}$$

where
$\quad M_k \quad$ = coupling torque at start $\quad$ (Nm)

M_i = internal motor torque (air-gap torque) at start (Nm)
J_{mo} = moment of inertia of motor (kgm²)
J_{ma} = moment of inertia of driven machine (kgm²)
t_o = motor starting time without load (s)
t = motor starting time with load (s)

The moment of inertia of a centrifugal pump can be given in relation to that of the motor; the following figures can be used as a guide:

2-pole motor $J_{ma} = 0.04$ to $0.2\ J_{mo}$
4-pole motor $J_{ma} = 0.25$ to $0.6\ J_{mo}$
6-pole motor $J_{ma} = 0.60$ to $3.2\ J_{mo}$

By inserting these figures in equation 8.1 and assuming that M_i is 6 to 10 times the rated torque, values for coupling torques at starting of up to 3.8 times the rated torque for 4-pole motors and 7.6 times for 6-pole motors are obtained. Thus it can be concluded that care must be taken when sizing couplings which are subjected to direct-on-line starting, especially when the driven machine has a large inertia.

8.6 Speed

Centrifugal forces increase with speed squared. The material of the coupling and the permissible peripheral velocities must be calculated. The maximum peripheral velocity for grey iron, for example, is 35 m/sec. To avoid vibrational damage it is necessary, for couplings which are not fully machined, to carry out both static and dynamic balancing at much lower speeds than is necessary for those which are fully machined.

The mass of the coupling is often quite small in relation to the rotating masses in the driving and driven machines. For a pump unit the relationship coupling/total rotor weight is approximately 0.02 to 0.08. It thus follows that out-of-balance in the coupling normally, has less effect on bearings and vibration than out-of-balance of the actual main components. However the actual position of the coupling relative to the bearings may change this. The following relationship applies:

$$F = m \cdot e \cdot \omega^2 \cdot 10^{-3} \qquad \text{Equ 8.2}$$

where
F = out-of-balance force (N)
m = out-of-balance mass (kg)
e = distance from centre of rotation to centre of gravity of out-of-balance mass (mm)
ω = angular velocity (rad/s)

For highly resilient rubber element couplings with a spacer, the out-of-balance can be further increased by whirling. It is also important that balancing is carried out using whole keys, half keys or without keys, depending upon the method of balancing the attached component.

Example:

A fully machined coupling can be assumed to have an inherent degree of balancing, without dynamic balancing, equivalent to VDI 2060 Ql6 to Q40, i.e. approximately 0.08 mm permissible centre line deviation at 3000 rev/min. If the concentricity tolerance for the shaft bore in the hub is 0.05 mm, the maximum centre line deviation can thus be 0.13 mm. This is in no way abnormal. In many cases the tolerance alone reaches this value. This centre line deviation generates an out-of-balance force of about 12 N per kg coupling weight at 3000 rev/min. A coupling for 50 kW can weigh 10 to 15 kg, which thus generates a rotational out-of-balance force of 120 to 180 N.

Most couplings have no components which can move radially to create out-of-balance forces. The gear coupling is different. The teeth on the hubs and the spacer must have clearance at the top and bottom; this can allow the spacer to move radially. In theory, the angle of the teeth flanks should provide a centralising force to counteract any tendency for the spacer to run eccentrically. Problems have been experienced with gear couplings and special attention should be paid to radial clearances and spacer weight.

The flexible spring coupling has a spring, which in theory, could move and run eccentrically. These couplings are usually used on positive displacement pumps at speeds which are low enough not to have balance problems.

8.7 Size and weight

The importance of small size and low weight so as to achieve as small a moment of inertia as possible, as well as reducing the out-of-balance forces, has been mentioned previously. In certain extreme cases light-alloy metal spacers and diaphragms are used to reduce weight. Apart from the necessity of maintaining a small size/transmitted torque ratio, it is also important, from the point of view of cost and standardisation that the coupling should be able to accommodate large variations in shaft diameter. Figure 8.9 shows the normal range of shaft diameters possible.

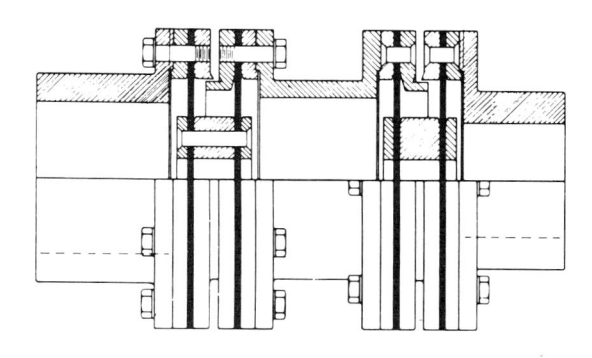

Figure 8.9 Non-sparking diaphragm coupling

8.8 Environment

Corrosive and abrasive environments can affect the service life of the coupling by causing abnormal wear to the component elements. Extremes of heat and cold can affect the strength and elasticity of the component materials. Oils, chemicals, sunlight and ozone can completely destroy a rubber element. A coupling made entirely of metal, diaphragm couplings or flexible spring, for example, are usually the only solution in such cases. The process industries offer a very poor environment from the point of view of coupling life. In certain types of industry, the petrochemical industry for instance, in refineries as well as oil and gas tankers, for example, it is necessary to use non-sparking couplings.

A non-sparking diaphragm coupling can be manufactured by making the diaphragm of monel and the remaining components of carbon steel or bronze. Non-sparking types are usually used in conjunction with flameproof electric motors in environments where there is risk of explosion, either continuously or normally during operation. Statutory regulations must be observed, see Section 9.2.8.

A flexible spring coupling has the important elements housed in a seal cover and coated with lubricant, grease. Environmental changes have little effect on the coupling. Cases of spring breakage are rare, but any parts which could create a spark are fully enclosed, see figure 8.10.

Another method of overcoming explosion risks, especially on board ship and with engine drivers, is by means of gas-tight bulkheads and bulkhead fittings consisting of two mechanical seals with barrier fluid between them, together with bellows which absorb misalignments. This type of gas-tight bulkhead fitting must be equipped with non-sparking shaft couplings.

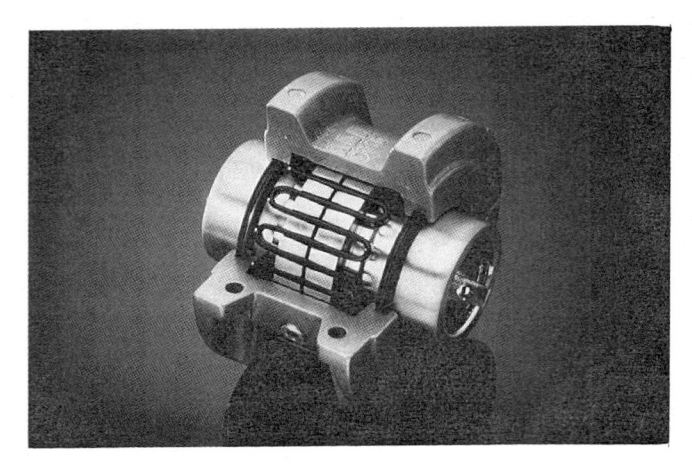

Figure 8.10 Cut-away view of a flexible spring coupling

8.9 Installation and disassembly

To maintain maximum operational reliability and to simplify assembly and service it is important that the machines connected are securely mounted, preferably on a common foundation and baseplate. Guards must be fitted to rotating parts according to safety requirements, see 8.13.

Alignment of couplings or, more correctly, alignment of the shafts which the coupling is to connect, should be carried out as accurately as possible. For pumps packaged on baseplates with their driver and other equipment, provisional alignment should be achieved by chocking the baseplate during levelling. After grouting, the alignment should be set correctly by adjusting the shims. A perfect alignment should be considered as an economic possibility, since alignment can considerably affect both service life and maintenance costs. See Section 8.11 with regard to methods of shaft alignment.

It is normal practice to bolt the pump directly to the baseplate. Other drive train equipment is shimmed to achieve correct alignment.

In the case of cardan shafts the angular deviation should be equally distributed between the two joints so as to avoid unequal rotational velocities. Furthermore, a universal coupling should always rotate with a slight amount of angular misalignment to promote lubrication.

The attachment of a coupling half to a shaft usually presents a dilemma. The hub should be securely attached and preferably absorb part of the torque, so as to reduce the load on the key, as well as being easy to detach. The practice of hub attachment is similar to that for motor shafts where the fit is usually H7/k6, light push fit up to 48 mm diameter. A push fit H7/m6 is preferred for diameters above 55 mm. Some pump manufacturers prefer a positive interference fit, typically 0.001 mm per mm of shaft diameter. These couplings are heated for mounting and dismounting. Large couplings become unwieldy. Oil injection on shallow taper shafts, without keys, can be very successful.

The tighter fit is necessitated by the fact that the height of the key is reduced from 12.5% of the diameter at 24 mm diameter to only 6% at 100 mm shaft diameter. This reduction should also be compensated for by increasing the length of the hub. In the case of electric motors the key does not normally extend right to the end of the shaft, which also increases the strain on the key. This must also be compensated for by increased hub length.

Assembly and disassembly of the coupling halves must be carried out carefully so as to avoid damage to the shaft ends and bearings. This operation could be simplified considerably if the motor, pump and coupling suppliers fitted their equipment with suitable lugs, etc., to facilitate the attachment of pullers. For electric motors a tapped hole in the end of the shaft, as shown in figure 8.11 can be supplied at extra cost. This ought to be standardised on all equipment.

Thread diameter mm								Shaft journal diameter
d_1	d_2	d_3	d_4	t_1 +2 0	t_2 min	t_3	t_4 ~	d_6
M 3	2.5	3.2	5.3	9	13	2.6	1.8	7-10
M 4	3.3	4.3	6.7	10	14	3.2	2.1	11-13
M 5	4.2	5.3	8.1	12.5	17	4	2.4	14-16
M 6	5	6.4	9.6	16	21	5	2.8	17-21
M 8	6.8	8.4	12.2	19	25	6	3.3	22-24
M 10	8.5	10.5	14.9	22	30	7.5	3.8	25-30
M 12	10.2	13	18.1	28	37.5	9.5	4.4	31-38
M 16	14	17	23	36	45	12	5.2	39-50
M 20	17.5	21	28.4	42	53	15	6.4	51-85
M 24	21	25	34.2	50	63	18	8	86-130

Figure 8.11 Tapped assembly hole in electric motor shaft

Other methods of attaching the coupling halves are shrink fits, bolted joints or some form of clamping sleeve. Taper bushes, figure 8.12, are used primarily for chain wheels and pulleys, but can be a useful alternative for couplings where space permits. Some manufacturers offer taper bushes as an alternative to parallel bores. The hydraulically loaded clamping sleeve shown is a relatively new innovation and is not used extensively in pumps.

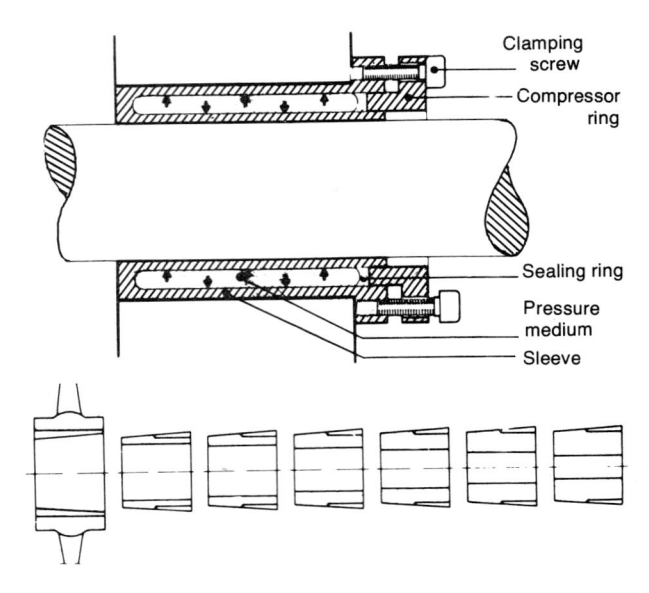

Figure 8.12 Examples of clamping sleeves

The resilient elements in the shaft coupling must be easy to purchase, replace or repair. That it must be possible to replace without disturbing the machines or coupling hubs, goes without saying.

8.10 Service life

The life of the coupling is influenced by many different factors, which vary according to the style of construction. One factor, which above all affects couplings with rubber elements, is the surrounding environment. The service life of a gear coupling is largely dependent upon regular lubrication using the correct type of lubricant according to the ambient temperature, etc. Flexible spring couplings are now available with special grease which lasts five years, routine maintenance is almost zero, and the effects of environment zero. Alignment affects the service life of all couplings irrespective of type or manufacturer.

For certain types of installations it can be desirable to use a coupling that allows a certain amount of emergency drive even in the event of failure of the flexible element. For other installations it may be necessary to use a limited torque coupling with overload protection.

It is important to carry out regular service and alignment checks according to the manufacturer's instructions, and equally important that these instructions are placed in the

Method	Shaft coupling type	Measuring device and location	Zero setting and notation rules**	Parallel misalignment mm	Inclination* mm per 100 mm measured length	Remarks
I	Short shaft coupling. Machined outer diameter. Machined on insides	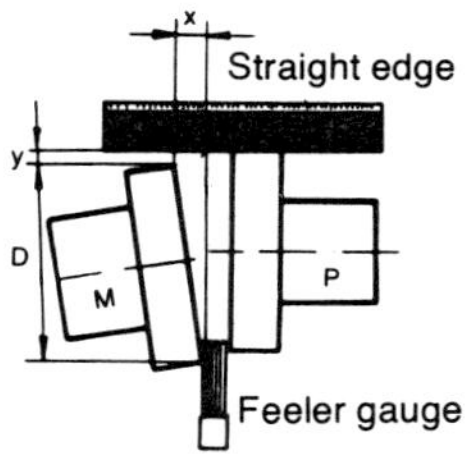 Straight edge; Feeler gauge	Misalignment according to the figure is positive i.e. the difference is measured above on the motor side.	Measured directly as dimension y	$L=\dfrac{100 \cdot x}{D}$	Make due allowance for bearing end float in the machines.
II	Short shaft coupling Requires at least a good surface at measuring pointer. Machined on insides	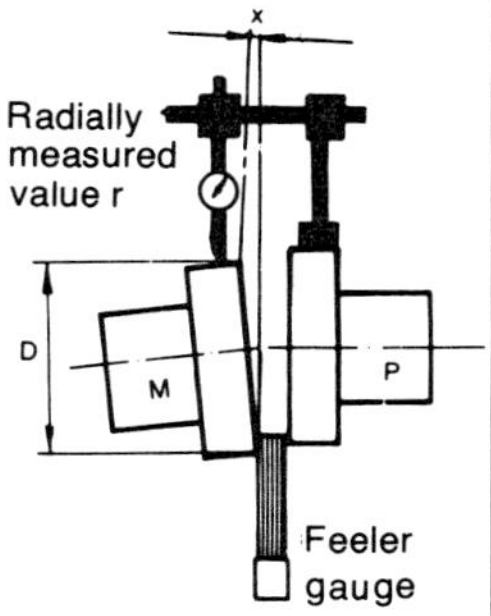 Radially measured value r; Feeler gauge	For vertical location zero set the dial above. Measured value is read after rotating one half turn.	$y=\dfrac{r}{2}$	$L=\dfrac{100 \cdot x}{D}$	Make due allowance for bearing end float in the machines. (Zero set the dial indicator underneath if the pointer is resting on the pump half.)
III	Short shaft coupling. Good surfaces at the measuring pointers	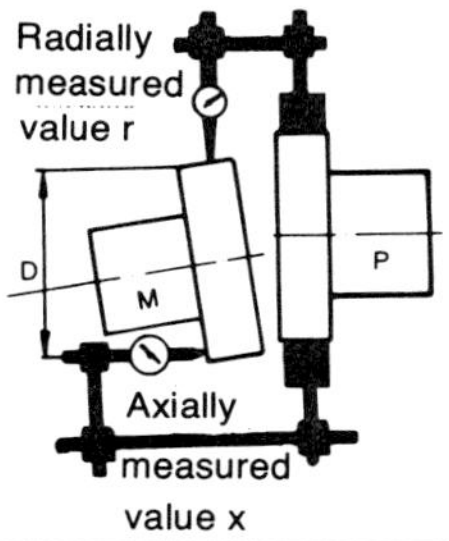 Radially measured value r; Axially measured value x	For vertical location zero set both dial indicators in the position shown, i.e. for radial deviation above and axial deviation underneath. The dials are read after rotating one half turn	$y=\dfrac{r}{2}$	$L=\dfrac{100 \cdot x}{D}$	Make due allowance for bearing end float in the machines. If both dial gauges are placed with their pointers on the pump half, then zero setting should be carried out from underneath.
IV	Long shaft couplings, i.e. couplings with a distance between the coupling halves. Good surfaces at the measuring pointers	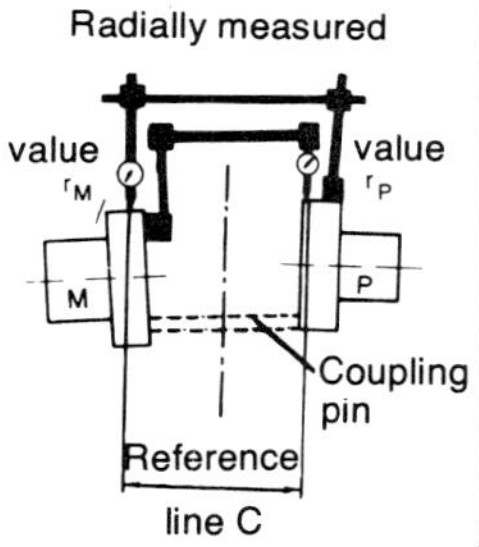 Radially measured value r_M, value r_P; Coupling pin; Reference line C	Zero set both dial gauges from above. The measurements r_M and r_P are read after rotating one half turn	$y=\dfrac{r_M - r_P}{4}$***	$L=\dfrac{r_M + r_P}{2 \cdot C} \cdot 100$	N.B. Measurement can also be carried out on "smooth" shaft ends.
V	Long shaft couplings, i.e. couplings with distance between the coupling halves. Good surfaces at the measuring pointers	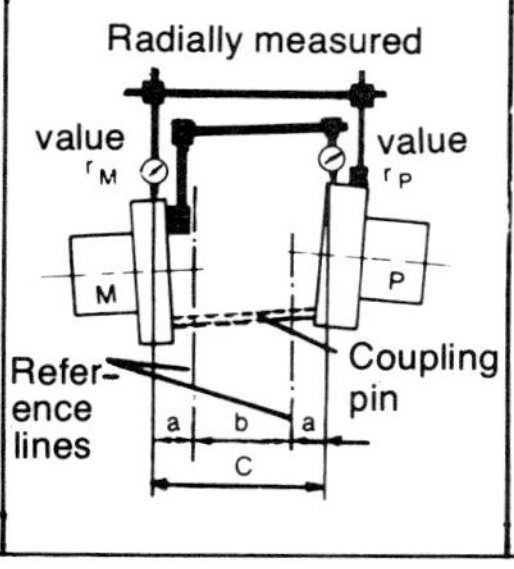 Radially measured value r_M, value r_P; Coupling pin; Reference lines	Zero set both dial gauges from above. The measurements r_M and r_P are read after rotating one half turn.	$y_M = \dfrac{r_M}{2}$*** $y_P = \dfrac{r_P}{2}$	$L=\dfrac{r_M + r_P}{2 \cdot C} \cdot 100$	Similar to method IV. Notice the position of the reference lines for calculating angular misalignment.

Figure 8.13 Shaft alignment methods

* Angular misalignment in degrees = $\dfrac{180}{\pi} \cdot \dfrac{L}{100} = 0.57 L$
** Dial gauge readings are reckoned to be positive when the pointer moves clockwise.
*** Parallel misalignment refers to a reference line as show in the figure.

hands of the personnel concerned. Unfortunately methods or regulations for assessing the degree of wear are often lacking.

8.11 Shaft alignment

8.11.1 General

Flexible shaft couplings are normally used to transfer torque between rotating shafts where the shafts are not necessarily in perfect alignment. It should be noted that a flexible coupling is not an excuse for poor alignment. Careful alignment is important for the purpose of achieving maximum operational reliability whilst reducing service and maintenance. When carrying out alignment, consideration must be given to relative movements of the respective machines due to thermal expansion and deformation caused by pipe forces/moments and setting of baseplates on foundations, etc. In certain cases, such as electric motors with plain bearings, regard must be taken of the electric motor's magnetic centre. Alignment should be carried out at various stages during installation. When alignment is carried out at cold temperatures, it is necessary to make allowances in order to compensate for the thermal expansion caused by the difference in temperature to that of the normal operating temperature of a pump, pipeline and driver. When possible, a final check should be made at operating conditions after a few weeks in service. Alignment checks should then be carried out at regular intervals. Misalignment, apart from being caused by any of the previously mentioned loads and deformations, can depend upon worn bearings and loose holding down bolts. An increase in vibration levels can often be caused by a change in alignment.

Within the petrochemical industry and refineries, reports are made with respect to alignment. The reports note the alignment prior to operation and after operation before removing the pump or dismantling for repairs. The same procedure is carried out to check alignment of hot pumps after warm running.

Correct alignment can be achieved in many ways depending upon the type of equipment and degree of accuracy required. Information regarding alignment requirements is usually to be found in the pump manufacturer's instructions. Never use the limiting values for the coupling as given by the coupling manufacturer since they greatly exceed the values for machines if smooth running and long service life are to be achieved. As a guide it can be stated that a final alignment check should not produce greater parallel misalignment than 0.05 to 0.1 mm or an angular misalignment exceeding 0.05 to 0.1 mm per 100 mm measured length. For the definition of misalignment see section 8.11.2.

Alignment is adjusted by means of brass or stainless steel shims, usually placed beneath the machine supports. Baseplates are generally machined so that a minimum number of shims are always required under the motor. Horizontal adjustment is performed by moving the machine sideways on its mountings. Large machines must have horizontal jacking screws fitted. Sometimes the pump and driver are fixed after final adjustment by means of parallel or tapered dowels.

8.11.2 Methods of alignment

Types of misalignment. Reference line

In principle alignment is based upon the determination of the position of two shafts at two points. Measurement or assessment can be made by straight edges, feeler gauges and dial indicators for the various radial and axial distances or run-out, see figure 8.13. Adjustment is continued until these deviations are zero, or nearly zero.

Two shafts in a vertical plane, for example, can display two deviations from their common centre line, namely parallel misalignment and angular misalignment, see figure 8.14. The amount of misalignment at the flexible section of the coupling is that which is of interest. It is therefore appropriate to use a reference line which passes through the flexible section. Par-

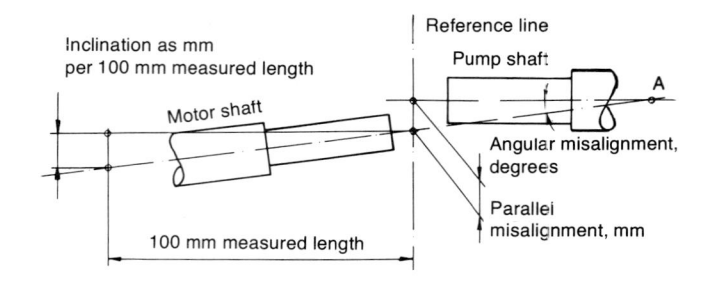

Figure 8.14 Misalignment of two shafts in a common plane

allel and angular misalignments are then referred to this reference line, figure 8.15. Notice in figure 8.14 that if the reference line were to be chosen at the intersection point of the two centre lines of the shafts, point A, then only angular misalignment would exist. From a practical point of view angular misalignment is best measured as an inclination expressed as mm per 100 mm measured length rather than as an angular measurement in degrees.

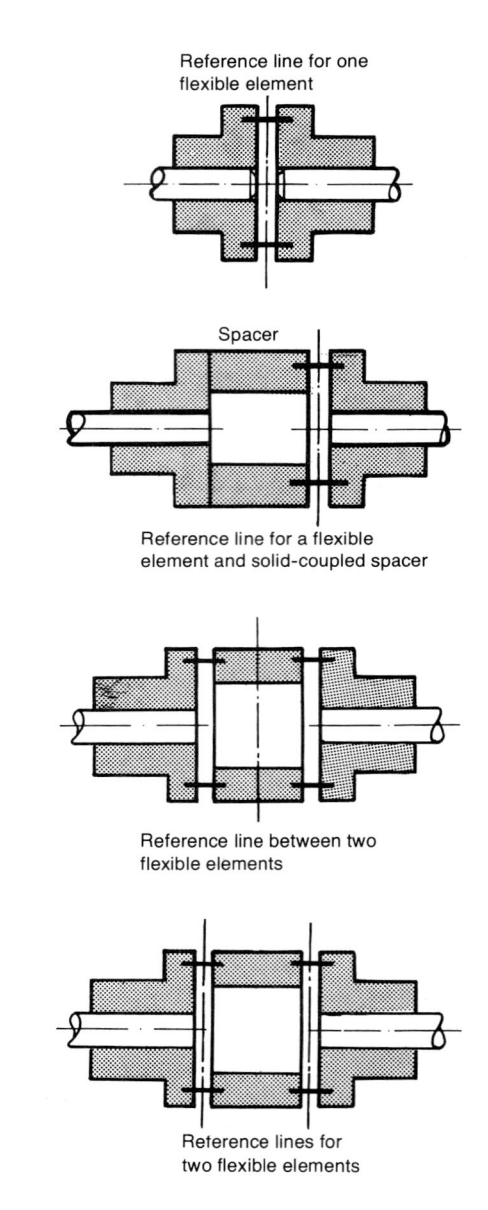

Figure 8.15 Location of reference lines for various types of coupling

The position of the reference line depends upon the type of coupling and should naturally always be located in relation to the flexible section of the coupling. For couplings with spacers and one or two flexible elements the position of the reference line is shown in figure 8.15. Unless otherwise stated by the coupling manufacturer the permitted misalignment is considered to be that which is measured from the reference line.

Alignment procedure

In the case of a horizontal unit, alignment is best carried out by first aligning in the vertical plane, followed by transverse alignment. For vertical units alignment is measured in two directions at 90° to each other. For a horizontal unit, alignment is carried out in the following steps:

1. Align the machines visually and check that the coupling is not crushed in any way.
2. Attach the measuring device(s) and check that the dial indicator(s) moves freely within the area to be measured.
3. Check possible distortion of the motor mounting or baseplate by tightening and loosening each holding down bolt individually. Shim the motor feet if distortion is present.
4. Set the dial indicator(s) to zero in the position shown in figure 8.13.
5. For methods II, III, and IV, rotate both shafts simultaneously through 180°, half revolution, thus eliminating the influence of run-out between shaft bores and the outer diameter of a coupling half. The coupling halves need not then be cylindrical. Determine the measured values according to figure 8.13. Note the measured values with plus or minus signs, see figure 8.13 for notation. Determine parallel and angular misalignments.
6. Determine shim thickness according to section 8.11.3 or 8.11.4 and adjust.
7. Carry out checks according to steps 4 and 5.
8. Carry out transverse alignment in the same manner as in the vertical plane.
9. Perform final alignment checks in both vertical and transverse directions and record for future reference remaining parallel or angular misalignments in both vertical and transverse directions. Also make note of operational conditions at the time of alignment, for example, cold motor with warm pump.

Choice of measuring method

Figure 8.13 shows the five most common measuring methods. From the point of view of accuracy it is difficult to compensate for manufacturing tolerances between the two halves of the coupling by using a straight edge and feeler gauge, method I. The difference in accuracy between method III and method IV is determined by the differences in the dimensions D and C respectively. Accuracy increases in both cases as each respective dimension increases, whereby method III is chosen if D is larger than C and method IV or V is chosen if C is larger than D. The choice of method is also determined, apart from accuracy, by the available measuring surface and by attachment facilities and space requirements of the measuring devices.

The difference between methods IV and V lies in the location of the reference lines. Method IV is universally applicable and suitable for smooth shafts or where it is sufficient to measure the total parallel misalignment and inclination. In the case of a coupling with two flexible elements method V is suitable if the angular misalignment for each element is first calculated individually.

Optical methods are now available. Light sources and mirrors are attached to each coupling half. The units are connected to a small dedicated portable computer which, when supplied with information regarding the feet position, will calculate the respective feet adjustments. Similar optical devices can be attached to machine casings to detect differential expansion when warming up.

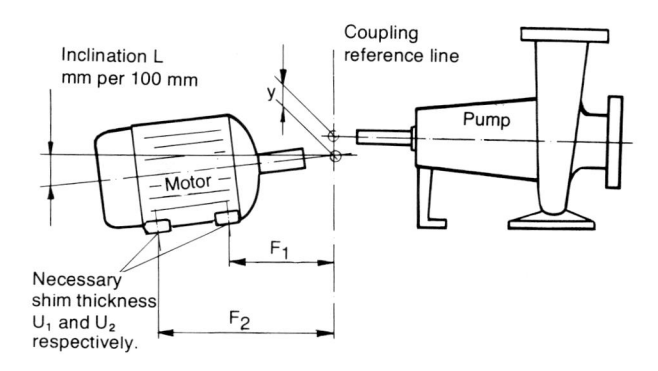

Figure 8.16 Positive misalignments y and L

8.11.3 Determination of shim thickness

Using the measured parallel and angular misalignment, the necessary shim thickness can be calculated directly. The misalignment is expressed as positive or negative, + or -, according to figure 8.16, which shows positive misalignment.

The shim thicknesses are calculated from the simple relationship:

$$U_1 = y + L \cdot \frac{F_1}{100} \qquad \text{Equ 8.3}$$

$$U_2 = y + L \cdot \frac{F_2}{100} \qquad \text{Equ 8.4}$$

where
- U_1 = shim thickness at foot 1 (mm)
- U_2 = shim thickness at foot 2 (mm)
- y = signed parallel misalignment (mm)
- L = inclination expressed as mm per 100 mm measured length
- F_1 & F_2 = distance in mm from coupling reference line to each respective foot, see figure 8.16. The coupling reference line usually passes through the middle of the coupling.

Example:

Indicator reading shows parallel misalignment y = +0.28 mm and inclination L = -0.06 mm/100 mm. The distances to the feet are F_1 = 300 mm and F_2 = 500 mm. The shim thicknesses required are:

$$U_1 = +0.28 - 0.06 \cdot \frac{300}{100} = 0.28 - 0.18 = 0.10 \text{ mm}$$

$$U_2 = +0.28 - 0.06 \cdot \frac{500}{100} = 0.28 - 0.30 = -0.02 \text{ mm}$$

Shims of thickness 0.1 mm are placed under foot 1. The calculated value of U_2 = -0.02 mm means that 0.02 mm should be removed from foot 2, but can probably be accepted as permissible misalignment.

Equations 8.3 and 8.4 can also be combined so that parallel and angular misalignments can be determined in cases where it is not possible to fit the calculated shim thickness, in which case:

$$y = \frac{U_1 + U_2}{2} \qquad \text{Equ 8.5}$$

$$L = \frac{U_2 - U_1}{\frac{F_2}{100} - \frac{F_1}{100}} \qquad \text{Equ 8.6}$$

where: y and L are residual misalignments

U_1 and U_2 respectively (with sign notation) are shim thickness deviations.

For the previous example, when the proposed correction has been carried out, the residual misalignment is:

$$y = \frac{0 - 0.02}{2} = -0.01 \text{ mm}$$

$$L = \frac{-0.02 - 0}{\frac{500}{100} - \frac{300}{100}} = -0.01 \text{ mm}/100 \text{ mm}$$

8.11.4 Graphical method of determining shim thickness

The required shim thickness can also be determined graphically by drawing the position of the shaft in respect of the measured values using a greatly enlarged vertical scale, 100:1 for example, and a reduced horizontal scale, 1:5 or 1:10 for example. The method is illustrated by the following example carried out according to measuring method IV or V with the various stages:

1. Fit the measuring device according to figure 8.13 method IV or V and take readings r_P and r_M on the dial gauge.

Example:

dial reading at pump half gives r_P = -1.40 mm

dial reading at motor half gives r_M = +1.20 mm

2. Determine the dimensions C, F_1 and F_2. Note that the reference line in this example has been chosen to pass through the measuring pointer as shown in figure 8.17

Example: measured results C = 180 mm

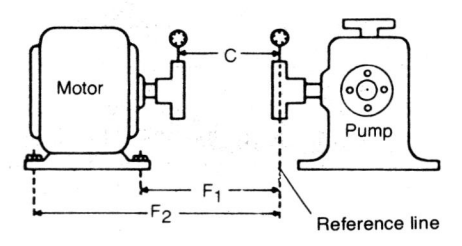

F_1 = 470 mm

F_2 = 890 mm

Figure 8.17 Length measurements and location of reference line

3. Draw up a diagram on squared paper as shown in figure 8.18. Mark in the dimensions C, F_1 and F_2 on the horizontal scale.

4. Mark half the measured value at the pump half, 0.5 r_P, on the vertical axis furthest to the right. The positive sign for r_P means that the motor shaft lies above the pump shaft and is marked upwards, whilst a minus sign is marked downwards. The reading r_P = -1.4 mm should thus be marked as -0.7 mm, i.e. downwards.

5. Mark half the measured value at the motor half, 0.5 r_M, at distance C. The reading's positive value means that the motor shaft lies below the pump shaft and should be marked as a minus value and vice versa for negative readings. The reading r_M = + 1.2 mm should thus be marked as - 0.6 mm, i.e. downwards.

6. Join both points and extend the line to the motor feet locations F_1 and F_2 respectively. The motor shaft shown in the example lies 0.44 and 0.21 mm too low at the respective foot locations and should be raised by shims of corresponding thickness, after which transverse alignment is carried out in the same manner.

7. The alignment can be checked simply by using the two measured values r_P and r_M and the distance 'b' between the two flexible elements. In the case of couplings with two flexible elements, only the total angular misalignment of each element should be calculated. Parallel misalignments are experienced as angular misalignments by the coupling.

In order to calculate angular misalignment, the parallel misalignment at the flexible element must be calculated first, i.e. calculated at both reference lines. These misalignments are:

$$h_P = \frac{r_P}{2} - \frac{a \cdot L}{100} \qquad \text{Equ 8.7}$$

$$h_M = \frac{r_M}{2} - \frac{a \cdot L}{100} \qquad \text{Equ 8.8}$$

The angular misalignment in the vertical plane is then determined from the relationship:

$$\alpha_M = \frac{h_P}{b} \text{ (radians)} = 57.3 \cdot \frac{h_P}{b} \text{ (degrees)} \qquad \text{Equ 8.9}$$

$$\alpha_P = \frac{h_M}{b} \text{ (radians)} = 57.3 \cdot \frac{h_M}{b} \text{ (degrees)} \qquad \text{Equ 8.10}$$

The angular misalignments in the horizontal plane β_M and β_P are calculated in the same way.

Thereafter, the total angular misalignment, θ, per flexible element is calculated from the relationships:

$$\theta_M{}^2 = \alpha_M{}^2 + \beta_M{}^2 \qquad \text{Equ 8.11}$$

or

$$\theta_P{}^2 = \alpha_P{}^2 + \beta_P{}^2 \qquad \text{Equ 8.12}$$

8.12 Choice of coupling

8.12.1 Costs

Generally speaking the cost per kW of a coupling is only a fraction of that of a pump or motor. A pump usually costing at least 40 times that of a coupling and a 4 pole electric motor at least 20 times. The cost varies according to the size and the type. The market for couplings is very competitive, the cost difference between manufacturers is usually small.

Gear couplings are the most costly. If a spray oil lubrication system is required this obviously increases the total cost considerably. Diaphragm and flexible spring couplings, together with the rubber buffer couplings, are about the same cost. Some of the rubber ring couplings are surprisingly costly.

A good way to compare the cost of couplings is to set the price in relation to the torque and range of shaft end sizes to which the coupling can be fitted. The same pump shaft can, for example, be used for a torque range of 1:20 which occasionally means that the shaft end dimension and not the torque is used when selecting the size of a coupling.

Furthermore, the motor shaft is often larger than the corresponding pump shaft. The motor shaft is dimensioned for bending stress to a greater degree than the pump shaft, a motor is often used for belt drive, for example. This can also contribute to the necessity of using a larger size coupling.

8.12.2 Factors influencing choice

Standardisation on Supplier

It is important, not least of all from an initial cost point of view but also cost and space required for spare parts, to establish a viable internal standard, if one does not already exist, by means of which a small number of type or style variations can cover the majority of coupling requirements within a company or plant. If no such standard exists, or if a new standard is to be developed, there is perhaps reason to modify the existing standard. The factors reviewed in the check-list, figure 8.19 should be considered.

For most centrifugal pumps, the diaphragm spacer coupling has become the standard. These couplings are very reliable and can easily cope with the loads and speeds encountered in most situations. For higher speed applications, pumps driven by gas turbines, the gear coupling is preferred by some

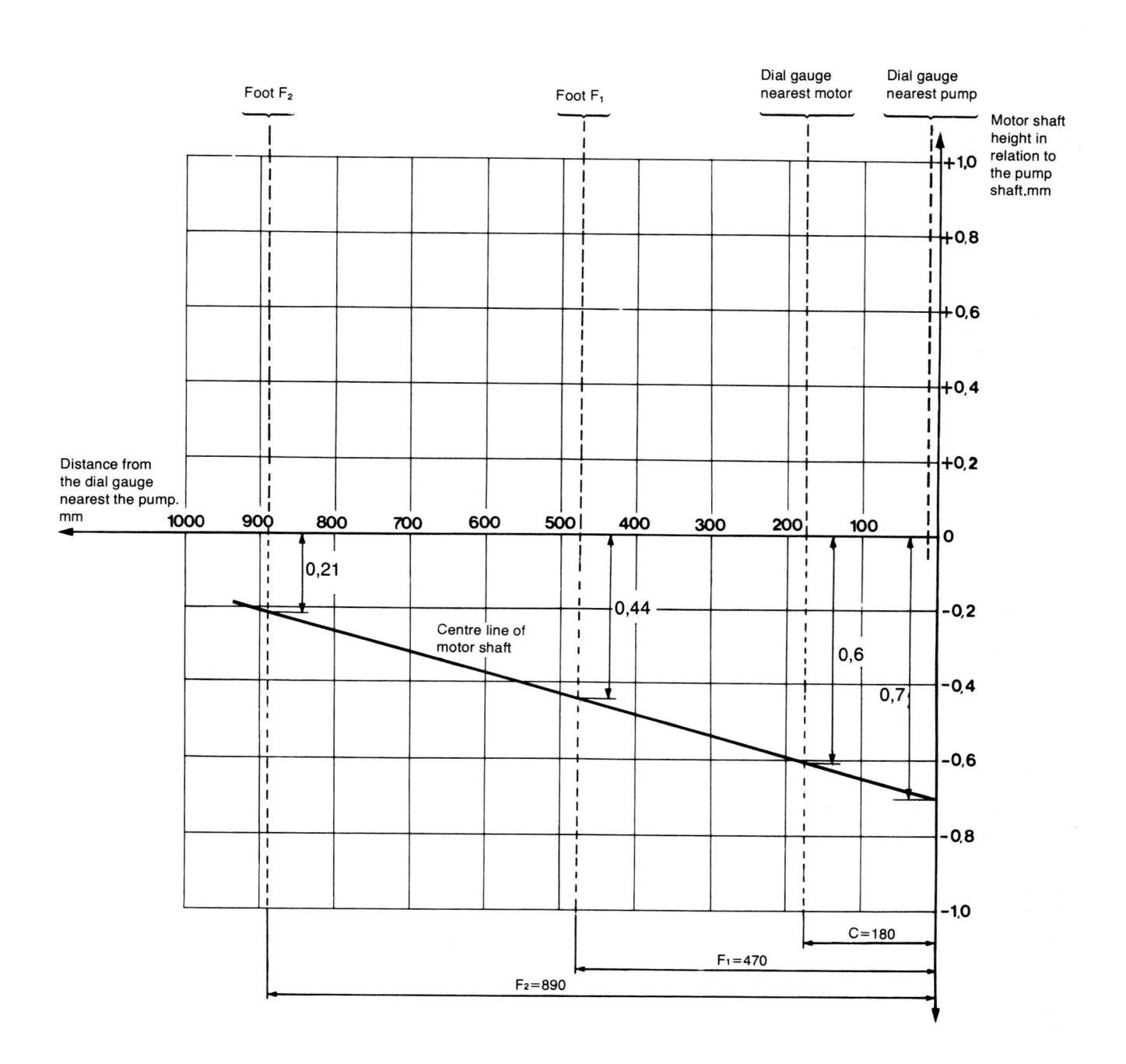

Figure 8.18 Graphical representation of method IV, scaled sketch of motor shaft position

users. Positive displacement pumps operate better with a torsionally flexible coupling; flexible spring and couplings with rubber cushioning are favourite.

Users who have a large number of pumps tend to adopt a single coupling manufacturer whenever possible. This philosophy increases the purchasing power of the user while reducing inventory requirements for spares.

Factor	Influencing parameters
Type of coupling	Non-disengaging Disengaging Torque limitations Torsional rigidity Torsional elasticity
Type of movement	Radial and axial deviation Angular deviation
Forces and moments	Torsional moment Bending moment Axial and radial forces
Operational factors	Frequency of starting Connection frequency Operating time Ambient temperature Moment of inertia Method of calculation
Speed	Balancing Strength Throw protection (safety flange)
Size, weight	Shaft bore Space requirements Spacer for disassembly
Environment	Corrosive Abrasive Temperature Explosive (spark-free, flameproof)
Installation and disassembly	Horizontal and vertical shafts Alignment Fit
Others	Attachment facilities etc. for alignment measuring device. Replaceable wear elements. Service life. Internal standard Costs Coupling safeguards

Figure 8.19 Check-list for shaft coupling selection

8.13 Guards

The pump manufacturer is normally responsible for machine guards. In the case of standard pumps, a distributor may package the pump with its driver and other equipment and it would become the distributor's responsibility to supply and fit guards.

Standard guards are generally painted steel. Sometimes aluminium is used because it is easier to bend and may not need painting. When pumps are to be installed in a potentially hazardous environment special motors are used to reduce the chances of the motor igniting any gas present. A steel coupling rubbing on a steel guard could cause a spark and is not appropriate. In these situations, onshore, an aluminium or bronze guard would be fitted. Offshore pumps in potentially hazardous atmospheres have bronze guards; the salt laden atmosphere offshore is not compatible with most aluminium alloys. Aluminium and bronze guards would be described as 'non-sparking' guards.

With high speed couplings; the distinction between high and low speed is subjective; there is a remote chance that the

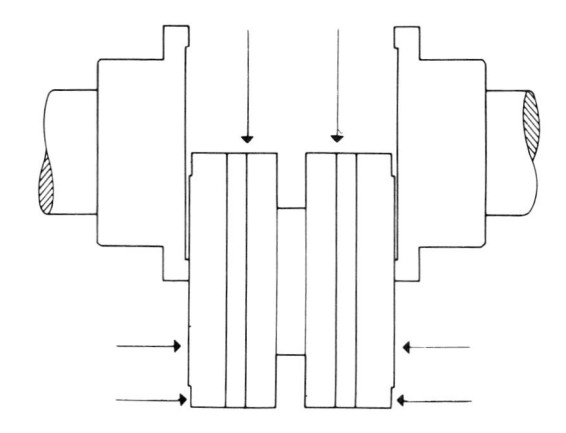

Figure 8.20 Burst-proof diaphragm coupling with spigotted spacer

coupling may fail physically and explode due to the centrifugal force acting on the pieces. It is generally thought that bolting is the weakness link and may be sheared due to an unforeseen overload. If the coupling is not 'burst-proof', see figure 8.20, then the guard must be capable of retaining any scattered material.

Within Europe, the safety of machinery in general is covered by the Machinery Directive which is implemented by EN 292, Safety of Machinery. The safety of pumps is covered by prEN 809. Guards are specifically regulated by EN 953, Safety of machinery; general requirements for the design and construction of guards (fixed, movable).

Other interesting safety standards worth reviewing include BS 5304, DIN 31001, ANSI B15.1 and OHSA coupling guard requirements.

8.14 Literature

ISO 1940, ISO 8821, ISO 5406, BS 6861 Pt1, VDI 2060, NFE 90600, AGMA 515 — Balance classification of flexible couplings

Mekanresultat 72003 — Shaft couplings, Product information issued by the Swedish Association for Metal Transforming, Mechanical and Electro-mechanical Engineering Industries.

DIN 740 — A coupling standard giving dimensioning methods, system of designation and dimensions. It is often used for pumps manufactured according to ISO 2858, BS 5257, and can form a basis for internal standardisation to simplify handling, storage, etc.

API 671 — Special purpose couplings

ISO/R 773:1969 — Rectangular or square parallel keys and their corresponding keyways (Dimensions in millimetres).

ISO/R 774:1969 — Taper keys with or without gib head and their corresponding keyways (Dimensions in millimetres).

ISO/R 775:1969 — Cylindrical and 1/10 conical shaft ends.

Addendum 1:1974 to ISO/R 775:1969 — Checking of the depth of keyways in conical shaft ends.

ISO 2491:1974 — The parallel keys and their corresponding keyways (Dimensions in millimetres).

ISO 2492:1974 — Thin taper keys with or without gib head and their corresponding keyways (Dimensions in millimetres).

ISO 3117:1977 — Tangential keys and keyways.

ISO 3912:1977 — Woodruff keys and keyways.

9 Electric motors for pumps

A pump requires external energy, this is supplied by a driver. Most pumps are driven by an electric motor. It is usual to employ a separate motor coupled by means of a suitable coupling, see Chapter 8. This is the simplest solution and gives the smallest energy loss and is always employed when the pump and motor speeds are the same. If the speeds of the motor and the pump are different or if the motion is reciprocating, as in the case of piston pumps, the transmission of energy becomes more complicated involving the application of gears, convertors or belt drives, see Chapters 5 and 10.

Drivers can be motors, engines or turbines. Engines and turbines are dealt with as special cases at the end of this chapter. Electric motors, however, are completely dominant, which is why this chapter deals predominantly with motor driven pumps. It must be pointed out, however, that relatively large numbers of pump types for intermittent use at construction sites, etc., drainage pumps for example, are equipped with some type of engine.

Contents

9.1 General electrical technology, basic principles
- 9.1.1 Electrical units
- 9.1.2 Electrical systems
- 9.1.3 Active, apparent and reactive power
- 9.1.4 Phase compensation
- 9.1.5 Speed
- 9.1.6 Torque
- 9.1.7 Voltage
- 9.1.8 Starting
- 9.1.9 Efficiency

9.2 Regulations and standards
- 9.2.1 Controlling authorities
- 9.2.2 Physical protection
- 9.2.3 Cooling categories
- 9.2.4 Mounting arrangements
- 9.2.5 Terminal markings and direction of rotation
- 9.2.6 Rated power, centre height and outline dimensions
- 9.2.7 Temperature classification
- 9.2.8 Potentially explosive atmospheres
- 9.2.9 Certification

9.3 Motor types
- 9.3.1 Constant speed AC motors
    - 9.3.1.1 Synchronous motors
    - 9.3.1.2 Squirrel cage motors
    - 9.3.1.3 Slipring motors
- 9.3.2 Variable speed AC motors
    - 9.3.2.1 Pole-changing motors
    - 9.3.2.2 Variable frequency motors
- 9.3.3 Variable speed DC motors

9.4 Motor starters
- 9.4.1 Direct-on-line
- 9.4.2 Star-Delta
- 9.4.3 Soft start
- 9.4.4 Resistance

9.5 Noise

9.6 Maintenance

9.7 Example

Special Note : Engines

Special Note : Turbines

Special Note : Power recovery turbines

9.1 General electrical technology, basic principles

9.1.1 Electrical units

The most common electrical units are reviewed in the table, figure 9.1

Unit		Designation, symbol
Current	I	Ampere (A)
Voltage	U	Volt (V)
Resistance	R	Ohm (Ω)
Active power	P	Watt (W) or kilowatt (kW) 1 kW = 1000 W 1 W = 1 J/s (see energy)
Apparent power	S	Voltampere (VA) or kilo voltampere (kVA) 1 kVA = 1000 VA
Reactive power	Q	Voltampere reactive power (VAr) or (kVAr)
Power factor	cos φ	Dimensionless, cos φ = p/S
Energy		Joule (J) or wattsecond (Ws) Kilowatt hour (kWh) is usually used for convenience 1 kWh = 3,600,000 J

Figure 9.1 Electrical units

Electric current is available in two distinct types; alternating current, AC, and direct current, DC. AC is the most popular type of electricity because it is much easier to distribute and control. AC is the type of electricity that power stations generate and that we use every day in the home. DC is type of electricity produced by batteries. The electrical system in cars is usually 12 V DC, larger commercial vehicles are 24 V DC. Both AC and DC can be generated by rotating machines. Also, one can be converted to the other electronically.

9.1.2 Electrical systems

Electric current is supplied to the consumer as either single phase or 3-phase AC with a frequency of 50 or 60 Hz, i.e. 50 or 60 oscillations per second. Single phase current is supplied to domestic consumers when the total load is less than 50 A. Three phase current is supplied to all consumers when the load is higher. The supply frequency of 50 Hz is used throughout Europe but not world-wide. America uses 60 Hz, as do some Middle East countries. Some offshore oil platforms also use 60 Hz supplies.

When describing the electric supply, a full description must be given, the following are typical of motor supplies:

110 V	1ph	50 Hz
240 V	1 ph	50 Hz
380 V	3 ph	50 Hz
415 V	3 ph	50 Hz
660 V	3 ph	50 Hz
3.3 kV	3 ph	50 Hz
6.6 kV	3 ph	50 Hz

The 110 V supply described is used for site work where increased electrical safety is assured by using a centre tapped transformer; the maximum voltage to earth is 55 V.

When pump packages are supplied with instrumentation, the user specifies electrical supplies for various other functions. Control supplies may be 24 V DC or 110 V 1ph AC. For critical installations, emergency power supplies will be available to allow equipment to operate during power cuts. Emergency supplies can be AC or DC depending upon the source.

Engine driven pumps may have electric auxiliaries and the engine may have electric starting. 24 V DC is the most popular.

When a user purchases a pump package with controls, the electric supplies for various functions will be specified as shown:

Electric motors, > 250 kW	3.3 kV 3 ph 50 Hz
Electric motors, ≤ 250 kW	380 V 3 ph 50 Hz
Motor heaters	220 V 1 ph 50 Hz
Controls	110 V 1 ph 50 Hz

AC electricity is generated as three phase in the power station. It is distributed at high voltage as a three phase supply. Before it arrives at the consumer its voltage is reduced to a suitable level. A small commercial consumer will have electricity delivered at 380 V or 415 V three phase. The consumer can then install equipment to produce a single phase supply. Figure 9.2 schematically illustrates a three phase system. Between the system voltage U_s, also referred to as mains voltage, and the phase voltage, U_f, the following relationship applies:

$$U_f = \frac{U_s}{\sqrt{3}} \qquad \sqrt{3} = 1.732$$

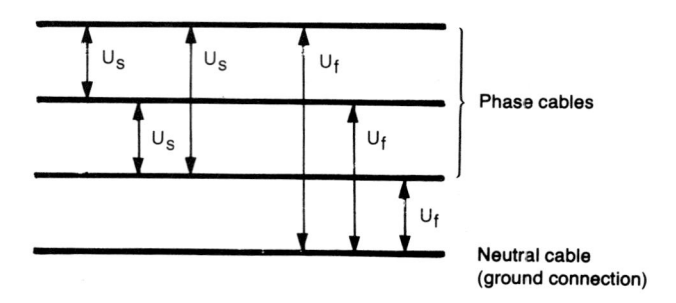

System designation 380/220 V (Alternatively 415/240 V)
System voltage U_s = 380 V (415 V)
Phase voltage U_f = 220 V (240 V)

Figure 9.2 Three phase electric system

The three phase supply shown in figure 9.2 has the neutral grounded. This is the normal situation on land. Offshore electrical systems may not have the neutral grounded and the neutral must be conducted individually to each device. This type of supply is called '4 wire three phase' and must be included in any specification.

In a three phase supply with system or mains voltage of 415 V a suitable electric motor can be connected to all phase cables. The motor is designated as a three phase motor. Smaller electric motors and other loads, up to several kW, can be constructed for connection between a single phase cable and the system's neutral cable, a normal wall socket. These types of motors are designated as single-phase motors.

For the units reviewed in figure 9.1 the following relationships apply:

For direct current, DC

$$U = I \cdot R \qquad \text{Equ 9.1}$$

$$P = U \cdot I = I^2 \cdot R = \frac{U^2}{R} \qquad \text{Equ 9.2}$$

Direct current motors are used for variable speed pumps and emergency pumps, see section 9.3.3 and 5.9.

For single phase AC current

$$U = \frac{I \cdot R}{\cos \varphi} \qquad \text{Equ 9.3}$$

$$P = U \cdot I \cdot \cos \varphi \qquad \text{Equ 9.4}$$

For three phase AC current

$$P = U \cdot I \cdot \sqrt{3} \cdot \cos \varphi \qquad \text{Equ 9.5}$$

If the required pump shaft power is P_p kW and the efficiency of the intended three phase motor is ηm, then the current required in amps is:

$$I = \frac{P_p \cdot 1000}{U \cdot \sqrt{3} \cdot \cos \varphi \cdot \eta m} \qquad \text{Equ. 9.6}$$

See also example calculations in section 9.7

9.1.3 Power; active, apparent and reactive

An electric motor for AC consumes not only active power, P, which is converted to mechanical work, but also a reactive power, Q, required to build up the necessary magnetic field in the motor. This reactive power does not perform any work and is not converted into heat but is nevertheless a load on the electricity supply. Electric power suppliers make a charge therefore for consumed reactive power if it exceeds a certain value.

This relationship is illustrated in figure 9.3. The active and reactive powers combine to form the apparent power, S. This trigonometrical relationship between P and S constitutes the power factor $\cos \varphi$.

The power factor can be determined by measuring an electric motor's power input, current and voltage, at its rated power. The tolerances in this respect are:

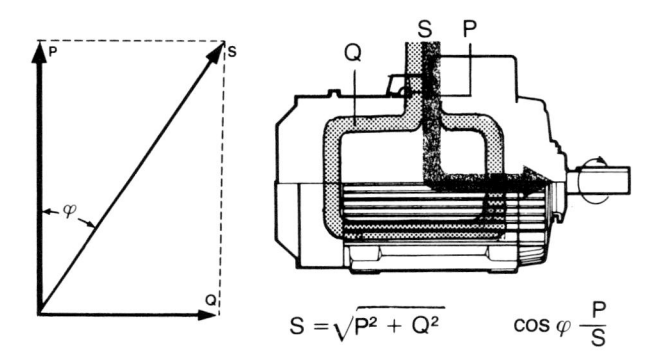

$$\frac{1 - \cos \varphi}{6}$$

$$S = \sqrt{P^2 + Q^2} \qquad \cos \varphi \frac{P}{S}$$

Figure 9.3 Power factor

For electric motors the power factor usually lies between 0.7 and 0.9. It is lower for small motors, see figure 9.4. The power factor is dependent upon the motor load, the table in figure 9.4 gives approximate values for 4 pole motors of various size at various loads.

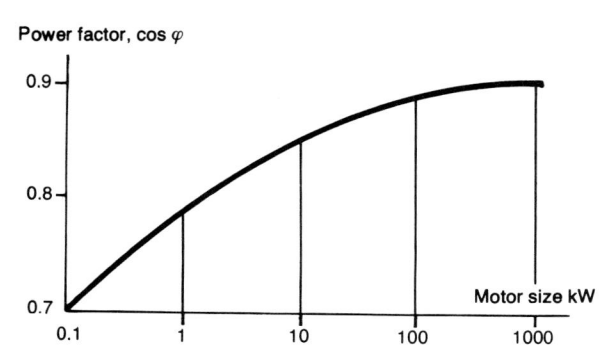

Power factor, $\cos \varphi$

Motor size kW	Degree of loading (Shaft output)			
	¼	½	¾	1/1
< 0.75	0.5	0.6	0.68	0.75
0.75 - 7.5	0.55	0.65	0.73	0.8
7.5 - 75	0.6	0.7	0.78	0.85
> 75	0.65	0.75	0.83	0.90

Figure 9.4 Variation in power factor with motor size and load

The power required by the pump determines the power which the electric motor supplies. This means that if a pump at certain operating conditions requires 11 kW then the motor supplies this power irrespective of whether the rated motor power is 15 or 7.5 kW. This will result in overloading in the case of the 7.5 kW motor, which can cause damage to the motor winding due to overheating. Some form of motor protection device would thus be necessary, whereby, after a certain period of time under overload conditions, the motor is automatically disconnected from the supply, causing the pump to stop, see section 9.4.

A pump motor must, of course, be capable of providing the power which the pump requires. A small margin, depending upon the pump size, is added to the calculated power required to account for small discrepancies in the calculations and minor overload situations which may occur. The motor size may be controlled by starting requirements rather than running requirements. Too large a motor should not, however, be chosen. The procurement costs are unnecessarily high, the efficiency and power factors are lower, whilst the starting loads imposed on the supply are unnecessarily high.

9.1.4 Phase compensation

If an installation contains many electric motors, large quantities of reactive power are consumed and the overall power factor is lowered. The problem can be aggravated by flow regulation systems such as load-unload when motors can be running at very low loads. A solution to this is to install capacitors, which generate reactive power and thereby increase the power factor. This procedure is usually called phase compensation or power factor correction. The capacitors can be applied to motors individually or to groups of motors. Extensive calculations must be performed for sizing phase compensation capacitors. The economics must be fully investigated to justify the extra initial investment. Modern systems include automatic compensation over a range of power factor values by switching various capacitors in and out of circuit.

Improvement of the installation's power factor results in a conservation of energy. It is carried out in order to:

- avoid excess consumption of reactive power involving extra cost,
- reduce loads on transformers and electric cables,
- maintain voltage potential within the system.

9.1.5 Speed

The speed of an AC electric motor is determined by the frequency of the supply and the number of poles in the motor stator according to the following relationship:

$$n = \frac{2 \cdot f \cdot 60}{p}$$

where
- n = speed (r/min)
- f = supply frequency (Hz)
- p = number of stator poles

From the above it can be seen that a change in frequency causes a change in speed. Electric motors can thus be speed regulated by means of varying the frequency, see Section 9.3.2.2.

In Europe the frequency is 50 Hz, thus the speed of electric motors is 6000 divided by the number of poles. At least two poles are required which produces a maximum speed of 3000 r/min; 4 poles gives 1500; 6 gives 1000; 8 gives 750 r/min, etc.

Electric motors which are constructed so that they exactly follow the stipulated frequency and number of poles are referred to as synchronous motors. For large sizes it is necessary, however, to have a special exciting winding, supplied with direct current, together with special starting equipment. This is not necessary for induction motors, squirrel cage motors, which are characterised by the fact that there is a speed differential between the frequency of the electrical supply, rotating magnetic field of the stator, and the motor speed. An induction motor thus has a 1 to 5% lower speed than a corresponding synchronous motor. This slip increases with increased motor loading and decreases with motor size. Because of its simple construction the induction motor is most popular for pump operation.

Most standard three phase squirrel cage motors are designed to operate satisfactorily when the supply frequency is within -5% and +2% of the nominal frequency.

Electric motors for AC can be equipped to enable the number of poles to be varied in operation, thereby obtaining changes in speed. Such motors are called pole-changing motors and to a certain extent can be used for speed regulation of pumps, see Section 9.3.2.1.

In most AC motors the rotor is short-circuited and thus has no connections with the stator winding or any external current supply. Such motors are called short-circuited or squirrel cage motors. In order to limit the starting current and to regulate the speed a variable resistance can be connected to the rotor whereby the rotor must be equipped with sliprings. This type of motor is referred to as a slipring or wound-rotor motor. Usually once the motor is up to running speed the sliprings are shorted to produce a squirrel cage motor.

The speed of a DC motor is a function of the motor design and the voltage applied. The DC motor stator does not have a rotating magnetic field which is dependant upon the supply frequency. The DC motor is somewhat similar to a slipring motor in that it has a wound rotor, an armature. But instead of sliprings, which supply current to all the windings all the time, the armature has a commutator which supplies current to some of the windings some of the time. The stator and armature need not necessarily be supplied at the same voltage. Up to the full power design speed, the motor speed is controlled by the armature voltage. Further speed increase is possible, at the expense of torque, by reducing the stator voltage, field weakening.

9.1.6 Torque

The torque of an electric motor is a measure of the motor's ability to rotate a shaft against resistance. The following relationship exists between torque, power and speed:

$$T = \frac{30000 \cdot P}{\pi \cdot n}$$

where
T = torque (Nm)
P = power (kW)
n = speed (r/min)

When starting an electric motor it is necessary to develop a greater torque than for the normal operating torque, so that the motor can accelerate to operating speed without overheating. Figure 9.5 illustrates a typical three phase induction motor speed/torque curve. When starting, the motor has a relatively high torque, usually 1.5 to 2 times the full load torque. As speed increases the torque first reduces followed by an increase to maximum torque and finally rapid reduction to the point at which the motor torque is equal to the load torque, equilibrium being reached at the designed duty point.

Electric motor catalogues usually state the starting torque in relation to the full load torque and full load current. It should be noted that full supply voltage is assumed. This is not always a valid assumption.

Maximum torque is a measurement of the electric motor's momentary overload capacity. This is usually required to be at least 1.6 times the full load torque, although most electric motors have values which are considerably higher.

The various critical torques, relevant to specific motor designs, have special names as defined below.

Locked rotor torque   The torque at the instant the current is applied before the rotor begins to rotate.

Pull-up torque   The minimum torque the motor will supply during run up to full load speed.

Pull-out torque   The maximum torque the motor will supply during run up to full load speed.

Some motor designs have a pull-up torque which is less than the full load torque. Depending upon the pump type and the starting conditions of the pump, this design feature can be unacceptable.

Single phase motors produce no torque when starting and require an auxiliary winding to facilitate starting and to ensure rotation in the correct direction. The following categories of motors are supplied:

- Motors with a resistance starter winding, which is disconnected automatically after starting.

- Motors with auxiliary winding in conjunction with a capacitor. For low starting torque requirements, the capacitor can be connected during operation, whilst for larger starting torque requirements the capacitor must be disconnected automatically after start-up. Other combinations are necessary to achieve optimum conditions for large starting torques and good maximum performance.

- Motors with commutators, 'universal current motors'.

9.1.7 Voltage

Generally small three phase motors, with the exception of pole changing motors, can be used for 380 V and 220 V. The motor in this case is marked Y/Δ 380/220 V and, depending upon the method of connection, can be used for either voltage.

Normally 380 V or 415 V is used for three phase motors up to about 250 kW. For larger motors, up to approx 450 kW or installations with large motor loadings, higher voltage may be motivated for both technical and economical reasons. According to present international standards for standard voltage, IEC 38, a system voltage of 660 V shall be used. This voltage has the advantage that it enables squirrel cage motors to be constructed for connection to either 660 or 380 V since $\sqrt{3} \cdot 380 = 660$, see section 9.1.2.

In the case of larger motors local voltage drops are often achieved by using transformers. For motors connected to high voltage supplies of 11 kV or 22 kV, transformers are normally used for a nominal motor voltage of 3.3 or 6.6 kV. Motors larger than 1000 kW can be connected directly to 11 kV supplies.

The choice of motor voltage must be made from technical and economical considerations. The determination of voltage involves not only the motor but the complete electrical equipment such as cabling, transformer requirements, switch gear, etc. Generally it can be stated that for larger motor loads it is advantageous to use a higher voltage than 380 V (415 V).

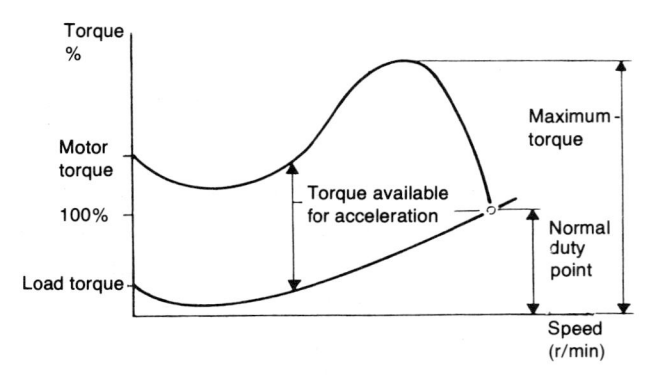

Figure 9.5 Induction motor speed/torque curve for direct on line starting

If the supply voltage reduces then the motor torque is also reduced, in theory. Standard low voltage motors are designed to operate correctly on the nominal voltage ± 6%. Low voltage producing reduced torque applies when starting as well as during normal operation. The torque reduces in proportion to the square of the voltage. It is therefore important to ensure that electric cables to the motor are suitably sized to avoid excessive voltage drop. The maximum acceptable figures usually being 10% when starting and 5% during normal operation. The voltage drop during starting is higher because of the increased current flowing, typically between 4 and 7 times full load current depending upon the starting method. To ensure that starting is not a problem, some user specifications require the motor to be able to start at 80% nominal supply voltage.

9.1.8 Starting

There is an obvious need to start and stop electric motors. The type of equipment required is dependant up on the type of motor, the starting torque required and the capabilities of the supply system. Three phase motors are manufactured in power ratings down to 0.06 kW, single phase motors are rarely used in sizes over 7.5 kW. Motors are not started by a switch like turning a light on. Motors are started by contactors, sometimes called circuit breakers. Under normal circumstances a contactor appears to be a switch. When the start button is pressed the motor starts. When the stop button is pressed the motor stops. The difference is seen when there is a power cut or an overload. When the power is restored after a power cut, motors which were running before do not start. When an overload occurs, the motor seems to switch itself off, after the overload condition has cleared the motor does not start again. A contactor reacts this way because a normal switch could be very dangerous, motors could start unexpectedly. When the start button is pressed, what actually happens depends upon the size and voltage of the motor. For small, low voltage motors the button may press the switch contacts together, making the circuit. Once the circuit is made, a small electric coil holds the contacts 'in'. On high voltage motors, pressing the start button energises a low voltage coil which closes the contacts, again a coil holds the contacts 'in'. Pressing the stop button breaks the circuit of the 'hold-in' coil and the contacts open or come 'out'. When a power cut occurs there is no power for the motor so it stops; there is also no power for the 'hold-in' coil so the contacts open. If power is restored the contactor is in the off position; the motor cannot start. When an overload occurs the power supply to the 'hold-in' coil is broken so the motor stops as if the stop button had been pressed. Motor control does get a little more complicated than this. Three phase motors can be started by various methods, see section 9.4. Single phase motors are started by a simple two pole contactor. Single phase motors, other than commutator motors, have the complicated starting control performed automatically by a centrifugal switch inside the motor.

Controlling motors by electrically operated contactors has the advantage of allowing more than one motor to be started by the same signal. Also automatic control of motors becomes easy. The control system produces a simple low voltage digital signal which can open or close the contactors. Control systems can be constructed from separate relays and timers which are wired to perform the necessary logic functions. Alternatively, one of the modern Programmable Logic Controllers, PLC, could be used. PLC's have digital outputs which can be rated at 110 V or 220 V to power the contactor hold-in coils.

Contactors can be extended to operate several circuits simultaneously. Not all circuits must close, it is possible to open some while closing others. The motor contactor can generate digital logic signals for the control system as well as operating indicator lamps.

When considering starting motors, the regularity of starting must be evaluated. Small motors, 0.06 kW, can be rated for 2300 to 5000 starts per hour. Larger motors, 132 kW, are rated for only 30 starts per hour. When the pump duty cycle requires frequent motor starting, the motor manufacturer should be consulted with all available data. The type of starting method and the run up time, together with environmental conditions, are crucial.

DC motors are not usually used as fixed speed drives, see section 9.3.3 about starting variable speed motors.

9.1.9 Efficiency

The efficiency of medium sized electric motors, 10 to 100 kW, is usually 85 to 90% at full load. Better values can be obtained for larger motors, whilst 75 to 85% is usual for smaller motors. Losses in motors consist of bearing and friction losses together with electrical and magnetic losses in the windings. Some manufacturers produce energy efficient motors, which are 1.5 to 3% more efficient than their standard motors. Whether the increased efficiency can be cost effective is dependant upon local energy costs and the constancy of the application. Figure 9.6 schematically shows speed, efficiency, power factor and current consumption of an electric motor. From the figure it can be seen that efficiency varies with motor load.

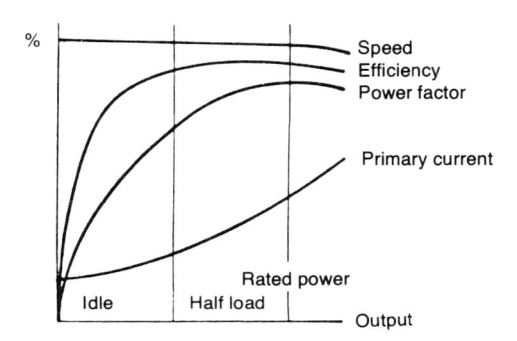

Figure 9.6 Variation in performance of electric motors with load

9.2 Regulations and standards

9.2.1 Controlling authorities

The same regulations apply to the installation of electric motors as for other electrical installations and equipment. Some of the authorities in the U.K. responsible for the issuing and enforcement of regulations are listed below:

Department of Energy

Department of the Environment

Health and Safety Executive (HM Factory Inspectorate)

Ministry of Defence

Post Office

British Coal

Regional Electricity Boards

Department of Trade and Industry

Within the U.K., national standards are issued by British Standards, BSi. Other European countries have national standards authorities as indicated below:

Belgium	IBN	Institut Beige de Normalisation
	BIN	Belgisch Instituut voor Normalisatie
Denmark	DS	Dansk Standardiseringsrad
Finland	SFS	Suomen Standardisoimisliitto r.y.
France	AFNOR	Association Francaise de Normalisation
Germany	DIN	Deutsches Institut für Normung

Germany	VDE	Verband Deutscher Elektrotechniker
Greece	ELOT	Hellenic Organization for Standardization
Luxembourg	ITM	Inspection du Travail et des Mines
Ireland	NSAI	The National Standards Authority of Ireland
Italy	UNI	Ente Nazionale Italiano di Unificazione
Netherlands	NNI	Nederlands Normalisatie Instituut
Norway	NSF	Norges Standardiseringsforbund
Portugal	IPQ	Instituto Portugues da Qualidade
Spain	AENOR	Asociacion Espanola de Normalizacion y Certificacion
Sweden	SIS	Standardiseringskommissionen i Sverige

The European Community has agreed to work together to establish European Standards. General standards are produced by CEN, European Committee for Standardisation. Electrical standards are issued by CENELEC, European Commission for Electrotechnical Standardization. CENELEC issues two types of standard. EN's, which are identical in principle to EN's issued by CEN; each member country must publish the EN and withdrawn any national standards which conflict. CENELEC's other standards are HD's, Harmonising Documents. Each national standard body must publicise the existence of the HD but it does not constitute a standard which requires compliance. The member countries agree to drift towards HD compliance at a future date when it will become an EN.

Other major industrial countries have national electrical standards. Compliance with these may be necessary when pump packages are built for export.

International organisations have established standards for motors and other electrical equipment with the hope that eventually all electrical equipment will be interchangeable. This is perhaps an ideal which will never be realised. The more parties involved in negotiating standards the weaker the standard tends to become and the longer it takes to finalise. (One international pump test standard has been in production for 12 years; not yet ratified.) IEC, the International Electrotechnical Committee is the most important international body. There are 15 countries covered by CENELEC standards, IEC standards cover about 44 countries.

Area of design	BS	CEN/CENELEC	IEC / ISO
Balance, (residual vibration)	4999 Pt 2	≡ HD 347	≠ 34-14 ≡ ISO 2373 Gr N
Cooling arrangements	4999 Pt 106		= 34-6
Dimensions	4999 Pt 141		≠ 72 & 72A
Hazardous area classification	5501 Pt 1		≠ 79-10
Insulation	2757		≡ 34-7
Mounting arrangements	EN 60034 Pt 7	EN 60034 Pt 7	≡ 85
Noise	4999 Pt 109	≡ HD 53.9	≠ 34-9
Performance	4999 Pt 101	≡ HD 53.1	≡ 34-1
Physical protection	EN 60034 Pt 5	EN 60034 Pt 5	≠ 34-5 = 529
Starting	4999 Pt 112	≡ HD 53.12	≠ 34-12
Temperature limitations	EN 50014 & 50018	EN 50014 & 50018	≠ 79-0 & 79-1
Terminals and rotation	4999 Pt 108	≡ HD 53.8	≠ 34-8
Thermal protection	EN 60730-2-2	EN 607302-2	≠ 730-2-2

Figure 9.7 British, European and International motor design standards

The important areas of electrical equipment design which require standardization and regulation are shown in figure 9.7. The relationship between British, European and International Standards are shown.

Equipment specifically for hazardous areas can be designed to reduce the attendant risks by utilising different philosophies. Figure 9.8 indicates the relationship between standards.

Construction		BS	CEN/CENELEC	IEC / ISO
General requirements		EN 50014	EN 50014	≠ 79-0, 79-1A, 79-4
Oil immersed	'o'	EN 50015	EN 50015	≠ 79-6
Pressurised	'p'	EN 50016	EN 50016	≠ 79-2
Powder filled	'q'	EN 50017	EN 50017	≠ 79-5
Flameproof	'd' 'd'	EN 50018 BS 4683 pt 2	EN 50018	≠ 79-1, 79-1A = 79-1
Increased safety	'e'	EN 50019	EN 50019	≠ 79-7
Intrinsically safe	'i'	EN 50020	EN 50020	≠ 79-11
Encapsulated	'm'	EN 50028	EN 50028	
Non-sparking	'N'	BS 6941		≠ 79-15
Std for 'i' systems	'i'	EN 50039	EN 50039	≠ 79-11

NOTE: 'i' can also be 'ia' and 'ib', similarly 'N' can be 'n'

Figure 9.8 Standards for hazardous area equipment

BS 4683 Part 2 and IEC 79-1 regulate the construction and testing of flameproof apparatus.

BS 5345, Code of practice for selection, installation and maintenance of electrical apparatus for use in potentially explosive atmospheres (other than mining applications or explosive processing and manufacture, is worth reviewing. There is no alternative, or close equivalent in European Standards. Parts 1 and 2 are similar to some IEC standards, IEC 79-10 and IEC 79-14, but the remainder of 5345 is unique. Figure 9.9 indicates the acceptable types of construction which can be used in the various zones, see section 9.2.8.

Zone	Type of protection
0	Ex 'ia' Ex 's' (specifically certified for use in Zone 0)
1	Any type of protection suitable for Zone 0 and Ex 'd' Ex 'ib' Ex 'p' Ex 'e' Ex 's'
2	Any type of protection suitable for Zone 0 or 1 and Ex 'N' or 'n' Ex 'o' Ex 'q'

Figure 9.9 Electrical apparatus construction for various zones (from BS 5345 Pt 1)

Commercial organisations also have standards which regulate motors in specific applications, take shipping for example. The following bodies have specific motor regulations for marine applications. Off-shore oil installations may be included.

    Lloyds Register of Shipping

    Germanischer Lloyd

    Det Norske Veritas

    Bureau Veritas

OCMA, Oil Companies Materials Association, has specifications relating to motor protection. EEMUA, Engineering Equipment and Materials Users Association and EEMAC,

Electrical and Electronics Manufacturing Association of Canada, both have motor specifications.

Specification compliance

In order to ensure motor manufacturers comply with all the relevant standards independent test houses have been established to validate specific motor designs. Test houses are mostly employed for equipment destined for hazardous areas, usually potentially explosive atmospheres. Motors are tested on load under laboratory controlled conditions to ensure that construction tolerances limit air gap sizes and that surface temperatures do not exceed prescribed values.

Test houses are located in several countries and each issues their own certification. The areas of expertise are not identical, some test houses are preferred to others for certain equipment. The test houses listed below are in alphabetical order and not in any order of superiority.

Australia	NSW	New South Wales Mines Approval
Canada	CSA	Canadian Standards Association
Canada	EMR	Energy, Mines and Resources Canada
Denmark	DEMKO	Dansk Elektrische Materialkontrol
France	INERIS	Institut National de l'Environment Industriel et des Risques
France	ISSeP	Institut Scientifique de Service Public
France	LCIE	Laboratoire Central des Industries Electriques
Germany	DMT-BVS	Bergbau-Versuchsstrecke Dortmund-Derne
Germany	PTB	Physikalisch-Technischen Bundesanstalt
Italy	CESI	Centro Elettrotecnico Sperimentale Italiano
Netherlands	N.V. KEMA	Keuring van Electrotechnische Materialen
Spain	LOM	Laboratorio Oficial J M Madariaga
United Kingdom	EECS	Electrical Equipment Certification Service, incorporating:
	BASEEFA	British Approvals Service for Electrical Equipment in Flammable Atmospheres
	MECS	Mining Equipment Certification Service
United Kingdom	SCS	SIRA Certification Service
U.S.A	UL	Underwriters Laboratory
U.S.A	FM	Factory Mutual

9.2.2 Physical protection

EN 60034 Part 5 classifies the types of enclosure for rotating electrical machines in respect of the degree of protection afforded to personnel against contact with live parts, and against ingress of solid foreign particles and liquids. The various types of classifications are designated by using the letters IP followed by two numbers. The same method is also applied in IEC 34-5 and IEC 529 but the design requirements are not identical. The table in figure 9.10 reviews the standard designations for various protection categories.

First digit		Second digit	
0	No protection	0	No protection
1	Protection against solids ≥ 50 mm	1	Protected against vertical drips
2	Protection against solids ≥ 12 mm	2	Protected against drips 15° from vertical
3	Protection against solids ≥ 2.5 mm	3	Protected against spray 60° from vertical
4	Protection against solids ≥ 1.0 mm	4	Protected against splashing water
5	Protection against dust	5	Protected against water jets
		6	Protected against heavy seas
		7	Protected against intermittent immersion
		8	Protected against continuous immersion

Figure 9.10 Physical protection categories for electric motors

The dust used for testing is talc, 75 µm nominal size at a concentration of 2 kg/m^3. The test for intermittent immersion is conducted with water for 30 minutes. The minimum immersion depth is 150 mm for the top on the motor, the minimum immersion for the bottom of the motor is 1 m.

Motors which are weather-proofed are designated by the addition of the letter 'W' between 'IP' and the number.

In certain cases electric motors must be equipped with condensation drainage facilities, usually by drilling a 6 to 8 mm diameter hole at the lowest point. IP 22 used to be the standard for motor protection, most motors now are IP 44 minimum. Motors for offshore platforms are usually IP W 55. Motors for hot, humid atmospheres are often required to be 'tropicalised'; this treatment has not be standardised.

European and International Standards recognize the effect of vibration on personnel. Vibration from motors can be transmitted through support structures to the ground. To limit personnel discomfort, and ensure proper bearing life, European Standard HD 347 has been incorporated in BS 4999 Part 142. IEC 34-14 has identical requirements to ISO 2373 Grade N, which are similar to European requirements.

Noise is becoming an increasing problem. Motor noise is specified in HD 53.9 and BS 4999 Part 109. IEC 34-9 is similar.

9.2.3 Cooling categories

In IEC 34-6 the categories for air cooling are designated by means of the letters IC followed by two figures. The designations used in BS 4999: Part 21 utilises the same two figures but omits the letters IC. The first character refers to the type of cooling and the second to the method of circulation. The most common methods of cooling electric motors and their designations are shown in figure 9.11. The IEC standard also contains a more detailed code for more complicated cooling systems. Standard air cooling designs are based on an ambient temperature of 40 °C and an altitude of 1000 m maximum.

BSI Standard	IEC Standard	Cooling Category
0/1	IC01	Free circulation, self circulation
0/6	IC06	Free circulation, independent components mounted on machine
1/1	IC11	Inlet duct ventilated, self circulation
1/7	IC17	Inlet duct ventilated, independent and separate device or coolant system pressure
2/1	IC21	Outlet duct ventilated, self circulation
2/7	IC27	Outlet duct ventilated, independent and separate device or coolant system pressure
3/1	IC31	Inlet and outlet duct ventilated, self circulation
3/7	IC37	Inlet and outlet duct ventilated, independent and separate device or coolant system
4/1	IC41	Frame surface cooled, self circulation
5/1	IC51	Integral heat exchanger (using surrounding air), self circulation

Figure 9.11 Motor air cooling categories

Motors for pumps in boreholes and similarly limited spaces usually have product lubricated bearings. Cooling is achieved by utilising the pumped product. The rotor chamber is filled with liquid and protected against freezing. The motor is connected to the pump by means of an intermediate section, which also serves as the pump inlet. Because of the limited space both the pump and motor must be specially designed.

Motors for pumps immersed in larger vessels can be connected directly to the pump casing. Shaft sealing is by means of two mechanical seals in an oil bath. The motor is normally cooled by means of the pumped liquid, which is effective and requires less motor power than for an equivalent air-cooled motor. The motor should be fitted with temperature detectors, thermistors, for 120 °C, integrated in the stator windings to protect against dry running. The temperature detectors control a contactor which can be incorporated in the terminal box. The following cooling systems are utilised.

- Direct cooling of the motor outer casing, which can be corrugated to increase the surface area, is used when the motor is permanently immersed in liquid which does not exceed 40 °C.

- Casing cooling using the pumped liquid. Part or all of the process liquid passes between the outer casing and the motor casing. The pump can be either 'wet' or 'dry', although dry-running is not permitted. At part flow cooling, the liquid must not contain solid particles. The temperature of the liquid must not exceed 40 °C.

- Oil cooling. Oil from the pump's oil case is circulated in the outer casing during operation.

- Casing cooling using external liquid which passes between the outer casing and the motor casing. Temperature of the pumped liquid can normally be up to 70 °C. This system is used in potentially explosive atmospheres. The motor should stop in the absence of positive pressure.

A 'wet motor' consists of a three phase or single phase cage induction motor, hermetically attached to the pump casing. The pump impeller is connected directly to the motor shaft. The rotor and bearing in the motor are lubricated and cooled by the pumped liquid. This construction constitutes a leak-free pump suitable for transportation of hazardous liquids and for use in potentially explosive atmospheres.

In a wet motor the pumped liquid completely fills the motor casing. Both the stator and the rotor are immersed in the liquid. For wet rotor motors the stator winding is kept dry by means of a thin stainless steel shroud which separates the stator from the liquid. This type of construction is used for particularly troublesome media and also very small pumps, such as domestic central heating circulators. Wet motors and wet rotor motors are also used for submersible pumps.

Motor cooling can be monitored by embedding temperature detectors in the motor windings. Thermistors are normally used but resistance temperature detectors, RTD's, can be used at extra cost. One detector per phase is the minimum requirement; two per phase for critical applications. See BS 4999 Part 111, EN 60730-2-2, IEC 34-11 and IEC 730-2-2.

9.2.4 Mounting arrangements

The design of electric motors with respect to bearings, shaft extension and methods of fixing is covered by BS 4999: Part 107, EN 60034 Part 7 and IEC 34-7. The motor construction allows for operation in all mounting positions, a motor which is normally intended for a particular mounting position is also

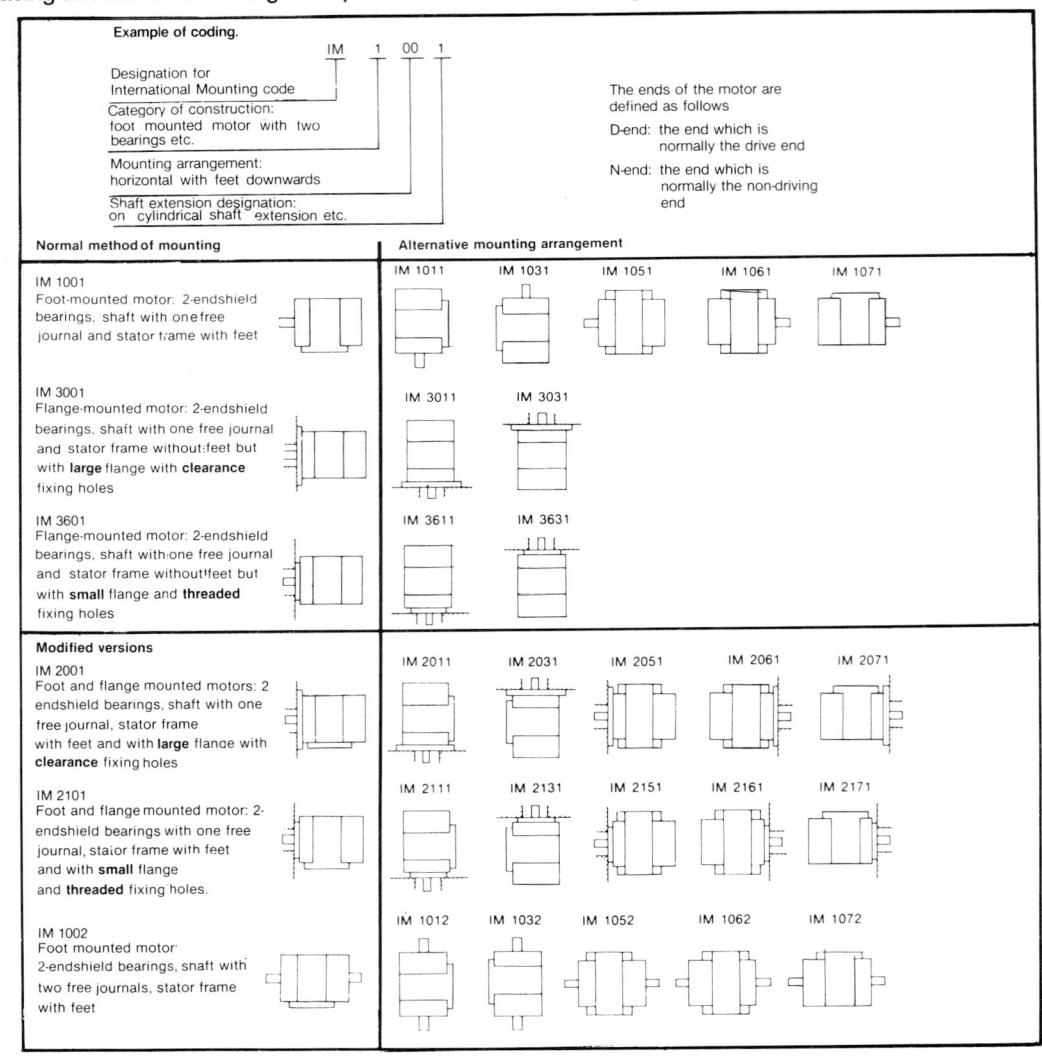

Figure 9.12 Mounting arrangements for electric motors

shown in alternative positions. Figure 9.12 shows examples of various types of mounting arrangement of motors together with their international standard designations. For practical reasons larger manufacturers of electric motors often have their own methods of designation.

9.2.5 Terminal markings and direction of rotation

Terminal markings and direction of rotation are covered by BS 4999: Part 108 which is directly equivalent to CENELEC HD 53.8. IEC 34-8 is a similar specification. According to these standards the phases in a three phase motor shall be designated by the letters U, V, W and the external earth or neutral connection shall be designated by the letter N. The normal direction of rotation in electric motors is clockwise, viewed on the shaft drive end, for phase sequence U, V, W.

Three phase electric motors can be reversed by changing the connection positions of any two phases. This operation is easily carried out by a contactor. **Not all** motors are suitable for reverse operation. Most standard motors are suitable. Motors, specifically designated as 'quiet', are probably not suitable. One method of reducing the motor noise is to fit a special cooling fan which is designed for one specific direction of rotation. Care must be exercised when reverse running is required.

Some pumps are not suitable for reverse running. Shaft driven oil pumps may not function causing bearing lubrication problems. Pumping rings in mechanical seals can be another problem. The direction of motor rotation must be checked during installation/commissioning. Start the motor when the coupling spacer is removed. Change two phases if necessary.

9.2.6 Rated power, centre height and outline dimensions

The IEC standards 72 and 72 A contain recommendations concerning power, centre heights, foot mounting dimensions, shaft end dimensions and flange dimensions for electric motors. The preferred series of electric motor output (kW) are as follows; 0.37, 0.55, 0.75, 1.1, 1.5, 2.2, 3, 4, 5.5, 7.5, 11, 15, 18.5, 22, 30, 37, 45, 55, 75, 95, 110, 132. Ratings and performance within Europe are covered by HD 53.1 which has been incorporated in BS 4999 Part 101. Motor output is structured in HD 231.

According to the relevant electric motor standards, dimensioning should be carried out in the manner shown in figure 9.13 using standardised dimensions. Interchangeability is thus facilitated between different manufacturers providing the non-standardised dimensions are reasonably similar. It should be pointed out that this standardisation is not yet fully implemented, so that it is prudent to carry out checks when purchasing replacement motors.

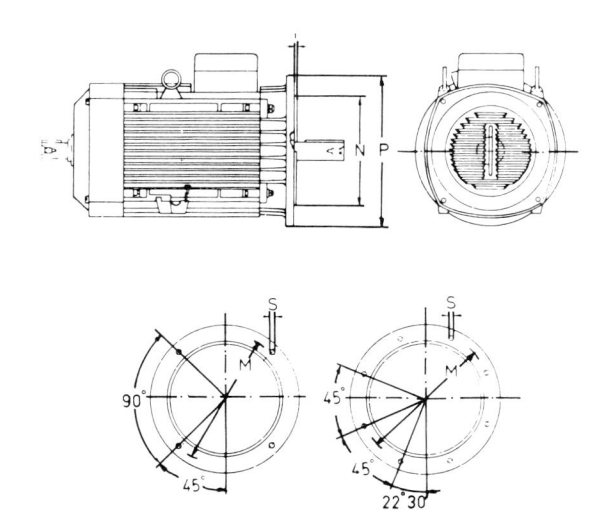

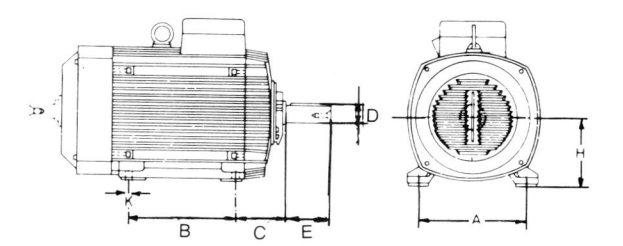

Designation for main dimensions
 A = Distance between fixing hole centres (end view)
 B = Distance between fixing hole centres (side view)
 C = Distance from fixing hole centres at drive end of motor to shoulder of shaft
 H = Shaft centre height
 K = Dia of holes in feet or mounting pads of machine
 E = Length of shaft extension from the shoulder
 D = Dia of shaft extension

Dimensioning for foot mounted motors with mounting arrangement IM 1001 and 1002
 M = Pitch dia of fixing holes
 N = Spigot dia
 P = Outside dia of flange
 T = Depth of spigot
 S = Fixing hole dia.

Fixing dimensions for flange-mounted motors with mounting arrangement IM 3001 and 3002

Figure 9.13 Standard dimensions for electric motors

For induction motors, which are the most common type of electric motor, there is a special standard, BS 4999 Part 141, which is very similar to IEC 72/72A.

9.2.7 Temperature classification

Winding insulation thickness is normally designed to cope with the hottest point for an ambient temperature of +40 °C at an altitude up to 1000 m. If the motor is exposed to excessively high temperatures then the service life of the insulation, and thus the motor, is considerably reduced. It is therefore necessary to use suitably designed insulation if the motor is to be operated at high temperatures. BS 2757, which is related to IEC 85, categorises insulation material according to temperature and insulation classes which are designated by means of a number and a letter. These represent the upper limit of use for normal operating conditions and acceptable service life for the particular insulating material see figure 9.14. For example, the 120 (E) indicates that the insulation can be used up to a maximum of +120 °C, ambient temperature of 40 + temperature rise 75 + safety margin 5.

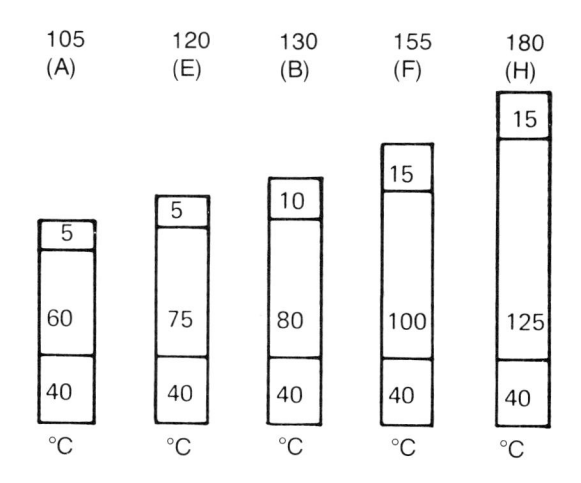

Figure 9.14 Temperature classification for insulating material

The temperature rise of an electric motor can be determined by means of direct temperature readings or by measuring the resistance of the motor windings when cold and hot, respectively, as follows:

$$\text{Temperature rise °C} = 225 - \left(\frac{\text{cold resistance}}{\text{hot resistance}} - 1\right) + 5 \pm 15$$

For extra safety, and increased insulation life, some purchasers will specify Class 'F' insulation with the temperature rise limited to Class 'B' specification. This means the insulation runs 20 to 25 °C cooler than designed.

Motors for potentially explosive atmospheres have additional restrictions placed on temperature rises. To avoid ignition of any gas present in the atmosphere, the motor temperature is restricted according to a grouping for the gas(ses) likely to be present. All flammable gasses have been allocated one of six categories. The maximum motor temperature, according to BS 4683, EN 50014 and EN 50018 for flame-proof motors, are shown in figure 9.15.

Temperature class	Maximum surface temperature °C
T1	450
T2	300
T3	200
T4	135
T5	100
T6	85

Figure 9.15 Maximum surface temperatures for flame-proof motors

9.2.8 Potentially explosive atmospheres

In areas where there is a potential hazard of fire or explosion because of the generation, handling or processing of flammable liquids, vapours, mists, etc. and in areas where there are other kinds of explosive materials, such as dust, special regulations apply for electrical equipment and installations. Fire hazardous liquids, vapours and gases are defined and classified according to their flash point, see section 1.1.7.

Electrical equipment should not be installed in potentially explosive locations if this can possibly be avoided. If this is unavoidable then the apparatus must conform to the relative standards and regulations.

The British Standard Code of Practice BS 5345 offers guidance in the selection and installation of electrical apparatus. The code corresponds to the IEC recommendations for electrical apparatus for use in explosive atmospheres, although the numbering systems are different. BS 4683: Part 2 = IEC 79-1, Electrical apparatus for explosive atmospheres, prescribes features for flameproof enclosures and groups them according to the type of atmosphere for which they are suitable and gives dimensions and permitted gaps.

Hazardous areas are classified in zones according to internationally accepted concepts, BS 5345 Part 2 and IEC 79-10, as follows:

Zone 0 - in which an explosive gas-air mixture is continuously present, or for long periods.

Zone 1 - in which an explosive gas-air mixture is likely to occur in normal operation.

Zone 2 - in which an explosive gas-air mixture is not likely to occur in normal operation, and if it occurs it will exist only for a short time.

In practice, physically defining a zone or zones around a piece of equipment may be a problem. American API 500 and NFPA 30 produces very small zones. RoSPA and the HSE CS5 regulations produce much larger zones. There are no agreed European or International regulations. The basic problem revolves around different approaches adopted by chemical and electrical engineers.

As a rule standard squirrel cage three-phase motors, mechanically and electrically adapted, or derated, are used for hazardous environments. To maintain effective protection it is important that both installation and servicing are carried out thoroughly and carefully.

Potentially explosive gasses are divided into groups. Group I is reserved for underground mining applications. Normal industrial and processing applications are covered by group II, which is subdivided into IIA, IIB and IIC. BS 5345 Part 1 and IEC 79-12 list gasses and vapours which have been classified, indicating their temperature class and group. A few gases and vapours cause problems and have not been classified; acetylene is an example. A motor environment should be described fully: Zone 1 Group IIA T3.

Potentially explosive atmospheres can be created by dust clouds. Layers of dust on equipment can be ignited by hot surfaces as well as by flame or spark, see BS 6467 and BS 7535. Some dust clouds can be ignited by temperatures as low as 280 °C and dust layers at 225 °C.

Zones for dust are designated Z, dust present under normal operating conditions, and Y, dust present only under abnormal operating conditions. Physical protection testing specifically for dust, IP rating, is conducted using French chalk with particle sizes down to 1 μm. IP 6x designates complete immunity to dust ingress. The minimum physical protection recommended for Zone Y or Z is IP 65.

9.2.9 Certification

Electrical equipment manufacturers read the standards relating to their equipment and design and manufacture in accordance with the requirements. In order to be certain that their equipment does comply with all the requirements, the equipment is submitted to an independent test authority, see section 9.2.1, who checks the equipment and certifies compliance. When equipment has been certified the type of certification is shown on the equipment nameplate. The following information will appear on the nameplate:

applicable standards,

test house identification,

test house certificate number.

As part of a project quality assurance programme, the purchaser may ask for a copy of the actual test certificate. In extreme cases, say a T5 or T6 temperature limitation, the purchaser may send an inspector to review the original test house certificate and confirm the photocopy to be a true copy.

European certification may change completely in the next couple of years. There is a proposal to issue an EC directive for hazardous area equipment. This is not thought to be a good idea by all of the electrical industry.

9.3 Motor types

The commonest type of electric motor used for driving pumps is probably the multi-speed, single phase canned motor with a wet rotor; this is the motor type used to drive water circulators in domestic central heating systems. Single phase cage induction motors are used for smaller pumps with power requirements up to several kW, for example, barrel emptying pumps which need to be portable and be fed from wall sockets. For industrial and process applications, the three phase, fixed speed squirrel cage motor is definitely the most popular.

9.3.1 Constant speed AC motors

9.3.1.1 Synchronous motor

This type of motor is characterised by the fact that speed is totally dependent upon the frequency of the electric supply and the number of poles, see section 9.1.5. The motor thus runs at constant speed irrespective of load.

The synchronous motor must be fitted with a special starter winding, which is disconnected during operation. Also necessary is a DC energised magnetic winding; the DC being supplied by a generator in the motor. The reactive power can be regulated by means of varying the magnetisation. A synchronous motor can, therefore, contribute to an increase of the power factor cos φ, see sections 9.1.3 and 9.1.4.

For pump applications synchronous motors are normally used in larger sizes and in cases where there are special requirements. The reason being that this type of motor is relatively complicated, costly and sensitive to frequency variations in the supply system.

9.3.1.2 Squirrel cage motor

Squirrel cage seems an odd name for a motor type. It is an old name. If the design was invented today it would be called 'Hamster wheel'. The important electrically parts of the rotor resemble the exercise wheel found in hamster cages. Squirrel cage motors are also called just 'cage' motors.

Three-phase squirrel cage induction motors are the predominant type for industrial and process pump applications. The construction is simple, requiring a minimum of maintenance and spare parts. Their starting qualities are generally good. There are usually two types of rotor to choose between: simplex and doublecage. The former is used for direct-on-line starting if the starting torque does not need to be high. The simplex rotor is the most common and is especially suited when frequent momentary overloading occurs.

Motors for submersible pumps are usually three phase motors. Although single phase motors are not unknown for this purpose.

Squirrel cage motors are constructed on the principle that a speed differential shall exist between the rotating stator magnetic field and its rotor, thereby maintaining induction in the rotor in a simple and reliable manner. The speed of an induction motor is therefore somewhat less than for a corresponding synchronous motor with the same number of poles, see section 9.1.5. The speed of an induction motor reduces with increased load, see the relationship between torque, power and speed in section 9.1.6. The direction of rotation can be changed by means of simple switching.

Squirrel cage motors are popular because the simple design can be adapted to suit many situations. Slipring and DC motors, because of their internal rubbing connections, are not as easy to protect for hazardous area situations. Also the internal rubbing connections are subject to wear and require maintenance.

9.3.1.3 Slipring motor

In this type of motor the stator construction is identical to the squirrel cage motor. The rotor windings are connected to rotating sliprings on the rotor. By connecting the rotor windings to an external power supply via resistances during starting, considerable limitation of the starting current can be achieved. The resistance can be adapted to suit the desired starting torque. During starting the resistance can be successively reduced as required as speed increases, this can be arranged as an automatic function. The resistance is completely disconnected at full load speed and the sliprings are shorted to create, in effect, a cage rotor, the principle is shown in figure 9.16. The resistance is successively reduced as speed increases. Figure 9.17 shows the effect of a starting sequence of a slipring motor with five resistance steps.

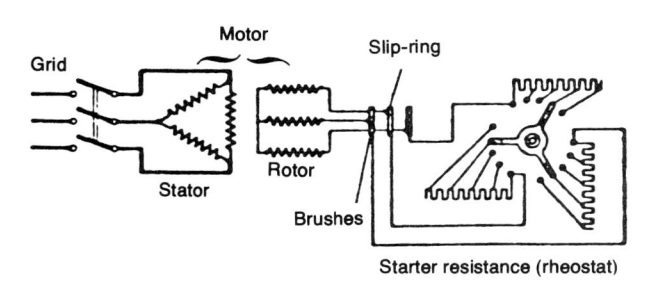

Figure 9.16 Wiring diagram of a slipring motor with starter resistance

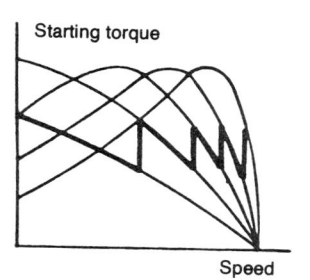

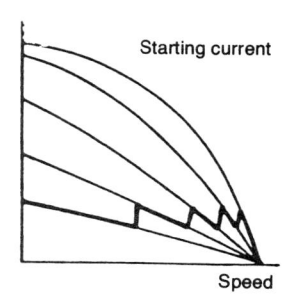

Figure 9.17 Starting torque and current for a slipring motor

The connection of starting resistance to the rotor windings is via brushes rubbing on the sliprings. Sliprings and brushes have been standardised in BS 4999 Part 147 which is identical to IEC 136. Slipring motors are used when there is a requirement for low starting current and in cases where the starting torque is relatively high or the run up time is long. This type of motor can also be used when speed regulation of the pump is required.

9.3.2 Variable speed AC motors

9.3.2.1 Pole-changing motors

By using pole-changing motors, see section 9.1.5, discrete speed regulation can be achieved. For pump operation it is common to use two speed motors whereby the combination of pole numbers determines the speed. The combination 2/4 gives speeds of 3000/1500 r/min, the combination 2/8 gives 3000/750 r/min. The speeds quoted are ideal without consideration of slip. Lower efficiencies are normally expected at low speeds. Three and four speed motors are available.

For the usual type of two-speed motor the stator is fitted with separate windings for each set of poles. The output of the motor will thus be relatively low when only half of the stator capacity is used for each set of poles. The same winding can, however, be used for both sets of poles by means of a special connection method, Dahlander connection, see figure 9.18. Motors with this type of winding cannot be Star-Delta started.

Pole-changing motors are constructed as cage electric motors. Starting can be achieved at either the low or high speed. In the latter case, however, it does not always follow that the starting current will be correspondingly lowered.

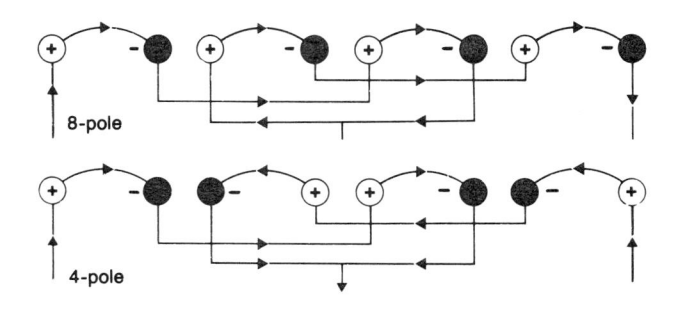

Figure 9.18 Principle of pole-changing stator winding for two-speed motors

9.3.2.2 Variable frequency motors

We have seen in section 9.1.5 that AC motor speed is a function of the supply frequency and the number of poles in the motor stator. Pole-changing motors achieve discrete speeds by changing the number of stator poles. Up until about 15 years ago it was very difficult and costly to change the frequency of AC power supplies. Then the development of power electronics made it possible to convert AC to DC and then DC to any frequency AC. The DC current was switched on and off at high speed to generate a sinusoidal waveform. The new solid state power electronic components allowed high currents to be switched at voltages up to about 600. Ideal for low voltage three phase systems and standard three phase squirrel cage motors. As frequency invertors developed from the original current and voltage source devices, additional features were added. The motor supply current was compensated, at low frequencies, to allow full load torque to be developed. The high speed switching technique was refined to allow varying 'switched on' periods, pulse width modulation. Modern invertors can drive multiple motors simultaneously. In theory, invertors can offer a speed range of 1% to 100%, in practice the minimum allowable pump speed will be 5 to 10%. Invertors are capable of supplying frequencies up to 400 Hz, useful if a high speed motor is required. Contactors for starting are built into the invertor cubicle.

The three phase supply produced by invertors is not as smooth as the standard current available from the grid. The distortion of the waveform can cause heating in the motor. Some motors may require derating to operate with an invertor. Some motors may make more noise when operating with an invertor. Motors which are required to run at low speeds, at full torque, for extended periods may need to be fitted with a separated motor driven cooling fan. The invertor cubicle may have to be mounted 'close' to the motor; the maximum cable length should be confirmed. With some invertor packages the low frequency performance is poor resulting in cyclic motor speed variations, 'cogging'. Invertor packages emit radio frequency interference, RFI. The interference is transmitted from the invertor into the electric supply system. Small invertors produce little interference. Large invertors can produce significant interference, enough to worry electricity supply companies. If an invertor drive over 500 kW is planned consult the local electricity supplier. The small problems which can occur are more than compensated for by the increased versatility of continuously variable speed over a wide range. In order to ensure compatibility between the motor and the invertor, it is advisable to buy both from one vendor.

The use of an invertor to supply motor power can invalidate the hazardous area certification of a motor. This problem can be overcome by certifying the motor and the invertor together. The German test house P.T.B. specialises in this type of testing. Certification problems, and overheating problems, can sometimes be remedied by fitting thermistors in the motor windings which are connected to the contactor 'hold-in' coil. Overheating problems with motor shafts have occurred; winding temperature detectors may not solve this particular difficulty.

Modern invertor circuits are sophisticated and can offer many useful features, the following are pertinent for pump applications:

- short circuit protection,
- earth fault protection,
- set minimum speed,
- set maximum speed,
- programmable acceleration,
- programmable deceleration,
- torque limit setting for starting,
- torque limit setting for normal running,
- automatic reset to minimum speed on trip,
- smooth recovery from momentary power failure,
- slip compensation (for accurate flow only),
- motor voltage/current indication,
- diagnostic fault indication.

Modern pulse width modulation drives produce an almost constant power factor, close to 1.0, irrespective of the motor duty. Depending upon the pump type, the invertor may be required to supply 150% torque for 10 or 20 seconds for starting. The method of speed control must be compatible with the process control system; manual, 0 to 10 V DC; 4 to 20 mA DC, RS 232, RS 422, RS 485, IEEE 488. Remote control and telemetry, via radio or microwave links, is possible for outlying installations.

Variable frequency supplies can be applied to synchronous motors to produce variable speed motors. Slip compensation is not required because the motor is locked to the supply frequency. Salient pole rotor machines are ideal for high speeds. These drive packages would be used for larger power requirements, 1 MW and larger. Designs are available for 5 MW at 8000 r/min. Motors are installed for 3.5 MW at 7720 r/min, 5.3 MW and 9.15 MW. Brushless motors are possible by using standard induction techniques for the rotor excitation current. The lagging power factor, which varies with speed, may need correction on large powers. The increased cost of these motor/invertor packages can be weighed against the increased efficiency. The modified invertor design can be up to 7% more efficient than conventional squirrel cage invertors.

9.3.3 Variable speed DC motors

Before the advent of power electronics, variable speed drives which required accurate speed and torque control would have been DC. Before power electronics, the DC power supply would have been supplied by a motor-generator set; a squirrel cage motor driving two dynamos. One dynamo to supply the field current, the other for the armature, commonly known as Ward-Leonard. By utilising two dynamos, the two current supplies could be controlled independently. Also by utilising dynamos, control was affected through the low power excitation currents, heavy current power cables were wired direct from the dynamos to the motor.

Power electronics changed DC motor control as significantly as it affected AC motor control. DC could now be derived electronically without motor-generators. DC motor control equipment includes two sets of rectifiers to allow independent supplies to the motor field and armature. DC motors can be designed for any voltage, up to about 600 V, and any speed. Small motors, up to 7.5 kW, have field voltages of 150 to 360 V and armature voltages of 180 to 300 V. Larger motors have armature voltages up to 500 V. Motor speeds are generally between 1500 and 3000 r/min. Common specifications for DC systems include:

- 20:1 speed range at constant torque,
- 150% torque for 10 seconds,
- 2% speed change for 100% load change,
- set minimum speed,
- set maximum speed,
- programmable acceleration,
- programmable deceleration,
- set current limit.

For accurate speed control tacho feedback can be added, speed change reduced to 0.5%. Small standard units are available with 0 to 10 V DC or 4 to 20 mA DC control signals. Larger units can have the same control facilities as AC invertors. For units over 7.5 kW, speed stability can be as good as 0.1% and overload current rating extended to 30 seconds.

Power electronic controllers for DC motors produce a lagging power factor which varies with speed. This will not be a problem on small drives.

Small variable speed DC systems are very economical but the motors can be larger than standard AC; AC variable frequency systems become more attractive as the drive power increases. There is no simple guide to the economic change-over. The particular application requirements will determine which drives are suitable. AC variable frequency drives can now do almost everything the DC drive was intended to do. DC motors can be a problem for hazardous areas. The commutator and brushes, like sliprings and brushes, require extra maintenance over simple squirrel cage motors.

9.4 Motor starters

We have discussed single phase starting in 9.1.8 and here we look at three phase starting in some detail. Motor starters or contactors, and wiring must conform to installation requirements and safety standards. Figure 9.19 shows how three phase motors are typically wired up. All electric installations must be safe and include protective devices for personnel, the wiring as well as the motor and any motor control equipment. IEC 947-4 of 1990 covers motor starters and contactors. A survey reported in 1987 that over 60% of motor failures were caused by: insulation failure, overloading, stalled rotors and single phasing. The various protective devices are described in the following paragraphs.

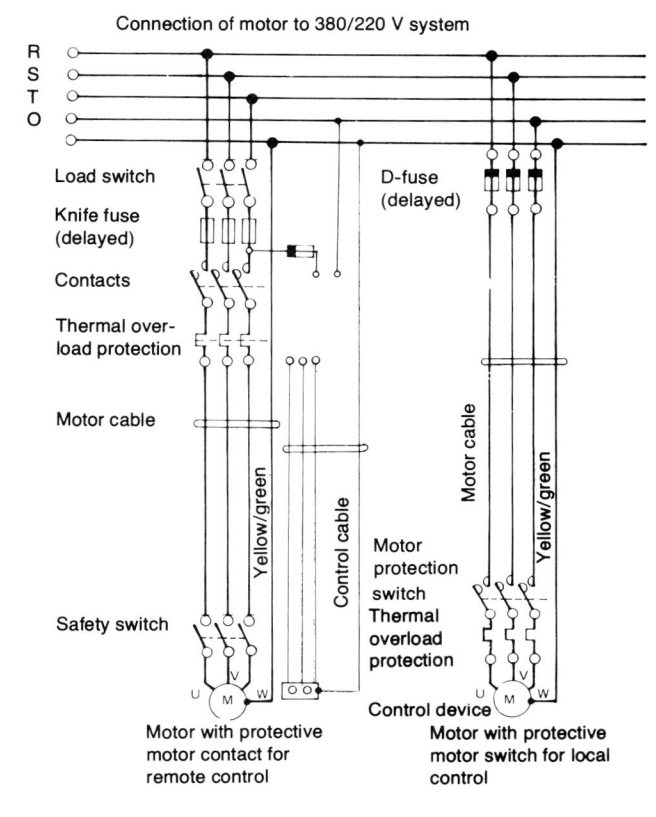

Figure 9.19 Electric motor wiring diagram

Short circuit protection

The purpose of the short circuit protective device is to quickly disconnect the motor, starter/contactor and any control equipment, from the supply in the event of a short circuit or earth fault. Protection can be provided by fuses or circuit breakers. Fuses are old-fashioned and slow to replace, but extremely reliable and fail-safe. Miniature plastic cased circuit breakers, using magnetic sensing, are available for the smallest loads. Resetting is very quick. Short circuits can be caused by broken cables and damaged insulation.

Overload protection

Overload protection is built into the starting contactor by means of a current detection device, an overload relay. Two popular methods are available; thermal and magnetic. Thermal protection is accomplished by passing the motor current through a bimetallic strip. Due the difference in expansion of the two metals the strip bends as it warms up. The temperature rise of the strip is related to the current passing through. When the strip bends passed a set limit, the 'hold-in' coil circuit is broken and the contactor opens. Magnetic protection is implemented by using the variation in magnetic strength of a coil as the flowing current varies. When a set magnetic pull is exceeded the 'hold-in' coil circuit is broken. Dampening of thermal bimetallic strips is inherent in the mass of the strip.

Dampening of the magnetic devices is provided by adjustable dash-pots. Motor stopping with overload symptoms can be caused by:

- pump running out on curve,
- liquid SG too high,
- pump running at relief valve accumulation,
- poor cable insulation,
- single phasing,
- bearing seizure,
- solids trapped in wear rings.

Single phasing is caused when the connection between one of the phases is broken while a motor is running. The motor will continue to run but at a slightly lower speed. The current drawn by the two working phases will increase by about 75% causing overheating in the windings. A motor with a broken phase connection cannot be started.

High temperature protection

We have already mentioned motor winding temperature detection in 3.9.2.2 and 3.2.3. Thermistors normally, or resistance temperature detectors, RTD's, are optionally embedded within the motor field windings. One per phase is standard, two per phase for critical applications. As the windings warm up during normal operation the temperature can be measured and indicated. A high temperature switch can be included in the circuit to alarm and/or trip at set temperatures. The motor will overheat under the following conditions:

- developed power too high (overload),
- too frequent starting,
- single phasing,
- earth fault in a winding,
- ambient air too hot,
- cooling system fault;
  fan problems,
  motor covered in dust,
- motor mechanical problem.

Thermistors and RTD's can work with the starter overload protection and allow current protection to be set a little higher for better starting performance.

Safety switch

When the motor controls are distant from the motor, in a control room for instance, it is necessary for safety reasons to fit some local control next to the motor. In the event of an emergency, personnel close to the motor can stop it before too much damage is done. The local safety switch must have a 'stayput' stop button to prevent the motor being restarted remotely. Also, it is a good idea to have a lockable isolator for maintenance purposes.

Ammeter

The ammeter is used for measuring the motor's operating current. The ammeter is connected to one phase only on the assumption that the phases are balanced. In the event of single phasing, the current will increase significantly or drop to zero. Regular readings from the ammeter should be logged, these can indicate trends in the motor or pump performance.

Electronic manufacturers are now producing motor protection devices which are completely independent of the starter/contactor. The devices measure parameters in all three motor phases. If a fault is detected the starter/contactor is tripped using the 'hold-in' coil. The following conditions can be monitored:

- unbalanced load,
- single phasing,
- short circuit,

- earth fault,
- thermal overload,
- repeated start,
- long run up time,
- locked rotor.

Review sections 9.1.8 and 9.3.1.3 before investigating squirrel cage motor starting in detail.

9.4.1 Direct-on-line

When starting a squirrel cage motor direct-on-line, DOL, there is a substantial inrush of current from the supply, about 600% full load current, FLC, which results in a local voltage drop. This affects the motor as well as other equipment connected to the supply. Suppliers of electricity therefore lay down certain restrictions in relation to DOL starting. Normally DOL starting is only permitted for motor sizes up to 7.5 kW. It is therefore normally necessary to take suitable precautions in order to limit the starting current of larger motors. Figure 9.20 shows the difference in torque and current for DOL and star-delta, Y-Δ starting.

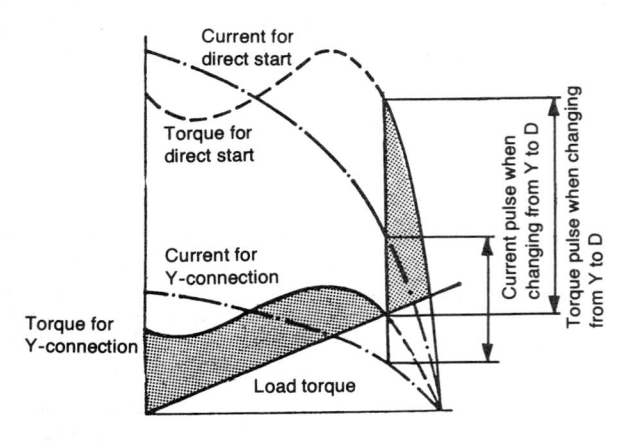

Figure 9.20 Squirrel cage motor starting torque and current

With DOL starting most motors will start most pumps. Most motors will have a locked rotor torque, see 9.1.6, of 130% and a pull-up torque of over 110%. With the safety margin between required power and installed power, most motors will accelerate the pump fairly quickly. However, some motors have very poor starting characteristics. Locked rotor torque can be as low as 80% and pull-up torques down to 70%. These motors may be motors which have been derated for hazardous area operation. Pumps with high inertias, long drive trains, or positive displacement pumps starting on load, may have serious problems when the starting torque reduces to less than 100%. In these situations a sealed fluid coupling may help.

9.4.2 Star-Delta

Star-Delta, Y-Δ, the most usual method for starting large motors. In principle, this means that when starting, the motor's windings are connected between the phase cables and the neutral cable, Y connection. Two windings are connected between each pair of phases. Each winding then receives a lower voltage than that for which they are intended, a three phase motor for 380 V only receives 220 V, see section 9.1.2, and the motor develops only a small part of its starting torque, which of course means that the acceleration time is extended. At approximately 90 to 95% of normal operating speed the stator winding connections are changed over automatically to the system phase cables, Δ connection, which then receives full voltage. Figure 9.20 illustrates the comparison between DOL and Y-Δ respectively for a squirrel cage motor. Star-Delta starting requires about 200% FLC.

If Y-Δ starting is to be used, care must be taken to ensure that the run up time is not so long as to allow the heat generated by the motor, prior to achieving full speed, to reach damaging proportions. Star-delta creates more starting heat than DOL. It is also necessary to check that the motor torque will be sufficient to facilitate starting i.e. it must be greater than the load torque at all times during run up. It should also be noted that changing from Y to Δ can cause a substantial torque and current pulse if performed too early. For this reason it is desirable to achieve a speed which is as near to maximum speed as possible using the Y connection. Y-Δ starting is often unsuitable for pump motors because of the rapid increase in load torque with speed. Positive displacement pumps starting on-line is usually impossible. If DOL is not possible and if the starting torque is too low when using Y-Δ starting then a soft start, variable frequency drive or slipring motor may be necessary. A sealed fluid coupling may help.

9.4.3 Soft start

Soft starts are an electronic method of controlling starting current. The soft starter is wired between the motor and a DOL contactor. By using thyristors to control the supply voltage to the motor, the starting current can be controlled. The supply voltage can be programmed to increase over a fixed period of time or the supply current can be limited during the starting sequence. Soft starters allow a very smooth acceleration of the motor unlike DOL and Y-Δ. Some sophisticated units allow the motor torque to be held at a preset value during starting. Newer versions have energy conservation optimisers which reduce the power consumption when the motor is running light. A soft start will probably still peak at a minimum of 200% FLC.

Because the soft start is electronic, a wide range of extra features can be built in without a large increase in cost. The following features are available on some units:

- adjustable thermal overload protection,
- unbalanced phase detection,
- single phasing,
- inch or jog function,
- internal diagnostics,
- adjustable soft stop.

Soft starters cost about twice as much as Y-Δ starters. The extra features and protection, plus possible savings on cabling, maximum demand charges and reduced surge problems, may make them worthwhile.

9.4.4 Resistance

Star-Delta starting reduces the starting current by reducing the starting voltage. For a 380 V system, the starting voltage would be reduced to 220 V. Starting current and torque is a function of the square of the applied voltage. By applying 58% voltage, the current and torque are reduced to 33% full load values. By using resistance connected in series with the windings, any suitable value of voltage can be applied. By decreasing the resistance value in steps, similar to a slipring motor rotor resistance, the motor torque and current can be gradually increased. This type of starting is similar to soft starts but without the electronics. Because of the reduced starting torque, and the lack of diagnostics associated with electronic methods, resistance starting must be monitored carefully to ensure the motor accelerates successfully.

Other methods of limiting the starting current are reactor and transformer starters. Both these methods involve reduction of both starting current and starting torque. It is therefore necessary to check that sufficient starting torque is available.

9.5 Noise

The quiet running of electric motors is becoming increasingly important and special construction modifications can be made in this respect. However, the noise level requirements of the motor should not be more stringent than for other components, which can often generate quite high noise levels. The noise level of the motor should not be taken in isolation. The noise level of the pump unit should be considered, if too high the various options available should be reviewed.

Noise from motors is generated by the cooling system, by the bearings and also by the magnetic circuit. The dominant source of noise depends upon the size of the motor, speed and the design and construction of the motor. Bearing and magnetic noise cause vibration in the motor and this can be transmitted through the baseplate/skid to other locations.

Small motors are generally quiet; as motor speed and size increases the noise level increases. Figure 9.21 illustrates the trends for squirrel cage motors with IP 54 enclosures.

		Sound pressure level dB(A)	
Motor power kW	Motor speed r/min	Normal design	Quiet option
0.75	2850	63	
	1400	48	
	920	65	
	700	48	
132	2965	93	84
	1480	84	81
	985	80	
	735	74	

Figure 9.21 Electric motor noise at 1 m

A standard requirement is for equipment packages to have noise levels below 84 dB(A). As pump packages increase in power, this noise restriction creates problems. Also a complex package may have multiple motors; oil pumps, circulation pumps, cooling water pumps. A partial solution is to site equipment in areas which are designated 'Ear protection zones'. Personnel in these areas must wear ear protection. This approach is much more cost effective, when possible, than fitting acoustic enclosures. Acoustic enclosures are very costly and can lower the overall efficiency of the pump unit if ventilating fans are necessary. Acoustic enclosures are disliked by maintenance departments because access to the machine is ruined. Electric motors are very reliable and routine maintenance, weekly or monthly, is low. Acoustic enclosures around motors cause less problems than enclosures around pumps.

9.6 Maintenance

An electric motor for a pump requires regular attention, as does the pump itself, in order to give good performance and reliability. The basic prerequisites for long service life of an electric motor are, apart from selecting the correct motor, careful alignment, mounting and proper lubrication. Regular balancing checks to prevent out-of-balance running should also be made. Vibration can be caused by faulty bearings, faulty windings or irregular air gaps and also by out-of-balance of the couplings. Motors fitted with accelerometers can be monitored continuously for vibration problems.

Most electric motors use rolling contact bearings and it is important to ensure that they are correctly lubricated. The service intervals between lubrication depends upon the type of bearing and the environment in which it operates. Lubrication instructions should form part of the motor documentation and must be followed carefully. Removal of bearings for inspection should be avoided whenever possible since they can be damaged when being withdrawn and reassembled. Checks should also be made by listening to bearing noise and observing the lubricant. See Chapter 12 for further information.

Motor control equipment must also be maintained. Starters which operate regularly must have their contacts examined and replaced if necessary. Poor connections at the starter can lead to single phasing and high resistance causing low motor volts.

9.7 Example

A clean water pump for a flow of 25 l/s at 65 m differential head is to be purchased and connected to a three phase system with 415 V supply, and the electric current requirement is to be calculated. The following method can then be used:

The manufacturer's literature shows the pump efficiency to be 70%. First determine the pump's power requirement, see Chapter 3, combine equations 3.3 and 3.5.

$$P = \frac{\rho \cdot Q \cdot g \cdot \Delta H}{\eta}$$

$$P = \frac{1000 \cdot 0.025 \cdot 9.81 \cdot 65}{0.7}$$

$$P = 22773 \text{ W} = 22.8 \text{ kW}$$

The pump is fixed speed so a three phase squirrel cage motor is appropriate. A safety margin of 10% is considered necessary so the minimum motor size acceptable would be 22.8 · 1.1 = 25 kW. Looking at the motor catalogues, a 30 kW motor must be used. The motor has a full load efficiency of 89%, power factor of 0.89 and current of 53 amps. At normal operating conditions the motor will deliver 76% power, when the efficiency will be 88% and the power factor 0.87. Using equation 9.6 from section 9.1.2 the motor current can be calculated:

$$I = \frac{22.8 \cdot 1000}{415 \cdot 1.732 \cdot 0.87 \cdot 0.88}$$

$$I = 41.4 \text{ A}$$

During normal operation of the pump unit the current consumption should, therefore, be approximately 41 A. An ammeter connected to the pump motor gives a good indication of normal operation. Any increase or decrease in the current consumption indicates that there is a fault somewhere or the pump duty has changed and that an investigation is necessary.

The power supply to which the pump is connected must be fitted with fuses or a circuit breaker. The motor also requires a starter. The fuses or circuit breaker, and the motor starter must be rated for the motor full load current, not the running current. When the motor is started, star-delta probably, the motor will absorb 2 to 3 times the full load current, not 2 to 3 times the operating current, see figure 9.20 in section 9.4.1. The fuses must therefore be of the delayed action type to prevent spurious tripping. The cable sizing can be based on the normal running current providing starting does not occur too often.

Special Note : Engines

This chapter has been all about electric motors. Engines are used to drive pumps in certain circumstances. The most popular occasions are for portable pumps and for fixed installations where electricity has insufficient power or is unreliable. In some cases a very good supply of fuel is much cheaper to use than electricity. See section 5.9.1.2 for information on the different engine types available.

Engines can be treated much like electric motors in most respects but there is one major difference which cannot be ignored. Motors supply the driving torque to the pump very smoothly, once started. Engines, on the other hand, do not have such a smooth torque characteristic. Engines produce a cyclic speed irregularity due to the firing of the individual cylinders. The magnitude of the irregularity is dependant upon the number of cylinders. Engines with six or more cylinders are fairly smooth. A large number of pumps, particularly small contractors site pumps, are driven by single or two cylinder engines. The nature of the power supply must be remembered when choosing and sizing couplings and gearboxes. Service factors must be increased. The torsional stressing of the pump shaft may need reconsidering depending upon the ability of the chosen coupling to dampen the cyclic speed changes.

The cyclic speed changes of engines can excite torsional vibrations in the drive train. In general, this is not a problem for rotodynamic pumps with few stages. It can be a severe problem for reciprocating pumps. Whenever a reciprocating pump is driven by an engine, a torsional vibration analysis must be performed. If the analysis highlights excessive amplitudes or stresses, or nodes at gear meshes, then corrective action must be taken. Changing coupling stiffnesses is the first modification.

Torsional vibrations caused by engines can almost certainly be eliminated completely by utilising a fluid coupling at the engine crankshaft.

Special Note : Turbines

Turbines are used to drive large pumps or small pumps when a free steam supply is available, see section 5.9.1.2. Gas turbines are used in situations where a large amount of power is required in a small space or high speed is required. API 611 and API 612 are steam turbine specifications for refinery applications. API 616 is a specification for gas turbines for refinery applications. Turbine driven pumps generally run at 3000 r/min and faster. Fast pumps can have problems with lateral shaft vibrations and critical speeds. Rotodynamic pumps are categorised as 'stiff shaft' or 'flexible shaft'. If the pump always operates below the first critical speed then it is 'stiff shaft'. If, during run up, the pump passes through the first critical speed then it is 'flexible shaft'. The problem is complicated because of the variables involved. The following aspects of pump design affect the critical speed:

- length of shaft between bearings,
- number of stages,
- type of bearings, plain or rolling,
- bearing housing support,
- type of coupling,
- design of driver,
- other drive train items, gearbox, etc.

The critical speed is further complicated by differences of opinion as to whether the critical speed should be 'dry' or 'wet'. The critical speed can be calculated for all the rotating masses running in air; this is easy. When the pump is operating the rotor is surrounded by liquid. Liquid in the wear rings, centre bush and throat bush(es) creates a bearing effect so that the 'wet' critical speed is higher than the 'dry'. The 'wet' can be difficult to calculate accurately as the bearing effects of some of the components cannot be predicted precisely. In theory, a pump handling a liquid with a viscosity of 100 cSt will derive more bearing stiffness from internal close clearances than a pump handling a 5 cSt liquid. It is worthwhile considering critical speed calculations and monitoring during test when the wear ring material combinations are marginal for the liquid, and high speed operation with lots of stages.

Special Note : Power recovery turbines

Some pumps do operate with power recovery turbines as the sole driver. Processes which operate continuously and have pressurised liquid streams which must be expanded to a lower pressure are ideal. However in most cases the turbine is used to augment the main driver and unload it. A main driver is usually required for start-up until the process has stabilised. Oil refining and reverse osmosis water purification plants are popular applications. Adding a power recovery turbine to a pump which has a main driver considerably lengthens the shaft system. Read the comments above in 'Turbines' regarding critical speeds. See section 5.9.1.2 for details of power recovery turbine designs.

Pumps with Built-In Reliability

For Acids Alkalis Corrosive and erosive fluids

Glandless Centrifugal Pumps - designed and manufactured to meet industry's highest standards

KESTNER
Engineering Co. Limited

Station Road, Greenhithe,
Kent, DA9 9NG England

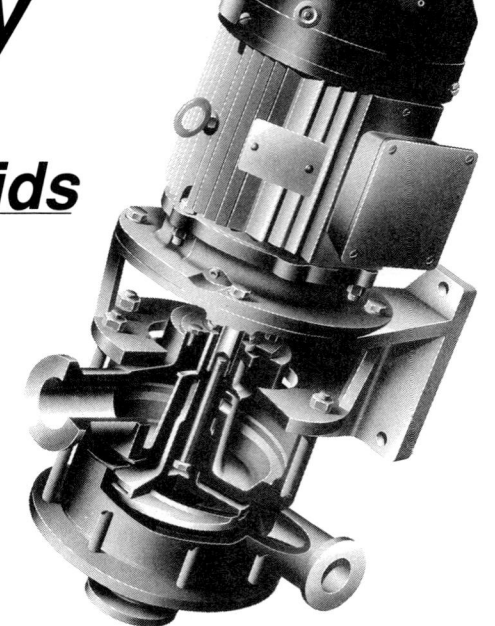

Telephone: 0322 383281
Fax: 0322 386684

10 Ancillary equipment

Small pumps can be installed without any additional equipment. This applies particularly to in-line pumps mounted in and supported by the pipework. Larger pumps tend to have other pieces of equipment to assist with the prime function, pumping liquid.

Chapter 10 looks at ancillary equipment which is frequently supplied as part of a complete package but which is not regulated by easily accessible standards. In this context it is worth explaining two definitions which form the basis of European pump standards.

A 'pump' is a single piece of equipment which terminates at its suction and discharge connections and its shaft end. Pump feet are not mentioned but these are another terminal point.

A 'pump unit' consists of a pump and any other associated equipment which is defined as being included in an assembly. Whenever the term 'pump unit' is applied, it must be defined to show the content intended. Equipment supplied loose, for mounting in the user's process pipework, would not constitute part of a pump unit. In practice, the smallest pump unit would be a pump plus a baseplate; when the purchaser was supplying the motor and coupling. A large pump unit could consist of several pumps, couplings and motors, pipework and controls all mounted on one baseplate.

Contents

10.1 Baseplates, skids and trailers

10.2 Belt drives

10.3 Gearboxes

10.4 Relief valves

10.5 Accumulators

10.6 Pulsation dampeners

10.7 Instrumentation

10.1 Baseplates, skids and trailers

Pump units are generally mounted on a structure to support and align the various components. The structure itself must perform additional functions depending upon its duty.

Baseplates

Pump units are mounted on baseplates to allow the unit to be bolted down. Unless otherwise stated, baseplates are designed for mounting on concrete foundations and to be supported over the whole length by grouting. The pump supplier must be aware if a baseplate is intended for mounting on structural steelwork with intermittent supports. The design of the baseplate side members may need adjusting to preserve alignment. If the baseplate is to be grouted in with epoxy grout rather than concrete, a special paint finish will be required on the contact surfaces. If the baseplate is to be welded down to structural steelwork the pump supplier should be informed.

The baseplate should form the perimeter of the pump unit. Only in exceptional circumstances should equipment extend beyond the baseplate outline. By allowing the baseplate to form a perimeter, delicate and sensitive equipment can be protected by a substantial static structure. If equipment must extend over the baseplate edge, provision should be made to protect the equipment by enlarging the concrete plinth. Problems with motor terminal boxes overhanging the side of the baseplate can be cured by top mounted terminal boxes.

When a pump unit is built to order, the purchaser likes to be in control and provides lists of national or international standards, and includes proprietary specifications for different aspects of the pump/pump unit. Baseplates are a problem. There are no good standards. Consequently, the purchaser has to use inappropriate standards, such as 'Structural steelwork for buildings'. Baseplates are not designed to the same sort of criteria as buildings. Deflections are much smaller and stressing is much lower. The pump supplier will normally have plenty of experience designing and fabricating baseplates and will just need a few little reminders about good practice.

All structural welds must be continuous. Main longitudinal members must be one continuous piece with no joints. Welds for main cross members to main longitudinals must be on all sides of the cross member. Pads for equipment must be machined and thicker and larger than the equipment foot. Pads for pumps and drive train must be fully supported. Drive train pads must be machined low, about 3 mm, so that shims are always fitted. Baseplates, with a fully assembled weight of over 2500 kg must have levelling bolts at each foundation bolt. Individual pieces of equipment over 500 kg must have provision for horizontal positioning bolts; these bolts can be detachable. Provision must be provided for a four point lift with equal length slings. If a lifting beam is necessary the pump supplier must state so and quote a price for its supply. Lifting arrangements must comply with insurance company regulations.

The simplest lifting arrangement is to incorporate full width tubes across the baseplate. The user inserts a longer bar through the tubes to accept the eyes of the slings.

Baseplates for larger pumps are designed to cope with transportation and installation without suffering permanent distortion. This is not always true for smaller pumps. Some small pumps may be delivered with temporary bracing or reinforcement to prevent permanent distortion.

Baseplates can be constructed with or without various desirable features; the options for these features must be specified at the inquiry stage.

Drain rim	The baseplate incorporates a full length sloping trip tray with one drain connection.
Full depth cross members	The structural cross members extend to the bottom of the main longitudinal members.
Shallow cross members	The cross members do not extend closer to the bottom of the main longitudinal members than a specified distance.
Lifting lugs	Four lugs, shaped to accept standard shackles, are welded on the outside of the outer longitudinal structural members.
Top covers	All accessible areas of the baseplate top are fitted with covers; solid, grid, permanent, removable.

In some instances, options may not be possible. Shallow cross members may not be viable if the pump supplier has agreed that the pump can accept very large nozzle forces and moments.

When a pump user purchases a special pump from a manufacturer the user is anxious to have a general arrangement drawing which shows the overall sizes, positions of nozzles and foundation bolts. The pump manufacturer may include a preliminary drawing with the quotation but the user needs a certified drawing to finalise pipework and foundations. We have seen in section 9.2.6 that some dimensions are standardised on standard motors. Special motors, and other types of drivers, need a drawing to confirm all the relevant dimensions. Manufacturers of motors and engines only issue certified drawings after a purchase order has been placed. The pump manufacturer cannot issue a certified pump general arrangement until receipt of the certified drawings for the drive train. At the beginning of a pump order, there appears to the user, to be a delay before anything happens. This time is taken up with issuing purchase orders and finalising certified drawings.

Skids

A skid is very much like a baseplate but it is not intended to be fastened down to concrete or structural steelwork. A skid is designed to rest on a reasonably flat surface. Consequently the longitudinal structural members are slightly deeper, and perhaps a heavier section, than a similar baseplate. Most of the comments for baseplates are valid for skids. The obvious exceptions are those requirements for baseplate levelling screws and an option for permanent distortion during transportation or installation. Skids are very rugged. A skid package will operate satisfactorily with four points of support.

Skids are constructed with angled or rounded ends on the main longitudinal members to assist movement. Towing points, to accept slings, are provided at both ends. Full depth cross members are not an option.

A special version of a skid is a lifting frame as fitted to lightweight pumps for site work and fire fighting, see figure 3.55.

Trailers

Trailers are a slightly more complicated version of a skid. When purchasing a trailer mounted pump one of the two major options must be specified before any other decisions can be made; off-road or road worthy.

Off-road trailers are designed specifically for rough terrain and cannot be towed on public roads in Europe. Trailers can have two or four wheels and the wheels may or may not have pneumatic rubber tyres. Four wheel trailers may have a steering axle, brakes are an option not mandatory. Depending upon the size and design of the trailer, stabilising jacks may be essential. There are no standards specifically for off-road trailers. Safety requirements inherent in the Machinery directive must be met, but there are no specific design requirements.

Road worthy trailers must comply with the current relevant Road Traffic Acts. These trailers must be complete with pneumatic rubber tyres, suspension, brakes and lights. It may be worthwhile specifying a maximum speed for towing. The following standards may be useful:

BS AU 177a ≠ ISO 3732
BS AU 194 ≠ ISO 4091
BS AU 197 = ISO 1185
BS AU 198 ≠ ISO 3731
BS AU 213 ≡ ISO 8703
BS AU 219 ≡ ISO 8755
BS AU 220 ≡ ISO 1102
BS AU 231 ≡ ISO 4141

Some pumps are packaged on a vehicle; the size of the vehicle being dependant upon the size of the pump and the nature of any accessories. The facilities required on such vehicles must be fully specified.

10.2 Belt drives

Belt drives are an economical method of speed increasing, transmitting and reduction. The speed does not necessarily have to change. Belt drives, 1:1, are used as an alternative to couplings. The driver and driven shafts can be moved to better locations not possible when utilising a coupling.

Belt drives should be considered when the following conditions exist:

- power rating up to 200 kW,
- speed ratio up to 7:1,
- belt linear speed up to 40 m/s,
- infrequent speed changes,
- ambient temperature -30 to 65 °C,
- fairly clean ambient air.

Belt drives are available for higher powers but these generally involve special pulleys incurring extra costs and causing spares availability problems. Belts can be used at higher speeds but this necessitates a specially designed drive. Pulleys with quick mounting hubs are standard for inch and metric sizes. Pulley sizes can be changed fairly quickly for irregular speed modifications.

Belts are available certified for underground mining applications and for potentially explosive atmospheres. Cast iron metric pulleys generally comply with ISO 4183 and belts ISO 4184.

Vee-belts with a service factor of 1.0, for use with rotodynamic pumps, would have an efficiency of about 96%. Best efficiency, 97.5%, occurs at a service factor of about 2.0. Three cylinder reciprocating pumps would normally have a service factor around 1.6.

Vee-belts do slip slightly and require occasional tension checking. These shortcomings can be overcome by using toothed belts. Higher powers can be transmitted in the same space requirements at slightly higher efficiencies. Toothed belts are not popular because of the noise they produce. Toothed belt drives with 1500 r/min motors are very noisy, 105 dB(A) is possible. Silencing belt guards causes problems with air circulation and belt cooling.

Belt drives cannot be fitted with continuous monitoring instrumentation. At present, to inspect the condition of a belt drive the drive must be stopped and the guard removed. Belt guards should be treated in a similar manner to coupling guards, see section 8.13.

10.3 Gearboxes

Gearboxes can be applied to any shaft speed and any ratio is possible. By selecting the correct service factor, working life can be long with minimal maintenance. The most popular style of gear box is the parallel shaft helical gearbox. Figure 10.1 shows the range of ratios possible with varying numbers of gear meshes or gear pairs.

Number of meshes	Range of ratios
1	1.0 to 6.4
2	6.6 to 21
3	18 to 100
4	90 to 525

Figure 10.1 Gear ratios of parallel shaft helical gearboxes

A pump unit can become quite long by the addition of a gearbox and an extra coupling. To reduce the overall length, a flange mounted motor could be bolted directly to the side of the gearbox. Another ploy would be to replace one of the helical ratios with a bevel ratio. The gearbox input shaft could be located vertically on top of the gearbox. Epicyclic gearboxes and in-line helical gearboxes, both with flange mounted motors are popular for sizes up to 100 kW to minimise package dimensions.

Gearboxes are available with more than one ratio. Parallel shaft helical gearboxes are built with up to four ratios. These gearboxes are manual change and require the pump to be stopped before a change can be made. Special provision, 'inching' or 'jogging' of the drive motor, may be necessary in order to engage the dog clutches. Small units can be turned by hand from the high speed coupling. This type of gearbox is suitable when discrete speed changes are acceptable and changes do not occur too often. Maintenance personnel may be required to change gear if operators are unhappy or lack confidence for technical intervention. Efficiency is high. Because the gearboxes are made to order, ratios are exactly as required.

Semi-automatic multi-speed gearboxes, similar to automotive transmissions, are available from a few manufacturers. These gearboxes designed for commercial vehicles, rail locomotives and marine applications, are normally used on portable engine driven pump packages. The number of gears varies from four to ten. A limited choice of ratios is possible but the multiplicity of ratios more than compensates for the lack of an optimum ratio. Speed changing can be manual or automatic via a process control system.

Epicyclic gearboxes are popular for larger and high speed applications; steam and gas turbine drivers. Very costly in comparison to parallel shaft helical boxes; but maybe more efficient. Epicyclic gearboxes can be used as a variable ratio transmission by adding a small control drive.

Gearboxes tend to rely on natural convection cooling. When necessary, fans can be added to the high speed shaft(s) to encourage forced convection. When fans are insufficient, an external cooling circuit can be added for the lubricating oil.

Most gearboxes have rolling contact bearings. Special versions can be built with plain bearings. API 613 for high speeds and API 677 for special purpose gearboxes for refinery applications both require plain bearings as standard. Gearboxes with rolling contact bearings can manage with a simple splash lubrication system. Gearboxes with plain bearings may need an external pressurised lubrication system. API 614 specifies the requirements for refinery lubrication systems. API 614 systems tend to be 'over-engineered' and require a lot of space; individual applications should be evaluated and designed accordingly.

Gearboxes are sized by using service factors. The service factor is dependant upon the smoothness of the drive and the smoothness of the load. The service factor adjusts the working tooth stresses to suit the application. Rotodynamic pumps handling clean liquids driven by an electric motor would have a service factor between 0.8 and 1.25. If the driver was a four cylinder engine, the service factor would increase to 1.0 to 1.5. For a reciprocating pumps the equivalent service factors would be 1.5 to 2.0 and 1.75 to 2.25.

There are no European Standards for service factors. National standards exist in U.K., Germany and the U.S.A. The American Gear Manufacturers Association is the most prolific for issuing standards. AGMA 2001 covers rating factors for spur and helical gears. British Standard 346 Part 3 covers service factors as does DIN 3990. The British Standard is considered to be the most conservative for through hardened gear designs. Most modern spur and helical gears tend to be case hardened; AGMA is considered to be the most conservative. DIN 3990 is thought to produce the most realistic designs for case hardened gears.

An International Standard is being prepared but this is going to complicate gear design and rating, not simplify it. The new proposed standard has five different calculation methods and many options for design factors. Service factors are going to be changed to Application Factors but the two are not identical or interchangeable. Service factors include a calculated life function. Application factors will not address the life problem.

The best way to specify a gearbox is to pick an old, well tried standard that has been used successfully for many years. AGMA 420 is an old standard for spur and helical gears. Rotodynamic pumps driven by standard squirrel cage motors for continuous operation, not frequent starting would be AGMA 420 SF= 1. For a three cylinder reciprocating pump driven by a squirrel cage motor the factor would be AGMA 420 SF= 2. For pumps with five or more cylinders the service factor can be reduced to 1.75. To be absolutely sure of what is purchased it would be advisable to specify the minimum acceptable life of teeth and bearings, i.e.

Minimum tooth life at rated conditions 85000 hours

Minimum bearing life at rated conditions 30000 hours

If pumps are subject to cyclic operation, including duty changes, on-off or load-unload flow regulation, then the full operating conditions should be discussed with the gear manufacturer for suitable recommendations.

Unlike belt drives, gearboxes can be fitted with instrumentation to monitor the health of the teeth and the bearings. Accelerometers can be fitted to each shaft and the gearcase itself. Harmonic analysis of the vibration signals can show how wear is developing and if bearings are working correctly. Analysis of oil samples can indicate the onset of oil degradation and the level of metal particles.

10.4 Relief valves

Relief valves are fitted to piping systems and vessels to prevent excessive internal pressures damaging the equipment. When it is possible for a pump to produce pressures in excess of the system or vessel design pressure a relief valve should be fitted. In the case of positive displacement pumps a relief valve must always be fitted to the discharge. Relief valves should be fitted as dedicated pieces of equipment. Modern designs allow relief valves to be combined with load-unload valves. Although this saves cost it is not a preferred method of construction. Relief valves are sometimes fitted in pump suction systems to prevent the pump experiencing excessive suction pressure leading to high axial thrust and subsequent bearing wear. In reciprocating pumps, high suction pressure can result in rapid wear of the crosshead pin bearings. Relief valves fitted to pumps must be sized to pass the total pump flow. The following terms are used in conjunction with relief valves.

Set pressure	The pressure at which the relief valve will begin to open.
Overpressure	The pressure which is required for the relief valve to pass 100% flow. Sometimes called accumulation. For liquid relief valves purchased without any special requirements the overpressure will be 25% of the set pressure. Overpressure can be set as low as 8% in some cases.
Reseat pressure	The pressure at which the valve will close and reseal. Normally between 80 and 97% of the set pressure.
Back pressure	The pressure at the relief valve outlet connection when piped into a system.
Modulating action	The lift of the valve is proportional to the flow.
Pop action	Once the valve opens the lift is 100% until it closes.

Relief valves are connected to a short branch on the process pipework. The branch must be short so that there is no significant pressure drop between the process pipe and the valve seat. When the valve opens the flow is piped to a low pressure system. If the pump differential pressure is less than 20 bar, the relief valve outlet may be piped into the pump suction although it is always preferable to pipe the valve back to the suction source. If the relief valve outlet is not at atmospheric pressure, a back pressure is present, the valve may require bellows compensation and/or the size increased.

A special application of relief valves, not directly involving pumps, is thermal relief. When a piping system is built it is possible to have a length of pipe with an isolating valve at each end. If the pipe is full of liquid when both valves are closed high pressures can be created when the liquid expands due to temperature changes. Small spring loaded relief valves are fitted to relieve the excess pressure.

Relief valves are built in three different styles to suit different applications and to provide different facilities. Two additional designs are available for special applications.

Spring loaded relief valves

Spring loaded valves are the basic and simplest design. The valve is held closed by a spring load. The liquid pressure under the valve seat has to overcome the spring load before the valve can open. As more liquid flows through the valve the lift increases. When the liquid stops flowing the valve reseats and seals. Manufacturers have proprietary nozzle and poppet designs, some with adjustable lift characteristics, to suit various applications. Spring loaded relief valves are suitable for gases and liquids; pumps require a valve with a liquid trim. High viscosity liquids, like some lubricating oils and crude oil, may require modification to prevent the poppet sticking.

Spring loaded relief valves are the first choice for most applications. Variable flow applications can create problems for spring valves. If the valve is required to pass flows less than 30%, stability and chattering can cause high wear rates and damage the seat resulting in poor sealing.

These valves are generally set at +10% with 10% overpressure. Running at overpressure is considered as intermittent. If longer running at overpressure conditions is necessary, this must be specified in the operating conditions. In some cases it is possible to use 25% overpressure to advantage; a smaller valve can be fitted. Low pressure applications and when the installed motor power is much larger than required are typical cases.

Pilot operated relief valves

A pilot operated relief valve uses liquid pressure to hold the valve closed. When the set pressure is reached the pilot releases all, or some, of the holding pressure and allows the valve to open. When the process pressure falls, the pilot re-energises and the valve is firmly closed. Pilot operated valves are more costly than spring loaded valves but they are more versatile.

A pilot operated valve can be set to have zero overpressure, a pop action. Once the set pressure has been reached, the valve opens 100%. A modulating pilot is also available. The set pressure can be fixed closer to the operating pressure because the valve is normally clamped closed hydraulically. The reseat pressure can be adjusted to be as close as 97% of the set pressure.

A set pressure of +5% and a 5% overpressure is a good compromise for system operation. A pop action valve can create pressure pulsations within the system, for this reason a modulating valve is preferable. Pilot operated valves do not suffer instability at low flows.

Air assisted relief valves

An air assisted valve is a modified spring loaded valve with pneumatic control. The poppet, as well as the spring load, is attached to a pneumatic diaphragm. The valve is held closed by the spring force and the air pressure above the diaphragm. A control valve senses the pressure in the process pipe. When the liquid pressure reaches the set pressure the control valve releases the air above the diaphragm. The valve opens as a normal modulating spring loaded valve. When the liquid pressure reduces to the reseat pressure the air is reapplied to the diaphragm, closing the valve. The addition of the air assistance allows the set pressure, overpressure and reseat pressure to be brought closer together. As with pilot operated valves, the operating pressures can be + 5%, 5% and 97%. A slightly higher reseat is possible. The benefit of an air assisted valve is in its back-up capability; it the pneumatics fail it will operate as a normal spring loaded valve. The drawback is the cost and need for an air or gas supply. The valve does not use gas continuously, only when the valve operates, bottled gas is adequate.

Shear pin and buckling pin relief valves operate slightly differently to other valves. A shear pin valve is held closed by a relatively weak pin restraining a plunger. For the plunger to move, and allow the valve to open, the pin must be broken by direct shear at two points. A buckling pin relief valve is held closed by a slender pin under direct compression at its ends. When the relief valve reaches set pressure, the axial load on the pin exceeds the critical load, the pin buckles and the valve opens. In both designs the valve does not reset after opening. These styles of relief valve are often used for liquid/solid mixtures.

10.5 Accumulators

Accumulators are used for pressurised liquid storage in systems which require variable flow rates or intermittent flow requirements. We have seen in sections 5.4, 5.5 and 5.6 that storage volumes are necessary when the pump flow does match the flow demand. Accumulators are suitable for pressurised applications up to 345 barg.

The accumulator consists of a pressure vessel which houses a bladder, rather like a balloon. The bladder is attached to the top of the pressure vessel by its gas valve. The bottom of the pressure vessel is piped to the process system. The bladder is precharged with nitrogen. When the accumulator is charged, the nitrogen in the bladder is compressed to a higher pressure as its volume is reduced. During the working cycle, the system pressure will fall when the pump is unable to match the demand. As the pressure falls, the nitrogen in the bladder expands and pushes some of the liquid out into the system. The next time the pump flow is greater than the system flow the accumulator will be recharged. The volume of liquid which can be stored is a function of the volume of gas. As we saw in section 5.4.6, it is the application of Boyle's Law for gases at constant temperature which controls the pressure and volume relationships. If more liquid is required, more gas must be provided. If a smaller pressure change is required, more gas must be provided. The manufacturers of accumulators have guide-lines which indicate the range of working pressures possible, some have design charts. Working pressure ranges of up to 3.3 are possible, but this is far too large for normal pump applications, see run-down accumulators later.

Accumulators are manufactured with gas volumes of 200 l as standard. If these are too small, accumulators can be collected in racks and piped together. The pressure/volume characteristic can be enhanced by adding more gas volume. Standard nitrogen bottles can be piped into the gas side of the bladders. The liquid volume per unit volume of pressure drop is thus increased.

Bladders are not completely gas tight. Just as with tyres on a car, the gas pressure must be checked regularly and topped up when necessary.

Another application for accumulators is in lube and seal systems on large machines. Large machines, particularly high speed machines, do not always stop when the power is switched off. Such machines are normally stopped by control logic so that after the power has been switched off to the main driver the auxiliaries; lube oil pumps, seal oil pumps, cooling water pumps; carry on running for 30 seconds or a minute to allow the main pump to decelerate. After a preset time the control logic switches off the auxiliaries. This type of controlled run-down works well in normal circumstances but can have problems in emergencies. If the main motor is high voltage and the auxiliaries are low voltage, it is possible to have a fault in the low voltage supply so that all auxiliaries are lost but the main motor is still running. Interconnection between the HV and LV systems could switch the main motor off as soon as the LV supply is lost. However, there are no auxiliary supplies for run-down. This type of problem is solved by fitting and charging accumulators. The auxiliaries would normally be started and allowed to stabilise before the main motor was started. This period can be used to charge the accumulators. Once charged, the system acts as though they are not there. When auxiliary power is lost the accumulators discharge their liquid into the auxiliary systems during the run-down. If the main motor was required to continue operating under these conditions it would be necessary to fit stand-by pumps to all the auxiliary services powered from an alternative supply, probably 24 V DC.

10.6 Pulsation dampeners

Reciprocating pumps, because of the nature of the piston/plunger motion, cause flow variations and pressure pulsations in the system pipework. Pulsation dampeners are fitted to the suction and discharge systems of reciprocating pumps in order to reduce the severity of the flow disturbances. Pulsation dampeners attempt to disconnect the pump from the system in terms of inertia. The flow variations produced by a pump cannot be changed. If the density of the liquid were zero, then there would be no pressure pulsations, no pipe vibration and the effect in the suction system, 'Acceleration Head Loss', would disappear. The pulsation dampener absorbs excess liquid and provides liquid when the pump is deficient. This is very similar to the action of accumulators but it happens much faster.

Flow variations are difficult to measure. Pressure pulsations are easy to measure and the effect of pressure pulsations is easy to observe. Pressure pulsations can be measured with a pressure transducer and displayed on an oscilloscope or captured for analysis by computer. The peak to peak pressure value is readily evaluated. The effect of pressure pulsations is pipework vibration. The product of the peak to peak pressure and the pipe bore area is a force; this force acts alternatively along the pipe. If the force is large in comparison to the mass of the pipe and the stiffness of the pipe supports the pipe will vibrate. By reducing the peak to peak pressure, the severity of the vibration will be reduced. Dampeners reduce the peak to peak pressure by reducing the flow variation. Pulsation dampeners are manufactured in three basic types; gas charged, inertial and acoustic.

Gas charged dampeners

A gas charged dampener is similar in many respects to the accumulators described in 10.5. A gas charged dampener relies on the increased compressibility of a volume of gas to absorb some of flow variation. Consider the discharge stroke of a single cylinder pump. When the piston/plunger accelerates towards maximum velocity the pressure at the pump increases due to liquid inertial loads and increased friction losses in the pipe system. If a pulsation dampener was placed close to the pump, some of the liquid could flow into the dampener, and increase the gas pressure, rather than flow up the discharge pipe. When the piston/plunger decelerated and the pressure at the pump decreased, the liquid would be forced out of the dampener due to the gas pressure. The flow variation into the discharge pipe would be reduced thereby reducing the pressure pulsations. The suction system would react in a similar manner.

The volume of liquid which flowed into the dampener would depend upon the gas volume and the nature of the gas. Nitrogen is nearly always used so the gas characteristics become constant. The volume of gas used in a dampener is the important sizing criterion. The pump user or manufacturer specifies the residual pressure pulsations acceptable and the dampener manufacturer calculates the gas volume required.

Gas charged dampeners can be constructed in various styles. A popular style for suction systems is a vertical vessel which contains approximately 75% liquid. The remainder of the vessel is charged with nitrogen. No bladder or diaphragm is used to separate liquid and gas. This type of dampener, also called a suction stabiliser, is very successful for low pressure and hot applications. At higher pressures the nitrogen is absorbed more quickly into the liquid. This style of dampener is a 'flow-through' dampener; all the liquid passes through the dampener. Flow through dampeners are also constructed with bladders or diaphragms to contain the gas charge, see figure 10.2. Some dampeners are constructed exactly like accumulators and are attached to a branch on the process pipe, figure 10.3.

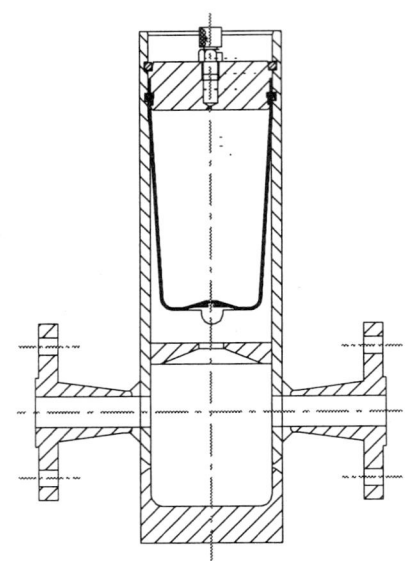

Figure 10.2 Flow through gas charged bladder dampener

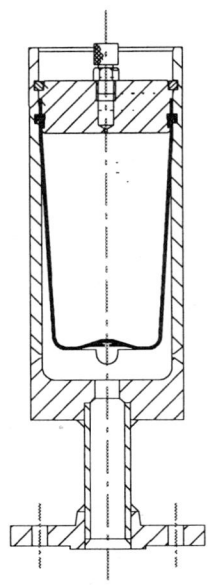

Figure 10.3 Single port gas charged bladder dampener

Gas charged dampeners must be pre-charged before use. The gas charge is usually 60 to 80% of the working pressure. Dampeners without bladders or diaphragms can suffer problems with gas loss when the gas expands during shut-downs. In general, these dampeners do not create a pressure loss. Gas charged dampeners are the most used type of dampener.

In some installations, spares inventory is reduced by using high pressure pulsation dampeners on low pressure suction systems. High pressure dampeners are constructed with fairly rugged bladders and diaphragms; some over 20 mm thick. Heavy, thick diaphragms do not work very well with low differential pressures trying to move them. Consequently high pressure dampeners do not realise the predicted pulsating pressure attenuation. High pressure dampeners are suitable for high pressures. Low pressure dampeners are essential for good performance on low pressure suction systems.

Inertial dampeners

Inertial dampeners rely on a moving mass of liquid to limit the velocity changes in discharge system. A mass of liquid, inside a pressure vessel, is made to rotate as the liquid flows through. The mass of liquid acts as a flywheel, in mechanical terms, and reduces the flow variations. Inertial dampeners are used for high temperature applications where elastomers are unsuitable. There is no pressure limit on inertial dampeners and they are unaffected by pump speed or discharge pressure. Inertial dampeners do create a pressure loss and this must be considered in the pump selection. Inertial dampeners are only used as discharge dampeners.

Acoustic dampeners

Acoustic dampeners are designed specially for each application. Pressure vessels consisting of connected volumes and choke tubes attenuate discrete frequencies. The dampeners are tuned to remove the worst pressure harmonics from the flowing liquid. Like inertial dampeners, there are no moving parts other than the liquid. Acoustic dampeners are unaffected by working pressure but very sensitive to changes of speed. Acoustic dampeners would be of little use on infinitely variable speed pumps. All of the pump flow passes through the dampener and a pressure loss is induced. Acoustic dampeners can be used on the suction and discharge sides of the pump. There are two major obstacles to popular use of acoustic dampeners. Because they are designed for each specific application the cost is relatively high. To work effectively they must be large. Acoustic dampeners are much bigger than any other dampener.

When choosing a pulsation dampener the initial cost must be weighed against running and maintenance costs. Gas charged dampeners may need gas every two weeks. If the gas charge is lost, can the process continue ?

At the beginning of this section we discussed the effect of the peak to peak pressure pulsation on the pipework. The pressure pulsation creates an axial force which shakes the pipework. Some shaking force is allowable as the inertia of the pipework plus the restraining forces of the supports restrict the pipe movement. Figure 10.4 indicates acceptable peak to peak pulsation levels as a function of the mean process pressure. These values are based on pipework mechanics only. The process or instrumentation connected to the pipewrok may require lower pulsation levels. Pressure pulsations in the pump suction pipework may require lower pulsation levels. Pressure pulsations in the pump suction pipework reduce the NPSHa, this must be considered when deciding acceptable levels.

The pressure pulsation radiated by a reciprocating pump is a complex signal which includes all the integer harmonics from $1 \cdot r/sec$ and greater. The power of the signal is in the lower harmonics, usually starting at r/sec $\cdot$ number of plungers. The pressure pulsation can cause lateral vibration of the pipework, and resonance if the natural frequency of the pipe span is very close to one of the integer harmonics. It is also possible to achieve acoustic resonance. A pipework system can be tuned

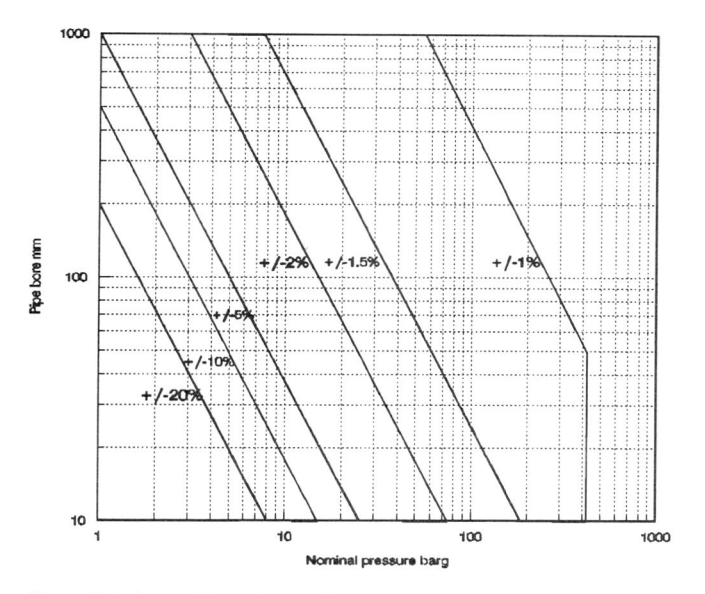

Figure 10.4 Acceptable peak to peak pressure pulsations for pipework

to various frequencies due to the pipe length and the arrangement of branches and other connections. If one of the pipework acoustic frequencies is very close to a pump integer harmonic very serious vibration problems may arise. The magnitude of the problem is dependant upon the operating conditions and the potential resonant frequency.

System mechanical and acoustic resonance can be predicted by computer simulation. The complete system must be analysed. The discharge system cannot be analysed without the suction system and vice versa. Complete, and accurate piping isometrics must be available for analysis. There is little point in analysing one system and building another. Analysis can either be analogue or digital. The electric analogue approach is a much more mature technique being an extension of reciprocating compressor technology. The following points should be raised with any prospective contractor:

- can your technique model pump run-up and run-down conditions,
- can your technique model process transient conditions,
- can all types of pulsation dampener be modelled,
- how easy is it to change parameters to evaluate modifications,
- can the interaction between multiple pumps be modelled,
- what are the limits of your simulation, system length, etc.

Simulations for a single pump system start at about £10000 and could cost as much as £40000. The analysis cost must be judged against the pump unit cost and the criticality of the service.

It is not possible to give generalised advice on when simulations should be performed but it is possible to state some obvious facts. Pressure pulsations are largely attenuated by liquid viscous effects. Pulsations will radiate further in low viscosity systems than in high viscosity. Pumps running fast, both r/min and m/s, will produce worse pulsations than pumps running slow. Pipework with poor supports will vibrate more easily than properly supported pipework. It is much better to have pipework for reciprocating pumps at ground level securely attached to concrete. Piping systems running at high liquid velocities will tend to give more problems than low velocity systems.

If it is difficult to decide whether a simulation should be performed seek independent advice. Good advice can only given after reviewing all the information; this is the information that would be supplied to allow the simulation to be performed.

10.7 Instrumentation
Integrated control & diagnostics

Many pump suppliers have developed standard packages, complete with instrumentation and starting equipment, for popular applications. These include:

- twin circulating pump packages for commercial and industrial heating,
- pressure maintaining packages for closed systems,
- trailer mounted engine driven site pumps,
- dosing pump packages complete with suction tank,

The ultimate in packaging tends to be the high pressure cleaner suppliers. Pumps are packaged in floor mounted coin operated booths for garages, on two and four wheel trailers, and truck mounted for complete portability.

Pump packages which are built to the user's specifications can incorporate almost anything. As pump motor sizes increase the motor starters are generally not included in the package but are housed remotely in motor control centres. The pump package would include control logic and local operating functions. The following list indicates typical equipment found in local control panels:

- pump start/stop buttons,
- hand/off/auto switches,
- lag/lead switches,
- indicator lamps,
- hours run counters,
- ammeters,
- fault/trip annunciators.

Annunciators consist of numerous small windows which are illuminated to reveal hidden messages. Not all the messages relate to faults or trips; it is possible to indicate normal functions. The annunciator is fitted with a test button to check all the lamps are functioning. Most annunciators have connections to allow remote signalling of fault and trip conditions. In automatic packages the following annunciator messages are typical of the range of information displayed:

- pump 1 running,
- pump 2 running,
- pump 1 tripped,
- pump 2 tripped,
- low suction pressure,
- low low suction pressure,
- high discharge pressure,
- high high discharge pressure.

When a pump motor trips the annunciator does not know the reason; the annunciator receives a signal from the motor starter to notify that the contactor has opened when it should have closed. The operator must go to the motor control centre to look at the motor starter. Motor control centres are becoming much more sophisticated and a group of motor starters may have something equivalent to an annunciator which indicates a wider range of motor faults.

Operating conditions designated low or high are fault or alarm conditions. The annunciator flashes the relevant lamp and sounds an audible signal. The alarm function is to warn operators that some operating condition is outside normal limits and someone should investigate and take corrective action. The pump continues to operate. Low low and high high are trip conditions, something is switched off. If, after the alarm signal, corrective action was not or could not be taken, the operating condition becomes further away from normal running, the control system takes over and shuts down the equipment. At the alarm condition, the control system can be

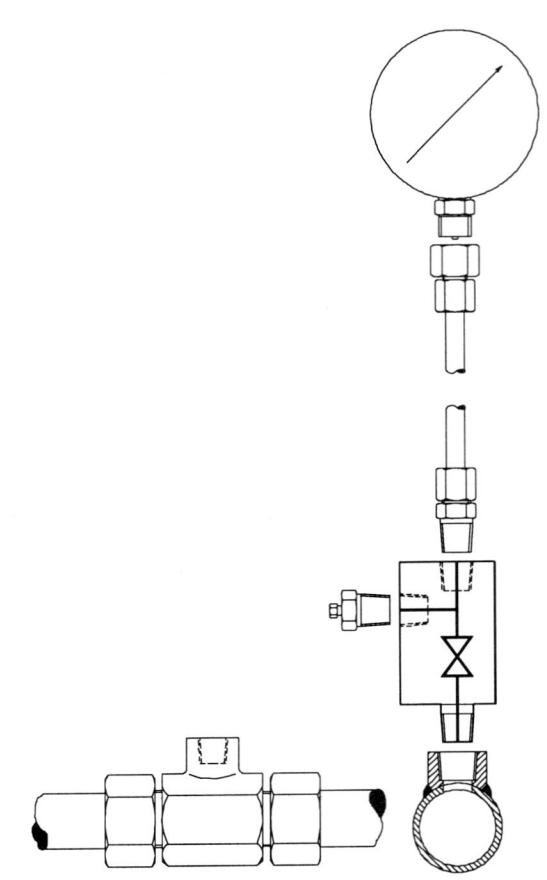

Figure 10.5 Pressure gauge for water or lube oil

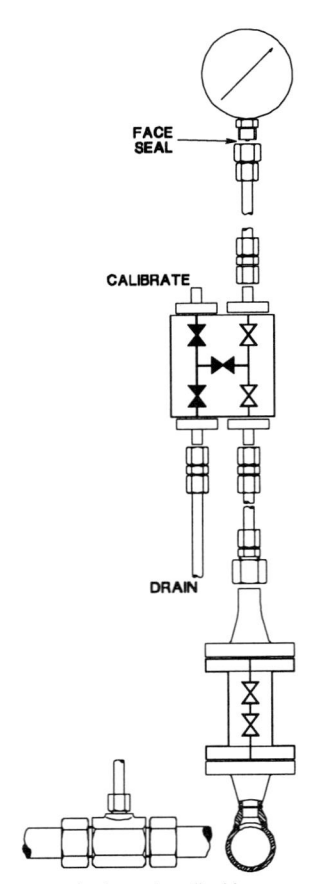

Figure 10.6 Pressure gauge for hazardous liquid

designed to start stand-by equipment when this is fitted. Modern control systems, using PLC's, can manage very complicated systems. The use of PLC's allows transducers to replace switches. A transducer can monitor the rate of change of a parameter and take pre-emptive action rather than wait for the parameter to reach a prescribed limit.

One important aspect of instrumentation which is sometimes not specified is how instruments should be connected to process pipework. This part of detail design becomes more important when the process liquid is hazardous. In some cases the user should specify that no screwed connections are permissible for sealing process liquids. Compression fittings, flanges and other connectors must be used. Valves are used to isolate instruments from the process to allow servicing and calibration. Bleed, vent and drain valves may also be required. For liquids such as water and lube oil, bleed and vent plugs could be acceptable, but not for propane. Figure 10.5 shows how a pressure gauge could be connected to water or lube oil systems. Figure 10.6 shows how a pressure should be connected to a hazardous liquid system. The user must be fully aware of the hazards posed by the process liquid. These hazards must be transmitted to the pump supplier to allow appropriate designs to be furnished.

When pressure switches are fitted, a pressure gauge should be fitted so that local operators can see the normal pressure and report on trends. If multiple switches are fitted, for separate alarm and trip functions, they must be connected to the same point on the process pipe.

Modern electronic instruments allow much more sophisticated control of pumps and equipment. However, qualified, experienced staff must be on hand to look after the equipment if high costs are not to be incurred using the supplier's site personnel.

11 Quality assurance and testing

All manufactured equipment is produced to some level of quality. The minimum quality level is initially set by the manufacturer. The rate of component rejections must be small enough to be an acceptable proportion of the equipment cost. For mass produced goods, the customer can only judge the quality from the final results. The only tests possible are initial appearance, performance and life. Some consumer goods are rejected by failing some or all of the tests. The consumer has no control over the manufacturing process directly. Indirectly the consumer can influence quality by not buying the goods.

When the equipment is manufactured specifically for the customer the question of quality is treated differently. Because the equipment does not exist, completely, at the time of ordering the customer is able to exert some control. The extent of control is dependant upon confidence, trust, legal requirements and insurance companies.

This chapter concentrates on the inspection of metals, sub-assemblies and pumps. Non-metallic materials are normally purchased against trade names, the registered owner being well known as is his quality reputation. It also deals with the various inspection functions and tests possible and what information the purchaser should expect to receive with a pump and what extra information should be available if the purchaser wishes to inspect it. Guidelines are given for purchaser quality requirements.

Contents

11.1 Introduction
    11.1.1 Physical properties
        11.1.1.1 Ultimate strength
        11.1.1.2 Limit of proportionality
        11.1.1.3 Elongation
        11.1.1.4 Reduction in area
        11.1.1.5 Hardness
        11.1.1.6 Impact strength
        11.1.1.7 Fatigue strength
        11.1.1.8 Creep resistance
    11.1.2 Heat treatment
    11.1.3 Chemical composition
    11.1.4 Corrosion resistance
    11.1.5 Non destructive testing
        11.1.5.1 Radiographic inspection
        11.1.5.2 Ultrasonic inspection
        11.1.5.3 Dye penetrant inspection
        11.1.5.4 Magnetic particle inspection
    11.1.6 Repairs
    11.1.7 Welding
    11.1.8 Inspection
    11.1.9 Assembly
    11.1.10 Packaging
    11.1.11 Pressure testing
    11.1.12 Running tests
    11.1.13 Painting
    11.1.14 Purchased equipment
    11.1.15 Functional testing
    11.1.16 Witnessing
    11.1.17 Clarification of specifications
    11.1.18 Certification
    11.1.19 Documentation

11.2 Mass produced standard pumps

11.3 Pumps built to purchaser's specification

11.4 Guidelines for documentation
    11.4.1 Rotodynamic pumps
    11.4.2 Positive displacement pumps

11.5 Bibliography

11.1 Introduction

The manufacture of pumps can be divided into logical steps. Each step can have unique inspection operations and tests. Each phase of production will be identified and the range of possible tests reviewed.

Raw material

The vast majority of pumps have the important components manufactured from metal, of which cast iron, carbon steel and stainless steel are the most used. Cast iron forms the basis of all steel products; even though some new steels are made from scrap steel. Iron ore can be converted to iron by three methods:

- blast furnace,
- sintering or pelletised/blast furnace,
- direct reduction.

In a blast furnace, iron ore reacts with hot coke to produce pig iron. The sintering or pelletising process prior to the blast furnace operation is added to allow blending of iron ores and also to control the size of the blast furnace feed. Sintering or pelletising improves the blast furnace operation and reduces energy consumption. Direct reduction produces sponge iron from iron ore pellets by using natural gas. Most iron is produced from sintered iron ore and coke. The steel maker controls the sintering process to produce a consistent iron quality. Modern blast furnaces are fitted with many instruments and, together with computer modelling, enable in-process control. Iron is taken from the blast furnace as finished material for iron foundries. Iron is transferred to the oxygen steel process for conversion to various grades of steel. Iron from direct reduction plants is mixed with scrap steel in an electric arc furnace to produce various grades of steel.

In general, the design engineer is not concerned which production method is used. For some alloy steels, for some arduous applications, the designer may specify vacuum remelted steel for reduced impurities or may require additives to promote fine grain structure. Principally the design engineer looks at the physical properties of the material to judge suitability. Standard tests are applied, solely, to assess compliance with the published specifications. Some materials are characterized only by their physical properties or chemical composition, others by both. Grey cast iron is specified by its physical properties. Some low grades of carbon steel are specified by their chemical composition, no physical properties are necessary. Most materials are described by both.

For the following physical properties, standard test pieces are stretched in a machine which simultaneously measures the increase in length and the applied load. There are several different test piece sizes which give slightly different results. One standard test piece is very small, this fits a machine called a Hounsfield Tensometer. Very small test pieces are useful when samples must be taken from castings or finished parts.

11.1.1 Physical properties

11.1.1.1 Ultimate strength

The strength of the material when it fractures. See section 6.2 for typical values.(*)

11.1.1.2 Limit of proportionality

The strength of the material when the relationship between stress and strain ceases to be linear. In low carbon steel this is classified as the yield point, the onset of plastic deformation, the material does not return to its original length when the load is removed. Most designs do not stress materials beyond the limit of proportionality.(*)

11.1.1.3 Elongation

How much the material has increased in length when it fractured. Different test pieces have different gauge lengths, each gauge length gives a slightly different result. Good elongation properties, 15 to 20%, are required for complex components which are highly stressed. Good elongation indicates ductility. Ductility is necessary so that components can deform very slightly to spread the load. A good cast iron may be 4%.(*)

11.1.1.4 Reduction in area

Ductile materials 'thin' slightly as they are stretched. When the material fractures, the cross-sectional area of the fracture is less than the original test piece. Reduction in area is reported in most American Standards but not used very much in Europe.(*)

11.1.1.5 Hardness

The ability of the material to withstand surface indentation. No special test piece is required, raw material and finished parts can be tested. Several scales of hardness are used; Brinell Hardness Number, Vickers Pyramid Hardness and Rockwell. Approximate conversions are available between scales. In carbon steels, the hardness is directly related to the strength.

11.1.1.6 Impact strength

The ability of the material to withstand shock or impact. A special test piece is required to fit the test machine. Most materials loose impact strength as the temperature reduces. Depending upon the material, impact properties should be checked when operating below 0 °C. Two different tests are used which give different results, very approximate conversions are available. Charpy is the most popular, Izod is little used. A popular bench-mark for offshore equipment is 27 J at the design temperature. It is normal to check three test pieces.(*)

(*) These mechanical properties are a function of the grain direction of the material. Unless specified otherwise all properties relate to the longitudinal direction. Properties in the transverse direction may be lower depending upon the physical treatment of the material and its grain structure.

11.1.1.7 Fatigue strength

All the tests defined so far can be performed fairly quickly; test pieces today, results tomorrow. Fatigue is different. A special test piece is either subjected to repeated tensile loads or repeated bending loads. For repeated tensile loads, the test piece experiences cyclic loads from 0 to + value. A bending test piece is loaded from -value to +value. To find the endurance limit the test piece must not fail. A test piece will look promising if it lasts five million cycles, 5,000,000. If the machine runs at 3000 r/min this will take 1667 minutes, 28 hours. Of course, it will not be possible to guess the correct stress so several tests must be run. Testing for fatigue in air is simple, testing for fatigue in contact with a specific liquid is more complicated. It is not usual for fatigue strength of materials to be checked on a contract basis. For critical pumps, fatigue testing of finished components may be necessary. Most pump designs are not based on fatigue. Exceptions include components for reciprocating pumps. The manufacturer should state if the life of any component is limit by fatigue when running at the rated conditions.

11.1.1.8 Creep resistance

Creep is the permanent distortion of the material after being subjected to a stress for a long period of time. Not normally a problem in pumps, although pumps built of GRP or PTFE may suffer at temperatures below 250 °C. Usually considered in steam and gas turbines, i.e. hot components. Creep testing is similar to fatigue testing but worse. Creep tests run for years. Published research data is used when necessary.

11.1.2 Heat treatment

Many materials require heat treatment to achieve the correct condition or strength. Carbon steels are hardened and tempered to achieve high strength, usually at the expense of ductility. Austenitic stainless steels are stress relieved, softened or solution annealed to modify the physical or chemical properties. The final condition is usually confirmed by taking hardness readings. When components are heat treated to achieve specific physical properties a test piece is heat treated as well. The necessary physical tests are conducted on the test piece.

11.1.3 Chemical composition

When a metallic material is produced as a raw material, its chemical composition is checked. When cast iron is converted to carbon steel in the oxygen process, all the relevant elements are weighed into the converter. Before the steel is poured, the chemical composition is checked. When the steel is poured a sample is cast. The sample is analysed and its chemical properties are the properties of the melt. Certificates will show the name of the steelmaker and the melt, cast or heat number. The chemical composition may show elements which are not called for in the specification. Low carbon steels may show traces of nickel, chromium and molybdenum. The trace elements are a welcome addition because they tend to enhance the physical properties of the material. Impurities, sulphur and phosphorous, will be shown very accurately. The chemical composition of specific components, when necessary, will traced back to the original melt.

On rare occasions, a sample will be taken from a component and analysed. Modern techniques only need very small samples. It is possible to analyse material without destroying it. Two devices are available which can analyse material without removal from the component. Neither method can detect carbon. However sufficient accuracy is present to differentiate between 304 and 316 stainless steel.

11.1.4 Corrosion resistance

Corrosion resistance of materials is judged from published research. Pump manufacturers do carry out long term research on corrosion and develop materials to cope with specific problems. If a pump user wishes to pump a new liquid, of which obviously no pump company has experience, the user should conduct basic corrosion testing.

11.1.5 Non destructive testing

Raw material, raw castings and completely finished components can be examined physically to determine the quality of certain aspects of the material. This type of examination falls into two categories: surface inspection and interior inspection. Surface inspection looks for discontinuities in the surface which could be detrimental to the service life of the component. Cracks in the surface create stress raisers which can lead to fatigue failures. Pinholes in the surface may indicate porosity. Internal examinations can show the integrity of the material and if there are any impurities, inclusions or voids in critical locations. Impurities, inclusions and voids detract from the cross-sectional area available for stressing and create stress raisers. Porosity can lead to problems of leakage.

When flaws are detected someone has to decide whether the flaw is serious, if it can be repaired or whether it should be repaired. Some National Standards, particularly pressure vessel standards, have categories for defects. The manufacturer's requirements may be more or less stringent than published standards. If the flaw is in raw material, a casting or piece of plate, it may be more cost effective to scrap it rather than expend more time and money on repairs. If the flaw is in a semi-finished piece there may be more incentive to repair. If the flaw is in a finished component there will be strong financial reasons for a repair.

11.1.5.1 Radiographic inspection

Radiography, X-ray, is accepted as the highest grade of internal inspection and the most costly. Radiography is good because a permanent record of the inspection is available at any time for review. It is used mostly for welds but also critical areas of castings. Components can be taken to fixed machines, large components or assemblies are radiographed using radioactive isotopes. Strict safety precautions must be enforced.

11.1.5.2 Ultrasonic inspection

Ultrasonic inspection is popular because it can be conducted with portable equipment. The display on a cathode ray tube indicates the position of flaws in respect to the thickness of the material. The size of the flaw must be assessed by an experienced operator. Very good for inspecting large flat plates prior to fabrication or forgings.

11.1.5.3 Dye penetrant inspection

'Dye pen' is a surface inspection method which is good for finding pinholes and hairline cracks. The surface is first sprayed with dye which is allowed to soak into any surface defects. After a specified time, the dye is washed off and the component cleaned. Chalk is finally sprayed onto the surface. If surface defects exist, the dye trapped in the defect is drawn into the chalk by capillary action. Defects are outlined by dye indications in the chalk. Generally used on finished machined surfaces. A skilled operator can judge the depth of the defect by the size of the 'bleed-out'.

11.1.5.4 Magnetic particle inspection

'Mag particle' is a surface inspection method which is popular for cast materials. 'Mag particle' inspection can only be conducted on materials which can be magnetised by an electric current. The surface is coated with a liquid bearing small magnetic particles. If the surface contains flaws, the magnetic flux is concentrated around them drawing the magnetic particles towards the flaw.

11.1.6 Repairs

When repairs are undertaken, usually by welding, the repair must be inspected. Initially the faulty material must be removed. This can be achieved by any suitable means. The component must be checked to ensure that all faulty material is removed, this is normally by dye penetrant or magnetic particle. The repair is effected. The component must then be inspected by the same method which found the original flaw. Pumps built specifically for a purchaser may have repairs categorised into 'major' and 'minor'.

11.1.7 Welding

Welding is a skilled occupation. There are many different types of welding and the welder doing any particular job must be proficient at that type of welding. Welders are graded by the types of components they weld, the positions of the welds and by the types of equipment used. Welding pipework is thought to be the most difficult type of welding. Welders for pressure vessels have to be qualified and this is usually assessed by an insurance company. Welding pipework to pump casings is difficult. Designing weld methods for exotic materials requires very special materials knowledge. All welding for process liquid parts should be supported by certification.

11.1.8 Inspection

Inspection of all components is carried out as a matter of course. The degree of inspection is dependant upon the nature and function of the component and the batch size. Mass produced parts are not scrutinised as closely, individually, as a small number of specially made parts. General dimensional inspection of parts does not normally form part of the purchaser's involvement. In special cases, the purchaser may wish to witness the measurement of specific dimensions.

11.1.9 Assembly

The basic process of pump assembly is the final dimensional check. If components do not have the correct fit or clearance it will be very apparent. Selective assembly of components is frowned upon unless the components are always supplied as a sub-assembly. Spare parts cannot be selectively assembled at site. Assemblies may have fits, clearances and alignment measured and recorded before testing or dispatch. This record is sometimes called a 'History Sheet'.

11.1.10 Packaging

In this context packaging refers to building the pump and any associated equipment into a pump unit. We have seen that a simple pump unit may be a pump and a baseplate when the purchaser supplies and mounts his own driver. At the other extreme, a pump unit may consist of several pumps and drivers with pipework and control equipment mounted on one baseplate or skid. When control equipment is supplied, the pump manufacturer is often required to wire all the control elements together. If a control panel is mounted on the package then switches and transducers can be wired to the panel. If the control panel is mounted separately, the pump manufacturer will probably wire the package instrumentation to a terminal box. The purchaser's inspector or instrument engineer may wish to witness the wiring process. Some cable glands are complicated to build up, wiring routes can be better assessed with a full size 3D package and wiring practices inside instruments and terminal boxes may be considered critical.

11.1.11 Pressure testing

Pressure testing of components and assemblies confirms the material integrity and the basic strength of the design. Porosity of castings can be detected easily without lengthy ultrasonic or costly radiographic inspections. Pressure testing does not always indicate thin walls. Whenever possible components should be pressure tested in a manner which closely represents the pressure effects during normal working. Clamping components in fixtures modifies the elastic behaviour of the component and does not highlight problems such as distortion of mating flanges.

Pressure testing does not automatically mean hydrotesting with water. Some tests may be conducted with oil, others with a detergent and water mixture. Some components may be pressurised with compressed air and submerged in a water tank. Critical components for very hazardous liquids may be tested with helium.

The pressure to which components may be subjected can be limited by standard flange designs. The purchaser and the manufacturer must ensure that each fully understands the requirements and the limitations of pressure testing.

11.1.12 Running tests

Running tests are thought to be the ultimate quality control test. This may or may not be the case. Many types of test are possible. The following list has been proposed to ISO for positive displacement pumps.

A confirmation test is a running test conducted to confirm the pump unit functions as required. No test measurements are made. The pump unit actually performs the specified functions. This type of test is limited to fully packaged equipment such as engine driven, trailer mounted pumps complete with suction and discharge hoses.

A performance test is a running test conducted to confirm the mechanical and volumetric efficiency and the mechanical soundness. The following test readings shall be taken and recorded:
   inlet pressure,
   inlet temperature,
   outlet pressure,
   outlet flow,
   pump speed,
   pump power input or torque.

The number of sets of readings will be confirmed by the manufacturer.

NOTE: Power input may be measured electrically at the input to a calibrated test motor or the contract motor when data on efficiency is supplied by the motor manufacturer. Driving torque may be measured directly or indirectly as a torque reaction when appropriate.

An NPIP test is a running test conducted to confirm the NPIPr is less than the NPIPa. The purchaser and the supplier shall agree the flow loss criterion to be used. All the readings required for a performance test shall be recorded.

NOTE: An NPIP test is not part of a performance test and must be specified separately. Test rig modifications may be necessary which have cost and delivery implications.

An MIP test is a running test conducted to confirm the Minimum Inlet Pressure is sufficient to support the rated operating conditions.

A linearity test is a running test conducted on dosing pumps to show that the variation in flow with stroke or speed adjustment conforms to a straight line relationship. All the readings required for a performance test shall be recorded.

A repeatability test is a running test conducted on dosing pumps to show that the pump delivery is consistent for specific stroke or speed settings. All the readings required for a performance test shall be recorded.

A steady state accuracy test is a running test performed on dosing pumps to show the precision of each dose under steady state operating conditions. All the readings required for a performance test shall be recorded.

A gas handling test is a performance test conducted with specific volumes of gas entrained in the inlet pipework. All the readings required for a performance test shall be recorded.

NOTE: This test should only be conducted after the normal performance test. This test can only produce useful results when the test is conducted at data sheet conditions with the data sheet liquid or a close approximation.

A self-priming test is conducted to show that the pump can evacuate gas from the inlet pipe and draw liquid into the pump and operate normally without external intervention. The pump casing shall be filled, according to the suppliers instructions, before the test.

Mass produced pumps shall be fitted with a standard inlet arrangement described by the supplier. Pumps tested for contract shall be fitted with pipework to provide a similar gas volume to the site arrangement. When necessary, the static lift can be increased by applying a partial vacuum. All the readings required for a performance test shall be recorded plus the time taken to evacuate the inlet.

Product data for mass produced pumps shall indicate all the test parameters including outlet pressure, liquid viscosity and temperature.

A mechanical run is a running test, at specified conditions, conducted to confirm mechanical soundness of the pump and any other auxiliary equipment. The following test data shall be recorded:
   inlet pressure,
   outlet pressure,
   pump speed.

A site simulation test is an extended performance test where the pipework local to the pump replicates the site pipework. An NPIP test, if required, must be specified separately. All the readings required for a performance test shall be recorded. The purchaser shall not specify the requirement to run a site simulation test until details of site process pipework are available for the supplier's review.

A strip inspection is a partial disassembly of agreed portions of the pump to confirm correct operation of components.

Retest after strip inspection is not required unless a fault is detected.

A string test is a running test which involves all, or part, of the contract drive train and any other equipment within the suppliers scope. The full extent of the test, including all equipment, operating conditions, test readings and whether mounted on the baseplate/skid/trailer, must be agreed.

An endurance run is a running test which involves all, or part, of the contract equipment at specified operating conditions and is intended to confirm the reliability of the equipment. The purchaser and the supplier shall agree the readings and frequency of readings to be taken and the total running time.

A relief valve operation test involves increasing the pump outlet pressure until the relief starts to open, then further increasing the pressure until all the pump flow is relieved. The pressure is then reduced until the relief valve closes and seals. All the readings required for a performance test shall be recorded. This test may be applied to integral and external relief valves.

NOTE: This test is only valid when the test liquid closely approximates the contract liquid.

A vibration test involves measuring the vibration severity, rms velocity, at agreed locations under agreed operating conditions.

NOTE: With some pump types, particularly reciprocating pumps, vibration levels experienced on the test bed, when baseplates are not grouted in or pumps are supported by temporary mounts, will be higher than those under normal operation at site. Under these circumstances, vibration data can only be considered as informative and not representative of the final installation.

A pulsation measurement test involves using pressure transducers and electronic instrumentation to measure the pressure pulsations, in inlet or outlet piping or within the pump cylinder or casing, over one or more cycles, during running tests. Pulsation measurement readings shall be taken at agreed operating conditions.

NOTE: When pulsation measurements are made on the manufacturer's test bed using test pipework the results can only be considered as informative. Such test results cannot confirm or guarantee the pulsations at site.

Noise testing was not included in this list because noise testing is sometimes very difficult and is the subject of a complete proposed standard in its own right. Most pump manufacturers do not have quiet rooms and noise testing is conducted on the normal testbed, sometimes adjacent to the machine shop. Background noise tends to be a problem. If the pump is tested with a slave motor, rather than the contract motor, the slave motor noise must be isolated because it cannot be treated as part of the package noise. Most manufacturers can offer a noise survey; a noise test not complying with any standard but reporting the background noise at each test point with the test reading.

The proposal to ISO stipulated that only fully assembled pumps could be tested. Pump rotors fitted to test casings would not comply with the testing requirements.

11.1.13 Painting

Painting is the last operation prior to dispatch. Some paint finishes, especially offshore requirements, are quite complicated and have many intermediate stages. The purchaser's inspector may wish to witness some operations.

Some paint specifications require the raw material to be blast cleaned to a bright finish. The surface has obviously to be dry and free from loose fragments. Blast cleaning may have to be performed in a controlled atmosphere. Some inspectors like to check the environment and to see the surface before the primer is applied. Individual coats of some paint finishes have specified thickness, the inspector may wish to measure the paint thickness as coats are applied. The total paint thickness and the finish of the top coat will be the final inspection.

Some components are machined on all surfaces. Pump manufacturers may not wish to blast clean a surface which is machined for design purposes. Paint systems are available for applying to bright machined surfaces, the only subsequent treatment being a thorough degrease.

11.1.14 Purchased equipment

When the pump manufacturer packages his pump to form a pump unit, the manufacturer will issue specifications and purchase orders for the bought out equipment. If the manufacturer is building his standard unit, all the specifications issued will be the pump manufacturer's. If the pump unit is built especially for a user the user may impose requirements on the bought out equipment. It is possible that conflicts may arise between the two sets of specifications. The pump manufacturer is responsible for supplying a complete working package, irrespective of what the purchaser's specifications require.

11.1.15 Functional testing

We saw earlier that pump testing is complicated. Current standards do not address the problem of package testing, that is how much of the contract equipment is used for the test. Another part of testing which is not covered by specifications is the functional testing of packaged instrumentation and control equipment. The following inspection functions are normal:

- wiring continuity,
- earth and leakage tests,
- circuit verification,
- calibration of instruments,
- verification of set points,
- simulation of control logic.

If the purchaser wishes to exercise control of these activities then proprietary specifications must be issued.

11.1.16 Witnessing

The purchaser, or his representative or the certifying authority, may wish to watch certain activities. The Certifying Authority is usually an insurance company who wishes to confirm that statutory requirements have been fulfilled. Viewing activities is approached in two ways. Activities which are interesting or complicated may be observed. The manufacturer informs the purchaser when the activity will take place, five to ten days notice is normal. The activity proceeds whether the purchaser is present or not. Critical activities are witnessed. The manufacturer informs the purchaser/Certifying Authority when the activity will take place. The activity cannot be performed if the inspector is not present. A hold is placed on that aspect of the contract. Both observing and witnessing cost money and take time. The purchaser has no automatic rights to either. The full extent of external participation in inspection functions must be agreed during the pre-contract negotiations. Changes to external inspection, or the unavailability of inspectors at specific times can lead to extended deliveries and increased costs.

The purchaser may wish to observe/witness activities at the pump manufacturer's sub-suppliers. This can only be done with prior agreement. Also it is important for the purchaser and the pump manufacturer to agree who has to pay for any third party inspection or the Certifying Authority.

11.1.17 Clarification of specifications

Pump users who have a large population of pumps and who purchase many pumps every year, will have proprietary specifications covering most aspects of design, operation and maintenance. The general trend is to take a National or International Standard and amend certain clauses. Some clauses are deleted as being irrelevant, others are edited to reduce option choices and some options are modified to become mandatory requirements. When the published standard is thought to be weak, the proprietary version includes new and rewritten clauses.

We have seen in Chapter 10 in baseplates, skids and trailers, pump users can specify standards which are inappropriate for the application. A similar problem can arise with proprietary specifications. When these problems arise it is essential that the pump manufacturer has access to the purchaser's engineers who understand why specific clauses were written. The manufacturer must have the opportunity for technical discussions with engineers who can change or delete parts of the specification. With modern business practices there is a great risk that the purchaser may buy the pump he specifies rather than buy the pump he needs. One way to visualise the consequences of this problem is the following edict;

> "Trouble-shooting is making the pump do what you specified.
> Making the pump do what you want is Up-grading!"

The difference between trouble-shooting and up-grading is, who pays for the work.

11.1.18 Certification

It is obvious from the range of testing and inspection operations discussed that the scope for certification is immense. The cost and delivery implications cannot be overlooked. Figure 11.1 lists, in the order of popularity, purchaser's favourite documentation.

Component/Assembly	Certification
Pump	Hydrotest Performance Material conformance Material physical Material chemical Material impact Welder qualification Weld procedures Vibration Noise
Pipework	Hydrotest Material physical Material chemical Welder qualifications Weld procedures Cleanliness
Motor	Hazardous area Type test Noise Performance Vibration

Figure 11.1 Popular certificates

A material conformance certificate is a declaration from the manufacturer that the materials used comply with the specifications. Certificates of conformance which begin "I .... certify that to the best of my knowledge......" are not acceptable.

When pumps are destined for cold climates, impacted tested material becomes very important. Even baseplate material is certified. Pipework does not really suffer impact certification problems. It is easier to use 304 or 316 stainless steel, for which many welders are qualified, rather than use a specialist nickel alloy.

As liquids become more hazardous and materials more exotic the requirements for non destructive testing and operator qualifications increase.

So far we have looked at the quality of the material. What about the quality of the certification? When a material certificate specifies physical properties or chemical composition, how accurate is it, how valid is it and which authority certified it? The first standard to grade certification was DIN 50049. However, this has now been superseded by a European Standard, EN 10204. The following grades of certification are classified:

2.1   A Certificate of Conformance.

> The manufacturer certifies the materials conform to the order specification.

2.2   A Works Report.

> The manufacturer certifies, on the basis of tests performed on the batch, the materials conform to the order specification. No actual certificates are supplied.

2.3   A Works Certificate.

> The manufacturer certifies, on the basis of tests performed on the contract material, the materials conform to the order specification. No actual certificates are supplied.

3.1.A   A certificate issued by the manufacturer's Quality Control or Inspection Departments; not someone from manufacturing, stating the test results of the actual material.

3.1.B   A certificate issued by an independent test house, employed by the manufacturer, stating the test results of the actual material.

3.1.C   A certificate issued by an independent test house, employed by the purchaser, stating the test results of the actual material.

Certification to 3.1.C is obviously the best from a pump purchaser's viewpoint. It is also the most costly and has the longest impact on delivery. On all but the most critical applications, certification to 3.1.B has become the standard for large pumps built to order. The purchaser's inspector can insist on positive material identification on random components to validate certification.

The pump manufacturer retains quality records for a considerable length of time, 15 years is not uncommon. The next edition of API 610 may ask for 20 years.

Manufacturer's quality could be considered as an extension of certification quality. How much control does the manufacturer have over the manufacturing process ? This aspect of Quality Assurance is standardised in BS 5750, EN 29000 and ISO 9000. The basis of manufacturing control is recognised as lying within strict operational instructions. Every employee has a written job description. Routine inspection and manufacturing procedures must be written down and strictly controlled. All of these documents should form part of the manufacturer's Quality Assurance Manual, QAM. The manuals are issued to all relevant departments. Personnel are not allowed to keep individual copies of often used procedures. These can become out-of-date and result in incorrect operations. Lines of communications are defined with critical decisions. Special emphasis is given to error reporting and feedback to the designer. Companies who advertise compliance with the standards have been audited by a third party. Routine inspections are performed at regular intervals to ensure standard compliance.

Compliance with the three standards mentioned is not essential for good manufacturing control. The basic principles can be incorporated into a manufacturing system without strict compliance with the standards. The critical items which are essential for effective control are a Quality Plan, for components or complete assemblies, and the power invested in the Quality Control Department. Inspection at every stage of manufacturing is essential.

Total Quality Management is a phrase which is used by management to boost the image of companies. If Quality Plans are not used, if an independent department does not check components/assemblies at each stage of production, what does Total Quality Management have to offer the purchaser?

Finally we must end on a note of warning. Quality Assurance is not intended as a means of achieving higher or better quality levels. Quality Assurance ensures goods are manufactured to the quality level specified. Quality Assurance is intended to produce consistency. Well regulated manufacturing systems will tend to refine designs and manufacturing techniques as the fault reporting system highlights component or assembly failures; this is an essential cost reduction feature. Improvement in product quality is usually driven by the purchaser, directly or indirectly.

11.1.19 Documentation

All pumps and pump units will be supplied with some documentation. The extent of the documentation will depend upon the nature of the pump and the method of purchase. However, some documentation is controlled by legislation and the purchaser must receive this documentation irrespective of the method of purchase. Safety is controlled in Europe by the Machinery Directive which is implemented by compliance with standards. It is the duty of the pump designer to design safe pumps. If complete safety cannot be incorporated in the design, documents must be provided which detail the areas of concern and what precautions must be taken by the user. In general these precautions will deal with the fitting of guards, protection against pressurised liquid and isolation of electric supplies prior to working on equipment. We shall see later what other documentation the purchaser would reasonably expect to receive.

Documentation can also apply to manufacturers' published literature. Typically manufacturers' show performance curves and list materials of construction either by generic term; carbon steel, stainless steel, bronze; or standard specification. We shall see later what information should be in publications to endorse the quality and reliability of the product.

11.2 Mass produced standard pumps

Mass produced pumps can be purchased by two methods; over-the-counter and by enquiry. A great many standard pumps are purchased over-the-counter from distributors. The use of distributors allows a manufacturer to have his products widely available at many outlets. Some distributors have technical knowledge and skills and can remove some of the skilled back-up required from the manufacturer. A purchaser can review published data, select a pump for an application and purchase a pump or unit over-the-counter. Alternatively, the purchaser can send the pump manufacturer or distributor an inquiry with the duty conditions. From the subsequent quotation, accompanied by standard literature, the purchaser can decide which pump to purchase.

Before purchasing a pump or unit the following questions should be answered, either from reading the published data or direct questions to the distributor or manufacturer:

- *Are the pumps designed and built to a standard?*
- *Are the pumps manufactured to a standard?*
- *Will my assembled pump be subjected to any running tests?*

Some manufacturers do not test all pumps. If not all pumps are tested, how many pumps in a batch are tested and what is a typical batch size. What liquid is used? Some manufacturers test rotating assembles in test casings rather than the contract casing. This does not fulfil the prime requirement of running tests, final assembly verification.

- *Will my assembled pump unit be subjected to any running tests?*
- *Will my assembled pump be pressure tested?*

If an assembled pump is pressure tested it is important to know the pressure and how the stuffing box was sealed.

- *To what standard are running tests conducted?*

- *What is the tolerance band on the published data for head, flow, NPSHr and efficiency?*

Some unscrupulous manufacturers publish data which is the best possible attainable. Tolerances will all be negative. Some pump designs relying on moulded non-metallic materials may have unexpectedly wide tolerance bands on flow.

- *What Quality Assurance documentation will be supplied with the pump or unit?*

The minimum documentation would be a hydrotest certificate and a final inspection release certificate. For a standard pump constructed of 316 stainless steel or a more exotic alloy, it would be reasonable to expect a Works Report, 2.2.

Other standard documentation would include:

- parts lists and assembly drawings,
- installation instructions and guide-lines,
- maintenance and safety instructions,
- trouble-shooting guide,
- telephone and fax number of service department.

11.3 Pumps built to purchaser's specification

When a pump is built to the purchaser's specification, the pump order will have been placed on the basis of quotations received from various manufacturers. The purchaser will have issued an enquiry detailing the operating conditions and any other requirements. The purchaser will have evaluated the quotations based on:

- pump duty point and efficiency,
- compliance with specifications,
- manufacturing schedule and delivery,
- manufacturer's experience,
- confidence in manufacturer,
- documentation,
- cost.

Cost is a difficult variable. If the quotation is evaluated by a contractor, initial cost plus a cursory look at the efficiency may suffice. A user will evaluate life-cycle costs. Documentation plays an important role in business and technical documentation plays a critical role in modern plants. The manufacturer should have listed all the documents which would form part of the technical contract. A Vendor Documentation Requirement form, VDR, should have been completed and returned. This form lists all the technical documents and at what time they will be submitted. A preliminary quality plan, QP, would have been submitted with the quotation to indicate the inspection included.

An important part of any quotation is a good installation list; this together with a good quality plan and a good VDR can instil confidence. A quality plan indicates the important inspection functions the manufacturer will conduct to assure an acceptable quality level. The QP also shows the areas where control is taken from the Quality Control Department, QC, and assumed by the manufacturer's Engineering Department or the purchaser's inspector or engineer. In some contracts a Certifying Authority may also have inspection rights; the CA would normally be an insurance company. The QC department is always in overall control of any contract. Some problems may be due to design rather than to manufacturing. If a pump casing fails hydrotest, and inspection confirms the machining to be correct, the problem will be referred to engineering. Some components are critical, any problems are referred directly to engineering. If the purchaser and the manufacturer have agreed to differentiate between minor and major repairs, major repairs will result in a purchaser's hold point. If the pump has difficulty in achieving the desired performance on test, both sets of engineers will be involved.

If the preliminary QP is not acceptable to the purchaser, the modifications must be agreed during the pre-contract negotiations. Extra certification and hold points not only affect the cost but also the delivery.

Quality Control Plans can take many forms. Early QP's were produced as drawings in the drawing office. The inspection operations were listed horizontally across the top of the drawing, the components were listed vertically down the left hand side. Different types of dots in the squares indicated various operations and hold points. Because of the style, these plans were often called 'pools plans' because of the similarity with football pools coupons. This type of QP is difficult to arrange initially and even worse to modify.

The Energy Industries Council, the largest UK trade association of diverse companies from various energy sectors, considered the problem of QP's and issued a proposal, 'Guideline for the generation of Quality Plans, QAG 01'. The proposal advocated a new style of QP, on A4 paper, which could be word processed. The basic idea was to have each component on a separate sheet, organised in a specific format. Complicated components could have their quality requirements extending over several pages. Figure 11.2 indicates a slightly modified format which has been used very successfully.

The references in the 'Q.C. Procedure Reference' column; ES and WIM; refer to the manufacturer's standard inspection and workshop instructions. Inspection to National, European or International Standards can be performed when appropriate. Components which are not listed in a QP are subject to the manufacturer's standard quality requirements. If the purchaser is considering placing an order, he may ask for a copy of the manufacturer's Quality Control Manual. This would normally be an 'Uncontrolled Document'. The prospective purchaser would not receive modified or new standards as they were released.

When a contract is subject to such structured quality requirements, it is usual for the purchaser and the manufacturer to have a meeting, very early after purchase order placement, to clarify any ambiguous points and present a contract manufacturing schedule. This meeting, sometimes called a 'Kick-off meeting' or a 'Vendor co-ordination meeting' takes place between the manufacturer's engineers and contract management and the purchaser's engineers and inspectors. If a Certifying Authority is imposed, they must be represented so that their involvement and hold points are correctly interpreted. This meeting confirms the quality requirements and the documentation requirements shown on the VDR. The major and minor repair categories may be finalised on a component by component basis. The manufacturer may agree to submit progress reports during the contract.

As the contract progresses the manufacturer's contract engineer and the purchaser's inspector can work through the plan and ensure all inspections are completed and that documentation is available.

One important document, which is only used occasionally, is a design review. Pumps for very critical applications, very hazardous liquids or subject to legislative control, usually through a Certifying Authority, have important areas of design reviewed by a third party. This review is considered as an additional safety measure and would evaluate important areas of construction such as:

- pump casing pressure stressing,
- effect of nozzle loads on casing stressing,
- shaft torsional and lateral vibration analysis,
- baseplate stressing.

It is quite normal for small pressure vessels, seal pots and pulsation dampeners, to have their pressure vessel calculations examined and approved.

The pump test is one of the most important aspects of the quality assurance programme. The preliminary QP with the quotation should have outlined the tests proposed and how they related to the rated operating conditions. The vendor co-ordination meeting should have clarified any missing data. Manufacturers' test facilities do have limitations and all rated operating conditions cannot be accommodated. Typical limitations are suction pressure, temperature and power. High suction pressure is difficult to provide, especially on large pumps. Rotodynamic pumps can be tested at the contract differential head at low suction pressure; the absorbed power will be very similar. Positive displacement pumps can be tested at the contract differential pressure or the contract discharge pressure if extra power is provided. Discussions with the manufacturer are essential. Pumps routinely operate at temperatures over 250 °C. Manufacturers of large pumps will have difficulty sustaining such high temperatures. A compromise of a temperature around 100 °C may be possible. Modern pumps can be quite large, over 50 MW. Most manufacturers do not have power supplies to test such large pumps. Pumps up to 5 MW should not be a problem.

The choice of test liquid can cause formidable problems. Most popular rotodynamic pump test specifications call for water tests. Many process and chemical pumps operate on liquids with viscosities up to 300 cSt. Centrifugal pumps operate at higher viscosities when running at low speeds. Surely, a good

CYLINDER

REF NO.	PROCESS DESCRIPTION OR Q.C. ACTIVITY	Q.C. PROC. REF	ACCEPTANCE CRITERIA			VERIFYING DOCUMENT	INSPECTION			
			IN HOUSE	CLIENT	ACTUAL		IN HOUSE		CLIENT	CERTIFYING AUTHORITY
							ENG	QC	INSP	ENG
1	VERIFICATION OF MATERIAL COMPLIANCE	NITRONIC 50	CONFORMS TO SPEC.		EN 10204 3.1.B	CERT. OF MECH. & CHEMICAL PROPS.	H		X	
2	ULTRASONIC INSPECTION	ES 509-005	CONFORMS TO SPEC.			CERT. OF U.T. INSPECTION	H		X	
3	GOODS INWARD INSPECTION	ES 509-018	CONFORMS TO P.O. & SPEC.			INSPECTION CARD	R			
4	INSPECTION AFTER MACHINING	TO DRAWING	CONFORMS TO DRAWING			INSPECTION CARD	R			
5	INSPECTION/Q.A. REVIEW & RELEASE FOR ASSY	WIM 9-200-81	INSPECTION/QC REQMTS. COMPLETE			INSPECTION CARD	R			

H ≡ hold point or witness point

R ≡ document retained at manufacturer's works

X ≡ document included in purchaser's data pack

Figure 11.2 Example of quality plan format for a component

performance test is one which closely approximates the operating conditions. Some positive displacement pumps rely on viscosity to operate correctly; there would be no point in specifying water as the test liquid. The test liquid must be agreed with the manufacturer, but one closely approximating the properties of the process liquid would obviously be better than an arbitrary choice based on dogma.

Depending upon the equipment being supplied by the manufacturer, the actual test set up should be clarified. Will the pump be tested on the baseplate with the contract motor? Will the pump be tested on temporary supports and driven by a calibrated slave motor? Will the pump performance test be conducted as part of a string test with all the contract equipment available? This whole aspect of testing is not covered by current standards and not by the proposed ISO standard for rotodynamic pumps.

Larger, more complex pump units would normally have two volumes of documentation supplied with the equipment, an installation and maintenance manual, and a data book. The maintenance manual would contain all the information listed for standard pumps. A typical data book would contain:

    quality plan,

    material certificates,

    weld procedures,

    welder qualifications,

    radiographs,

    non destructive test reports,

    ndt operator qualifications,

    hydrotest certificates,

    pump test procedures,

    pump test results,

    motor certification,

    motor test results,

    instrument certification,

    wiring check results,

    functional test results,

    paint procedure,

    paint inspection certificate,

    overall dimension and weight certificate,

    inspector's release certificate.

If the purchaser's inspector had carried out random positive material identification, the material chemical composition certificate would marked as such next to the inspector's stamp.

11.4 Guidelines for documentation

11.4.1 Rotodynamic pumps

Speed r/min	Discharge press barg	Stages	Materials	Pump assy hydro	Performance test	String test	Cert of conform	Casing material	Impeller material	Shaft material	Rubbing parts	Torsional analysis	Lateral analysis
≤ 3600	≤ 40	1	Cast iron casing/cast iron impeller Carbon steel casing/carbon steel impeller Bronze casing/bronze impeller	■	□		2.1						
≤ 3600	≤ 100	1	Carbon steel casing/carbon steel impeller Bronze casing/bronze impeller	■	■		2.2						
≤ 3600	≤ 100	1	Stainless steel casing/stainless steel impeller	■	■			3.1.A	3.1.A		3.1.A		
> 3600	≤ 40	1	Any materials	■	W	W		3.1.A	3.1.A	3.1.A	3.1.A	■	■
> 3600	> 40	1	Any materials	W	W	W		3.1.B	3.1.B	3.1.B	3.1.A	■	■

■ ≡ required   □ ≡ required if pump modified, i.e. impeller diameter changed   W ≡ witnessed

Figure 11.3 Documentation for single stage overhung pumps

Speed r/min	Discharge press barg	Stages	Materials	Pump assy hydro	Performance test	String test	Cert of conform	Casing matrial	Impeller material	Shaft material	Rubbing parts material	Torsional analysis	Lateral analysis
≤ 3600	≤ 100	≤ 6	Cast iron casing/cast iron impeller Carbon steel casing/carbon steel impeller Bronze casing/bronze impeller	■	■		2.2						
≤ 3600	≤ 100	≤ 6	Stainless steel casing/stainless steel impeller	■	■		2.3						
3600 to 6000	≤ 100	≤ 6	Carbon steel casing/carbon steel impeller	■	W	W	2.3					■	■
3600 to 6000	≤ 100	≤ 6	Stainless steel casing/stainless steel impeller	■	W	W		3.1.B	3.1.B	3.1.B	2.3	■	■
> 6000	≤ 100	≤ 6	Any materials	■	W	W		3.1.B	3.1.B	3.1.B	3.1.B	■	■
≤ 3600	≤ 100	> 6	Carbon steel casing/cast iron impeller Carbon steel casing/carbon steel impeller	■	n		2.2			3.1.B	3.1.B		■
≤ 3600	> 100	> 6	Any materials	W	W			3.1.B	3.1.B	3.1.B	3.1.B	■	■
> 3600	> 100	> 6	Any materials	W	W	W		3.1.B	3.1.B	3.1.B	3.1.B	■	■

Figure 11.4 Documentation for multi-stage pumps

11.4.2 Positive displacement pumps

Speed r/min	Discharge press barg	Power KW	Materials	Pump assy hydro	Performance test	String test	Cert of conform	Casing material	Rotor material	Torsional analysis	Lateral analysis
≤ 250	≤ 50	≤ 10	Cast iron casing/cast iron rotor Carbon steel casing/cast iron rotor Carbon steel casing/carbon steel rotor Bronze casing/bronze rotor	■			2.1				
≤ 250	≤ 50	≤ 10	Stainless steel casing/stainless steel rotor	■			2.2				
≤ 250	≤ 160	≤ 10	Cast iron casing/cast iron rotor Carbon steel casing/cast iron rotor Carbon steel casing/carbon steel rotor Bronze casing/bronze rotor	■			2.2				
≤ 250	≤ 160	≤ 10	Stainless steel casing/stainless steel rotor	■			2.3				
≤ 250	> 160	≤ 10	Any materials	■	■		3.1.A				
≤ 250	≤ 50	≤ 100	Cast iron casing/cast iron rotor Carbon steel casing/cast iron rotor Carbon steel casing/carbon steel rotor Bronze casing/bronze rotor	■	■		2.2				
≤ 250	≤ 50	≤ 100	Stainless steel casing/stainless steel rotor	■	■		3.1.A				
≤ 250	≤ 160	≤ 100	Any materials	■	■		3.1.B	3.1.B			
≤ 250	> 160	≤ 100	Any materials	W	W		3.1.B	3.1.B			
≤ 250	> 160	> 100	Any materials	W	W	W	3.1.B	3.1.B			

Figure 11.5 Documentation for rotary pumps

Torsional and lateral vibration analysis should be considered for powers over 30 kW when engines are used as drivers.

Reciprocating positive displacement pumps

Mean piston or plunger speed r/min	Discharge press barg	Power KW	Materials	Pump assy hydro	Performance test	String test	Cert of conform	Crankshaft material	Liquid end material	Rubbing parts material	Torsional analysis	Lateral analysis
≤ 0.5	≤ 160	≤ 25	Any materials except stainless steel	■			2.1					
≤ 0.5	≤ 160	≤ 25	Stainless steel liquid end	■			2.2					
≤ 0.5	≤ 435	≤ 25	Any materials except stainless steel	■	■		2.1					
≤ 0.5	≤ 435	≤ 25	Stainless steel liquid end	■	■			3.1.A				
≤ 0.5	> 435	≤ 25	Any materials	■	■			3.1.B				
0.5 to 1.0	≤ 160	≤ 100	Any materials except stainless steel	■	■			3.1.A				
0.5 to 1.0	≤ 160	≤ 100	Stainless steel liquid end	■	■			3.1.A	2.1			
0.5 to 1.0	≤ 435	≤ 100	Any materials	■	■		3.1.A	3.1.B	3.1.A			
0.5 to 1.0	> 435	≤ 100	Any materials	W	n		3.1.A	3.1.B	3.1.B			
1.0 to 1.5	≤ 160	≤ 200	Any materials	W	■		3.1.A	3.1.B	3.1.B			
1.0 to 1.5	> 160	≤ 200	Any materials	W	W	W	3.1.A	3.1.B	3.1.B			
> 1.5	> 160	> 200	Any materials	W	W	W	3.1.B	3.1.B	3.1.B			

Figure 11.6 Documentation for reciprocating pumps

Torsional and lateral vibration analysis must be conducted for engine drivers.

11.5 Bibliography

API RP 11S2 Recommended practice for electric submersible pump testing

API 610 (references the Hydraulic Institute Centrifugal Pump Test Code)

BS 599 Methods of testing pumps, performance & efficiency

BS 1394:Pt 2 Stationary circulating pumps for heating and hot water service systems. Specification for physical and performance requirement. Power driven circulators, domestic glandless pumps, up to 300W

BS 4617 = ISO 4409 Methods for determining the performance of pumps and motors for hydraulic fluid power transmission. Defines test installation & procedures & presentation of results

BS 5316:Pt 1 ≡ ISO 2548 Specification for acceptance tests for centrifugal, mixed flow and axial pumps, Class C Performance & efficiency on cold water

BS 5316:Pt 2 ≡ ISO 3555 Specification for acceptance tests for centrifugal, mixed flow and axial pumps, Class B Performance & efficiency on cold water

BS 5316:Pt 3 ≡ ISO 5198 Specification for acceptance tests for centrifugal, mixed flow and axial pumps, Precision Class Tests Performance & efficiency on cold water, including thermodynamic method (Laboratory testing)

BS 5750 ≡ EN 29000 ≡ ISO 9000 Quality systems

BS 5994:Pt 1 ≡ ISO 4412.1 Measurement of airborne noise from hydraulic fluid power systems and components. Method of test for pumps for sound power level

BS 5944:Pt 2 ≡ ISO 4412.2 Measurement of airborne noise from hydraulic fluid power systems and components. Method of test for motors for sound power levels

BS 5994:Pt 3 Measurement of airborne noise from hydraulic fluid power systems and components. Guide to application of Pt 1 & Pt 2

BS 5944:Pt 5 Measurement of airborne noise from hydraulic fluid power systems and components. Simplified method for pumps using anechoic chamber

BS 5944:Pt 6 ≡ ISO 4412.3 Measurement of airborne noise from hydraulic fluid power systems and components. Method of test for pumps using a parallel piped microphone array

BS 6001 = ISO 2858 ≠ IEC 410 Sampling procedures and tables for inspection by attributes

BS 6002 = ISO 3951 Specification for sampling procedures and charts for inspection by variables for percent defective

BS 7201:Pt 1 Hydro-pneumatic accumulators for fluid power purposes. Specification for seamless steel accumulator bodies above 0.5l water capacity. Materials, design, construction and testing

BS 7250 ≡ ISO 8426 Methods of determining the derived capacity of hydraulic fluid power positive displacement pumps and motors. Test installations, procedures, presentation of results

BS M65 ≡ ISO 8278 Specification for pressure compensated variable delivery hydraulic pumps. Quality assurance and acceptance tests

EN 287-1 Approval testing of welders for fusion welding: Steels

EN 287-2 Approval testing of welders for fusion welding: Aluminium and aluminium alloys

EN 288-1 Specification and approval of welding procedures for metallic materials: General rules for fusion welding

EN 288-2 Specification and approval of welding procedures for metallic materials: Specifications and procedures for arc welding

EN 288-3 Specification and approval of welding procedures for metallic materials: Welding procedure test for arc welding of steels

EN 288-4 Specification and approval of welding procedures for metallic materials: Welding procedure test for arc welding of aluminium and aluminium alloys

EN 10045-1 Charpy impact test on metallic materials, V and U notches, Test method

EN 10204 = DIN 50049 Metallic products. Types of inspection documents

HI 1.6 Centrifugal pump test standards

HI 6.6 Reciprocating pump test standards

IEC HD 289 S1 Routine production tests for electrical equipment

ISO 83 Methods for notched bar tests: The Charpy U-notch impact test on metals

ISO/R 148 Methods for notched bar tests: The Charpy V-notch impact test on metals

ISO/DIS 9906 Centrifugal, mixed flow and axial pumps - code for hydraulic performance tests for acceptance - engineering classes I and II

A record that speaks for itself.

100 000's of IMO pumps silently performing their work on the seven seas.

Not only there but also in factories, powerplants, elevators and many other applications.

The first IMO pump was manufactured in 1931. The brilliant principle of the IMO pump is still unsurpassed. Only three moving parts – three intermeshing screws. No gears, no valves, no vanes.

The reliability of an IMO pump has been proved over and over again.

Its efficiency and the elimination of pulsation, turbulence and vibration problems are other special IMO qualities.

These qualities meet present and future requirements for modern pump technology.

IMO

IMO AB, P.O. Box 42090, S-126 12 Stockholm, Sweden.
Tel +46 8 19 01 60, Telex 11548 IMO S, Telefax +46 8 645 15 09.

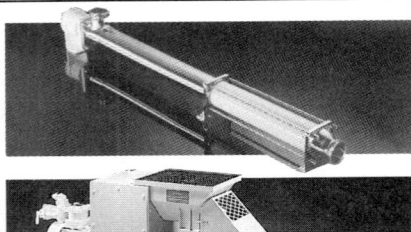

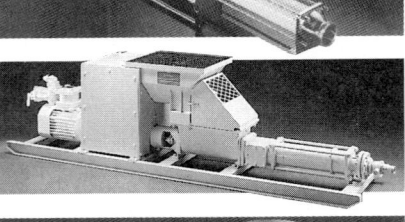

Our expertise is just as extensive.

For over half a century, Mono Pumps Ltd. have been at the forefront of progressing cavity pump design. A position based upon our commitment to service, quality and innovation.

Innovations such as the Flexishaft, a unique development in pump design and one which is indicative of our ability to solve your problems. Whatever the application we have a pump that suits it exactly.

And whatever the pump, you will maximise its efficiency by only using genuine Mono parts, – available 24 hours a day through an international distribution network that represents only one part of our total package of pre and post sales services.

Sevices which also include Computer Aided Pump Selection (CAPS), a stator manufacturing plant and the centralised production of all critical items.

With such extensive expertise it's no wonder that Mono pumps are used across the globe.

Mono®

Mono Pumps Limited,
P.O. Box 14, Martin Street,
Audenshaw, Manchester M34 5DQ
Tel: 061 339 9000 Fax: 061 344 0727 Telex: 667733 Mono AG

Mono Pumps *DRESSER*

12 Installation and maintenance

Pumping installations always require a more or less extensive system of additional equipment such as valves, measuring devices, control systems, etc. The aim of this chapter is to provide advice and information on this subject as well as on the installation of pumps and pipes.

The design of pump sumps is dealt with in some detail in this chapter.

The sections on the care and maintenance of pumps includes commissioning, and have been written to provide guidance for operating staff at pump installations as well as maintenance staff unfamiliar with pumps. A special section is devoted to trouble-shooting.

Contents

12.1 Installation

12.2 Foundations

12.3 Tanks and sumps
12.3.1 General
12.3.2 Submersible pumps

12.4 Pipe systems for pumps

12.5 Care and maintenance
12.5.1 General considerations
12.5.2 Preventative maintenance
12.5 3 Stocking spare parts
12.5.4 Trouble-shooting guide

12.1 Installation

The installation of pumps can be carried out by the pump manufacturer, the driver manufacturer or the contractor building the installation. Large pumps are usually installed as early as possible while access at site is at its best. Small pumps can usually be fitted in at any time. The pump order should be placed so that its delivery occurs at the correct time in the overall site programme. Also consider the pump general arrangement drawing. The certified drawing must be available to allow foundations or structural steelwork to be designed and manufactured before the pump is delivered.

The space allocated for the pump must be sufficient to permit installation, care, disassembly and maintenance. Transport routes and lifting facilities must be available. Sufficient space must be available around the installed pump for personnel access. Operators may need frequent access for running adjustments. Maintenance personnel will require access and space to remove components and assemblies. Some pump designs require withdrawal space at the non-drive end. The motor may need space to withdraw the rotor. The building or plot must have drainage facilities for leakage, priming, liquid used for flushing sealing devices and water if used for quench. The risks of flooding the pump unit, in addition to the electrical and control systems, which can be very costly and time consuming to dry out, must also be taken into account.

The pump should be sited as low as possible in relation to the liquid level in the suction system and in a place permitting the shortest possible suction pipe with the minimum number of bends.

The ventilation of the pump site should be further studied. The electric motor of the pump must receive the necessary cooling. If hazardous liquids, harmful to the environment or highly flammable, are to be pumped, special ventilation requirements must be observed. The electrical and control systems for the pump must also be protected against damage. It should be noted here that the heat generated by the pump's electric motor can be considerable.

Consideration must also be given to the noise caused by the pump, its motor and drive train. More and more attention is being devoted to noise from pump units and their pipe systems, and special measures may have to be taken in this respect. In order to reduce vibration transmission, which may be further conveyed through the building structure, it may be necessary for the pump foundation and parts of the piping system to use isolation mountings. Mounting the pump unit on isolation mountings can lead to piping and bearing problems. Piping close to the pump nozzle can fracture if the pump vibrates but the piping does not. This problem can be solved by using flexible pipes adjacent to the pump. Short bearing life can be a problem if the pump does move appreciably on its isolation mountings. It is better to use an isolated foundation block.

Acoustic enclosures may have to be erected around the pump unit. Permissible noise levels are often controlled by legal, safety and environmental requirements, at places of work. In the case of noise exceeding 75 to 85 dB(A), measures must usually be taken. Staff can be protected by declaring a 'Ear protection zone'.

If the pump unit has been on site for some time, it may have been necessary to carry out some preventative maintenance operations. Shafts may have been turned to maintain an oil presence in the bearings or to prevent rolling bearings from 'Brinelling'. It is the user's responsibility to ensure the pump unit is stored in suitable conditions and that the factory preservation instructions have been followed.

Unpack the pump unit and check it thoroughly. Look for any damage which may have occurred in transit. Also check the packing list to ensure all parts are present. Some delicate instruments may have been removed and packed separately. If there is damage or shortages, fax or telex the manufacturer immediately. If delicate parts have been packed separately, repack them temporarily. Remove any temporary bracing or locking clamps; replace parts as instructed by the manufacturer. Before placing the pump on its foundation, thoroughly clean the top of the foundation, remove any thin ridges and roughen the top to provide a good key for the grout. Prepare enough shims to level the baseplate, two sets for each foundation bolt. The shims should be longer than the width of the bottom flange. A packer should be placed either side of each foundation bolt, about 20 to 35 mm thick. Remove coupling spacers or driving pins. Lift the pump over the foundation block and fit the foundation bolts into the baseplate holes before lowering onto the packers. Level the baseplate, by adding shims, using an accurate spirit level on the machined pads. Long baseplates may be fitted with targets for laser or optical alignment.

Horizontal pumps and motors are usually supplied complete and aligned on a common baseplate, see later for separate pumps and motors. Whether the pump has been tested or not, the pump and motor will have been accurately aligned prior to dispatch from the manufacturer's works. Check the coupling alignment, see Chapter 8, section 11. Adjust the shimming until the coupling alignment is close, read the manual for thermal growth correction. Protect the tops of the foundation bolts and fit shuttering around the block in preparation for grouting. A good mixture for grout is one part of cement to two parts of sharp sand. The final consistency should be able to flow easily.

Pour the grout and ensure the top of the foundation is covered evenly. If full depth cross members are fitted, ensure complete support. If a cast iron baseplate is used, fill it completely. Remove the shuttering after two days, but allow the grout to harden for at least 10 days. Do not allow the grout to dry out too quickly, protect from direct sunlight. In hot environments it may be necessary to cover the grout with damp sacks. Also, protect the grout from frost if the temperature is low. Sacks, covered with polythene sheets will be adequate unless the temperature is very low. Tighten up the foundation bolts and smear with grease. Check the coupling alignment, remember thermal growth corrections, adjust the shimming under the motor feet if necessary. Record all the alignment settings. Spacer couplings are not an excuse for poor alignment. Diaphragm coupling life will be short with poor alignment. Pumps with cardan shafts should have some radial misalignment, read the manufacturer's instructions.

The suction and discharge process pipework should be finalised after the grout has hardened and final alignment has been completed. The pipework must line up naturally with the pump connections. Do not force pipework into alignment. Remove the piping sections adjacent to the pump and clean out the pipework, as best as possible. Fit a temporary suction strainer when pipework is rebuilt. Recheck coupling alignment with recorded figures. Modify pipework if necessary. Fit coupling spacers or drive pins. It is important that coupling bolts are tightened with the correct torque and that the appropriate quality of bolt is used. Information on the subject should be obtained from the instruction book. Fit any loose equipment which was removed for transportation. Wire up unit and 'lock-off' all local isolators.

Pumps driven by electric motor over 5 MW, or steam or gas turbines, will probably be delivered in two sections, pump and driver. Start with the biggest section; level and grout in as described above. After the grout has hardened, align, level and grout in the second section. When the second section grout has hardened, then proceed with final alignment and assembly.

12.2 Foundations

With the exception of pumps directly incorporated into the piping, as well as some submersible pumps and site pumps fitted with skids or trailers, a pump requires a solid foundation. This may be of concrete or a steel structure. Standard pump baseplates are designed to be grouted to a continuous concrete foundation block. If the pump is to be installed on a steel structure, the pump manufacturer should be informed of the size and position of the steelwork. Baseplate modifications may be necessary.

It is the user's responsibility to provide adequate support for the pump unit. The size of a foundation block will depend upon the nature of the sub-soil and the magnitude of the vibrations produced by the pump. Increasing the mass of the foundation block will reduce the amplitude of radiated vibration. A block isolated by proprietary elastomer mats may be necessary. Structural steelwork will be much more flexible than a concrete foundation block. Consideration must be given to the natural frequency of the support structure.

Further, the foundation should be sufficiently high to facilitate the connection of pipes and to ensure adequate space for drain lines, oil drains from reservoirs and that there is space for the fitting and maintenance of filters in the pipework. The foundation block should be longer and wider than the baseplate to provide extra physical protection, from such items as forklift trucks and barrows.

Horizontal pumps are usually supplied with pump and motor mounted on a common baseplate of cast iron or fabricated steel, tested and aligned. Large pumps may be delivered with the driver separate from the pump package. It may be necessary to provide a split level foundation.

Vertical pumps with flange mounted motors are supplied with a base or mounting plate, tested and aligned. In the sizing of foundations regard must be paid to the relationship between the height of the pump and the surface of the base or mounting plate. If the pump unit is divided, for example, so that the electric motor can be mounted above an overflow level, separate foundations must be designed for both the motor and the pump. Large vertical pumps can suffer structural vibration problems. Sound foundation designs may help to alleviate such problems.

12.3 Tanks and sumps

12.3.1 General

Many pump installations possess an open vessel, a tank or pump sump, from which the pump receives the liquid to be transported. The pump performance stated by the manufacturer is based on an uninterrupted flow to the pump inlet. Defective design of the pump sump or tank and suction pipe will result in inferior pump performance, possible damage and increased running costs. The commonest faults affecting the function of a pump due to defective suction design are:

- entrainment of air,
- foam formation,
- uneven velocity distribution in the pump inlet,
- formation of undesirable turbulence,
- suction pressure too low.

All these faults are dependent on the properties of the process liquid and lead to different degrees of reduced pump performance and damage for various types of pumps. Viscous liquids and liquid-solid mixtures demand, as a rule, the utmost care with regard to the design of the suction system of the pump installation.

In order to prevent the entrainment of air the end of any pipe filling a tank or sump should extend below the surface of the liquid. In certain cases, however, the pipe end must be located above the liquid surface. Figure 12.1 illustrates different tank arrangements.

The volume of the tank or pump sump represents a compromise between, on the one hand, economic factors associated with low initial cost for the plant and, on the other hand, technical criteria such as maximum permissible number of starts per hour in the case of on-off regulation or adequate

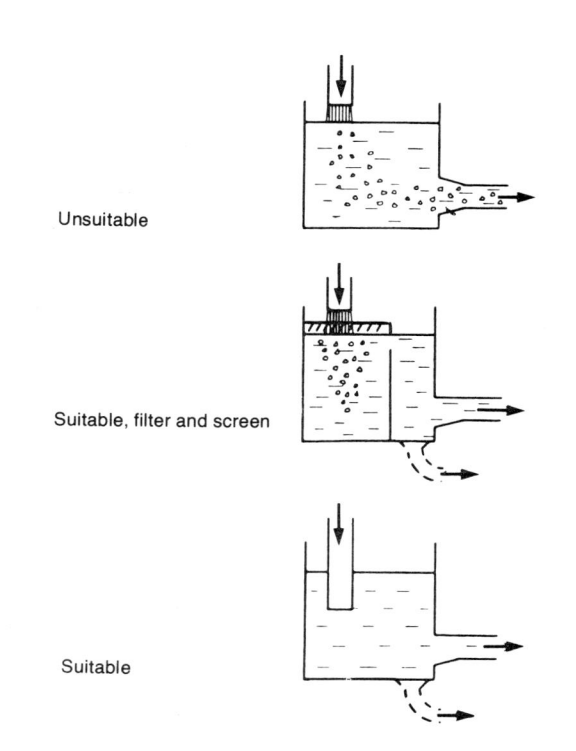

Figure 12.1 Pump suction tanks

retention time for degassing. An example of the effect of liquid properties in this respect, is the different rates of ascent of air-bubbles in oil and water.

The design of the pump sump or tank should be such as to ensure even low flow velocity of the liquid. Local high velocities increase the risk of the formation of turbulence, vortexes and the associated problems of air entrainment. Areas of very low velocity have a tendency to collect particles which may be present in the liquid. However, this is not a serious problem. The local velocity cannot drop below the critical velocity so the flow area cannot be reduced progressively to zero. A sensitive place is the connection of the suction pipe to the tank. The risk of unfavourable flow conditions is considerable. Figures 12.2 and 12.3 illustrate examples of unsuitable and suitable designs.

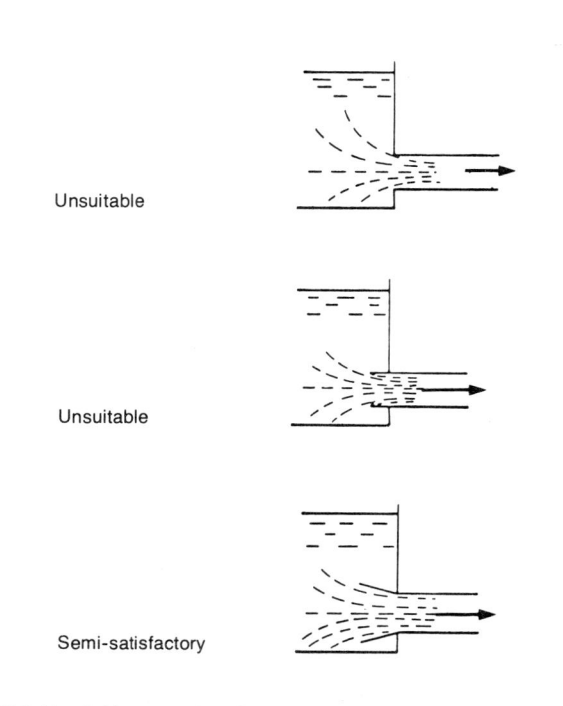

Figure 12.2 Unsuitable connections for suction pipes

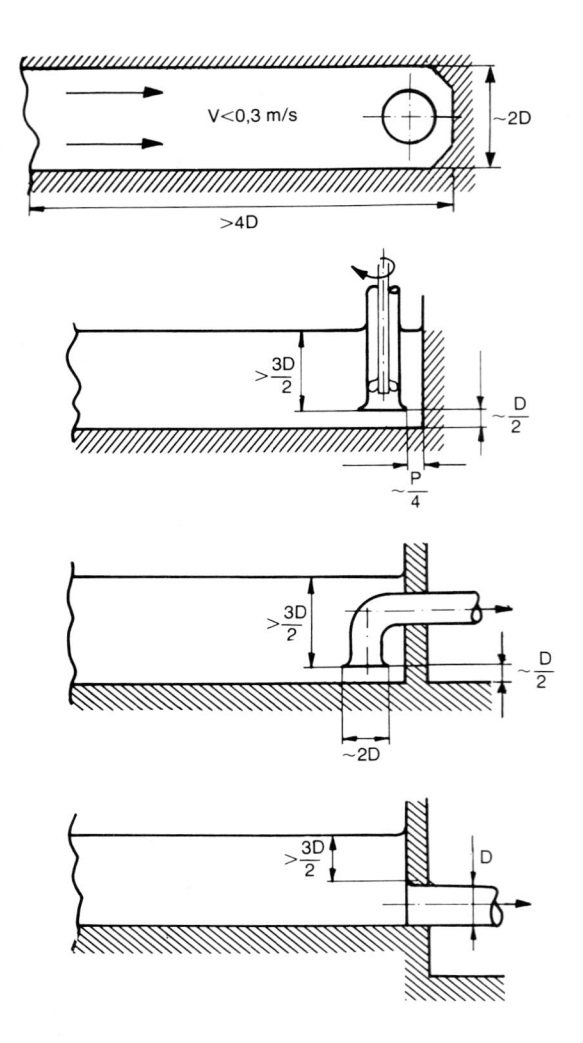

Figure 12.3 Recommended designs for the connection of suction pipes. The numerical values stated refer to water and a single pump

Although the numerical values stated in figure 12.3 refer to water, the design of the connections are also applicable to other liquids. The unsuitable connections shown in 12.2 can be converted to acceptable arrangements by reducing the velocity at the pipe entrance. Rotodynamic pumps should be less than 1.5 m/s and reciprocating pumps less than 0.3 m/s.

Low pressure at the pump suction connection, with consequent risk of cavitation damage, can be avoided by means of small losses in the suction pipe and restricted suction lift, see section 3.6 on suction performance. Whenever possible, the pump is placed to advantage below the level of the liquid in the suction tank, thereby permitting the liquid to flow down into the pump with positive head.

12.3.2 Submersible pumps

We are mainly concerned here with pump sumps for sewage pumps with flows exceeding 100 to 200 l/s, i.e. for medium to large pumps. To a considerable extent these guidelines also apply to pump sumps for storm water and rain water. In these pump sumps we are not confronted with problems of sewage sediment but rather with problems of sand and soil.

A pump sump must have a sufficiently large volume to avoid too frequent starting and stopping of the pumps, otherwise there is a risk of the pump motors over-heating. The problem of air being entrained into the water, the waterfall effect, arises if the inlet to the sump is located above the surface of the liquid. In such cases it is necessary to ensure that the pump sump is designed so that the water is degassed before it reaches the pumps. Otherwise there is a risk that the pumps will run unevenly, vibrate and even lose prime if the amount of air in the water is too great. Vortexes are a problem which can also cause pumps to operate unevenly and create vibration. Often these eddies originate at the pump intakes and can be reinforced into a vortex if the pump sump is incorrectly designed. In the case of totally calm liquid surfaces it is, for example, necessary to have greater submersion depths in order to avoid eddies other than those required when the liquid surface is disturbed.

The inlet to the pump sump must be located in such a way that the flow of liquid to the pumps may be evenly distributed. If the flow in to the sump discharges immediately in front of a pump, there is considerable risk of disturbance to the pump intake. This can increase or reduce the power consumption of the motor at the same time as the capacity of the pump is changed. In extreme cases this can cause the motor protection to be activated. The formation of sludge and scum must also be borne in mind when designing a pump sump. It must be designed to prevent the formation of dead zones with stagnant water. Similarly it must be possible to pump away scum at regular intervals.

Pump sump volumes for on-off flow regulation are dealt with in detail in Chapter 5.

Four alternative solutions are given below for pump sumps with different locations for inlets. The dimensions of the pump sump are calculated according to the flow per pump. The dimensions A to E for the different designs are determined by using the nomogram shown in figure 12.4.

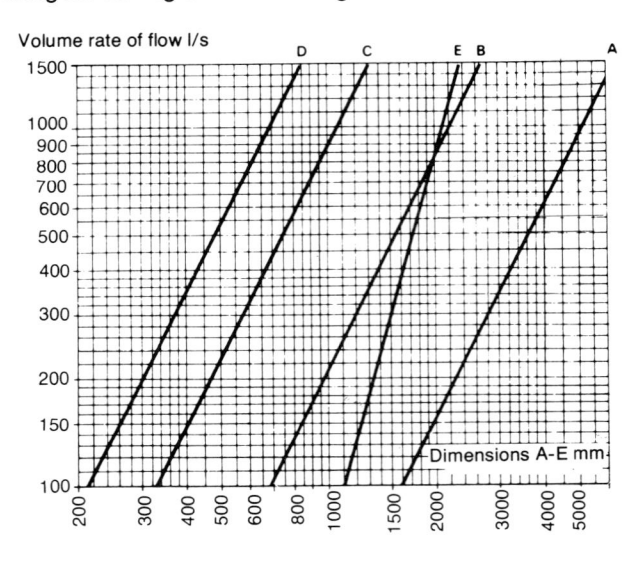

Figure 12.4 Nomogram for determining the dimensions of a pump sump

If the available height is restricted, the volume required is most easily obtained by increasing dimension A.

General comments

1) Dimension B: The clearance between two pump casings should not be less than 200 mm.

2) Dimension C: The clearance between a side wall and pump casing must not be less than 150 mm.

Alternative 1.

Central, inline inlet at high level, see figure 12.5.

Due to the high location of the inlet, much air is entrained into the incoming water. The sizing of the inlet trough allows time for the water to degas before it is led into the pump chamber through holes in the bottom of the trough. The side wall, section X-X, which is extended at least half way up the motor housing, prevents the formation of eddies between the outer pump and the side wall. The continual movement of the water means that the risk of sedimentation is minimal, provided that the minimum dimensions are not too greatly exceeded.

The trough wall immediately opposite the inlet must be high enough to catch the inflowing water. A weir can be arranged at the side so that floating waste can be removed by pumping.

Alternative 2.

Side inlet with inlet trough, see figure 12.6.

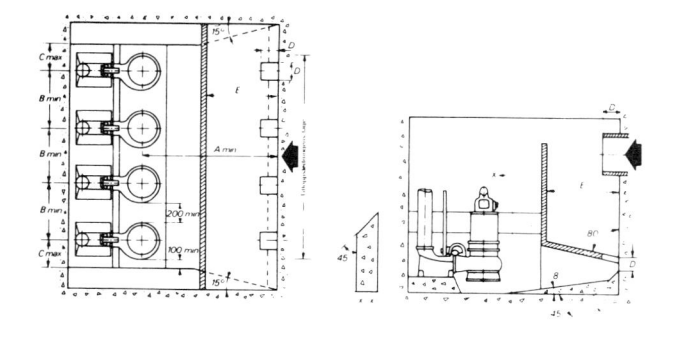

Figure 12.5 Alternative 1

This is a variant of Alternative 1. The square openings in the inlet trough are necessary to spread the flow evenly through the entire sump and to allow time for degassing any air which may still become whipped into the incoming water. Compared with Alternative 1, the flow conditions in the pump chamber are somewhat more uneven and turbulent. The breadth of the inlet trough and the height of the baffle are calculated with respect to the diameter of the inlet, the former 1.25 x the diameter and the latter 0.75 x the diameter.

Alternative 3.

Side bottom, inlet, see figure 12.6.

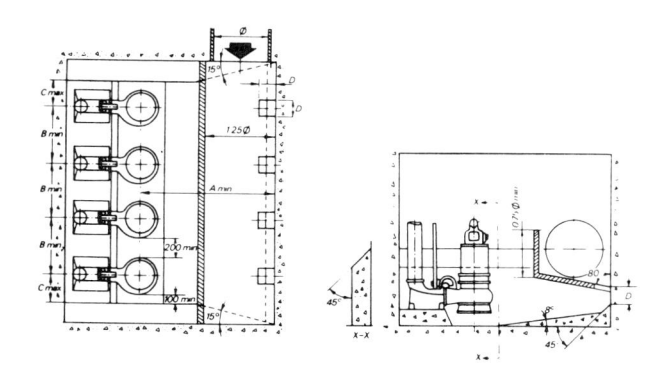

Figure 12.6 Alternative 2

A partition wall separates the pump sump from the inlet and forms a kind of inlet chamber. From this chamber water flows through square openings with measurements 2D x D located immediately in front of every pump. The openings are dimensioned so that the flow velocity through them does not exceed 1 m/s. In this way the water is led without risk of eddies forming or of sedimentation at the pump intakes. The partition wall should be at least high enough to reach up to a point on a level with half way up the pump motor housing. At the same time

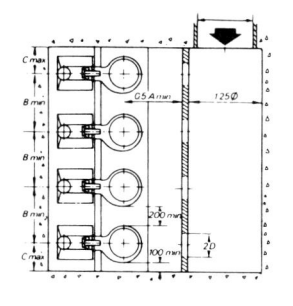

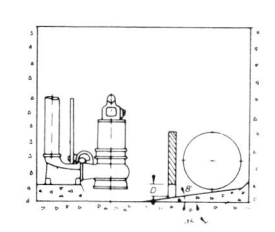

Figure 12.7 Alternative 3

its height should not exceed a level situated a little below the highest starting level, thereby making possible the removal of floating waste by pumping. The breadth of the intake chamber shall be 1.25 x the diameter of the inlet pipe, provided that the maximum flow velocity in the inlet pipe does not exceed 3 m/s. The distance between the pump centre lines and the wall can be as little as 0.5 x A.

Alternative 4.

Central inline inlet at low level, see figure 12.8.

This is a simple and functional arrangement which gives a smooth turbulent-free flow to the pumps. The tapered inlet section should have an included angle of 30 to 60°. The distance between the pump centre lines and the end of the inlet section should not exceed 0.5 x A.

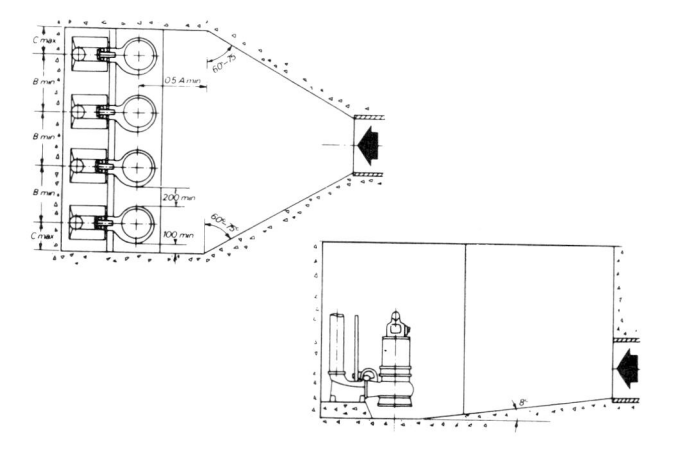

Figure 12.8 Alternative 4

Sometimes there is a demand that it should be possible to empty the inlet pipe completely. Alternatives 1 and 2 meet this requirement. In the case of Alternatives 3 and 4, connection may be made according to figure 12.9. If we do not follow this arrangement but allow the inlet pipe to discharge directly above the lowest stop level in the pump sump, we are confronted in the case of major flows with the problem of air being entrained due to the "waterfall" effect.

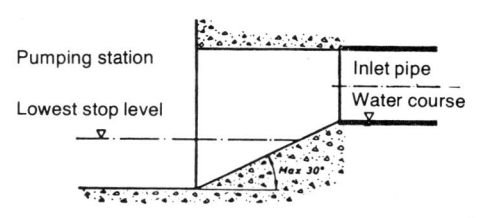

Figure 12.9 Recommended arrangement of low level inlet pipe

12.4 Pipe systems for pumps

Pumps with back-pull-out features, axially-split pumps and radially-split pumps are designed so that rotating elements can be removed without disturbing the pipework. Not all pumps have these facilities. The pipework must be designed to accommodate the style of pump used. A pipe system should be designed so that the pump unit can be dismantled without involving extensive work on the pipes. This may mean that spool pieces have to be incorporated in the pipe adjacent to the pump. It should be noted that it is very difficult to fit in a pipe section with flanges as well as gaskets unless it is possible to displace neighbouring sections. This problem is eliminated if a bend is removed. The need for flexible and expansion joints to compensate for pipe movement and to accommodate pipe expansion in the event of changes in temperature must also be borne in mind. Necessary supports and mountings for the pipe must also be provided. The pump should not be considered as a pipe support or anchor. Considerable forces are released by flow surges and pressure pulsations in the case of rapid valve operation or during the starting and stopping of pumps, see Chapter 4. Figure 12.10 illustrates an example of normal mounting requirements of pipes adjacent to a pump. Notice the suction pipe is larger than the pump suction connection. Pipes should be provided with rigid clamps at the supports.

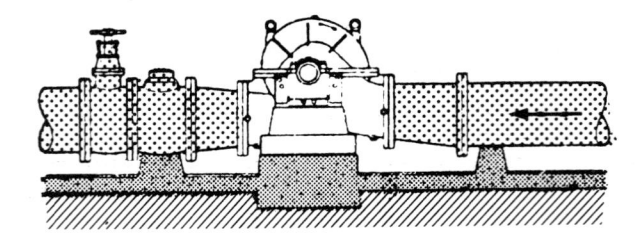

Figure 12.10 Correct mounting of suction and delivery pipes

When attaching pipes to the pump, care must be taken to ensure that the pipes align correctly and do not impose stresses in the casing. Stresses of this kind can cause cracks in the pump casing and the distortion cause galling between the impeller and wear rings. Checking the coupling alignment before and after the pipes are fitted can quantify distortion produced by the pipes.

The pump suction pipe should be designed to rise or fall evenly with few bends. If a valve must be incorporated, it should be of suitable type, see later under 'Essential equipment in pipework'. If the pump is lifting on the suction then the suction pipe must be completely free of air leaks. Also the suction pipe must be designed without changes of section which could trap air pockets.

Bends should not be fitted directly to the suction connection of the pump. This can lead to uneven loads which can result in both a deterioration in performance as well as abnormal axial loads. Figure 12.12 shows the worst case of this problem. A bend fitted close to the suction of a single stage double suction

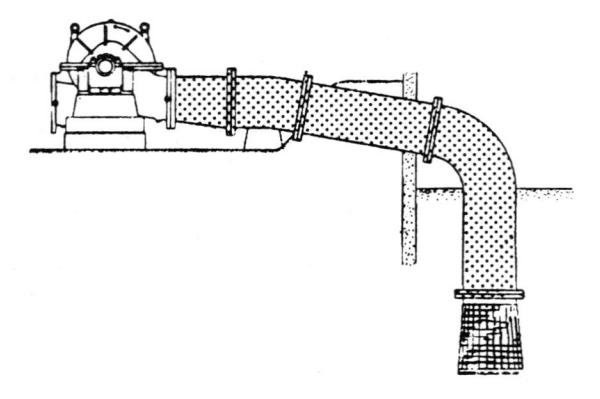

Correct routing of suction pipe

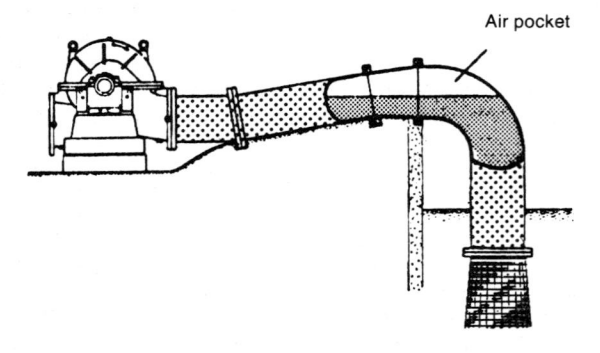

Incorrect routing of suction pipe

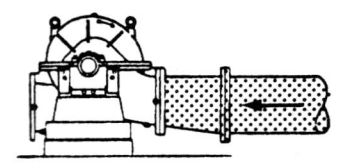

Correct installation with eccentric reducer

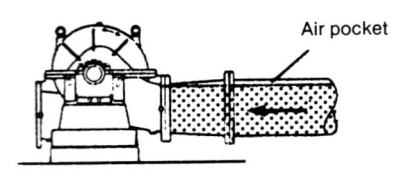

Formation of air pocket due to use of concentric reducer

Figure 12.11 Examples of correctly and incorrectly designed suction pipes

impeller. The bend disturbs the flow pattern which results in uneven flow distribution to the two sides of the impeller. This type of installation leads to short bearing life and vibration problems. With this type of pump bends close to the pump can only be in the vertical plane.

Figure 2.25 lists typical values for velocities in pipes. In multiple pump installations, pumps for one duty may be supplied by a header. The header should be larger in diameter than the branches supplying each pump. Maintain the header diameter constant over the full length. Do not attempt to save costs by reducing the header diameter after pump branches. In theory this may be attractive, in practice it makes plant modifications and expansion difficult in the future. Square tees

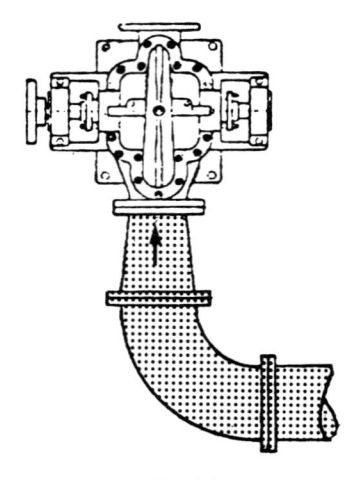

Figure 12.12 Incorrect installation of a bend close to a pump

in suction headers are difficult to assess for pressure losses. A full size swept tee, followed by an eccentric reducer, is preferred. In suction systems, 'pulled' bends are preferred to welded pressed bends. For reciprocating pumps, a full size swept 45° branch, followed by 3 x D of straight and then a 45° set followed by the eccentric reducer has provided the best results. For reciprocating pumps, bends should be avoided in suction and discharge; pairs of 45° sets separated by 3 pipe diameters give much less pipework vibration.

Some text books suggest the fitting of bends to attenuate noise and vibration transmission by the liquid. This is a fallacy. Consider what the bends in a trumpet or French horn do to the noise. The best pipework design from a vibration viewpoint is straight, it is the bends which cause most problems.

Methods for calculating flow losses in pipe systems are given in Chapter 2. For standards requirements on pipework and installations see sections 3.7.3 to 3.7.8.

Essential equipment in pipework

Valves of various types must normally be installed in the pipe system associated with the pump. When choosing valves, allowance must be made for the pressure drop which always takes place. The difference in this respect between various kinds of valves intended for the same application can be considerable. Information on this subject should be requested from possible suppliers and evaluated, see also Chapter 4. Manually operated isolating valves with position indicators, e.g. with rising screw, are suitable for use when it is necessary to rapidly establish the position of the valve. In the case of larger sizes, geared types of valve are necessary.

Non-return valves of flap type with external levers are often of practical use. In this way it is possible to achieve functional control in addition to enabling the valve to open manually. The disadvantage is difficulty in efficiently sealing the shaft bushing to the external lever. Non-return valves should be provided with inspection covers so that the operation can be tested and flaps and balls replaced without dismantling the valve. If a non-return valve is required for the system, it should be located between the pump discharge connection and an isolating valve, so that it is not necessary to drain the delivery pipe when repairing the non-return valve. When pumps are operating in parallel, the delivery pipe of every pump must have its own non-return valve so that one pump cannot pump backwards through another. In long piping systems it may be necessary to fit damped non-return valves, which cannot slam shut, to prevent or reduce pressure and flow surges.

When pneumatically or hydraulically operated valves are used, a special operating and control system is required for the purpose. This can be relatively complicated. As a rule small electrically controlled solenoid valves are used in systems of this kind. It is usually necessary to insist on comparatively clean operating media in such cases to avoid the risk of blockages and subsequent functional defects. If pneumatic or hydraulic valves must close in the event of pump stoppage caused by an electrical power failure, the solenoid valves must fail 'closed' and the operating medium must be independent of the electric circuit. This generally means the incorporation of an accumulator. Pneumatically and hydraulically operated valves must include adjustable orifices or needle valves in the fluid circuits so the speed of valve closure can be adjusted.

Valves operated by electric motors are relatively expensive, but afford very good possibilities of regulation. Naturally they do not function in the event of a power failure.

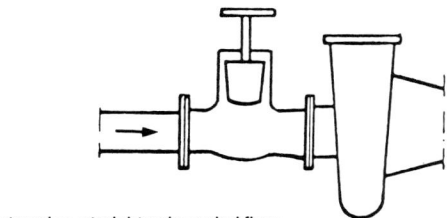

Gate valve, straight unimpeded flow

Globe valve, turbulence

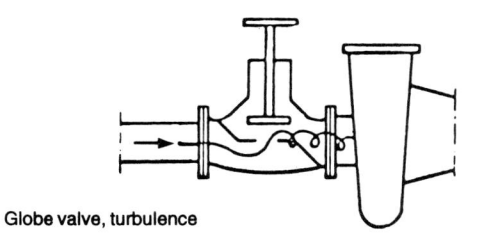

Figure 12.13 Valves in suction pipes

If throttle regulation is applied by means of a valve, the throttle valve must always be located in the delivery pipe. Valves in the suction pipe must only be used for isolation. They should be of a type permitting unimpeded flow to ensure the minimum losses and undisturbed flow into the pump. Globe valves must not be installed in a suction pipe immediately before a pump. See figure 12.13.

Best results for low pressure drop and minimum flow disturbance are achieved by full bore ball valves. Relief valves are fitted in positive displacement pump systems to prevent harmful increases in pressure, see section 10.4.

In the suction system of pumps, strainers or filters of various kinds may be necessary to provide protection against objects which can cause damage within the pump. These must be accessible for cleaning or replacement. The pressure drop across strainers or filters must be measured as this increases with the degree of blockage. It will be necessary to install a differential pressure indicator to check conditions. Standard screens are designated by width of mesh or diameter of holes as well as number of meshes or holes per unit of surface. For this purpose there are National Standards, see figure 12.14. Pump sumps for water containing sand should be equipped with a sand and gravel trap, in which the rate of flow should be reduced to about 0.3 m/s so that sedimentation may take place. Sand traps of this kind must be emptied at suitable intervals if they are to function efficiently.

DIN 4188	Former Designation Meshes		Width of Mesh		USA Standard	Tyler	British Standard	Afnor
Meshes mm	Every cm	Every cm^3	Micr.	Inch	Meshes per inch			Module
			45	.0018	325	(325)	(350)	
0.063	100	10000	63	.0025	230	(250)	240	19
0.09	70	4900	90	.0036	170	170	170	
0.125	50	2500	125	.0049	120	115	120	22
			180	.0071	80	80	85	
0.25	34	576	250	.0098	60	60	60	25
			355	.0139	45	42	44	
0.5	12	144	500	.0197	35	32	30	28
			710	.0280	25	24	22	
1	6	36	1000	.0394	18	16	16	31
			1400	.0552	14	12	12	
2	3	9	2000	.0787	10	9	8	34
			2800	.1102	7	7	6	
4	(1.5)	(2.25)	4000	.1570	5	5	34	37
			5600	.2200	3.5	3.5		

Figure 12.14 Comparison of National Standards for screen mesh

Pumping stations for sewage are sometimes provided with grills to trap rags, pieces of wood, etc. and thereby prevent blockage of the pumps. The cleaning of these grills is a complicated process and is usually carried out by mechanical means in large-scale systems. Reverse flushing through the pump during every shut-down period can reduce the risk of blockage considerably. If the delivery pipe is very short, its entire volume can be used for reverse flushing; in other cases an arrangement of valves according to figure 12.15 is also effective.

The volume of liquid required for reverse flushing is 50 to 150 litres, depending on the size of the pump. A check should be made that the pump is able to tolerate a certain degree of reverse rotation. This does not, however, need to involve any high speed operation. Care must be taken to ensure that reverse flushing of this kind does not cause damage to the pump. It may be necessary to install devices for filling and tapping, air separation and sampling, etc. in the pump piping system.

Accurate flow measurement is generally required for metering pumps in order to check the amount of dosage. Such pumps are often supplied with graduated vessels which can be sealed off with valves as shown in figure 12.16. In dosage systems it is always necessary to take into account the risk of over-dosage due to defects in the control system, ejector effects, etc. It may therefore be necessary to incorporate special devices to counteract such risks, etc.

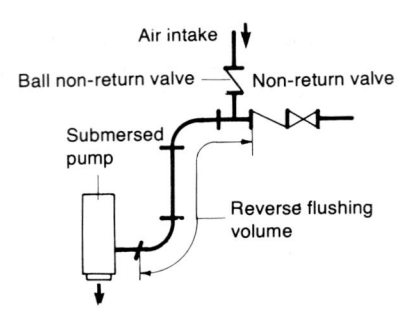

The ball in the air valve should be of a suitable weight to achieve good functional effect. When the pump is started a small amount of liquid passes through the air intake, which should therefore be located in the pump sump or other drained well

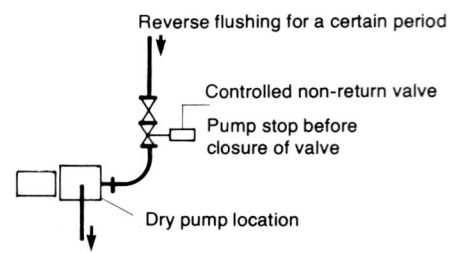

Here the non-return valve is a pneumatically or hydraulically operated shut-off valve

Figure 12.15 Reverse flushing systems for sewage pumps

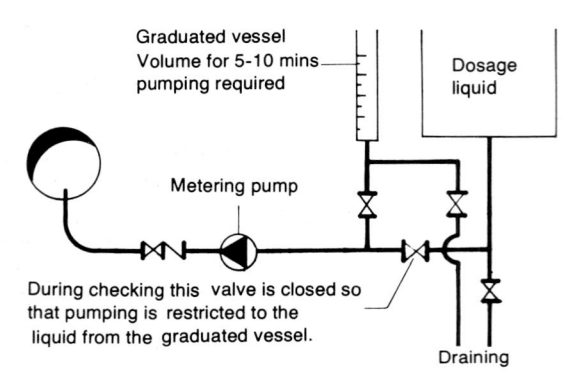

Figure 12.16 Graduated vessel for flow measurement of metering pump

12.5 Care and maintenance

Modern pump units are often supplied with vast quantities of paperwork and manuals. It is not unusual for 10 sets to be provided; 20 on occasions. However, when a service engineer from the factory visits the site he generally cannot find a manual for reference. Paperwork for pump units is costly to produce. It is full of interesting and essential information. Process operators and site maintenance staff are the people who need the information on a day-to-day basis.

It is essential that site staff are given original data produced by the pump manufacturer. Some contractors extract 'essential' data from the pump manual and produce process manuals. Costly errors have been made which were avoidable.

12.5.1 General considerations

The first prerequisite for long service life and reliable operation of a pump unit is naturally, as applies to all types of machinery, correct installation. Secondly, once the pump is operating it should not be subjected to working conditions other than those for which it was specified and designed. The third is that the unit should receive the necessary care and maintenance. This is not generally carried out to preserve the economic value but to maintain reliability of operation.

Commissioning

The transition from installation to operation is commissioning. Commissioning can be performed by site staff or the manufacturer's staff. If site staff are unfamiliar with some aspects of the pump unit, the manufacturer's personnel can combine commissioning with staff training. The following list can be used as a commissioning guide. The manufacturer's help must be sought if clarification is required.

1. Read the manufacturer's instructions regarding preparation for starting.

   Read all of this list after the manufacturer's instructions.

   The manufacturer's instructions take priority.

   If in doubt, consult the manufacturer.

2. Fit temporary suction and discharge pressure gauges if necessary. Fit extra gauges to measure pressure drop across temporary suction strainer.

3. Check minimum flow by-pass is fitted correctly on rotodynamic pumps.

   Check by-pass is fitted to positive displacement pumps.

4. Check mechanical seal or packing is fitted.

5. Check all external nuts and bolts are tightened correctly.

6. Check relief valve, if fitted, for correct setting. The relief valve manufacturer should have sealed and tagged the valve with contract settings.

7. Check gas precharge on pulsation dampeners and accumulators for run-down systems. If pump is to be commissioned at lower pressures than normal adjust dampener precharge pressures. Precharge to 70% of nominal liquid pressure unless the manufacturer quotes alternative values.

8. Check external cooling, heating and quench systems are piped up correctly and supplies are available. Heating and cooling may be connected to the pump casing as well as the stuffing box(es), check thoroughly.

9. Flush out all equipment which has been treated with rust preventatives.

10. Check specifications for lubricants and other liquids which must be used in reservoirs. Fill reservoirs to correct levels.

    Do not use close approximations of liquids as a temporary measure.

11. Turn all rotating equipment by hand. If not free, investigate before proceeding.

12. Check all power supplies are available to control system. Test all indicator lamps.

    Ensure all systems are set to 'OFF' except system to be started which should be set to 'HAND'.

    Check the resistance of all motor winding insulation with a Megger, log results, switch on space heaters if condensation appears to be a problem.

    NOTE: See IEEE Recommended practice for testing insulation resistance of rotating machines, Publication No. 43.

13. Spin motors very briefly to check direction of rotation, adjust if necessary.

14. Start the smallest auxiliary system first. Check if any pumps on the system require venting or priming, do this before switching on. Some systems are equipped with hand pumps for priming.

    Check and record operating conditions, make allowances for system being cold, check with manufacturer's figures.

    Inspect all pipe runs and connections for leaks.

If pump does not develop correct discharge pressure or flow seems low, stop immediately and investigate. See trouble-shooting chart in section 12.5.4.

NOTE: Venting is removing all the air from the pump casing. Some pumps are self venting so that when the suction pipework is vented, and the discharge pipe up to the isolating valve, the pump will vent. Priming is filling the pump casing with liquid. If the pump is not self venting there will be one or more vent valves on top of the casing.

NOTE: Do not start motors too frequently if problems occur. Do not restart motors while the shaft is rotating.

15 Simulate fault conditions and check instrument set points.

Check operation of annunciator and control logic for starting stand-by equipment.

16 Allow system to run for at least one hour, then stop and inspect filters.

If more than one pump is fitted in a system, start the other pump(s) and run each for one hour.

17 If pumps have stand-by's, set controls to 'AUTO' and check correct operation.

18 Start all the auxiliary systems in turn by repeating steps 14 to 17.

19 Complicated drivers, such as engines and gas turbines, will have their own auxiliary systems, start these first. Remove coupling spacers or driving pins and start the engine or turbine, according to the manufacturer's instructions, and run without the pump. When the driver is running properly, commission the pump.

20 Vent and prime the pump if necessary. If the process liquid is much hotter or colder than ambient conditions, the pump temperature must be adjusted slowly without thermal shock. The pump casing temperature should not change faster than 4 °C per minute. Large high pressure pumps may need much slower changes. If the casing is heated or cooled by auxiliary systems, ensure even temperature distribution through the whole casing.

21 Check the manual for starting instructions for the pump. Small rotodynamic pumps should be started with the discharge valve closed. As pump size increases, the pump may be started with discharge valve partially open. Pumps fitted with minimum flow by-passes will probably start with the discharge valve closed.

Positive displacement pumps must be started with the discharge valve open unless a starting by-pass is fitted. If a by-pass is fitted, close the discharge valve and open the by-pass valve.

22 Turn on external quench and external heating or cooling circuits.

23 Start the pump. Rotodynamic pumps should have the discharge valve opened slowly as soon as the pump has reached rated speed. The valve does not have to be opened fully but opened sufficiently to allow a reasonable flow through the pump.

Positive displacement pumps can be run at low discharge pressure until the pump unit has been checked for correct operation. After successful operation at low pressure the discharge pressure can be increased in stages by throttling the by-pass valve. Over a period of about one hour the pressure should be increased to rated discharge pressure. If all systems operate correctly, the discharge valve should be opened and the by-pass valve closed completely.

24 Pumps fitted with torque convertors may experience difficulties running the pump at very low loads. The discharge pressure or flow may have to be increased until the convertor can stabilise.

25 While the pump is running check the stuffing box(es) and bearings for overheating. Check the leakage at the stuffing box(es), adjust soft packing carefully, see Chapter 7.

Record operational values from all gauges fitted. Monitor progress of cooling water and heating medium temperatures. Monitor the condition of all filters, including the temporary suction strainer.

Stop the pump immediately if the discharge pressure falls unexpectedly.

26 When the pump is seen to operate correctly, simulate auxiliary pump failures and monitor the automatic changeover to stand-by pumps.

27 After 12 hours running remove and clean any gearbox filters.

If pump operation is satisfactory, take readings and calculate overall pump unit efficiency. Take vibration readings, at marked locations on all bearing housings and suction/discharge pipework, and analyse harmonics. For reciprocating pumps, take pressure pulsation readings and analyse harmonics.

After 750 running hours, remove all oil filters and wash out oil filter housings. Drain all lubricating oil and replace. Fit new oil filters.

Remove coupling spacers or drive pins, check coupling alignment.

28 After an appropriate period of successful running remove all the temporary equipment, including the suction strainer, and hand over to production.

Give production operators log sheets to record operating conditions.

NOTE: If the equipment is not put into service immediately, then run at least once a month until warmed up.

NOTE: In the event of operational problems, the log sheets are a good guide to the changing condition of the pump. The pump manufacturer's service engineers will ask to see the log sheets if they are ever called to site.

Care of equipment

In the selection of pumps, the planning of installations, etc., it is essential to take account of all the maintenance aspects. The assessment of a particular pump should be based on technical grounds connected with maintenance; ease of dismantling, availability of spare parts, trade skills required. Further, steps must be taken to ensure that adjacent pipe spools can be loosened easily, that working space, lifting facilities and adequate access routes are available. See Section 12.1. The distinction between servicing and care, on the one hand, and preventive maintenance, on the other, is difficult to draw. However, in the text which follows, these activities will be dealt with. The aim of care and maintenance is, of course, to prevent as far as possible breakdowns, with costly loss of production and subsequent process plant operational problems.

The necessary amount of care and maintenance is difficult to estimate. Factors such as the need of reliability of operation, the properties of the liquid pumped, the working environment of the pump, etc., exercise a marked influence in this respect. In general it can be recommended that care, as well as maintenance, should be incorporated into regular routines and that some form of database of faults noted and measures taken should be initiated and up-dated. Modern computers allow the storage and retrieval of vast quantities of information. Software is available which permits the storage of machinery data and each machine's service history. With only a cursory glance, computer records do not appear to offer any advantage over card files or machine wallets. Consider some typical questions which are easy for a computerised system but virtually impossible, unless you examine every machine wallet, for a manual system:

- how many machines use bearing 32010X?
- how many machines use ISO 320 oil?
- how many times has 'Joe' checked pump alignments?
- how many machines went off-line when No.2 boiler blew a tube?

A properly organised computerised system can offer information unimaginable when only using cards and wallets. It also provides the basis for good cost control. It is estimated that the annual maintenance costs of a pump unit including spare parts amount to between 10% and 20% of the purchase cost of the unit.

Before we consider how pumps fail, let us consider how often pumps cause major problems as opposed to other equipment. Data published in 1988 in Hydrocarbon Processing revealed the following facts regarding the 100 worst incidents world-wide over a 30 year period. Figure 12.17 lists the type of plant involved and figure 12.18 lists the type of equipment involved.

Type of plant	Number of incidents
Refineries	42
Petrochemical plants	16
Plastic-rubber plants	11
Chemical plants	10
Gas processing	7
Bulk terminals	7
Pipelines	2
Underground storage	2
Special	2
Gas oil production	1
Drilling rigs	0

Figure 12.17 Types of plant involved in major incidents

Type of equipment	Number of incidents
Piping systems	29
Miscellaneous or unknown	22
Storage tanks	17
Reactors	10
Process holding tanks	5
Heat exchangers	4
Valves	4
Process towers	3
Compressors	2
Pumps	2
Gauges	2

Figure 12.18 Types of equipment involved in major incidents

The worst incident in the 30 year period was caused by a piping problem and cost $114,000,000 in property damage.

We can see from figure 12.18 that pumps created 2% of the problems overall and only 7% of the problems caused by pipework. No one would deny that pumps have problems, but those problems must be placed in context.

The tables in figures 12.19, 12.20 and 12.21 review some practical cases of pump faults and their frequency of occurrence. These give us some idea as to the types of functional troubles to be anticipated.

Component	Relative frequency of breakdown %
Centrifugal pump	88
Coupling	8
Electric Motor	4

Figure 12.19 Frequency of breakdown of a horizontal centrifugal pump

Sub-component	Number of items	Relative frequency of breakdown %
Sundry small items	96	41
Soft packings	-	16
Mechanical seals	3	10
Control valves for sealing water	2	7
Roller bearings	2	8
Gland halves	4	3
Shaft	1	10
Impeller	1	4
Pump casing	1	1

Figure 12.20 Frequency of breakdown of an axially-split pump with a capacity of 5,200 l/min @ 17.8 m

Type of fault	Frequency %
Defective bearings	24
Blockage by foreign matter and formation of plugs	14
Gland packings, shaft sleeves	12
Packing	11
Worn out and damaged impellers	11
Broken shafts	7
Corrosion of components	4
Motor defects	4
Erosion and wear	4
Worn out pump casing	3
Wear on side covers	3
Loss of prime and insufficient NPSHa	3

Figure 12.21 List of faults experienced by a large number of pumps of different types used in the cellulose industry

A general conclusion to be drawn from the tables in figures 12.19, 12.20 and 12.21 may be that faults in the pump are much more common than in couplings or motors. This should not be surprising. Faults affecting bearings and packings are of relatively high frequency in the pump, as are also blockages and damage to impellers when troublesome liquids are pumped. A control study of a sewage disposal plant with about 250 pumps in operation in pumping stations and purification plants yielded much the same results. The second most common fault, 'Blockage by foreign matter and formation of plugs' is not really a pump problem. It is, more correctly, a pump application problem or a system failure. The term 'foreign matter' suggests that material which should not have been there blocked the pump; this is not the pump's fault. Formation of plugs suggests the pump was running too fast, an application problem, or the process liquid was 'off-spec', a system problem.

12.5.2 Preventative maintenance

The care of a pump unit should be implemented by routine inspection performed according to a definitive schedule. The timing intervals required are dependent on the working conditions and environment of the pump and the demands of operational reliability. Operational reliability should have been a major part of the pump specification and evaluation criteria. Good maintenance practices cannot reverse deficiencies of pump selection. Some pumps may require attention every day; topping up oil reservoirs for example. Periods exceeding one week are inadvisable, especially in the case of pumps in distant pumphouses and lacking alarm systems. Observations such as listening and feeling for vibrations, and checks on pressure, flow and power consumption should be performed in connection with every inspection in addition to the checking of packings and mechanical seals. More detailed investigation should, of course, be undertaken if deviations from normal operation are noted. Section 12.5.4 includes a trouble-shooting guide which may be of assistance.

If several pumps are included in the system, the starting sequence should be adjusted with the 'lag-lead' switch so that the running time is divided between all the units. Automatic and alarm devices should also be regularly tested.

Dosing pumps often require special checks with regard to flow measurement, necessitating the use of graduated vessels or other measuring equipment, and also the testing of control equipment, see Section 12.4.

The concept of care must include the external cleaning of pump and motor in addition to the cleaning of screens and filters on the suction side and in pump sumps. The necessary cleaning of pump casings and impellers may also be classed under the same heading. This work may be extensive, depending on the type of pump and the nature of the product. The possibility of regular reverse flushing can considerably reduce trouble caused by blockage, see Section 12.4. This kind of cleaning of the casing and impeller means that the unit must be able to tolerate a certain amount of low speed reverse rotation, which should be checked with the supplier, and that care should be exercised.

Reciprocating positive displacement pumps used in sewage and solid handling applications frequently have easy access to valves for inspection and cleaning. These should be inspected a regular intervals.

Pumps used in hygienic applications; food processing, dairy products, etc.; must be cleaned and sterilised in accordance with the prevailing regulations.

Timing	Activity
Daily	Check and record process operating conditions Check and record values of all indicators on the unit Check oil levels; add oil as required (1) Check levels in any external reservoirs; add liquid as required Check liquid level in gas charged suction stabilisers; add gas as required Check water in radiators of engines Check bearings and stuffing boxes, by hand, for temperature Check for leakage from mechanical seals Assess leakage from soft packed stuffing boxes Check vibration levels by hand Test annunciator lamps and audible alarm Listen for unusual noises or change in noise
Weekly	Pumps with valve covers, inspect valves Pumps with soft packing, back-off gland slightly, oil and reset Switch over gearbox filters and clean used filter Switch over fuel filters and clean used filter Vent water side of coolers Vent filters Check for leaks on pipework Check for leaks at shaft seals Check glands on control valves, back-off and reset Check air filters on engines and gas turbines Check water in batteries Assess pressure pulsations from pressure indicators Check gas pre-charge pressure in dampeners on stand-by units Adjust 'lag-lead' switches so that stand-by equipment runs 1 week in 4
Monthly	Vent all gauge connections Drain off water from all oil reservoirs Smear grease on important external nuts and exposed stud threads Take temperature readings on bearings and stuffing boxes Take vibration readings on pump and driver and at marked locations on the process pipework
Quarterly	Clean all air breathers Clean all cooling fins on housings Clean motor fan and any air filters Check vee-belt tension Check coupling bolts Check battery connections, smear with grease Take oil samples and analyse Take vibration readings and analyse harmonics
Biannually	Check torque settings of important fasteners + holding down bolts Check clearances of sliding bearings, wear rings, balance disc Check stuffing box clearances Check for general wear and corrosion Check seals on all drive shafts Check battery cell voltages individually Take readings and calculate overall pump unit efficiency
Annually	Drain all oil reservoirs Clean reservoirs internally Clean any internal oil passages Check and clean any orifices Check and clean any strainers Check oil pumps internally for wear Check condition of any rubber hoses Check seals on auxiliary pumps Check reciprocating pump valves, change springs Clean water side of coolers Refill with new oil Grease motor bearings (2) Check motor insulation resistance Inspect relief valve seats, poppets and springs Check calibration and set points of instruments
2 years	Thoroughly inspect all pressure vessels Re-hydrotest all pressure vessels (3)

(1) Do not mix different grades of oil. Do not mix different brands of oil.
(2) Check motor manual, regreasing period is based on motor size.
(3) Local regulations may be more stringent

Figure 12.22 Schedule of routine maintenance.

Maintenance in this context refers to preventive maintenance intended to reduce the number of breakdowns and the resulting unscheduled shut-downs. Attempts have been made to work out optimum maintenance statistics for pumps, but the results are uncertain and the systems are difficult to handle. Generalisation of pump maintenance is a waste of time. Pumps operate in widely differing circumstances in a vast array of materials. The best policy is to initiate a strict, very regular, routine inspection of the equipment and continue this until a pattern of equipment behaviour is apparent. At this stage, it may be possible to relax the inspection routine in some areas. It is important to remember that the pump cannot be held responsible for the effects of process upsets. If the pump requires an overhaul after a serious process upset this should not affect the pump's normal inspection regime. The cause of the upset must be traced and rectified; this is the area where the normal inspection routine should be modified. Figure 12.22 lists the type of routine inspections and maintenance functions which should be considered. Not all pumps will have all the systems mentioned. The timings of these activities may be relaxed if the pump unit performance is stable and acceptable. Oil changes may need to be more frequent if operating conditions are severe. Oil must be changed at least every 18 months irrespective of operational time or apparent condition.

Checking the gas pre-charge pressure is the best indicator of a healthy bladder in a dampener or an accumulator. If the bladder is chemically attacked it will lose gas faster.

Small handheld instruments are available which are capable of measuring; temperature, speed, vibration, bearing condition and cavitation assessment. Devices like these greatly improve the necessary data collection to improve equipment reliability.

Overhaul implies taking the pump to pieces and examining all parts. In the case of pumps working with abrasive liquid-solid mixtures or other difficult liquids, overhauls must be carried out relatively frequently, possibly every year. Routine inspection will indicate the necessity of a complete or partial overhaul. Standard process pumps and sewage pumps usually require overhauls, etc., at intervals of three to five years. For pumps handling clean water the intervals can be still further extended. During overhauls the pump is cleaned internally and externally. Bearings and other items subject to wear are checked and where necessary replaced. Wear ring clearance should be checked. This should not exceed the manufacturer's dimension, as the mechanical efficiency of the pump is adversely affected if the clearance is large. The impeller and the pump casing are checked for deposits, cavitation and corrosion damage. If the impeller is damaged it must be repaired or replaced. An impeller in poor condition means a drastic reduction of pump performance. Pump overhauls are expensive and it is necessary to consider whether they should be carried out more than a few times. In the case of small and simple pumps it is usually more profitable to replace the pump than to perform a complete overhaul.

12.5.3 Stocking spare parts

When buying pumps it is necessary to enquire fully into the supplier's resources with respect to stock spare parts. The pump supplier should be able to supply a quotation for:

- commissioning spares,
- spares for 1 year,
- spares for 2 years,
- insurance spares.

Spares quotations will include delivery times. The spares quoted will be the parts which the supplier knows from experience will wear or become damaged. Insurance spares may be required if the pump is on a critical service and space is not available for stand-bys. Spares are costly. Some manufacturers offer a discount if the spares are purchased with the complete pump. If the quotation does not mention discounts, ask. The assessment of spares requirement is dependent on many factors; the process liquid, the operating conditions, the number of hours of operation per year, delivery times for spare parts from the manufacturer, storage costs, etc. Figure 12.23

shows a proposal for stocking spares for centrifugal pumps based on the German VDMA Standard 24296. Remember: the VDMA is a trading association of German manufacturers.

Spare Parts	Number of Pumps				
	2	3	4	5	6	8
	Number of spare parts				
Impellers	1	1	1	2	2	3
Wear rings	1	1	1	2	2	3
Sealing rings	2	2	2	3	3	4
Shafts with impeller attachment	1	1	2	2	2	3
Shaft sleeves	2	2	2	3	3	4
Bearings	1	1	2	2	3	4
Mechanical seals, sets	4	6	8	8	9	12
Other seals, sets	4	6	8	8	9	12
Stuffing box packings	16	16	24	24	24	32
Sealing rings	2	3	4	4	4	6
Seats	2	3	4	4	4	6
O-rings	2	3	6	8	8	10
Seat packings	2	3	6	8	8	10
Springs	1	1	1	1	2	2

Figure 12.23 Proposal for stocking spare parts for centrifugal pumps

The pump supplier should also be asked about the supply of spare parts if a pump model is made obsolete and current production ceases.

When failure of the pump involves risk of damage, for example the possibilities of flooding or vital cooling systems being put out of action, a stand-by unit with an automatic starting system should always be installed if space allows. When a number of pumps of the same size share a service, it is a prudent to have one complete unit in reserve for change-over in the event of a breakdown or in the case of a planned overhaul. Naturally the pipe connections, electrical installation, etc. should permit really rapid change-over with the minimum of inconvenience and effort.

If overhauls and repairs are to be performed by the user's staff, it is also important that the tools required for the purpose should be obtained. Pump suppliers can make the necessary recommendations. If special tools are necessary, the pump supplier should quote costs for the tools when quoting for the pump. The use of unsuitable tools frequently results in more serious damage than that for which measures were to be taken.

12.5.4 Trouble-shooting guide

Trouble-shooting pumps is somewhat taken for granted; it may not be what you think. Trouble-shooting is making the pump do what it was specified to do. The first step is to check the actual operating conditions against the data-sheet conditions. If the conditions are different then it is perhaps not surprising that the pump does not perform correctly. If the plant operating conditions have changed since the pump was specified and purchased then the pump may require 'up-grading'. Review figure 12.24, mainly applicable to rotodynamic pumps, but certain parts also apply to other types of pumps.

Symptom	Possible cause	Remedial action
Pump won't start	Power failure Blown fuses Motor single phasing	Check all power supplies Check all fuses Check motor wiring
No flow	Pump not primed	Vent and prime pump
	Air leaks in suction pipes	Check suction pipework, replace gaskets Check stuffing box
	Excessive suction lift	Check calculations, check liquid temperature
	Suction pipe blocked or closed valve	Check suction valve, check suction pipework
	Impeller blocked	Pump vibrates badly; strip pump
	Differential head too high	Check pressure gauges, check calculations
	Pump speed low	Check pump speed, check motor frequency, check motor voltage
	Wrong direction of rotation	Check rotation

Symptom	Possible cause	Remedial action
Low flow	Differential head too high	Check pressure gauges, check calculations
	Pump speed low	Check pump speed, check motor frequency, check motor voltage
	Impeller partially blocked	Pump vibrates badly, strip pump
	Insufficient NPSHa	Noisy pump, check calculations Small air leak through stuffing box Gas evolving from liquid
	Pump wear or damage	Excessive wear ring clearance Excessive balance drum/disc clearance Damaged impeller
Low head	Pump running out on curve	Check flow, throttle discharge
	Pump speed low	Check pump speed, check motor frequency, check motor voltage
	Impeller too small	Check flow, check calculations, check pump curve
	Pump wear	Excessive wear ring clearances Excessive balance drum/disc clearance
High power consumption	Pump power high	Liquid SG high, check liquid Liquid viscosity high, check liquid Differential head low, flow high, check operating conditions Pump bearings/stuffing boxes binding Pump speed high, check speed
	Electrical faults	Voltage and frequency low, check supply Motor problem, check motor
Pump vibration	Poor alignment	Check alignment Check coupling assembly Check baseplate grouting Check process pipework
	Insufficient NPSHa	Check operating conditions, check gauges
	Damaged impeller	Check vibration harmonics, strip pump
	Bearing problem	Check bearing clearance Check vibration harmonics
Motor vibration	Poor alignment	Check alignment Check coupling assembly Check baseplate grouting
	Bearing problem	Check bearing clearance Check vibration harmonics
	Motor rotor fault	Loose rotor bars

Figure 12.24 Trouble-shooting guide for rotodynamic pumps

The following points should be remembered when trouble-shooting positive displacement pumps:

- flow is almost directly proportional to speed, low speed will always produce low flow,

- flow can be low because of pump wear or low viscosity liquid,

- low NPIPa or suction pressure below MIP will produce low flow,

- the discharge pressure generated is the pressure required to pump the flow produced into the discharge system,

- pump and process pipework vibration can be caused by operating below the pulsation dampener gas pre-charge pressure,

- discharge pipework vibration can be caused by low NPIPa, low MIP or broken valves in reciprocating pumps,

- motor overheating problems can be caused by running at relief valve overpressure for too long.

Harmonics have been mentioned during commissioning as well as trouble-shooting. Vibration and pressure pulsation harmonics can be analysed by specialists and the cause(s) of the problem diagnosed. Bench-mark readings must be taken immediately after commissioning for comparison later if problems arise.

For soft packed stuffing box problems see Sections 7.3.1.6 and 7.5.6, for mechanical seals see Section 7.3.2.9.

13 Pump economics

The chapter begins with a general discussion on the purpose and extent of economic optimisation, both in connection with the planning of new plant and also with respect to existing plant. It should be emphasised that it is important that the complete liquid transport system, including the secondary costs are considered in the optimisation. Usually it is piping and energy costs which are the two most dominant factors in pumping plant life-cycle cost. Energy consumption costs are heavily dependent upon the choice and methods of control and regulation. Economic optimisation thus involves additional aspects, which are dealt with in more detail in Chapter 5, Flow regulation.

The section dealing with economic assessment criteria reviews the various methods of economic evaluation of alternative pumping plant proposals. It is important, when assessing the profitability of an investment, to consider the rate of increase of energy costs in comparison to general inflation, together with the investment returns which can be gained by introducing energy conservation improvements.

An important characteristic of pumping plant, from an economic point of view, is its efficiency factor, i.e. the efficiency with which it converts the absorbed energy, usually electrical energy, into hydraulic power. This is especially pertinent, since efficiency factors normally have extremely low values. Reasons for low efficiency factors are discussed and suggestions for improvement are presented. The secondary costs for the main process, of which the pump unit constitutes a part, can be considerable if factors such as process adaptation and available capacity are not considered when designing the plant. The most precious form of energy in pumping is hydraulic energy, which warrants attention because of the parameters which, not only determine the hydraulic power requirements, but also the energy consumption.

Certain components in a pumping plant can be optimised individually, which greatly assists the optimisation procedure. For existing plant the choice of feasible optimisations are limited.

Obviously there are different prerequisites for economic optimisation depending upon the type and size of process for which the pumping plant is intended. Chapter 13 is generalised, offering in certain sections, practical advice on cost reductions for the majority of pumping applications.

Contents

13.1 Economic optimisation

    13.1.1 Introduction
    13.1.2 New and existing plant

13.2 Economic assessment criteria

    13.2.1 Investment calculation - new plant
    13.2.2 Investment calculation - existing plant
    13.2.3 Estimated profits and service life
    13.2.4 Energy costs

13.3 Important system characteristics

    13.3.1 Pumping efficiency
    13.3.2 Demand variations
    13.3.3 Availability
    13.3.4 Hydraulic power

13.4 Partial optimisation

    13.4.1 Economic pipe diameter
    13.4.2 Component efficiency
    13.4.3 Existing plant

13.1 Economic optimisation

13.1.1 Introduction

The basic requirement of every pump installation is to transport liquid. The object of the economic optimisation is to enable the liquid conveyance to be carried out at the lowest possible cost. The actual liquid transportation, however, often constitutes a part of a larger process. An evaluation of the liquid transport costs must, therefore, also embrace the costs of adaptation to the main process, or if adaptation is not carried out, the resultant costs of a less effective main process.

Since the required liquid transportation cannot be performed without the contribution of all components in the pumping plant, the complete liquid transportation system should be considered when carrying out economic optimisation. Only when the remaining costs are not affected, can partial optimisations give completely accurate results. Partial optimisations do, however, have the advantage that they are easier to evaluate and are therefore applied extensively, despite the risk of certain errors.

The cost which should be minimised is the plant life-cycle cost, i.e. the summation of all the costs which occur during the total economic life span of the plant. This begins with planning and drafting and finishes when the plant is written off or otherwise disposed of. Life-cycle costs are traditionally divided into investment and operating costs. The investment costs consists of the sum of the following:

- planning,
- design and drafting,
- equipment,
- installation and commissioning,
- process adaptation,
- staff training,
- disposal, writing off.

Operating costs consist of the sum of:

- maintenance, parts and labour,
- energy consumed.

If pumping plant is critical to a process and space is not available for stand-by capacity, the cost of lost production can be considered as a running cost.

The investment costs are basically fixed price in character, i.e. in principle they are independent of the extent to which the plant is used. Whilst the operating costs increase with the number of hours of the operation.

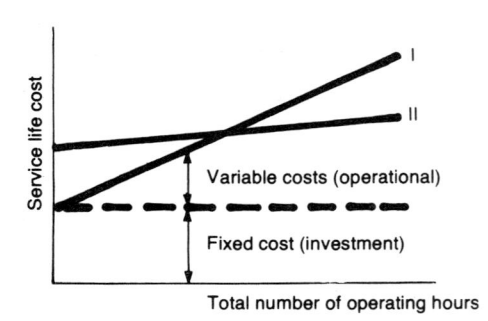

Figure 13.1 Graphical representation of life-cycle costs

The two alternatives in figure 13.1 show the influence which the number of operating hours has in respect of plant life-cycle costs. Plant I, which is characterised by its simple construction with low investment costs, gives the lowest total cost despite its higher operating costs, low system efficiency, when the total number of operating hours is low. For longer operating times, however, the relatively expensive plant II, with its high efficiency, shows the lowest life-cycle cost. In practice, the operating costs will not be linear. As the plant ages, efficiency will be reduced and maintenance costs will increase.

The distribution of costs for the various items of expense are naturally different for each case. An example of the items, together with their cost distribution, is shown in figure 13.2. These figures are derived from estimates made within the cellulose and paper industry.

Expense item	Cost %	Cost %
Pumps including mounting	4	
Temporary liquid storage	8	
Electrical equipment	12	Investment
Instrumentation	8	60 %
Pipework	20	
Administration	8	
Maintenance including operational administration	11	Operating
Energy	29	40 %
Total	100	100

Figure 13.2 Distribution of life-cycle costs for a liquid transportation system within the cellulose and paper industry

It is worth mentioning that according to figure 13.2, the cost for pumps represents the smallest part of the fixed costs, whilst energy and maintenance are the largest part of the total costs. A doubling of the pump investment costs would only increase the total cost by 4%. A reduction of the pump energy consumption by 25%, which in many cases is quite feasible would reduce the life-cycle costs by as much as 7%. It can also be seen from the table that energy costs, 29%, together with the investment costs for the pipework, 20%, are the two dominant cost items and account for half of the total cost of the system.

Even though the values in figure 13.2 are taken from a special branch of industry, similar distribution trends are to be found in many pump installations. Within sewage systems, water supply, heating, ventilation, sanitation and district heating, the cost of pipes and fittings usually show higher values but, together with energy costs, they still constitute the largest items. On the other hand, pumps driven by large engines, steam or gas turbines, will be much more costly than electric motor driven pumps. For small, simple pumping plant, however, the costs associated with labour and administration can often predominate. In such cases it is important that the plant is of a simple construction being easy to install and requiring little maintenance.

13.1.2 New and existing plant

The design and layout of new plant offers a greater freedom of choice with respect to economic optimisation than is the case for existing plant. For new plant the conditions are more favourable for a total optimisation, since all the plant components can be chosen freely. Normally, economic optimisation is carried out for a number of technically feasible alternatives. The costs for the various items are determined and summated as in the previous section. The alternative selected is that which fulfils the liquid transportation requirements whilst incurring the lowest life-cycle costs. The opportunities presented by a new plant should be especially utilised to achieve good adaptation to the process. Since the major costs are often associated with the main process, of which the liquid transportation is only a part, considerable cost reductions for the main process can be achieved by suitably designing the pumping plant. The following examples are given in order to explain what is meant by the term 'process adaptation'.

- The transfer of process liquid between two process operations without synchronised flow requirements; the process availability is increased if temporary storage is included in the liquid transportation system.

- The pumping of sewage water to a sewage works regularly; constant flow assists the purification function of the sewage works.

- The pumping of drinking water to a water tower; the matching of supply to consumption results in a smaller water tower.
- The pumping of hazardous liquids; careful selection of the pump size and design of the shaft seals or use of glandless pumps reduces the costs of environmental protection.
- The gentle pumping of food products; reduces product damage and costs.
- The correct choice of low speed pumps for abrasive liquid-solid mixtures; reduces wear, costs of spare parts and labour costs.
- The scheduled adjustment of pump configuration, adding a stage, changing impeller diameter, changing piston/plunger diameter; matches flow/pressure requirements to system changes reducing energy consumption.

Optimisation of existing plant should also, of course, consider the costs of process adaptation. Here, however, many parameters are already established which imposes limitations. Optimisation of existing plant assumes therefore, a partial optimisation function. The question posed then, is whether an extra investment can reduce the operating costs sufficiently to make the improvement profitable. Examples of modifications proposed for existing plant are:

- Replacement of pump, replacement of impeller with larger diameter, reduce impeller diameter.
- Change method of regulation or control.
- The addition of temporary liquid storage.
- Pipework or system modifications to reduce pressure losses.
- The introduction of improved shaft sealing arrangements.

All modifications are introduced with the intention of reducing future maintenance and energy costs.

Another type of optimisation situation occurs when the transportation requirements can be performed by alternative existing pumping units. Such situations appear, for example, in large pipe networks, which are supplied by several pumping stations. Since the fixed costs for the plant in this situation cannot be changed, the rule is to choose the pump unit which results in the lowest operational costs. The operational costs in these situations are complicated by differing pump efficiencies and varying system pressure losses due to distance and pipe size.

13.2 Economic assessment criteria

13.2.1 Investment calculation - new plant

Present capitalised value method

The pumping plant life-cycle cost K_L can be calculated as the sum of the investment cost K_I and the capitalised operational costs K_{Dcap}. The operational costs must be paid annually and must therefore be summated to a capitalised present value. This is carried out with the aid of a present capitalised value factor F. The life-cycle cost becomes:

$$K_L = K_I + K_{Dcap} = K_I + F \cdot K_D \qquad \text{Equ 13.1}$$

where
  K_L = life-cycle cost (currency)
  K_I = investment cost (currency)
  K_{Dcap} = capitalised operational cost (currency)
  F = present capitalised value factor
  K_D = annual operating costs (currency/annum)

The present capitalised value factor is calculated from the relationship:

$$F = \frac{1 - (1+r)^{-n}}{r} \qquad \text{Equ 13.2}$$

where
  F = present capitalised value factor
  r = estimated annual profits (decimal percentage)
  n = number of years

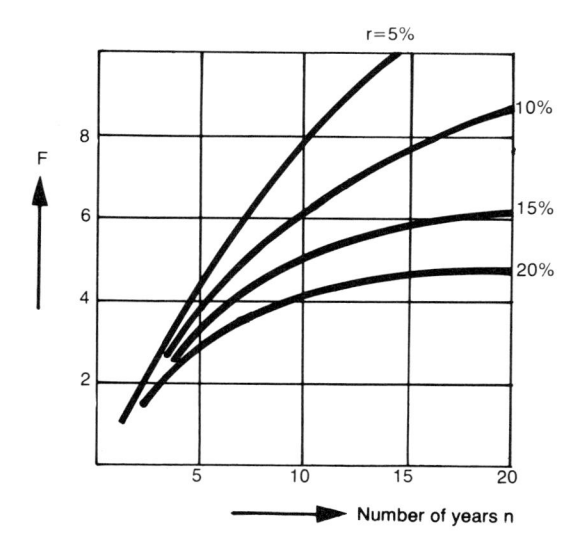

Figure 13.3 Present capitalised value factor

Example:

Two alternative designs of pumping plant have been proposed. Their cost review is as follows:

I investment = £10,000, annual operational costs = £1,000

II investment = £7,500, annual operational costs = £1,500

Which plant is preferable from an economic point of view, if the estimated profits are 15% and the service life is 10 years?

If r = 15% and n = 10 years, the present capitalised value factor F = 5, from figure 13.3. Thus:

I K_L = £10000 + 5 · 1000 = £15000

II K_L = £7500 + 5 · 1500 = £15000

Both plants are similar in this particular instance. If it is probable that the service life will not exceed 10 years, then alternative II would normally be preferred since it requires a smaller initial capital investment.

Annuity method

For the annuity method the annual plant costs are calculated and minimised. The annual costs consist of the sum of the annuity interest and repayments for a proposed loan which covers the investment cost and the annual operational costs.

$$K = A + K_D = a_F \cdot K_I + K_D \qquad \text{Equ 13.3}$$

where
  K = annual cost (currency/annum)
  A = annuity (currency/annum)
  K_D = annual operating costs (currency/annum)
  a_F = annuity factor
  K_I = investment (currency)

The annuity factor is identical to the reciprocal of the present capitalised value factor

$$a_F = \frac{1}{F} \qquad \text{Equ 13.4}$$

and is determined by service life and estimated profit. Applying this method, for the two alternatives in the previous example:

I $K = \dfrac{10000}{5} + 1000 = £3000$

II $K = \dfrac{7500}{5} + 1500 = £3000$

The two alternatives are equivalent, as before.

13.2.2 Investment calculation - existing plant

Present capitalised value method

For existing plant the question is often asked if an improvement of the plant can reduce the operational costs in such a way as to reduce the life-cycle costs. According to the present capitalised value method the summation of the savings during the service life will be expressed at today's current value.

$$B_L = K_{Dcap} - K_I = F \cdot K_D - K_I \qquad \text{Equ 13.5}$$

where

B_L	=	total savings during working life	(currency)
K_{Dcap}	=	capitalised reduction of operating costs	(currency)
F	=	present capitalised value factor	
K_D	=	annual operating cost reduction	(currency/annum)
K_I	=	investment cost	(currency)

Example:

By investing £3,000 it is possible to reduce the annual costs for liquid transportation by £1,000. The plant is designed to operate for 15 years and the estimated profits to be a minimum of 15%. How great are the savings during the service life of the plant at today's currency value?

According to figure 13.3 the present capitalised value factor F = 5.9 for n = 15 years and r = 15%.

$$B_L = 5.9 \cdot 1{,}000 - 3{,}000 = £2{,}900$$

Over and above the repayment of investment, including estimated profits, an extra £2,900 is obtained.

If several alternatives exist to achieve a similar technical improvement, then it is normal to choose the alternative which produces the greatest savings.

Annuity method

The annuity method answers the question as to how great the first year's saving will be at today's currency value.

$$B = K_D - A = K_D - A_F \cdot K_I = K_D - \dfrac{K_I}{F} \qquad \text{Equ 13.6}$$

where

B	=	annual saving	(currency/annum)
K_D	=	annual operating cost reduction	(currency/annum)
A	=	annuity	(currency/annum)
a_F	=	annuity factor	
K_I	=	investment cost	
F	=	present capitalised value factor	(currency)

Using the values from the previous example,

$$B = 1000 - \dfrac{3000}{5.9} = £492$$

For otherwise similar alternatives, the size of the annual saving is decisive when making the selection.

Pay-off method

The pay-off method utilises an imaginary "repayment time" defined by the relationship:

$$T_p = \dfrac{K_I}{K_D} \qquad \text{Equ 13.7}$$

where

T_p	=	pay-off time	(years)
K_I	=	investment cost	(currency)
K_D	=	annual operating cost reduction	(currency/annum)

The shorter the pay-off time the more profitable the investment. By comparing with Equation 13.5, it is found that for $B_L = 0$ then $T_p = F$, i.e. the pay-off time and present capitalised value factor have the same numerical value when the total saving is equal to zero.

Using the values from the earlier example

$$T_p = \dfrac{3000}{1000} = 3 \text{ years}$$

With an estimated profit requirement of 15% and F = 3, the actual repayment time from figure 13.3 is approximately 4.3 years.

Investment grant

Grants for energy saving investments may apply. For up-to-date information it is best to seek the advice of the relevant local or central government authority.

13.2.3 Estimated profits and service life

Estimated profits

Profit estimates should, in principle, correspond to the interest on capital which would otherwise be realised from an alternative investment. It is a measure of a company's profitability and is higher than the current bank rate. The profit estimate increases with reduced capital resources, since it is necessary to be more particular when investment capital is limited. Profit estimates are rarely considered at less than 15%.

The methods of calculation previously reviewed assume that the annual operating cost reductions are of the same magnitude from year to year. The majority and largest savings on the operational side are achieved by reducing energy consumption. Energy costs, currency per kWh, can also be expected to rise more quickly than other costs, which means that energy savings will become more profitable with time.

One way of considering energy cost increase when making economic calculations is to use a corrected profit estimation.

$$r_k = r - e + i \qquad \text{Equ 13.8}$$

where

r_k	=	corrected profit estimate	(percentage)
r	=	uncorrected profit estimate	(percentage)
e	=	rate of increase of energy costs	(percentage)
i	=	general inflation rate	(percentage)

A more rapid rate of increase of energy costs can in this way be transferred to a reduced profit estimation requirement for energy saving investments.

Service life

The economic service life is determined by factors such as write-off rules, the technical service life of components and the planned period of use of the plant. As with other parameters for pumping plant, the economic service life is dependent upon the size and type of industry. As an approximation with the exception of small plant the following applies:

Buildings	40 years
Pipelines, underground	50 years
Other pipelines	20 years
Machines	15 years
Control equipment	10 years
Instrumentation	10 years

13.2.4 Energy costs

Energy costs depend upon the amount of energy consumed, the prevailing energy price scale and fixed costs for the supply installation. Premiums may be levied if the 'maximum demand' is exceeded. In the most usual cases of electric motor operation, energy prices are determined by the relevant electricity tariffs.

Tariffs

The basis of most forms of tariff is a fixed charge dependent on the 'maximum demand' taken by the consumer and designed to cover:

a) the costs dependent upon maximum load, e.g. interest and depreciation of generating plant, rates, taxes, insurance, salaries; and

b) the costs incurred for each consumer, e.g. transformers, meters, meter reading labour, service cabling. And a running charge depending on the energy supplied, e.g. fuel, losses and maintenance of the supply plant, equipment, etc.

For industrial consumption two-part tariffs are usual with the fixed charge proportional to maximum kW or maximum kVA demand; and a running cost, kWh, which may be dependent on the time of day and/or year, i.e. peak and off-peak periods, and which may also include, for example, a fuel cost variation clause. A kVA maximum demand is preferable since it takes into account the effect of low power factor. It involves, however, more expensive metering equipment. The cost of metering maximum demand makes it uneconomic, in any case, for loads of less than 20 to 50 kW.

If supplies are taken at a high voltage instead of the usual 415V/380V for distribution, the maximum demand charge may be less since the consumer then has the option to provide his own transformer.

Industrial Tariff (annual)

The energy costs are a product of energy consumption and cost per unit. The annual energy cost will thus be:

$$K_E = k_e \cdot E \qquad \text{Equ 13.9}$$

where

K_E = annual energy cost (currency)

k_e = energy cost (currency per unit)

E = annual energy consumption (units)

or capitalised for the service life of the pumping plant

$$K_{Ecap} = F \cdot k_e \cdot E \qquad \text{Equ 13.10}$$

The Restricted Tariff, see Figure 13.5, is only available up 50 kW maximum load. The Catering Tariff is only available to consumers who already have an ordinary electricity supply.

The electricity supply business is very competitive. The UK has two areas where the consumer can negotiate an agreement. For loads between 100 kW and 1 MW the consumer negotiates with the regional supply company. For loads over 1 MW the consumer negotiates directly with the generator.

In Chapters 5 and 9 we have talked about power recovery turbines and engines and gas turbines using local fuel supplies. Every energy supply has a cost. In the case of high pressure liquid, which is supplied to power recovery turbines, the pipework and the control valves will have fixed and running costs. The same type of costs will apply to waste process steam. Local fuel supplies, such as crude oil or LNG, will have a value either as refining stock or sale to a consumer. These costs must be considered when evaluating the potential of site energy supplies.

	Abbreviation	Tariff Name	Upper Availability Limits	Quarterly Charge	Charges per unit used in:- Weekday Daytime	Charges per unit used in:- Evening and Weekend	Charges per unit used in:- Night-time
1	Q.1	Ordinary general purpose	50kW OR 60000 units per annum	£11.74 (£11.23)	8.83p	(2.36p)	
2	Q.1.T.	Economy 7 general purpose	50kW OR 60000 units per annum in daytime	£14.67 (£13.97)	9.23p (2.53p)		2.60p (0.24p)
3	E.W.	Evening and weekend	50kW OR 60000 units per annum	£14.67 (£13.97)	11.47p (3.09p)	5.40p	(1.24p)

Figure 13.4 Example of General Purpose Industrial Tariff

	Abbreviation	Tariff Name	Availability		Charges per unit used in:- Weekday Daytime	Charges per unit used in:- Evening and Weekend	Charges per unit used in:- Night-time
4	E.W.S.	Evening and weekend small user	Up To 1000 units per annum	Quarterly Charge £11.04 (£10.43)	7.67p	(1.78p)	
5	C.7	Catering tariff	For Heavy Duty Cooking Equipment	Minimum Charge £11.04 (£10.43) per quarter	7.67p	(1.78)	
6	S.T.B.	Super tariff	For Approved Storage Heating Equipment	Quarterly Charge £4.93 (£3.69)	Designated Periods 2.39p	(0.26p)	

Figure 13.5 Example of Restricted Industrial tariff

The charges include amounts, shown in brackets ( ), for Use of the Distribution System	Q2 Quarterly maximum demand	Q2T Economy 7 maximum demand
1. A Quarterly Charge of	£22.88 (£21.75)	£22.88 (£21.75)
2. An Availability Distribution Charge for each kilovolt ampere of agreed distribution capacity subject to a minimum of 15kVA	£3.77/kVA/qtr (3.42)	£3.77/kVA/qtr (3.42)
3. Maximum Demand Charge Rate per kW per quarter	£7.53	£7.53
4. Unit charges **either** Tariff Q2 'Any Time' unit charges For each unit taken	5.21p (1.06)	
or Tariff Q2T "Day/Night" unit charges Night Units: for each unit taken in the Night Period		2.60p (0.25p)
Day Units: for each unit taken at any other time		5.47p (1.07p)
5. Reactive Power Charge: For each lagging kVArh (if any) supplied by the Company in excess of 0.5 times the number of kWh supplied in each quarter	0.62p (0.33p)	0.62p (0.33p)

Figure 13.6 Example of Maximum Demand Tariff for larger Industrial Users

13.3 Important system characteristics

13.3.1 Pumping efficiency

The cost of energy is the dominant cost item in most pumping plants. The energy, for which the pump user must pay, is the quantity measured by the electrical supplier or equivalent. The annual energy consumption consists of the product of the power used and the period of actual power consumption. See equation 13.9 for constant load applications and equation 13.11 for variable load applications.

$$E = \int_0^{t_0} P \cdot dt = \sum_0^{t_0} P \cdot \Delta t = P_m \cdot t_0 \qquad \text{Equ 13.11}$$

where
- E = annual energy consumption (kWh/annum)
- P = actual power used (kW)
- t = time (hours)
- P_m = mean annual power used (kW)
- t_0 = operational hours (hours/annum)

The power used depends partly upon the useful hydraulic power required in order to maintain the necessary flow in the pipe and partly upon the efficiency of converting the electrical power, or equivalent, into useful hydraulic power. This efficiency is called the pumping efficiency factor and is defined as:

$$\eta_E = \frac{P_h}{P} = \eta_r \cdot \eta_p \cdot \eta_{tr} \cdot \eta_m \cdot \eta_{others} \qquad \text{Equ 13.12}$$

where
- η_E = pumping efficiency factor (decimal percentage)
- P_h = required hydraulic power (kW)
- P = power absorbed (kW)
- η_r = flow regulation efficiency (decimal percentage)
- η_p = pump efficiency (decimal percentage)
- η_{tr} = drive train efficiency (decimal percentage)
- η_m = motor efficiency (decimal percentage)
- η_{others} = other component efficiencies (decimal percentage)

The flow regulation efficiency takes into account the power losses which are caused as a direct result of the method of flow regulation. Such power losses are caused by throttle regulation in the discharge pipe, by-pass control and on-off or load-unload control.

The expression for regulation efficiency for on-off control consists of the quotient of energy requirements. Whereas the others are represented by the quotient of power requirements.

The pump efficiency will obviously be determined for the actual operating point and not the rated duty point efficiency, maximum flow point or any other arbitrary operational point.

The transmission efficiency represents losses in gears, couplings and speed variators. It is important to consider the transmission efficiency when speed regulation is used. Losses in rectifiers, frequency convertors, regulation resistances, additional motor losses, etc., i.e. losses when utilising electrical methods of speed regulation, are traditionally calculated as transmission losses.

The motor efficiency should be taken at the actual operating point. A common misconception is to assume that the motor efficiency is still high even at part load. At part load the percentage of reactive power is increased, this may be costed as a separate item.

The efficiency of other components takes into consideration, for example, losses for phase compensation and protection against supply disturbances, power requirements for supplementary ventilation, quench water, cooling, heating or other power consuming arrangements which can be directly associated with energy conversion in the pumping plant when working.

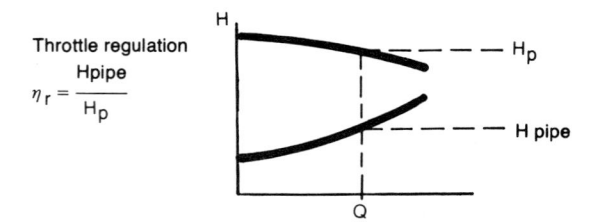

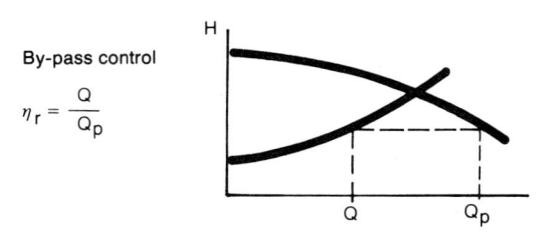

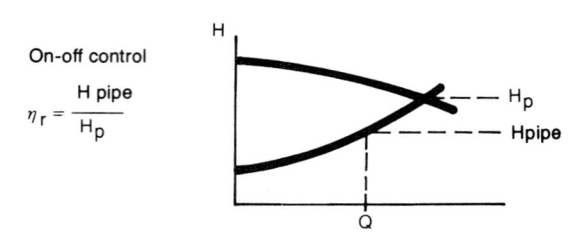

Figure 13.7 Illustration of flow regulation efficiency

It is extremely important to remember that it is the pumping efficiency factor and not the efficiency of isolated components which is the characteristic unit for energy conversion efficiency. If all the component efficiencies are high at the actual operating point then a high pumping efficiency factor will obviously be obtained. However, it only requires one of the component efficiencies to be low to cause a poor total pumping efficiency.

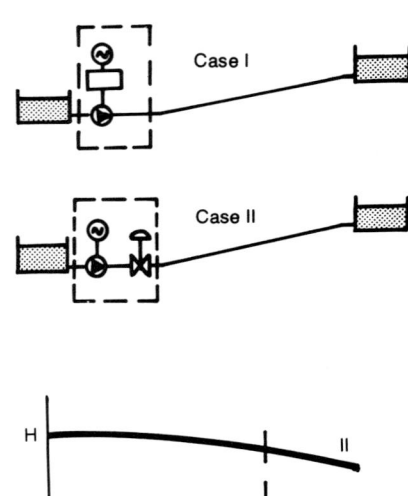

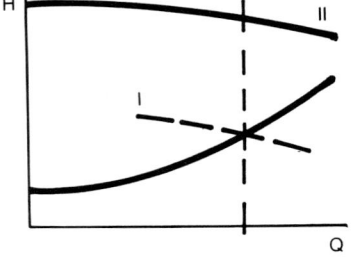

Figure 13.8 Illustration of pump efficiency factor

Example:

Determine the pumping efficiency factor for a flow of 70% of maximum flow for the installation shown in the figure, for (I) speed regulation and (II) throttle regulation.

Using estimated component efficiency values:

Case I			Case II		
η_r	= 1	(from figure)	η_r	= 0.44	(from figure)
η_p	= 0.75	(estimated)	η_p	= 0.75	(estimated)
η_{tr}	= 0.75	(estimated)	η_{tr}	= 1	(estimated)
η_m	= 0.90	(estimated)	η_m	= 0.90	(estimated)
η_{others}	= 1	(not included)	η_{others}	= 1	(not included)
η_E	= 0.51		η_E	= 0.30	

The efficiency factor for on-off control becomes evident, if it is assumed, for example, that the pump operates at full flow, 100% flow, for 75% of the time. During 30% of the time the pump is then at rest. In this case higher pipe losses are obtained during the time the pump operates and the efficiency factor becomes:

$$\eta_E = 0.75 \cdot 1 \cdot 0.90 \cdot 0.60 = 0.41$$

For multi-pump systems the efficiency factor can be determined in a similar way. Here, as always, it is only the total efficiency for the actual flow and not the maximum efficiency of individual components which is characteristic for energy conversion efficiency. For on-off control of multi-pump systems there are a number of pertinent operating points. The efficiencies for these flows are greatly influenced by the sizes of the pumps selected. An example of this is shown in figure 13.9. In the case of three identical pumps having maximum efficiency at Q_{max}, poor part load efficiencies, shown as circles on the graph are obtained. By choosing larger pumps, with η_{max} at a somewhat higher Q value, better part load efficiencies can be obtained, shown as crosses on the graph. Only by choosing three different sizes of pump can the highest efficiency be achieved for all three operating points. Three different pumps, however, involve the stocking of additional spare parts and more expensive maintenance. Also the pumps cannot be used as stand-by for each other.

The basic pump efficiency is very important. Pump efficiency is greatly influenced by size and choice of pump. Figure 13.10 shows typical efficiencies for single stage overhung impeller centrifugal pumps, the most popular rotodynamic pump. Figure 13.11 shows typical efficiencies for reciprocating plunger pumps.

Absorbed power kW	Efficiency %
3	37 to 59
7.5	38 to 74
15	50 to 78
30	50 to 79
55	55 to 83
110	60 to 85

Figure 13.10 Typical best efficiencies for single stage centrifugal pumps

Absorbed power kW	100% Pressure Efficiency %	50% Pressure Efficiency %	25% Pressure Efficiency %
18.5 to 45	88	76	65
45 to 200	90	78	67
200 to 600	92	80	68
600+	94	84	72

Figure 13.11 Typical efficiencies for reciprocating plunger pumps

The efficiency of plunger pumps is closely related to discharge pressure, not speed. Suction pressure does influence the efficiency when high. It can be seen from figures 13.10 and 13.11 that plunger pumps are much more efficient than centrifugal pumps. Reciprocating pumps are more costly than centrifugal pumps from an initial investment, but the running costs can be much lower. Reciprocating pumps can be operated at reduced speeds, without loss of efficiency, to extend the life of parts such as packing and valves.

13.3.2 Demand variations

The primary cause of a low pumping efficiency factor is often due to the variation, with respect to time, of the desired flow through the pipes, insufficient consideration having been given to this aspect at the design stage. Normally the mean flow per annum, total supplied volume of liquid divided by a calendar year, is about 10% to 15% of the installed flow capacity, Q_{max}. The efficiency factor calculated as a mean efficiency over the year, theoretical hydraulic energy consumption/actual electrical energy consumption, usually gives a value in the region 5% to 40%, which should then be compared with the normal maximum momentary value of 55% to 75%.

The reasons for production or demand variations and the low mean flow are many, and will not be dealt with here in great detail. Some examples can, however, be mentioned:

- Seasonal variations in industry production
- Variations in supply of sewage water
- Variations in quantities of rainfall/snow
- Variations in water consumption
- Variations in heating requirements for central heating
- Component and calculation tolerances
- Margins of safety over and above the normal tolerances

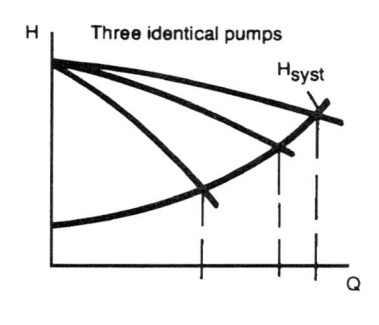

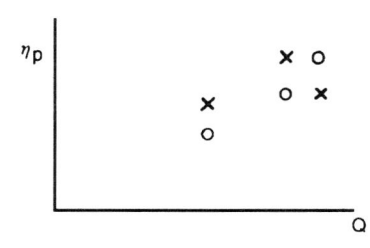

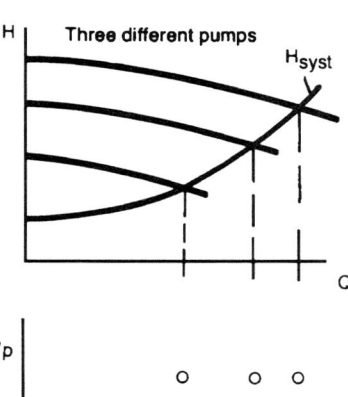

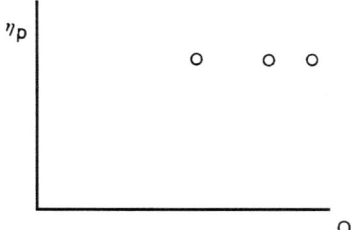

Figure 13.9 Example of pumping efficiency for three pumps connected in parallel

A survey of flow variations in respect of time is a necessary basic requirement for all economic optimisations of liquid transportation. It is desirable that both progressive curves and constancy diagrams are produced.

Modifications for the improvement of the efficiency factor and thus the costs can be of two types:

- To improve the part load efficiency by means of suitable flow control and regulation equipment.
- To reduce extreme flow peaks.

The latter modification involves the addition of a liquid reservoir of some kind in the liquid transport system. Examples of liquid reservoirs are:

- Sumps and reservoirs
- Water towers
- Accumulators

Sumps and reservoirs can offer the largest storages volumes at atmospheric pressure. Water towers have limited capacity although they can be substantial. Accumulators are small in comparison, they can be mounted in parallel, but they are the only option for pressures up to 350 bar. Unfortunately, there is no general method for assessing the optimal effect of various modifications and it is necessary to make estimates for each individual case. There is no doubt, however, that considerable cost reductions can be made by improving the degree of demand variation.

13.3.3 Availability

Pumps

Availability is generally defined by the relationship:

$$A = \frac{t_{do} - t_{db}}{t_{do}} \qquad \text{Equ 13.13}$$

where
- A = availability (decimal)
- t_{do} = total operating available time in a fixed period (hours)
- t_{db} = operating time lost due to malfunctions (hours)

The loss of operational time includes both planned shut-down for maintenance as well as unscheduled shut-downs. The availability for different pumps varies considerably depending upon the pump type, speed, operating conditions and environment. Apart from the pump design, materials, seals, properties of the process liquid, etc., availability is also affected by the success of preventive maintenance, spare parts availability, repair time, etc. Centrifugal pumps within the process industry, for well-tried, normal operating conditions, achieve values of availability in the order of $A = 0.99$ to 0.9999, i.e. pump shut-down due to pump malfunction is between 1 and 80 hours per annum for continuous operation. These values also include the squirrel cage induction motor but not other electrical equipment. Availability calculations are applied mainly to multi-pump systems. The percentage of operating time which a given number of pumps in a pumping station can be expected to be available for operation, is calculated from the formula:

$$P = \frac{N!}{K!(N-K)!}(1-A)^K \cdot A^{(N-K)} \qquad \text{Equ 13.14}$$

where
- N = number of pumps installed (integer)
- K = number of pumps unavailable (integer)
- A = availability of individual pumps (decimal)
- $!$ = factorial ($4! = 24$, $0! = 1$)

All the pumps are thus assumed to have the same availability values. By using Equation 13.14 the resultant availability of a pumping station can be calculated and the effect of a stand-by pump can be demonstrated.

Without reserve pump		With reserve pump	
Function requirement	Resultant availability	Function requirement	Resultant availability
1 of 1	0.9900	1 of 2	0.9999
2 of 2	0.9801	2 of 3	0.9997
3 of 3	0.9703	3 of 4	0.9994
4 of 4	0.9606	4 of 5	0.9990
5 of 5	0.9510	5 of 6	0.9985

Figure 13.12 Availability of pumping station with and without stand-by pump. Function requirement 3 of 4 means that 3 pumps out of 4 will be available for operation. The individual pump availability has been taken as 0.99.

From figure 13.12 it can be seen that a pumping station's availability is considerably improved by the installation of a stand-by pump. For this also to apply in practice it is necessary to maintain the stand-by pump to ensure that it will operate when required. In cases where it is possible to calculate the cost of a non-functioning pump, for example, in the form of lost production, availability calculations offer a direct economic optimisation possibility. Stand-by pumps usually run a limited number of hours regularly, one week in four say, to ensure the pump is functioning correctly.

Liquid transport system

The liquid transport system constitutes a link in a more extensive main process. The design of the liquid transport system affects the availability of the main process, see figure 12.18. The pipework may be 14 times more likely to create serious problems than a pump. The type of problem posed can be illustrated by a typical practical example. A process consists of two sub-processes for liquid treatment connected by a liquid transport system, see figure 13.13.

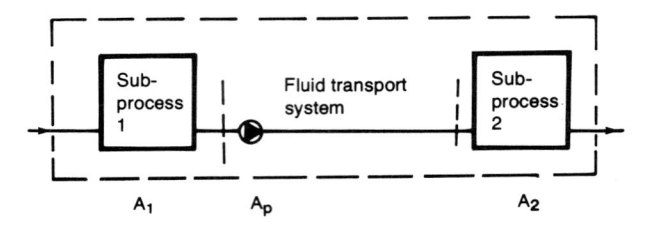

Figure 13.13 Schematic arrangement of a simple liquid system

The liquid transport system comprises the connection between the two liquid processes and in figure 13.13 it can be characterised as a rigid connection. If the sub-process 1 must be shut-down because of an internal malfunction, then the liquid transportation and sub-process 2 must also be shut-down. The resultant availability for the process will thus be:

$$A = A_1 \cdot A_p \cdot A_2 \qquad \text{Equ 13.15}$$

The sub-process can consist of a large number of complicated and sophisticated components and thus have a considerably lower availability than the liquid transport system. For this reason it is often unsuitable to make the sub-processes rigidly dependent upon each other.

By introducing a tank into the liquid transport system it is possible to loosen the connection between the two sub-processes. If the tank is sufficiently large, then the two sub-pro-

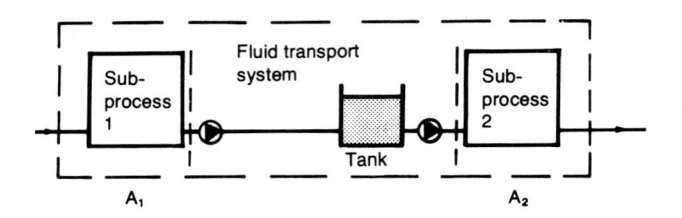

Figure 13.14 Schematic arrangement of a flexible liquid system

cesses can become totally independent of each other. Sub-process 1 can deliver liquid to the tank while sub-process 2 is being repaired and vice versa. If $A_1 = A_2$ and $A_{p1} = A_{p2}$ the resultant process availability becomes:

$$A = A_1 \cdot A_{p1} = A_2 \cdot A_{p2} \qquad \text{Equ 13.16}$$

The higher value of availability facilitates a lowering of the mean capacity during operation for the same production.

The required installed capacities of all components in the sub-processes and the liquid transport system are thus reduced. Lower mean flow also results in lower power and energy consumption if the efficiency of the smaller components is as high as the larger components. These possible cost reductions should be weighed against the investment costs of the tank. For numerical values of $A_1 = A_2 = 0.90$ and $A_{p1} = A_{p2} = 0.99$ for mean capacity and mean flow during operation, the following results are obtained:

without buffer $A = 0.9 \cdot 0.99 \cdot 0.9 = 0.802$

with buffer $A = 0.9 \cdot 0.99 = 0.891$

With a tank large enough to completely isolate sub-process 1 from sub-process 2 the availability is increased by 11%.

13.3.4 Hydraulic power

General

The useful hydraulic power from the pump unit is used to satisfy the pipe system differential head requirements at the actual flow.

$$P_h = \rho \cdot Q \cdot g \, (H_{stat} + H_{fpipe}) \cdot 10^{-3} \; (kW) \qquad \text{Equ 13.17}$$

In order to achieve an optimal economic result, it is important that the hydraulic power is kept as low as possible. A limitation of the required hydraulic power is worthwhile, especially when bearing in mind the low pumping efficiency factors, normally 5% to 40%. A reduction of P_h by 1 kW reduces the electric power consumption by 2.5 to 20 kW because of the low efficiency of converting electricity to hydraulic power.

Static delivery head

The static delivery head is determined when planning the plant and is often difficult to influence. It should be remembered that for installations where liquid is first pumped up to a tank in order to be subsequently gravity fed to a consumer, the energy required is always greater than if the liquid were pumped direct to the consumer. Such cases should be redesigned, if possible, at the planning stage.

Pipe flow losses

The magnitude of pipe friction losses in straight pipes is described by the relationship:

$$h_{fpipe} = \lambda \cdot \frac{l}{d} \cdot \frac{Q^2}{\left[\frac{\pi \cdot d^2}{4}\right]^2} \cdot \frac{1}{2g} \qquad \text{Equ 13.18}$$

$$h_{fpipe} = \frac{\text{constant}}{d^5} \qquad \text{Equ 13.19}$$

See also section 2.2. The pipe friction losses for a given flow in a 200 mm pipeline are 3.05 times greater than for the same flow in a 250 mm pipeline, i.e. $(250/200)^5 = 3.05$. This calculated example illustrates the importance of correct pipe sizing. Pipe losses are not only friction losses. Pipe fitting, flanges and bends, also make a contribution. The selected pipe velocity must consider the quality of the assembled pipe system and any problems caused by turbulence at poorly aligned connections.

The corresponding expression for losses in valves and fittings, etc., is:

$$h_{fvalves \, and \, fittings} = \frac{\text{constant}}{d^4} \qquad \text{Equ 13.20}$$

See also section 2.2. Here also there is a large dependence upon size. For laminar flow:
h_{fpipe} = constant $\cdot$ d^{-4} and $h_{fvalves \, and \, fittings}$ $\cdot$ constant d^{-3}.

Demand and production variations do not only lead to a low pumping efficiency factor, but also increase the hydraulic power requirement and the hydraulic energy consumption. Hydraulic energy consumption is relative to the conversion from pressure to heat, pressure drop, caused by the effects of pipe friction. For pump installations where $H_{stat} = 0$, which is assumed here solely for the purpose of simplifying the illustration of the importance of flow variations, then:

$$P_h = \rho \cdot Q \cdot g \cdot h_{fpipe} = \text{constant} \cdot Q^3 \qquad \text{Equ 13.21}$$

The installed hydraulic power, which is required in a system with mean flow Q_m and demand/production variation $\pm \Delta Q$ is:

$$F_h = \text{constant} \cdot (Q_m + \Delta Q)^3$$

$$= \text{constant} \cdot Q_m^3 (1 + \frac{\Delta Q}{Q_M})^3$$

$$= P_{hm} \cdot (1 + \frac{\Delta Q}{Q_m})^3 \qquad \text{Equ 13.22}$$

The hydraulic energy consumption becomes dependent upon the shape of the flow demand/supply curve and is illustrated below for three simple special cases.

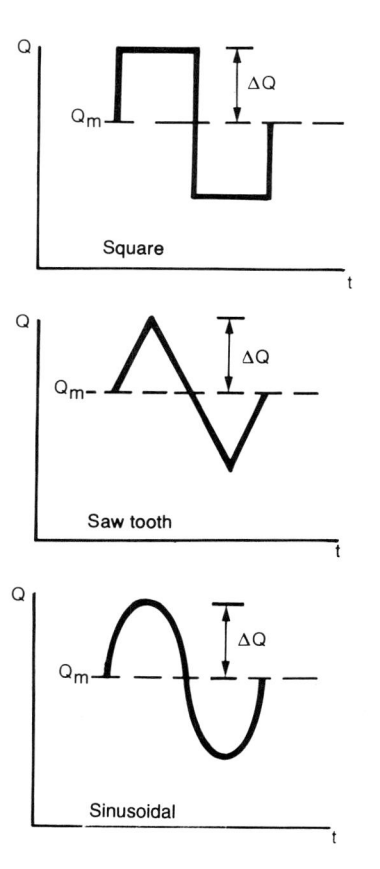

Figure 13.15 Three special cases of demand/supply variations

The hydraulic energy consumption becomes:

$$E_h = [1 + 3 \, (\frac{\Delta Q}{Q_m})^2] \cdot P_{hm} \cdot t \quad \text{Square}$$

$$E_h = [1 + (\frac{\Delta Q}{Q_m})^2] \cdot P_{hm} \cdot t \quad \text{Saw tooth}$$

$$E_h = [1 + \frac{3}{2} (\frac{\Delta Q}{Q_m})^2] \cdot P_{hm} \cdot t \quad \text{Sinusoidal} \qquad \text{Equ 13.23}$$

$\Delta Q/Q_m$	P_h/P_{hm}	$E_h/P_{hm} \cdot t$		
		Square	Saw tooth	Sinusoidal
0	1	1	1	1
0.1	1.33	1.03	1.01	1.02
0.2	1.73	1.12	1.04	1.06
0.5	3.37	1.75	1.25	1.38

Figure 13.16 Hydraulic installed power and hydraulic energy consumption for different types of demand/supply variation

As shown by the table the demand/supply variations have a considerable influence upon both the installed hydraulic power and the energy consumption. From the economic point of view, reducing the flow peaks and troughs can give good returns.

Suspensions

In the case of suspensions, liquid-solid mixtures, yet another variable must be considered, the solids concentration. The adaptation of the quantity of solid material transported per unit of time can be made by varying the flow, changing the concentration or a combination of both. Valve regulation is not usually suitable because of the risk of blockage and high wear rates. Other feasible alternatives are illustrated by the example in figure 13.17.

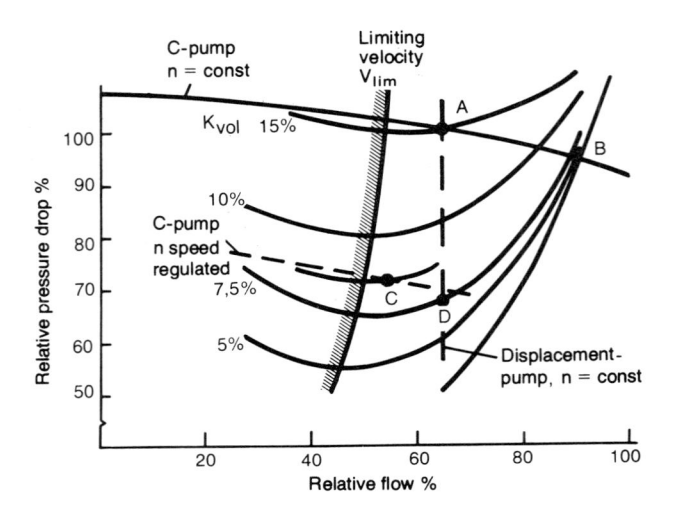

Figure 13.17 Approximate effects of pumping heterogeneous suspensions with nominal capacity, A, and at half capacity solid flow at B, C and D

At nominal capacity, operating point A at 65% centrifugal pump flow, a certain quantity of solid material is transported through the pipeline. The relative required hydraulic power is set at point A as being equal to unity. It is desired to reduce the quantity of solid material per unit time to approximately half the nominal capacity. For a constant speed centrifugal pump this can be carried out by reducing the volume concentration K_{vol}. The liquid flow increases simultaneously and the relative hydraulic power becomes approximately 1.3 at point B. The liquid velocity increases because the pressure loss through the pipe reduces as the solids concentration is reduced, see section 2.3. In the case of a speed regulated centrifugal pump the operating point C is obtained with relative power requirement of 0.6. Alternatively, a positive displacement pump can be used for the transportation. At half capacity, operating point D is obtained with relative hydraulic power 0.7. The original power of the positive displacement may have been less due to better pump efficiency. The risk of unstable operation and blockage of the pipe is also illustrated by figure 13.17. The centrifugal pump curve can maintain equilibrium with the pipe loss curve at two points, of which the operating point at the lower flow leads to blockage in the pipeline. Since the curves intersect each other at nearly the same angle then the risk of instability is relatively great. Corresponding risks do not occur when using a positive displacement pump. The actual power consumptions can only be confirmed by evaluating complete pump data.

Considering positive displacement pumps for long pipeline applications introduces yet another variable, the number of pumping stations. Positive displacement pumps can produce high pressures without high speeds or high liquid velocities. The pipe friction losses can consequently be higher before the mixture must be re-pressurised. Positive displacement pumps can handle solid mixtures up to 70% by weight. Pump efficiency is not necessarily a function of solids concentration.

13.4 Partial optimisation

13.4.1 Economic pipe diameter

The two major items for many pumping plants are the costs of pipelines on the investment side and the energy costs on the operational side. The investment costs increase with increased pipe diameter whilst at the same time operational costs decrease, see figure 13.18.

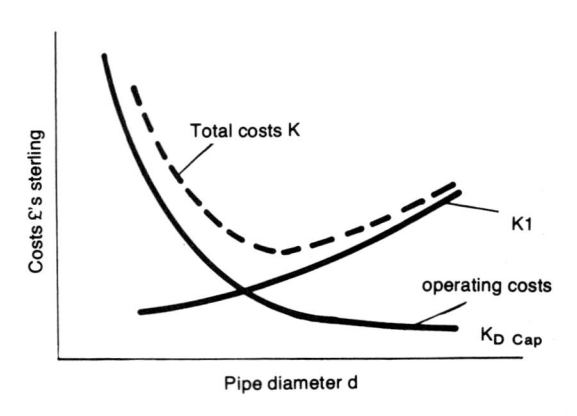

Figure 13.18 Effects on costs of pipe diameter

The total cost curve represents a minimum for a certain pipe diameter. As indicated in the figure the shape of the total cost curve is such that it rises more rapidly with reduced diameter, to the left of the economic diameter, than in cases of increased diameter, to the right of the optimum value. In cases of doubt choose, from the feasible alternatives, the one having the largest pipe diameter.

By calculating the total costs for a number of different pipe diameters it is possible to establish the most economical pipe diameter. The economic diameter varies for different situations. Short operating periods and costly pipe sections, for example, stainless steel, tend to reduce the economic pipe diameter and to increase the economic flow velocity.

By carrying out a partial optimisation, i.e. if it is assumed that all costs are independent of the pipe diameter except for the costs of the pipeline and energy, the following relationship is obtained.

$$Q_1 = \sqrt[3]{\frac{k_2 - k_1}{1/d_1^5 - 1/d_2^5} \cdot \frac{\eta_E \cdot \pi^2 \cdot a_F \cdot 10^3}{8 \cdot \rho \cdot \lambda \cdot t \cdot k_e}} \quad \text{Equ 13.24}$$

where

Q_1	=	the mean flow for which pipe diameters d_1 and d_2 give the same annual cost	(m³/h)
k	=	cost/m of pipeline	(currency/m)
d	=	pipe diameter	(m)
η_E	=	pumping efficiency factor	(decimal)
a_F	=	annuity factor	(decimal)
ρ	=	liquid density	(kg/m³)
λ	=	pipeline loss co-efficient	
t	=	operating time per year	(hour/annum)
k_e	=	cost of energy	(currency/kWh)

For this Q_1 value the total energy costs are the same. If the mean annual flow for the plant is greater than Q_1 then it is economically feasible to choose pipe diameter d_2. If Q_m is much greater than Q_1 then the procedure must be repeated using diameters d_2 and d_3.

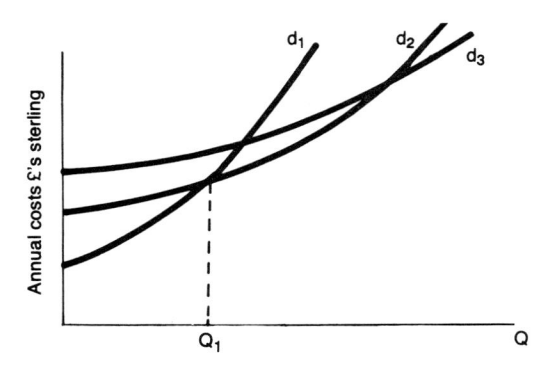

Figure 13.19 Annual costs for a pumping plant using various pipe diameters

The procedure for valves and other fittings is carried out in a similar manner.

$$Q_1 = \sqrt[3]{\frac{k_2 - k_1}{1/d_1^4 - 1/d_2^4} \cdot \frac{\eta_E \cdot \pi^2 \cdot a_F \cdot 10^3}{8 \cdot \rho \cdot \zeta \cdot t \cdot k_e}} \quad \text{Equ 13.25}$$

where k = valve cost including connectors for pipe diameter d.

It should be noted that an improved pumping efficiency factor tends to reduce the economic pipe diameter.

13.4.2 Component efficiency

When purchasing new equipment, it is often possible to choose between a cheap pump having low efficiency and an expensive pump with high efficiency. The comparative costs for the two alternatives are:

$$P_{1m} \cdot t \cdot k_e \cdot F + K_{I1} \Leftrightarrow P_{2m} \cdot t \cdot k_e \cdot F + K_{I2} \quad \text{Equ 13.26}$$

where
- P_m = consumed mean electrical power (kW)
- t = operating hours per year (hours/annum)
- k_e = energy cost (currency/kWh)
- F = present capitalised value factor
- K_I = investment cost (currency)

It is important to remember that the energy costs are debited according to readings taken from the electricity meter and not from the pump shaft. Because of cable losses, the electric power used is always greater than the pump shaft power. The capitalised value of 1 kW mean electric power is $t \cdot k_e \cdot F$.

The efficiency of other components can be tested in the same way. Conversion to electrical power used must always be made, however. The closer the power consumption lies to the useful hydraulic power in the conversion chain, the greater the energy saving and the greater the motivation for additional investment.

13.4.3 Existing plant

In the case of existing plant there are certain limitations regarding energy saving possibilities. Changes to the lay-out and pipework are often difficult to make. Possible improvements are replacements or the addition of supplementary components which do not require too great a disturbance to the plant.

All reductions of the hydraulic power, by means of improvements which reduce the static differential head or pressure drop, in the case of throttle-regulated plant, only result in a greater pressure drop across the control valve and are thus worthless from an energy-conservation point of view. In order to realise a saving of energy some form of improvement must also be carried out on the pump side. For speed-regulated plant all pressure reductions are automatically utilised.

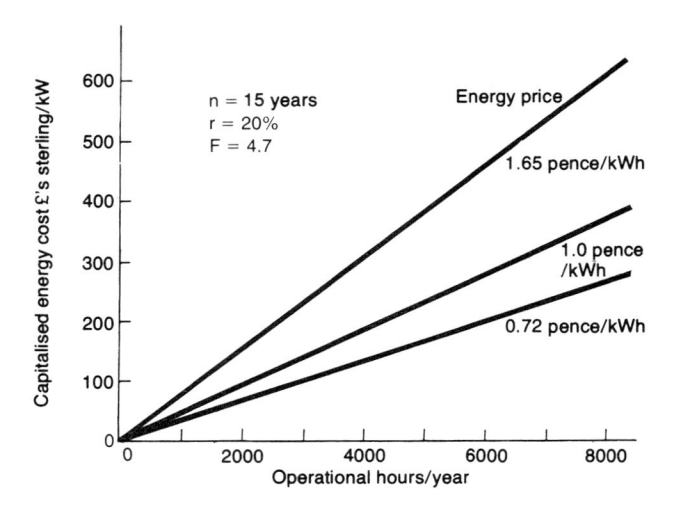

Figure 13.20 Capitalised value in £'s sterling/kWh for a reduction of the mean annual electric power used

When planning a project, at the time of determining pump specifications, it is often the case that not all of the plant details are completely known. It is not therefore unnatural to choose pump sizes with a certain safety margin to be on the safe side. For this reason many pumps are over-sized and result, if adjustments are not made, in unnecessarily high energy costs. In the case of pure over-sizing, the following corrections to the pump may be considered.

- Machining the pump impeller
- Fitting a smaller piston/plunger
- De-staging a multi-stage pump
- Fitting a smaller motor
- Changing the gear ratio of a gearbox

It may also be advantageous to review the process(es) to optimize or de-bottleneck to use the excess flow or pressure.

A more common basic reason, other than over-sizing, for unnecessarily high energy costs is, however, the occurrence of demand/supply variations. In which case the above improvements cannot be applied. Attention should instead be directed to corrections such as:

- energy efficient methods of control, the most usual being speed regulation
- energy efficient methods of control together with reduction of pressure drop and reduced flow peaks and troughs.

The fact that energy conservation proposals and suggestions cannot be realised in practice unless adaptation is carried out on the pumping side, is a strong argument for speed regulation or some other form of energy efficient method of control. Since the investment costs for the pump unit are usually a small proportion of the energy costs, the addition of energy conservation equipment is nearly always profitable. The secondary costs of an over-sized pump are of a very limited nature when using energy conservation methods of control and the excess pump capacity can be considered as a reserve in the event of possible future production increases.

Fewer engineering hours are used when designing a project and during operation, since the secondary costs of over-sizing are automatically limited.

Bran+Luebbe. World Leaders for Control Volume Pumps...

- Any combination of pumpheads:
- Packed plunger up to 1000 bar
- Hydraulically actuated PFTE or Stainless Steel diaphragm up to 500 bar
- PTFE mechanical diaphragm
- Accuracy better than ± 0.5%
- Single and multi head dosing pumps.

...introduce High Pressure Process Pump Range

NOVAPLEX
- All benefits of existing range
- Up to 7 head configuration
- Leak free design
- High volume - High Pressure
- Solids and suspensions
- 20,000 hrs operation life - minimum

Bran+Luebbe Ltd, Scaldwell Road, Brixworth, Northants NN6 9UD
Tel: 0604 880751 Fax: 0604 880145

BRAN+LUEBBE
MPL PUMPS

Stainless Steel Centrifugal Pumps Vent Valves & Strainers for hygienic applications

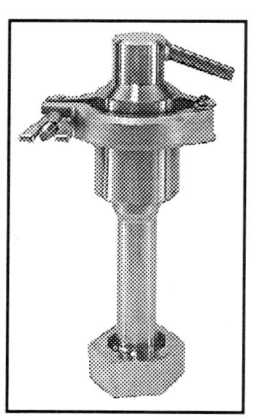

AE and AP
Hygienic Polished Stainless Steel Air Venting Valve

F.D.A. Approved Hygienic Stainless Steel Centrifugal Pumps

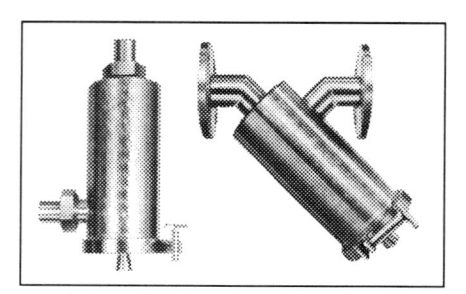

K2 Stainless Steel Pipeline Strainers

MDM PUMPS LTD
Malvern, Worcs.
England. WR14 1BP
Tel: 0684 892678
Fax: 0684 892841

Certificate No. FM 20542
ISO 9001 EN 29001 BS 5750: Part 1

14 Pump selection

This chapter is concerned with the general choice of pump which can subsequently be made more precisely with the help of the manufacturers and suppliers listed in Chapter 16. The choice is made with the assistance of categories, consistently applied throughout the entire book, with a numbered column for every type of pump. The choice can also be made according to other points of reference. In Sections 14.2 and 14.3 the choice is based on pump designation and direct choice tables or according to the operating conditions and rapid choice tables. In the following sections the pump is chosen on the basis of discharge pressure, differential head, flow, temperature, viscosity and the ability to handle various solids.

It should be pointed out that centrifugal pumps are the most popular choice for most applications, but not necessarily the best choice. When all relevant information is reviewed the basic type of pump, rotodynamic or positive displacement, should become clear. The pump characteristic compared to the system characteristic may be the clue. The type of pump can be selected to optimise other operational requirements.

The basis for purchase consists mainly of a number of check lists for pumping plants and of a presentation of typical purchasing criteria with regard to quality and specification.

Contents

14.1 General operating conditions

14.2 Selection of pump according to pump designation
    Direct selection table

14.3 Selection of pump according to duty and capabilities
    Rapid selection table

14.4 Selection of pump according to hydraulic performance
    14.4.1 Pumps for low viscosity liquids
    14.4.2 Pumps for viscous liquids
    14.4.3 Pumps for highly viscous liquids

14.5 Pumps for liquid solid mixtures
    14.5.1 Pumping non-abrasive solids
    14.5.2 Pumping abrasive solids
    14.5.3 "Gentle" pumping
    14.5.4 Pumping waste water, sewage

14.6 Check list for pump purchase specification

14.7 Purchasing

14.1 General operating conditions

Liquid properties uncertain

Operating conditions uncertain

In order to consider any type of pump for an application, the following information must be known:

- flow,
- temperature,
- suction pressure,
- NPSHa/NPIPa,
- differential head or pressure,
- discharge pressure,
- constituents in the liquid,
- the properties of the liquid.

Constituents in the liquid, in this context, mean solids, gases and other liquids. Pumps selected for clean liquids may only last 20% of the time predicted if abrasive solids are present. Entrained gases may reduce the effective NPSHa by over 3.5 m; cavitating pumps do not always last very long. Liquids as trace elements can have a serious impact on the corrosion properties of the bulk liquid.

The properties of the liquid are of critical importance for selecting the correct type of pump. The first obvious property is the vapour pressure. The liquid temperature plus the vapour pressure plus the suction system arrangement provide the NPSHa and the NPIPa. The liquid density varies with temperature; this will affect the conversion to pressure in rotodynamic pumps. The liquid viscosity varies considerably with temperature. Perhaps the liquid is non-Newtonian, a pulp or emulsion. If the liquid is extremely hazardous, then a pump which cannot leak may be required. The corrosive properties of the liquid may be important for the type of pump. Normally, liquid corrosion would decide the choice of material for the pump. If the pump must be manufactured from a material which cannot be cast or welded, the choice of pump types is severely restricted.

Any variation in the operating conditions must be quantified. For any specific variation, the other conditions must be stated. The duration of expected running must also be specified. Changes in operating conditions are always assumed to be relatively slow; slow enough not to cause instability or surges. Rapid changes in temperature can create distortion; for instance it is possible to crack thick casings due to thermal shock. Any rapid changes in operating conditions must be identified. For pumps running continuously, it is usual to specify the Mean Time Between Overhaul, MTBO. Chemical processing and oil refining plants can operate for 25 or 50 weeks without a shut-down. The following operating descriptions can be useful:

continuous over 8 hours running in any 24 hour period,

light 3 to 8 hours running in any 24 hour period,

intermittent up to 3 hours running in any 24 hour period,

irregular the pump operates for differing times with various periods of extended rest between operation,

cyclic the pump operates with a set pattern of rest or unloaded running followed by a period on-load.

More costly, slower equipment can, through higher reliability and availability, pay for itself very quickly from increased plant output.

If starting or stopping the pump results in significant changes to the liquid properties or system conditions, these effects must be clarified. High suction pressure during start-up for example, can create tremendous bearing wear problems. Pumps starting cold, at high viscosity, may not achieve the desired performance. Routine maintenance operations on the pump, such as steam or chemical cleaning, may result in modified material selections.

Variations in flow determine whether the application should be shared by several pumps. The system curve, flow/pressure characteristic, then determines whether the division should be between units of equal or different sizes. Great variations in the viscosity of the liquid may mean that, for example, a centrifugal pump should be used alternatively with a positive displacement pump in cases of low and high viscosity respectively. A positive displacement pump may be able to cope with the complete viscosity range. Viscosity is extremely important. Some pumps rely on viscosity for lubrication. Other pumps rely on viscosity for sealing. Some pumps are unaffected by viscosity.

Variations in flow demand and the consequential change, or otherwise, of differential head or pressure determine the type of flow regulation to be used. In this context the working time of the pump is of considerable interest. The efficiency of the pump unit, at all its operating conditions must be used to total the energy requirement, kWh/annum. The energy costs for pumping can be as great as the purchase cost of the pump after only a few months of operation. The cost of the energy supply must be considered carefully. Can alternative energy sources by utilised? Can the energy total be reduced by power recovery?

Flow should also be considered both from the point of long term changes and also from short term fluctuations. Different regulation methods, with different initial costs and running efficiencies can be applied depending upon the frequency of the changes.

The effect of system pressure changes on the pump flow may be important. A positive displacement pump can offer almost constant flow over a wide range of pressures. This feature may be important and can simplify flow control methods.

Operating conditions can change as the installation wears, corrodes or fouls. Changes to the pump unit, from as new, to progressive ageing may be used to optimise performance at different duty points.

The location of the pump within the site also has a marked influence on the choice of style. The location determines the NPSHa and NPIPa and evacuation and also dry pit, wet pit or submersible. Requirements for space can also lead to a choice between horizontal and vertical designs. The environment at the installation site may also raise special demands with regard to the driver for the pump.

The size, nature and concentration of any particles are also, of course, important factors. Hard abrasive solids will generally have a much more serious effect on pump life and efficiency than soft deformable solids like wax. The abrasive properties of hard solids should be quantified by testing. The Miller Number test is used extensively for reciprocating pumps. Wax however, may coat pipe walls and impeller vanes and progressively choke a pump and its system.

When employing positive displacement pumps it is important to know the liquid compressibility. As differential pressures increase, the compressibility becomes more essential to evaluate the volumetric efficiency. Few liquids are less compressible than water. Liquified gases, such as CO_2, can be up to six times more compressible than water with consequential reductions in volumetric efficiency.

The hazards posed by the liquid are an important factor. Hazards to personnel, the site and the environment, and the time period involved, must be considered carefully. The location of the site may influence decisions. An oil production platform in the North Sea must be treated differently to a remote site in the heart of the French countryside. Local installation and statutory regulations may also require specific actions to be implemented. The pump may be situated in a hazardous area caused by potentially explosive gases. Surface temperatures imposed on electrical equipment will apply to mechanical equipment, even if not stated. Pneumatically or hydraulically operated pumps may be necessary.

When selecting pumps for a range of applications for one site, select the pump which is best for the application providing life-cycle costs are considered. Selecting the correct pump will optimise efficiency and reduce running costs. The cost of extra spare parts for different machines will be outweighed by the reduced energy consumption and extended running times of appropriate equipment. The skills of local personnel must also be considered. Most pumps are relatively simple and can be maintained with standard tools using established techniques. Some pumps, and/or drivers, may have complicated systems requiring specialist knowledge. In these cases the costs of factory staff for maintenance must be evaluated.

For choice of pump, the auxiliary systems necessary for the pump should also be taken into consideration. The pump itself must often be furnished with protection against dry running, against over-heating when running against a closed valve and, in the case of positive displacement pumps, against unintentional throttling in the suction or discharge. Shaft seals require in many cases sealing or cooling liquids at the correct pressure and with a guaranteed flow.

In some cases the properties of the liquid may be uncertain. When this problem arises it is best to discuss the effects of variations on pump performance, and perhaps life and reliability, with pump manufacturers or consultants. Values must be placed on uncertain parameters. If this is left to pump manufacturers, they will chose values which are best for their particular pump. The user must analyse the problem and select safe values for the limits of operation. Figure 14.1 lists some typical important ranges commonly encountered.

Parameter	Range
Density	500 to 1700 kg/m^3
Solids content	up to 65% by wt
Solids size	up to 100 mm
Viscosity	0.15 to 20000 cSt
Vapour pressure	negligible to 130 barg
Compressibility	1.96E-7 to 1.95E-5

Figure 14.1 Typical liquid, liquid-solid mixture, properties
Compressibility units; reduction in unit volume per bar differential pressure

Just as with liquid properties, operating conditions can be doubtful, figure 14.2 lists typical ranges of operating conditions.

Condition	Range
Temperature	-100 to 350 °C
Suction pressure	0.15 to 140 barg
NPSHa	0.75 m and greater
Discharge pressure	0.25 to 1000 barg
Flow	0.002 to 5000 m^3/h
Power	0.05 kW to 5 MW

Figure 14.2 Typical pump operating conditions

14.2 Selection of pump according to pump designation

Direct selection table

For many applications there are special designs or adaptations and these, almost without exception, are the most suitable. Heating, water and sanitation pumps as well as foodstuff pumps are typical examples. Accepted pump designations are often helpful in connection with the choice of pumps. Such designations can be derived from the liquid or from the function of the pump. Figures 14.3 and 14.4 are direct choice tables referring the pump designation to the numbered categories consistently applied in this book to different types of pumps. These are also to be found in the pump specifications in Chapter 3 and in the Manufacturers and Suppliers List in Chapter 16. However, in some cases pump designations alone sometimes result in too many alternative categories and must in such cases be supplemented with more details, above all by flow, differential head, discharge pressure and the viscosity of the liquid.

When using the direct choice tables a check should be made of the category number obtained according to:

- Specifications of different pump types and designs, for more details see list of contents for Chapter 3.
- Classification guide to the different category numbers, for more see Section 16.2.
- Technical features, for more information see rapid choice table, Section 14.5.

Pump designation	Category Number	Prime consideration
Acid pump	095, 100, 105, 110, 120, 130, 235, 240, 250, 270, 280, 285	material choice, seals
Ammonia pump	235, 240, 270, 280	Q, H, NPSHa
Ash pump	050, 145, 235, 250, 270, 280	abrasion resistance
Ballast pump	045	
Beer pump	150, 235, 240	
Boiler circulation pump	095, 100, 105, 110, 125	high suction pressure
Boiler feed water pump	090, 165, 170, 270	water purity, temperature
Carbamate pump	095, 110, 155, 160, 270	material choice, seals
Clean water pump	005, 010, 015, 020, 025, 055, 060, 195, 200, 235	Q, H, installation
Concrete pump	145, 235, 240, 250, 270	Q, H
Condensate pump	060, 085, 095, 175	Q, H, NPSHa
Cooling water pump	005, 010, 025, 055, 060, 085, 090, 195	Q, H, NPSHa
Crude oil pump	095, 100, 105, 110, 115, 165, 230, 235, 270, 275	Q, H, NPSHa, viscosity, sand, H_2S
Drinking water pump	005, 010, 015, 020, 150	
Emergency boiler circulation pump	270	
Fertiliser pump	030	Q, H
Fish pump	135, 140, 240	
Foodstuffs pump	135, 140, 145, 150, 240	
Fresh water pump	005, 010, 015, 025, 085, 090	
Fuel oil pump	095, 100, 105, 110, 120, 220, 225, 230, 235, 240	Q, H, viscosity, temperature
Grease pump	220, 225, 235, 240, 250, 255	
Ground-water pump	035, 050, 070, 075	
Heat media transfer pump	085, 090, 095, 100, 105, 110, 220, 225, 230	Q, H, temperature
Heavy fuel oil pump	220, 225, 235, 240	
High temperature water pump	085, 090, 095, 100, 105, 165	
Hot oil pump	095, 100, 105, 110, 115, 120, 165, 220, 225, 270, 280	
Hot water pump	025, 060, 085, 090, 165	
Hydraulic oil	220, 225, 245, 260	
Liquid metals pump	120, 125	
Liquid salts pump	120, 125	
Liquor (lye) pump	095, 100, 105, 120	
Lubricating oil pump	220, 225, 230, 245	
Milk pump	150, 235, 240	Q, H
Mud pump	275	
Paint pump	095, 100, 105, 235, 240, 250, 255	
Pesticide pump	030	
Petrol pump	095, 100, 105, 110, 120	
Pulp pump	135, 140, 180, 185, 190	

Pump designation	Category Number	Prime consideration
Raw water pump	005, 010, 025, 035, 050, 055, 070, 075, 085, 090	Q, H, NPSHa
Salt leach pump	095, 100, 105, 130	
Salt water pump	045, 055, 060, 085, 090, 095, 100, 105, 165, 170, 175, 195, 200, 235, 270	Q, H, NPSHa, temperature
Solids pump	050, 065, 135, 140, 145, 180, 185, 190, 220, 230, 235, 240, 250, 270, 275, 280, 290, 295, 305	Q, H, solid size and nature
Sewage pump	140, 180, 185, 190, 235, 270	Q, H, NPSHa
Sludge pump	145, 180, 185, 190, 235, 240, 270	Q, H, NPSHa
Water pump	Most pumps except those that rely on the process liquid for lubrication	Q, H, NPSHa, temperature

Figure 14.3 Direct choice table for pumps designated according to liquid (See also Figure 14.4).

Pump designation	Category Number	Prime consideration
Building, Construction		
Building site pump	050, 065, 140, 185, 190	
Concrete pump	145, 235, 240, 250, 270	
Dredging pump	140	
Excavation pump	050, 065, 140, 145	
Gravel pump	140, 145	
Ground water pump	050, 065	
Submersible pump	050, 065	
Sump pump	050, 055, 065	
Chemical Industry		
Acid pump	095, 100, 105, 120, 130, 250, 255, 270, 280	material choice, seals
Bottom pump	095, 100, 105, 145, 235, 240, 270, 280	viscosity, solids, NPSHa
Chemical pump	095, 100, 105, 120, 125, 130, 220, 225, 235, 240, 250, 255, 270, 280	material choice, seals, viscosity
Laboratory pump	120, 125, 130, 245, 250, 255, 280, 285	
Liquor (lye) pump	095, 100, 105, 120	
Metering pump	280, 285	
Standard pump	005, 025, 085, 090	
Vertical chemical pump	060, 105, 125, 130, 175	
Petrochemical refining		
Acid pump	110, 120, 130, 250, 255, 270, 280	material choice, seals
Bottom pump	110, 115, 235, 240, 270, 280	viscosity, solids, NPSHa
Export pump	060, 165, 170, 175, 270	Q, H
Gas scrubber pump	110, 115, 165	entrained gas
Laboratory pump	120, 125, 130, 245, 250, 255, 280, 285	
Liquor (lye) pump	110, 120	
Metering pump	280, 285	
Process pump	110, 115, 155, 160, 165, 170, 175	
Standard pump	005, 010, 085, 090, 270	
Vertical process pump	060, 155, 175	
Crude oil production		
Carbon dioxide injection	270	suction pressure, temperature
Ethylene glycol injection	270, 280, 285	suction pressure
Gas drying	110, 115, 270	viscosity, discharge pressure, sand
LNG re-injection	270	
Methanol injection	270, 280, 285	
Mud pump	275	

Pump designation	Category Number	Prime consideration
Steam injection	165, 170, 270	
Viscous crude oil pipelines	270, 275, 280	
Viscous crude oil transfer	230, 235, 240, 245, 270, 275, 280	
Water injection	165, 170, 270	
District Heating		
Central heating pump	005, 015, 025, 125	
Circulation pump	005, 015, 025, 125	
District heating pump	005, 015, 025, 085, 090	
Fuel oil pump	095, 100, 105, 110, 120, 220, 225, 230, 235, 240	viscosity, discharge pressure
Oil-burner pump	220, 225, 230, 235	viscosity, discharge pressure
Service hot water pump	025, 125	
Transfer pump	085, 090, 165	
Drainage and Dewatering		
Ditch drainage pump	030, 050, 065	
Drainage pump	005, 010, 025, 030, 050, 055, 060, 065, 190, 195	
Flood pump	005, 010, 025, 030, 050, 055, 060, 065, 190, 195	
Ground-water pump	035, 050, 070, 075	
Reclamation drainage pump	030, 050, 065	
Sump pump	035, 050, 055, 060, 065	
Fire Fighting Services		
Fire pump	005, 045, 085, 090, 175	
Motor fire spray	005, 065	
Sprinkler pump	005, 025, 085	
Irrigation and Agriculture		
Irrigation pump	005, 010, 025, 030, 035, 055, 060	
Fertiliser pump	030	Pumps driven by tractor power take-off can be centrifugal, diaphragm or roller vane
Pesticide pump	030	
Spray pump	030	
Tractor-powered pump	030	
Mining		
Coal sludge pump	140, 145, 270	
Dressed ore pump	140, 145	
Flotation pump	140, 145	
Gallery pump	050	
Mine dewatering pump	085, 090, 270	
Solids pump	140, 145	
Shaft pump	075	
Waste pump	140, 145, 270	
Paper and Cellulose Industry		
Bagasse (cane trash) pump	135	
Chips pump	180, 185, 190	
Condensate pump	060, 085, 095, 175	Q, H, NPSHa
Couch pump	135, 195	
Dense pulp pump	235, 240	
Digester circulation pump	095, 100, 105	
Fan pump	010, 095, 100, 105	
Filter pump	135	
Linters (brown cotton fibre) pump	135	
Liquor (lye) pump	095, 100, 105, 120	
Pulp pump	135	
Sludge pump	140, 145, 180, 185, 190	
Waste pulp pump	135	
White water pump	095, 100, 110, 120	
Power Stations		
Accumulator pump	(*)	

Pump designation	Category Number	Prime consideration
Boiler circulation pump	095, 100, 105, 110, 125	high suction pressure
Boiler feed water pump	090, 165, 170, 270	water purity, temperature
Condensate pump	060, 085, 095, 175	
Hot-well pump	055, 060, 175	
Pump/turbine	(*)	
Reactor pump	120, 125	
Return condensate pump	060, 085, 095, 175	
Water storage pump	(*)	

(*) Very large machines which run as pumps or 'power recovery turbines' coupled to motors which also function as generators. Built by turbo-generator manufacturers rather than pump manufacturers.

Sewage Treatment Plant (including sewage transport)		
Decantation pump	180, 185	
Free flow pump	180, 185, 190	
Non-clogging pump	180, 185, 190, 305	
Mud pump	180, 185, 190, 270, 275	
Sand pump	140, 145, 185, 295	
Self-cleaning pump	180, 185, 190	
Sludge pumps raw sludge active sludge chemical sludge digested sludge	180, 185, 190, 270, 275 180, 185, 190, 270, 275 180, 185, 190, 295 180, 185, 190, 295 235, 270, 275	
Storm water pump	180, 185, 190, 195, 200, 315	
Submersible pump	185	
Vortex pump	180, 185, 190	
Waste water pump	180, 185, 190, 270, 275	

Shipping and Shipbuilding Industries		
Ballast pump	045	
Bilge pump	045, 275	Q, H, viscosity
Bilge and fire pump	045	
Butterworth pump	045	
Deck flushing pump	045	
Dock pump	200	
Fresh water pump	045	
Oil cargo pump	045, 275	Q, H
Pitching pump	200	
Post-bilge pump	275	
Residual oil pump	275	
Salt water pump	045, 055, 085, 095, 100, 165, 175, 195, 200	Q, H
Salvage pump	045, 050, 065, 195, 200	Q, H
Sanitary pump	180	
Slinger pump	200	
Trimmer pump	045	

Sugar Industry		
Beet pump	190	
Beet pulp pump	135	
Beet root pump	190	
Chip pump	135	
Circulation pump for raw juice	095, 100, 105	
Clear juice pump	095, 100, 105, 220, 225, 230, 235, 240, 245, 250	
Dense sludge pump	255, 270, 275, 280	
Digesting pulp pump	135, 220, 225, 230, 235, 240, 245	
Filter sludge pump	095, 100, 105	
High green syrup	235, 240	
Juice pump	095, 100, 105	
Lime juice pump	095, 100, 105	
Milk of lime pump	095, 100, 105, 140, 145	
Molasses pump	220, 225, 230, 235, 240, 245, 250, 255, 270, 275, 280	

Pump designation	Category Number	Prime consideration
Raw juice pump	095, 100, 105, 220, 225, 230, 235, 240, 245, 250, 255, 270, 275, 280	
Treacle pump	095, 100, 105, 220, 225, 230, 235, 240, 245, 250, 255, 270, 275, 280	
Sludge juice pump	095, 100, 105	
Water Supply Industry		
Automatic water sets	020	
Borehole pump	070, 075, 195, 315	
Clean water pump	005, 010, 015, 020, 025, 055, 060, 195, 200, 235	
Deep-well pump	070, 075, 200	
Excavation pump	005, 195, 200	
Household pump	015, 020, 070, 075, 210, 215, 315	
Raw water pump	005, 010, 070, 075, 095, 100, 105	
Tubular well pump	075, 195	
Water ring pump	010, 195, 200	
Well pump	005, 010, 070, 075, 195, 200, 210	

Figure 14.4 Direct choice table for pumps designated by industry

14.3 Selection of pump according to duty and capabilities

Rapid selection table

Decisive factors for the choice of pump are the properties of the process liquid, the desired pump performance, flow and differential head or pressure, as well as certain specific requirements in connection with the pump installation. Typical technical features have been listed in the rapid choice table, see figure 14.5, for types of pumps with the category numbering. Naturally it has not been feasible to list the thousands of possible combinations of pump performance and liquid properties. In the rapid choice table the critically important properties are taken into account one after another. The suitable main type of pump is obtained when all the features and properties are most suited to the application. In some cases, however, it is necessary to accept a choice of pump which is not ideal with respect to all features. Then the choice must be made entirely on the basis of the requirements of the specific case, i.e. a compromise must be reached between technical features such as wear capacity or resistance to corrosion. The table does not consider economic effects or the difference in characteristics between rotodynamic pumps and positive displacement pumps. Centrifugal pumps are the most suitable types for about 80% of all pump requirements. Positive displacement pumps are commonly mistaken for high pressure, small flow and for viscous liquid applications. Some positive displacement pumps can be used to advantage for arduous low pressure applications; pumping sewage and viscous tank bottom sludge with sediment. We have seen from Section 13.11 that small plunger pumps running at 25% rated pressure can be more efficient than centrifugal pumps.

The following data can be used to supplement the rapid choice table:

- Rate of flow is stated in m^3/h. Conversion from other units is calculated as follows:

  in the case of m^3/s, multiply by 3600

  in the case of l/s, multiply by 3.6

  in the case of l/min, multiply by 0.06

- Differential head, pressure head, in metres liquid column is always used to indicate performance of rotodynamic pumps. Differential pressure, on the other hand, is normally referred to in the case of positive displacement pumps, the units being MPa, kPa or bar. For liquids having

Figure 14.5
Rapid selection table for pumps according to duty and capabilities

Based on 50 Hz pump speeds
Rotodynamic pump flows @ BEP
✳ Geared twin screw pumps
★ Can be hygienic quality
◆ Diaphragm pumps up to 20 barg

● Especially suitable
○ Suitable
▲ Suitable with reservations

Rotodynamic Pumps

Pumps in C.I., C.S. and Bronze for water, mildly corrosive non-hazardous liquids

1 and 2 stage:
- 005 Standard water pumps
- 010 Double suction axially-split pumps
- 015 Heating, water and sanitation pumps
- 020 Automatic water supply packages
- 025 Standard pumps to prEN 733
- 030 Agricultural pumps without driver
- 035 Fixed irrigation pumps
- 040 Machine tool coolant pumps
- 045 Marine pumps
- 050 Electrically driven submersible pumps
- 055 Vertical wet pit pumps
- 060 Vertical dry pit pumps
- 065 Portable self-priming pumps and submersible pumps not electrically driven

Multi-stage:
- 070 Deep well pumps with ejector
- 075 Submersible deep well pumps
- 080 Wash water pump packages
- 085 Multi-stage segmental pumps < 300 m
- 090 Multi-stage segmental pumps > 300 m

Pumps in various materials for food, corrosive and hazardous liquids

1 and 2 stage:
- 095 Standard pumps to ISO 2858/ISO 3069/ISO 3661
- 100 End suction pumps to ASME/ANSI B73.1
- 105 Inline pumps to ASME/ANSI B73.2
- 110 End suction pumps to API 610
- 115 Double-suction pumps to API 610
- 120 Magnetic drive pumps
- 125 Canned motor pumps

Property		005	010	015	020	025	030	035	040	045	050	055	060	065	070	075	080	085	090	095	100	105	110	115	120	125
Max discharge pressure	10 barg	●	●	●	●	●	●	●	●	●	●	●	●	●	●	●	●	○	○	○	●	●	○	●	●	○
	16 barg	●	●	●	●		◆			●	●	●	●			●	●	○	○	●	●	○	●	○	●	●
	40 barg			●							●					●		●	●				●	●	●	●
	50 barg																		●				●	●	●	●
	100 barg																		●				●		●	●
	150 barg																		●							●
	240 barg																		●							●
	430 barg																		●							●
	> 430 barg																		●							
Differential head	< 30 m	●	●	●	●	●	●	●	●	●	●	●	●	●	○						●	●	●		●	●
	30 to 100 m	●	●	●	●	●	●	●		●	●	●	●	●		●	●	●	●	●	●	●	●	●	●	●
	100 to 300 m	●	●	●	●		◆	●			●	●	●	●	▲	●		●	●	●	●	●	●	●	●	●
	300 to 1000 m										●					●		●	●				●	●	●	●
	1000 to 3000 m																		●							
	> 3000 m																		●							
Flow	< 1 m³/h			●			◆		●		●				●										●	
	1 to 10 m³/h	●		●	●		◆		●		●	●	●	●			●			●	●	●			●	
	10 to 100 m³/h	●	●	●	●	●	●	●		●	●	●	●	●		●	●	●		●	●	●	●		●	●
	100 to 1000 m³/h	●	●	●	●	●		●		●	●	●	●	●				●		●	●	●	●	●	●	●
	1000 to 10000 m³/h	●	●					●			●	●												●		
	> 10000 m³/h		●									●	●													
Temperature ≤ 120 °C		●	●	●	●	●		●	●		●	●	●	●		●	●	●	●	●	●	●	●	●	●	●
Temperature 120 to 250 °C		▲	▲			●											●	●		●	●	●	●	●	●	●
Temperature > 250 °C																				●	●	●	●	●		●
Non corrosive		●	●	●	●	●	●	●		●	●	●	●	●	●	●	●	●	●	○	○	○	●	●	●	●
Corrosive		▲		▲		▲	▲				●	●	●	▲		●		▲	●						●	●
Hazardous					▲	▲				●		●	●						▲						●	●
Entrained gas ≤ 15%		▲	▲	▲	▲	▲	▲			▲	▲	▲		●		▲		▲	▲	▲	▲	▲	▲	▲	▲	
Entrained gas > 15%																										
Size	≤ 0.1 mm	▲	▲	▲	○	▲	●	▲		▲	●	●	●	▲	▲	●		▲	▲	▲	▲	▲	▲	▲	▲	○
	0.1 mm to 1 mm						●	▲			●	●	▲	●									▲	▲		○
	1 mm to 10 mm							▲			●	●	▲	●												○
	> 10 mm										●			○												○
Nature	Deformable	▲	▲	▲		▲	●	▲		▲	▲	▲	▲		▲	▲		▲	▲	▲	▲	▲	▲	▲	▲	○
	Friable	▲	▲	▲		▲	▲	▲		▲	▲	▲	▲		▲	▲		▲	▲	▲	▲	▲	▲	▲	▲	○
	Abrasive				○			▲			●	●	●	●		●								▲	▲	
Conc	< 1% by wt	▲	▲	▲	○	▲	●	▲		▲	●	●	●	▲	▲	●		▲	▲	▲	▲	▲	▲	▲		○
	1 to 30% by wt							▲			●	●		○												
	> 30% by wt																									
Thin	< 10 cSt	●	●	●	●	●	●	●	●	●	●	●	●	●	●	●	●	●	●	●	●	●	●	●	●	●
	10 to 300 cSt	●	●	●		●	▲	●		●	●	●	●	●		●		●	●	●	●	●	●	●	●	●
Thick	300 to 1000 cSt	▲	▲	▲		▲				▲	▲	●	●	○		▲		▲	▲	▲	▲	▲	▲	▲	○	○
	1000 to 10000 cSt											●	●												○	
Paste	> 10000 cSt																								○	
Delicate handling/low shear																										

a density similar to that of water, 1000 kg/m³, the following conversions apply:

1 MPa ≈ 100 m

1 kPa ≈ 0.1 m

1 bar ≈ 10 m

1 kgf/cm² ≈ 10 m

NOTE: The bar and kgf/cm² are the most useful units in process applications. The MPa is too big, the kPa is too small. Pump manufacturers think it unwise to have suction gauges calibrated in kPa and discharge gauges in MPa. The pump industry has approached ISO with the request that the bar be retained permanently.

- Liquid properties such as viscosity, resistance to chemical attack, variations in the rate of the chemical reaction with temperature and contamination are especially important not only for choosing the type of pump but also for the specifications intended to serve as a basis for purchasing. Some chemicals will not react at one temperature but will cause substantial damage if the temperature rises above a certain point. Variations in pH can make what is sometimes a harmless solution into a highly dangerous and corrosive solution. Some pumps may be resistant to chemical compounds over a short period but may be attacked over a long period. This factor is important in cleaning, where a corrosive chemical may be used for a short period to remove foreign matter but would damage the pump if it was left in contact for a long period.

- The various grades of suitability are based on the pump type's ability to operate under the specified conditions. When more than one pump type appears suitable, check the efficiency and whether an identical pump will be used elsewhere on the same site. Purchasing pumps from one manufacturer may limit pump choice. Penalties for lower than possible efficiency must be evaluated.

- Suitability symbols are not shown in some places in the rapid choice table. This means that the type of pump in question is altogether technically or economically unsuitable.

When using figure 14.5 one should start with the most important parameter; then the second, then the third. Make a list of all the categories fulfilling the parameters. Remember the difference between rotodynamic and positive displacement pump characteristics, delete those not suitable. Evaluate the additional benefits which could be utilised from the remaining categories. Review the relevant categories in Chapter 3. Efficiency may be the final test.

14.4 Selection of pump according to hydraulic performance

Guidance is given for the rapid selection of pump types for various operating conditions. When considering viscosity, the limitations given for any pump design are complicated by pump size. Small pumps have higher parasitic losses, lower efficiency, due to friction losses in the small passages. As pump size increases these losses become smaller and less important. Increased viscosity, in rotodynamic pumps, increases the parasitic losses and reduces pump performance and efficiency. As a general rule, a viscosity of 300 cSt is a good bench-mark for evaluating viscosity effects. The debilitating effect of viscosity on rotodynamic pumps can be mitigated by running pumps at reduced speed. In the following sections it is important to remember that viscosity limitations are not 'cast-in-stone' but indicate areas of concern where the transition from low viscosity to viscous and viscous to highly viscous are blurred.

14.4.1 Pumps for low viscosity liquids

Suitable ranges of operation for flows and differential heads can be indicated on the basis of popular rotodynamic pump availability for different types of pumps for low viscosity liquids according to figure 14.6.

Double suction axially-split pumps and axial propeller pumps can have flows over 50000 m³/h. Figure 14.7 indicates the flow and discharge pressure capabilities of positive displacement pumps. In some cases it is necessary for the liquid to have certain lubrication properties.

Plunger pumps are shown with a maximum pressure of 2000 barg; pumps are available for pressures up to 10000 barg. Very few applications use pressures over 1000 barg. Performance requirements, purchasing cost, running costs, estimated service life and reliability of operation must be considered when choosing pumps. Rotodynamic pumps of various designs cover nearly 80% of all pump requirements. However, looking at figures 14.6 and 14.7, for flows from 1 to 100 m³/h, and pressures up to 200 m or 20 barg, there are plenty of pump types to choose from.

In some instances the choice of pump is very simple, based on the duty of a single pump:

- flow over 1860 m³/h ∴ rotodynamic
- flow less than 0.001 m³/h ∴ positive displacement
- differential head over 4500 m ∴ plunger pump

In some special applications there are dedicated designs which almost without exception are the most suitable. Some examples of these are heating, water and sanitation pumps.

In certain borderline cases for rotodynamic pumps it is possible to choose between single-stage and multi-stage pumps. An analysis of the efficiency and running costs on the one hand and of operational reliability on the other will provide sufficient grounds for a decision in such cases.

The speed of the pump is a factor which is often discussed. In principle the choice should be in favour of a speed sufficiently high to ensure that the NPSHa in the plant, less a suitable margin, is utilised. Possible fears of a reduction in operational reliability in the case of higher speeds should not be allowed to result in the choice of a lower speed, but rather in better quality and testing requirements when purchasing. Requirements of this kind can refer to critical speeds or bearing life or sealing problems, for example. It should, however, be observed that a higher speed may result in higher noise levels requiring acoustic treatment.

In the borderline cases between rotodynamic and positive displacement pumps the form of the pump characteristics and method of regulation often influence the choice. Both types of pump have their distinct advantages, see Sections 3.2 to 3.5 with regard to different pumps designs and characteristics.

A difficult problem for all pumps is the presence of air or gas in the liquid. The effect on the effective vapour pressure must not be overlooked. Rotodynamic pumps suffer from a reduction of efficiency when the gas content on the suction side is in excess of about 1% by volume. Some rotodynamic pumps, centrifugals for example, can deal with up to 15% without losing prime. Some positive displacement pumps, particularly geared twin screw, can cope with much larger volumes. Dry running must be specified as an operating condition when required. The ability of the pump to handle slugs of gas or liquid must be evaluated.

14.4.2 Pumps for viscous liquids

The viscosity of the liquid can greatly affect the choice of pump. The effect is dependent on the flow and differential head and therefore the size of the pump. The effect of viscosity on the efficiency of centrifugal pumps is shown in figure 14.8.

Figure 14.6 Operating ranges of popular rotodynamic pump types
(Pump category number bold in brackets)

(1) Single stage pumps to ISO 2858 (**095**)
(2) High speed single stage pumps & multi-pump packages (**155, 160**)
(3) Multi-stage radially-split pumps (**170**)
(4) Multi-stage axially-split pumps (**165**)
(5) Double suction pumps to API 610 (**115**)
(6) Double suction axially-split pumps (**010**)
(7) Axial propeller pumps (**200**)
(8) Electrically driven submersible pumps (**050**)
(9) Standard pumps to prEN 733 (**025**)
(10) End suction pumps to API 610 (**110**)
(11) Multi-stage segmental pumps (**085, 090**)

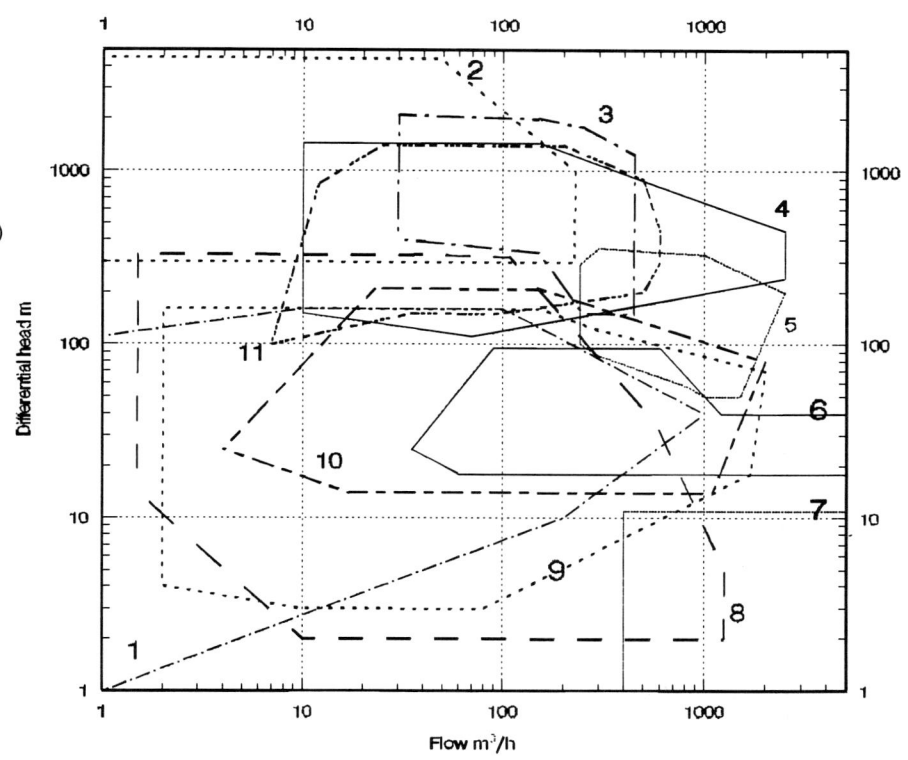

Figure 14.7 Operating ranges of positive displacement pump types
(Pump category number bold in brackets)

(1) Peristaltic pumps (**250, 255**)
(2) Metering pumps (**285**)
(3) Plunger pumps (**270**)
(4) Screw pumps, 2 geared screws (**230**)
(5) Screw pumps, 3 screws (**230**)
(6) Diaphragm pumps, air operated (**280**)
(7) Piston pumps (**275**)
(8) Lobe and rotary piston pumps (**240**)
(9) Progressing cavity pumps (**235**)
(10) Gear pumps (**220, 225**)

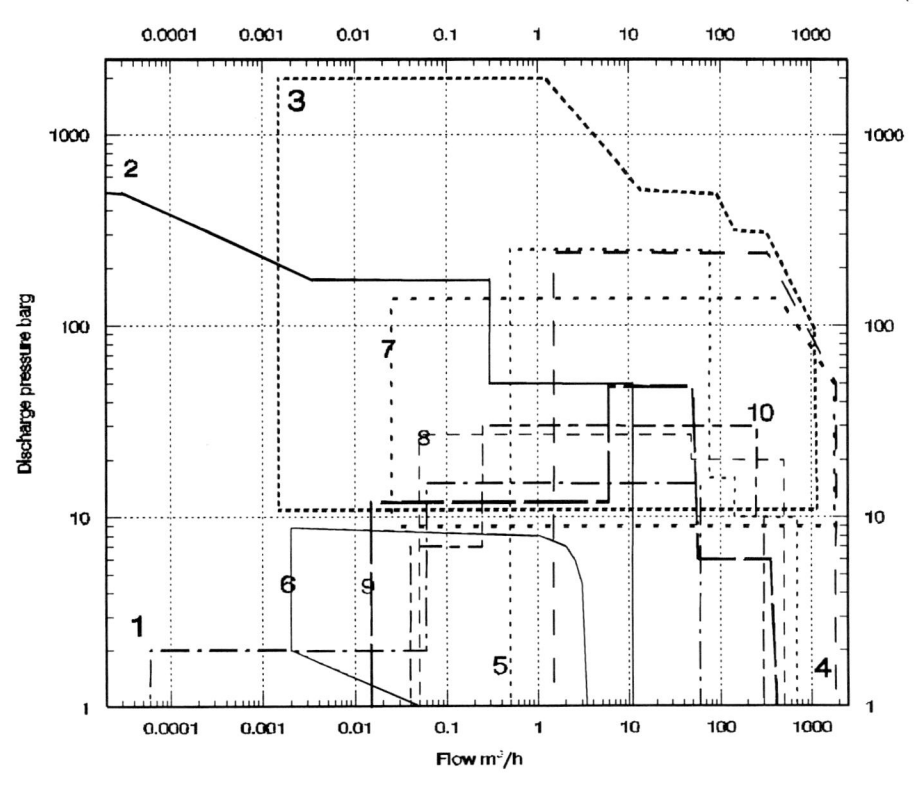

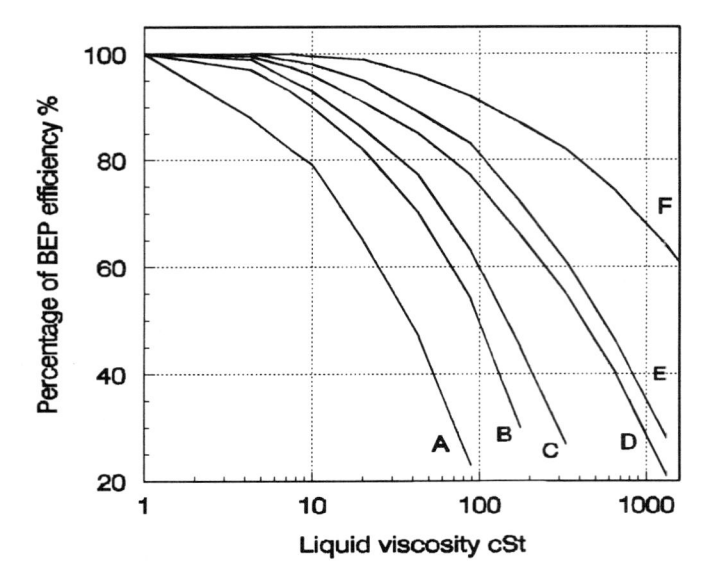

(A) BEP duty 2 m³/h at 11 m differential
(B) BEP duty 11 m³/h at 30 m differential
(C) BEP duty 25 m³/h at 30 m differential
(D) BEP duty 110 m³/h at 45 m differential
(E) BEP duty 225 m³/h at 45 m differential
(F) BEP duty 2500 m³/h at 185 m differential

Figure 14.8 The viscous effects on efficiency of centrifugal pumps

The five pumps from A to E are taken from the range outlined by ISO 2858 for chemical pumps; F indicates the effects on a very large pump. This data was prepared by using the viscosity correction graphs shown in figures 3.31 and 3.32. Reduced efficiency leads to increased power consumption which results in higher shaft stresses. It may be the pump shaft stressing which limits viscous operation.

In rotary positive displacement pumps the increase in viscosity reduces slip and improves volumetric efficiency, hence greater flow per revolution. But increased viscosity can also lead to increased fluid friction losses. In triple screw pumps for example, the differential pressure is limited by the viscosity. On one range of pumps, used mainly for lubricating oil duties, the differential pressure capabilities vary from 21/40 bar at 2 cSt to 160 bar at 40 to 200 cSt. The volumetric efficiency improves by about 10% which leads to a slight improvement in mechanical efficiency. Minimum inlet pressure will tend to increase with viscosity.

Depending upon the valve design, reciprocating pumps are not affected very much by medium viscosity operation. An increase in NPIPr will be apparent.

Intelligent economic pump selections for viscous liquid operations should be based on life-cycle costing. We know that energy consumption is likely to be more important than initial pump cost. Review pump efficiencies at operating conditions to assess actual energy consumptions.

14.4.3 Pumps for highly viscous liquids

As we have seen the problems caused by viscosity are related to pump size. In general, for high viscosities only positive displacement pumps are suitable. The choice is then to select the most suitable type of positive displacement pump.

At high viscosities the NPIPr and MIP will increase. The speed of the positive displacement pump may need to be reduced. When choosing pumps for highly viscous liquids the design of the suction system must be studied in great detail. NPIPa and MIP must evaluated very carefully. If the liquid cannot be supplied at the correct flow rate the pump will never function correctly. Severe problems may be experienced on the discharge side of the pump. Large pumps with large connections and low speeds are to be preferred. For extremely viscous liquids, and solids such as wax or lard, feed hoppers and augers may be necessary. Those positive displacement pumps which can handle the maximum pumpable viscosities, approximately 10000 Poise or Stoke, are progressing cavity pumps, lobe and rotary piston pumps and reciprocating plunger/piston pumps. Figure 14.9 indicates the recommended speed of gear pumps for viscous operation.

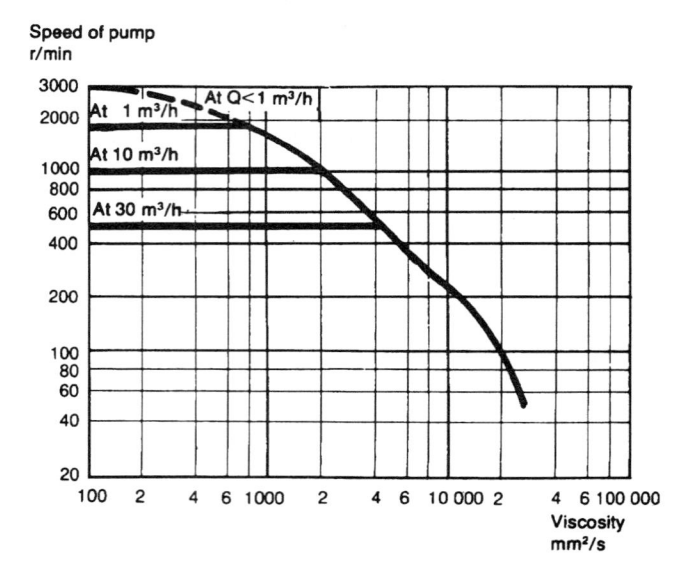

Figure 14.9 Speed of gear pumps for viscous liquids

Again, ultimate pump choice should be made based on energy consumption and life-cycle costs.

14.5 Pumps for liquid solid mixtures

When pumping solids the distinction must be made between abrasive and non-abrasive solids. The Miller Number test is used for reciprocating pumps. Values below 50 are considered non-abrasive. Review figure 1.25 for the effects of very low concentrations of solids. Figure 1.27 indicates some typical Miller Number values while figure 1.26 shows the solid size capabilities of pumps. It is possible to pump solids concentrations up to 70% by weight.

The Miller Number test is conducted using a high chrome cast iron test block. Most pump applications will use a different material. Stauffer, of Escher Wyss, conducted metal tests with a sand-water mixture. The bench-mark material was a case hardened carbon steel. Some interesting data, adjusted for 27 Cr iron to be unity is shown in figure 14.10.

Material	Relative abrasion resistance
Hard chrome plate	2.06 / 2.28
Tungsten carbide	1.39 / 4.12
Stellite 6	0.83 / 3.31
Ni-Hard	0.89 / 1.11
27 Cr cast iron	1
13 Cr steel, 441 BHN	0.32
316L stainless steel	0.26
Carbon steel, 195 BHN	0.22
Al bronze	0.13
Ni Al bronze	0.12
Cast iron	0.09

Figure 14.10 Relative abrasion resistance to sand in water

There is no consistent method of selecting pumps for solids-handling applications. This is because all the pump types which are used rely on different techniques and the manufacturers adopt different philosophies for derating and viscous corrections depending upon their experience. Centrifugal pumps designed specifically to handle solids react differently to centrifugal pumps designed for clean liquids. This problem is due to the different design methods used. The rheology of the mixture can be a problem; assuming Newtonian behaviour can lead to poor pump selections. Rheology has very serious implications for the pipeline design and predicting the pressure drop. However, rheology is not a problem to all pump types. Plunger and piston pumps can have the plunger/piston diameter optimised, at site after commissioning, without too much trouble. The purchaser would have to admit to being uncertain about the Q-H characteristic. The solids concentration can be adjusted to modify the mixture rheology. Chemical additives are also available for the same purpose.

Efficiency and predicted life of wearing parts are the essentials for life-cycle costing. The manufacturer should be able to advise the life of any rapidly wearing parts and the Mean Time Between Overhauls. The time taken to replace rapidly wearing parts is also very important.

14.5.1 Pumping non-abrasive solids

Already, at solids concentrations of 0.5%, the service life of a pump can be severely affected. For light particles, e.g. paper pulp, refer to Sections 2.3.4, 2.4.4 and 3.2.4 on pulp pumps. Liquid-solid mixtures which can be considered as non-abrasive are:

- aragonite (calcium carbonate)
- bauxite
- chalk
- clay
- coal
- gilsonite (natural asphalt)
- gypsum (selenite)
- lignite
- limestone
- mud, drilling
- potash
- rutile (titanium dioxide)
- sewage

The critical element of pump selection is the size of the particles and whether any stringy or fibrous material will be present.

One particular application of non-abrasive pumping that may have immense possibilities, is liquid coal fuels. Crushed coal is mixed with water and some trace chemicals to produce a liquid-solid mixture which can be handled by piston pumps. The mixture acts as an emulsion and can be stored for reasonable periods without separation. The liquid coal fuel is pumped from the mine direct to the boiler, usually at a power station to produce electricity, and is injected into the boiler. The solid-liquid mixture is not de-watered. The liquid-solid mixture is a complete fuel. Large scale testing commenced in 1989. Boilers were operated on liquid coal fuel and produced 650 tonne/hour of steam. Combustion was stable and complete. It is believed that the testing was successfully concluded and production commenced. However, the coal mine has since closed and the pump operations have ceased.

14.5.2 Pumping abrasive solids

Liquid-solid mixtures which can be considered as abrasive are:

- alundum (Al_2O_3)
- copper concentrate
- fly ash
- iron ore
- limonite (Fe_2O_3)
- microsphorite
- magnetite
- phosphate
- pyrites (FeS_2, $CuFeS_2$)
- sand
- shale
- serpentine (magnesium silicates)
- tailings
- tar sand

Again the particle size and concentration is critical in pump selection.

Some users have manufactured their own jet pumps. The basic jet pump principle has been adapted in various ways to reduce the effects of wear from the solids. Mine operators seem to be the most versatile in this respect.

14.5.3 "Gentle" pumping

"Gentle" pumping can be redefined as low shear or delicate handling pumping. The process liquid is not subject to surfaces which travel at high velocities. "Gentle" pumping is used when the solids to be transported by the liquid are to be damaged as little as possible or the liquids are not to be emulsified. For this purpose the "goods" can be anything from delicate bacteria cultures to foodstuffs and root crops. The basis for a systematic choice of pump is seldom available. In the case of rotodynamic pumps, the differential head per stage is a measurement of the stirring of the liquid. Front and back shrouds of an impeller create shear in an almost stationary liquid. In order to avoid damage to the transported particles the passages through the pump should be large. This means that among rotodynamic pumps the categories **180**, **185** and **190** should be used, and among positive displacement pumps in the first instance lobe and rotary piston pumps are of special interest. For higher pressures plunger and piston pumps can be used.

Figure 14.11 shows a comparison between damage to goods caused by two different types of non-clogging pump. The figure demonstrates that in this case the channel impeller type pump is superior to the free-flow pump. On the other hand if bacteria cultures are pumped as active sludge in a purification plant other test results indicate that the free-flow pump is

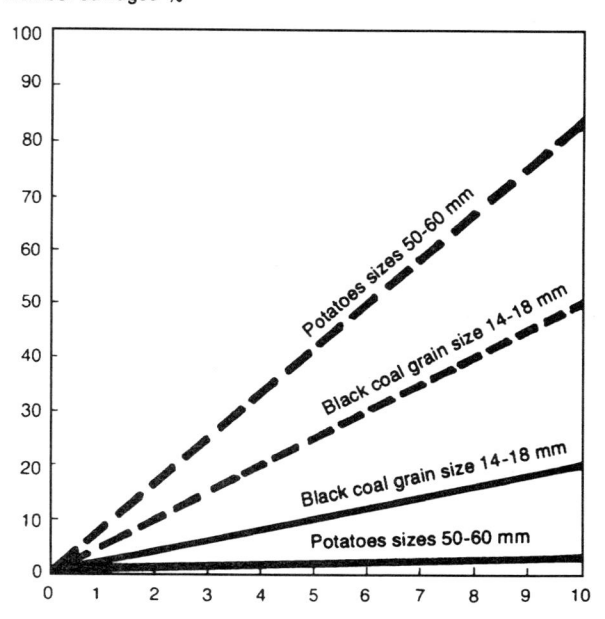

Figure 14.11 Results of pumping tests with a free-flow pump - - - - and a channel impeller pump ———. For black coal the damage criterion was when the size of grain was reduced to less than 12 mm

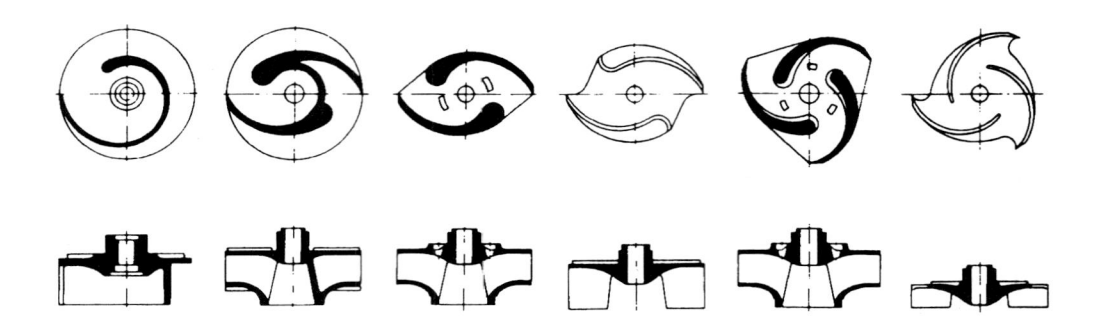

Figure 14.12 Various designs of channel impeller

clearly superior to the channel impeller type. These examples illustrate the difficulties in a general assessment of the "gentle handling" property.

14.5.4 Pumping waste water, sewage

Low pressure vertical plunger pumps

The properties of the pump must be formulated in terms of the waste water to be pumped. An up-to-date waste water pump must be designed to pump, if possible without trouble, waste water of the consistency found today and likely to be found during the next 10 years. It is possible to grade the water to some extent. The extreme cases range from very rural areas to big city centres. Whatever the scope of a definition of waste water, the following requirements apply:

- Minimum risk of blockage (non-clogging)
- High degree of operational reliability (mechanical)
- Low running costs (energy consumption and service)
- Simple, quick maintenance in the event of breakdown

In recent years the definition self-cleaning, non-clogging, has been used less and less to describe the ideal pump. This was a pump equipped with a through-flow impeller of channel type. Previously this type of pump could be considered, with regard to the waste water of that time, to have satisfied the special requirements for which it was designed. However, the waste water produced today and still more likely, that to be produced in the future, make it increasingly difficult for pumps with through-flow impellers, channel impellers, to meet the non-clogging requirement.

The dimensions and design of the channels in relation to the type of contaminant found in the liquid are of critical importance to the non-clogging capability of the pump. The design of the cross-sectional area of a channel impeller varies from rectangular to circular. It is not known for certain whether the design of the channel has any real effect on the non-clogging properties of the impeller. A circular through-flow area may be assumed to offer superior characteristics for the pumping of spherical objects than an impeller with a rectangular cross-section. In the case of textile contaminants a rectangular design may be preferable. Available statistics concerning breakdowns due to the blockage of channel impellers give little guidance to the assessment of different designs of impeller, see figure 14.12.

Impellers for waste water pumps are usually designed with one or two channels. Apart from reasons connected with the hydraulic design, which can dictate different numbers of channels, there is also a lack of reliable assessment of the respective through-flow capacities of one and two channel impellers. In general it may be said that two outlets from the impeller are better than one, provided that the accessible area of each channel is of an acceptable size. The acceptable size of a channel area is of course dependent on the size and character of the contaminants in the waste water. Certain guide values have established themselves in purchase specifications.

Often the area is indicated as corresponding to a spherical through-flow of 75 mm for small pumps and of 100 or 125 mm for larger pumps. In some cases there is a requirement that the through-flow area should be as large as that of the discharge pipe. In such cases it may be advantageous to have suction and discharge connections the same size. The specification of these values probably has its basis in the fact that the magnitude of the H-Q curve in relation to the motor speeds in current use does not allow scope for larger impellers.

The speed of the impeller has a definitive effect on the flow of the through-flow impeller and thereby on the size of the through-flow area. With the aid of the laws of affinity, it can be demonstrated that a given pump at a certain speed will, at double the speed, double its flow, the delivery head will increase four times while the power requirement is increased eight times, neglecting all system effects. For pumps with small flow and moderate pressure the speed must be determined on the basis of the minimum permissible through-flow area. The speed greatly influences the dimensioning of the through-flow area of a through-flow impeller. On the other hand the speed in itself is without significance with regard to ability of the pump to handle contaminants.

Thanks to the hydraulic design of the interior passages, the flow through a free-flow pump, see figure 3.84, is not dependent on speed. The only function of the impeller is to supply energy to the liquid and it can be dimensioned without regard to the through-flow area. The low symmetrical impeller creates negligible deflection to the shaft. This results in increased service life for shaft seals and bearings. For normal waste water the free-flow pump may be regarded as non-clogging. The choice of a non-clogging pump represents a compromise between non-clogging properties and energy consumption. On the grounds of experience it may be claimed that free-flow pumps are the best choice up to capacities of about 5 kW and up to about 10 kW at remotely located pumping stations. Travelling costs for cleaning and maintenance greatly affect the optimum. Waste water pumps of the future will probably take the form of submersible high-speed free-flow pumps with capacities of up to 30 kW. In the case of larger pumps, through-flow types will probably continue to dominate the field.

For duties up to 45 m on systems with a large H_{stat} component, a low pressure vertical plunger pump may be better. These pumps run at low speeds and combine "gentle" pumping with solids handling. Built for quick, simple maintenance, the low speed ensures long operating cycles. Positive displacement principles means easy, accurate flow control.

For waste water pumps the manner of installation deserves attention. Submersible pumps are easy to install and replace for maintenance. Dry pit pumps are usual in the case of high power capacities and generally operate at a higher efficiency. Their speed can be easily regulated and they can be more reliably supervised. They are, however, more costly to install. Due to maintenance problems and industrial safety requirements, the current trend is away from submersible to dry pit pumps.

14.6 Check list for pump purchase specification

Many problems and unnecessary costs can be avoided if the following check lists are used. Irrespective of the method of purchase, over-the-counter or by enquiry, the following data should be collated and reviewed by all interested parties. Although more than 100 check items are included under more than 15 main groups, they cannot be regarded as entirely exhaustive, because in certain special fields attention must be paid to some additional factors. Above all, check lists should be used when planning and purchasing pumping equipment for unusual, critical or difficult liquids. For normal applications in a particular branch of industry, e.g. a dairy, or for a certain type of pump, e.g. submersible sewage pumps, a special check list or specification can easily be drawn up. Terms described as 'rated' relate to the pump's normal operating conditions; the conditions to be reproduced during testing. Items designated with (pd) apply particularly to positive displacement pumps. It must be remembered in the context of constant speed positive displacement pumps that there can only be one flow, rated flow. If maximum and minimum flows are specified the method of achieving the flow range must also be specified.

The data collated must be accurate. When accuracy is not possible, the data presented must be suitably qualified. There must be no ambiguity. The data collected applies to the pump, not the system. Some system data may be necessary, H-Q characteristic for variable flow systems in some cases, but this must be described as 'system data'. It is important that the data does not require interpretation. A pump manufacturer or distributor must not be required to 'read-between-the-lines' to extract important information.

When users or contractors work on large installations some decisions taken early in the project are forgotten. All relevant data should be reviewed when preparing the final pump unit specification.

Liquid properties
See figure 6.31

Solids properties
See figure 6.31

Pump capacity
Suction head, pressure; minimum, rated, maximum
Suction temperature; minimum, rated, maximum
Rate of change of suction temperature
Flow; minimum, rated, maximum
Flow duration; constancy
Differential head; at minimum, rated, maximum flow
Suction and discharge pressure; at minimum, rated, maximum flow (pd)
Suction and discharge conditions at starting
Differential head for filling pipe system with consideration to possible syphon effects
Differential head affected by possible starting inertia of liquid
Discharge pressures for relief valve operation (pd)
Pump characteristic; rotodynamic or positive displacement
NPSHa, NPIPa; minimum, rated, maximum
Suction system; self priming, evacuation time
Flow regulation; throttling,
    by-pass,
    on-off,
    load-unload,
    adjustable guide vanes,
    infinitely variable speed,
    discrete speed steps,
    manual or automatic control.
    Require system characteristic !
    Possibility of liquid reverse flow
    Also review figure 6.31

Pump, mechanical requirements
Shaft deflection at seals at rated conditions
Service life of bearings at rated conditions
Suction pressure rating
Discharge pressure rating
Suction hydrotest pressure
Discharge hydrotest pressure
Suction connection allowable forces and moments
Discharge connection allowable forces and moments
Thermal growth of pump
Mean time between failures
Mean time between overhauls
Starting torque requirements
Permissible number of starts per hour
Permissible number of starts per year
Internal clearances
Changes in efficiency resulting from clearance variations
Dry running capabilities
Heat tracing or cooling
Warm-up time or cool down time
Hot stand-by
Extent of inspection without removing pipework
Extent of maintenance without removing from pipework
Access for cleaning out residue, if required
Effect of pump running backwards
Balancing
Vibration
Noise level
Possibility of modifying pump for duty changes; increased impeller, plunger, piston diameter, de-staging

Shaft seal
See figures 7.16, 7.32, 7.33 and 7.42

Material
See figure 6.31

Pipes and pipe connections
Type of connection; screwed, flanged, special
Special surface finish on flanges
Pressure ratings
Pipe movement and thermal strain

Flexible pipe connections
Pipe spools adjacent to pump
Suction pipe routing
Location of pump process connections
Minimum distance from pump to bends or valves
Location of pump auxiliary service connections
Isolating valves
Non-return valves
Relief valve (pd)
Tapping points for measuring pressure and/or temperature
Possibility of water hammer during normal start
Possibility of water hammer during normal stop
Possibility of water hammer if power failure
Allowable axial shaking forces for pressure pulsations (pd)
Insulation against heat and/or cold
Heat tracing
Design and facilities for mechanical cleaning or inspection by 'pig'
Will pump manufacturer review and advise on pipework (pd)
Acoustic and mechanical analysis (pd)

Driver

Electric motor: type,
speed(s),
AC / DC,
speed regulation,
direction of rotation,
voltage, phases, frequency,
method of starting,
starting torque characteristics,
number of starts per hour,
area classification,
certification,
type of physical protection,
cooling system,
power, minimum, rated, maximum,
efficiency and power factor,
bearings and lubrication,
space heaters,
location of terminal box,
electrical protection/monitoring equipment,
thermal growth,
balancing,
vibration,
noise level.
See Chapter 9.

Engine: speed,
speed/torque characteristic,
naturally aspirated, turbo-charged,
direction of rotation,
governor,
fuel, calorific value,
fuel consumption,
efficiency,
method of staring,
cooling system,
time to warm-up,
controls and instruments,
auxiliary equipment,
waste heat boiler,
hazardous area modifications or mount in safe area,
thermal growth,
balancing,
vibration,
noise level.
See sections 5.9.1.2 and the end of Chapter 9.

Turbine: speed,
speed/torque characteristic,
direction of rotation,
condensing/pass-out,
steam conditions,
governor,
over-speed trip,
steam consumption,
efficiency,
bearings and lubrication,
barring system,
time to warm-up,
steam valves,
controls and instrumentation,
thermal growth,
balancing,
vibration,
noise level.
See sections 5.9.1.2 and the end of Chapter 9.

Drive train

Couplings: service factor,
maximum speed,
spacer length,
long spacer for fire wall,
capacity for misalignment,
mounting,
balancing,
lubrication.
See Chapter 8.

Gearbox: service factor,
service life of gears and bearings at rated conditions,
exact ratio,
efficiency,
cooling,
bearings and lubrication,
minimum speed,
thermal growth,
balancing,
vibration,
noise level.
See section 10.3

Vee-belts: service factor,
ratio,
efficiency,
cooling,
pulley balancing,
noise level.
See section 10.2

Auxiliary equipment
Baseplate, skid, trailer; see section 10.1
Relief valve; see section 10.4 (pd)
Accumulators; see section 10.5
Pulsation dampeners; see section 10.6 (pd)
Instrumentation; see section 10.7

Environment
Stand-by facilities
Electricity supplies
Auxiliary services supplies
Industrial safety standards
Overall pump package noise level
Allowable transmitted vibration level
Leakage in the form of liquid
Leakage in the form of vapour
Safety with respect to hazardous gasses
Radiation of heat
External condensation on cold surfaces, possible formation of ice on pump
Permissible gas concentration in premises (ppm) and necessary ventilation
Atmospheric pollution
Biological attack
Altitude
Weather variations
Indoor, outdoor, onshore, offshore, sheltered, sunshade, attended, unattended

Erection and installation
Horizontal or vertical shaft
Submerged pump
Space and transport routes for erection and maintenance
Lifting facilities for installation
Lifting facilities for maintenance
Foundation; concrete, structural steelwork
Dynamic loads on foundation
Static loads imposed by pipework
Space for future developments
Drainage for safe leakage
Drainage for hazardous leakage
Drainage of the pump contents
Drainage of lubricants and cooling liquid
Vents for hazardous vapour
Vents for steam

Maintenance
Parts lists and cross-sectional drawings
Lubrication schedule
Routine maintenance instructions
Repair instructions
Trouble-shooting guide for all equipment
List of rapidly wearing parts with predicted life
Quotation for commissioning spares
Quotation for 1 years operations spares
Quotation for 2 years operations spares
Quotation for insurance spares
Quotation for special tools
Pump curves
Locations of measuring points for preventive maintenance (vibration, temperature, electric current, etc.)
Hours-run meter
Staff training
Location of manufacturer's nearest service centre
Location of manufacturer's nearest spares stockist

NOTE: Manufacturers will not supply drawings of spare parts. Drawings contain proprietary information which is confidential. Using spare parts manufactured by others will invalidate any guarantee or warranty. The pump manufacturer has no responsibility for the damage caused or production lost due to pirate parts.

Quality Assurance
Pumps designed and built to a standard
Pumps manufactured to a standard
Fully assembled pump subjected to a running test
Fully assembled pump subjected to a pressure test
Typical batch size of mass produced pumps
Number of pumps in a batch tested
Tolerance band on published pump performance data
Fully assembled pump unit tested
Which standard is used for testing?
What quality assurance documentation supplied?

Purchase conditions
Quotation validity
Length of guarantee for workmanship and materials
Delivery time
Point of delivery
Price
Currency
Drawings
Manuals
Certification
Erection at site
Commissioning

NOTE: Manufacturers are unlikely to give a guarantee regarding the performance of the pump at site or how the performance will degrade with time. The pump performance and life are greatly affected by the quality of the installation and particularly by the pipework design. Pump manufacturers are not normally involved in the foundation design or the pipework design. Bearing life can be shortened dramatically by poor foundations, misalignment and pipe strains. Pump performance can be severely impaired by poor pipework design. The pump manufacturer shows how the pump can operate by testing in the factory. The factory tests should be conducted as close to the site operating conditions as possible. A site simulation test may be possible. Except in the cases of fully packaged equipment such as portable high pressure cleaners, factory testing cannot reproduce site conditions. Problems such as acoustic resonance or pipework vibration can only be detected by computer simulation. The pump manufacturer should be prepared to work at site to analyse and isolate problems. The pump user must be prepared to pay the pump manufacturer for solving problems outside his sphere of control.

14.7 Purchasing

Section 3.7 deals with the criteria and standards which are of prime importance in connection with specification. Chapter 11 deals with the implementation of testing and quality controls. These are concerned, amongst other things, with guarantees of performance and quality. Criteria for service life and reliability are lacking, however. Over and above the technical requirements, suitable general terms of delivery should be drawn up. In the processing industry and at refineries API 610, 674, 675 and 676 are usually applied as requirement criteria as they specify in detail how a pump should be designed, manufactured and tested. Reference may be made where applicable to API standards for guidance for pumps to meet these requirements. However, it must be remembered that an inspector looking at a pump cannot certify that it will be capable of operating without interruption for three years.

The purchase documents should contain references to legislative regulations and requirements affecting design, to the properties of the liquid as well as to operating conditions and function of the pump in order that the manufacturer may be able to determine the necessary material and the design of the pump. There is no point in referencing obscure regulations. Manufacturers do not have easy access to all standards and legislation. If particular, local regulations must be observed, a copy should be supplied with the enquiry.

It is important that the specifications detail the pump operations. Safety margins should be shown separately as such.

Remember the site cost-saving edict:

Trouble-shooting is making the pump do what you specify.

Making the pump do what you want is up-grading !

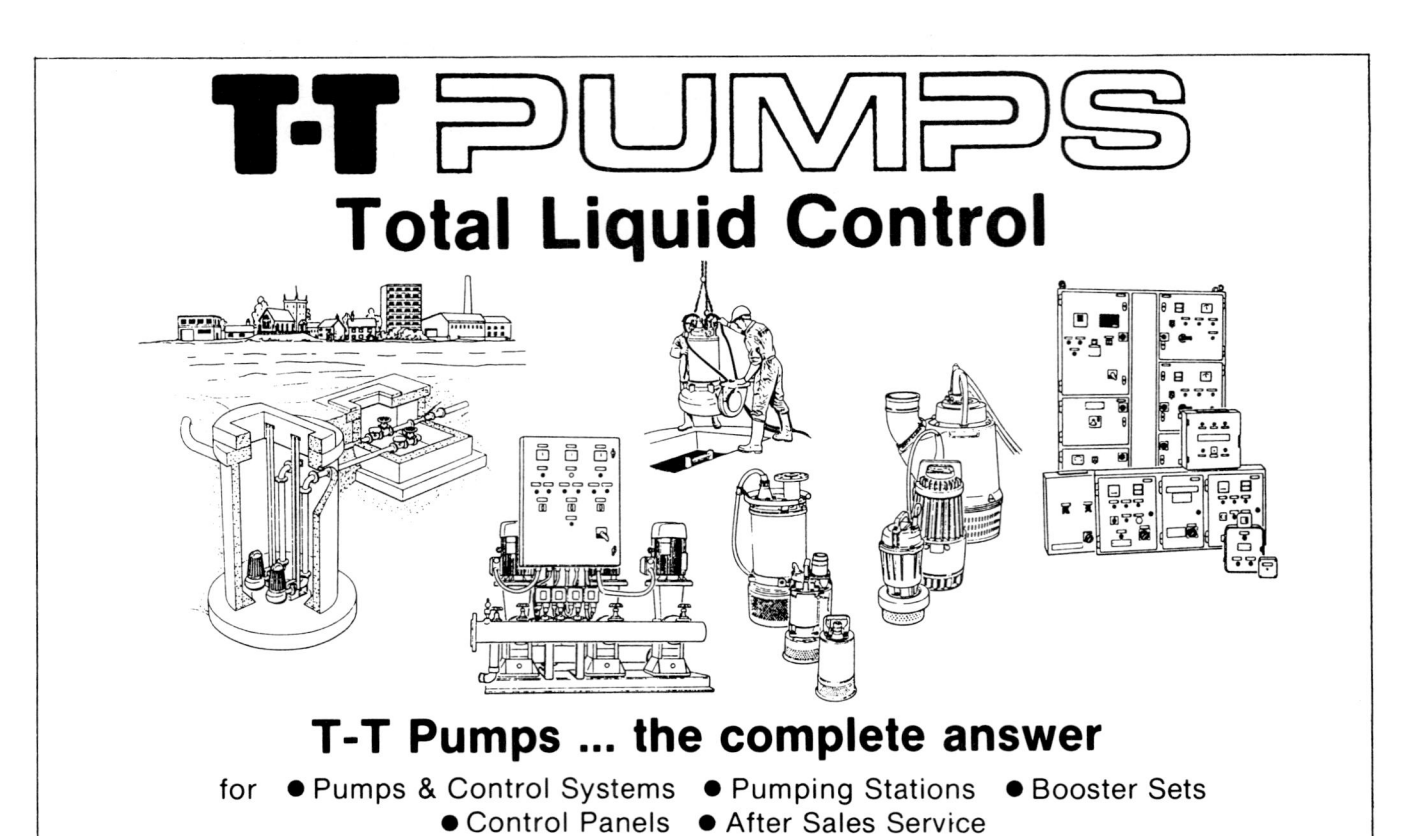

15 Case studies

The case studies compiled in this chapter are intended to illustrate some of the problems encountered in the practical use of pumps. Since pumps are used widely as auxiliary components in various process systems, each individual application is a special case. The various case studies are not therefore universally applicable, but should nonetheless be of interest to both the specialist in the subject as well as to anyone who wishes to form his own views on the practice in a particular field.

Contents

15.1 Pumping fresh water to a high reservoir

15.2 A small sewage pumping station

15.3 A fresh water booster station

15.4 Circulation pump for domestic central heating

15.5 Adjustable jet pump

15.6 Modernisation of water supply for industrial use

15.7 Contractors' use of electric powered submersible sump pumps

15.8 Engine driven self-priming pumps

15.9 Land reclamation pump

15.10 Cargo pump for tankers
    15.10.1 Large tankers for crude oil
    15.10.2 Smaller tankers for finished products
    15.10.3 Tankers for LNG

15.11 Liquid detergent manufacture - pump installation

15.12 Positive displacement pumps - variable discharge conditions

15.13 Energy recovery turbines

15.14 Economic aspects of energy utilisation

15.15 Choosing non-return valves

15.1 Pumping fresh water to a high reservoir

Operating conditions

The clean water collected in a low reservoir, is to be pumped up to a high reservoir. The high and low water levels in the two reservoirs and the altitude of the site for the location of the pump are shown in figure 15.1. The pipe linking the reservoirs is long, about 20 km, and during much of the day water is drawn from the pipe by consumers with consequent loss of pressure.

The daily consumption of water varies considerably. Furthermore, other water resources are operated simultaneously. For these reasons and also with regard to economy it must be possible to adjust the pumping from the water works in question from between about 50 to 115 litres per second. In cases of emergency, it should be able to pump about 170 litres per second.

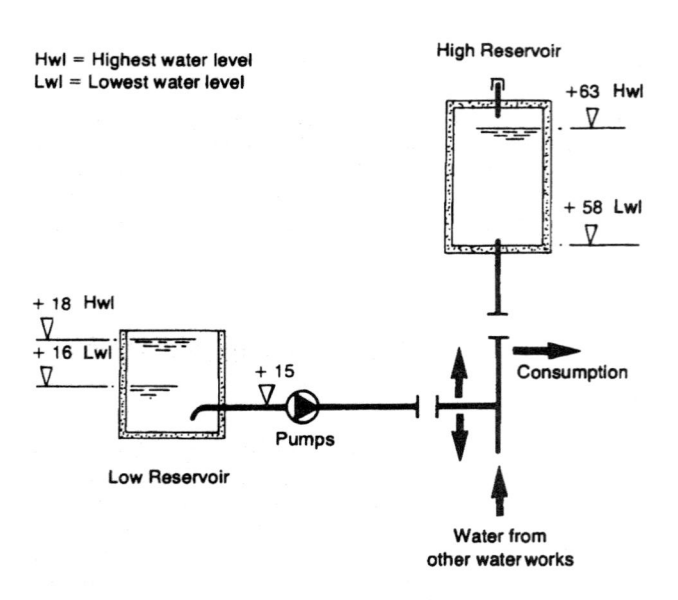

Figure 15.1 The system

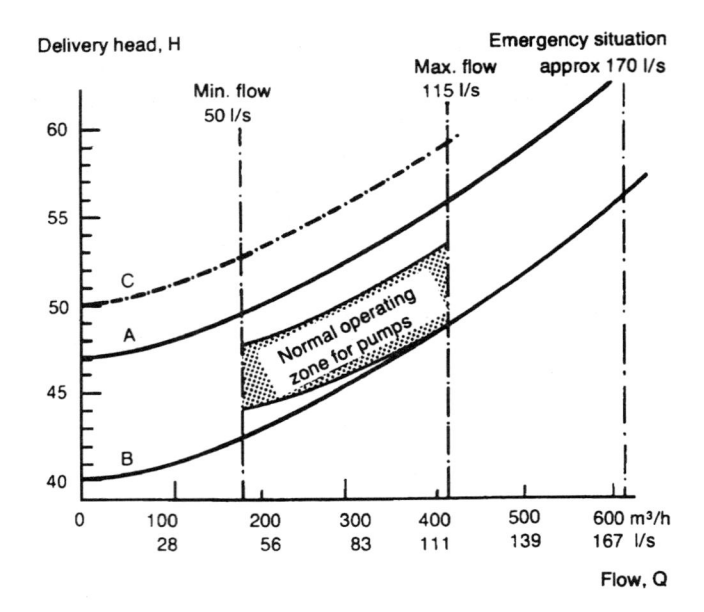

Figure 15.2 System curves

System curves

The system curves are shown in figure 15.2. It proved possible to install the pumps 1 metre below the lowest water level in the low reservoir so that a positive head of 1 to 3 metres was obtained on the suction side. At the high reservoir the difference in height between high water and low water is 5 metres. The static delivery head in the system therefore varies between 47 m (+63 - +16) and 40 m (+58 - +18).

The losses in the pipe system are shown by curves A and B. From time to time the high reservoir may be disconnected for cleaning purposes or as the result of a breakdown. Pumping must then be directed to another higher reservoir and this results in an increase of static losses by about 3 m. Curve C illustrates the upper limit of the head requirement in this situation. The normal operating zone for the pumps is clearly indicated in the figure. The highest operating efficiency should therefore lie within this zone.

Choice of pumps

On the basis of the great variations in flow demand, and respective head requirements, found in this case a combination of pumps was deemed necessary. However, it was scarcely possible to find space for more than three. High reliability of operation required that a stand-by unit should, at least to some extent be available. The necessary delivery heads implied that virtually only centrifugal pumps could be considered. From the running and maintenance points of view it was desirable that all the pumping units should be identical.

For the best possible reliability, company policy was to use pumps running at 4 pole motor speeds. Standard, double suction axially-split pumps would only fit the duty when running at 2 pole speeds. The use of three pumps would mean a motor power of 55 to 75 kW depending upon the pump efficiency. It had been decided to use a variable speed DC motor on one

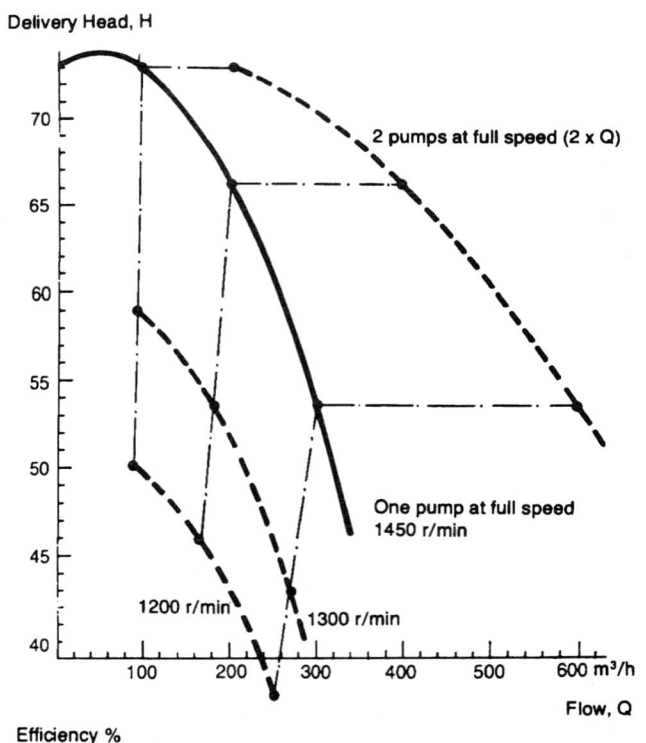

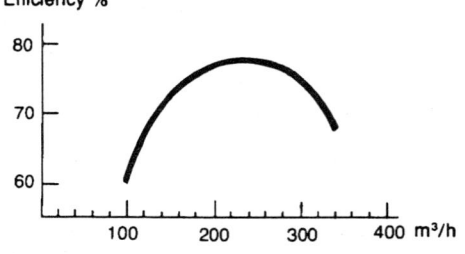

Figure 15.3 Pump curves

pump for supply adaptation. Mass produced DC motors of the size necessary only ran up to 2200 rpm. Eventually a two stage axially-split pump was found; the first stage being double suction for good suction performance.

The pump curve is shown in figure 15.3. Curves calculated according to the affinity laws for 1200 and 1300 r/min are also shown in the figure, as well as a combined curve for two pumps at full speed of 1450 r/min. The efficiency curve is also included in the figure.

In figure 15.4 the pump curves have been overlaid on the system curves. As may be seen, the speed regulated pump provides the desired minimum flow at a speed between 1200 and 1300 r/min. Maximum flow was obtained with a constant speed plus the speed regulated pump. One unit was then kept in reserve. An emergency temporary demand of 170 litres per second can be met with all three pumps in operation.

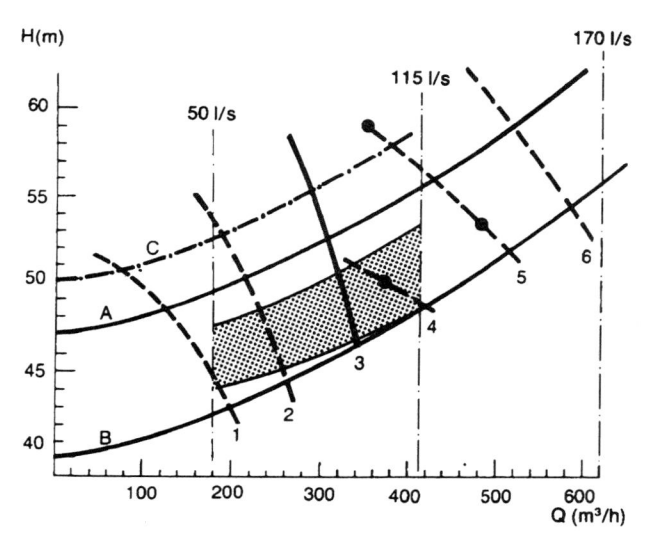

Pump curves:
1: Speed regulated pump, 1200 r/min
2: Speed regulated pump, 1450 r/min (full speed)
3: Speed regulated pump, 1300 r/min
4: Constant speed pump + speed regulated pump 1200 r/min
5: Constant speed pump + speed regulated pump 1300 r/min
6: Constant speed pump + speed regulated pump full speed

Figure 15.4 Pump and system curves

The efficiency of the constant speed pumps is not the best in their normal operating zone, only about 70%, but it is better than that obtainable with faster pumps having cut-down impellers.

The maximum power requirement for the pumps selected occurs at H = 46 m and Q = 340 m³/h, see figure 15.3, and is 61 kW. The efficiency of the electric motor is estimated to be 90% and the pumping units are supplied with 75 kW standard 4 pole squirrel cage motors.

Installation of pumps

Figure 15.5 indicates schematically how the pumps were installed. Thus the pumps work in parallel with the speed regulated pump as lead unit. When this reaches full speed, one of the two constant speed pumps is started, the constant speed pumps operate alternatively controlled by switched sequencing, and the speed regulated pump reduces to its minimum speed. In the case of increasing flow requirements it increases speed again. The procedure is the reverse in the case of decreasing water requirements.

Control of the variable speed as well as the starting and stopping of the constant speed pumps is regulated by a theoretical daily demand which in turn is constantly corrected by input from the water level in the high reservoir.

P1 = Speed-regulated pump
P2 = Constant speed pump
P3 = Constant speed pump

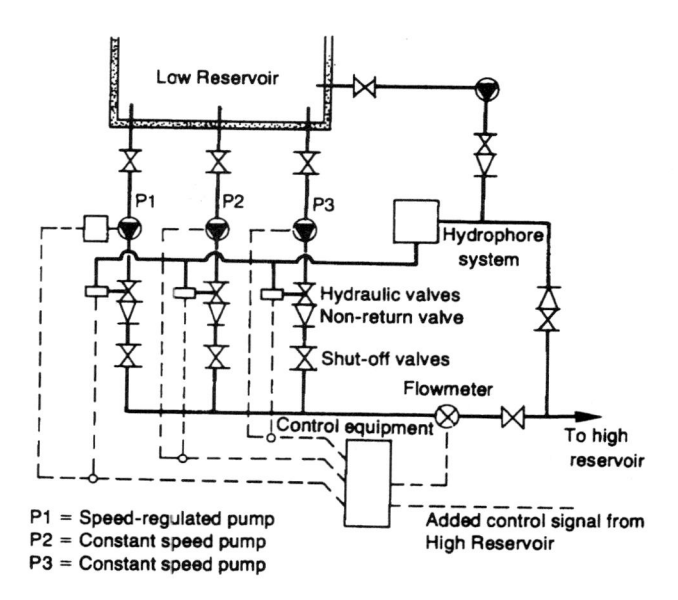

Figure 15.5 Schematic arrangement of pump installation

The delivery pipes of the pumps are fitted with hydraulically controlled slow operating valves to reduce water hammer when starting and stopping. Pumps are thus always started with the valve closed and when stopping the valve closes first. For safety reasons the pumps are also supplied with non-return valves.

15.2 A small sewage pumping station

Requirements

The pumping station serves a small housing estate. The supply of sewage during the initial period is a minimum of approximately 1 litre/s and a maximum of approximately 5 litre/s. It is estimated that gradually the minimum flow will increase to 2 litre/s and the maximum will increase to about 8 litre/s. The positive elevations at the pumping station are shown in figure 15.6. The rising main from the station is about 700 m long and discharges into a well at level +55 m. The diameter of the pipe is 150 mm and the volume is thus about 12.4 m³.

As may be seen in the figure, pump stop occurs at + 28 m. The station has an emergency outlet, overflow, at + 31 m. The static delivery head will be maximum 55-28 = 27 m and minimum 55-31 = 24 m. The stop level of the pumps has been placed as low as possible, approximately 250 mm above the floor in the pump sump, in order to avoid the formation of sludge deposits.

The station consists of two units, the pumping station itself and a valve chamber. The pumping station is constructed of pre-cast concrete rings with rubber joints. The internal diameter is 2 m and the volume per metre of height is thus about 3.14 m³. The station is equipped with two identical submersible centrifugal pumps with quick release pipe couplings to allow quick pump removal. The pumps are started and stopped via level rockers, one of which is also used to transmit a signal to the control centre when the water level in the station is too high. Level rockers are proprietary level switches which change attitude when floating on the liquid surface or are submerged.

In the valve chamber, which is also constructed of pre-cast concrete rings, there is a system of valves which ensures reverse-flushing of the pumps at every stop to prevent clogging. During pump shut-down, air is introduced into the system via the non-return ball valve, the ball should have a specific gravity of 0.9 to function efficiently and the water in the rising main in front of the flap non-return valve flows back to the

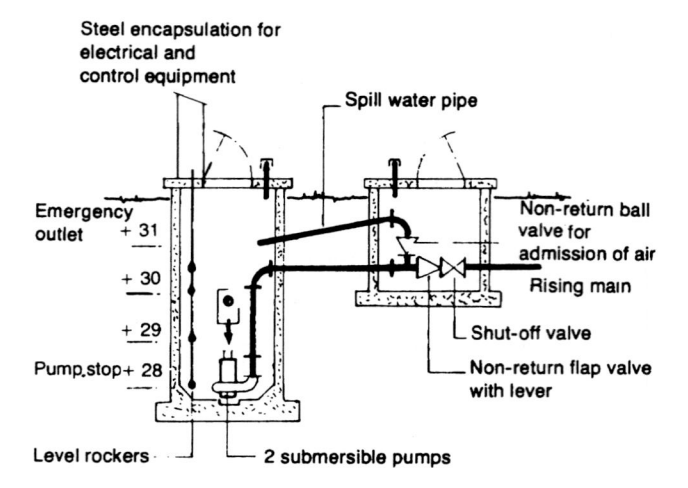

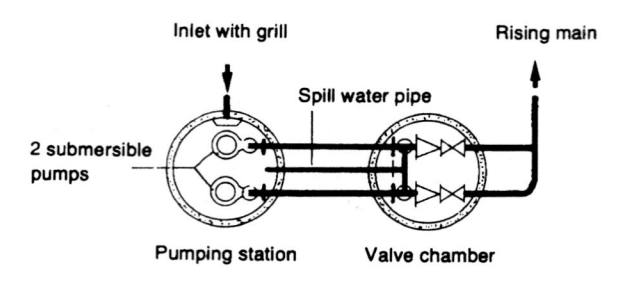

Figure 15.6 Arrangement of a small sewage pumping station

pumping station and flushes the pump clean. The amount which flows back in this particular case is about 90 litres. Thus for a short time the pump rotates backwards.

When the pump starts up it is impossible to prevent a slight jet of liquid being expelled through the non-return ball valve as this cannot be closed instantaneously. A spill water pipe returns the liquid to the pumping station.

Pump and system curves

These are illustrated in figure 15.7. As stated above, the static rising delivery head can vary between 27 m and 24 m. The losses in the rising main and its valves are shown by the system curves. The size of pump chosen has flows from 9 to 12 litre/s, depending on the differential head. Figure 15.6 also shows the curve for both pumps in operation. The efficiency of the pumps in the operating zone is about 70%. The size of the motor is 9.6 kW, the normal output being about 6 kW, corresponding to a current of about 12 A. The pumps are started direct-on-line.

Operating conditions

As mentioned in the introduction, the flow of water to the station in the initial period is 1 to 5 litre/s. The volume of the pump sump is about 3140 litre/metre height. Pumping should occur at least once every hour to ensure water replacement. The start level of the first pump has therefore been provisionally adjusted to level +29 m, see figure 15.6, i.e. when 1 m of the sump is utilised. The operating sequences will then be as shown in figure 3. At minimum supply flow the sump used is filled in about 52 minutes, 3140/(1 · 60) and pumping takes place for about 6 minutes, output approximately 10 litre/s at $H_{mean} = 26.5$ m, see figure 15.8, minus in-flow, i.e. 3140/((10-1) · 60). As already mentioned, the volume of the rising main is 12.4 m^3 and after 4 hours the water in the pipe has been replaced. At maximum in-flow the sump level rises sufficiently in 10.5 minutes and pumping out takes equally long, pump start occurring every 21 minutes. The pump motor is rated at 15 starts per hour.

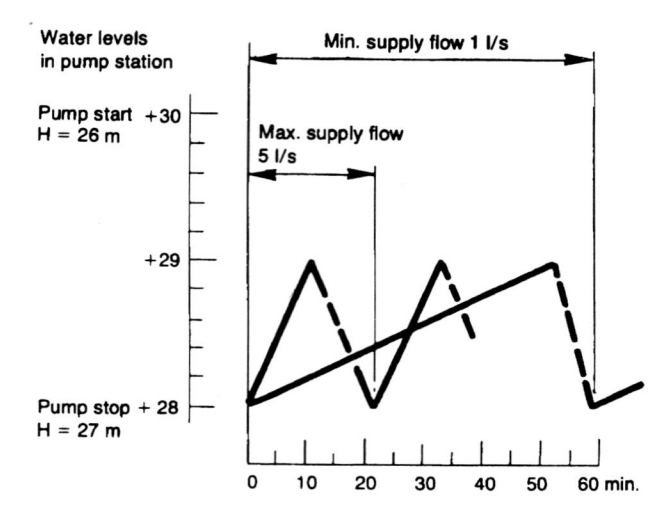

Figure 15.8 Operating sequence with 1 m working level

Gradually as the minimum and maximum rates of flow are increased, the sump volume will also be increased. Figure 15.9 illustrates operating conditions at full supply flow and when 2 m of the sump, volume = 2 x 3140, is used. This figure also includes two curves, I and II, showing the time required for the emptying of the station from the emergency outlet level, e.g. after a power failure, with one pump in operation and minimum in-flow, I, and both pumps in operation and maximum in-flow, II.

In calculating the above times no allowance has been made for the reverse-flushing flow volumes. These, however, would be a marginal effect only.

As may be seen from the above, normally only one pump is required to operate. Change over takes place at every pump start so that both units are used. The second pump therefore acts as a stand-by and is started when the level in the sump exceeds the permissible value, +30.5 m. This situation can occur when:

- lead pump fails,
- lead pump partially blocked,
- high rain water inflow.

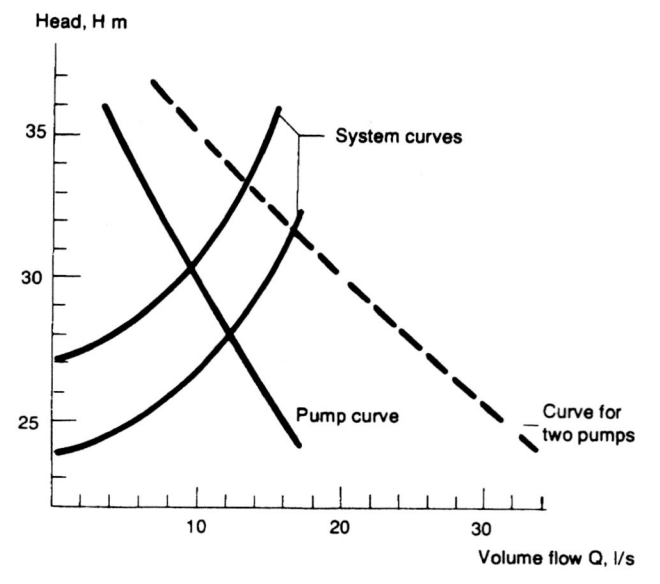

Figure 15.7 Pump and system curves

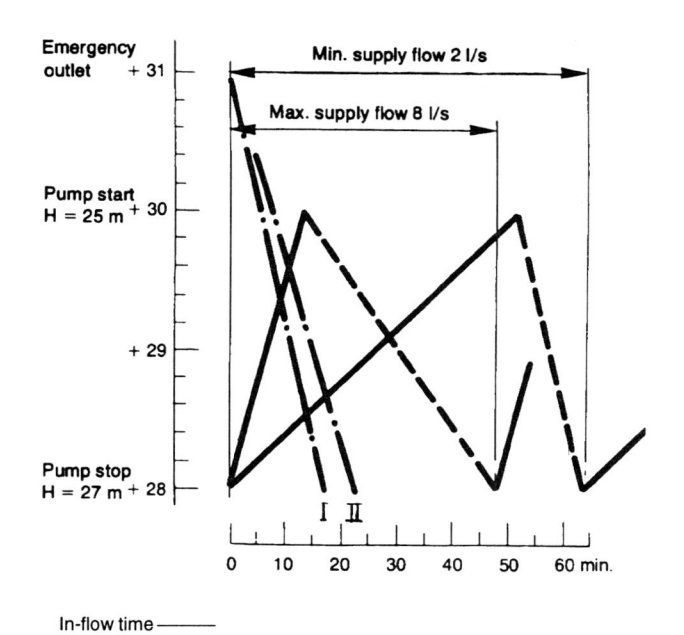

Figure 15.9 Operating sequence with 2 m working level

The station is monitored by an alarm system via telemetry and is visited for inspection only once a week, when the sump is flushed clean and functional checks, etc., are carried out.

15.3 A fresh water booster station

The function of the booster station is to create the necessary increase in pressure for the distribution of water in the higher areas of an urban district. Figure 15.10 illustrates the schematic arrangement of the booster station and its control system.

Method of operation

Normally, the pump operation will be automatically adjusted to water consumption in the high area in such a way that the pressure in the distribution pipe is maintained more or less constant. Adjustment to the actual pressure is affected partly by means of changes in speed of one of the electrically driven pumps and partly by changing the number of pumps in operation. Variable speed is obtained by the use of a DC motor.

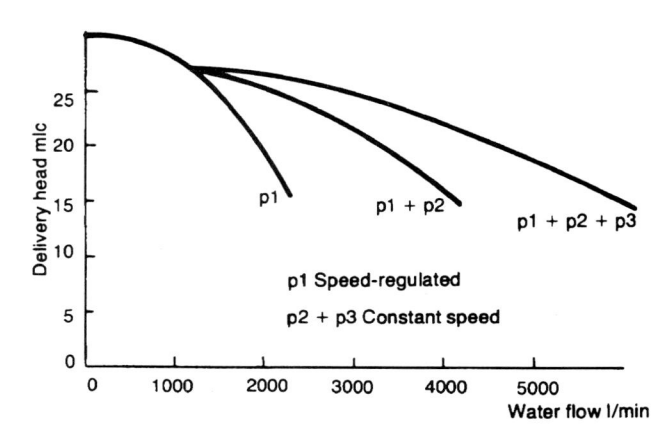

Figure 15.11 Combined pump curves

Pump curves

The effects of running the pumps in combination are shown in figure 15.11.

Valves

Each pump has a manual isolation valve located in the suction pipe to the common suction header. Each pump is fitted with a non-return valve and an isolating valve in the discharge connection to the common header. The gate valves are of the full bore rising stem design. A mechanical indicating, totalising flow meter is fitted in the booster station supply pipe.

Control

A pressure transducer, fitted to the distribution pipe, measures the changes in pressure caused by variations of consumption and transmits signals to the logic controller. This initiates the necessary changes in speed and number of pumps operating so that the pressure is restored to the predetermined value.

The pumping station operates without local personnel supervision. Control is performed by the local logic controller by monitoring the water pressure on the delivery side of the pumping station. In addition, certain fault condition signals are transmitted to the central control station.

Operating sequence

When water consumption is low, the speed regulated pump, p^1, operates alone. Speed is increased as consumption increases. When consumption surpasses the capacity of p^1, either p^2 or p^3 is brought into operation. Which of the two pumps, p^2 or p^3, is to be started up is decided manually by a

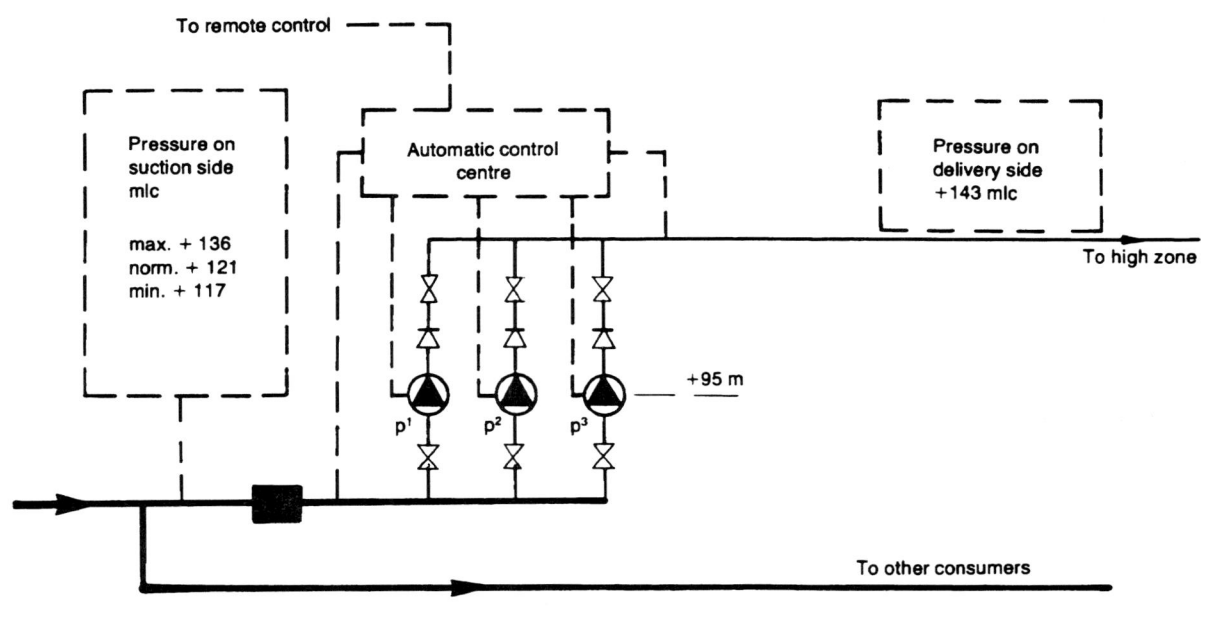

Figure 15.10 Schematic arrangement of booster station and control system

lag-lead switch or automatically by a timer. With the intention of reducing variations in pressure experienced when starting a constant speed pump, the speed of pump p^1 is reduced as another pump comes into operation. A similar procedure is followed when a constant speed pump is withdrawn from operation. At peak consumption all three pumps are in operation.

Limitations of operation

Pumping station operation is suspended in the case of insufficient pressure on the suction side or excessive pressure on the delivery side. Not until the pumping station has been inspected by staff is it possible to start up the pumps again. The speed regulated pump can only be started at minimum speed.

Operational experiences

The booster station has been in operation for 20 years.

15.4 Circulation pump for domestic central heating

The piping and instrumentation around the boiler of a typical domestic central heating system are shown in figure 15.12.

Boiler

The outlet water temperature from the boiler, 90 °C, is controlled by an integral thermostat which switches the burner on and off.

Expansion vessel

The expansion vessel accommodates changes of volume of the water and maintains a suitable static head in the system in order to avoid cavitation.

Pipes

Pipes are sized in accordance with the pressure drop and the length of the runs. Pipes are normally copper but steel and stainless steel are also used on occasions. Figure 15.12 shows a double pipe system. Single pipe systems are also used but are more difficult to control.

Radiators

Transfer heat from the hot water to the air in the rooms. Suitable distribution of water flow to the radiators is ensured by means of adjustable throttle valves. Fitting thermostatic control valves to the radiators removes the need for preset throttle valves.

Hot water for taps

An independent water circuit is shown for heating domestic hot water for taps. In some designs, the hot water heating is controlled by an electrically driven divertor valve which changes the priority between hot water heating and room heating.

Mixing valve

Regulates the system water temperature by admitting return water. Often located close to, or built into, the boiler.

Clock-thermostat

Reduces the system water temperature at night or at other desired times when a lower room temperature is acceptable. Can be used as energy conservation through the day when the house is not occupied or as frost protection.

Automatic control centre

Receives signals from the system water temperature and outdoor temperature sensors. Adjusts the thermal load via the mixing valve by a suitable outward pipeline temperature as a function of the outdoor temperature. In this way a more even room temperature is obtained and a lowering of the room temperature at night is made possible.

Thermostat valves

Throttle the flow to radiators depending upon the local air temperature, thereby permitting individual room temperature adjustment. Also make allowance for other heat sources, from the sun, electric sources, etc.

Design temperature

Forward pipe temperature approximately 80 °C, return pipe approximately 60°, $\Delta t = 20$ °C. The heat required is adjusted in accordance with outdoor temperature.

Pump

Circulates a flow which is nearly constant depending upon the pressure drop created by the thermostatic valves and the mixing valve. Pumps can be too large and can cause flow noise. This is avoided by using pumps with built-in flow adjustment by throttling or by-pass or speed control. Choose a wet rotor pump to avoid shaft seals.

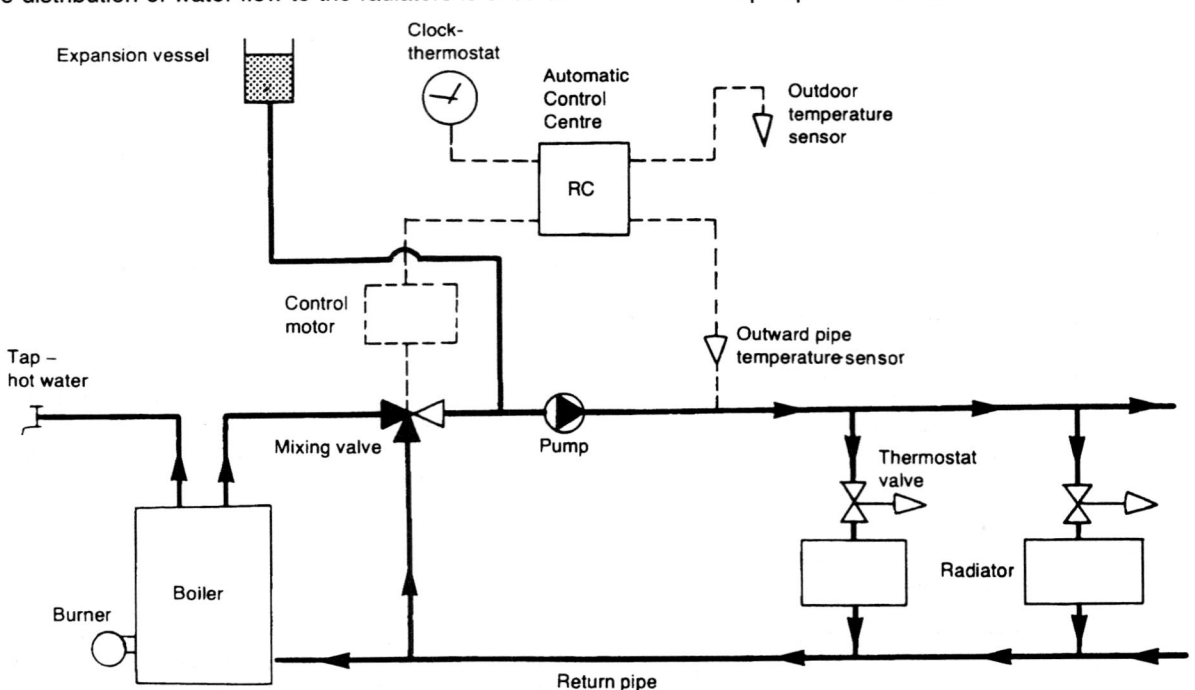

Figure 15.12 Boiler and controls for domestic central heating

Pitfalls

Check the rotation of three phase pumps when installing. The ingress of contaminants from pipes during installation and corrosion products from boiler and pipes causes wear to the pump bearings. Heat up the system and drain it before finally putting into operation. Clean out regularly or use chemical dosing. In the summer when the system is not used, run the pump for 15 minutes every week to prevent any sediment collecting and causing possible damage to the pump.

15.5 Adjustable jet pump

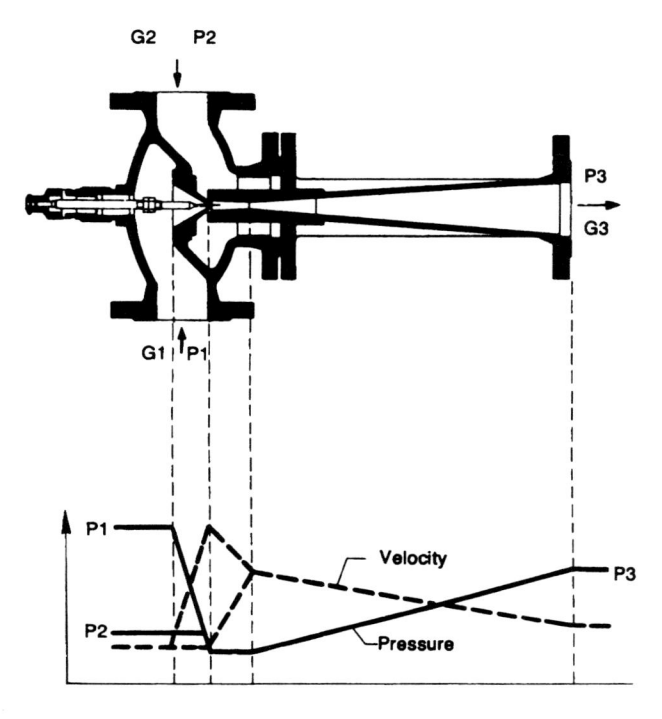

Figure 15.13 Cross-sectional arrangement and pressure/velocity distribution for a jet pump

Description

Figure 15.13 shows an infinitely variable jet pump with corresponding pressure and velocity distributions. Regulation is affected with the aid of an adjustment needle valve. In the figure:

- G1 = motive mass flow with inlet pressure P1
- G2 = pumped mass flow with inlet pressure P2
- G3 = G1 + G2 with outlet pressure P3

Typical application: central heating

Two central heating schemes are shown in figure 15.14. When comparing the upper conventional plant with the lower embodying the adjustable jet pump, it can be seen that the jet pump replaces both the mixing valve and the circulation pump. In order to drive the jet pump the necessary pressure must be generated elsewhere. Both plants can be controlled by the outdoor temperature and can be fitted with a timer to provide for reduced indoor temperatures at night.

Regulation

For varying heating requirements caused, for example, by variations in outdoor temperature, the setting of the adjustment needle valve is changed. This results in changes in the mixing ratio G2/G1, the temperatures t2 and t3 and the flow G3 through the radiator circuit. In a comparison with the conventional plant at reduced heating requirements the result is:

- lower return pipe temperature,
- reduced flow in the radiator circuit,
- greater temperature difference in t3 and t2.

Properties of the system

The reduced flow in the radiator circuit results in reduced pressure drop and thereby in reduced pump energy consumption. In the case of larger heating systems the necessary pressure for the jet pump is supplied by a centrally located carefully sized centrifugal pump. The adjustable jet pump can in theory replace the circulation pump and the mixing valve in all existing sub-circuits of the heating system.

Operational experience

Some 5000 adjustable jet pumps have been installed in heating plants, mainly in Germany.

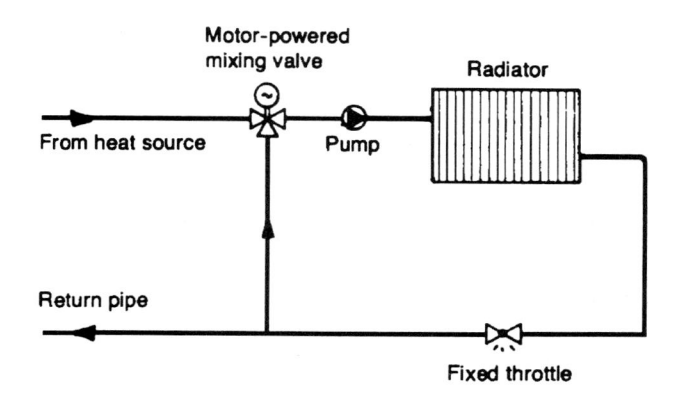

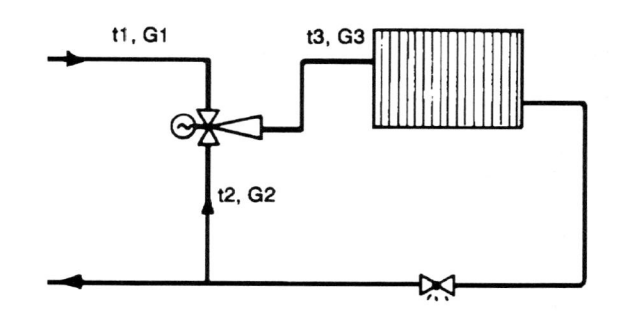

Figure 15.14 Central heating; typical and with jet pump

15.6 Modernisation of water supply for industrial use

Figure 15.15 shows the schematic arrangement of a water extraction plant feeding a booster station after modification to variable speed control.

Control before modernisation

Before modernisation both pumps were driven at a constant speed. The flow was controlled by a manual throttle valve, set on the basis of experience, at various times. Variations in water consumption led to varying throttle valve positions and energy wastage. The system could not react to unexpected changes in demand.

Control after modernisation

After modernisation both pumps were driven at variable speeds by thyristor controlled DC motors. The speed of the pump in Pumping Station 1 is controlled by the water level in the sump at Pumping Station 2. The speed of the pump in Pumping Station 2 is controlled by the water level in a storage tower at the end of the main distribution pipeline. Throttling is no longer required.

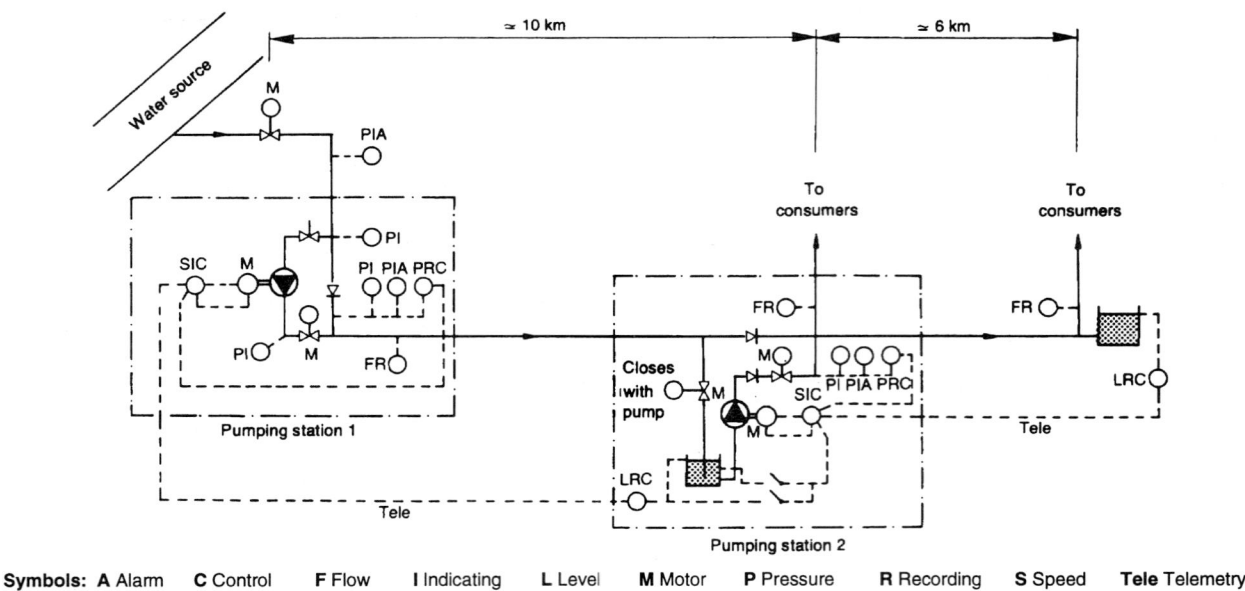

Symbols: **A** Alarm   **C** Control   **F** Flow   **I** Indicating   **L** Level   **M** Motor   **P** Pressure   **R** Recording   **S** Speed   **Tele** Telemetry

Figure 15.15 Schematic arrangement of modernised water supply system

The speed of the pumps cannot be regulated to zero. In the case of extremely low water demand the pumps are switched off. The storage volumes in the sump and storage tower being adequate to prevent rapid on-off cycling.

Alternative control

In the event of faulty transmission of remote control signals it is possible to control the speed of the pumps in accordance with the pressure on the discharge side of the respective pumps. If Pumping Station 1 is out of service, Pumping Station 2 can be controlled by the level in the sump at the station. If Pumping Station 2 is out of service, Pumping Station 1 can pump directly to the storage tower at slightly lower flow rates.

Some facts

Installed motor capacity 960 kW per pumping station.

Flow approximately 3400 m³/h.

Diameter of pipelines 800 mm

Operational results after modernisation

Modernisation has resulted in a reduction in water pumping for the industrial plant of about 40%, i.e. water savings of about 2160 m³/h. The consumption of electricity has been reduced by about 60% corresponding to energy savings of about 10 GWh per year.

15.7 Contractors' use of electric powered submersible sump pumps

Pumping units

Electrically driven sump pumps for contractors are described in Section 3.2.4, Category **050**. A number of manufacturers and distributors maintain stocks of various sizes from 1 kW up to 20 kW both for sale and leasing. Standard pumps of up to 60 kW can usually be supplied on short delivery from the manufacturer. Pumping of this kind is primarily intended for temporary draining work in connection with building and construction activities on sites. In the case of major construction works, e.g. hydro-electric power plants, the installations may be of a permanent nature. Pumps constructed of aluminium may be restricted to certain water grades.

Three types of drainage

The nature of the drainage is characterised by the type of civil engineering activity. Industrial and housing construction does not normally require involvement with rivers or lakes and drainage is therefore mainly concerned with ground-water, storm and rain water etc. The drainage period is short in relation to the construction period and the pumping requirement is often of a limited nature. Sizes 1 to 5 kW easily meet the requirements.

Harbour and bridge construction works involve drainage factors of their own, specially characterised by large flows and moderate to low delivery heads. Discharge pipes can often be short, further limiting the total delivery head. Typical of works of this kind is drainage and drying out of coffer-dams. This type of draining is often simple to plan and supervise. Pumps of low head of 10 to 30 kW are required to yield flows of 180 to 600 m³/h.

The greatest pump investment, in percentage terms of the total costs, is required by hydro-electric power plant construction works and mining. The investment is particularly large in the case of long deep tunnel sections. Sophisticated dewatering techniques involving severe demands on planning and maintenance are mainly concerned in the case of tunnel driving.

A description follows of a case study involving a four year power plant construction job.

General facts

Headrace tunnel 4200 m long, area 24 m²

Two transport tunnels, each 400 m, area 15 m²

Tailrace tunnel 5600 m, area 30 m²

Three transport tunnels, each 350 m, area 15 m²

Maximum head 62 m

Differences in level in the transport tunnels between ground level, open excavation, and lowest level, i.e. at entrance to main tunnel.

Transport 1   41 m

Transport 2   40 m

Transport 3   34 m

Transport 4;5   36 m

The transport tunnels are driven at a gradient of 1:10.

Transport tunnel 1 may be used as typical for drainage techniques for tunnels and rock chambers.

Pump sizing data

Total estimated flow of water Q_b comprises:

- Estimated leakage of water due to poor rock.

- Estimated penetration of melt and rain water through open excavation.
- Estimated amount of drilling water.

Total difference in elevation levels between lowest water course and discharge pipe outlet = H_{stat}

Total length of pump pipeline = L_{tot},

Pipe standard: quick-coupling pipes 100 mm diameter and 150 mm diameter.

Two drilling units are used with a total water requirement of 8.4 m³/h when in operation. Leakage from the rock is estimated to be 6 m³/h and melt/rain water to be 12 m³/h, Q_b = 26.4 m³/h. A safety margin added to ensure pumps could always cope with water flow brought the flow to Q_{max} = 36 m³/h.

An estimate of the pressure loss in 100 mm and 150 mm pipes respectively with a design flow of 36 m³/h indicates 100 mm pipe to be acceptable. Pressure loss equals 9 m. Total differential head:

$H_{stat} + H_f = 41 + 9 = 50$ m

The drainage is divided into two phases.

Phase 1.

Driving of the tunnel from tunnel face and open excavation to level at -41 m at the main tunnel. During this phase, use is made of pumps capable of pumping full flow but about half the differential head, i.e. 36 m³/h x 25 m. The intention is that the pump should be light and well-matched to the actual duty. When about half of the length of the tunnel has been driven, i.e. about 200 m and H_{stat} = 20.5 m, the first pumping station will be constructed and can be made permanent for the entire construction phase.

After the transport tunnel has been put to use, water flows from the main tunnel now under construction. Admittedly, the 8.4 m³/h drilling water from the transport tunnel itself has now been eliminated, but the four units in the main tunnel produce 16.8 m³/h and this must be added to the leakage from the main tunnel.

A rough estimate indicates that Q_b through the transport tunnel will be 42 m³/h.

This means that:

Q_b = 42 m³/h

H_{stat} = 41 m

Q_{max} estimated to be 60 m³/h

H_f then becomes 20 m for 100 mm pipe, which is too high a loss. If the pipe size is changed to 150 mm, H_f at 60 m³/h will be 4 m. Thus:

$H_{stat} + H_f = 41 + 4 = 45$ m

This station may be constructed to two designs: a pump chamber blasted out of the rock or a steel pump tank.

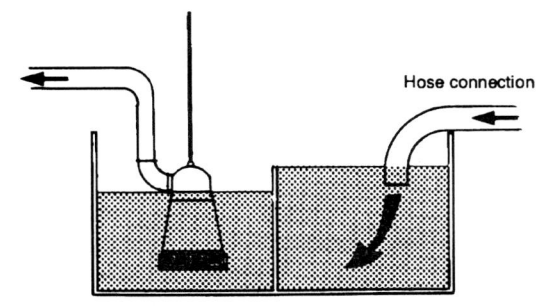

Figure 15.16 Steel tank option

The advantage of the steel tank, see figure 15.16, is that it can be easily moved and can therefore follow the advance of the tunnel downwards. This enables the waste pump on the drilling car to work constantly with the same length of hose and delivery head.

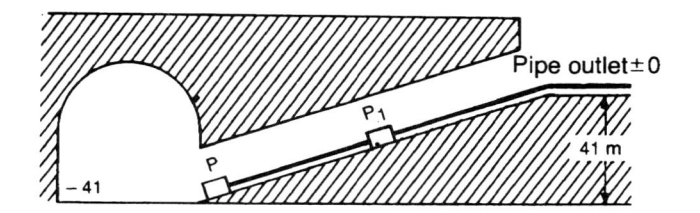

Figure 15.17 Drainage with two pumping stations

On reaching level -20 m a complete pumping station is established at that level and an additional similarly equipped station follows the working face down to the beginning of the main tunnel at level -41 m.

Phase 2.

When the transport tunnel joins up with the main tunnel two possibilities for the continued drainage of the transport tunnel may be considered, figure 15.17

Alternative 1

Retention of two pumping stations each with a total delivery head of 22.5 m and maximum flow of 60 m³/h.

Alternative 2

Installation of one central pumping station at level -41 m, the pumping duty then being 60 m³/h x 45 m.

Alternative 1 is preferable because the pump size used permits greater range of application and reduces the risk of water hammer in the case of power failure. The advantage of Alternative 2 is that inspection can be limited to a single station.

The total pumping requirement for the transport tunnel until the commencement of Phase 2 is shown in figure 15.18.

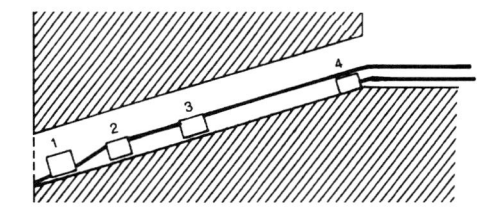

Figure 15.18 Drainage with four pumping stations
1. Drilling car with air-powered diaphragm pump.
2. Pumping station with submersible electric pump 60 m³/h x 22.5 m.
3. Pumping station with submersible electric pump 60 m³/h x 22.5 m.
4. Submersible electric pump for rain and melt water with its own pipe. Size about 30 m³/h x 5 m.

General installation instructions for drainage

Determination of size: Difference in elevation

                             Length of pump pipe

                             Flow

The difference in elevation between the liquid surface and the pipe outlet is always known. The length of the pump pipe can always be determined.

The flow of water to be pumped can usually only be estimated and this must be done on the basis of experience. When this has been done, refer to the manufacturer's instructions to calculate the pipe friction loss and add this to the difference in elevation.

$H_{stat} + H_f = H_{tot}$

When using hose as a pump pipe, a 20% addition should be made for unavoidable kinks. Use the shortest possible route.

In open shafts and pits always place the delivery connection at right angles to the pump shaft to avoid kinks.

In the case of sharp edges, e.g. rebates, use pipe bends and connector pipes. See figure 15.19.

Figure 15.19 Correct routing of flexible pipes

Always select a pump giving consideration to the wear factor. This is especially important in the case of the pumping of tunnels. Drilling waste is very abrasive for impellers and rubber components, see Miller Number in Chapter 1, figure 1.27, causing rapid reduction in capacity. See figure 15.19.

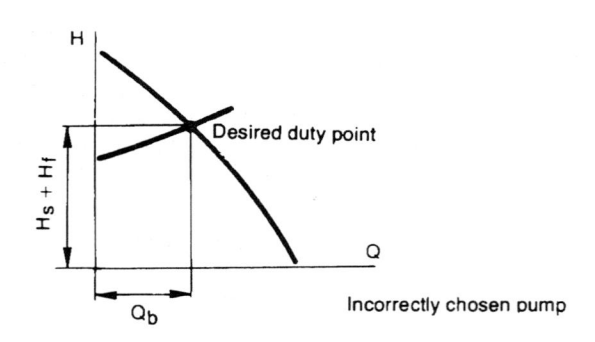

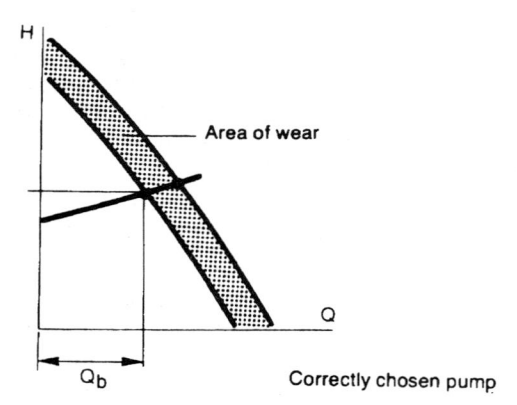

Figure 15.20 Pump curves corrected for wear

15.8 Engine driven self-priming pumps

Pumping Unit

Figure 15.21 shows a self-priming site pump with a petrol or diesel engine mounted on skid. Suction hose with strainer. Discharge hose is normally about 10 m long. Extra lengths can be added when necessary. Skid fitted with lifting and towing eyes.

Size of pumps

Pumps in good condition can lift 8 m. Figure 15.22 indicates the flow ranges of popular pump sizes at differential heads between 10 and 20 m.

This type of pump can handle solid particles. It is not a solids handling pump in the sense of solid-liquid mixtures but a rugged pump capable of passing occasional, fairly large solid plus mud and soil. The size of the maximum solid increases with pump size. The mesh of the suction strainer is selected on the basis of the pump inner passage size. In practice it is accepted that all solids capable of passing through the strainer are pumpable.

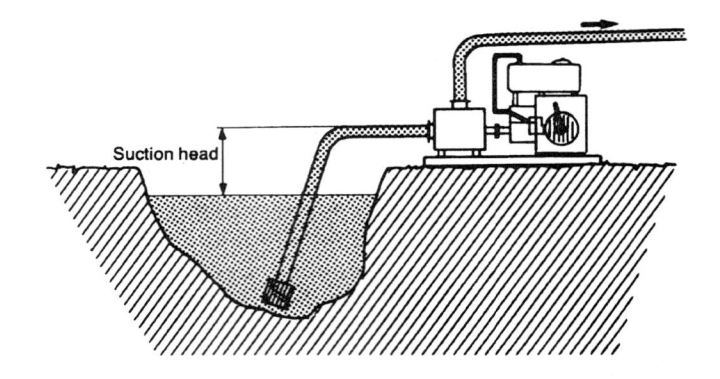

Figure 15.21 Skid mounted self-priming pump

In the case of a discharge hose which is more than about 30 m long, it is advisable to check the head drop of the hose. This is easily evaluated with the aid of the nomograms in Chapter 2, starting at figure 2.30. The head drop has the effect of reducing the capacity of the pump.

Suction connection mm	Flow range m³/h
40	10 to 20
50	10 to 30
75	20 to 60
100	30 to 100
150	30 to 300
200	30 to 500

Figure 15.22 Capacities of popular engine driven pumps

Site installation

Place the pump on a flat, level and if possible rigid pad. Most engines and pumps are splash lubricated so levelling is quite important. The difference between the water level and the centre-line of the suction connection should not exceed 8 m. The suction hose should ascend uninterruptedly to the pump. Position the suction strainer near the bottom without burying it in mud, etc. In the case of small flows the strainer can be placed in a shallow pit approximately 0.5 x 0.5 m.

Pumping

Fill the pump with water through the filler. The volume of water necessary is 2 to 5 litres.

Start up and evacuate the suction pipe. If the time for evacuation exceeds 2 to 3 minutes, there is a fault, e.g. too great a suction lift, leaky suction hose, worn-out pump, etc.

As the level in the pit reduces, move the strainer to greater depths. As the suction lift increases the flow will reduce. If there is plenty of water, it is only necessary to check the pump every quarter of an hour. A full fuel tank usually suffices for 2 to 3 hours, but try to avoid running at very low fuel levels. Sediment in the fuel tank can block filters and injectors. If fuel runs out, the fuel system may have to be vented and primed. If the water flow is small, reduce the engine speed to a suitable level.

Each day when pumping is finished, wash the pump and suction strainer out with clean water. Dried mud and setting concrete can seriously affect performance. In cold weather, drain the pump completely each night to prevent freezing. Before storing between jobs, protect in accordance with the manufacturer's instructions.

State when purchasing or hiring:

Flow and corresponding delivery head.

Suction lift.

Maximum particle size.

Desired power source, e.g. 4-stroke petrol engine.

Maximum weight which can be handled at site.

Maximum noise level.

Requirements for skids or trailers.

Pitfalls

Leaky suction hose results in reduced pump performance. An air leak on the suction side can be detected by placing the discharge hose under water. Air bubbles indicate air leakage in the suction hose or the pump.

Petrol-oil mixtures should only be used for two-stroke engines. Wrong fuel causes long delays.

A blocked suction strainer reduces the capacity of the pump. In such cases check the strainer regularly and clean out as required.

Kinks in the hoses cause unnecessary throttling.

15.9 Land reclamation pump

Drainage

Low-lying swampy areas can be changed into agricultural land by drainage. The catchment area determines the size of the pump. In round figures the capacity is usually about 1.5 to 2 litre/s for every hectare, every 10000 m^2. The material for the dyke is taken from the neighbouring ground and should be of clay to seal off the water. If good quality clay cannot be found in the vicinity, plastic sheeting provides good protection against leakage through the dyke.

Pumping stations

Pumping stations constructed before the 1960s were cast in situ and were large and costly. The pumps were of the wet pit type, see Chapter 3, figures 3.51, 3.52 and 3.53. During the 1960s submersible pumps with capacities from 180 to 3600 m^3/h were developed for drainage purposes. The stations were usually constructed as wells made of pre-cast concrete rings. See figure 15.23.

Pump

The pump is usually driven by an electric motor, or in exceptional cases, by a diesel engine with hydraulic power transmission. The pump is an axial-flow propeller or mixed-flow type. The delivery heads are between 1 and 3 m and with capacities currently up to 7.2 m^3/h. Since the water flow varies during the year, sometimes two or more pumps are installed and put into operation when required.

Pipes and valves

The pipes are of welded steel construction or made of plastic reinforced with glass fibre. The valves are reverse-acting flap types and capable of handling a variety of solids.

Control

The pump or pumps are started and stopped by level-control. The levels are sensed by rocker or float switches or electrodes. Axial-flow propeller pumps can be flow-regulated by automatic adjustment of the impeller blades. Mixed-flow pumps can be flow-regulated by change of speed or by inlet guide vanes in the water before the impeller.

Pitfalls

From the power consumption point of view, differential head is critical. Because the pump differential head is low, small changes in suction or discharge head have a dramatic effect on power requirements. The regulation of levels must be carefully adjusted. If the stopping level is too low, the pump can suck in air at constant water levels without stopping. Too low a starting level may lead to rapid cycling because the water rises rapidly when it flows backwards through the outlet pipe. Excessive switching on and off causes fatigue of the reverse-flap valve and there is a risk that it may work loose. Contactors in the electric control, seals and bearings are subjected to more rapid wear. It is therefore an advantage to construct large water storages in the form of drainage ditches or large diameter feed pipes. Alternatively pump flow can be regulated so that they run continuously.

15.10 Cargo pump for tankers

15.10.1 Large tankers for crude oil

Large tankers are fitted with a small number of oil cargo pumps, usually four. The pumps are located in a separate pump room. The cargo space on board the vessel is divided into 10 to 15 individual tanks. With the aid of a relatively complicated pipe system every individual pump can be used to unload any of the tanks. Figure 15.24 shows a simplified diagram of how an oil cargo pump system works:

The suction pipe to every tank is connected to a suction funnel close to the bottom of the tank so that the tank can be completely emptied.

The separator tank, which is connected to a vacuum unit, separates gases drawn into the suction funnel during the final stage of unloading.

The pump is normally driven by a steam turbine, but electric motors can also be used. Due to the risk of fire or explosion due to the gases given off from the crude oil the pump driver

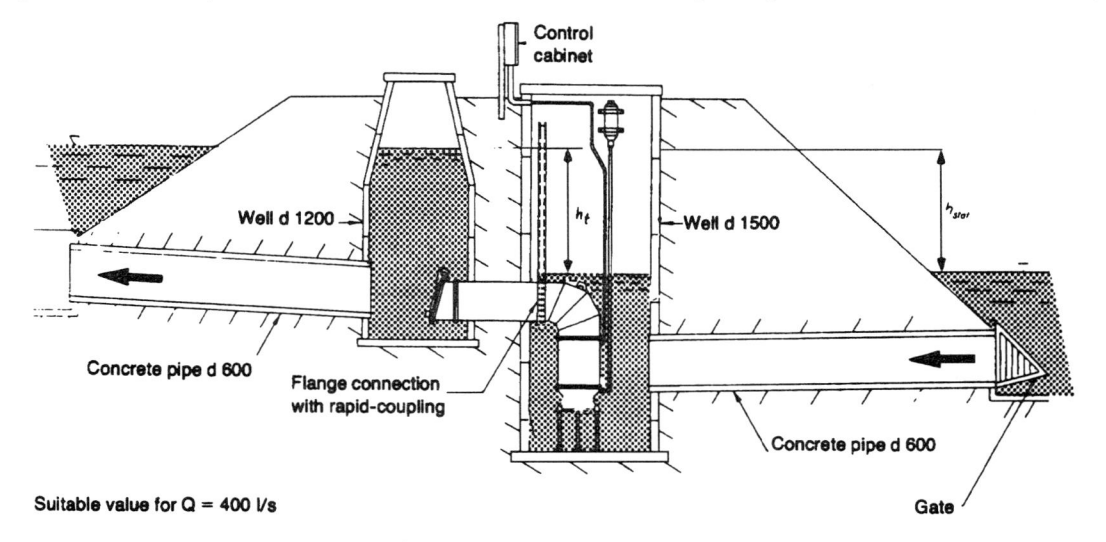

Figure 15.23 Pump station constructed from pre-cast concrete rings

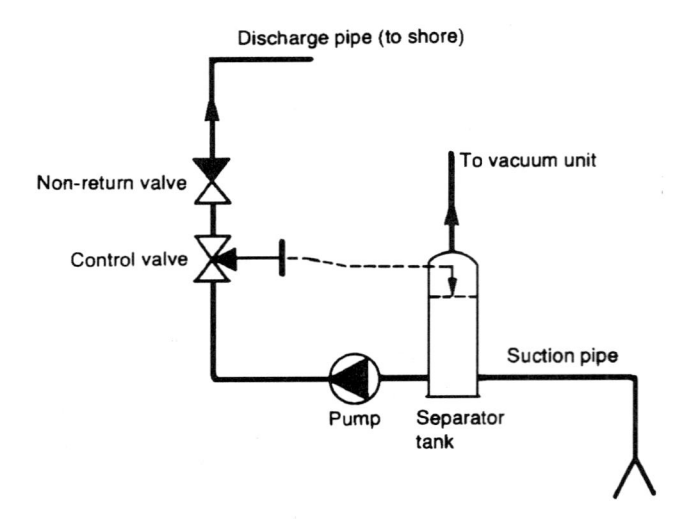

Figure 15.24 Cargo pump pipework

must not be located in the pump room. Figure 15.25 illustrates a horizontal shaft pump installation.

The control valve is used to reduce the flow during the final unloading stage, which is necessary to ensure the tank is completely emptied. The valve is automatically controlled by the level in the separator tank.

The non-return valve prevents oil from flowing backwards through the pump when unloading is stopped.

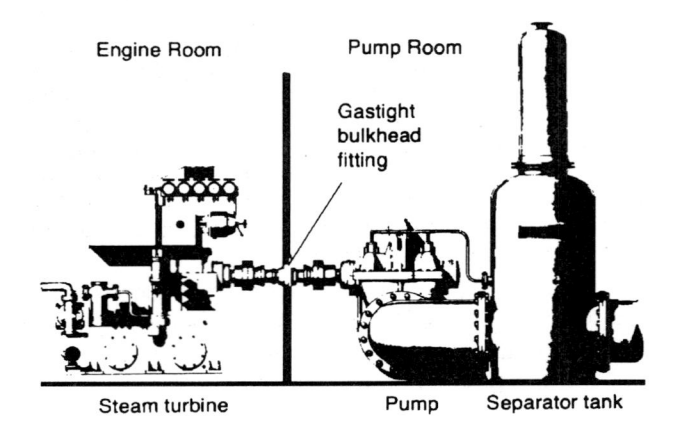

Figure 15.25 General arrangement of steam turbine driven cargo pump

The bulkhead acts as a firewall between the two compartments. A spacer coupling is used with a mechanical seal to prevent gas leakage.

The pumps are normally sized to unload the vessel completely in about 24 hours. The delivery head required to pump oil to the onshore terminal is 120 to 180 m. This results in large pumps with heavy power consumption, up to 7 MW per pump on the largest vessels. Oil cargo pumps are also used for the loading and unloading of water ballast.

Observations

The risk of explosion makes it essential that all equipment capable of generating sparks must be excluded from the pump room and the tank spaces. Pumps should be fitted with mechanical seals, since stuffing boxes can become overheated. Classification societies check that safety regulations are observed.

During installation, allowance must be made for the movements which take place in the hull. The shaft coupling should be able to absorb minor misalignments between pump and driver.

15.10.2 Smaller tankers for finished products

A characteristic of vessels of this type is their ability to transport a great number of different kinds of liquid cargoes. The cargo space in smaller tankers is also divided into small tanks, the number varying in relation to the design and size of the vessel. As the cargo often consists of a number of different liquids which must not be mixed, a common unloading system cannot be used. Instead every tank is provided with a pump of its own, which, bearing in mind the risk of explosion, should be hydraulically powered. Also, hydraulic equipment can be easier to protect against the ravages of a marine environment. In order to simplify installation, the pumping unit is submersed in the liquid. The outline drawing in figure 15.26 shows the pump installation in a tank.

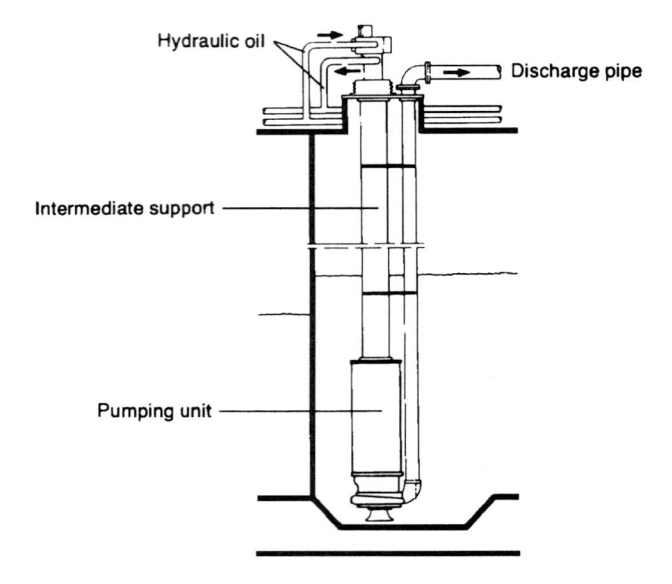

Figure 15.26 Hydraulically driven vertical submersible pump

The pumping unit consists of a centrifugal pump and a hydraulic motor. The suction connection of the pump is bell-mouthed and its entrance is close to the bottom of the tank. The capacity of the pump is determined by the volume of the tank. The delivery head is 70 to 150 m.

The delivery pipe and intermediate support or column are cross braced to each other to increase rigidity and reduce vibration. On larger vessels of this kind the distance between the deck and the bottom of the tank is about 15m. In addition to feed and return pipes for hydraulic oil, the intermediate support also contains pipes for sealing liquid.

A constant flow valve maintains the flow of oil to the hydraulic motor, which means that the pump operates at a constant speed independently of the load. The hydraulic motor also contains devices for the remote control of the pump speed. The pumping unit is supplied with hydraulic oil from a power pack which can supply oil to several pumps. The oil pressure in the hydraulic motor feed pipe is normally 160 to 240 barg.

Observations

Since the cargo may consist of highly corrosive media, the material of parts in contact with liquids must be given special consideration. The sealing of the pump shaft, which must function in liquids with varying properties, also deserves special attention.

The breakdown of a pump with a tank full of liquid raises difficulties with respect to rapid repair. For such eventualities portable pumps must be available for emptying and flushing out tanks.

15.10.3 Tankers for LNG

In the last few years the world market for liquified natural gas, LNG, has enlarged considerably. There is now a requirement to transport considerable quantities of LNG from producers to consumers. LNG usually consists mostly of methane so the liquified gas must be stored at temperatures around -160 °C at atmospheric pressure. The material requirements for operating at such low temperatures are well known. The main problem with pumping such cold liquids is ensuring the pump unit is cooled sufficiently for the liquid gas to remain liquid.

This problem can be solved by using submerged pumps in the tanks. Specially adapted canned motor pumps, category **125**, have been developed. As the complete pump unit is surrounded by the liquid thorough cooling is practically assured. Great care must be taken with the motor connections into its terminal box. Submerged pumps, of the wet pit type, category **055**, have also been used. Double mechanical seals with a suitable barrier liquid, such as propanol or methanol, must be employed.

When spherical storage tanks are mounted on the ship's deck it may not be possible to fit submerged pumps. In these situations, a canned pump built in an external can may be suitable. The external can must be fitted with vapour vents and liquid level controls to ensure the can is cooled and filled properly. Adequate insulation must be applied to the suction pipe and the external can to prevent ambient conditions vaporising the LNG. Very slight temperature rises cause considerable loss of NPSHa.

LNG pumps may seem to 'rattle' slightly when first started. This can be due to mild cavitation while vapour is clearing. The 'rattling' should stop quite quickly; if not, shut-down and investigate.

15.11 Liquid detergent manufacture- pump installation

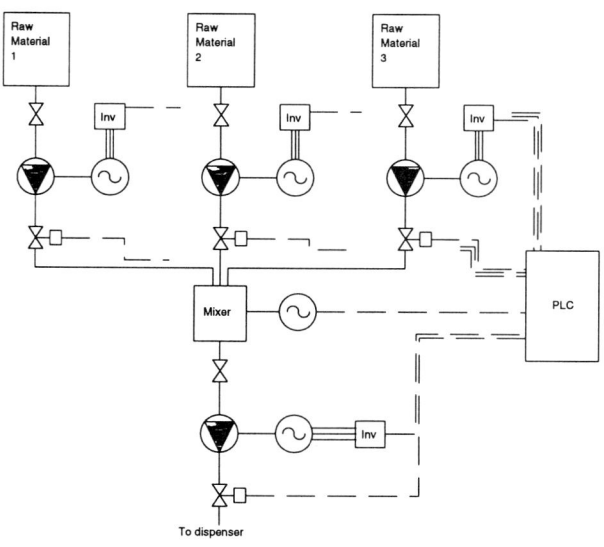

Figure 15.27 shows the schematic arrangement of a detergent manufacturing system

The product

The finished detergent is manufactured from three raw materials which are mixed in various proportions, depending upon the grade of the finished detergent, before being dispensed into five or 10 litre containers.

Raw material tanks

The raw material tanks are elevated above the pumps to provide plenty of NPIPa. Each tank is fitted with a tapered bottom leading to a central connection which is twice the diameter of the pump suction connection. The manual isolating valves are full-bore quarter turn ball valves.

Raw material pumps

The pumps are triple lobe pumps operating at low speed to maintain the volumetric efficiency close to unity. All three pumps operate together controlled by the programmable logic controller, PLC. The speeds of the three pumps are adjusted by the PLC, via solid state variable frequency invertors, to conserve the raw material proportions fed to the mixer. After a preset time the three pumps stop and the electrically actuated valves in the discharge pipes close to prevent 'dribbling'.

The mixer

Once the mixer has been filled to the required level, the PLC starts the mixer motor which runs for a preset time. The mixer tank also has a tapered bottom and a large connection fitted with a full-bore ball valve.

Finished detergent pump

The finished detergent pump is a low speed triple lobe pump. Volumetric efficiencies close to unity are assured by adequate NPIPa due to low loss pipework. The pump speed is controlled in response to signals from the dispenser.

The PLC

The PLC is a small unit which stores the sequence of events, the programme, on an EEPROM, electrically eraseable programmable memory. The unit can be programmed in several European languages as well as a choice of programming languages. The EEPROM can be programmed, validated and debugged with a small hand-held unit.

15.12 Positive displacement pumps- variable discharge conditions

The problem

The new Severn Bridge will have a cable-stayed mid-estuary section. The steel cables of 27 individual strands are essential for the support of the centre span bridge deck. To avoid corrosion the steel cables are shrouded with polythene cable ducting. A small radial clearance is provided sufficient to allow removal of the cables for inspection. It was deemed essential to protect the cables insitu by coating them with an anti-corrosive wax. The wax was obviously highly viscous and the maximum cable length was 250 m. The problem was compounded by the fact that the polythene cable duct could only be pressurised to 6 barg maximum.

The solution

The pump manufacturer, Stork Pumps Ltd, was able to predict the pressure losses for various flows, viscosities and cable lengths by using the standard equation for laminar flow in an annular orifice.

$$\Delta P = \frac{12 \cdot \mu \cdot L \cdot Q}{b \cdot x^3}$$

where

μ =	liquid viscosity	(cP)
L =	length	(m)
Q =	flow	(m³/s)
B =	breadth of orifice	(m)
x =	height of orifice	(m)

Calculations indicated that if the wax viscosity was reduced to 100 cP it could be pumped along the longest cable.

Rotodynamic pumps were not suitable for the high viscosity wax during the warming up stage. Flow control, when warm at constant pressure, would require an air operated control valve; no air was available.

An internal gear pump, capable of handling viscosities up to 80000 cP, was selected for the application. Driven by a variable speed motor from 290 to 765 r/min the pump was capable of flows from 10 to 26 m³/h at pressures up to 16 barg. In order to reduce the wax viscosity the pump had optional heating jackets fitted.

The pump system was fitted with a full-flow by-pass and a full-flow relief valve, set at 5 barg, in the discharge line. The pump was started on by-pass and continued to run this way until the suction tank of wax was completely heated. When flow was switched to the discharge line and directed down the cable shroud, pump speed was regulated automatically by a pressure sensor.

15.13 Energy recovery turbines

Introduction

Pumps are usually purchased to pressurise a fluid. This then enables flow to take place by overcoming resistance in the piping system. In most cases, the pressure generated is entirely absorbed into the system.

In some cases the pump also has to generate additional pressure to start a process. Boiler feed pumps are one example of this. Here the system frictional resistance might only account for 30% of the generated pressure.

Reverse Osmosis, R.O., is another example. Again, the frictional resistance only forms a small percentage of the generated pressure. Most of the pressure is used to stimulate the osmotic process of water purification. About half of the pressurised saline water fails to pass through the R.O. membrane. This is then discarded. Since this fluid is typically at 45 to 90 barg it must be depressurised before it can be dumped.

In a scrubber plant, impure gas passes through absorbent fluid in the absorber tower. Impurities are absorbed into this carrier fluid. This takes place at relatively high pressure. The fluid then passes to the stripper tower where the impurities are released. To do this the pressure then has to be reduced, or "let-down".

In the refining of oil to gasoline, the product passes through a hydro-cracking process. Again the charge product is pressurised to finalise the process. After reaction, this pressure has to be dissipated before the next part of the process.

In all these cases, liquid has been pressurised to permit some process to take place. It then has to be depressurised for some reason. A simple way of depressurising the fluid is to pass it through an orifice plate. In this approach, all the energy embedded in the liquid is lost as heat and noise.

Another approach is to pass the liquid through a control valve. This is not much different from an orifice plate but at least it is variable.

Yet another approach is to pass the fluid through a turbine. This might appear to be an expensive solution. However, it has the advantage that not only is the liquid depressurised, but the energy embedded in the liquid is largely recovered. The value of this power can then be offset against the turbine cost. Experience has shown that this recovered power can repay the cost of the turbine in 9 to 18 months. After that the energy recovered is almost free. On this basis, it is almost irresponsible not to explore all possible uses of energy recovery turbines. How can the energy be used?

- It can be used as the sole power source for another machine.
- It can be used to contribute power to anther machine and reduce the power consumed by the motor.

Conventional turbines are expensive when compared with pumps of the same power. They are not so readily available, neither is the choice so great. The material options are also more limited. For special applications, such as high temperature refinery service the choice is very limited. Fortunately, all forms of centrifugal pump can also be run in reverse to perform as very effective turbines. The disadvantages of commercial turbines are then eliminated. The only shortcoming of reverse running pumps is that their efficiency is lower than a true turbine. However, this is typically only two or three per cent and hardly impacts on the pay-back period. The notation of reverse running pump is as shown in figure 15.28.

One apparent disadvantage of any turbine is that it needs some form of speed control. Fortunately this can easily be

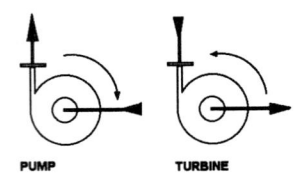

Figure 15.28 Schematic notation of pumps and turbines

overcome in most cases. This depends upon the turbine being connected into a train that also utilises an electric motor. Electric motors lock into the mains frequency and act as excellent governors. This establishes the speed control the turbine needs.

Fundamentals

The performance characteristics of centrifugal pumps are well known. Those for a turbine are less well known. They are compared below.

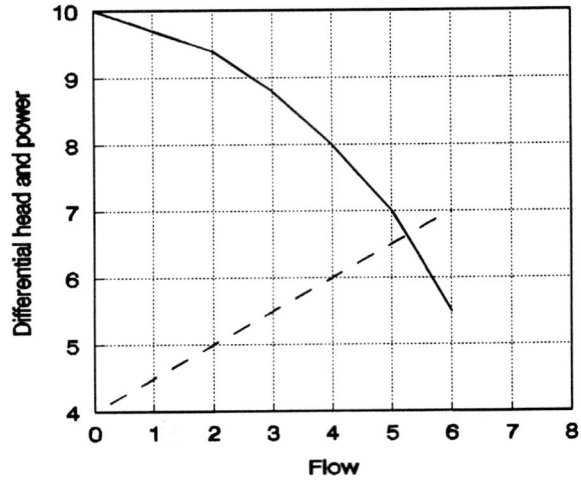

Figure 15.29 Typical centrifugal pump charcteristic

Figure 15.29 shows typical pump performance characteristics. Generated pressure decreases as pump flow increases. On the other hand, absorbed power increases with flow

Figure 15.30 shows the characteristics of a pump used as a turbine, i.e. the flow direction and direction of rotation are reversed but the pressure differential is the same. Note that the differential pressure increases as the flow increases. Also that the recovered power increases with flow. However, below a certain flow, the turbine actually absorbs power.

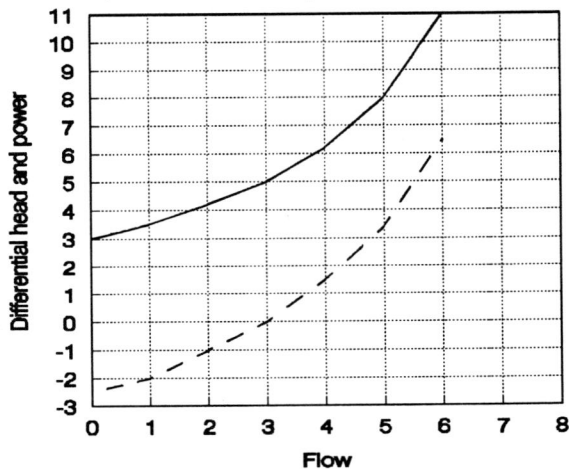

Figure 15.30 Typical inward flow radial turbine

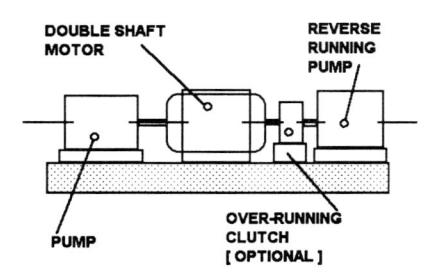

Figure 15.31 Typical arrangement of energy recovery turbine with a motor as the main driver

Figure 15.31 shows the most popular way of connecting a pump and energy recovery turbine. The clutch is optional and its effect is discussed below. (Figures 15.30 & 15.32.)

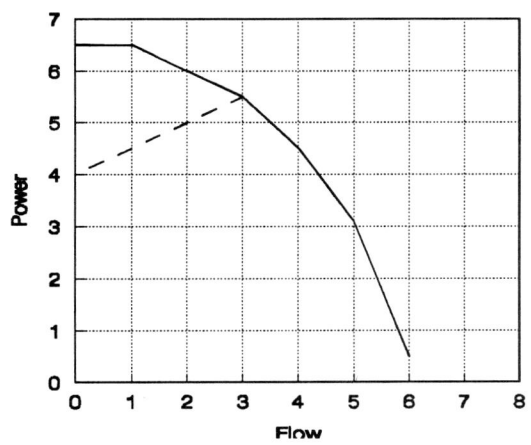

Figure 15.32

Figure 15.32 shows the effect of coupling an energy recovery turbine in-train with the pressurising pump as shown above. This shows the nett power delivered by the motor. It is calculated by subtracting the power recovered in the second chart from that absorbed in the first chart.

This nett power curve should be compared with the unassisted power curve shown in figure 15.29. Note that at low flow, the motor is actually overcoming the power absorbed by both the turbine and the pump. As flow increases, the turbine begins to contribute and reduces the motor power requirements.

One way of reducing the low flow power demand is to install an overrunning clutch between the turbine and the motor. In this way the turbine is uncoupled from the train during starting at flows where it is absorbing power. It only becomes connected when power is being contributed.

15.14 Economic aspects of energy utilisation

Influence on choice of circulation pumps in heating systems

When selecting circulation pumps for heating systems, due consideration must be given not only to the relative merits of a particular type of pump, its efficiency and serviceability etc., but also to the overall economy of the heating system as a whole.

Depending upon the size and type of heating system and the relative costs of heating fuel and electrical power sources, it can be shown that, in spite of its lower efficiency, a wet rotor circulation pump can be more economical in terms of total energy costs, than a dry motor pump.

Figure 15.33 Circulation pump with standard motor and seal, single and twin head versions

Figure 15.34 Wet rotor circulation pump, direct coupled in-line versions

In most instances the final choice of circulation pump depends upon factors other than optimum energy utilisation. In small and medium plants it is advantageous to use wet rotor circulation pumps, figure 15.34, since they can, to a large extent, operate without service or maintenance. In large plants, which are often supervised by operating personnel, it is usual to utilise pumps driven by standard motors, dry motor pumps, figure 15.33, because of their higher efficiency.

In border line cases between small and large plants it may be necessary to carry out a comparative evaluation of the two types. In this comparison the question of total energy utilisation should directly influence the choice of pump. The dry pump motor losses are radiated and convected to the surrounding air, whilst the wet pump motor losses are almost entirely given up to the heating system media.

The following describes methods of evaluating the economic aspects related to the use of wet or dry motor pumps respectively.

Definitions and assumptions

When comparing wet rotor and dry motor pumps, it is usual to refer to pump efficiency, η_P, which is defined as the relationship between the input and output power, equation 15.1, the difference in the two values being equal to the pump losses, equation 15.2.

For pumps, the input energy is defined as the required shaft input power, P, or as the input electrical power supplied from the electric supply system for the pump unit as a whole, the motor losses being also taken into account. The output is usually defined as work done, P_q, equation 15.3.

The motor efficiency is defined as mechanical power output divided by electrical power output, the difference between the two being mechanical and electrical losses, see equation 15.4. In close coupled pumps the pump power input and the motor power output must be equal.

The overall efficiency of the pump unit, η_U, is obtained by multiplying the motor efficiency, η_M by the pump efficiency, η_P, equation 15.5.

All the losses in a pump unit, and even pump work is converted into heat and can be utilised by the heating system. The pump output power, P_q, is dissipated by friction which results in heat. The pump losses, P_{VP}, are converted into heat within the pump. If the pump unit is mounted within the heated space, the motor losses, P_{VM}, are given up directly to the surroundings.

In order to make a proper comparison between various types of pump unit, account must be taken of the economic relationship of the heating fuel and the electrical energy supplied to the pump motor. In the following analysis the term 'economic efficiency', η_E, is used to represent the relationship between the output and input energy, with due consideration to the specific energy costs of the losses.

The cost per heat unit extracted from the fuel, allowing for the efficiency of the heating system, η_H, is defined by 'a' in equation 15.6 and the cost for electrical energy supplied by 'b' in equation 15.7, both values in currency/kJ.

Standard motor pump units

For explanation of symbols see equations

Figure 15.35 illustrates the energy flow for a pump unit with a standard motor and seal. The motor takes energy from the electric supply system and supplies the pump with mechanical energy via the shaft coupling. The motor energy losses are given up to the surroundings, see equation 15.4.

The mechanical power is converted into hydraulic energy, equation 15.3, and the pump losses, equation 15.8, are converted into heat and transferred to the liquid.

When calculating the 'economic efficiency' of a heating system using a 'dry' pump unit as in Figure 15.33, then the unit efficiency must include the pump losses, equation 15.9. The energy costs related to the pump losses being the only losses which can be usefully utilised by the heating system.

Wet rotor motor pump units

For explanation of symbols see equations

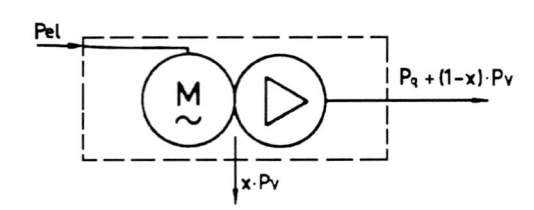

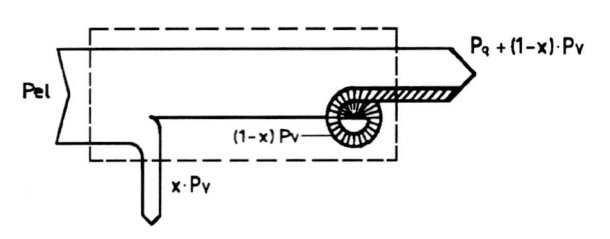

Pel = Supplied electrical power
P_q = Hydraulic pump output
P_V = Pump aggregate's energy losses
X = Losses to surroundings

Figure 15.36 Energy flow diagram for a wet rotor motor pump unit

Figure 15.36 illustrates the energy flow for a pump unit with a wet rotor motor. A large proportion of the energy consumed by the motor is utilised by the heating system.

The greater part of the motor heat losses and the pump friction losses are utilised by the plant as additional heat. The heat losses dissipated to the surroundings are negligible and need not, therefore, be included in the calculations. The 'economic efficiency' of a wet rotor motor pump unit is obtained by considering the pump unit efficiency and the value of the additional heat supplied, equation 15.10.

Equations

$$\eta_P = \frac{\text{pump output power}}{\text{pump input power}} = \frac{P_q}{P} \qquad \text{Equ 15.1}$$

$$P_{VP} = \text{pump power input} - \text{pump power output}$$

$$= P - P_q \qquad \text{Equ 15.2}$$

where

η_P = pump efficiency (decimal)
P_q = pump output power (W)
P_{VP} = pump losses (W)
P = pump input power (W)

$$P_q = Q \cdot \Delta H \cdot \rho \cdot g \qquad \text{Equ 15.3}$$

where

Q = flow (m^3/s)
ΔH = differential head (m)
ρ = liquid density (kg/m^3)
g = gravitational acceleration (9.81 m/s^2)

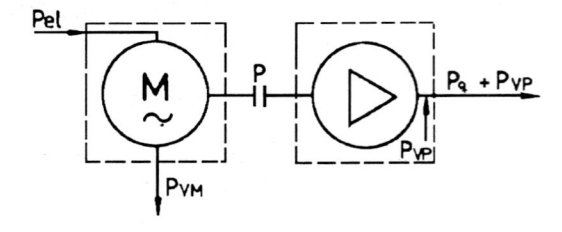

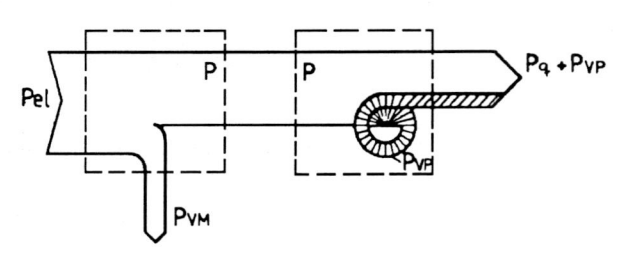

Pel = Electrical energy supplied
P = Mechanical shaft power
P_{VM} = Motor energy losses
P_q = Hydraulic pump output
P_{VP} = Pump energy losses

Figure 15.35 Energy flow diagram for a standard motor pump unit

$$\eta_M = \frac{P_{el} - P_{VM}}{P_{el}} \qquad \text{Equ 15.4}$$

where
- μ_M = motor efficiency (decimal)
- P_{el} = electric power input (W)
- P_{VM} = motor losses (W)

$$\eta_U = \eta_P \cdot \eta_M \qquad \text{Equ 15.5}$$

where
- η_U = unit overall efficiency (decimal)

$$a = \frac{\text{heating fuel cost}}{\text{calorific value} \cdot \eta_H} \qquad \text{Equ 15.6}$$

$$b = \frac{\text{electrical energy cost}}{3600} \qquad \text{Equ 15.7}$$

where
- a = heating fuel energy cost (currency/kJ)
- b = electrical energy cost(currency/kJ)
- η_H = heating system efficiency (decimal)
- heating fuel cost (currency/kg)
- calorific value (kJ/kg)
- electrical energy cost (currency/kWh)

$$P_{VP} = Q \cdot \rho \cdot C_p \cdot \Delta t \qquad \text{Equ 15.8}$$

where
- C_p = specific heat of liquid (kJ/kg/K)
- Δt = temperature rise across pump (°C)

$$\eta_{EDry} = \frac{P_q + ((P_{el} - P_q - P_{VM}) \cdot a/b)}{P_{el}}$$

substituting $(P_{el} - P)$ for P_{VM}

we obtain

$$\eta_{EDry} = \frac{P_q + ((P_{el} - P_q - (P_{el} - P)) \cdot a/b)}{P_{el}}$$

which can be simplified to

$$\eta_{EDry} = \frac{P_q + ((P - P_q) \cdot a/b)}{P_{el}}$$

which can be written as

$$\eta_{EDry} = \eta_{UDry} + ((\eta_{MDry} - \eta_{UDry}) \cdot a/b) \qquad \text{Equ 15.9}$$

where
- η_{EDry} = economic efficiency for a pump unit with a standard motor (decimal)

$$\eta_{EWet} = \frac{P_{el} + ((1 - x) \cdot P_v \cdot a/b)}{P_{el}}$$

which can be rewritten as

$$\eta_{EWet} = \eta_{UWet} + ((1 - x - \eta_{UWet} + x \cdot \eta_{UWet}) \cdot a/b)$$

realising $x \cdot \eta_{UWet} \approx 0.03$

it is considered neglible

$$\therefore \eta_{EWet} = \eta_{UWet} + ((1 - \eta_{UWet} - x) \cdot a/b) \qquad \text{Equ 15.10}$$

- η_{EWet} = economic efficiency for a pump unit with a wet rotor motor (decimal)

The critical ratio of a/b is given by

$$a/b_{crit} = \frac{\eta_{UDry} - \eta_{UWet}}{1 - \eta_{UWet} - x - \eta_{MDry} + \eta_{UDry}} \qquad \text{Equ 15.11}$$

$$\text{oil cost} = \text{calorific value} \cdot \rho \cdot a/b_{crit} \cdot b \cdot \eta_H \qquad \text{Equ 15.12}$$

where
- oil cost (currency/m³)
- calorific value (kJ/kg)
- ρ = density (kg/m³)
- a/b_{crit} = value from Equ. 15.11 (decimal)
- b = electrical energy cost (currency/kJ)
- η_H = heating system efficiency (decimal)

Comparison

Two extreme cases have been chosen to demonstrate the relative differences in total economy when considering similar heating systems.

Figure 15.37 Wet rotor motor pumps in a heating system

1. Very low fuel costs and high electricity costs, i.e. the ratio a/b approaching zero,
2. The cost of fuel and the cost of electricity is approximately the same, i.e. the ratio a/b approaching unity.

The expression 'economic efficiency', defined previously, is used here to calculate the effect of energy prices and should not be considered as a value of efficiency in the normal sense.

Case 1. In Holland where gas is very cheap and electricity is relatively expensive, the situation is very similar to case 1 above.

According to equation 15.9 and 15.10 and with a/b = 0 the 'economic efficiency' for both dry and wet rotor motor pump units will be equal to the unit efficiency, η_U.

Since the unit efficiency for a dry motor pump is always higher than for a wet rotor motor pump, then it follows that the 'economic efficiency' for a dry motor pump is always greater than for a wet rotor motor pump, i.e.

$$\eta_{EDry} > \eta_{EWet}$$

Case 2. In Sweden, for example, electrical energy is relatively cheap by international standards. The relationship between fuel and electrical energy costs, a/b, is almost unity, if the normal heating system efficiency for larger installations of 0.75 to 0.80 is taken into account.

If a/b is 1, then from equation 15.9 for dry motor pump units, the 'economic efficiency' is equal to the motor efficiency. For wet rotor motor pump units operating under the same conditions, a/b = 1-x, where 'x' is the losses to the surroundings, then the 'economic efficiency' is determined by applying equation 15.10.

Figure 15.38 Dry motor pumps in a heating system

Measurements have shown that the losses 'x' to the surroundings are normally between 4 and 7%, for wet rotor pumps with motors up to 2.2 kW, which is less than the corresponding dry motor pump losses of about 20%.

It can therefore be concluded that, for heating systems in these circumstances, better running economy will be achieved using wet motor circulation pumps than by using dry motor pumps, i.e.

$\eta_{EWet} > \eta_{EDry}$

Calculated example for heating systems in Britain at current fuel and electricity prices.

The following values have been used for a typical small industrial system:

Pump output	about 2 kW
Pump efficiency	$\eta_P = 0.7$
Motor efficiency for dry pump	$\eta_{MDry} = 0.8$
Motor efficiency for wet pump	$\eta_{MWet} = 0.5$
Losses to surroundings for wet pumps	$x = 0.06$
Cost of fuel oil	£130 /m^3
	14.4 p/kg
Density	900 kg/m^3
Calorific value	41000 kJ/kg
Average efficiency of heating system	$\eta_H = 0.78$
Electricity charge rates	8.06 p/kWh

According to equation 15.5, the unit efficiencies become:
for dry motor pumps $\quad \eta_{UDry} = 0.56$
for wet motor pumps $\quad \eta_{UWet} = 0.35$

By using equations 15.6 and 15.7, the specific energy costs corresponding to the above values become:
a = 4.5 · 10^{-6} £/kJ
b = 22.4 · 10^{-6} £/kJ
i.e. a/b = 0.2

By applying equations 15.9 and 15.10, the 'economic efficiencies' for the two respective pump types can be calculated:

for dry motor pumps $\quad \eta_{EDry} = 0.608$
for wet motor pumps $\quad \eta_{EWet} = 0.468$

This suggests that in Britain the 'economic efficiency' of a heating system utilising a dry motor pump unit is higher than that of a wet rotor motor pump unit for a particular heating system and assuming the above conditions.

Dry pump units remain more economically efficient, even if the ratio a/b increases to 0.6, at which value the economic efficiencies for wet and dry motor pump units are equal. This corresponds to an increase in fuel oil price to 387 /m^3, as calculated by equation 15.12.

In a gas heating system, the ratio a/b would be less which would approach the conditions discussed in case 1, where dry motor pumps would prove to be even more economic.

Conclusions

The heat output of a heating plant is provided by the heat extracted from the fuel and the power supplied to the circulating pump.

The electrical energy contribution, although being relatively small, is important when considering the optimum economy of the complete heating system.

A comparison of heating system energy costs for similar systems in different countries shows that in countries like Sweden where electrical energy is relatively inexpensive, the ratio a/b approaches 1 thus making wet rotor motor circulation pumps, despite their lower efficiency, more economical in service than dry motor pumps.

In countries like Holland where electricity is expensive, the ratio a/b approaches 0 and dry motor circulation pumps are more 'economically efficient'.

15.15 Choosing non-return valves

General remarks

Non-return valves are incorporated in the pipe system to prevent, undesirable reverse flow. Normally the valve cannot be actuated by external sources, it is self-actuating, i.e. the flow alone determines its function. A non-return valve should be able to operate under varying conditions with respect to media, pressure, temperature, flow and viscosity. Non-return valves selected for clean liquid operation should not be expected to function as well when handling abrasive solids. A desire common to both suppliers and consumers is that it should be possible to use the same type of non-return valve under different operating conditions, i.e. its application should be universal.

In many pipe systems non-return valves have performed indifferently, a chattering non-return valve is extremely irritating, if not to say, hazardous. The valve is subject to considerable wear. The chattering releases forces which work on the pipes and supports, pumps and the valve itself. The risk of a breakdown of the system is therefore very great. Chattering valves indicate a change in flow velocity and perhaps the use of an incorrect valve type. For increased system reliability the true cause should be investigated.

Water hammer and pressure surge in pipes

Following a pipe fracture or shut-down of a pump in a pipe system, the forward flow is reduced and after some time it may be succeeded by a reverse flow. If this remains in operation for some time, the result can be that the pumps reverse direction of rotation and attain high speeds. This usually entails damage to pumps, couplings and motors. Efforts are made to prevent reverse flow with the aid of automatic non-return valves.

Automatic non-return valves rely on forward flow to hold the valve open. The amount of valve opening or lift is dependant upon the flow velocity. When the flow velocity reduces to zero the valve closes. Due to the inertia of the moving components within the valve there can be a short delay between the flow stopping and the valve closing. The delay period can allow reverse flow to start. When the reverse flow is arrested, pressure surge takes place in the pipe system both upstream and downstream of the non-return valve.

Choice of type of non-return valve

For non-return valves the most important requirements are:

- Controlled closure in the event of reverse flow — reduced risk of line shock
- Low opening pressure — reduced risk of line shock
- Low flow resistance — low energy consumption
- Low maintenance requirements — low operating costs
- Reliable — low operating costs

The range of non-return valves is considerable. Figure 15.39 lists popular types of non-return valve with their respective merits.

Type	Flow resistance	Risk of water hammer	Maintenance	Temperature restrictions
Swing-check valve	low	yes	low	no
Poppet valve	high	slight	medium	no
Disc valve	high	yes	medium	no
Piston valve	high	slight	medium	no
Ball check valve	medium/high	slight	low	no
Rubber tube valve	high	yes	medium	as low as 80 °C
Butterfly check valve	high	yes	low	no

Figure 15.39 Types of non-return valves

Sizing non-return valves

It is important that the valve is sized so that it is fully open under normal steady state running conditions. If there is a flow variation the valve must be selected very carefully. At maximum flow the valve should not be wide open and up against the 'stops'. Minimum flow should not correspond to less than 50% opening. The inertia of the moving parts must be minimised and the movement damped.

In the case of non-return butterfly valves, operating under steady-state conditions, sizing is carried out in accordance with the procedures below and figures 15.40 and 15.41.

For water

Use figure 15.40:

1. Enter the figure along the 'x' axis at the minimum normal flow
2. Move vertically until the pipe diameter line is intersected
3. Check valve operation by noting relative positions of curves A, B, C, D and E
4. If valve not fully open, select a smaller diameter
5. Check pressure drops at all flows.

For liquids of similar viscosity to water but differing densities

Use figure 15.41:

1. Calculate the flow velocity in the pipe for the minimum normal flow
2. Calculate ρv^2
3. Enter the figure along the 'x' axis at the ρv^2 value
4. Move vertically until the valve lines are intersected

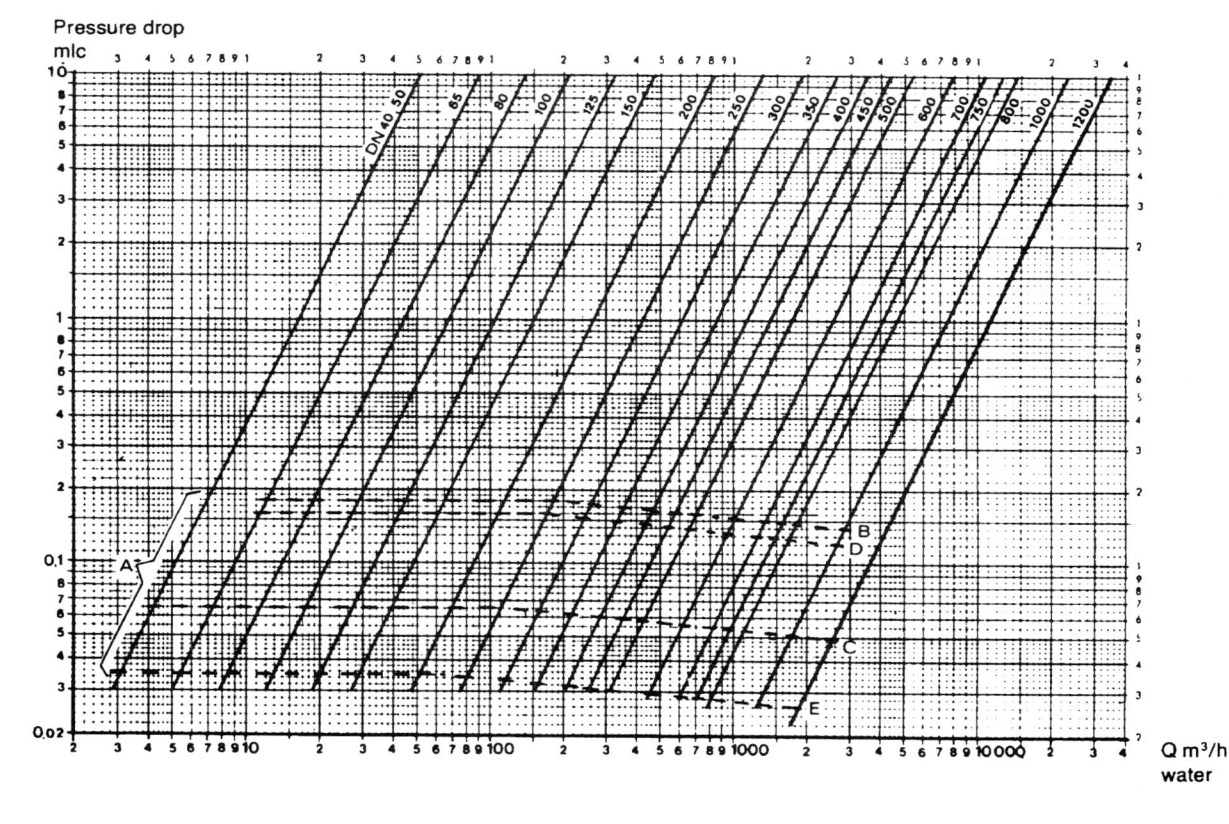

A. Before the valve is fully open the pressure drop is greater than indicated on the curve, see figure 15.41

B. Valve is fully open - applies to designs with springs and horizontal pipes.

C. Valve fully open - applies to designs without springs and horizontal pipes.

D. Valve fully open - applies to designs with springs and vertical pipes.

E. Valve fully open - applies to designs without springs and vertical pipes.

Figure 15.40 Butterfly check valves for water

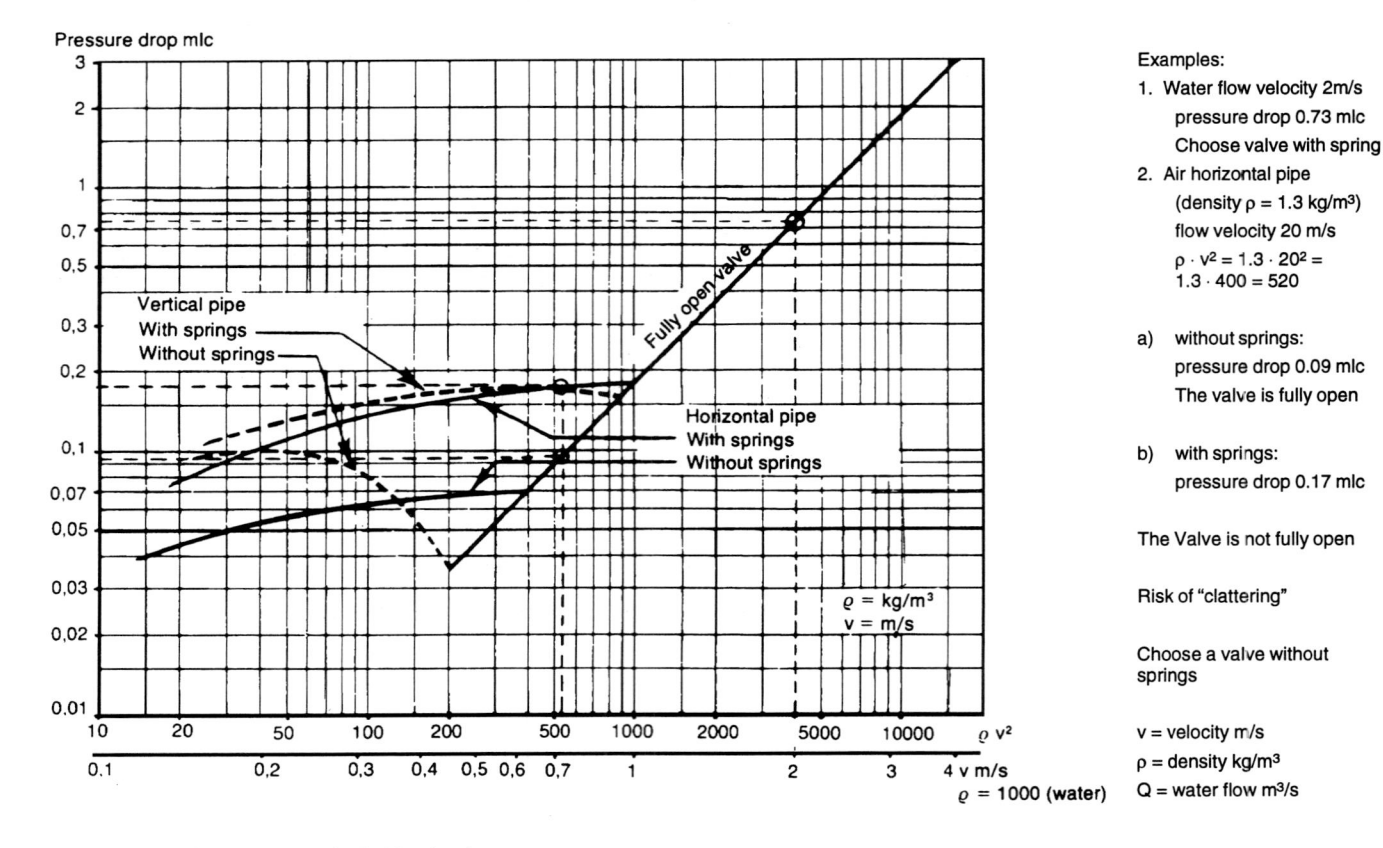

Figure 15.41 Butterfly check valves for liquids other than water

Examples:
1. Water flow velocity 2 m/s
   pressure drop 0.73 mlc
   Choose valve with spring
2. Air horizontal pipe
   (density $\rho = 1.3$ kg/m³)
   flow velocity 20 m/s
   $\rho \cdot v^2 = 1.3 \cdot 20^2 = 1.3 \cdot 400 = 520$

   a) without springs:
      pressure drop 0.09 mlc
      The valve is fully open

   b) with springs:
      pressure drop 0.17 mlc

   The Valve is not fully open

   Risk of "clattering"

   Choose a valve without springs

v = velocity m/s
ρ = density kg/m³
Q = water flow m³/s

5 If valve not fully open, select a smaller diameter at higher velocity
6 Check pressure drops at all flows

Pressure drop

The pressure drop across the valve according to figure 15.41 is based on the resistance figure $Z = 3.6$ for valves up to and including DN 250. For larger valves the drop in pressure is lower and the pressure drop values in the diagram should therefore be reduced by the factors stated below.

DN 300 - 350   factor 0.89
    400 - 700   factor 0.83
    750 - 1000 factor 0.78
          1200 factor 0.70

The pressure drop is dependent on density and flow velovcity.

Pitfalls

Centrifugal pumps operating in parallel

Pumps for parallel operation are usually installed as shown in figure 15.42. To avoid problems in cases when a second pump is to be started while the first is in operation, it is necessary for the pumps to have stable characteristics, see Chapter 3, figure 3.25. The characteristics should be steep rather than flat, see Chapter 3, figure 3.26.

The suction side of centrifugal pumps

A centrifugal pump requires a certain pressure above the vapour pressure on the suction side, NPSHr, in order to operate correctly and to avoid cavitation. The non-return valve should therefore be located on the delivery side.

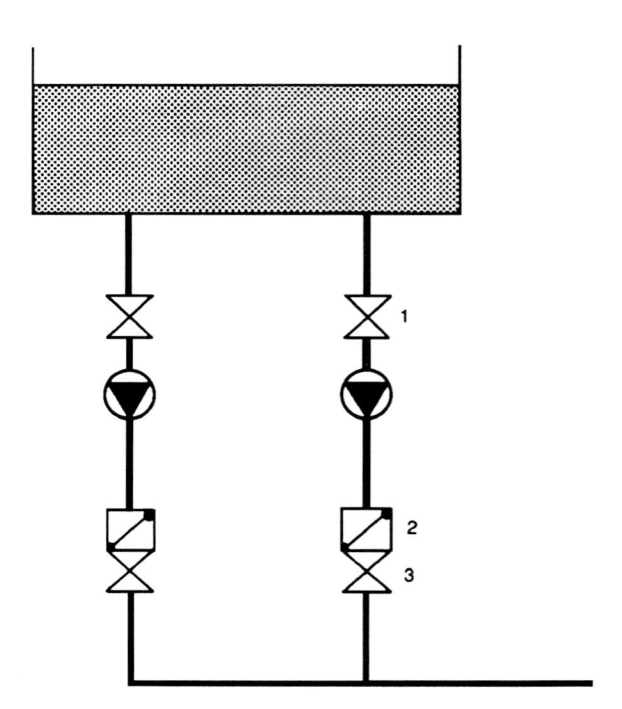

1. Suction isolating valve for pump. Select a full-bore valve with a low pressure drop to avoid loss of NPSHa and the risk of cavitation. Full-bore ball valve or wedge gate valve.

2. A non-return valve with low pressure drop is selected to avoid losses of energy and with low risk of water hammer, i.e. poppet or ball check valve or in the case of small sizes a piston valve.

3. Discharge isolating valve after the pump with low pressure drop to avoid great losses of energy. Preferably a ball valve or wedge gate valve.

Figure 15.42 Pumps connected in parallel

16 Pump classification guide and guide to manufacturers and suppliers

The chapter is introduced with an explanation of the various names, designations and limits of various types of pumps. The classification guide summarises the various types of pump according to the categories which have consistently been applied throughout this book, especially in the Direct Selection Tables, figures 14.3 and 14.4, the Rapid Selection Table, figure 14.5 and the detailed technical descriptions throughout Chapter 3.

Contents

16.1 Introduction

16.2 Pump classification guide

16.3 Pump manufacturers and suppliers guide

16.4 Manufacturers and suppliers of ancillary products

16.5 Names and addresses of European manufacturers and suppliers

16.6 Trade Names

16.1 Introduction

General

The classification of various pumps into categories has been made according to their differing styles and sizes and basic principles of operation. Within the pump industry there are certain established practices by which pumps are sometimes named according to their design or construction, sometimes according to the pumped liquid and sometimes according to the field of use or particular application.

Despite the fact that the methods of designation and descriptions are not always strictly logical, it is relatively obvious to both user and manufacturer as to what is intended by a particular main group. In certain cases, however, it is difficult to establish precise boundaries between two different main groups. In Section 16.2, therefore, specific points of recognition have been given for certain exceptions and with reference to other similar types.

One reason for the classification guide being grouped into basic types has been to impose tight boundary limits, with the intention of simplifying the choice of supplier from the point of view of purchasing.

Classification has been based primarily upon the design principles with rotodynamic pumps and positive displacement pumps constituting the main groups. Subdivision has also followed designations according to construction principles and only the final break-down of groups has led to designation according to the pumped liquid and application.

European pump manufacturers and suppliers guide, Section 16.3

This important section has been compiled by means of a questionnaire sent to all European manufacturers and suppliers of pumps, ancillary equipment and services inviting them to submit details of their products. The guide lists the names of suitable manufacturers and suppliers and their country of origin, against specific pump categories. Discussions were held, where necessary, with individual companies to make certain their activities were correctly interpreted. There will, however, inevitably be some overlapping due to limitations of category descriptions and the information is for guidance only.

The guide is cross-referenced to Chapter 3, enabling the various pump types to be compared and potential manufacturers and suppliers to be identified.

European manufacturerers and suppliers of ancillary products, Section 16.4

An alphabetical guide to manufacturers and suppliers of ancillary products, such as couplings, motors and packing materials has also been compiled from the questionnaire.

Names and addresses of European pump manufacturers and suppliers, plus Trade Names, Sections 16.5 and 16.6

This section has also been compiled by means of the same questionnaire. Full company details are listed alphabetically, by country.

16.2 Pump classification guide

	Category	Description	Characteristics and/or Distinctive Features Approximate operational limits	Exceptions/Remarks	See also Category
Rotodynamic pumps	005	Standard water pumps	Overhung impeller, single stage pumps for general service on clean, mildly or non-corrosive, non-hazardous liquids. Pumps can be close-coupled or baseplate mounted, inline or end suction. Max pressure 16 barg Max temperature 80/120 °C	For pump interchangeability use prEN 733 std	025, 095
	010	Double suction axially-split pumps	Between bearings single stage pumps of larger capacity than 005 for general service on clean, mildly or non-corrosive, non-hazardous liquids. Generally inline connections. Max pressure 16 barg Max temperature 80/120 °C	Size for size can run faster than **005**. Better NPSHr than **005**.	055, 060
	015	Heating, water and sanitation pumps	Small pumps for domestic and commercial service. Heating circulation pumps, 'wet' and 'dry' motors. Hot, cold and drinking water booster pumps. Max pressure 40 barg Max temperature 120 °C	Single and multi-stage pumps available. Can be twin pump installations with one set of pipe connections.	
	020	Automatic water supply packages	Self-priming pumps for shallow wells with packaged storage tank and pressure or level sensing for on-off control.	When self-priming not required plunger or piston could be used for greater pressure.	070, 075, 210, 215, 270, 275
	025	Standard pumps to prEN 733	Horizontal end suction single stage pumps with specified performance and dimensions for clean liquids. Metric flanges. Max pressure 10 barg Max temperature 80/120 °C		005, 095
	030	Agricultural pumps without driver	General purpose pumps, usually geared single stage, for mounting on tractor power take-off. Diaphragm pumps are also available. Max pressure 20 barg Max temperature 80 °C	Centrifugal pumps can be self-priming and cope with solids	
	035	Fixed irrigation pumps	Vertical suspended wet pit or borehole pumps. Max pressure 10 barg Max temperature 80 °C	Pumps designed to cope with some small solids.	070, 075, 085
	040	Machine tool coolant pumps	Special pumps for machine tool water/oil emulsions. Usually vertical suspended pumps in reservoir. Max pressure 10 barg	Complete packaged systems available See BS 3766 See DIN 5440 See NF E44301	
	045	Marine pumps	Special pumps for use onboard ships. Used in engine rooms, cargo and ballast pumps. Piston pumps used for some applications.	Bronze used extensively. Insurance company certification necessary.	275

16.2 Pump classification guide

	Category	Description	Characteristics and/or Distinctive Features Approximate operational limits	Exceptions/Remarks	See also Category
Rotodynamic pumps	050	Electrically driven submersible pumps	Usually water pumps suitable for complete immersion. Intended for cellar or foundation drainage. Can be used for lifting water from streams, rivers and excavations. Max pressure 40 barg Max temperature 80 °C	Special materials available for corrosion resistance. Special designs available for solids and sludge. Non-metallic designs available.	055, 060, 065
	055	Vertical wet pit pumps	Vertical pumps for suspension in tanks or pits. Available as axial, mixed or centrifugal pumps. Max pressure 16 barg Max temperature 120 °C	Special materials available including non-metallic	060, 175
	060	Vertical dry pit pumps	Vertical suspended pumps with the liquid piped to the suction connection, similar to **055**. Max pressure 16 barg Max temperature 120 °C		055, 175
	065	Portable self-priming pumps and submersible pumps not electrically driven	Contractors site pumps, usually engine driven. Compressed air submersible pumps most popular after electric. Max pressure 10 barg Max temperature 80 °C	Can handle occasional fairly large solids	050
	070	Deep well pumps with ejector	A jet pump driven by a self contained pump package for water extraction from levels below 5 to 7 m. Max pressure 10 barg Max temperature 80 °C		020, 075
	075	Submersible deep well pumps	Multi-stage pumps from very small sizes for installation in bore holes for water extraction. Max pressure 100 barg Max temperature 100 °C		020, 070
	080	Wash water pump packages	Self-contained fully assembled packages for supplying medium pressure water for car washing or car washers and similar applications. Max pressure 16 barg Max temperature 100 °C	Usually multi-stage segmental centrifugal pumps but piston and plunger pump versions also built.	270, 275
	085	Multi-stage segmental pumps < 300 m	Horizontal and vertical types available in cast iron and bronze. Max pressure 25 barg Max temperature 180 °C	Clean liquids, not hazardous. Can be used as power recovery turbines.	090
	090	Multi-stage segmental pumps > 300 m	Usually horizontal in cast iron, bronze, stainless steel and super stainless steel. Max pressure > 431 barg Max temperature 220 °C	Popular for boiler feed water pumps and sea water. Can be used as power recovery turbines.	
	095	Standard pumps to ISO 2858/ ISO 3069/ISO 3661	Single-stage over-hung impeller end suction pump with back pull-out facilities. 16 barg chemical pump. Stuffing box and seal cavities to ISO 3069. Package dimensions to ISO 3661. Max pressure 16 barg Max temperature 250 °C	Lighter version of API 610 pumps. Similar to ANSI B73.1. Some manufacturers make a 25 barg version.	100, 105, 110, 130
	100	End suction pumps to ASME/ANSI B73.1	Single-stage over-hung impeller end suction pump with back pull-out facilities. Inch version of ISO 2858. Chemical pump in various materials. Max pressure 12/20 barg Max temperature > 250 °C	Size and duty ranges specified. Non-metallic versions available.	095, 105, 110, 130
	105	Inline pumps to ASME/ANSI B73.2	Single-stage vertical over-hung impeller pump. Chemical pump in three styles of construction. Max pressure 12/20 barg Max temperature 250 °C	Size and duty ranges specified. API 610 versions available.	095, 100, 110, 130
	110	End suction pumps to API 610	Single-stage over-hung impeller with end suction and back pull-out facilities. Centre-line mounted heavy duty refinery pump in most materials. Steel bearing brackets. Cast iron casings not popular. Max pressure > 100 barg Max temperature > 250 °C	Not a dimensional standard or a duty range standard.	095, 100, 105, 130
	115	Double suction pumps to API 610	Single-stage double entry impeller between bearings with rotor removal facilities. Centre-line mounted heavy duty refinery pump in most materials. Cast iron casings not popular. Max pressure > 50 barg Max temperature > 250 °C	Not a dimensional standard or a duty range standard.	
	120	Magnetic drive pumps	Sealed pumps without mechanical seals or packing. Usually single-stage centrifugal pumps. Many materials available. Max pressure 50 barg Max temperature 250 °C	Non-metallic pumps available. Gear pumps also available.	125
	125	Canned motor pumps	Pumps sealed with electric motor making all dynamic seals unnecessary. Vertical and horizontal types. Max pressure > 431 barg Max temperature > 250 °C	Designs available for special applications such as molten metals. Also cryogenic applications and high suction pressures.	120
	130	Non-metallic pumps	Usually single-stage over-hung end suction centrifugal pumps. Chemical pumps for corrosive applications. Max pressure ≈ 7 barg Max temperature 100 °C	Similar in construction to categories **095** and **100**.	
	135	Pulp pumps	Horizontal single-stage over-hung centrifugal pumps with open impellers of rugged construction. Normally in cast iron or stainless steel. Max pressure 7.5 barg Max temperature 110 °C	Similar to non-clogging pumps and solids handling pumps.	140, 180, 185, 190

16.2 Pump classification guide

	Category	Description	Characteristics and/or Distinctive Features Approximate operational limits	Exceptions/Remarks	See also Category
Rotodynamic pumps	140	Pumps for solids > 10 mm	Heavy duty horizontal single-stage over-hung centrifugal impeller pumps. Generally in cast iron or chrome alloy cast iron. Max pressure ≈ 12 barg Max temperature 80 °C	Maximum solids size dependant upon pump size. Electrically driven submersible pumps should also be considered.	050
	145	Pumps for solids < 10 mm	Heavy duty horizontal single-stage over-hung centrifugal impeller pumps generally with a rubber lining. Max pressure ≈ 7 barg Max temperature 80 °C	Electrically driven submersible pumps should also be considered.	050
	150	Hygienic quality pumps	Usually horizontal single stage over-hung impeller pumps in stainless steel. Pumps suitable for 'cleaning in place'. Max pressure ≈ 10 barg Max temperature 120 °C	Progressive cavity, lobe and jet pumps may be suitable.	235, 240, 290
	155	High speed single-stage pumps	Horizontal and vertical over-hung centrifugal impeller pumps. Max pressure 150 barg Max temperature 250 °C	Chemical and refinery applications.	
	160	High speed multi-pump packages	Two or three high speed single-stage pumps driven by one motor. Horizontal versions only. Max pressure 310 barg Max temperature 250 °C	Chemical and refinery applications. Piston or plunger possible alternatives.	270, 275
	170	Multi-stage radially-split pumps	The most rugged multi-stage pump design for the most arduous duties. Usually centre-line mounted. Max pressure ≈ 430 barg Max temperature > 250 °C		
	175	Vertical multi-stage pumps	Built as axial, mixed flow and centrifugal versions. Used when NPSHa is low or space is restricted. Max pressure 50 barg Max temperature > 250 °C		
	180	Non-clogging pump with standard motor	Mainly for sewage with free-flow or recessed impeller design. Max pressure 10 barg Max temperature 80 °C	Efficiency may be low. Capable of 'gentle pumping'.	185, 190
	185	Submersible non-clogging pump	Mainly for sewage with free-flow or recessed impeller design. Max pressure 10 barg Max temperature 80 °C	Efficiency may be low. Capable of 'gentle pumping'.	180, 190
	190	Other non-clogging pumps	Centrifugal pumps with auger inducers or pumps with maceraters built-in. Max pressure 10 barg Max temperature 80 °C	Low speed. Capable of 'gentle pumping'.	180, 185
	195	Mixed flow pumps	Usually vertical with pump suspended in the liquid. High flows at low differential heads. Max pressure 10 barg Max temperature 80 °C	Can be multi-stage. Can be used for bore-holes.	055, 175
	200	Axial propeller pumps	Can be horizontal or vertical, usually for large flows. Can be built into heavy pressure casings for high suction pressure applications. Max pressure 100 barg Max temperature > 250 °C	Easy to build in large sizes. Can have inlet guide vanes for flow control.	
	205	Power recovery turbines	Can be single or multi-stage. Pumps running in reverse more popular than 'proper' turbines.	Very cost effective.	600
	210	Liquid ring pumps	Single and multi-stage versions. Can be self priming once filled. Max pressure 40 barg Max temperature 120 °C	Usually small. Steep characteristic.	
	215	Peripheral pumps	Higher heads than centrifugal pumps, can be single and multi-stage. Max pressure 40 barg Max temperature 250 °C		
Positive displacement pumps	220	External gear pumps	Capable of high pressures and handling high viscosity liquids. Limited solids handling capabilities. Many material combinations. Max pressure > 430 barg Max temperature > 250 °C	Liquid must have some lubricating qualities. Can be magnetic drive.	225
	225	Internal gear pumps	Preferred for some process applications, built in larger sizes than external gears. Max pressure 16 barg Max temperature > 250 °C	Liquid must have some lubricating qualities. Can be magnetic drive.	
	230	Screw pumps with two or more screws	Three and five screw pumps are ideal for clean lubricating liquids such as lubricating oil. Geared twin screw pumps are good for liquids which have small solids and gases. Max pressure ≈ 240 barg Max temperature > 250 °C	Geared twin screw pumps good with gases up to 97%.	235, 240
	235	Progressing cavity pump	Good for all liquids. Good solids handling capabilities. Max pressure 40 barg Max temperature 120 °C	Can be hygienic quality.	
	240	Lobe pumps and rotary piston pumps	Good with low and high viscosity liquids. Depending upon lobe profiles can be very good with solids. Max pressure 40 barg Max temperature 250 °C	Can be hygienic quality. Capable of 'gentle pumping'.	
	245	Rigid vane and flexible vane pumps	Vane pumps can handle a wide range of liquids including quantities of gas and small solids. Low latent heat liquids not a problem. Hydraulic oil pumps capable of high pressures. Max pressure ≈ 200 barg Max temperature 80 °C	Maximum temperature dependant upon vane material.	220, 225

16.2 Pump classification guide

	Category	Description	Characteristics and/or Distinctive Features Approximate operational limits	Exceptions/Remarks	See also Category
Positive displacement pumps	250	Peristaltic hose pump	Available from very small to quite large. Liquid compatibility good because of wide range of hose materials. Suitable for high viscosities and solids. Max pressure ≈ 10 barg Max temperature ≈ 80 °C	Hygienic and pharmaceutical quality. Capable of 'gentle pumping'.	
	255	Rotary peristaltic pump	Restricted choice of sizes compared to hose pumps. Max pressure ≈ 10 barg Max temperature ≈ 80 °C		
	260	Piston pumps for hydraulic power - radial and axial pistons	Radial and axial piston pumps for hydraulic oil and water-oil emulsions. Generally very clean liquids. Variable capacity and variable speed. Max pressure > 430 barg Max temperature ≈ 120 °C	Some pumps are capable of pumping in both directions. New designs for sea water.	
	265	Portable plunger pump packages for high pressure descaling/cleaning	High pressure descaling and cleaning packages. Mostly water, some with chemical dosing. From 7.5 kW.		
	270	Plunger pumps - horizontal and vertical	Horizontal and vertical, with 3 to 9 cylinders. Standard versions for clean water. Chemical, sewage and solid handling versions available. Max pressure > 2000 barg Max temperature > 250 °C	Design affected by liquid properties. Performance not greatly affected. Capable of 'gentle pumping'.	
	275	Piston pumps	Horizontal with 2 or 3 cylinders. Standard versions for water and 'mud' pumps. Max pressure 100 barg Max temperature 250 °C	Cast iron and chrome cast iron standard. Capable of 'gentle pumping'.	
	280	Diaphragm pumps	Horizontal with 2 or more cylinders. Mechanical, compressed air or hydraulic drive. Good with most liquids and solids. Max pressure ≈ 240 barg Max temperature 120 °C	Some designs have very complicated hydraulic drive systems. Capable of 'gentle pumping'.	
	285	Metering pumps	Can be piston, plunger or diaphragm design depending upon the liquid and the pressure. Very low accurate flows possible. Max pressure > 430 barg Max temperature ≈ 120 °C	Constant speed and variable speed possible.	
Other pumps	290	Jet pumps	Also called ejectors and injectors depending upon the process. Good solids and viscous handling capabilities. Need a hydraulic source of power rather than electricity. Max pressure 10 barg for commercial units	Simple, robust, low efficiency. Can be 'home-made' for special applications.	
	295	Air lift pumps	Injected air causes a liquid column to rise. Simple, robust, low efficiency.	'Gentle pumping'.	
	300	Barrel emptying pumps	Pumps designed to pass through caps of standard barrels. Pumps can be electric or compressed air powered. Most pumps are low pressure but high pressure models are available. Max pressure ≈ 240 barg Max temperature 80 °C		
	305	Rigid screw pumps	Sometimes called Archimedean screw pumps. High flow, low head for safe liquids. Very good for all types of contaminants in liquids.	'Gentle pumping'.	
	310	Pitot tube pumps	Also called Prandtl tube pumps. Generally small flows at relatively high heads. Max pressure 50 barg Max temperature ≈ 250 °C	More efficient than an equivalent centrifugal pump.	
	315	Miscellaneous types	Hydraulic ram Contraction pump	'Self-powered' and hand-operated pumps for water	

14.3 Pump manufacturers and suppliers guide

Rotodynamic Pumps

- Pumps in C.I., C.S. and Bronze for water, mildly corrosive non-hazardous liquids
  - 1 and 2 stage
- Pumps in various materials for food, corrosive and hazardous liquids
  - Multi-stage
  - 1 and 2 stage

Column code	Description
005	Standard water pumps
010	Double suction axially-split pumps
015	Heating, water and sanitation pumps
020	Automatic water supply packages
025	Standard pumps to prEN 733
030	Agricultural pumps without driver
035	Fixed irrigation pumps
040	Machine tool coolant pumps
045	Marine pumps
050	Electrically driven submersible pumps
055	Vertical wet pit pumps
060	Vertical dry pit pumps
065	Portable self-priming pumps and submersible pumps not electrically driven
070	Deep well pumps with ejector
075	Submersible deep well pumps
080	Wash water pump packages
085	Multi-stage segmental pumps < 300m
090	Multi-stage segmental pumps > 300m
095	Standard pumps to ISO 2858/ISO 3069/ISO 3661
100	End suction pumps to ASME/ANSI B73.1
105	Inline pumps to ASME/ANSI B73.2
110	End suction pumps to API 610
115	Double-suction pumps to API 610
120	Magnetic drive pumps
125	Canned motor pumps
130	Non-metallic pumps

Manufacturer	Country Code	005	010	015	020	025	030	035	040	045	050	055	060	065	070	075	080	085	090	095	100	105	110	115	120	125	130
A de Backer & Co, BVBA	B																●						●				
APV Rosista	DK	●															●				●						
Abel GmbH	D	●																									
ABS Pumpen AG	D							●								●											
ABS Pumps Ltd	UK		●								●																
AGI Pompe Srl	I												●										●				
A Ahlstrom Corporation	SF	●	●																	●							
Air Pumping Engineering Services	UK																										●
The Albany Engineering Co Ltd	UK								●																●		●
Alfa-Laval Pumps	UK																										
Alldos Eichler GmbH Dosing Pumps	D																										
Allis Mineral Systems (UK) Ltd	UK																				●		●				
Allweiler AG	D																								●		
Alpha Pompe Srl	I	●														●	●						●				
AMB Rateau SA	B	●		●												●			●					●			
Andrews Sykes Ltd	UK					●								●													
Apex Fluid Engineering Ltd	UK	●	●	●													●		●								
Apollowerk Gössnitz GmbH	D				●		●										●										
Appliance Components Ltd	UK																										
APV Gaulin GmbH	D																										
APV Howard Pumps Ltd	UK																										
APV Industrial Pump & Mixer Division	UK									●															●	●	●
APV Rannie A/S	DK																										
Aqua Hydraulics Ltd	UK																										
Argal Srl	I	●				●															●	●	●				
Armstrong Pumps	UK	●		●													●				●						
The ARO Corporation (UK) Ltd	UK																										
Aro, N.V.	B																										
Arsopi SA	P	●																									
ASTI	F																										●
Aturia Pompe SpA	I	●															●				●		●				
Audoli & Bertola Co	I					●					●										●						
Charles Austen Pumps Ltd	UK	●																									
Autoclude Ltd	UK																										
Autometric Pumps Ltd, Lowara (UK) Ltd	UK																										
Baric Pumps Ltd	UK	●		●													●				●						
BBC Elettropompe srl	I																										
The Bedford Pump Company Ltd	UK									●																	
Heinrich Behrens Pumpenfabrik GmbH & Co	D								●																		
Bellin SpA	I																										
Berendsen Fluid Power	UK																										

	Rotodynamic pumps																Positive displacement pumps													Other pumps					
	Pumps in various materials for food, corrosive and hazardous liquids								Self-cleaning								Rotary								Reciprocating										
	1 and 2 stage				Multi-stage																														
Pulp pumps	Pumps for solids > 10mm	Pumps for solids < 10mm	Hygienic quality pumps	High speed single-stage pumps	High speed multi-stage pump packages	Multi-stage axially-split pumps	Multi-stage radially-split pumps	Vertical multi-stage pumps	Non-clogging pumps with standard motor	Submersible non-clogging pump	Other non-clogging pumps	Mixed flow pumps	Axial propeller pumps	Power recovery turbines	Liquid ring pumps	Peripheral pumps	External gear pumps	Internal gear pumps	Screw pumps with 2 or more screws	Progressing cavity pump	Lobe pumps and rotary piston pumps	Rigid vane pumps and flexible vane pumps	Peristaltic hose pump	Rotary peristaltic pump	Piston pumps for hydraulic power - radial and axial pistons	Portable plunger pump packages for high pressure descaling/cleaning	Plunger pumps - horizontal and vertical	Piston pumps	Diaphragm pumps	Metering pumps	Jet pumps	Air lift pumps	Barrel emptying pumps	Rigid screw pumps	Pilot tube pumps	Miscellaneous types
35	140	145	150	155	160	165	170	175	180	185	190	195	200	205	210	215	220	225	230	235	240	245	250	255	260	265	270	275	280	285	290	295	300	305	310	315

14.3 Pump manufacturers and suppliers guide

Manufacturer	Country Code	005 Standard water pumps	010 Double suction axially-split pumps	015 Heating, water and sanitation pumps	020 Automatic water supply packages	025 Standard pumps to prEN 733	030 Agricultural pumps without driver	035 Fixed irrigation pumps	040 Machine tool coolant pumps	045 Marine pumps	050 Electrically driven submersible pumps	055 Vertical wet pit pumps	060 Vertical dry pit pumps	065 Portable self-priming pumps and submersible pumps not electrically driven	070 Deep well pumps with ejector	075 Submersible deep well pumps	080 Wash water pump packages	085 Multi-stage segmental pumps < 300m	090 Multi-stage segmental pumps > 300m	095 Standard pumps to ISO 2858/ISO 3069/ISO 3661	100 End suction pumps to ASME/ANSI B73.1	105 Inline pumps to ASME/ANSI B73.2	110 End suction pumps to API 610	115 Double-suction pumps to API 610	120 Magnetic drive pumps	125 Canned motor pumps	130 Non-metallic pumps
Beresford Pumps Ltd	UK																								●		
Bevi Wright Ltd	UK			●				●																			
BGA International	UK							●																			
Bieri Pumpenbau AG, Biral International	CH			●												●							●				
Biwater Pumps	UK										●	●	●														
Bizzi & Tedeschi snc DT Bizzi G & G	I														●	●											
Boehler GmbH	A																										
Bomba Elias SA	E	●			●															●			●				
Bombas Azcue SA	E	●		●												●											
Bombas Grundfos (Portugal) Lda	P			●							●					●											
Bombas Ideal SA	E			●									●						●	●							
Bombas Itur-Manufacturas Aranzabal SA	E	●		●				●												●			●				
J H Bornemann GmbH & Co KG	D																										
Bran + Luebbe (GB) Ltd	UK																										
Bran + Luebbe GmbH	D																										
E Braude Ltd	UK																			●					●		●
Brimotor Ltd	UK	●						●						●													
David Brown Pumps Ltd	UK																						●	●			
Burdosa UK Ltd	UK																										
BW/IP International BV	NL															●							●	●			
Calpeda Ltd	UK	●									●																
Calpeda SpA	I	●									●																
Capital Engineering Services Ltd	UK		●																								
Caprari SpA	I					●	●				●		●			●											
Cardo Pump AB	S			●							●									●					●		
CAT Pumps (UK) Ltd	UK																										
CDR Pompe Motori Srl	I	●				●														●	●		●		●		
Ceetak Fluid Handling	UK																								●		
Centrilift	UK															●											
Chem-Resist	UK																								●		
Claessen Pumps Ltd	UK										●																
Clasal N.V.	B	●														●											
Cleghorn Waring & Co (Pumps) Ltd	UK									●															●		
Clextral	F																										
Combimac BV	NL														●												
Componenta International Ltd	UK	●									●																
Cornell Pumps (Europe)	UK	●					●	●																			
Wm Coulthard & Co Ltd	UK																										
CP Instrument Co Ltd	UK																								●		
Crane Fluid Systems	UK	●		●	●	●					●																
Crest Pumps Ltd	UK													●													

Pump Selection Chart

Category	Subcategory	Pump type	Page
Rotodynamic pumps	Pumps in various materials for food, corrosive and hazardous liquids — 1 and 2 stage	Pulp pumps	135
		Pumps for solids > 10mm	140
		Pumps for solids < 10mm	145
		Hygienic quality pumps	150
		High speed single-stage pumps	155
	Multi-stage	High speed multi-stage pump packages	160
		Multi-stage axially-split pumps	165
		Multi-stage radially-split pumps	170
		Vertical multi-stage pumps	175
	Self-cleaning	Non-clogging pumps with standard motor	180
		Submersible non-clogging pump	185
		Other non-clogging pumps	190
		Mixed flow pumps	195
		Axial propeller pumps	200
		Power recovery turbines	205
		Liquid ring pumps	210
		Peripheral pumps	215
Positive displacement pumps	Rotary	External gear pumps	220
		Internal gear pumps	225
		Screw pumps with 2 or more screws	230
		Progressing cavity pump	235
		Lobe pumps and rotary piston pumps	240
		Rigid vane pumps and flexible vane pumps	245
		Peristaltic hose pump	250
		Rotary peristaltic pump	255
	Reciprocating	Piston pumps for hydraulic power - radial and axial pistons	260
		Portable plunger pump packages for high pressure descaling/cleaning	265
		Plunger pumps - horizontal and vertical	270
		Piston pumps	275
		Diaphragm pumps	280
		Metering pumps	285
Other pumps		Jet pumps	290
		Air lift pumps	295
		Barrel emptying pumps	300
		Rigid screw pumps	305
		Pilot tube pumps	310
		Miscellaneous types	315

351

14.3 Pump manufacturers and suppliers guide

		Rotodynamic Pumps																								
			Pumps in C.I., C.S. and Bronze for water, mildly corrosive non-hazardous liquids														Pumps in various materials for food, corrosive and hazardous liquids											
			1 and 2 stage												Multi-stage						1 and 2 stage							
	Country Code	Standard water pumps	Double suction axially-split pumps	Heating, water and sanitation pumps	Automatic water supply packages	Standard pumps to prEN 733	Agricultural pumps without driver	Fixed irrigation pumps	Machine tool coolant pumps	Marine pumps	Electrically driven submersible pumps	Vertical wet pit pumps	Vertical dry pit pumps	Portable self-priming pumps and submersible pumps not electrically driven	Deep well pumps with ejector	Submersible deep well pumps	Wash water pump packages	Multi-stage segmental pumps < 300m	Multi-stage segmental pumps > 300m	Standard pumps to ISO 2858/ISO 3069/ISO 3661	End suction pumps to ASME/ANSI B73.1	Inline pumps to ASME/ANSI B73.2	End suction pumps to API 610	Double-suction pumps to API 610	Magnetic drive pumps	Canned motor pumps
		005	010	015	020	025	030	035	040	045	050	055	060	065	070	075	080	085	090	095	100	105	110	115	120	125
Alan Dale Pumps Ltd	UK														●											
Datronix	UK																									
Dawson Downie Lamont Ltd	UK																									
A/S De Smithske (DESMI)	DK		●							●											●	●				
Delasco Pumps	F																									
Delaval-Stork v.o.f.	NL																									
Deloule Espanola SA	E																						●			
Denco Ltd	UK																									
Denver Process Equipment Ltd	UK							●				●								●						
Denver Sala, Sala International AB	S										●															
DIA Pumpenfabrik	D		●		●						●										●					
Dia-Meter Pumps Ltd	UK																									
Dickow Pumpen KG	D	●		●																	●					
Dosapro Milton Roy	F																									
DP Pumps	NL			●												●			●			●				
Drakos-Polemis Inc	GR			●		●						●										●	●			
Dual Pumps Ltd	UK					●												●								
Dualway Metering Pumps Ltd	UK																									
Durco Europe	B																				●	●	●		●	
Durco Process Equipment Ltd	UK																				●	●	●		●	
Dymatic Industri AB	S																									
Ebara Italia SpA	I				●														●		●					
Ebara UK Ltd	UK				●														●							
Edur Pumpenfabrik	D	●																	●			●				
Edwards and Jones Ltd	UK																									
Edwards High Vacuum International	UK																									
Emile Egger & Cie.	CH							●				●		●												
Eheim GmbH & Co KG	D																								●	
EMU-Unterwasserpumpen GmbH	D										●	●	●			●										
Ensival SA	B												●					●			●		●			
EnviroTech Pumpsystems BV	NL																									
Envirotech Pumpsystems UK	UK																									
FA-SA Pompe Srl	I																									
Fapmo	F							●																	●	
Feluwa Pumps, Schlesiger & Co KG	D									●																
Ferguson & Timpson Ltd	UK																									
FIPS Srl	I			●								●														
Flotronic Pumps Ltd	UK																									
Flotronics GmbH	D																									
Fluid Pumps (Bristol) Ltd	UK	●								●	●									●						
Flux-Geräte GmbH	D			●										●	●	●									●	

14.3 Pump manufacturers and suppliers guide

Rotodynamic Pumps

Pumps in C.I., C.S. and Bronze for water, mildly corrosive non-hazardous liquids

1 and 2 stage / Multi-stage

Pumps in various materials for food, corrosive and hazardous liquids

1 and 2 stage

	Country Code	Standard water pumps (005)	Double suction axially-split pumps (010)	Heating, water and sanitation pumps (015)	Automatic water supply packages (020)	Standard pumps to prEN 733 (025)	Agricultural pumps without driver (030)	Fixed irrigation pumps (035)	Machine tool coolant pumps (040)	Marine pumps (045)	Electrically driven submersible pumps (050)	Vertical wet pit pumps (055)	Vertical dry pit pumps (060)	Portable self-priming pumps and submersible pumps not electrically driven (065)	Deep well pumps with ejector (070)	Submersible deep well pumps (075)	Wash water pump packages (080)	Multi-stage segmental pumps < 300m (085)	Multi-stage segmental pumps > 300m (090)	Standard pumps to ISO 2858/ISO 3069/ISO 3661 (095)	End suction pumps to ASME/ANSI B73.1 (100)	Inline pumps to ASME/ANSI B73.2 (105)	End suction pumps to API 610 (110)	Double-suction pumps to API 610 (115)	Magnetic drive pumps (120)	Canned motor pumps (125)
Fluxinos Italia Srl	I	●				●																	●			
Foras Pumps Srl	I					●						●										●	●			
Fordwater Pumping Supplies Ltd	UK	●										●											●		●	
Forward Industries Ltd	UK																									
Fraccarolo Pompe	I																									
Friatec Ltd	UK									●										●					●	
Fristam Pumpen F Stamp KG (GmbH & Co)	D															●										
Fristam Pumps Ltd	UK	●																							●	
GEC Alsthom Fluids Group	UK									●									●							
GEA Wiegand GmbH	D																									
GEA Wiegand Kestner	F										●	●														
GEC Alsthom Bergeron SA	F		●			●									●										●	
Gilbert Gilkes & Gordon Ltd	UK													●												
Girdlestone Pumps Ltd	UK	●										●													●	●
Godiva Ltd	UK															●										
R Goodwin International	UK									●	●	●														
Goulds Pumps Europe BV	NL													●	●					●				●		
Graco UK Ltd	UK																									
Graham Precision Pumps Ltd	UK	●		●																●		●				
Willi F Grassel	D																									
Grindex AB	S										●															
Grosvenor Pumps Ltd	UK																									
Grundfos (Ireland) Ltd	IRL			●	●						●					●										
Grundfos A/S	DK	●		●	●		●	●		●						●										
Grundfos Pumps Ltd	UK			●				●								●										
H₂O Waste-Tec	UK															●										
Halberg Maschinenbau GmbH	D					●																			●	
Paul Hammelmann Maschinenfabrik GmbH	D																									
Hamworthy Allweiler Ltd	UK																								●	
Hamworthy Hydraulics Ltd	UK																									
Hamworthy Pumps & Compressors Ltd	UK	●														●					●		●			
Harrison Group Ltd	UK													●												
Haskel Energy Systems Ltd	UK																									
Hauke GmbH & Co KG	A																							●	●	
Hayward Tyler Fluid Dynamics Ltd	UK																									
Hayward Tyler Sumo	UK										●					●								●	●	
Hermetic-Pumpen GmbH	D																			●					●	●
A.C. Herstal SA	B					●		●																		
Hick Hargreaves & Co Ltd	UK																									
Hidrostal Process Engineering Ltd	UK				●	●															●					
Hilge Pumps Ltd	UK																			●						

14.3 Pump manufacturers and suppliers guide

Rotodynamic Pumps

Pumps in C.I., C.S. and Bronze for water, mildly corrosive non-hazardous liquids

1 and 2 stage / Multi-stage

Pumps in various materials for food, corrosive and hazardous liquids

1 and 2 stage

Manufacturer	Country Code	005 Standard water pumps	010 Double suction axially-split pumps	015 Heating, water and sanitation pumps	020 Automatic water supply packages	025 Standard pumps to prEN 733	030 Agricultural pumps without driver	035 Fixed irrigation pumps	040 Machine tool coolant pumps	045 Marine pumps	050 Electrically driven submersible pumps	055 Vertical wet pit pumps	060 Vertical dry pit pumps	065 Portable self-priming pumps and submersible pumps not electrically driven	070 Deep well pumps with ejector	075 Submersible deep well pumps	080 Wash water pump packages	085 Multi-stage segmental pumps < 300m	090 Multi-stage segmental pumps > 300m	095 Standard pumps to ISO 2858/ISO 3069/ISO 3661	100 End suction pumps to ASME/ANSI B73.1	105 Inline pumps to ASME/ANSI B73.2	110 End suction pumps to API 610	115 Double-suction pumps to API 610	120 Magnetic drive pumps	125 Canned motor pumps
Richard Hill Pumps Ltd	UK	●						●						●				●	●							
HMD Seal/Less Pumps Ltd	UK																			●	●	●		●		
Holden and Brooke Ltd	UK	●		●							●															●
H R Holfeld (Pumps) Ltd	UK				●											●					●					
Homa Pumpenfabrik GmbH	D	●											●													
Houttuin Holland BV	NL																									
Huedig GmbH & Co	D					●									●											
Hydrainer Pumps Ltd	UK	●												●												
Hydrair Ltd	UK																									
Hydrobel SA	B																								●	
IHC Holland NV	NL																									
IMO AB	S																									
Imovilli Pompe Srl	I																									
Ingersoll-Dresser Pumps (UK) Ltd, Gateshead	UK																	●			●	●				
Ingersoll-Dresser Pumps (UK) Ltd, Newark	UK	●	●															●							●	
Ingersoll-Rand Sales Company Ltd	UK																									
INOXPA SA	E	●																			●					
Inoxpa UK (Euro-Industrial Engineering Ltd)	UK	●																								
Interdab Ltd	UK			●	●											●	●									
ITT Flygt AB	S							●			●	●		●												
ITT Jabsco	UK									●																
Jacuzzi Europa SpA	I	●		●	●										●											
Jesco Dosiertechnik GmbH	D																									
JLC Pumps & Engineering Co Ltd	UK								●	●																
Johnson Pump (UK) Ltd	UK	●				●															●				●	
Jung Pumpen GmbH & Co	D									●	●	●														
Jurop SpA	I					●																				
Kamat-Pumpen GmbH & Co KG	D																									
Emmanuel N Kazis S.A.	GR	●									●															
Kecol Industrial Products Ltd	UK																									
Kestner Engineering Co Ltd	UK																			●						●
Kinder-Janes Engineers Ltd	UK																									
Klaus Union	D	●			●															●	●		●			
Oy Kolmeks AB	SF																									
KOVO Konice, v.d.	CZ															●										
Kracht GmbH	D																									
Kraeutler GmbH & Co	A								●	●																
KSB Aktiengesellschaft	D	●														●				●					●	
KSB Ltd	UK	●			●											●										
Kvaerner Eureka AS	N		●																		●					
Lederle GmbH	D																			●					●	

356

	Rotodynamic pumps																	Positive displacement pumps													Other pumps					
	Pumps in various materials for food, corrosive and hazardous liquids								Self-cleaning									Rotary								Reciprocating										
	1 and 2 stage				Multi-stage																															
Pulp pumps	Pumps for solids > 10mm	Pumps for solids < 10mm	Hygienic quality pumps	High speed single-stage pumps	High speed multi-stage pump packages	Multi-stage axially-split pumps	Multi-stage radially-split pumps	Vertical multi-stage pumps	Non-clogging pumps with standard motor	Submersible non-clogging pump	Other non-clogging pumps	Mixed flow pumps	Axial propeller pumps	Power recovery turbines	Liquid ring pumps	Peripheral pumps	External gear pumps	Internal gear pumps	Screw pumps with 2 or more screws	Progressing cavity pump	Lobe pumps and rotary piston pumps	Rigid vane pumps and flexible vane pumps	Peristaltic hose pump	Rotary peristaltic pump	Piston pumps for hydraulic power - radial and axial pistons	Portable plunger pump packages for high pressure descaling/cleaning	Plunger pumps - horizontal and vertical	Piston pumps	Diaphragm pumps	Metering pumps	Jet pumps	Air lift pumps	Barrel emptying pumps	Rigid screw pumps	Pilot tube pumps	Miscellaneous types
35	140	145	150	155	160	165	170	175	180	185	190	195	200	205	210	215	220	225	230	235	240	245	250	255	260	265	270	275	280	285	290	295	300	305	310	315

14.3 Pump manufacturers and suppliers guide

Rotodynamic Pumps

Pumps in C.I., C.S. and Bronze for water, mildly corrosive non-hazardous liquids

1 and 2 stage

Pumps in various materials for food, corrosive and hazardous liquids

Multi-stage | 1 and 2 stage

Manufacturer	Country Code	005 Standard water pumps	010 Double suction axially-split pumps	015 Heating, water and sanitation pumps	020 Automatic water supply packages	025 Standard pumps to prEN 733	030 Agricultural pumps without driver	035 Fixed irrigation pumps	040 Machine tool coolant pumps	045 Marine pumps	050 Electrically driven submersible pumps	055 Vertical wet pit pumps	060 Vertical dry pit pumps	065 Portable self-priming pumps and submersible pumps not electrically driven	070 Deep well pumps with ejector	075 Submersible deep well pumps	080 Wash water pump packages	085 Multi-stage segmental pumps < 300m	090 Multi-stage segmental pumps > 300m	095 Standard pumps to ISO 2858/ISO 3069/ISO 3661	100 End suction pumps to ASME/ANSI B73.1	105 Inline pumps to ASME/ANSI B73.2	110 End suction pumps to API 610	115 Double-suction pumps to API 610	120 Magnetic drive pumps	125 Canned motor pumps	130 Non-metallic pumps
Leistritz AG	D																										
LEWA Herbert Ott GmbH & Co	D																										
Linatex Ltd	UK																										●
Lincoln GmbH	D																										
Lowara (UK) Ltd	UK							●				●			●	●				●							
Lowara SpA	I							●				●			●	●				●							
Lumatic (GA) Ltd	UK																										
Lutz (UK) Ltd	UK																										●
Lutz-Pumpen GmbH & Co KG	D																										
Mackley Pumps	UK	●	●													●		●	●		●	●		●			
Macro Precision Pumps Ltd	UK																										
Majmar Srl	I			●								●				●						●					
Mako Oy	SF	●																									
Mamec Oy	SF											●	●	●													
March May Ltd	UK																								●		●
Maschinenfabrik Andritz AG	A		●																				●				
Matra SpA	I			●																							
Maximator (Deeweld)	UK																										
MDM Pumps Ltd	UK																										
Megator Pumps & Compressors Ltd	UK	●								●																	
George Meller Ltd	UK																									●	
Melotte BV	NL										●					●											
Mencarelli Pompe e Valvole Srl	I	●													●												
Mercury Srl	I									●						●											
Milton Roy (UK) Ltd	UK																										
Mono Pumps Ltd	UK																										
Moret	F							●								●											
Mouvex	F																										
MPL Pumps	UK																										
MSA a.s.	CZ	●	●									●	●					●									
Multiphase Pumping Systems Ltd	UK																										
Munster Simms Engineering Ltd (Whale Pumps)	UK	●								●	●			●													
Nash Engineering Co (GB) Ltd	UK																										
Naybuk Pumps Ltd	UK					●	●			●																	
Nijhuis Pompen BV	NL															●											
Nuovo Pignone	I																						●	●			
Ochsner GmbH	A																										
Oilgear Towler Ltd	UK																										
Ondrejovicka strojirna spol. sro.	CZ																										
Osna Pumpen J Hartlage GmbH & Co KG	D															●											
Packo Diksmuide	B																						●				

358

	Rotodynamic pumps												Positive displacement pumps													Other pumps					

14.3 Pump manufacturers and suppliers guide

Rotodynamic Pumps

Pumps in C.I., C.S. and Bronze for water, mildly corrosive non-hazardous liquids

Pumps in various materials for food, corrosive and hazardous liquids

Manufacturer	Country Code	005 Standard water pumps	010 Double suction axially-split pumps	015 Heating, water and sanitation pumps	020 Automatic water supply packages	025 Standard pumps to prEN 733	030 Agricultural pumps without driver	035 Fixed irrigation pumps	040 Machine tool coolant pumps	045 Marine pumps	050 Electrically driven submersible pumps	055 Vertical wet pit pumps	060 Vertical dry pit pumps	065 Portable self-priming pumps and submersible pumps not electrically driven	070 Deep well pumps with ejector	075 Submersible deep well pumps	080 Wash water pump packages	085 Multi-stage segmental pumps < 300m	090 Multi-stage segmental pumps > 300m	095 Standard pumps to ISO 2858/ISO 3069/ISO 3661	100 End suction pumps to ASME/ANSI B73.1	105 Inline pumps to ASME/ANSI B73.2	110 End suction pumps to API 610	115 Double-suction pumps to API 610	120 Magnetic drive pumps	125 Canned motor pumps	13
Patay Pumps Ltd	UK																										
PCM Pompes	F																										
Pedrollo SpA	I	●													●	●											
Pegson Ltd	UK		●								●			●													
PEME	F										●					●											
Perfecta Pump AB	S			●																							
F Peroni & C SpA	I																										
Plenty Mirrlees Pumps	UK																										
Pollution Control Engineering Ltd	UK						●																				
Pompe Gabbioneta SpA	I											●	●														
Pompe Travaini SpA	I									●								●									
Pompes Grosclaude	F																								●	●	
Pompes Grundfos Distribution	F			●																							
Pompes Guinard KSB	F	●														●					●		●			●	
Pompes Industrie SA	F														●								●				
Precision Dosing Pumps Ltd	UK																										
Production Techniques Ltd	UK																										●
Pullen Pumps Ltd	UK	●		●	●					●							●										
Pump Engineering Ltd	UK	●																									
Pumpenbau G Schulte	D	●									●						●			●							
Pumpenfabrik Ernst Vogel GmbH	A	●		●	●		●	●	●		●	●	●			●	●	●	●						●		
Pumpenfabrik Oddesse	D														●		●										
Pumpenfabrik Wangen GmbH	D																										
Pumpex AB	S										●																
Putzmeister Ltd	UK																										
Putzmeister-Werk Maschinenfabrik GmbH	D																										
Reda (UK) Ltd	UK														●	●			●								
The Reiss Engineering Co	UK																										
RHL Hydraulics Ltd	UK																										
Richter Chemie-Technik GmbH	D																								●		
Ritz Pumpenfabrik GmbH & Co KG	D	●														●	●										
Riva Hydroart SpA	I															●	●										
Roach Pumps Ltd	UK										●																
Roban Engineering Ltd	UK										●																
Robbins & Myers Europe	B														●												
Robbins & Myers Ltd	UK																										
Robot Pompen BV	NL								●	●																	
Robuschi & C. SpA	I																										
Roper Industries (Europe) Ltd	UK																										
Rotant Srl	I																										
Rotos SpA	I															●					●		●		●		

14.3 Pump manufacturers and suppliers guide

Rotodynamic Pumps

Pumps in C.I., C.S. and Bronze for water, mildly corrosive non-hazardous liquids — 1 and 2 stage | Multi-stage | Pumps in various materials for food, corrosive and hazardous liquids — 1 and 2 stage

Manufacturer	Country Code	005 Standard water pumps	010 Double suction axially-split pumps	015 Heating, water and sanitation pumps	020 Automatic water supply packages	025 Standard pumps to prEN 733	030 Agricultural pumps without driver	035 Fixed irrigation pumps	040 Machine tool coolant pumps	045 Marine pumps	050 Electrically driven submersible pumps	055 Vertical wet pit pumps	060 Vertical dry pit pumps	065 Portable self-priming pumps and submersible pumps not electrically driven	070 Deep well pumps with ejector	075 Submersible deep well pumps	080 Wash water pump packages	085 Multi-stage segmental pumps < 300m	090 Multi-stage segmental pumps > 300m	095 Standard pumps to ISO 2858/ISO 3069/ISO 3661	100 End suction pumps to ASME/ANSI B73.1	105 Inline pumps to ASME/ANSI B73.2	110 End suction pumps to API 610	115 Double-suction pumps to API 610	120 Magnetic drive pumps	125 Canned motor pumps	130 Non-metallic pumps
Rovatti A. & Figli Pompe SpA	I	●					●				●					●											
Rutschi Pumpen AG	CH		●									●									●					●	
SPP Group Ltd	UK				●										●						●		●				
SPP LaBour Pump Company	UK										●									●					●		●
S.I.G.M.A. SpA	I	●		●							●																
SACEM Srl	I															●									●		
Saer Elettropompe Srl	I	●		●							●					●											
Oy E Sarlin AB	SF										●																
Sarlin Ltd	UK																										
Sauer-Sundstrand Ltd	UK																										
Savino Barbera snc	I											●													●		●
Schabaver	F											●	●	●													
Ernst Scherzinger GmbH	D																										
Schmalenberger GmbH & Co	D	●									●					●											
Kranz Schmidt & Co GmbH	D																										
Schuco International London Ltd	UK																								●		
Seddons (Plant & Engineers) Ltd	UK	●									●																
Seepex Seeberger GmbH & Co	D																										
Seepex UK Ltd	UK																										
Seko SpA	I																										
Selwood Pumps	UK																										
Serfilco Europe Ltd	UK																								●		●
Sero Pumpenfabrik	D															●	●	●									
Shurflo Ltd	UK																										
SIGMA Hranice a.s.	CZ	●																									
Sigma Lutin	CZ		●				●				●	●	●					●	●								
SIGMA Olomouc s.p.	CZ	●		●			●													●					●		
SIGMA Trading Ltd	CZ	●		●			●			●																	
Sihi Pumps (UK) Ltd	UK																			●					●	●	
Sihi-Halberg	D		●												●						●	●			●		
Sihi-Maters BV	NL																			●	●				●		
Simaco Elettromeccanica Srl	I		●				●			●																	
Simon-Hartley Ltd	UK																										
SLD Pumps Ltd	UK										●			●													
T Smedegaard A/S	DK	●			●						●									●		●					
Michael Smith Engineers Ltd	UK																								●	●	
Spandau Pumpen	D			●														●			●						
Speck-Kolbenpumpenfabrik	D			●																	●	●					
Speck-Pumpen, Daniel Speck & Sohne	D	●							●																		●
Stanhope Barclay Kellett	UK							●																			
The Sterling Pump Co Ltd	UK	●				●								●						●							

362

14.3 Pump manufacturers and suppliers guide

Rotodynamic Pumps

- Pumps in C.I., C.S. and Bronze for water, mildly corrosive non-hazardous liquids
  - 1 and 2 stage
  - Multi-stage
- Pumps in various materials for food, corrosive and hazardous liquids
  - 1 and 2 stage

Code	Pump type
005	Standard water pumps
010	Double suction axially-split pumps
015	Heating, water and sanitation pumps
020	Automatic water supply packages
025	Standard pumps to prEN 733
030	Agricultural pumps without driver
035	Fixed irrigation pumps
040	Machine tool coolant pumps
045	Marine pumps
050	Electrically driven submersible pumps
055	Vertical wet pit pumps
060	Vertical dry pit pumps
065	Portable self-priming pumps and submersible pumps not electrically driven
070	Deep well pumps with ejector
075	Submersible deep well pumps
080	Wash water pump packages
085	Multi-stage segmental pumps < 300m
090	Multi-stage segmental pumps > 300m
095	Standard pumps to ISO 2858/ISO 3069/ISO 3661
100	End suction pumps to ASME/ANSI B73.1
105	Inline pumps to ASME/ANSI B73.2
110	End suction pumps to API 610
115	Double-suction pumps to API 610
120	Magnetic drive pumps
125	Canned motor pumps
130	Non-metallic pumps

Manufacturer	Country	005	010	015	020	025	030	035	040	045	050	055	060	065	070	075	080	085	090	095	100	105	110	115	120	125	130
Stork Pompen BV	NL																					●			●		
Stork Pumps Ltd	UK									●								●				●					
Stuart Turner Ltd	UK	●		●	●					●																	
Sulzer (UK) Pumps Ltd	UK	●	●	●				●							●			●	●								
Sulzer Weise GmbH	D															●						●					
Gebr. Sulzer AG	CH																			●				●			
Sundstrand International Corporation SA	B																					●					
Svenska Tapflo AB	S																										●
T-T Pumps Ltd	UK	●								●																	
Termomeccanica Italiana SpA	I		●	●														●				●					
Thrige Pumper A/S	DK			●						●												●					
Thyssen Maschinenbau GmbH Ruhrpumpen	D	●	●					●			●	●										●	●	●			
Torres Engineering & Pumps Ltd	UK									●																	
Totton Pumps Ltd	UK	●																							●		●
Treadway Flow Control	UK																										
Otto Tuchenhagen GmbH & Co KG	D																					●					
Tuthill UK Ltd	UK																								●		
Uraca Pumpenfabrik GmbH & Co KG	D																										
Vanton Pumps Ltd	UK										●									●	●						●
Varisco Pompe Srl	I							●						●													
Vergnet SA	F														●	●											
J M Voith GmbH	D		●	●																							
JM Voith Aktiengesellschaft	A							●																			
Wallace & Tiernan Ltd	UK																										
Wanner International Ltd	UK					●																●					
Warman International Ltd	UK										●	●															
Warsop Power Tools	UK										●			●													
Watson-Marlow Ltd	UK													●													
Waukesha Bredel Fluid Handling BV	NL																										
Weda Pump AB	S															●											
Weir Pumps Ltd	UK	●	●					●								●											
Welland Engineering Co Ltd	UK	●			●													●									
Wemco Pumps	UK																										
Wepuko-Hydraulik GmbH & Co	D																										
Wernert Pumpen GmbH	D																								●	●	
Wilo GmbH	D			●														●				●					
Wilo Salmson Pumps Ltd	UK			●										●				●	●								●
Wirth Maschinen- und Bohrgeräte-Fabrik GmbH	D																										
WOMA Apparatebau GmbH	D																										
Yamada Europe BV	NL																										
Albert Ziegler GmbH & Co KG	D		●												●		●								●		

364

Pump Selection Chart

Category	Subcategory	Pump Type	Code
Rotodynamic pumps	Pumps in various materials for food, corrosive and hazardous liquids — 1 and 2 stage	Pulp pumps	35
		Pumps for solids > 10mm	140
		Pumps for solids < 10mm	145
		Hygienic quality pumps	150
		High speed single-stage pumps	155
Rotodynamic pumps	Multi-stage	High speed multi-stage pump packages	160
		Multi-stage axially-split pumps	165
		Multi-stage radially-split pumps	170
		Vertical multi-stage pumps	175
Rotodynamic pumps	Self-cleaning	Non-clogging pumps with standard motor	180
		Submersible non-clogging pump	185
		Other non-clogging pumps	190
Rotodynamic pumps		Mixed flow pumps	195
		Axial propeller pumps	200
		Power recovery turbines	205
		Liquid ring pumps	210
		Peripheral pumps	215
Positive displacement pumps	Rotary	External gear pumps	220
		Internal gear pumps	225
		Screw pumps with 2 or more screws	230
		Progressing cavity pump	235
		Lobe pumps and rotary piston pumps	240
		Rigid vane pumps and flexible vane pumps	245
		Peristaltic hose pump	250
		Rotary peristaltic pump	255
Positive displacement pumps	Reciprocating	Piston pumps for hydraulic power - radial and axial pistons	260
		Portable plunger pump packages for high pressure descaling/cleaning	265
		Plunger pumps - horizontal and vertical	270
		Piston pumps	275
		Diaphragm pumps	280
		Metering pumps	285
Other pumps		Jet pumps	290
		Air lift pumps	295
		Barrel emptying pumps	300
		Rigid screw pumps	305
		Pilot tube pumps	310
		Miscellaneous types	315

16.4 Manufacturers and suppliers of ancillary products

ACCUMULATORS
Fawcett Christie Hydraulics Ltd   (UK)
Flowguard Ltd   (UK)
Liquid Dynamics   (UK)
Shurflo Ltd   (UK)

ANTI-VIBRATION MOUNTINGS
Appliance Components Ltd   (UK)
Edwards and Jones Ltd   (UK)
Linatex Ltd   (UK)
James Walker & Co Ltd   (UK)

BEARINGS
Bearings (Cardiff) Ltd   (UK)
Fife Bearing & Transmissions Ltd   (UK)
Glacier Industrial Bearings   (UK)
Headland Engineering
   Developments Ltd   (UK)
Inverness Bearing &
   Transmissions Ltd   (UK)
Yarmouth Bearing &
   Transmissions Ltd   (UK)

EJECTORS AND INJECTORS
GEA Wiegand GmbH   (D)
ITT Flygt AB   (S)

ELECTRIC MOTORS - POLE CHANGING
Mackley Pumps   (UK)

ELECTRIC MOTORS - STANDARD
Bevi Wright Ltd   (UK)
Grundfos A/S   (DK)
Mackley Pumps   (UK)
Magnetek Universal Electric Ltd   (UK)

ELECTRIC MOTORS - STARTING EQUIPMENT
T-T Pumps Ltd   (UK)

ELECTRIC MOTORS - SUBMERSIBLE
Bevi Wright Ltd   (UK)
Bombas Grundfos (Portugal) Lda   (P)
Franklin Electric Europa GmbH   (D)
Grundfos A/S   (DK)
KOVO Konice, v.d.   (CZ)
Mercury Srl   (I)
Ziehl-Ebm (UK) Ltd   (UK)

ELECTRIC MOTORS - VARIABLE FREQUENCY
Allspeeds Ltd   (UK)
Bevi Wright Ltd   (UK)
Grundfos A/S   (DK)

FILTERS - HIGH PRESSURE
Airpel Filtration Ltd   (UK)
Fawcett Christie Hydraulics Ltd   (UK)
Forward Industries Ltd   (UK)
Haskel Energy Systems Ltd   (UK)
Hydrair Ltd   (UK)
Pall Industrial Hydraulics Ltd   (UK)
Simon-Hartley Ltd   (UK)

FILTERS - LIQUID
Airpel Filtration Ltd   (UK)
E Braude Ltd   (UK)
Fawcett Christie Hydraulics Ltd   (UK)
GEC Alsthom Fluids Group   (UK)
Pall Industrial Hydraulics Ltd   (UK)
Serfilco Europe Ltd   (UK)
Simon-Hartley Ltd   (UK)

FILTERS - WATER
Airpel Filtration Ltd   (UK)
Pall Industrial Hydraulics Ltd   (UK)
Shurflo Ltd   (UK)
Superior Industrial Products Ltd   (UK)

GASKETS AND JOINTS
A.E.S. Engineering Ltd   (UK)
Advanced Products
   (Seals & Gaskets) Ltd   (UK)
Beldam Crossley Ltd   (UK)
Clough (Croydon) Ltd   (UK)
Du Pont de Nemours
   International SA   (CH)
F.T.L. Fluid Seals Ltd   (UK)
Ferguson & Timpson Ltd   (UK)
Fluorocarbon Co Ltd   (UK)
Hofland (UK) Ltd   ()
Hughes Wynne Ltd   (UK)
Nicholsons Seals Ltd   (UK)
James Walker & Co Ltd   (UK)
Whitby & Chandler Ltd   (UK)

HOSE COUPLINGS AND FITTINGS
ASTI   (F)
Inverness Bearing &
   Transmissions Ltd   (UK)
Norgren-Martonair Ltd   (UK)
Pollution Control Engineering Ltd   (UK)
Shurflo Ltd   (UK)
Uniclip Ltd   (UK)
Warsop Power Tools   (UK)

INSTRUMENTATION
Advanced Energy Monitoring
   Systems Ltd   (UK)
Chromalock Ltd   (UK)
Danfoss Ltd   (UK)
Dosapro Milton Roy   (F)
Edwards High Vacuum International   (UK)
Hawker Electronics Ltd   (UK)
Roban Engineering Ltd   (UK)
Seko SpA   (I)

LIQUID LEVEL CONTROLLERS
Appliance Components Ltd   (UK)
E Braude Ltd   (UK)
Danfoss Ltd   (UK)
Gentech International Ltd   (UK)
R Goodwin International   (UK)
Hawker Electronics Ltd   (UK)
Hydrair Ltd   (UK)
ITT Flygt AB   (S)
Sarlin Ltd   (UK)

LUBRICATORS
Baric Pumps Ltd   (UK)
Wm Coulthard & Co Ltd   (UK)
Durametallic UK   (UK)
Lumatic (GA) Ltd   (UK)
Mato Curt Matthaei GmbH & Co KG   (D)

PACKING - METALLIC
A.E.S. Engineering Ltd   (UK)
Beldam Crossley Ltd   (UK)
John Crane UK Ltd   (UK)
F.T.L. Fluid Seals Ltd   (UK)
Nicholsons Seals Ltd   (UK)

PACKING - SOFT
A.E.S. Engineering Ltd   (UK)
Beldam Crossley Ltd   (UK)
Clough (Croydon) Ltd   (UK)
John Crane UK Ltd   (UK)
Durametallic UK   (UK)
F.T.L. Fluid Seals Ltd   (UK)
Ferguson & Timpson Ltd   (UK)
Fluorocarbon Co Ltd   (UK)
James Walker & Co Ltd   (UK)
Whitby & Chandler Ltd   (UK)

POWER TRANSMISSION - BELT DRIVES
Aberdeen Transmissions Ltd   (UK)
Fife Bearing & Transmissions Ltd   (UK)
Headland Engineering
   Developments Ltd   (UK)
Renold Gears   (UK)
Yarmouth Bearing &
   Transmissions Ltd   (UK)

POWER TRANSMISSION - FIXED RATIO GEARBOXES
Aberdeen Transmissions Ltd   (UK)
Bearings (Cardiff) Ltd   (UK)
Couplings and Drives (UK) Ltd   (UK)
Fife Bearing & Transmissions Ltd   (UK)
Hamworthy Hydraulics Ltd   (UK)
Inverness Bearing &
   Transmissions Ltd   (UK)
Mackley Pumps   (UK)
Renold Gears   (UK)
Yarmouth Bearing &
   Transmissions Ltd   (UK)

POWER TRANSMISSION - INFINITELY VARIABLE RATIO
Aberdeen Transmissions Ltd   (UK)
Allspeeds Ltd   (UK)
Bearings (Cardiff) Ltd   (UK)
Couplings and Drives (UK) Ltd   (UK)
Renold Gears   (UK)

POWER TRANSMISSION - MULTI-RATIO GEARBOXES
Bearings (Cardiff) Ltd   (UK)
Couplings and Drives (UK) Ltd   (UK)
Mackley Pumps   (UK)

POWER TRANSMISSION - SHAFT COUPLINGS - FLEXIBLE
Aberdeen Transmissions Ltd   (UK)
Couplings and Drives (UK) Ltd   (UK)
Flexibox Ltd   (UK)
Hamworthy Hydraulics Ltd   (UK)
Headland Engineering
   Developments Ltd   (UK)
Inverness Bearing &
   Transmissions Ltd   (UK)
Renold Gears   (UK)
Yarmouth Bearing &
   Transmissions Ltd   (UK)

POWER TRANSMISSION - SHAFT COUPLINGS - RIGID
Aberdeen Transmissions Ltd   (UK)
Bearings (Cardiff) Ltd   (UK)
Couplings and Drives (UK) Ltd   (UK)
Fife Bearing & Transmissions Ltd   (UK)
Headland Engineering
   Developments Ltd   (UK)

Renold Gears (UK)
Yarmouth Bearing &
  Transmissions Ltd (UK)

PULSATION DAMPENERS
ASTI (F)
Burgess-Manning Ltd ()
Dosapro Milton Roy (F)
Edwards and Jones Ltd (UK)
Fawcett Christie Hydraulics Ltd (UK)
Flotronic Pumps Ltd (UK)
Flotronics GmbH (D)
Flowguard Ltd (UK)
Liquid Dynamics (UK)
Milton Roy (UK) Ltd (UK)
Waukesha Bredel Fluid
  Handling BV (NL)

PUMP REPAIR
A.E.S. Engineering Ltd (UK)
Advanced Energy Monitoring
  Systems Ltd (UK)
Clasal N.V. (B)
Edwards and Jones Ltd (UK)
Edwards High Vacuum International (UK)
Fluid Pumps (Bristol) Ltd (UK)
Forward Industries Ltd (UK)
Friatec Ltd (UK)
Hayward Tyler Fluid Dynamics Ltd (UK)
Hick Hargreaves & Co Ltd (UK)
Houttuin Holland BV (NL)
ITT Flygt AB (S)
KOVO Konice, v.d. (CZ)
Mackley Pumps (UK)
Megator Pumps & Compressors Ltd (UK)
RHL Hydraulics Ltd (UK)

Roban Engineering Ltd (UK)
Sundstrand International
  Corporation SA (B)
Torres Engineering & Pumps Ltd (UK)

RELIEF VALVES
Forward Industries Ltd (UK)
Inoxpa UK
  (Euro-Industrial Engineering Ltd) (UK)
Linatex Ltd (UK)
Megator Pumps & Compressors Ltd (UK)
Mono Pumps Ltd (UK)
Oilgear Towler Ltd (UK)
Richter Chemie-Technik GmbH (D)

SEALING RINGS
BT Tenute Meccaniche Rotanti SpA (I)
BW/IP BW Seals (NL)
Clough (Croydon) Ltd (UK)
Du Pont de Nemours
  International SA (CH)
Ferguson & Timpson Ltd (UK)
Fluorocarbon Co Ltd (UK)
Forsheda Ltd (UK)
James Walker Britco Ltd (UK)
Whitby & Chandler Ltd (UK)

SEALS - BELLOWS
BT Tenute Meccaniche Rotanti SpA (I)
BW/IP BW Seals (NL)
Clough (Croydon) Ltd (UK)
John Crane UK Ltd (UK)
Durametallic UK (UK)
EG & G Sealol (UK)
Fluorocarbon Co Ltd (UK)
James Walker Britco Ltd (UK)

SEALS - MECHANICAL
A.E.S. Engineering Ltd (UK)
Advanced Products
  (Seals & Gaskets) Ltd (UK)
Beldam Crossley Ltd (UK)
BT Tenute Meccaniche Rotanti SpA (I)
BW/IP BW Seals (NL)
John Crane UK Ltd (UK)
Durametallic UK (UK)
Edwards High Vacuum International (UK)
F.T.L. Fluid Seals Ltd (UK)
Flexibox Ltd (UK)
Fluorocarbon Co Ltd (UK)
Headland Engineering
  Developments Ltd (UK)
Pioneer Weston Ltd (UK)
James Walker & Co Ltd (UK)
James Walker Britco Ltd (UK)
Whitby & Chandler Ltd (UK)

SEALS - OIL
Beldam Crossley Ltd (UK)
Clough (Croydon) Ltd (UK)
Durametallic UK (UK)
Edwards High Vacuum International (UK)
F.T.L. Fluid Seals Ltd (UK)
Ferguson & Timpson Ltd (UK)
Fife Bearing & Transmissions Ltd (UK)
Forsheda Ltd (UK)
Forward Industries Ltd (UK)
Inverness Bearing &
  Transmissions Ltd (UK)
Pioneer Weston Ltd (UK)
James Walker & Co Ltd (UK)
Whitby & Chandler Ltd (UK)

PULSATION DAMPERS

Flowguard is the U.K.'s leading manufacturer of pulsation dampers and offers units in **316 Stainless** and **Carbon Steel, Polypropylene, PVC, Hastelloy, Titanium, Monel** with separating membranes in **Nitrile, EPDM, Viton and PTFE**. Available with **Flanged, Graylock, NPT & BSP** connections and desinged in accordance with **BS 5500 and ASME 8**, units are available from **0.1 to 100 litres volume** and up to **2000 bar design pressure**. Flowguard application engineers can be contacted by telephone to assist with damper selection and advise on materials compatibitliy.

Applications

- BLENDING & PROPORTIONING
- SPRAYING
- DOSING
- BULK TRANSFER
- MIXING
- METERING

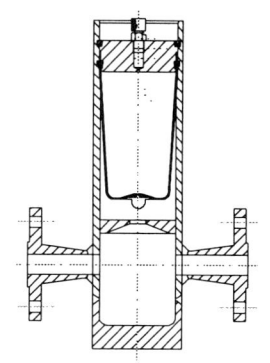

Fitted to

- METERING PUMPS
- MULTI PLUNGER PUMPS
- PERISTALTIC PUMPS
- AIR OPERATED
  DIAPHRAGM PUMPS

FLOWGUARD

Flowguard Limited, Watford Bridge, New Mills, Stockport, Cheshire SK12 4HJ
Tel: (INT 44 + 663) 0663 745976 Fax: 0663 742788

16.5 Names and addresses of European manufacturers and suppliers

AUSTRIA
(Country Code: A)

Boehler GmbH
High Pressure Division
PO Box 96
Mariazellerstrasse 25
A-8605 Kapfenberg
Austria
Tel: 03862 208118
Fax: 03862 207582

Hauke GmbH & Co KG
Cumberlandstrasse 46-50
Postfach 103
A-4810 Gmunden
Austria
Tel: 07612 4133
Fax: 07612 413385

Kraeutler GmbH & Co
Industrie - Nord
Bildgasse 40
A-6893 Lustenau
Austria
Tel: 05577 86644-0
Fax: 05577 88433

Maschinenfabrik Andritz AG
Statteggerstrasse 18
A-8045 Graz
Austria
Tel: 0316 69022514
Fax: 0316 6902413

Ochsner GmbH
Oberfeldstrasse 8
A-4204 Linz
Austria
Tel: 0732 47288
Fax: 0732 46340

Pumpenfabrik Ernst Vogel GmbH
Ernst Vogel-Strasse 2
A-2000 Stockerau
Austria
Tel: 02266 604
Fax: 02266 65311

JM Voith Aktiengesellschaft
Linzer Strasse 55
PO Box 168
A-3100 St Polten
Austria
Tel: 02742 806
Fax: 02742 71876

BELGIUM
(Country Code: B)

A de Backer & Co, BVBA
Nieuwsstraat
B-9230 Wetteren
Belgium
Tel: 09 3693496
Fax: 09 3695752

AMB Rateau SA
Leuvensesteenweg 474
B-2812 Muizen
Belgium
Tel: 015 41298
Fax: 015 423337

Aro, N.V.
Eurolaan 3
B-9140 Temse
Belgium
Tel: 03 7710921
Fax: 03 7715584

Clasal N.V.
Ruddervoordsestraat 82
B-8210 Zedelgem
Belgium
Tel: 050 278253
Fax: 050 275494

Durco Europe
Rue de Geneve 6
B-1140 Bruxelles
Belgium
Tel: 02 242 6610
Fax: 02 242 7342

Ensival SA
Rue Hodister 44
B-4860 Pepinster
Belgium
Tel: 087 460111
Fax: 087 460441

A.C. Herstal SA
Rue Hayeneux 148
B-4040 Herstal
Belgium
Tel: 041 640840
Fax: 041 640848

Hydrobel SA
Rue Libeau 36
B-4682 Houtain-Saint-Simeon
Belgium
Tel: 041 864481
Fax: 041 864487

Packo Diksmuide
Cardijnlaan 10
B-8600 Diksmuide
Belgium
Tel: 051 501606
Fax: 051 501778

Robbins & Myers Europe
Leuvensesteenweg 41
B-1800 Vilvoorde
Belgium
Tel: 02 253 2166
Fax: 02 253 2093

Sundstrand International Corporation SA
Woluwelaan 140A
Box 1
B-1831 Diegem
Belgium
Tel: 02 717 5045
Fax: 02 725 9820

CZECH REPUBLIC
(Country Code: CZ)

KOVO Konice, v.d.
9. kvetna 11
CZ-798 52 Konice
Czech Republic
Tel: 0508 98531
Fax: 0508 98530

MSA a.s.
Hlucinska 41
CZ-747 22 Dolni Benesov
Czech Republic
Tel: 0653 54111
Fax: 0653 51236

Ondrejovicka strojirna spol. sro.
CZ-793 76 Zlate-Hory Ondrejovice
Czech Republic
Tel: 0645 4233
Fax: 0645 4462

SIGMA Hranice a.s.
Tovarni 605
CZ-753 33 Hranice
Czech Republic
Tel: 0642 2461
Fax: 0642 3477

Sigma Lutin
CZ-783 50 Lutin
Czech Republic
Tel: 068 475 1111
Fax: 068 31930

SIGMA Olomouc s.p.
Barakova 15
CZ-772 43 Olomouc
Czech Republic
Tel: 068 301111
Fax: 068 33564

SIGMA Trading Ltd
Trida Svobody 39
CZ-772 00 Olomouc
Czech Republic
Tel: 068 23955
Fax: 068 32364

DENMARK
(Country Code: DK)

APV Rosista
Ternevej 61-63
DK-8700 Horsens
Denmark
Tel: 075 643777
Fax: 075 643868

APV Rannie A/S
Roholmsvej 8
DK-2620 Albertslund
Denmark
Tel: 042 649300
Fax: 042 640330

A/S De Smithske (DESMI)
Tagholm 1
PO Box 226
DK-9400 Norresundby
Denmark
Tel: 098 178111
Fax: 098 175499

Grundfos A/S
Paul Due Jensens Vej 7
DK-8850 Bjerringbro
Denmark
Tel: 086 681400
Fax: 086 682224

T Smedegaard A/S
Sydvestvej 57-59
DK-2600 Glostrup
Denmark
Tel: 043 961028
Fax: 043 631766

Thrige Pumper A/S
Tolderlundsvej 2
DK-5000 Odense C
Denmark
Tel: 066 111315
Fax: 065 913415

FINLAND
(Country Code: SF)

A Ahlstrom Corporation
Pump Industry
PO Box 18
SF-48601 Karhula
Finland
Tel: 052 291111
Fax: 052 63958

Oy Kolmeks AB
PO Box 27
SF-14201 Turenki
Finland
Tel: 017 82071
Fax: 017 81286

Mako Oy
Linjatie 4
SF-01260 Vantaa
Finland
Tel: 0 8751 700
Fax: 0 8751 701

Mamec Oy
SF-08450 Hormajärvi
Finland
Tel: 012 39151
Fax: 012 39189

Oy E Sarlin AB
PO Box 750
SF-00101 Helsinki
Finland
Tel: 0 504441
Fax: 0 5633989

FRANCE
(Country Code: F)

ASTI
13 Avenue de la Republique
F-92400 Courbevoie
France
Tel: 01 43 33 17 70
Fax: 01 43 34 06 66

Clextral
Zone Industrielle de Chazeau
BP 10
F-42700 Firminy
France
Tel: 077 40 31 31
Fax: 077 40 31 23

Delasco Pumps
19 rue Ernest Laval
F-92170 Vanves
France
Tel: 01 46 45 63 73
Fax: 01 46 45 65 60

Dosapro Milton Roy
BP 5, 10 Grande Rue
F-27360 Pont Saint Pierre
France
Tel: 032 68 30 00
Fax: 032 68 30 93

Fapmo
35/37 Rue Roger Salengro
F-62230 Outreau
France
Tel: 021 31 72 00
Fax: 021 80 46 88

GEA Wiegand Kestner
BP 44
F-59003 Lille
France
Tel: 020 15 39 00
Fax: 020 15 39 77

GEC Alsthom Bergeron SA
10-12 Avenue des Olympiades
Le Niemeyer
F-94132 Fontenay Sous Bois
France
Tel: 01 43 94 87 12
Fax: 01 43 94 08 50

Moret
Chemin des Ponts et Chausees
F-02100 Saint-Quentin
France
Tel: 023 64 05 05
Fax: 023 62 02 30

Mouvex
5 rue du Sahel
F-75012 Paris
France
Tel: 01 44 73 12 53
Fax: 01 43 41 24 58

PCM Pompes
17 rue Ernest Laval
F-92170 Vanves
France
Tel: 01 46 45 21 88
Fax: 01 46 42 90 06

PEME
51 rue de la Liberation
F-69920 Gonnehem
France
Tel: 021 57 11 00
Fax: 021 01 08 80

Pompes Grosclaude
29 rue du 35eme Regiment d'Aviation
Zone Industrielle du Chene
F-69500 Bron
France
Tel: 072 37 88 91
Fax: 078 26 38 64

Pompes Grundfos Distribution
Parc D'Activités de Chesnes
57 rue de Malacombe
F-38290 St Quentin-Fallavier
France
Tel: 074 82 15 15
Fax: 074 94 10 51

Pompes Guinard KSB
179 Boulevard Saint Denis
F-92400 Courbevoie
France
Tel: 049 04 83 00
Fax: 047 88 23 78

Pompes Industrie SA
BP 30
F-33810 Ambes
France
Tel: 056 77 08 78
Fax: 056 77 10 16

Schabaver
20 rue de Melou
F-81100 Castres
France
Tel: 063 59 51 19
Fax: 063 59 06 92

Vergnet SA
6 Henri Dunant
F-45140 Ingre
France
Tel: 038 43 36 52
Fax: 038 88 30 50

GERMANY
(Country Code: D)

Abel GmbH
Feldstrasse 10
PO Box 1220
D-21511 Buchen
Germany
Tel: 04155 80020
Fax: 04155 2075

ABS Pumpen AG
Scheiderhoehe
D-5204 Lohmar 1
Germany
Tel: 02246 130
Fax: 02246 13200

Alldos Eichler GmbH Dosing Pumps
Reetzstrasse 85
D-7507 Pfinztal-Solingen
Germany
Tel: 07240 610
Fax: 07240 61177

Allweiler AG
Werk Radolfzell
Postfach 11 40
D-78301 Radolfzell
Germany
Tel: 07732 86 440
Fax: 07732 86 436

Apollowerk Gössnitz GmbH
Walter-Rabold Strasse 26
Postfach 61/62
D-04639 Gössnitz
Germany
Tel: 03 449370
Fax: 03 44937210

APV Gaulin GmbH
Postfach 160 164
Mecklenburger Strasse 223
D-2400 Lübeck 16
Germany
Tel: 0451 693010
Fax: 0451 690866

Heinrich Behrens Pumpenfabrik GmbH & Co
Von-Thünen-Strasse 7
D-28307 Bremen
Germany
Tel: 0421 481513
Fax: 0421 483857

J H Bornemann GmbH & Co KG
PO Box 11 80
Bornemannstrasse 1
D-3063 Obernkirchen
Germany
Tel: 05724 3900
Fax: 05724 4344

Bran + Luebbe GmbH
Werkstrasse 4
Postfach 1360
D-22803 Norderstedt
Germany
Tel: 040 522020
Fax: 040 52202444

DIA Pumpenfabrik
Postfach 10 20 62
D-40011 Düsseldorf
Germany
Tel: 0211 310050
Fax: 0211 3100565

Dickow Pumpen KG
Siemenstrasse 22
D-8264 Waldkraiburg
Germany
Tel: 08638 6020
Fax: 08638 5520

Edur Pumpenfabrik
Eduard Redlien, Hamburger Chausee 148-152
Postfach 1949
D-2300 Kiel
Germany
Tel: 0431 688071
Fax: 0431 642683

Eheim GmbH & Co KG
Postfach 1180
D-7301 Deizisau
Germany
Tel: 07153 700201
Fax: 07153 700274

EMU-Unterwasserpumpen GmbH
Heimgartenstrasse
Postfach 3309
D-8670 Hof/Saale
Germany
Tel: 09281 974-0
Fax: 09281 9 65 28

Feluwa Pumps, Schlesiger & Co KG
Beulertweg
D-5537 Murlenbach/Eifel
Germany
Tel: 0659 41000
Fax: 0659 41640

Flotronics GmbH
Mönchsbergstrasse 97
D-7000 Stuttgart 40
Germany
Tel: 0711 870 13 74
Fax: 0711 87 46 41

Flux-Geräte GmbH
Talweg 12
D-75433 Maulbronn
Germany
Tel: 07043 1010
Fax: 07043 10133

Franklin Electric Europa GmbH
Rudolf-Diesel-Strasse 20
Postfach 12 80
D-5560 Wittlich
Germany
Tel: 06571 105-0
Fax: 06571 105-55

**Fristam Pumpen F Stamp KG
(GmbH & Co)**
Kampchaussee 55
D-21033 Hamburg
Germany
Tel: 040 725 56-0
Fax: 040 725 56-166

GEA Wiegand GmbH
Eisteinstrasse 9-15
D-76275 Ettlingen
Germany
Tel: 07243 7050
Fax: 07243 705-330

Willi F Grassel
Braasstrasse 6
D-3260 Rinteln
Germany
Tel: 05751 41001
Fax: 05751 44400

Halberg Maschinenbau GmbH
Halbergtrasse 1
D-67061 Ludwigshafen
Germany
Tel: 0621 56120
Fax: 0621 5612209

**Paul Hammelmann
Maschinenfabrik GmbH**
Zum Sundern 13-21
Postfach 3309
D-59302 Oelde
Germany
Tel: 02522 760
Fax: 02522 76444

Hermetic-Pumpen GmbH
Postfach 1220
D-71919 Gundelfingen
Germany
Tel: 0761 58300
Fax: 0761 5830280

Homa Pumpenfabrik GmbH
Industriestrasse 1
D-5206 Neunkirchen-Seelscheid 2
Germany
Tel: 02247 7020
Fax: 02247 70244

Huedig GmbH & Co
Heinrich-Huedig Strasse 2
D-3100 Celle
Germany
Tel: 051 4188450
Fax: 051 4186918

Jesco Dosiertechnik GmbH
Postfach 10 01 64
D-30891 Wedemark 1
Germany
Tel: 05130 58020
Fax: 05130 580268

Jung Pumpen GmbH & Co
Industriestrasse 4-6
D-33803 Steinhagen
Germany
Tel: 05204 170
Fax: 05204 80368

Kamat-Pumpen GmbH & Co KG
Salinger Feld 10
D-5810 Witten 6
Germany
Tel: 02302 89030
Fax: 02302 801917

Klaus Union
Blumenfeldstrasse 18
D-4630 Bochum 1
Germany
Tel: 0234 459040
Fax: 0234 432387

Kracht GmbH
Postfach 1420
D-58774 Werdohl
Germany
Tel: 02392 540
Fax: 02392 54231

KSB Aktiengesellschaft
Johann Klein Strasse 9
D-6710 Frankenthal
Germany
Tel: 06233 860
Fax: 06233 863401

Lederle GmbH
Gewerbestrasse 53
Postfach 1169
D-79190 Gundeldingen
Germany
Tel: 0761 58300
Fax: 0761 5830180

Leistritz AG
Markgrafenstrasse 29-39
D-8500 Nurenberg 40
Germany
Tel: 0911 43060
Fax: 0911 4306490

LEWA Herbert Ott GmbH & Co
Ulmer Strasse 10
D-7250 Leonberg
Germany
Tel: 07152 14-0
Fax: 07152 14 303

Lincoln GmbH
Heinrich Hertz Strasse
D-6909 Waldorf
Germany
Tel: 06227 330
Fax: 06227 33259

Lutz-Pumpen GmbH & Co KG
Erlenstrasse 5-7
PO Box 14 62
D-6980 Wertheim
Germany
Tel: 09342 879-0
Fax: 09342 879404

Mato Curt Matthaei GmbH & Co KG
Biebererstrasse 215-217
D-63071 Offenbach/Main
Germany
Tel: 069 850040
Fax: 069 8500485

Osna Pumpen J Hartlage GmbH & Co KG
Postfach 22 40
D-4500 Osnabrück
Germany
Tel: 0541 12110
Fax: 0541 1211220

Pumpenbau G Schulte
Kruppstrasse 1A
D-41469 Neuss-Norf
Germany
Tel: 02137 4077
Fax: 02137 12453

Pumpenfabrik Oddesse
Anderslebener Strasse 56-60
D-3230 Oschersleben
Germany
Tel: 037949 520
Fax: 037949 2431

Pumpenfabrik Wangen GmbH
Simoniusstrasse 17-19
D-7988 Wangen/Allgäu
Germany
Tel: 07522 503134
Fax: 07522 3971

**Putzmeister-Werk
Maschinenfabrik GmbH**
Postfach 2152
D-7447 Aichtal
Germany
Tel: 07127 599-0
Fax: 07127 599520

Richter Chemie-Technik GmbH
Otto Schott Strasse 2
PO Box 69
D-4152 Kempen 1
Germany
Tel: 02152 1460
Fax: 02152 146190

Ritz Pumpenfabrik GmbH & Co KG
Postfach 1780
D-7070 Schwäbisch Gmünd
Germany
Tel: 07171 609-0
Fax: 07171 609287

Ernst Scherzinger GmbH
Bergstrasse 23
D-7743 Furtwangen
Germany
Tel: 07723 65060
Fax: 07723 650540

Schmalenberger GmbH & Co
Postfach 23 80
Im Schelmen 9 - 11
D-72013 Tübingen
Germany
Tel: 07071 7008-0
Fax: 07071 7008-10

Kranz Schmidt & Co GmbH
Hauptstrasse 123
D-5620 Velbert 11 (Langenberg)
Germany
Tel: 02052 3011
Fax: 02052 3016

Seepex Seeberger GmbH & Co
Scharnholzstrasse 344
Postfach 10 15 64
D-46215 Bottrop
Germany
Tel: 02041 9960
Fax: 02041 96200

Sero Pumpenfabrik
Postfach 69
D-74907 Meckesheim
Germany
Tel: 06226 92010
Fax: 06226 8091

Sihi-Halberg
Lindenstrasse 170
D-25524 Itzehoe
Germany
Tel: 04821 77101
Fax: 04821 771274

Spandau Pumpen
Motzener Strasse 35/37
D-12277 Berlin
Germany
Tel: 030 720020
Fax: 030 72002111

Speck-Kolbenpumpenfabrik
Elbestrasse 8
Postfach 12 40
D-8192 Geretsried 1
Germany
Tel: 08171 62007
Fax: 08171 61517

Speck-Pumpen, Daniel Speck & Sohne
Postfach 12 09
Lohbachstrasse 6
D-91156 Hilpoltstein
Germany
Tel: 09174 9720
Fax: 09174 97249

Sulzer Weise GmbH
PO Box 3028
D-7520 Bruchsal
Germany
Tel: 07251 760
Fax: 07251 6329

Thyssen Maschinenbau GmbH Ruhrpumpen
Stockhumer Strasse 28
Postfach 6320
D-58449 Witten
Germany
Tel: 02302 661-01
Fax: 02302 661666

Otto Tuchenhagen GmbH & Co KG
Postfach 1140
D-2059 Büchen
Germany
Tel: 04155 49-0
Fax: 04155 2035

Uraca Pumpenfabrik GmbH & Co KG
Sirchinger Strasse 5
D-72563 Bad Urach
Germany
Tel: 07125 12-0
Fax: 07125 12202

J M Voith GmbH
St Poltener Strasse 43
D-7920 Heidenheim
Germany
Tel: 07321 37-0
Fax: 07321 373000

Wepuko-Hydraulik GmbH & Co
Max Eyth Strasse 31
Postfach 11 62
D-72542 Metzingen
Germany
Tel: 07123 1805-0
Fax: 07123 41231

Wernert Pumpen GmbH
Postfach 10 21 53
D-45421 Mülheim an der Ruhr
Germany
Tel: 0208 400011
Fax: 0208 407472

Wilo GmbH
Nortkirchenstrasse 100
D-44263 Dortmund
Germany
Tel: 0231 4102-0
Fax: 0231 4102363

Wirth Maschinen- und Bohrgeräte-Fabrik GmbH
Kölner Strasse 71-78
D-41812 Erkelenz
Germany
Tel: 02431 83405
Fax: 02431 83215

WOMA Apparatebau GmbH
Werthauser Strasse 77-79
D-4100 Duisburg 14
Germany
Tel: 02065 3040
Fax: 02065 304200

Albert Ziegler GmbH & Co KG
Postfach 1680
D-7928 Giengen/Brenz
Germany
Tel: 09722 9510
Fax: 07322 951211

GREECE
(Country Code: GR)

Drakos-Polemis Inc
Kryoneri-Attiki
PO Box 51024, Kifissia
GR-14510 Athens
Greece
Tel: 01 8161402
Fax: 01 8161262

Emmanuel N Kazis S.A.
Kryoneri-Attica, N. Erytrhea
GR-14610 Athens
Greece
Tel: 01 8161702
Fax: 01 8134765

IRELAND
(Country Code: IRL)

Grundfos (Ireland) Ltd
Unit 34, Stillorgan Industrial Park
Blackrock
Co. Dublin
Ireland
Tel: 01 295 4926
Fax: 01 295 4739

ITALY
(Country Code: I)

AGI Pompe Srl
5 Via Amundsen
I-20148 Milano
Italy
Tel: 02 48704924
Fax: 02 48700377

Alpha Pompe Srl
Via Molino Emili
PO Box 10
I-25030 Maclodio BS
Italy
Tel: 030 978661
Fax: 030 978663

Argal Srl
Via Averolda 31
I-25039 Travagliato BS
Italy
Tel: 030 686 3629
Fax: 030 686 3684

Aturia Pompe SpA
Piazza Aturia 9
I-20060 Gessate MI
Italy
Tel: 02 95382701
Fax: 02 95382161

Audoli & Bertola Co
Corso Vercelli 251
I-10091 Torino
Italy
Tel: 011 263333
Fax: 011 203420

BBC Elettropompe srl
Via S Martino Del Piano
I-61034
Italy
Tel: 0721 716590
Fax: 0721 716518

Bellin SpA
Via Carbon 8
Orgiano
I-36040 Vicenza
Italy
Tel: 0444 874900
Fax: 0444 874742

Bizzi & Tedeschi snc DT Bizzi G & G
Via Brodolini 35/A
I-42040 Campegine
Italy
Tel: 0522 677209
Fax: 0522 677633

BT Tenute Meccaniche Rotanti SpA
Via Leornardo Da Vinci 7/13
I-36057 Arcugnano VI
Italy
Tel: 0444 566399
Fax: 0444 588076

Calpeda SpA
Via Roggia di Mezzo 25
I-36050 Montorso Vicentino VI
Italy
Tel: 0444 685066
Fax: 0444 685011

Caprari SpA
Via Emilia Ovest n 900
I-41100 Modena
Italy
Tel: 059 330533
Fax: 059 828153

CDR Pompe Motori Srl
Via Togliatte 26/A
Senago
I-20030 Milano
Italy
Tel: 02 99010900
Fax: 02 9980606

Ebara Italia SpA
Via Pacinotti 32
I-36040 Brendola VI
Italy
Tel: 0444 401145
Fax: 0444 400018

FA-SA Pompe Srl
Via Costituzione 20
Correggio
I-42015 Reggio Emilia
Italy
Tel: 0522 642840
Fax: 0522 694466

FIPS Srl
Viale Toscana 46
I-20089 Rozzano MI
Italy
Tel: 02 8258923
Fax: 02 57512095

Fluxinos Italia Srl
Via Genova 8
I-58100 Grosseto
Italy
Tel: 02 610 1731
Fax: 02 610 1376

Foras Pumps Srl
Via Mazzini 2
I-36040 Brendola VI
Italy
Tel: 0444 401188
Fax: 0444 401190

Fraccarolo Pompe
Via Valtellina 31
Rescaldina
I-20027 Milano
Italy
Tel: 0331 577710
Fax: 0331 465554

Imovilli Pompe Srl
Via Masaccio 1
I-42010 Reggio Emilia
Italy
Tel: 0522 516595
Fax: 0522 514735

Jacuzzi Europa SpA
S.S. Pontebbana Km. 97,2
I-33098 Valvasone PN
Italy
Tel: 0434 85141
Fax: 0434 85278

Jurop SpA
Via Crociera di Corva 27
Tiezzo di Assano Decimo
I-33088 Pordenone
Italy
Tel: 0434 632847
Fax: 0434 632252

Lowara SpA
Via Dott. Lombardi 14
I-36075 Montecchio Maggiore VI
Italy
Tel: 0444 698555
Fax: 0444 694471

Majmar Srl
Via Apelle 43
I-20128 Milano
Italy
Tel: 02 255 0301
Fax: 02 255 3020

Matra SpA
Via Papa Giovanni XXIII, 33
I-41100 Modena
Italy
Tel: 059 250407
Fax: 059 251548

Mencarelli Pompe e Valvole Srl
Via Campestre 19
I-20091 Bresso MI
Italy
Tel: 02 6101 731
Fax: 02 6101 376

Mercury Srl
Via Mantova 2
I-20020 Lainate
Italy
Tel: 02 93570827
Fax: 02 93572007

Nuovo Pignone
Via Felice Matteucci 2
I-50127 Firenze
Italy
Tel: 055 423211
Fax: 055 4232800

Pedrollo SpA
Via E Fermi 75/0
I-37047 San Bonafacio VR
Italy
Tel: 045 7611766
Fax: 045 7614663

F Peroni & C SpA
Via Forze Armate 310/5
I-20152 Milano
Italy
Tel: 02 489401
Fax: 02 48910287

Pompe Gabbioneta SpA
Viale Casiraghi 68
I-20099 Sesto S. Giovanni MI
Italy
Tel: 02 2470451
Fax: 02 2424550

Pompe Travaini SpA
Via per Turbigo - Z.I.
I-20022 Castano Primo MI
Italy
Tel: 0331 880127
Fax: 0331 880511

Riva Hydroart SpA
Via Stendhal 34
I-20144 Milano
italy
Tel: 02 41461
Fax: 02 425749

Robuschi & C. SpA
Via San Leonardo 71/A
PO Box 8 Succ.8
I-43100 Parma
Italy
Tel: 0521 771811
Fax: 0521 771242

Rotant Srl
Via Stella Diana 11/13
I-22040 Civate CO
Italy
Tel: 0341 550513
Fax: 0341 551651

Rotos SpA
Via Bergamo 2
I-20060 Gessate MI
Italy
Tel: 02 95382002
Fax: 02 95382601

Rovatti A. & Figli Pompe SpA
Via Trento 22/24
I-42042 Fabbrico RE
Italy
Tel: 0522 665000
Fax: 0522 665020

S.I.G.M.A. SpA
Viale Cassala 28
I-20143 Milan
Italy
Tel: 02 89401090
Fax: 02 8372456

SACEM Srl
SS Valsugana 19
I-35010 San Giorgio PD
Italy
Tel: 049 9450022
Fax: 049 9450044

Saer Elettropompe Srl
Via Parma 8
Guastalla
I-42016 Reggio Emilia
Italy
Tel: 0522 830941
Fax: 0522 826948

Savino Barbera snc
Via Torino 12
I-10032 Brandizzo TO
Italy
Tel: 011 9139063
Fax: 011 9137313

Seko SpA
Via Salaria Km. 92,200
I-02010 S. Rufina di Cittaducale RI
Italy
Tel: 0746 607070
Fax: 0746 607072

Simaco Elettromeccanica Srl
S.S. 235 N. 16
Cadilana
I-20070 Corte Palasio MI
Italy
Tel: 0371 420567
Fax: 0371 420566

Termomeccanica Italiana SpA
Via del Molo 1
I-19100 La Spezia
Italy
Tel: 0187 532111
Fax: 0187 532267

Varisco Pompe Srl
Terza Strada 9
Zona Industriale Nord
I-35129 Padova
Italy
Tel: 049 8294111
Fax: 049 772168

THE NETHERLANDS
(Country Code: NL)

BW/IP BW Seals
Postbus 1300
4700 BH Roosendaal
The Netherlands
Tel: 01650 81400
Fax: 01650 52622

BW/IP International BV
Pump Division
Parallelweg 6
4878 AH Ettenleur
The Netherlands
Tel: 01608 28200
Fax: 01608 28487

Combimac BV
Kapt. Grantstr. 5
7821 AP Emmen
The Netherlands
Tel: 05910 11531
Fax: 05910 19730

Delaval-Stork v.o.f.
Lansinkesweg 1
Postbus 329
7550 AH Hengelo
The Netherlands
Tel: 074 402000
Fax: 074 427525

DP Pumps
Postbus 28
2400 AA Alphen aan den Rijn
The Netherlands
Tel: 01720 88325
Fax: 01720 33137

EnviroTech Pumpsystems BV
Postbus 249
5900 AE Venlo
The Netherlands
Tel: 077 895200
Fax: 077 824844

Goulds Pumps Europe BV
Bronsstraat 12
1976 BC IJmuiden
The Netherlands
Tel: 02550 37574
Fax: 02550 35444

Houttuin Holland BV
Postbus 76
Sophialaan 4
3500 AB Utrecht
The Netherlands
Tel: 030 484611
Fax: 030 411845

IHC Holland NV
Postbus 50
2960 AA Kinderdijk
The Netherlands
Tel: 01859 10556
Fax: 01859 10439

Melotte BV
Fregatweg 50
Postbus 85
6200 NZ Maastricht
The Netherlands
Tel: 043 632666
Fax: 043 631360

Nijhuis Pompen BV
Parallelweg 4
7102 DE Winterswijk
The Netherlands
Tel: 05430 47474
Fax: 05430 47475

Robot Pompen BV
Postbus 140
2400 AC Alphen aan den Rijn
The Netherlands
Tel: 01720 31541
Fax: 01720 42452

Sihi-Maters BV
Havenstraat 22 - 28
1948 NP Beverwijk
The Netherlands
Tel: 02510 63232
Fax: 02510 26309

Stork Pompen BV
Lansinkseweg 30
Postbus 55
7550 AB Hengelo
The Netherlands
Tel: 074 404000
Fax: 074 425696

Waukesha Bredel Fluid Handling BV
Sluisstraat 7
7491 GA Delden
The Netherlands
Tel: 05407 62605
Fax: 05407 61175

Yamada Europe BV
Topaasstraat 46
7554 TH Hengelo
The Netherlands
Tel: 074 422032
Fax: 074 421055

NORWAY
(Country Code: N)

Kvaerner Eureka AS
Joseph Kellers Vei
Tranby, PO Box 38
N-3401 Lier
Norway
Tel: 03 859000
Fax: 03 852193

PORTUGAL
(Country Code: P)

Arsopi SA
Apartado 10
P-3731 Vale de Cambra
Portugal
Tel: 056 422511
Fax: 056 422592

Bombas Grundfos (Portugal) Lda
Praceta Simones D'almeida Junior 13-14
P-2745 Queluz Ocidental
Portugal
Tel: 01 437 5039
Fax: 01 437 0143

SPAIN
(Country Code: E)

Bomba Elias SA
Ctra. Molins de Rey a Rubi, Km 8,700
E-08191 Rubi (Barcelona)
Spain
Tel: 03 6996004
Fax: 03 6971609

Bombas Azcue SA
PO Box 34
E-20750 Zumaia (Guipuzcoa)
Spain
Tel: 043 147047
Fax: 043 147440

Bombas Ideal SA
Av Navarro Reverter 10
E-46004 Valencia
Spain
Tel: 0334 6208
Fax: 0374 7102

Bombas Itur-Manufacturas Aranzabal SA
PO Box 41
E-20800 Zarauz (Guipuzcoa)
Spain
Tel: 043 131320
Fax: 043 134278

Deloule Espanola SA
Ctr. N-11, Km 759
E-17600 Figueras (Gerona)
Spain
Tel: 072 503766
Fax: 072 508525

INOXPA SA
Carrer Dels Telers 54
E-17820 Banyoles (Girona)
Spain
Tel: 0972 575200
Fax: 0972 575502

SWEDEN
(Country Code: S)

Cardo Pump AB
Box 2056
Krokslätts Parkgata 4
S-431 02 Mölndal
Sweden
Tel: 031 836300
Fax: 031 184906

Denver Sala, Sala International AB
S-733 25 Sala
Sweden
Tel: 0224 57000
Fax: 0224 16969

Dymatic Industri AB
Stenhuggarvägen 3
Box 54
S-132 22 Saltsjöbaden
Sweden
Tel: 08 715 0270
Fax: 08 715 5859

Grindex AB
Box 538
S-136 25 Haninge
Sweden
Tel: 08 606 6600
Fax: 08 745 5328

IMO AB
Box 42090
S-126 14 Stockholm
Sweden
Tel: 08 190160
Fax: 08 6451509

ITT Flygt AB
Box 1309
S-171 25 Solna
Sweden
Tel: 08 627 6500
Fax: 08 627 6900

Perfecta Pump AB
Box 3024
Smedjegatan 20
S-350 33 Växjö
Sweden
Tel: 0470 10475
Fax: 0470 13762

Pumpex AB
Box 5207
S-121 16 Johanneshov
Sweden
Tel: 08 649 2580
Fax: 08 659 3314

Svenska Tapflo AB
Filaregatan 4
S-442 34 Kungälv
Sweden
Tel: 0303 63390
Fax: 0303 19916

Weda Pump AB
Box 621
S-151 27 Södertälje
Sweden
Tel: 0755 32550
Fax: 0755 31050

SWITZERLAND
(Country Code: CH)

Bieri Pumpenbau AG, Biral International
Südstrasse 10
CH-3310 Münsingen
Switzerland
Tel: 031 720 9000
Fax: 031 721 5644

Du Pont de Nemours International SA
Specialty Elastomers
2 Chemin Du Pavillon
CH-1218 Le Grand Saconnex, Geneva
Switzerland
Tel: 022 717 5111
Fax: 022 717 5411

Emile Egger & Cie.
Route de Neuchatel 36
CH-2088 Cressier
Switzerland
Tel: 038 481122
Fax: 038 472290

Rutschi Pumpen AG
Herzogstrasse 11
CH-5200 Brugg
Switzerland
Tel: 056 410455
Fax: 056 411331

Gebr. Sulzer AG
PO Box 21
CH-8404 Winterthur
Switzerland
Tel: 052 2621155
Fax: 052 2620040

UNITED KINGDOM
(Country Code: UK)

A.E.S. Engineering Ltd
(Sealtec Division)
Mangham Road
Barbot Hall Industrial Estate
Rotherham
South Yorkshire S61 4RJ
United Kingdom
Tel: 0709 369966
Fax: 0709 374919

Aberdeen Transmissions Ltd
Wellington Road
Aberdeen
AB1 4DP
United Kingdom
Tel: 0224 898191
Fax: 0224 894812

ABS Pumps Ltd
Station Road
Horley
Surrey RH6 9HN
United Kingdom
Tel: 0293 821975
Fax: 0293 821976

Advanced Energy Monitoring Systems Ltd
The Energy Centre
Finnimore Industrial Estate
Ottery St Mary
Devon EX11 1NR
United Kingdom
Tel: 0404 812294
Fax: 0404 812603

Advanced Products (Seals & Gaskets) Ltd
Unit 25A
No 1 Industrial Estate
Consett
County Durham DH8 6SR
United Kingdom
Tel: 0207 500317
Fax: 0207 501210

Air Pumping Engineering Services
PO Box 239
East Ham
London
E6 3SG
United Kingdom
Tel: 081-470 8721
Fax: 081-470 4617

Airpel Filtration Ltd
Chiltern Hill
Chalfont St Peter
Gerrards Cross
Buckinghamshire SL9 9TN
United Kingdom
Tel: 0753 889251
Fax: 0753 887251

The Albany Engineering Co Ltd
Lydney
Gloucestershire GL15 5EQ
United Kingdom
Tel: 0594 842275
Fax: 0594 842574

Alfa-Laval Pumps
SSP Division
Birch Road
Eastbourne
East Sussex BN23 6PQ
United Kingdom
Tel: 0323 725151
Fax: 0323 730495

Allis Mineral Systems (UK) Ltd
Netherton Road
Wishaw
Lanarkshire ML2 0EJ
United Kingdom
Tel: 0698 355921
Fax: 0698 351376

Allspeeds Ltd
Royal Works, Atlas Street
Clayton le Moors
Accrington
Lancashire BB5 5LP
United Kingdom
Tel: 0254 235441
Fax: 0254 382899

Andrews Sykes Ltd
Coneygre Industrial Estate
Coneygre Road
Tipton
West Midlands DY4 8XD
United Kingdom
Tel: 021-557 2021
Fax: 021-557 5011

Apex Fluid Engineering Ltd
The Badminton Centre
Station Road, Yate
Bristol
BS17 5HT
United Kingdom
Tel: 0454 273464
Fax: 0454 273465

Appliance Components Ltd
Cordwallis Street
Maidenhead
Berkshire SL6 7BQ
United Kingdom
Tel: 0628 72121
Fax: 0628 75062

APV Howard Pumps Ltd
Fort Road
Eastbourne
East Sussex BN22 7SE
United Kingdom
Tel: 0323 722804
Fax: 0323 648955

APV Industrial Pump & Mixer Division
41-43 Glenburn Road
College Milton North, East Kilbride
Glasgow
G74 5BJ
United Kingdom
Tel: 03552 25461
Fax: 03552 63496

Aqua Hydraulics Ltd
Star Trading Estate
Partridge Green
Horsham
West Sussex RH13 8RA
United Kingdom
Tel: 0403 710855
Fax: 0403 710067

Armstrong Pumps
Peartree Road
Stanway
Colchester
Essex CO3 5JX
United Kingdom
Tel: 0206 579491
Fax: 0206 760532

The ARO Corporation (UK) Ltd
Walkers Road
North Moons Moat Industrial Estate
Redditch
Worcestershire B98 9HE
United Kingdom
Tel: 0527 61931
Fax: 0527 592299

Charles Austen Pumps Ltd
100 Royston Road
Byfleet
Surrey KT14 7PB
United Kingdom
Tel: 0932 355277
Fax: 0932 351285

Autoclude Ltd
Victor Pyrate Works
Arisdale Avenue
South Ockenden
Essex RM15 5DP
United Kingdom
Tel: 0708 856125
Fax: 0708 857366

Autometric Pumps Ltd, Lowara (UK) Ltd
Turkey Court
Ashford Road
Maidstone
Kent ME14 5PP
United Kingdom
Tel: 0622 677659
Fax: 0622 677650

Baric Pumps Ltd
Baltic Road
Felling Industrial Estate
Gateshead
Tyne & Wear NE10 0QF
United Kingdom
Tel: 091-469 8621
Fax: 091-469 2069

Bearings (Cardiff) Ltd
Cowbridge Road West
Cardiff
CF5 5XR
United Kingdom
Tel: 0222 592345
Fax: 0222 596132

The Bedford Pump Company Ltd
Brooklands
Woburn Road Industrial Estate
Kempston
Bedfordshire MK42 7UH
United Kingdom
Tel: 0234 852071
Fax: 0234 856620

Beldam Crossley Ltd
PO Box 7, Hill Mill
Temple Road
Bolton
BL1 6PB
United Kingdom
Tel: 0204 494711
Fax: 0204 493203

Berendsen Fluid Power
Sandy Way
Amington Industrial Estate
Tamworth
Staffordshire B77 4DS
United Kingdom
Tel: 0827 69369
Fax: 0827 65165

Beresford Pumps Ltd
Carlton Road
Foleshill
Coventry
CV6 7FL
United Kingdom
Tel: 0203 638484
Fax: 0203 637891

Bevi Wright Ltd
Pellon Lane
Halifax
West Yorkshire HX1 5QL
United Kingdom
Tel: 0422 360201
Fax: 0422 351092

BGA International
Swift House
Cosford Lane
Rugby
Warwickshire CV21 1QN
United Kingdom
Tel: 0788 543366
Fax: 0788 560733

Biwater Pumps
Clay Lane
Oldbury
Warley
West Midlands B69 4TF
United Kingdom
Tel: 021-544 2740
Fax: 021-544 2741

Bran + Luebbe (GB) Ltd
Scaldwell Road
Brixworth
Northamptonshire NN6 9UD
United Kingdom
Tel: 0604 880751
Fax: 0604 880145

E Braude Ltd
Liberta House
Sandhurst
Surrey GU17 8JR
United Kingdom
Tel: 0252 876123
Fax: 0252 875281

Brimotor Ltd
Orchard Centre
North Farm Road
Tunbridge Wells
Kent TN2 3DR
United Kingdom
Tel: 0892 537588
Fax: 0892 527724

David Brown Pumps Ltd
Green Road
Peniston
Sheffield
S30 6BJ
United Kingdom
Tel: 0226 763311
Fax: 0226 766535

Burdosa UK Ltd
Bellway House
270 Field End Road
Eastcote
Middlesex HA4 9NL
United Kingdom
Tel: 081-868 3377
Fax: 081-868 2929

Burgess-Manning Ltd
Rankin House
West Street
Ware
Hertfordshire SG12 9EE
United Kingdom
Tel: 0920 461268
Fax: 0920 466177

Calpeda Ltd
Wedgwood Road Industrial Estate
Bicester
Oxfordshire OX6 7UL
United Kingdom
Tel: 0869 241441
Fax: 0869 240681

Capital Engineering Services Ltd
Mailmech House
83 Copers Cope Road
Beckenham
Kent BR3 1NR
United Kingdom
Tel: 081-658 2243
Fax: 081-658 3853

CAT Pumps (UK) Ltd
Fleet Business Park, Sandy Lane
Church Crookham
Fleet
Hampshire GU13 0BF
United Kingdom
Tel: 0252 622031
Fax: 0252 626655

Ceetak Fluid Handling
Napier Road
Bedford
MK41 0QR
United Kingdom
Tel: 0234 327901
Fax: 0234 327909

Centrilift
Howe Moss Place
Kirkhill Industrial Estate, Dyce
Aberdeen
AB2 0ES
United Kingdom
Tel: 0224 723989
Fax: 0224 771020

Chem-Resist
Process Technology Division
Britannia House, Lockway
Ravensthorpe Industrial Estate
Dewsbury
West Yorkshire WF13 3SX
United Kingdom
Tel: 0924 499466
Fax: 0924 490334

Chromalock Ltd
Falkland Close
Coventry
CV4 8HQ
United Kingdom
Tel: 0203 466271
Fax: 0203 465298

Claessen Pumps Ltd
Horton Close
West Drayton
Middlesex UB7 8JA
United Kingdom
Tel: 0895 445261
Fax: 0895 441274

Cleghorn Waring & Co (Pumps) Ltd
Icknield Way
Letchworth
Hertfordshire SG15 6UD
United Kingdom
Tel: 0462 480380
Fax: 0462 482422

Clough (Croydon) Ltd
85 Manor Road
Wallington
Surrey SM6 0DH
United Kingdom
Tel: 081-395 8787
Fax: 081-647 9235

Componenta International Ltd
5 Ennis Close
Floats Road Industrial Estate, Wythenshawe
Manchester
M23 9LE
United Kingdom
Tel: 061-998 0717
Fax: 061-998 7972

Cornell Pumps (Europe)
Western Way
Bury St Edmunds
Suffolk IP33 3SZ
United Kingdom
Tel: 0284 706922
Fax: 0284 760256

Wm Coulthard & Co Ltd
Durranhill Trading Estate
Carlisle
CA1 3NS
United Kingdom
Tel: 0228 21418
Fax: 0228 511310

Couplings and Drives (UK) Ltd
Transmission House, 1 Lord Byron Square
Stowell Technical Park
Salford
M5 2XH
United Kingdom
Tel: 061-736 0760
Fax: 061-745 8041

CP Instrument Co Ltd
PO Box 22
Bishop's Stortford
Hertfordshire CM23 3DX
United Kingdom
Tel: 0279 757711
Fax: 0279 755785

Crane Fluid Systems
Nacton Road
Ipswich
Suffolk IP3 9QH
United Kingdom
Tel: 0473 270222
Fax: 0473 270221

John Crane UK Ltd
Crossbow House
Liverpool Road
Slough
Berkshire SL1 4QX
United Kingdom
Tel: 0753 531122
Fax: 0753 573677

Crest Pumps Ltd
163 Stourvale Road
Southbourne
Bournemouth
BH6 5HQ
United Kingdom
Tel: 0202 425000
Fax: 0202 425577

Alan Dale Pumps Ltd
Vitesse House
Ashford Road
Ashford
Middlesex TW15 1XE
United Kingdom
Tel: 0784 421114
Fax: 0784 421092

Danfoss Ltd
Perivale Industrial Park
Horsenden Lane South
Greenford
Middlesex UB6 7QE
United Kingdom
Tel: 081-991 7000
Fax: 081-991 7171

Datronix
Chaul End Lane
Luton
LU4 8EZ
United Kingdom
Tel: 0582 490437
Fax: 0582 598036

Dawson Downie Lamont Ltd
Elgin Works
Clydebank
Dunbartonshire G81 1YF
United Kingdom
Tel: 041-952 2271
Fax: 041-952 9088

Denco Ltd
PO Box 11
Holmer Road
Hereford
HR4 9SJ
United Kingdom
Tel: 0432 277277
Fax: 0432 268005

Denver Process Equipment Ltd
Stocks House
9 North Street
Leatherhead
Surrey KT22 7AX
United Kingdom
Tel: 0372 379313
Fax: 0372 379504

Dia-Meter Pumps Ltd
Unit 22, Bridge Industries
Broadcut
Fareham
Hampshire PO16 8SX
United Kingdom
Tel: 0329 233542
Fax: 0329 823209

Dual Pumps Ltd
47 Norman Way
Melton Mowbray
Leicestershire LE13 1NF
United Kingdom
Tel: 0664 67226
Fax: 0664 410127

Dualway Metering Pumps Ltd
Highfield Works
Spring Gardens, Washington
Pulborough
West Sussex RH20 3BS
United Kingdom
Tel: 0993 892358
Fax: 0903 892062

Durametallic UK
Unit 13B, United Trading Estate
Old Trafford
Manchester
M16 0RJ
United Kingdom
Tel: 061-848 7061
Fax: 061-872 6772

Durco Process Equipment Ltd
28 Heathfield
Stacey Bushes
Milton Keynes
MK12 6HR
United Kingdom
Tel: 0908 318212
Fax: 0908 320037

Ebara UK Ltd
Ebara House, 2 Ariel Way,
Heathrow Causeway
Great Southwest Road
Hounslow
Middlesex TW4 6JS
United Kingdom
Tel: 081-577 1331
Fax: 081-577 0492

Edwards and Jones Ltd
Whittle Road
Meir
Stoke-on-Trent
ST3 7QD
United Kingdom
Tel: 0782 599000
Fax: 0782 599001

Edwards High Vacuum International
Manor Royal
Crawley
West Sussex RH10 2LW
United Kingdom
Tel: 0293 528844
Fax: 0293 533453

EG & G Sealol
Coronation Road
Cressex Industrial Estate
High Wycombe
Buckinghamshire HP12 3TP
United Kingdom
Tel: 0494 451661
Fax: 0494 452425

Envirotech Pumpsystems UK
Swift House
Cosford Lane
Rugby
Warwickshire CV21 1QN
United Kingdom
Tel: 0788 546996
Fax: 0788 560733

F.T.L. Fluid Seals Ltd
Howley Park Road
Morley
Leeds
LS27 0BN
United Kingdom
Tel: 0532 521061
Fax: 0532 522567

Fawcett Christie Hydraulics Ltd
Chester Road
Sandycroft
Clwyd CH5 2QP
United Kingdom
Tel: 0244 535515
Fax: 0244 533002

Ferguson & Timpson Ltd
Thistle House
Selinas Lane
Dagenham
Essex RM8 1TB
United Kingdom
Tel: 081-593 7611
Fax: 081-595 9241

Fife Bearing & Transmissions Ltd
Whitehill Road
Whitehill Industrial Estate
Glenrothes
Fife
United Kingdom
Tel: 0592 630448
Fax: 0592 630488

Flexibox Ltd
Nash Road
Trafford Park
Manchester
M17 1SS
United Kingdom
Tel: 061-872 2484
Fax: 061-872 1654

Flotronic Pumps Ltd
Ricebridge Works
Brighton Road
Bolney
West Sussex RH17 5NQ
United Kingdom
Tel: 0444 881871
Fax: 0444 881860

Flowguard Ltd
Watford Bridge
New Mills
Stockport
Cheshire SK12 4HJ
United Kingdom
Tel: 0663 745976
Fax: 0663 742788

Fluid Pumps (Bristol) Ltd
Unit 1, Avonside Road
Avonside Ind Estate
Bristol
BS2 0UQ
United Kingdom
Tel: 0272 724102
Fax: 0272 724399

Fluorocarbon Co Ltd
Caxton Hill
Hertford
SG13 7NH
United Kingdom
Tel: 0992 550731
Fax: 0992 584697

Fordwater Pumping Supplies Ltd
49-53 Stratford Road
Birmingham
B11 1RQ
United Kingdom
Tel: 021-772 8336
Fax: 021-771 0530

Forsheda Ltd
Station Road
Bakewell
Derbyshire DE45 1GF
United Kingdom
Tel: 0629 814381
Fax: 0629 814658

Forward Industries Ltd
South Marston Park
Swindon
Wiltshire SN3 4RA
United Kingdom
Tel: 0793 823241
Fax: 0793 828474

Friatec Ltd
Friatec House, Old Parkbury Lane
Colney Street
St Albans
Hertfordshire AL2 2ED
United Kingdom
Tel: 0923 857878
Fax: 0923 853434

Fristam Pumps Ltd
Unit 11, Apex Business Centre
Diplocks Way
Hailsham
East Sussex
United Kingdom
Tel: 0323 849849
Fax: 0323 849438

GEC Alsthom Fluids Group
Cambridge Road
Whetstone
Leicester
LE8 3LH
United Kingdom
Tel: 0533 750750
Fax: 0533 750891

Gentech International Ltd
Grangestone Industrial Estate
Girvan
Ayrshire KA26 9PS
United Kingdom
Tel: 0465 3581
Fax: 0465 4974

Gilbert Gilkes & Gordon Ltd
Canal Iron Works
Kendal
Cumbria LA9 7BZ
United Kingdom
Tel: 0539 720028
Fax: 0539 732110

Girdlestone Pumps Ltd
Melton
Woodbridge
Suffolk IP12 1ER
United Kingdom
Tel: 0394 383777
Fax: 0394 386733

Glacier Industrial Bearings
Kirkstyle
Riccarton
Kilmarnock
Ayrshire
United Kingdom
Tel: 0563 39999
Fax: 0563 71426

Godiva Ltd
Charles Street
Warwick
CV34 5LR
United Kingdom
Tel: 0296 411666
Fax: 0926 402075

R Goodwin International
Goodwin House
Leek Road, Hanley
Stoke-on-Trent
ST1 3NR
United Kingdom
Tel: 0782 208040
Fax: 0782 208060

Graco UK Ltd
Wednesfield Road
Wolverhampton
West Midlands WV10 0DR
United Kingdom
Tel: 0902 351924
Fax: 0902 351365

Graham Precision Pumps Ltd
The Forge
Forge Lane
Congleton
Cheshire CW12 4HQ
United Kingdom
Tel: 0260 274721
Fax: 0260 276965

Grosvenor Pumps Ltd
8 Shaftesbury Industrial Centre
The Runnings
Cheltenham
Gloucestershire GL51 9NH
United Kingdom
Tel: 0242 227400
Fax: 0242 227404

Grundfos Pumps Ltd
Grovebury Road
Leighton Buzzard
Bedfordshire LU7 8TL
United Kingdom
Tel: 0525 850000
Fax: 0525 850011

H_2O Waste-Tec
Horsfield Way
Bredbury Park
Stockport
SK6 2SU
United Kingdom
Tel: 061-406 7111
Fax: 061-406 7222

Hamworthy Allweiler Ltd
Fleets Corner
Poole
Dorset BH17 7LA
United Kingdom
Tel: 0202 665566
Fax: 0202 660397

Hamworthy Hydraulics Ltd
Fleets Corner
Poole
Dorset BH17 7LB
United Kingdom
Tel: 0202 665566
Fax: 0202 665666

Hamworthy Pumps & Compressors Ltd
Fleets Corner
Poole
Dorset BH17 7LA
United Kingdom
Tel: 0202 665566
Fax: 0202 665444

Harrison Group Ltd
Pumps Division, Blagdon Pumps
Lambert Road
Washington
Tyne & Wear NE37 1QP
United Kingdom
Tel: 091-417 7475
Fax: 091-417 5435

Haskel Energy Systems Ltd
North Hylton Road
Sunderland
SR5 3JD
United Kingdom
Tel: 091-549 1212
Fax: 091-549 0911

Hawker Electronics Ltd
43 Melchett Road
Kings Norton Business Centre
Birmingham
B30 3HP
United Kingdom
Tel: 021-459 8911
Fax: 021-433 3041

Hayward Tyler Fluid Dynamics Ltd
Kimpton Road
Luton
Bedfordshire LU1 3LD
United Kingdom
Tel: 0582 31144
Fax: 0582 402563

Hayward Tyler Sumo
Howard Avenue
Nerston Industrial Estate
East Kilbride
G74 4PZ
United Kingdom
Tel: 03552 21301
Fax: 03552 33571

Headland Engineering Developments Ltd
Unit 5B, Navigation Drive
Hurst Business Park
Brierley Hill
West Midlands DY5 1YF
United Kingdom
Tel: 0384 76767
Fax: 0384 481040

Hick Hargreaves & Co Ltd
Crook Street
Bolton
BL3 6DB
United Kingdom
Tel: 0204 23373
Fax: 0204 395261

Hidrostal Process Engineering Ltd
Unit 5, The Galloway Centre
Express Way, Hambridge Lane
Newbury
Berkshire RG14 5TL
United Kingdom
Tel: 0635 550560
Fax: 0635 40525

Hilge Pumps Ltd
Hilge House
Pelham Court
Crawley
RH11 9AZ
United Kingdom
Tel: 0293 514433
Fax: 0293 519100

Richard Hill Pumps Ltd
Brooke Road
Ridlington
Uppingham
Leicestershire LE15 9AN
United Kingdom
Tel: 0572 823385
Fax: 0572 821660

HMD Seal/Less Pumps Ltd
Hampden Park Industrial Estate
Eastbourne
East Sussex BN22 9AN
United Kingdom
Tel: 0323 501241
Fax: 0323 503369

Hofland (UK) Ltd
Bull Lane Industrial Estate
Acton
Sudbury
Suffolk CO10 0BD
United Kingdom
Tel: 0787 880588
Fax: 0787 376486

Holden and Brooke Ltd
Wenlock Way
Manchester
M12 5JL
United Kingdom
Tel: 061-223 2223
Fax: 021-220 9660

H R Holfeld (Pumps) Ltd
65 Carshalton Road
Sutton
Surrey SM1 4LH
United Kingdom
Tel: 081-661 0714
Fax: 081-642 0454

Hughes Wynne Ltd
68 Guildford Street
Chertsey
Surrey KT16 9BB
United Kingdom
Tel: 0932 569700
Fax: 0932 569652

Hydrainer Pumps Ltd
Norwood Industrial Estate
Rotherham Close, Killamarsh
Sheffield
S31 8JU
United Kingdom
Tel: 0742 484868
Fax: 0742 510136

Hydrair Ltd
Berry Hill Industrial Estate
Droitwich
Worcestershire WR9 9AB
United Kingdom
Tel: 0905 772302
Fax: 0905 770309

Ingersoll-Dresser Pumps (UK) Ltd
Queensway
Team Valley Trading Estate
Gateshead
NE11 0QB
United Kingdom
Tel: 091-487 5051
Fax: 091-482 0504

Ingersoll-Dresser Pumps (UK) Ltd
Newark Operations
PO Box 17
Newark
Nottinghamshire NG24 3EN
United Kingdom
Tel: 0636 705151
Fax: 0636 705991

Ingersoll-Rand Sales Company Ltd
PO Box 2
Chorley New Road, Horwich
Bolton
BL6 6JN
United Kingdom
Tel: 0204 690690
Fax: 0204 691093

Inoxpa UK
(Euro-Industrial Engineering Ltd)
Lancashire House
133 Grosvenor Street
Manchester
M1 7HE
United Kingdom
Tel: 061-274 4477
Fax: 061-274 4001

Interdab Ltd
The Maltings Industrial Estate
Southminster
Essex CM0 7EQ
United Kingdom
Tel: 0621 773081
Fax: 0621 772223

Inverness Bearing & Transmissions Ltd
5 Harbour Road
Inverness
IV1 1SY
United Kingdom
Tel: 0563 243528
Fax: 0563 225039

ITT Jabsco
Bingley Road
Hoddesdon
Hertfordshire EN11 0BU
United Kingdom
Tel: 0992 467191
Fax: 0992 467132

JLC Pumps & Engineering Co Ltd
PO Box 225
Barton-le-Clay
Bedfordshire MK45 4BY
United Kingdom
Tel: 0582 881946
Fax: 0582 881951

Johnson Pump (UK) Ltd
Highfield Industrial Estate
Edison Road
Eastbourne
East Sussex BN23 6PT
United Kingdom
Tel: 0323 509211
Fax: 0323 507306

Kecol Industrial Products Ltd
Faraday Drive
Bridgnorth
Shropshire WV15 5BJ
United Kingdom
Tel: 0746 764311
Fax: 0746 764780

Kestner Engineering Co Ltd
Station Road
Greenhithe
Kent DA9 9NG
United Kingdom
Tel: 0322 383281
Fax: 0322 386684

Kinder-Janes Engineers Ltd
Porters Wood
St Albans
Hertfordshire AL3 6HU
United Kingdom
Tel: 0727 844441
Fax: 0727 844247

KSB Ltd
Thameside House
Grove Road
Northfleet
Kent DA11 9AX
United Kingdom
Tel: 0474 564359
Fax: 0474 564381

Linatex Ltd
Stanhope Road
Camberley
Surrey GU15 3BX
United Kingdom
Tel: 0276 63434
Fax: 0276 691734

Liquid Dynamics
Hurstfield Industrial Works
Hurst Street, Reddish
Stockport
Cheshire SK5 7BB
United Kingdom
Tel: 061-442 6222
Fax: 061-443 1486

Lowara (UK) Ltd
Millwey Rise Industrial Estate
Axminster
Devon EX13 5HU
United Kingdom
Tel: 0297 33374
Fax: 0297 35238

Lumatic (GA) Ltd
Ponswood
Hastings
East Sussex TN34 1YS
United Kingdom
Tel: 0424 436343
Fax: 0424 429926

Lutz (UK) Ltd
Gateway Estate
West Midlands Freeport
Birmingham
B26 3QD
United Kingdom
Tel: 021-782 2662
Fax: 021-782 2680

Mackley Pumps
Clarke Chapman Marine
Clarke Chapman Ltd
Victoria Works
Gateshead
Tyne & Wear NE8 3HS
United Kingdom
Tel: 091-477 2271
Fax: 091-477 1009

Macro Precision Pumps Ltd
Unit 33/41
Station Road Industrial Estate
Hailsham
East Sussex BN27 2ER
United Kingdom
Tel: 0323 842331
Fax: 0323 842980

Magnetek Universal Electric Ltd
PO Box 8
Gainsborough
Lincolnshire DN21 1XU
United Kingdom
Tel: 0427 614141
Fax: 0427 617513

March May Ltd
Eaton Works, Howard Road
Eaton Socon
Huntingdon
Cambridgeshire PE19 3NZ
United Kingdom
Tel: 0480 214444
Fax: 0480 405336

Maximator (Deeweld)
Sealand Industrial Estate
Chester
CH1 4NZ
United Kingdom
Tel: 0244 375375
Fax: 0244 377403

MDM Pumps Ltd
Spring Lane
Malvern
Worcestershire WR14 1BP
United Kingdom
Tel: 0684 892678
Fax: 0684 892841

Megator Pumps & Compressors Ltd
Hendon Industrial Estate
Hendon
Sunderland
SR1 2NQ
United Kingdom
Tel: 091-567 5488
Fax: 091-567 8512

George Meller Ltd
Orion Park
Northfield Avenue, Ealing
London
W13 9SJ
United Kingdom
Tel: 081-579 2111
Fax: 081-579 7326

Milton Roy (UK) Ltd
Oaklands Park
Fishponds Road
Wokingham
Berkshire RG11 2FD
United Kingdom
Tel: 0734 771066
Fax: 0734 771198

Mono Pumps Ltd
PO Box 14
Martin Street, Audenshaw
Manchester
M34 5DQ
United Kingdom
Tel: 061-339 9000
Fax: 061-344 0727

MPL Pumps
Ironstone Way
Brixworth
Northamptonshire NN6 9UD
United Kingdom
Tel: 0604 882525
Fax: 0604 882227

Multiphase Pumping Systems Ltd
Unit 1N, Farrington Fields
Farrington Gurney
Bristol
BS18 5UU
United Kingdom
Tel: 0761 453746
Fax: 0761 453761

Munster Simms Engineering Ltd (Whale Pumps)
Old Belfast Road
Bangor
Northern Ireland BT19 1LT
United Kingdom
Tel: 0247 270531
Fax: 0247 466421

Nash Engineering Co (GB) Ltd
Road One
Industrial Estate
Winsford
Cheshire CW7 3PL
United Kingdom
Tel: 0606 594242
Fax: 0606 863496

Naybuk Pumps Ltd
Thatched Folly, Lindow End
Mobberley
Knutsford
Cheshire WA16 7BA
United Kingdom
Tel: 0565 872222

Nicholsons Seals Ltd
Hamsterley
Newcastle-upon-Tyne
NE17 7SX
United Kingdom
Tel: 0207 560505
Fax: 0207 561004

Norgren-Martonair Ltd
PO Box 22
Eastern Avenue
Lichfield
Staffordshire WS13 6SB
United Kingdom
Tel: 0543 414333
Fax: 0543 268052

Oilgear Towler Ltd
Oaklands Road
Rodley
Leeds
LS13 1LG
United Kingdom
Tel: 0532 577721
Fax: 0532 559537

Pall Industrial Hydraulics Ltd
Europa House
Havant Street
Portsmouth
PO1 3PD
United Kingdom
Tel: 0705 753545
Fax: 0705 297647

Patay Pumps Ltd
Trevoole
Praze
Camborne
Cornwall TR14 0PJ
United Kingdom
Tel: 0209 831937
Fax: 0209 831939

Pegson Ltd
Mammoth Street
Coalville
Leicester
LE67 3GN
United Kingdom
Tel: 0530 510051
Fax: 0530 510041

Pioneer Weston Ltd
Douglas Green
Pendleton
Salford
Lancashire M6 6FT
United Kingdom
Tel: 061-736 5811
Fax: 061-736 5107

Plenty Mirrlees Pumps
Hambridge Road
Newbury
Berkshire RG14 5TR
United Kingdom
Tel: 0635 42363
Fax: 0635 49758

Pollution Control Engineering Ltd
PO Box 617, Chapel Hill
Longridge
Preston
PR3 3BU
United Kingdom
Tel: 0772 784444
Fax: 0772 784082

Precision Dosing Pumps Ltd
Alma Park Road
Grantham
Lincolnshire NG31 9SE
United Kingdom
Tel: 0476 77677
Fax: 0476 75233

Production Techniques Ltd
15 Kings Road
Fleet
Hampshire GU13 9AU
United Kingdom
Tel: 0252 616575
Fax: 0252 615818

Pullen Pumps Ltd
58 Beddington Lane
Croydon
CR9 4PT
United Kingdom
Tel: 081-684 9521
Fax: 081-689 8892

The Pump Centre
Building RD1, AEA Technology
Birchwood Science Park
Warrington
WA3 6AT
United Kingdom
Tel: 0925 252185
Fax: 0925 253576

Pump Engineering Ltd
Riverside Industrial Estate
Littlehampton
West Sussex BN17 5DF
United Kingdom
Tel: 0903 730900
Fax: 0903 730234

Putzmeister Ltd
Carrwood Road
Chesterfield Trading Estate, Sheepbridge
Chesterfield
Derbyshire S41 9QB
United Kingdom
Tel: 0246 260033
Fax: 0246 260077

Reda (UK) Ltd
Yiewsley Grange
High Street
Yiewsley
Middlesex UB7 7QP
United Kingdom
Tel: 0895 422115
Fax: 0895 448599

The Reiss Engineering Co
2 Dalston Gardens
Stanmore
Middlesex HA7 1BQ
United Kingdom
Tel: 081-204 7155
Fax: 081-206 0459

Renold Gears
Holroyd Gear Works
Milnrow
Rochdale
Lancashire OL16 3LS
United Kingdom
Tel: 0706 47491
Fax: 0706 42515

RHL Hydraulics Ltd
St Peters
Newcastle-upon-Tyne
NE6 1BS
United Kingdom
Tel: 091-276 4444
Fax: 091-276 4462

Roach Pumps Ltd
55 Monument Industrial Park
Chalgrove
Oxfordshire OX44 7RW
United Kingdom
Tel: 0865 400076
Fax: 0865 400078

Roban Engineering Ltd
18 Avenue Road
Belmont
Surrey SM2 6JD
United Kingdom
Tel: 081-643 8070
Fax: 081-642 1104

Robbins & Myers Ltd
2 West Links
Tollgate
Chandlers Ford
Hampshire SO5 3TG
United Kingdom
Tel: 0703 650610
Fax: 0703 650611

Roper Industries (Europe) Ltd
Western Way
Bury St Edmunds
Suffolk IP33 3SZ
United Kingdom
Tel: 0284 760406
Fax: 0284 760256

SPP Group Ltd
Theale Cross
Calcot
Reading
RG3 7SP
United Kingdom
Tel: 0734 323123
Fax: 0734 323302

SPP LaBour Pump Company
500 Wellingborough Drive
Northampton Business Park
Northampton
NN4 7YT
United Kingdom
Tel: 0604 700338
Fax: 0604 700064

Sarlin Ltd
High Cliffe Road
Hamilton Industrial Park
Leicester
LE5 1TY
United Kingdom
Tel: 0533 461527
Fax: 0533 460813

Sauer-Sundstrand Ltd
Cheney Manor
Swindon
Wiltshire SN2 2PZ
United Kingdom
Tel: 0793 530101
Fax: 0793 528301

Schuco International London Ltd
Woodhouse Road
London
N12 0NE
United Kingdom
Tel: 081-368 1642
Fax: 081-361 3761

Seddons (Plant & Engineers) Ltd
Hot Lane Industrial Estate
Burslem
Stoke-on-Trent
ST6 2AU
United Kingdom
Tel: 0782 575111
Fax: 0782 575241

Seepex UK Ltd
10 Oxford Road
Pen Mill Trading Estate
Yeovil
Somerset BA21 5HR
United Kingdom
Tel: 0935 72376
Fax: 0935 79836

Selwood Pumps
Bournemouth Road
Chandlers Ford
Hampshire SO5 3ZL
United Kingdom
Tel: 0703 266311
Fax: 0703 360906

Serfilco Europe Ltd
Broadoak Industrial Park
Ashburton Road West, Trafford Park
Manchester
M17 1RW
United Kingdom
Tel: 061-872 1317
Fax: 061-873 8027

Shurflo Ltd
The Old Forge
36 West Street
Reigate
Surrey RH2 9BX
United Kingdom
Tel: 0737 242290
Fax: 0737 242282

Sihi Pumps (UK) Ltd
Bridgewater Road
Altrincham
Cheshire WA14 1NB
United Kingdom
Tel: 061-928 6371
Fax: 061-928 3022

Simon-Hartley Ltd
Etruria
Stoke-on-Trent
ST4 7BH
United Kingdom
Tel: 0782 202300
Fax: 0782 261494

SLD Pumps Ltd
Portobello Trading Estate
Chester le Street
Co Durham DH3 2RY
United Kingdom
Tel: 091-410 4611
Fax: 091-410 7000

Michael Smith Engineers Ltd
Oaks Road
Woking
Surrey GU21 1PH
United Kingdom
Tel: 0483 771871
Fax: 0483 723110

Stanhope Barclay Kellett
Crown Division of Albany Engineering
Richter Works
Garnett Street
Bradford
BD3 9HB
United Kingdom
Tel: 0274 725351
Fax: 0274 742467

The Sterling Pump Co Ltd
Water Lane
Spalding
Lincolnshire PE11 2TH
United Kingdom
Tel: 0775 722404
Fax: 0775 710501

Stork Pumps Ltd
Meadow Brook Industrial Centre
Maxwell Way
Crawley
West Sussex RH10 2SA
United Kingdom
Tel: 0293 553495
Fax: 0293 524635

Stuart Turner Ltd
The Market Place
Henley-on-Thames
Berkshire RG9 2AD
United Kingdom
Tel: 0491 572655
Fax: 0491 573704

Sulzer (UK) Pumps Ltd
Manor Mill Lane
Leeds
LS11 8BR
United Kingdom
Tel: 0532 701244
Fax: 0532 772397

Superior Industrial Products Ltd
31a Cleveland Road
South Woodford
London
E18 2AE
United Kingdom
Tel: 081-989 1171
Fax: 081-530 1150

T-T Pumps Ltd
Lingard Street, Burlsem
Stoke-on-Trent
ST6 1EE
United Kingdom
Tel: 0782 812231
Fax: 0782 826751

Torres Engineering & Pumps Ltd
28 Sanderson Street
Sheffield
S9 2TW
United Kingdom
Tel: 0742 433353
Fax: 0742 425885

Totton Pumps Ltd
Rushington Business Park
Rushington Lane, Totton
Southampton
SO4 4AN
United Kingdom
Tel: 0703 666685
Fax: 0703 666880

Treadway Flow Control
26-30 Cubitt Street
London
WC1X 0LR
United Kingdom
Tel: 071-833 0395
Fax: 071-837 5673

Tuthill UK Ltd
Birkdale Close
Manners Industrial Estate
Ilkeston
Derbyshire DE7 8YA
United Kingdom
Tel: 0602 325226
Fax: 0602 324816

Uniclip Ltd
100 Royston Road
Weybridge
Surrey KT14 7PB
United Kingdom
Tel: 0932 355277
Fax: 0932 351285

Vanton Pumps Ltd
Unit 6
Radnor Park Industrial Centre
Congleton
Cheshire CW12 4XL
United Kingdom
Tel: 0260 277040
Fax: 0260 280605

James Walker & Co Ltd
Lion Works
Woking
Surrey GU22 8AP
United Kingdom
Tel: 0483 757575
Fax: 0483 755711

James Walker Britco Ltd
Britco Works, Orgreave Crescent
Dore House Industrial Estate, Handsworth
Sheffield
S13 9NQ
United Kingdom
Tel: 0742 690776
Fax: 0742 540455

Wallace & Tiernan Ltd
Priory Works
Tonbridge
Kent TN11 0QL
United Kingdom
Tel: 0732 771777
Fax: 0732 771800

Wanner International Ltd
Riverview House
Weyside Park, Catteshall Lane
Godalming
Surrey GU7 1XE
United Kingdom
Tel: 0483 861234
Fax: 0483 860209

Warman International Ltd
Halifax Road
Todmorden
Lancashire OL14 5RT
United Kingdom
Tel: 0706 814251
Fax: 0706 815350

Warsop Power Tools
Hever Road
Edenbridge
Kent TN8 5DL
United Kingdom
Tel: 0732 863081
Fax: 0732 865680

Watson-Marlow Ltd
Falmouth
Cornwall TR11 4RU
United Kingdom
Tel: 0326 373461
Fax: 0326 376009

Weir Pumps Ltd
179 Newlands Road
Cathgart
Glasgow
G44 4EX
United Kingdom
Tel: 041-637 7141
Fax: 041-637 7358

Welland Engineering Co Ltd
Water Lane
Spalding
Lincolnshire PE11 2TH
United Kingdom
Tel: 0775 722535
Fax: 0775 710501

Wemco Pumps
Swift House
Cosford Lane
Rugby
Warwickshire CV21 1QN
United Kingdom
Tel: 0788 546996
Fax: 0788 560733

Whitby & Chandler Ltd
Green Road
Sheffield
S30 6PH
United Kingdom
Tel: 0226 370380
Fax: 0226 767138

Wilo Salmson Pumps Ltd
Ashlyn Raod
West Meadows Industrial Estate
Derby
DE21 6XE
United Kingdom
Tel: 0332 385181
Fax: 0332 44423

Yarmouth Bearing & Transmissions Ltd
7 James Court, Faraday Road
Gapton Hall Industrial Estate
Great Yarmouth
Norfolk NR31 0NF
United Kingdom
Tel: 0493 655505

Ziehl-Ebm (UK) Ltd
17-19 Richmond Road
Dukes Park Industrial Estate
Chelmsford
Essex CM2 6TL
United Kingdom
Tel: 0245 468555
Fax: 0245 466336

16.6 Trade Names

ABS
Cardo Pump AB   (S)
ARO
Air Pumping Engineering Services   (UK)
ARTUS
Hughes Wynne Ltd   (UK)
BEGEMANN PUMPS
EnviroTech Pumpsystems BV   (NL)
CHEM-GARD
Vanton Pumps Ltd   (UK)
DEPA
Air Pumping Engineering Services   (UK)
DESMI
A/S De Smithske (DESMI)   (DK)
ELECTRO-SUB
Biwater Pumps   (UK)

FLEX-I-LINER
Vanton Pumps Ltd   (UK)
GEHO PUMPS
EnviroTech Pumpsystems BV   (NL)
HAGGLUNDS-DENISON
Hydrobel SA   (B)
HMD SEAL/LESS PUMPS LTD
HMD Seal/Less Pumps Ltd   (UK)
INSIGHT
HMD Seal/Less Pumps Ltd   (UK)
KALREZ
Du Pont de Nemours International SA   (CH)
PRACTI-SHIM
Hughes Wynne Ltd   (UK)
ROPER
Hydrobel SA   (B)

SANDPIPER/WARREN RUPP
Air Pumping Engineering Services   (UK)
SCANPUMP
Cardo Pump AB   (S)
SHURFLO
Shurflo Ltd   (UK)
SUMP-GARD
Vanton Pumps Ltd   (UK)
VARLEY
Hayward Tyler Fluid Dynamics Ltd   (UK)
WALLWIN
Biwater Pumps   (UK)
WILDEN
Air Pumping Engineering Services   (UK)
ZALAK
Du Pont de Nemours International SA   (CH)

17 Units and conversions

The modern SI-system has been adopted in legislation by practically every country in the world. There are, however, extensive changeover problems and many different kinds of data are still only available in units which do not conform to the SI-system. Chapter 17 attempts to bridge the difficulties experienced when dealing with quantities expressed in older units.

Almost without exception it is recommended that the quantities be converted directly to SI units; calculation can then be carried out using the coherent SI-system. However, it is important that all results of calculations should be reviewed by someone fully conversant with the system of units and the practical values encountered.

The SI-system is a logical and progressive system of inter-related units. However, the units of pressure, the Pascal, are the wrong order of magnitude for most engineering purposes. For safety reasons, and also to avoid confusion, it is better to use consistent units for suction and discharge pressures. The pump industry has proposed the 'bar' as a superior unit to the Pascal. Suction and discharge pressures can be expressed in 'bar' without recourse to subdivision or multiples. Of course all pressures, except differential pressures, and NPIP is a differential, must be qualified by 'absolute' or 'gauge'. Strict compliance with SI rules precludes the modification of unit names to 'bara' and 'barg'. The correct procedure is to qualify the quantity specified, not the unit.

Special Note: Some American publications have great difficulties with SI units. Treat all SI values quoted with caution.

Conversion factors for hardness and toughness testing, which are not defined by SI are also given.

Contents

17.1 SI, The International System of Units

17.2 Conversion factors for SI units

    17.2.1 Plane angle
    17.2.2 Length
    17.2.3 Area
    17.2.4 Volume
    17.2.5 Time
    17.2.6 Linear velocity
    17.2.7 Linear acceleration
    17.2.8 Angular velocity
    17.2.9 Angular acceleration
    17.2.10 Mass
    17.2.11 Density
    17.2.12 Force
    17.2.13 Torque
    17.2.14 Pressure, stress
    17.2.15 Dynamic viscosity
    17.2.16 Kinematic viscosity
    17.2.17 Energy
    17.2.18 Power
    17.2.19 Flow
    17.2.20 Temperature

17.3 Other conversion factors

    17.3.1 Hardness
    17.3.2 Material toughness

17.4 Normal quantities and units used within pump technology

17.1 SI, The International System of Units

SI, Systeme Internationale d'Unites, the international measurement unit system, is not a completely new system. It is based on an earlier metric system and is coming more and more into world-wide use. The SI-system is now systematically constructed to cover in practice all scientific, technical and daily requirements and is subject to international agreement. This means that it is now possible to apply the SI-system uniformly throughout the world. A measurement system which is suitable for all technical and scientific purposes has to fulfil many requirements. Some of the basic requirements which SI satisfies are consistency, consequential applicability, coherence, the convenient expression of multiples and sub-multiples over a wide range of numerical values and accuracy.

- Consistency means that each unit shall represent one, and only one, quantity.

- Consequential applicability means that each quantity shall be measured in one, and only one, unit.

- Coherence means that all units for every quantity shall be compatible so as to eliminate the need for arbitrary conversion factors in calculations involving related quantities.

- Convenient expression of multiples and sub-multiples means the convenient multiplication of units to enable the use of practical numerical values within a particular application.

- Accuracy means that the base units shall be precisely derived and defined. Six of the seven base units are thus determined from distinct precisely defined physical phenomena, the seventh, the kilogram, is determined by one standard body which is held in Paris.

In 1971 the Council of Ministers of the EEC ratified a Directive, 71/354/EEC, on units which committed all member states to amend legislation to authorise SI units within 18 months of that date and to implement all provisions of the Directive within a further five years. An amending Directive, 76/770/EEC, legislates the obligations. The Units of Measurement Directives place non-SI units into four chapters A to D.

Chapter A prescribes those units which are for permanent use and member states are obliged to authorise them in their laws by 21 April, 1978.

Chapter B contains a list of all units which member states have undertaken to cease to authorise in their laws with effect from 31 December, 1977.

Chapter C contains a list of units which member states have undertaken to cease to authorise in their laws with effect from 31 December, 1979.

Chapter D covers remaining units and some other units and will be reviewed before 31 December, 1979.

The formal content of the SI is determined and authorised by the General Conference of Weights and Measures (CGPM) and, for more detailed descriptions of the System reference should be made to BS 3763 and SI - The International System of Units published by HMSO. However, the basic advice to industry on the use of SI is now contained in ISO 1000, BS 5555.

The SI system includes three classes of units: 1) base units 2) supplementary units 3) derived units.

The system also includes a number of Non-SI units, the 'bar' for example, which are retained for use together with the SI units and their multiples because of their practical importance or because of their specialized fields. Multiples and sub-multiples of SI units and recognised Non-SI units are formed by attaching a prefix, micro, milli, kilo, Mega etc., to the unit symbol.

Base Units. The SI is founded on seven base units, which are independent of each other. All other types of unit can be expressed in terms of these.

The seven base units are:

length	metre	(m),
thermodynamic temperature	Kelvin	(K),
mass	kilogram	(kg),
amount of substance	mole	(mol),
time	second	(s),
luminous intensity	candela	(cd),
electric current	ampere	(A).

The base unit for temperature, absolute temperature, temperature differences and temperature changes is K, not °K. The unit K is identical in magnitude to the degree Celsius, centigrade, i.e. 1 K = 1 °C, which is why the term degree Celsius and the symbol °C is retained and recognised. The numerical values for temperature expressed in K and °C are not, however, the same. The numerical value expressed in K = the numerical value expressed in °C plus 273.15.

Supplementary units. Certain basic units have not yet been classified as either base or derived units and may be regarded as either.

These units are:

plane angle	radian	(rad)
solid angle	steradian	(sr)

Derived units are expressed in terms of base units and/or supplementary units by multiplication and division according to the laws of physics relating the various quantities, see figure 17.1.

Quantity	Name of derived SI unit	Symbol	Expressed in terms of base or supplementary units
frequency	Hertz	Hz	1 Hz = 1/s
force	Newton	N	1 N = 1 kg m/s^2
pressure, stress	Pascal	Pa	1 Pa = 1 N/m^2
energy, work, heat	Joule	J	1 J = 1 Nm
power	Watt	W	1 W = 1 J/s
electric charge, quantity of electricity	Coulomb	C	1 C = 1 A s
electric potential	Volt	V	1 V = 1 J/C = 1 W/A
electric capacitance	Farad	F	1 F = 1 C/V
electric resistance	Ohm	Ω	1 Ω = V/A
electric conductance	Siemens	S	1 S = 1/Ω
magnetic flux	Weber	Wb	1 Wb = 1 V s
magnetic flux density	Tesla	T	1 T = 1 Wb/m^2
inductance	Henry	H	1 H = 1 Wb/A
luminous flux	lumen	lm	1 lm = 1 cd sr
illuminance	lux	lx	1 lx = 1 lm/m^2
radioactivity	Becquerel	Bq	1 Bq = 1/s
absorbed dose	Gray	Gy	1 Gy = 1 J/kg

Figure 17.1 Some derived SI units having special names

Non SI units. There are certain units not included in SI which cannot, for a variety of reasons, be eliminated, despite the fact that these can, in principle, be expressed in SI units. Some of the non-SI units which may be used together with the SI units and their multiples and are recognised by the CIPM, Comité International des Poids et Mesures, are shown in figure 17.2.

Quantity	Name of unit	Unit symbol	Definition
time	minute hour day	min h d	1 min = 60 sec 1 h = 60 min 1 d = 24 h
plane angle	degree minute second	° ' "	1° = (π/180) rad 1' = (1/60)° 1" = (1/60)'
volume	litre	l	1 l = 1 dm³
mass	tonne	t	1 t = 10³ kg
pressure of fluid	bar	bar	1 bar = 10⁵ Pa

Figure 17.2 Non-SI technical units

There is a further group of specialised units which are primarily for use within astronomy and physics.

Multiples of SI units. The prefixes in figure 17.3 are used to form names and symbols of multiples and subdivisions of the SI units. The symbol of a prefix is considered to be combined with the unit symbol for the base unit, supplementary unit or derived unit to which it is directly attached, forming with it a symbol for a new unit which can be provided with a positive or negative exponent and which can be combined with other unit symbols to form symbols for compound units.

Factor by which the unit is multiplied	Prefix name	symbol	Example
10¹⁸	exa	E	
10¹⁵	peta	P	
10¹²	tera	T	1 terrajoule = 1 TJ
10⁹	giga	G	1 gigawatt = 1 GM
10⁶	mega	M	1 megavolt = 1 MV
10³	kilo	k	1 kilometre = 1 km
10²	hecto	h	1 hectogram = 1 hg
10¹	deca	da	1 decalumen = 1 dalm
10⁻¹	deci	d	1 decimetre = 1 dm
10⁻²	centi	c	1 centimetre = 1 cm
10⁻³	milli	m	1 milligram = 1 mg
10⁻⁶	micro	μ	1 microgram = 1 μm
10⁻⁹	nano	n	1 nanohenry = 1 nH
10⁻¹²	pico	p	1 picofarad = 1 pF
10⁻¹⁵	femto	f	1 femtometre = 1 fm
10⁻¹⁸	atto	a	1 attosecond = 1 as

Figure 17.3 Multiples of SI units

The new units of the international system of units can be viewed as follows:

```
Base units                              ⎫
Supplementary units    ⎫ SI units       ⎪
Derived units          ⎭                ⎬ Values in   ⎫ The only
Multiples of SI units                   ⎪ SI units    ⎬ units which
Non-SI units recognised by CIPM         ⎪             ⎭ should be used
Multiples of Non-SI units recognised by CIPM  ⎭
```

17.4 Conversion factors for SI units

When calculating convert all units to their basic values, i.e. remove any prefixes and insert the required number of zeros to show the correct magnitude.

17.2 Conversion Factors for SI units

17.2.1 Plane angle

Quantity designation: α, β, γ

SI unit: radian, rad.

Normal multiple units: mrad, μrad.

Example: 2 rad = 2 · 57.2958 = 114.5916°

rad	...ᵍ gon, grade	...° degree	...' minute	..." second	angular mil
1	63.6620	57.2958	3.43775 · 10³	0.206265 · 10⁶	1.00268 · 10³
15.7080 · 10⁻³	1	0.9	54	3.24 · 10³	15.75
17.4533 · 10⁻³	1.11111	1	60	3.6 · 10³	17.5
0.290888 · 10⁻³	18.5185 · 10⁻³	16.6667 · 10⁻³	1	60	0.291667
4.84814 · 10⁻⁶	0.308642 · 10⁻³	0.277778 · 10⁻³	16.6667 · 10⁻³	1	4.86111 · 10⁻³
0.997331 · 10⁻³	63.4921 · 10⁻³	57.1429 · 10⁻³	3.42857	205.714	1

grade (g) (or gon), 1 g = 1 gon = μ/200 rad.

1° = π/180 rad.

For some purposes, the angular mil is taken to be 10⁻³ rad. The figures shown here are based on the concept that an angular mil is equal to 360/6400 degrees.

NOTE: 1 grade (...ᵍ) = 1/100 of a right angle

17.2.2 Length

Quantity designation l.

SI unit: metre (m).

Normal multiple units: km, cm, mm, μm.

Example: 3 in = 3 · 25.4 · 10⁻³ = 76.2 · 10⁻³ m

metre m	inch in	foot ft	yard yd	mile
1	39.370	3.2808	1.0936	0.62137 · 10⁻³
25.4 · 10⁻³	1	83.333 · 10⁻³	27.778 · 10⁻³	15.783 · 10⁻⁶
0.3048	12	1	0.33333	0.18939 · 10⁻³
0.9144	36	3	1	0.56818 · 10⁻³
1.6093 · 10³	63.36 · 10³	5.28 · 10³	1.76 · 10³	1
1.852 · 10³	72.913 · 10³	6.0761 · 10³	2.0254 · 10³	1.1508

1 nautical mile = 6080 ft = 1853.184 m

1 Å, 1 Ångström = 10⁻¹⁰ m

1 astronomic unit = 0.1496 · 10¹² m

1 light year = 9.4605 · 10¹⁵ m

1 parsec = 30.857 · 10¹² m

17.2.3 Area

Quantity designation: A.

SI unit: square metre (m²).

Normal multiple units: km², dm², cm², mm².

Example: 4 ft² = 4 · 92.903 · 10⁻³ = 0.371612 m²

m²	in²	ft²	yd²	acre	square mile mile²
1	1.5500 · 10³	10.764	1.1960	0.24710 · 10⁻³	0.38610 · 10⁻⁶
0.64516 · 10⁻³	1	6.9444 · 10⁻³	0.77161 · 10⁻³	0.15942 · 10⁻⁶	0.24910 · 10⁻⁹
92.903 · 10⁻³	144	1	0.11111	22.957 · 10⁻⁶	35.870 · 10⁻⁹
0.83613	1.296 · 10³	9	1	0.20661 · 10⁻³	0.32283 · 10⁻⁶
4.0469 · 10³	6.2726 · 10⁶	43.56 · 10³	4.84 · 10³	1	1.5625 · 10⁻³
2.5900 · 10⁶	4.0145 · 10⁹	27.878 · 10⁶	3.0976 · 10⁶	640	1

1 acre = 100 m²

1 hectare = 100 acres = 10000 m²

17.2.4 Volume

Quantity designation: V.

SI unit: cubic metre (m³).

Normal multiple units: dm³, cm³, mm³.

Non SI unit: litre (l): 1 l = 0.001 m³ = 1 dm³.

Normal multiple units: cl, ml.

Example:

5 US gallon = 5 · 3.7854 · 10⁻³ = 18.927 · 10⁻³ m³ = 18.927 l

m³	in³	ft³	yd³	UK gallon	US gallon
1	61.024 · 10³	35.315	1.3080	219.97	264.17
16.387 · 10⁻⁶	1	0.57870 · 10⁻³	21.434 · 10⁻⁶	3.6046 · 10⁻³	4.3290 · 10⁻³
28.317 · 10⁻³	1.728 · 10³	1	37.037 · 10⁻³	6.2288	7.4805
0.76456	46.656 · 10³	27	1	168.18	201.97
4.5461 · 10⁻³	277.42	0.16054	5.9461 · 10⁻³	1	1.2010
3.7854 · 10⁻³	231	0.13368	4.9511 · 10⁻³	0.83268	1

gross (register) tonnage used in shipping
1 ton = 100 ft³ = 2.83168 m³

1 UK fluid ounce, fl oz = 28.4131 cm³

1 US fluid ounce, fl oz = 29.5735 cm³

17.2.5 Time

Quantity designation: t.

SI unit: second (s).

Normal multiple units: ms, μs, ns.

Non SI units: day (d), hour (h), minute (min).

Example: 100000 s = 100000/3600 = 27.778 h

s	min	h	d (day)	week
1	16.6667 · 10⁻³	0.277778 · 10⁻³	11.5741 · 10⁻⁶	1.65344 · 10⁻⁶
60	1	16.6667 · 10⁻³	0.694444 · 10⁻³	99.2063 · 10⁻⁶
3.6 · 10³	60	1	41.6667 · 10⁻³	5.95238 · 10⁻³
86.4 · 10³	1.44 · 10³	24	1	0.142857
604.8 · 10³	10.08 · 10³	168	7	1

1 tropical year = 31556925.974 s = 365.24219878 d

1 sidereal year = 31558150 s

1 calendar year = 365 d = 8760 h

17.2.6 Linear velocity

Quantity designation: v.

SI unit: metre per second (m/s).

Normal multiple units: km/h.

m/s	km/h	ft/s	mile/h
1	3.6	3.2808	2.2369
0.27778	1	0.91134	0.62137
0.3048	1.0973	1	0.68182
0.44704	1.6093	1.4667	1
0.51444	1.852	1.6878	1.1508

1 knot = 1 nautical mile per hour = 1.853 km/h

17.2.7 Linear acceleration

Quantity designation: a.

SI unit: metre per second squared (m/s²).

m/s²	cm/s²	ft/s²	in/s²	g
1	100	3.2808	39.37	0.10197
10 · 10⁻³	1	32.808 · 10⁻³	393.7 · 10⁻³	1.0197 · 10⁻³
0.3048	30.48	1	12	31.081 · 10⁻³
25.4 · 10⁻³	2.54	83.33 · 10⁻³	1	2.59 · 10⁻³
9.80665	980.665	32.174	386.09	1

17.2.8 Angular velocity

Quantity designation: ω.

SI unit: radian per second (rad/s).

The SI and Imperial units are identical.

Angular velocity is normally calculated from N revolutions/s by $2\pi N$

17.2.9 Angular acceleration

Quantity designation: α.

SI unit: radian per second squared (rad/s²)

The SI and Imperial units are identical.

17.2.10 Mass

Quantity designation: m.

SI unit: kilogram (kg).

Normal multiple units: μg, g, mg.

Non SI unit: tonne (t) = 1000 kg.

kg	lb (pound)	oz (ounce)	hundred-weight cwt (UK)	ton (UK)
1	2.2046	35.274	19.684 · 10⁻³	0.98421 · 10⁻³
0.45359	1	16	8.9286 · 10⁻³	0.44643 · 10⁻³
14.594	32.174	514.79	0.28727 · 10⁻³	14.363 · 10⁻³
28.350 · 10⁻³	62.5 · 10⁻³	1	0.55804 · 10⁻³	27.902 · 10⁻⁶
50.802	112	1.792 · 10³	1	50 · 10⁻³
1.0161 · 10³	2.24 · 10³	35.84 · 10⁻³	20	1

oz = ounce, also called ounce avoirdupois

1 ounce troy = 31.1035 · 10⁻³ kg

cwt = hundredweight

USA cwt = 100 lb   USA ton = 2000 lb

17.2.11 Density

Also called specific weight.

Quantity designation: ρ.

SI unit: kilogram per cubic metre (kg/m³)

Non SI units: kg/dm³, g/cm³.

kg/m³	g/cm³	lb/in³	lb/ft³
1	10⁻³	36.127 · 10⁻⁶	62.428 · 10⁻³
10³	1	36.127 · 10⁻³	62.428
27.680 · 10³	27.680	1	1.728 · 10³
16.019	16.019 · 10⁻³	0.57870 · 10⁻³	1

The term specific gravity or relative density is also used and is the ratio of the mass of a given volume of substance to the mass of an equal volume of water at temperature of 4 °C and a pressure of 101.325 kPa absolute. The density of water at 4 °C and 101.325 kPa absolute is 1000.02 kg/m³.

17.2.12 Force

Quantity designation: F.

SI unit: newton (N).

Normal multiple units: MN, kN.

N	dyn	Kilogram-force, kgf kilopond, kp	pound-force lbf
1	$0.1 \cdot 10^6$	0.10197	0.22481
$10 \cdot 10^{-6}$	1	$1.0197 \cdot 10^{-6}$	$2.2841 \cdot 10^{-6}$
9.8066	$0.98066 \cdot 10^6$	1	2.2046
4.4482	$0.44482 \cdot 10^6$	0.45359	1

17.2.13 Torque

Quantity designation: T.

SI unit: Newton metre (Nm).

Normal multiple units: MNm, kNm.

Nm	kgf m	lbf · in	lbf · ft
1	0.10197	8.8508	0.73756
9.8066	1	86.796	7.2330
0.11299	$11.521 \cdot 10^{-3}$	1	$83.333 \cdot 10^{-3}$
1.3558	0.13826	12	1

Torque, power and speed are related by the formula: $P = 2\pi NT$

17.2.14 Pressure, stress

Quantity designation: p, σ, τ.

SI unit: Pascal (Pa), 1 Pa = 1 N/m².

Normal multiple units: GPa, MPa, kPa and for stress: MN/m², N/m², N/mm².

Pa	bar	kgf/cm² technical atmos	kgf/mm²	torr (≈ mm Hg)	standard atm	lbf/in² (psi)
1	$10 \cdot 10^{-6}$	$10.197 \cdot 10^{-6}$	$0.10197 \cdot 10^6$	$7.5006 \cdot 10^{-3}$	$9.8692 \cdot 10^{-6}$	$0.14504 \cdot 10^{-3}$
$100 \cdot 10^3$	1	1.0197	$10.197 \cdot 10^{-3}$	750.06	0.98692	14.504
$98.066 \cdot 10^3$	0.98066	1	$10 \cdot 10^{-3}$	735.56	0.96784	14.223
$9.8066 \cdot 10^6$	98.066	100	1	$73.556 \cdot 10^3$	96.784	$1.4223 \cdot 10^3$
133.32	$1.3332 \cdot 10^{-3}$	$1.3595 \cdot 10^{-3}$	$13.595 \cdot 10^{-6}$	1	$1.3158 \cdot 10^{-3}$	$19.337 \cdot 10^{-3}$
$101.32 \cdot 10^3$	1.0132	1.0332	$10.332 \cdot 10^{-3}$	760	1	14.696
$6.8948 \cdot 10^3$	$68.948 \cdot 10^{-3}$	$70.307 \cdot 10^{-3}$	$0.70307 \cdot 10^{-3}$	51.715	$68.046 \cdot 10^{-3}$	1

The preferred pressure unit for pump applications is the 'bar'.

1 mm H₂O ≈ 9.81 Pa

1 in H₂O ≈ 249.09 Pa

1 in Hg ≈ 3386.4 Pa

1 ata = 1 technical atmosphere (absolute)

1 atu = 1 technical atmosphere (gauge)

17.2.15 Dynamic viscosity

Quantity designation: η.

SI unit: Pascal second Pa s.

Normal multiple units mPa s (= cP).

N s/m²	N s/mm²	P (poise)	cP m Pa s
1	10^{-6}	10	10^3
10^6	1	$10 \cdot 10^6$	10^9
0.1	$0.1 \cdot 10^{-6}$	1	100
10^{-3}	10^{-9}	$10 \cdot 10^{-3}$	1

17.2.16 Kinematic viscosity

Quantity designation: μ.

SI unit: square metres per second (m²/s).

Normal multiple units: mm²/s.

For conversion to other units of viscosity see nomogram below.

m²/s	mm²/s cSt	St (Stoke)
1	10^6	$10 \cdot 10^3$
10^{-6}	1	$10 \cdot 10^{-3}$
$0.1 \cdot 10^{-3}$	100	1

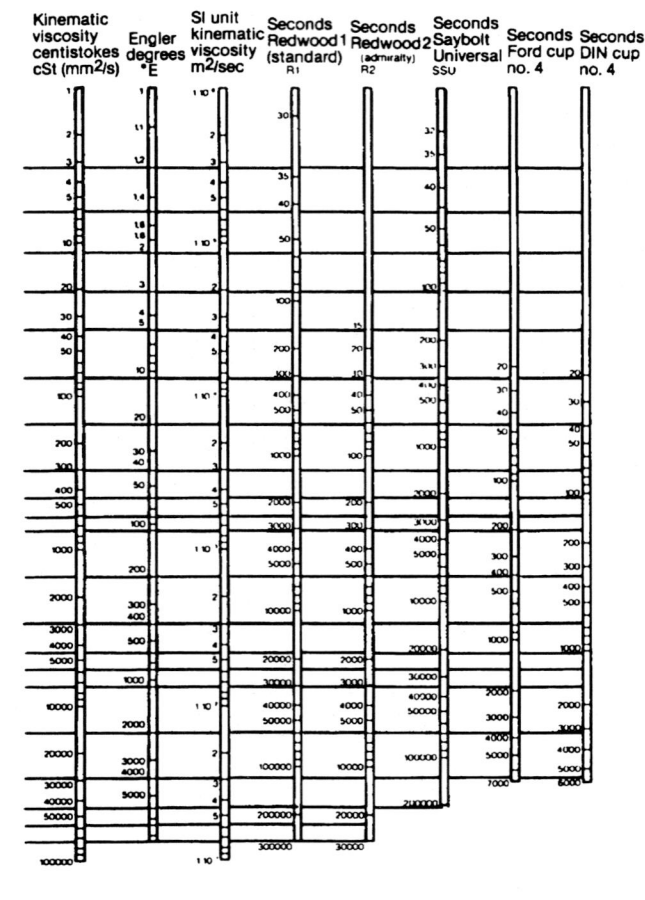

	cSt x	°E x	R1 x	SSU x
Kinematic viscosity cSt	1	7.58	0.247	0.216
Engler degrees °E	0.132	1	0.0326	0.0285
Seconds, Redwood 1, R1	4.05	30.7	1	0.887
Seconds, Saybolt Universal SSU	4.62	35.11	1.14	1

The above factors apply for values above 60cSt

17.2.17 Energy

Quantity designations: E, W, Q, L, U depending upon the type of energy

SI unit: Joule (J).

Normal multiple units: TJ, GJ, MJ, kJ, mJ.

Since 1 J = 1 Nm = 1 Ws then Nm and Ws can also be used for all types of energy. The unit Joule should, however, be used for expressing all types of energy.

$1 \text{ erg} = 0.1 \cdot 10^{-6}$ J

J Joule	kWh kilowatt hour	kgf m kilogram-force metre	kcal kilo-calorie	ch h metric horse-power hour	ft. lbf (foot pound-force)	Btu (British thermal unit)
1	$0.27778 \cdot 10^{-6}$	0.10197	$0.23885 \cdot 10^{-3}$	$0.37767 \cdot 10^{-6}$	0.73756	$0.94782 \cdot 10^{-3}$
$3.6 \cdot 10^{6}$	1	$0.36710 \cdot 10^{6}$	859.85	1.3596	$2.6552 \cdot 10^{6}$	$3.4121 \cdot 10^{3}$
9.8066	$2.7241 \cdot 10^{-6}$	1	$2.3423 \cdot 10^{-3}$	$3.7037 \cdot 10^{-6}$	7.2330	$9.2949 \cdot 10^{-3}$
$4.1868 \cdot 10^{3}$	$1.163 \cdot 10^{-3}$	426.94	1	$1.5812 \cdot 10^{-3}$	$3.0880 \cdot 10^{3}$	3.9683
$2.6478 \cdot 10^{6}$	0.73550	$0.27 \cdot 10^{6}$	632.42	1	$1.9529 \cdot 10^{6}$	$2.5096 \cdot 10^{3}$
1.3558	$0.37662 \cdot 10^{-6}$	0.13826	$0.32383 \cdot 10^{-3}$	$0.51206 \cdot 10^{-6}$	1	$1.2851 \cdot 10^{-3}$
$1.0551 \cdot 10^{3}$	$0.29307 \cdot 10^{-3}$	107.59	0.25200	$0.39847 \cdot 10^{-3}$	778.17	1

17.2.18 Power

Quantity designation: P.

SI unit: Watt (W).

Normal multiple units: GW, MW, kW, mW, μW.

W	kgf m/s	kcal/s	kcal/h	ch metric horse-power	hp horse-power	ft lbf/s	Btu/h
1	0.10197	$0.23885 \cdot 10^{-3}$	0.85985	$1.3596 \cdot 10^{-3}$	$1.3410 \cdot 10^{-3}$	0.73756	3.4121
9.8066	1	$2.3423 \cdot 10^{-3}$	8.4322	$13.333 \cdot 10^{-3}$	$13.151 \cdot 10^{-3}$	7.2330	33.462
$4.1868 \cdot 10^{3}$	426.94	1	$3.6 \cdot 10^{3}$	5.6925	5.6146	$3.0880 \cdot 10^{3}$	$14.286 \cdot 10^{3}$
1.163	0.11859	$0.27778 \cdot 10^{-3}$	1	$1.5812 \cdot 10^{-3}$	$1.5596 \cdot 10^{-3}$	0.85779	3.9683
735.50	75	0.17567	632.42	1	0.98632	542.48	$2.5096 \cdot 10^{3}$
745.70	76.040	0.17811	641.19	1.0139	1	550	$2.5444 \cdot 10^{3}$
1.3558	0.13826	$0.32383 \cdot 10^{-3}$	1.1658	$1.8434 \cdot 10^{-3}$	$1.8182 \cdot 10^{-3}$	1	4.6262
0.29307	$29.885 \cdot 10^{-3}$	$69.999 \cdot 10^{-6}$	0.25200	$0.39847 \cdot 10^{-3}$	$0.39302 \cdot 10^{-3}$	0.21616	1

17.2.19 Flow

Quantity designation: q_v.

SI unit: cubic metre per second (m³/s).

Non SI units: l/s, ml/s, m³/h.

l gallon/ min Igpm	US gallon/ min USgpm	barrel/ day bpd	m³/s	m³/h	l/s	l/min
1	1.2009	41.175	$75.768 \cdot 10^{-6}$	$272.76 \cdot 10^{-3}$	$75.768 \cdot 10^{-3}$	4.5461
0.83268	1	34.286	$63.09 \cdot 10^{-6}$	$227.12 \cdot 10^{-3}$	$63.09 \cdot 10^{-3}$	3.7854
$24.286 \cdot 10^{-3}$	$29.167 \cdot 10^{-3}$	1	$1.84 \cdot 10^{-6}$	$6.6244 \cdot 10^{-3}$	$1.84 \cdot 10^{-3}$	$110.41 \cdot 10^{-3}$
$13.198 \cdot 10^{3}$	$15.85 \cdot 10^{3}$	$543.44 \cdot 10^{3}$	1	3600	1000	60000
3.6662	4.4029	150.95	$277.78 \cdot 10^{-6}$	1	$277.78 \cdot 10^{-3}$	16.667
13.198	15.85	543.44	$1.0 \cdot 10^{-3}$	3.6	1	60
$219.97 \cdot 10^{-3}$	$264.17 \cdot 10^{-3}$	9.0573	$16.667 \cdot 10^{-6}$	$60 \cdot 10^{-3}$	$16.667 \cdot 10^{-3}$	1

1 barrel = 42 US gallon

17.2.20 Temperature

Absolute temperature: Quantity designation: T,

SI unit: Kelvin (K).

Temperature: Quantity designation: t,

unit degree Celsius (Centigrade) (°C).

	Kelvin*** scale	Celsius scale	Rankine scale	Fahrenheit** scale	Physical relation-ship
Relative temper-ature value	0 K	-273.15 °C	0 °R	-459.67 °F	Absolute zero
	273.15 K	0 °C	491.67 °R	32 °F	Melting point of ice*
	273.16 K	0.01 °C	491.688 °R	32.018 °F	Triple point of water*
	373.15 K	100 °C	671.67 °R	212 °F	Boiling point of water*
Relative temper-ature intervals (diffs)	1 K 0.55556 K	1 °C 0.55556 °C	1.8 °R 1 °R	1.8 °F 1 °F	

* For defined conditions

** Value in °C = $\frac{1}{1.8}$ (value in °F -32) = (value in K -273.15)

*** Value in K = $\frac{5}{9}$ · (value in °R)

17.3 Other conversion factors

17.3.1 Hardness

Hardness is not defined within the SI system. The following table can be used for conversion between the popular systems used.

Tensile strength	Vickers hardness (F≥ 98 N)	Brinell hardness	Rockwell hardness	
N/mm²	HV	BHN	HR$_B$	HR$_C$
200	63	60		
210	65	62		
220	69	66		
225	70	67		
230	72	68		
240	75	71		
250	79	75		
255	80	76		
260	82	78		
270	85	81	41	
280	88	84	45	
285	90	86	48	
290	91	87	49	
300	94	89	51	
305	95	90	52	
310	97	92	54	
320	100	95	56	
30	103	98	58	
335	105	100	59	
340	107	102	60	
350	110	105	62	
360	113	107	63.5	
370	115	109	64.5	
380	119	113	66	
385	120	114	67	
390	122	116	67.5	
400	125	119	69	
410	128	122	70	
415	130	124	71	
420	132	125	72	
430	135	128	73	
440	138	131	74	
450	140	133	75	
460	143	136	76.5	
465	145	138	77	
470	147	140	77.5	
480	150	143	78.5	
490	153	145	79.5	
495	155	147	80	
500	157	149	81	
510	160	152	81.5	
520	163	155	82.5	
530	165	157	83	
540	168	160	84.5	
545	170	162	85	
550	172	163	85.5	
560	175	166	86	
570	178	169	86.5	
575	180	171	87	
580	181	172		
590	184	175	88	
595	185	176		
600	187	178	89	
610	190	181	89.5	
620	193	184	90	
625	195	185		
630	197	187	91	
640	200	190	91.5	
650	203	193	92	
660	205	195	92.5	
670	208	198	93	
675	210	199	93.5	
680	212	201		
690	215	204	94	
700	219	208		
705	220	209	95	
710	222	211	95.5	
720	225	214	96	
730	228	216		
740	230	219	96.5	

Tensile strength	Vickers hardness (F≥ 98 N)	Brinell hardness	Rockwell hardness	
N/mm²	HV	BHN	HR$_B$	HR$_C$
750	233	221	97	
755	235	223	97.5	
760	237	225		
770	240	228	98	
780	243	231		21
785	245	233		
790	247	235	99	
800	250	238	99.5	22
810	253	240		
820	255	242		23
830	258	245		
835	260	247		24
840	262	249		
850	265	252		
860	268	255		25
865	270	257		
870	272	258		26
880	275	261		
890	278	264		
900	280	266		27
910	283	269		
915	285	271		
920	287	273		28
930	290	276		
940	293	278		29
950	295	280		
960	299	284		
965	300	285		
970	302	287		30
980	305	290		
990	308	293		
995	310	295		31
1000	311	296		
1010	314	299		
1020	317	301		32
1030	320	304		
1040	323	307		
1050	327	311		33
1060	330	314		
1070	333	316		
1080	336	319		34
1090	339	322		
1095	340	323		
1100	342	325		
1110	345	328		35
1120	349	332		
1125	350	333		
1130	352	334		
1140	355	337		36
1150	358	340		
1155	360	342		
1160	361	343		
1170	364	346		37
1180	367	349		
1190	370	352		
1200	373	354		38
1210	376	357		
1220	380	361		
1230	382	363		39
1240	385	366		
1250	388	369		
1255	390	371		
1260	392	372		40
1270	394	374		
1280	397	377		
1290	400	380		
1300	403	383		41
1310	407	387		
1320	410	390		
1330	413	393		42
1340	417	396		
1350	420	399		
1360	423	402		43
1370	426	405		
1380	429	408		
1385	430	409		
1390	431	410		
1400	434	413		44
1410	437	415		
1420	440	418		

Tensile strength	Vickers hardness (F≥ 98 N)	Brinell hardness	Rockwell hardness	
N/mm²	HV	BHN	HR$_B$	HR$_C$
1430	443	421		
1440	446	424		
1450	449	427		45
1455	450	428		
1460	452	429		
1470	455	432		
1480	458	435		46
1485	460	437		
1490	461	438		
1500	464	441		
1510	467	444		
1520	470	447		
1530	473	449		47
1540	476	452		
1550	479	455		
1555	480	456		
1560	481	457		
1570	484	460		48
1580	486	462		
1590	489	465		
1595	490	466		
1600	491	467		
1610	494	470		
1620	497	472		49
1630	500	475		
1640	503	478		
1650	506	481		
1660	509	483		
1665	510	485		
1670	511	486		
1680	514	488		50

Values based on DIN 50150.

17.3.2 Material toughness

Material toughness is not defined by SI. Most materials are assessed by conducting impact testing. Two differing test methods can be used with various sizes and styles of test specimen. The following table can be used as a **guide** to the relative toughness of the various tests:

Charpy V notch kgm/cm²	Charpy V notch ft lb	Charpy V notch Joule	Izod ft lb
0.4	2.3	3.1	2.5
0.9	5.2	7	6.4
1.5			10.8
2.2			16
3.1	18	24.4	21.5
4.1	23.8	32.2	27.8
5.2	30	40.6	34.1
6.5	37.7	51	40.4
8.0	46.4	62.9	46.7
9.4	54.5	73.9	53
10.9	63	85.4	59.3
12.6			65.6
14.1	82	111	71.9
15.8			78.2
17.7	102	138	84.5
19.4			90.8
21.1	122	165	97.1
23.0	134	182	103.4

17.4 Normal quantities and units used within pump technology

Quantity		Recommended unit	
Name	Symbol	Name	Units
Area	A	square metre square centimetre square millimetre	m² cm² mm²
Compressibility	χ	reduction in unit volume per bar differential pressure	bar^{-1}
Density	ρ	kilogram per cubic metre gram per cubic decimetre	kg/m³ g/dm³
Energy	E	Joule kiloJoule MegaJoule	J kJ MJ
Flow (1)	q$_v$	litre per second litre per minute litre per hour cubic metre per hour	l/s l/min l/h m³/h
Force	F	Newton kiloNewton MegaNewton	N kN MN
Frequency	f	Hertz kiloHertz MegaHertz	Hz kHz MHz
Head (suction, discharge, differential)	H	metre	m
Length	l	metre millimetre micron	m mm μm
Mass (weight)	m	tonne kilogram gram	tonne kg g
MIP (Minimum inlet pressure)	MIP	bar (gauge)	bar
Moment of inertia	J	kilogram metre squared	kgm²
NPIP (Net positive inlet pressure)	NPIP	bar	bar
NPSH (Net positive suction head)	NPSH	metre	m
Power	P	Watt kiloWatt MegaWatt	W kW MW
Pressure (absolute or gauge)	p	bar	bar
Shaft speed	n	revolutions per second revolutions per minute revolutions per hour	r/s r/min r/h
Stress	σ τ	MegaNewton per square metre	MN/m²
Temperature	T	degree Celsius	°C
Time	t	second minute hour	s min h
Torque	T	Newton metre kiloNewton metre MegaNewton metre	Nm kNm MNm
Velocity	v	metre per second	m/s
Viscosity, dynamic	η	Poise centiPoise	P cP
Viscosity, kinematic	μ	Stoke centiStoke	St cSt

(1) Pumps handle liquid by volume. The **user** must convert mass flow to volume flow.

18 Reference index

The reference index contains a large number of key words used within the pump industry. The reference index lists the page numbers on which the key words are used. The list of contents preceding every chapter also provides a useful guide. Where categories of pumps or ancillary equipment are listed the category reference is shown bold in parenthesis.

A

Abbreviations
- names of authorities etc.   18, 24, 145, 251-253

Abrasion
- abrasive liquids   127
- abrasive solids   26, 122, 127, 214, 316-317
- choice of materials   214, 316
- Miller Number   26, 106, 119, 308
- review   26
- testing   26
- values   26-27
- velocity   28

Acceleration head loss   128, 141

Accumulators (500)   267
- automatic water sets   94
- description   267
- run-down protection   267
- sizing for load-unload control   177
- sizing for on-off control   177

Acid resistant steel
see Stainless steel   204

AC motor
- description   256-257
- speed   250
- speed regulation   187

Acoustic resonance   15, 119, 165, 169, 268, 322

Adaptation   295-297, 304, 305, 325

Agricultural pump without driver (030)
- description   96

Air chambers   167, 175

Air separation
- capacity loss   116, 142
- effects on pump performance   116, 143
- in pipe system   97, 142, 288
- self-priming rotodynamic pumps   99, 114, 142, 332

Air-lift pump (295)
- description   134

Air motor   187

Alarm   269

Alignment
- checking   241-244
- during commissioning   291
- during installation   284, 288
- shaft couplings   236-238, 242, 284
see also Thermal growth   237-238

Aluminium alloys
- construction sump pumps   97
- groundwater pumps   97
- review   206
- shaft couplings (625)   239
- shaft coupling guards   246

Aluminium bronze
- cavitation resistance   215
- review   206
- salt water   23, 97
see also Nickel aluminium bronze   206

Ancillary equipment - for pumps
see Accumulators (500)   267
see Baseplates   264
see Belt drives (605)   265
see Gearboxes (610, 615)   265
see Instrumentation (570)   269, 284
see Pulsation dampeners (635)   267-269
see Relief valves (645)   266
see Skids   264
see Trailers   264
- maintenance   292-293

Angular misalignment
- definition   237
- shaft couplings (625, 630)   237, 242, 284
- universal joints   124

Annunciator   269

Annuity   297-298

ANSI B73.1
- end suction pumps (100)   103

ANSI B73.2
- inline pumps (105)   103, 106

API 610
- end suction pumps (110)   104
- double suction pumps (115)   104

Archimedian pump
see Rigid screw pump (305)   135

Automatic water supply packages (020)   94

Auxiliary equipment   309

Auxiliary pumps   220

Availability   302

Axial force
see Axial thrust   86

Axial piston (260)   117, 127

Axial propeller pumps (200)
- cavitation   140
- description   112
- minimum flow   163
- practical case   333
- pump selection graph   315
- types of performance curve   88
- use as heeling pump on ships   97
- variable blade settings, effects on performance   88
- vertical   98, 111, 112, 333

Axially-split pumps (010, 165)   93, 109

Axial thrust
- balancing   86, 106, 109
- bearings   93, 101, 103
- calculation   86
- from shaft coupling   238, 246
- multi-stage pumps   101
- pressure testing considerations   274
- pumps in series   108, 159

B

Back-pull-out design   95, 102, 103, 104

Balancing
- couplings   238-239
- impellers   111, 214
- motors   252
- rotors   111

Baseplates
- description   264
- design requirements   104, 264
- options   104, 264

Bearing materials   200

Belt drives (605)   265

BEP - Best efficiency point
- definition   83, 138
- relative to duty point   85, 104, 152, 162
- relative to curve shape   90, 136, 138, 308
- values   301

Bibliography
- quality assurance   280-281
- selection of materials   217
- shaft couplings   246
- shaft seals   234
- water hammer   169

Blades, adjustable
see Pre-setting   88

Boiling point
- definition   12
- liquids, others   32
- oils   24, 139
- water   16

Booster pumps   159

Brass
- cavitation resistance   215
- dezincification   211
- review   206

Bronze
- cavitation resistance   215
- marine pumps   97
- non-sparking shaft couplings   239
- non-sparking coupling guards   246
- review   206
- salt water   23

Bulk modulus
- definition   14
- liquids   15, 164

Bulk-head sealed fittings
- cargo oil pump   334
- description   239, 334
- fire wall   239, 334

By-pass regulation
- liquid ring pumps   113
- overflow control   184
- positive displacement pumps   120, 184
- temperature rise   114, 163
- with control valves   156, 161, 183

C

Canned motor pumps (125)   105, 256, 335
see also Magnetic drive pumps (120)   104, 123

Capitalisation
- energy costs   299
- present capitalised value   297-298
- operating costs   297-298
- when determining safety factors   301, 305

Cast iron
see Grey cast iron   201

Cast steel   202, 203

Cavitation
- calculated example   143-144
- damage   138-141, 214
- definition, calculation   137
- detection   138
- jet pumps (290)   100, 133
- material, cavitation resistant   214-215
- pipelines   168
- protective films, effects on   214
- pumps   138, 139
- relationship to N_{ss} 85,
- vapour pressure   12, 138

Central heating pumps
see Heating, water and sanitation pumps (015)   94

Centre-line mounted pumps   104, 109

Centrifugal pumps
- general review   82-91
- grouping   92
- vertical   98, 99, 100, 110

Certification
- electric motors   253, 276
- materials   276
- NDT inspection   210, 273, 276
- pipework   276
- pump conformance   276
- pump hydrotest   210, 274, 276
- pump MIP   274
- pump NPIPr   274
- pump NPSHr   274
- pump performance   274, 276
- QC operators   279
- welder qualifications   210, 273, 276
- welding procedures   201, 210, 273, 276

Channel impeller pumps
- description   318
- 'gentle' pumping   111, 112, 317
- waste water   318

Check-list
- liquid properties   215
- materials selection   215-217
- pump, mechanical   319
- pump selection   319
- shaft couplings   246
- shaft seals   229, 233
- solids properties   215

Chemical pump
see ANSI pump (100, 105)   103
see API pump (110, 115)   104
see Canned motor pump (125)   105, 256, 335
see Diaphragm pump (280)   131
see ISO pump (095)   102
see Magnetic drive pump (120)   104, 123
see Metering pump (285)   127, 131, 290, 292
see Non-metallic pump (130)   106
see Peristaltic hose pump (250)   127
see Piston pump (275)   130
see Plunger pump (270)   128

Chemicals
- chemical formulae   32
- liquid properties   31, 52, 60, 230
- trade names   31

Chokeless pump
see Non-clogging pump (180, 185, 190)   111, 112
see Pulp pump (135)   106

CIP - Cleaning in place   107, 308

Classification, of pumps, by design
- classification guide   81
- general   81
- other pumps   81
- positive displacement pumps   81, 114, 121
- rotodynamic pumps   81, 92

Classification guide   92, 121

Classification Societies
- for electric motors   252

Close coupled pump   93, 94, 98, 100, 105, 108

Closed valve head

- definition	89
Commissioning	**290**
Compact pump	93, 94, 98, 100, 105, 108
Compressed air driven pump	
- air-lift pump (**295**)	134
- barrel emptying pump (**300**)	134
- diaphragm pump (**280**)	119, 131
- jet pump (**290**)	132
- submerged pump (**050**)(**065**)	97, 99, 133, 134
Compressibility	
- definition	14
- displacement pumps	115, 116, 129
- liquids	15, 116, 117, 308
Concentration	
- concentration in H_2O	31
- corrosion	212
- effects on pipe flow	56-58
- liquid-solid mixtures	26, 27, 28, 214, 316
- suspensions	26, 28
Constancy (duration)	**173**
Contactor (motor)	**251**
Continuous	
- definition	308
Contraction pump	
- description	136
Control efficiency	
see Regulation efficiency	
Control methods	
see Flow regulation	
Control of pumps	**325, 326, 327, 329**
Control of pump packages	**327, 329**
Control valves	
- characteristics	157-158
- flow coefficient	157
- for by-pass regulation	161, 163, 184
- for throttle regulation	156, 181
- sizing	157, 182, 305
Conversion tables	**384-389**
Coolant pumps	
see Machine tool coolant pumps (**040**)	96
Cooling	
- electric motors	252-254
- mechanical seals	103, 227-228
- stuffing boxes	223
Copper alloys	
- review	206
- sea water	23, 206
Corrosion	
- allowance	210
- cast iron	201, 211
- crevice corrosion	210
- duplex stainless steel	205
- erosion process	212
- galvanic (electrolytic)	211-214
- general	200, 211, 215
- in sea water	22
- in water	19
- intercrystalline	205, 211
- lead	206
- monel	206
- pitting	211
- pitting resistance	23, 205
- protective films	204, 206, 211
- rate of corrosion	211
- resistance of various materials	216-217
- selective	211
- spheroidal or nodular cast iron (SG iron)	202
- stainless steel	204, 211
- stress corrosion	212
- stress corrosion cracking	205
- tantalum	207
- testing	213-214, 274-275
- titanium	207
- types of corrosion	211
Corrosion fatigue	
- fatigue strength	212
Cos φ	
see Electrical units	248
Costs	
- administration	296
- assessment criteria	297-297
- capitalisation	297-298
- energy consumption	178, 182, 195, 297-299
- flow-regulation	182, 196
- instrumentation	296
- investment	297-298
- life-cycle	105, 296
- material combinations in pumps	210, 215-216, 229
- operational	296
- pipework	296
- pump maintenance	296
- pump plant	330
- shaft couplings	246
- shaft seals	234
- speed-regulation	180-181, 188, 195-196
- when applying safety factors	301, 305
Coupling guards	**246**

Couplings	
see Shaft couplings (**625, 630**)	236
Creep	
- metal	272
- non-metallic pumps	200
- plastics	272
Crevice corrosion	**211**
Critical speed	
- definition	262
- dry	262
- long shafts	109, 112
- wet	262
Crude oil burning engine	**186**
Cyclic	
- definition	308
D	
Dammed point	
see Closed valve head	89
Dampers	
see Pulsation Dampeners (**635**)	267-269
DC motor	
- description	258
- speed	250
- speed regulation	188, 258, 324
Deep draw	
see Jet pumps (**290**)	132
Deep well pumps (070, 075)	
- contraction pump	137
- submersible	100
- with ejector	100
Delivery head	
see Differential head	82
Demand variation	
- causes, examples	173, 178, 301, 305, 324
- constancy curves	173
- efficiency factor	300
- hydraulic power, energy	303
- positive displacement pumps	335
- running curve	173, 308
Dense pulp	**106**
Density	14, 26, 29, 31, 43, 44, 45, 46, 47, 389
De-staging of rotodynamic pumps	**305**
Diagonal pump	
see Mixed flow pump (**195**)	112
Diaphragm shaft couplings (625)	
- description	237, 244
- non-sparking	239
Diaphragm pumps (280)	
- air operated	119
- description	119, 131
- as metering pump	131-132
Diesel engine	**186**
Differential head	
- conversion from pressure rise	82
- definition for pump	82
- for various pump types	214
- guarantee	90, 143
- head loss	152
- per stage	92, 108, 112, 113, 114, 115, 136, 214
- pump rated duty point	152, 156
- reduction due to cavitation	100
- reduction due to viscosity	90, 91
- static	152-156, 303
- system	152-156
Direct selection graphs	**315**
Direct selection tables	**308-311**
Disc coupling (630)	**236**
Discharge head	
see Differential head	
Discharge pressure	**290**
Displacement pump	
see Positive displacement pump	114
Dosing pump	
see Metering pump (**285**)	131
Double seal	
- double mechanical seal	103, 112, 228-229, 233
- double soft packing	118, 221-223, 232
Double suction pump (010, 115)	
- description	93, 96, 104
- first stage	109, 110
- pump selection graphs	315
- pump selection tables	312-313
Drivers	
- air motors	187
- check-list	320-321
- engines	99, 186, 261, 332
- motors	187, 248-261
- power recovery turbines (**205**)	112, 187, 262, 336
- turbines	109, 110, 187, 262
"Dry" running	
- during evacuation	142-143
- mechanical seals	226, 227, 228, 231
- positive displacement pumps	120
- protection	97
- soft packing	222

Dual fuel engine	**186**
E	
Eccentric screw pump	
see Progressing cavity pump (**235**)	123
Economy	
- assessment criteria	297
- economic pipe diameter	304
- economic service life	296, 298
- important system characteristics	300
- optimisation	296
- partial optimisation	304
- pumping	300
- when speed-regulating	301
Eddy current coupling	**186, 197**
Efficiency	
- belt drives	265
- dependence on specific speed	92, 158
- effects of differential head	85, 92
- effects of discharge pressure	118, 136, 191
- effects of flow rate	189
- effect of number of stages	92
- effects of size	92
- effects of system characteristic	152-154, 191
- effects of viscosity	90, 91, 120, 156, 314-316
- effects of wear	59, 90, 115, 119, 153-154
- efficiency factor	300
- electric motor	251, 338
- pumps	301
- regulation efficiency	300
- transmission efficiency	189, 300
Efficiency factor	
- definition	300
- examples	300
Ejector	
see Jet pump (**290**)	97, 132
Electric current requirements	
- calculated example	261
- calculation	249
- motor starting	251, 252, 259-260
Electric motors	
- AC motor	249
speed	250
starting current	251, 261
- commutator motor	258
- construction	252
- cost	188
- DC current	248
- DC motor	188
speed	250, 258, 325
- dimensions	103, 252, 255
- direction of rotation	255
- efficiency	251, 300, 338
- frequency regulated	257
- hazard protection	252
- hygienic	107-108
- immersed	252
- induction motor	
definition	257
description	257
slipring	257
speed	249
squirrel cage	257
- insulation	252, 255
- maintenance	261
- methods of cooling	252-254
- motor types	256
- mounting arrangements	252, 254
- noise	252, 260-261
- performance	252
- pole-changing	250
definition	257
pump performance	88, 89, 179
speed regulation	257
- physical protection	252, 253
- power rating	252, 255
- protection	259-260
- regulations	252-253
- single phase	
definition	248
description	248
- sizes	252
- slipring motor	
definition	257
speed regulation	188
starting	252, 257
- soft starts	260
- squirrel cage	
definition	257
speed regulation	187
starting	252, 259-260
- standards	251-256
- starters	259
- starting methods	251, 259-260
- starts per hour	251
- starting problems	259-260
- submersible	97
- synchronous motor	
definition	256
description	256
- temperature classification	252, 255
- terminals and rotation	252, 255
- thermal protection	252

- three phase
  - definition 248
  - description 248
- torque
  - during starting 250, 257
  - running 250
- voltage 250
- wet rotor 256, 337

Electrical frequency
- definition 248
- frequency regulation 257
- relationship to speed 88, 250

Electrical power
- active 249
- apparent 249
- calculated example 261
- power factor 249
- reactive 249

Electrical units
- relationships 248-249
- review 248

Electricity supply system
- potential drop 248
- three phase alternating current 248

Electricity tariffs 299

Electrolytic corrosion
see Galvanic corrosion 211

Electrolytic series
see Galvanic series 212

Emulsion
- definition 12
- water/glycerine 127
- water/glycol 127
- water/oil 127

Energy
- consumption
  - with by-pass regulation 184
  - with load-unload 175
  - with on-off control 175
  - with speed regulation 89, 180, 181
  - with throttle regulation 182
- electrical 248-249
- energy costs 299
- energy equation 52
- hydraulic 52, 303
- internal energy 52
- kinetic energy 50, 52
- mechanical energy 52
- potential energy 50
- pressure energy 52

Engines
- crude oil 96, 186
- diesel 96, 186
- dual fuel 186
- efficiency 189, 300
- gas 96, 187
- petrol 187
- speed 186, 187
- torque 186, 187, 261

Environment
see Health and environment
see Site conditions 321

Erosion
- abrasive liquids 127
- abrasive solids 26, 122, 127, 214, 316
- choice of materials 214, 316
- Miller Number 26, 106, 119, 308
- review
- testing

Erosion corrosion 212

Euler's equation 83

Evacuating pumps
see Venting and priming 142

Evacuating suction systems
see Venting and priming 142, 288

Evacuation time 99

Explosion risk
see also Hazardous area
- classification of gas zones 256
- classification of dust zones 256
- electric motors 252
- non-sparking couplings 239
- non-sparking guards 246
- temperature classes 252, 255

F

Face seals
see Mechanical seals (655 & 660) 221, 227

Fatigue
- component life 87, 130, 139-141, 143, 200, 209

Fault finding
see Trouble-shooting

Faults
- frequency in plants 292
- frequency in pumps 292

Filters (550 & 555)
- creating suction problems 289, 290-291
- in pump sumps 285
- in suction lines 289

Fire hazard class
- explanation and definitions 16, 252
- liquids etc. 26
- oils 24, 127, 334
see also Explosion risk 256

Fixed irrigation pump (035)
- description 96

Flow
- definition 50
- for different types of pump 312-313, 315
- laminar 53
- Reynolds Number 53, 152
- turbulent 53, 152
- units 387

Flow control
see Chapter 5

Flow measurement
- accuracy 73, 75
- checking 77, 289
- current meter 76
- displacement type 73
- electromagnetic flowmeter 73
- general 72
- hot wire anemometer 77
- laser-doppler 77
- measuring flumes 75
- measuring tanks 76
- methods of checking flow 77
- orifice plate, nozzle 73
- Pitot tube 76
- measuring range 75
- standards 77
- swirl meter 74
- turbine meter 74
- ultrasonic-meter 74
- vane flowmeter 76
- variable orifice flowmeter 75
- venturi flowmeter 74
- vortex shedding meter 74
- weirs and notches 75

Flow regulation
see Chapter 5
- continuous 120, 174, 185-192, 195, 329
- discrete 174-181
- economy 195, 195-196
- overview 173, 194, 197

Flow variations
- description 50, 119
- positive displacement pumps 50, 119, 128, 131, 132, 267

Flow velocity
- basic equations 50
- Bernoulli's equation 50
- cavitation, rotodynamic pumps 85
- economic velocity 304
- effects on corrosion 23, 213
- Euler's equation 83
- Jet pump (290) 132, 329
- optimum velocity 84
- Pitot tube pump (310) 135
- sedimentation of solids 58
- typical pipe 61

Fluid coupling
- efficiency 190
- speed range 89, 186

Fluid transportation
see Liquid transportation

Foodstuffs pump
see Hygienic quality pump (150) 107

Forces and moments
- permitted on pipe connections 103, 288
- transmitted by couplings 238

Free-flow, recessed impeller, pump 111, 318
see Non-clogging pump (180, 185 & 190) 111, 112

Fuel pump 96

G

Galvanic corrosion 23, 211, 213

Galvanic series, metals 212

Gas engine 186

Gas separation
see Air separation

Gas turbine 262

Gearboxes (610 & 615) 265, 305

Gear couplings (625)
- balancing 238-239
- maintenance, service life 240
- types 237

Gear pumps (220 & 225)
- description 115, 121, 122
- effects of pump speed 115, 120, 316
- effects of viscosity 115, 120, 156, 316, 335

Gentle pumping 111, 112, 125, 127, 317

Glandless pump
see Canned motor pump (125) 105, 256, 335
see Diaphragm pump (280) 131
see Magnetic drive pump (120) 104, 123
see Metering pump (285) 131

see Peristaltic hose pump (250) 127
see Rotary peristaltic pump (255) 127

Grain boundary corrosion
see Intercrystalline corrosion 211

Grey cast iron
- alloyed 202, 203
- protective films 201, 211
- review 201, 272

Ground-water pump
- for deep wells 98, 100
- for drainage 97-98, 101-102

Groupings
see Classification by design 81

GRP - Glass reinforced plastic 207

Grouting 284

Guide vanes
- effect on performance 88

Gunmetal
- description 206
- pumps 97
see also Copper alloys 206

H

Hardness
- conversion chart 388
- materials 214, 216
- water 20-22

Hastelloy
see Nickel alloys 206

Hazard
- definition 16

Hazardous area 16, 18, 252, 308

Hazardous gas 16, 18, 308

Hazardous liquid 16, 18, 26, 31, 230, 270, 308

Hazardous vapour 16, 18, 26, 31, 308

Head losses
see Pressure losses 53, 303-304

Health and environmental, dangers
- check-list 321
- general 321
- liquids 215, 230
- noise 100, 284, 321
- vibration 321

Heat treatment
- for corrosion resistance 272
- for grain structure restabilisation 204
- for strength 204, 272

Heating circulation pumps
see Heating, water and sanitation pumps (015) 94

Heating, water and sanitation pumps (015) 94

Hermetically sealed pump
see Canned motor pump (125) 105, 256, 335
see Magnetic drive pump (120) 104, 123

High speed single stage pump (155)
- description 108

High speed multi-pump packages (160)
- description 108

High suction pressure
- problems 106, 130, 308

Hose diaphragm pump 131

Hose pump
see Peristaltic hose pump (250) 127

Hot water pumps (015)
- domestic supply 94

Hours run meter 269

Hover seals 220

Hydraulic coupling
see Fluid coupling
see Torque convertor

Hydraulic diameter 54

Hydraulic gradient 164, 167

Hydraulic ram
- description 136

Hydraulic tolerances
see Performance tolerances

Hydraulic turbines 112, 187, 262, 336

Hydrodynamic transformers
- hydraulic ram 136
- Jet pump (290) 132

Hydrostatic coupling 185, 334

Hygienic quality pump (150)
- description of positive displacement type 108
- description of rotodynamic type 107, 114

I

Imo-pump
see Screw pump (230) 123

Impeller
- reduced diameter
  effect on performance 85, 88, 138, 153-154, 305

Inclination deviation
see Alignment 237

Inertia
- of liquid 117, 267

392

Inflation	298
Injector	
see Jet pump (290)	132
Inspection	
- by Certifying Authority	275
- by manufacturer	273, 275
- by purchaser	210, 273, 275
- by third party	275
Installation	284
Instrumentation	
- annunciator	269
- flow	72-77
- manometer	76
- pressure	290
Insulation	252, 255
Intercrystalline corrosion	205, 211
Intergranular corrosion	211
Intermittent	
- definition	308
Internal combustion engines	
see Engines	
Investment	
- calculation	297-298
- for various methods of speed regulation	188, 195
- grants	298
Irregular	
- definition	308
Irrigation pump	
see Agricultural pumps without driver (030)	96
see Fixed irrigation pump (035)	96
ISO pumps (095)	
- description	102

J

Jet pump (290)	
- adjustable	329
- applications	133, 134
- as deep-well pump (070)	100
- description	132
- for liquid-solid mixtures	133, 317
- for venting and priming	133, 142

L

Lateral vibration	
see Critical speed	
Lead and lead alloys	
- review	206-207
Leakage	
- canned motor pump (125)	105
- dangers to the environment	104, 233
- exposure of personnel	104, 233
- in positive displacement pump	222, 233
- in rotodynamic pumps	101, 225, 233
- leakage rates from shaft-seals	222, 233
- magnetic drive pump (120)	104, 123
- non-contact seals	220
- porosity of materials	104
- soft packings	222, 233
- vapour leakage, mechanical seals	230, 233
- vapour leakage, soft packing	222, 225, 232-233
see also Trouble-shooting	
Life cycle cost	
- choice of control method	194-196
- definition	296
- distribution	296
- energy	296
- investment	296-298
- maintenance	296
Light (duty)	
- definition	308
Line shock	
see Water hammer	
Lip seals	
- choice of seal	220
- construction	220
Liquid flow	
see Chapter 2	
Liquid properties	
see Chapter 1	
- for pump selection	215, 308
- pulps	60
Liquid ring pump (210)	
- automatic water supply packages (020)	94
- description	113
- for venting and priming	142
- performance curves	113
Liquids	
- barrier liquids for shaft seals	222, 230
- buffer liquids for shaft seals	220, 222, 232
- check-list for material selection	215, 224
- corrosion of metals	23, 211, 216
- flow	50
- flush liquids for shaft seals	220, 223, 228, 232
- for testing	278-279
- liquid table	32, 233
- lubricants	223, 232
- properties	12, 52, 60, 233
- quench liquids for shaft seals	222-223, 228, 232

Liquid-solid mixtures	
- abrasion	26, 214, 317
- concentration	26, 214
- density	28, 29, 107
- Miller Number	26, 106, 119
- pressure loss	56-59, 304
- sedimentation velocity	58
Liquid transportation	
- economy, Chapter 13	
- system considerations	152-156
Liquified gases	117
Load-unload control	
- operating principles	174
Lobe pump (240)	
- description	116, 124
- efficiency	120
- pump curves	120
- rotor profiles	125
Locked rotor torque	
- electric motors	250
Loss coefficients	
- calculation, pipe	54
- tables, pipe	53
Losses in bends and fittings	
- definition	53, 54
- diagrams for pulp	70-72
- equivalent pipe length	61
- loss of coefficient	54, 55

M

Macerater	112, 137
Machine tool coolant pump (040)	
- description	96
- standards	96
Magnetic drive pump (120)	
- description	104, 123
- limitations	105
Maintenance	
- alignment of shaft couplings	237, 241
- calibration	270
- check-list of routine inspections	291, 293
- electrical equipment	270
- electric motors	261
- pumps	291, 293
- shaft seals	224, 230
- spare parts inventory	293-294
Marine pumps (045)	
- description	96
see also Cargo pump for tankers	334
Material combinations	
- galvanic (electrolytic) corrosion	23, 211, 213
- price relationship for pumps	216
- pump	215-216
- shaft seals	220, 224, 229
Materials	
see Chapter 6	
- applications	216-217
- abrasion resistance	214
- cavitation resistance	214-215
- corrosion	23, 211-212, 216, 273
- costs	210, 216, 229
- elastomers for mechanical seals	229
- lip seals	220
- bibliography	217
- mechanical seals (655, 660)	229
- protective films	204, 206, 211
- review	200-201
- shaft couplings (625, 630)	239
- soft packings (585, 590)	224, 232
- strength	200, 202-207, 209, 272
- toughness	272, 390
Material testing	
- chemical composition	273
- corrosion	213, 273
- creep resistance	272
- elongation	272
- fatigue endurance	272
- hardness	272
- impact strength	272
- limit of proportionality	272
- reduction in area	272
- ultimate strength	272
see also NDT	
Mean time between overhauls	308
Mechanical seals (655, 660)	
- construction	221, 233
- functional principle	225
- leakage	233
- materials	229
- standards	229
- temperature considerations	211, 227
- trouble-shooting	140, 230
- seal cavities	102, 103
- types	225-229
Meehanite	201
Melting point	
- definition	12
- oils	24
- other liquids	31
- water	19

Mesh sizes	289
Metering pumps (285)	
- checking	289, 292
- description	131
- flow regulation	197
- speed regulation	197
Miller Number	
- definition	26
- values	27
Minimum flow	
- devices	161
- reasons for	161
- temperature rise	161, 163, 211, 227
MIP - Minimum inlet pressure	290
- definition	139
- test	139, 274
Mixed flow pumps (195)	
- deep-well pumps	100, 112
- land drainage	112, 333
- vertical	98, 110, 111, 112, 333
Moments	
see Forces and moments	
Momentum	51
Monel	
- cavitation resistance	215
- review	206
- shaft coupling	239
Mono pump	
see Progressing cavity pump (235)	123
Multi-phase flow	30, 123, 143
Multi-pump packages (160)	108
Multi-stage pumps	
- efficiency	92
see Deep well pumps (070)	100
see High speed multi-pump packages (160)	108
see Multi-stage axially-split pumps (165)	109
see Multi-stage radially-split pumps (170)	109
see Segmental pumps 300 m (085)	101, 110
see Segmental pumps 300 m (090)	102, 110
see Submersible deep well pumps (075)	100
see Vertical multi-stage pumps (175)	110
see Wash water pump packages (080)	101

N

NACE	145
NDT - Non destructive testing	225, 273
Newtonian liquids	
- definition	13
- pipe flow losses, calculation	53-56
- pipe flow losses, diagrams	62-69
Nickel alloys	
- review	206
Nickel aluminium bronze	
- review	206
- sea water	23
see also Aluminium bronze	
Noise	
- electric motors	252, 260-261
- noise levels, water, heating and sanitation pumps	94, 328
- standards	147, 252
- test	275
Non-clogging pumps (180, 185, 190)	
- description	111, 112
- "gentle" pumping	111, 112, 317
- other pumps	112
- positive displacement pumps	124-125, 128-131
- pump selection tables	309-311
Non-contact seals	220
Non-metallic materials	207, 214, 224, 229
Non-metallic pump (130)	
- creep strength, plastic materials	200
- Grp	106
- types	106, 104, 122-123
Non-Newtonian liquids	
- by-pass regulated pd pumps	184
- definition	13
- examples	13, 32
- main groups	13
- pulp	59, 60, 156
- pulp loss curves	70-72
- suspensions, calculation diagrams	60
- suspensions, examples at part load	304
- suspensions, pipe flow calculations	56, 59
- suspensions, properties	56-60
NPIP - Net positive inlet pressure	
- cavitation damage	130, 139, 214
- definition	139
- effect on displacement pump performance	139
- test	139, 274
NPIPa - Net positive inlet pressure available	
- definition	139
NPIPr - Net positive inlet pressure required	
- datum	52
- definition	139, 140
NPSH	
- calculated examples	143-144
- cavitation damage	140, 214

- definition	138
- effect on rotodynamic pump performance	85, 138
- pump choice	308, 319
- suction capacity	137

NPSHa
- definition	138
- in pump system	138

NPSHr
- datum	52, 138
- definition	138
- suction specific speed	85

N_s
see Specific speed

N_{ss}
see Suction specific speed

O

Oblique disc pump	**112**

Oils
- liquid properties	24
- pipe losses, diagram	66-69

On-off control
- principles and methods	174
Overflow control	**174, 184**

Overhauls
- electrical equipment	261
- electric motors	261
- pumps	293
- spare part stocks	293-294

P

Packing material
- soft packing	224, 232
Parallel misalignment	**237, 242**

Parallel operation
- rotodynamic pumps	108, 324-325, 327
efficiency	159
unstable curves	89, 159, 308
pump curves	308
- positive displacement pumps	165

Particles
- effects on pipe flow losses	56, 57, 304
- properties, other	56-60
- pump selection	27, 97, 107, 127, 131, 135, 214, 308
- rapid selection table	312-313
- sedimentation in pipelines	58
- sizes, examples of	26, 27, 214
- sizes, grouping	57
Pay-off method (investment)	**298**
Penetration	**14**

Performance tolerances
- general	161
- guarantee	90, 143
- positive displacement pumps	162
- rotodynamic pumps	161

Peripheral pump (215)
- automatic water sets	94
- description	114
- for venting and priming	142

Peristaltic pump (250, 255)
- description	127
- as metering pump	127, 131
Petrol engine	**186**

pH-value
- choice of material	216, 224
- definition	15
- grey cast iron	15-16, 201
- liquids, etc.	17, 29, 31
- packing material for stuffing boxes	224
- water	17, 20
Pig, for piping cleaning or inspection	**320**

Pipe flow
- basic equations	53, 54, 303
- diagrams for loss of head	62-72
- losses in bends, valves and fittings, etc.	54, 61
- pipe flow losses	53-56, 303
- pulp	59, 156
- straight pipe losses	53, 56, 57, 59, 303
- suspensions	56-59, 304
- typical velocity	61

Pipe systems
- branched	154, 160
- single	152

Pipelines
- check-list	319
- economic diameter	304
- evacuating	142
- forces and moments	288
- hydraulic gradient lines	164, 167
- hydraulic losses (see Pipe flow)	303-304
- insulation	320
- noise	61, 328
- pipe bends at pump	288
- pipe entrance losses	61
- regulations	146
- spools for maintenance	284
- standards	146

- suction lines	61, 285-286, 288
- supporting	287
- types	61
- typical velocity	61
- water hammer	156, 165

Piston pumps (275)
- acceleration of liquid in suction line	128, 141
- applications	96, 101, 130, 317
- description	130
- operating region	315
- venting and priming	142

Pitot tube pump (310)
- description	135

Pit pumps
- dry pit (060)	99, 110
- wet pit (055)	98, 110, 335
Pitting	**23**
Pitting resistance	**23, 205**

Plastic
- creep	200, 201
- Grp	207
- material review	207-209
- plastic pipe, pressure losses	53
- plastic pipe, speed of wave propagation	15
- plastic pipe, surface roughness	53

Plastic pumps
see Non-metallic pumps (130)   106

PLC
see Programmable logic controller

Plunger pumps (270)
- acceleration of liquid in suction line	141
- applications	96, 101, 128, 130
- as metering pumps	132
- description	128
- operating region	315
- venting and priming	142

Pole-changing electric motors (520)
- description	257
- methods of control	257
Pollution	**29, 127**

Positive displacement pump
- basic principles	114
- cavitation	130, 214
- classification guide	121
- curves	82, 119, 120, 179, 308
- description of various types	121-132
- flow regulation	120, 197
- general review	114-120
- grouping	121

Power
- electric power	248
- hydraulic	83, 127, 303
- pump performance curves	84, 87
- pump shaft power	83
- requirement for various methods of flow regulation	178, 180-182, 184, 190, 191
Power factor	**249**

PR
see Pitting resistance   23

Pre-setting (Pre-whirl)
- regulation	88
Present capitalised value	**297-298**

Pressure
- control	161
- dynamic	50
- pressure energy	52
- pressure loss	53, 303-304
- static	50, 52, 152, 303
- total	51

Pressure boosting
- control	161
- energy conservation	161
- installation	160

Pressure losses, diagrams
- oils	66-69
- paper pulp	70-72
- review	60
- water	62-65

Pressure maintenance
- function	161
- methods of control	161

Pressure measurement
- gauges	290
- instrument piping	270
- pressure tapping points	290
- transducers	270
Pressure pulsations	115, 119, 128, 131, 132, 142, 164, 168, 275, 294

Pressure rating
- flanges	209
- pressure testing	210, 274
- pump choice	95, 102, 103, 104
- safety standards	145
- series pump connection	108, 158
- strength of materials	200, 389
- water hammer	166

Programmable logic controller   **177, 251, 270, 327**

Profits

- estimation of	298
- investment calculation	296-298
Progressing cavity pump (235)	
- description	123

Protection equipment
- check-list when purchasing	321
- flow variations	119, 128, 131, 132, 267
- maximum pressure	266, 290
- minimum flow	161
- pressure pulsations	119, 128, 131, 132, 166, 175, 267
- water hammer	166-167, 175, 267

Pull out torque
- electric motors	250

Pull up torque
- electric motors	250

Pulp pumps (135)
- dense pulp pump	106
- centrifugal pumps	106, 123-125
- performance and air content	106

Pulp suspensions
- air content	29, 106
- liquid properties	29, 32, 60
- pipe losses, calculations	59
- pipe losses, diagrams	62-72, 156, 304
- pump choice	106
- pumps	106, 123-125

Pulsation dampeners (635)
- for reciprocating pumps (270, 275)	119, 141, 169, 267-269

Pump
- definition	263

Pump efficiency
- mechanical	83, 92, 115, 117, 134, 136, 300, 301, 305
- volumetric	83, 116
- wire to water	83, 300, 338
Pump for cleaning	**128**

Pump selection
see Chapter 14   307

Pump curves
- air lift pump	134
- as turbine	336
- axial pump	88, 140
- centrifugal pump	84, 87, 88, 140, 156, 179, 190, 192, 326, 327
- effects of cavitation	85, 140
- effects of entrained gas	20, 123, 142
- effects of viscosity	90, 91, 120, 123, 156, 316
- guarantee	90, 143
- hydraulic ram pump	136
- jet pump	100, 133
- in parallel	159, 308, 325-327
- in series	158, 308
- liquid ring pump	113
- mixed flow pump	88
- positive displacement pump	120, 140, 156, 179
- pulp	156
- pump and turbine	165, 166, 336
- range charts	89, 101
- rigid screw pump	135
- stable	89, 159, 308
- unstable	89, 159, 308

Pump foundations
- concrete	284-285
- requirements for baseplates	264
- requirements for skids	100, 264
- requirements for trailers	100
- structural steelwork	264

Pump impeller
- angles	83
- axial thrust	86, 101
- non-clogging	111
- radial load	87
- specific speed	85
- suction specific speed	85
- various types	91
- velocities	83

Pump installation
- alignment	241, 284, 288
- foundation	100, 284-285
- hazardous area	16, 252
- lifting facilities	284
- noise	94, 100, 284, 328
- pump selection	119, 308-318
- space requirement	284
- ventilation	284

Pump names and categories
see the beginning of Chapter 3
- classification guides	81, 92
- direct selection tables	308-311
- trade names	381

Pump tanks and sumps
- determination of size	285-287, 325
- for submerged sewage pumps	286-287, 325
- general	285
- suction pipe connection	285-286
- supply lines	285

Pump unit
- definition	263

394

Purchasing
- check-lists 319-321
- conditions 321

PV-factor
- seal selection 222
- wear 115

Q

QC
see Quality control department 277

QP
see Quality plan

Quality Assurance
- certified design control 149
- certified manufacturing system 149
- check-list 321
- design integrity 210
- design to National Standard 146-148
- documentation 149, 275-276
- inspection 210, 273, 276
- mature design 211
- operator qualification 210, 275
- quality plan 277-280
- testing 148, 274-275
- witnessing 210, 273, 275

Quality control department **277**

Quality plan **277-280**

R

Radial load
- balancing 87
- bearings 93
- calculation 87

Radial piston (260) **117, 127**

Radially-split pump **104, 109-110**

Rapid selection tables **312-313**

Reciprocating motion **117-118**

Refinery pumps
see API pumps (**110, 115**) 104, 310
see Piston pumps (**275**) 130, 310
see Plunger pumps (**270**) 128, 310

Recess impeller pump
see Free-flow pump 111
see Non-clogging pump 111-112

Red metal
see Gunmetal 206

Regenerative pump
see Peripheral pump (**215**) 114

Regulation efficiency
- definition, example 300
- for by-pass regulation 183
- for on-off control 178
- for overflow control 184
- for pole-changing motor 180
- for throttle regulation 182
- for variable speed control 189-191
- pumping efficiency factor 300

Regulation methods
- positive displacement pumps
  review 120, 174, 184
- rotodynamic pumps
  review 174

Relative density
- definition 14, 387

Relief valves (645)
- air assisted 267
- pilot operated 116, 266
- spring loaded 266
- tested with pump 275

Repairs **210, 273**

Reverse flushing **289, 325**

Reynolds number
- definition 53
- flow measurement 73, 75
- pipe flow 53, 152
- with respect to corrosion 212

Rheology **13, 29, 308**

Rigid Screw pump (305)
- description 135

Rotary piston pumps (240)
see Lobe pumps (**240**) 124

Rotodynamic pumps
- classification guide 92
- curves 82, 84, 87, 88, 179, 325-327
- general review 92
- grouping 92

Rubber
- abrasive solids pumps 107, 127, 214, 316
- contraction 137
- mechanical seals 229
- positive displacement pumps 131
- protective linings 214
- review 207-209
- seals 220
- shaft couplings 237, 239

Rubber element shaft couplings **237**

S

Safety
- suppliers legal obligation 146

Safety valves
see Relief valves 266

Sandtrap **289**

Saturation pressure
see Vapour pressure 20-26

Screens
- in pump sumps 285
- temporary for commissioning 284, 290

Screw pumps (230)
- description 123
- rigid screw pump (**305**) 135

Sea water
- analysis 22
- choice of materials 23, 217, 224, 229

Seals
- oil seals 220
- plunger seals 118, 231, 233
- shaft seals 102, 103, 220, 225, 233

Selective corrosion **211**

Self-clearing pumps
see Non-clogging pumps (**180, 185, 190**) 111, 112

Self priming
- engine driven pumps, capacity 332
- evacuation 99, 142
- starting 114, 122, 126, 130, 142, 332

Self venting **142**

Semi-rotary hand pumps **136**

Semi-axial pumps
see Mixed flow pumps (**195**) 112

Series connection of pumps
- efficiency 158
- pressure boosting 158
- pump curves 108, 158

Service and maintenance
- electrical equipment 261
- electric motors 261
- instrumentation 267
- pumps 224, 230
- routine 177, 291, 293, 327, 332

Service factors
- belt drives 265
- couplings 238
- gearboxes 265

Sewage
- properties 29, 30
- pump selection 310-311

Sewage pumps
see Non-clogging pumps (**180, 185, 190**) 111, 112

Shaft alignment
- checking 241-244
- commissioning 291
- installation 284, 288

Shaft couplings
see Chapter 8
- check-list 320
- installation 240
- service factors 238
- service life 240
- types 236-237, 241

Shaft seals
- back to back mechanical 229, 233
- balanced mechanical 226
- bellows mechanical 229
- bulkhead through-fitting 239, 334
- cartridge 112, 232
- check-list 224, 229, 233
- chemical 102, 103, 233
- double mechanical seals 103, 112, 228-229, 233
- lip seals 220, 231
- non-contacting 220
- overheating 211
- pressure testing together with pump 274
- seals in oil-bath 112
- single mechanical 221, 225, 228, 233
- trouble-shooting 225, 230
- types 231
- unbalanced mechanical seals 225

Shear thickening **13, 308**

Shear thinning **13, 308**

Ships' pumps (045)
- applications 96, 333-335
- description 96

Shut-off head
see Closed valve head 89

SI system
see Chapter 18 383

Similarity
- centrifugal pumps 85, 189
- selection of pump for
  viscous liquids 314-316, 335

Site conditions **321**

Skid
- description 264
- design requirements 264
- foundations 100
- options 264

Sleeve coupling **236**

Sliding-shoe pump **126**

Sludge
- liquid properties 29
- pipe flow 56-57
- pump selection 97, 310-311

Soft packings
- cooling 223
- costs 234
- function 117, 222, 232
- leakage 222
- trouble-shooting 225
- types 117, 231-232

Solids pumps
- agricultural (**030**) 96
- adjustment 56, 59
- air lift (**295**) 134
- centrifugal (**140, 145**) 106, 107
- diaphragm (**280**) 131
- effect of wear 56, 59, 90, 119, 154, 214, 223, 316
- engine driven 99, 332
- gear (**220, 225**) 122
- lobe rotor (**240**) 125
- materials 107, 214, 224, 229
- peristaltic hose (**250**) 127
- piston (**275**) 130
- plunger (**270**) 129, 214
- progressing cavity (**235**) 124
- pump selection 27, 214, 310
- rigid screw (**305**) 135
- screw (**230**) 123
- vane (**245**) 126

Solubility
- gas in liquids 30
- solubility in water 31, 48
- solutions, definition 12

Sound
see Noise

Spacer coupling
- description 237
- for pump construction 95, 102, 103, 104, 334
- reference line for alignment 241-242

Specific gravity
- definition 14

Specific speed
- definition 85
- for various pump impeller designs 85

Speed
- effects on balancing 239
- effects on NPIP 129
- effects on NPSH 138
- effects on performance
  rotodynamic pumps 89, 138, 325
  positive displacement pumps 129
- electric motors 88
- pump selection for viscous liquids 314-316, 335
- pump selection with regard to NPSH 308

Speed-regulation
- description 179-181, 185-191
- economy 180, 190-191

Spheroidal or nodular cast Iron (SG Iron)
- alloyed 202, 203
- pressure containment 202
- review 202

Split-muff coupling **236**

Stainless steel
- cavitation 215
- corrosion 23, 211
- pipe, pressure losses 53
- pipe, roughness 53
- protective film 211
- review 19, 20, 23

Standard pumps
- construction/design standards
  API 610 (**110, 115**) 103, 109
- dimensional/performance standards
  prEN 733 (**025**) 95
  ISO 2858 (**095**) 102
  ANSI B73.1 (**100**) 103
  ANSI B73.2 (**105**) 103

Standards
- bearing housing 104
- bearing life 103, 104
- belt drives 265
- chemical pump 102, 103
- components 146, 148
- dimensions 102, 103, 146, 252, 255
- domestic heating circulators 146
- electric motors 103
- flanges 102, 103, 104
- forces and moments on connections 103, 147
- gearboxes 265
- gearbox service factors 266
- ISO pump (**095**) 102
- issuing authorities 144-145, 251-252
- lube and seal systems 265
- machine tool coolant pumps (**040**) 96
- manufacturing 149

- materials	103, 104, 146, 202
- mechanical seals	102, 103, 104, 234
- mesh size	289
- motors	103, 251-256
- noise	147, 252
- operational life	102, 103, 146
- performance	146, 252
- pressure ratings	95, 102, 103, 146, 207, 210
- pressure relief systems	146
- pressure vessels	146, 200
- refinery pumps	104
- reliability	104, 146, 324
- safety	109, 145-146
- seal cavities	102, 103, 104, 234
- sealless pumps	106
- shaft couplings	95, 102, 104, 238
- steam tables	139
- stuffing boxes	102, 103, 104
- stuffing box systems	229-230, 234
- systems	146
- tanks	146
- testing	104, 148-149, 274-275
- units	383
- vibration	103, 104, 147
- water pump (025)	95
Standard water pumps (005)	**93**
Starting torque	
- electric motors	250, 257
- loads on shaft couplings	238
Steam turbine	**262**
Steel pipe	
- speed of wave propagation in pipe	15, 164
Steel spring coupling	**237, 239**
Storage volumes	
- for discrete flow increments	180-181
- for flow smoothing, buffer	177, 180, 181
- for load-unload control	177
- for on-off control	177
Stress corrosion	**23, 205, 212**
Structural steelwork	
- pipework supports	287
- pump supports	264
Stuffing box	
- circulation	129, 228-229
- flush	129, 222, 228-229
- lubrication	129, 223, 229
- quench	129, 223, 228-229
Submersible pumps	
- building drainage	97
- deep wells	100
- evacuating	97
- groundwater	97, 330
- non-clogging	111
- pump selection	310-311
Suction capacity	
- cavitation	137, 139-140
- evacuation	142
- gas separation	30
Suction lift	
- calculated examples	143-144
- definition	138
- jet pumps	133
- piping	288
- positive displacement pumps	121, 122, 124, 126, 131
- self-priming centrifugal pumps	99, 142, 333
Suction line	
- evacuation	142, 288
- venting and priming	124, 142, 288
Suction specific speed	
- definition	85
- relation to cavitation	85
Sump pumps	
- for building and construction purposes	97, 99
- for groundwater	97, 99
- non-electrically driven	99
- ships	96-97
Surge tower	**167, 175**
Suspensions	
- abrasion	26, 214, 316
- heterogeneous	57
- homogeneous	56
- liquid properties	13
- materials in abrasive mixtures	27, 316-317
- part-load, example	56-57
- pipe losses, calculations	56
- pipe losses, diagrams	62-72, 304
- pulp pumps (135)	29, 106, 156
- pump selection	97, 214, 310-311, 316
- rheology	13
- solids pumps (140, 145)	97, 106, 107, 214
- sludge	29
- types of shaft seals	220, 228
- typical velocities	58
System curves	
- branched system	154, 324
- control parameters	173
- single pipe system	152, 173, 326
- speed-regulation	179, 325
- suspensions	56-57, 156, 304
- variable	153, 324-325, 335

- viscosity, effects of	156, 335
System simulation	
- computer	119, 169, 269
T	
ΔT	
- parameter for shaft sealing	227-228
- pump minimum-flow	163
Tanker pumps	
- LPG	334-335
- oil	333-334
Tantalum	**207**
Test Houses	
- for electric motors	253
Testing	
- alignment of shaft couplings	241
- corrosion testing	213, 273
- flow variations	119
- functional	275
- hydrotest	210, 274
- leakage, pressure testing	274
- noise	275
- NPIP	274
- NPSH	274
- pressure pulsations	119, 169, 275
- pump performance	274
- site simulation	274
- vibration	99, 111
Thermal growth	**237-238**
- see also Alignment	237, 284
Throttle regulation	
- energy consumption	182
- minimum pressure drop	182
Through flow	
- of non-clogging pumps (180, 185, 190)	111, 112
Titanium	
- cavitation resistance	215
- cost	207
- protective films	211
- review	207
Torque	
- electric motors starting	250, 252, 259-260
Torque convertor	
- efficiency	189
- speed range	186
Torsional vibration	**262**
Toxicity	
- general	16, 215, 233
- liquids	32
Tractor-driven pump (030)	**96**
Trailers	**264**
Transmission efficiency	
- efficiency factor	300
- speed-regulation	189
Trip	**269**
Trouble-shooting	
- mechanical seals	230
- motors	259-260
- pumps	294
- soft packing	225
Turbo-charged engines	**187**
Turbo-expander	**187**
Twin pumps	**94**
U	
Units	
- SI	383
- units for pump technology	389
Upgrading	**59, 225, 276, 294, 322**
V	
Vacuum pump	
NOTE: a compressor	
- for venting and priming	142
Valves	
- characteristics	157
- control valves	156
- diaphragm pumps (280)	131
- electrically operated	289
- hydraulic	289
- isolating valves	61, 156, 289
- metering pumps (285)	132, 290
- non-return valves	61, 156, 165, 166, 289, 340-342
- piston pumps (275)	117, 130
- plunger pumps (270)	117, 129
- pneumatic	288
- positioning	289
- pressure drop	157, 289
- relief valves	116, 121, 163, 266, 275
- sizing	157, 305
- throttle valves	156, 182
Vane pumps (245)	
- description	126
Vapour pressure	
- cavitation	12, 138
- definition	139
- entrained gas	20, 142, 143

- jet pumps	133
- liquids, others	26, 31, 42
- oils	24, 139
- water	20, 139
Variable ratio transmissions	**185**
- efficiency	189-190
Variable speed drivers	**186**
- efficiency	189-190
VDR - Vendor documentation requirements	**277**
Velocity head	**51, 52, 82**
Venting and priming	
- methods	142
- pumps	142
- suction systems	142, 288
Vibration	
- balancing couplings	238-239
- balancing rotors	111
- harmonics	294
- in pipework	288
- minimum flow	163
- reciprocating pumps	119, 262
- standards for pumps	103, 104, 147
- vertical pumps	99, 111
Viscosity	
- alkydes	40
- black liquor	41
- conversion table	386
- definition	13
- effect on centrifugal pumps	90, 91, 156, 314-316
- effect on positive displacement pumps	120, 123, 124, 125, 126, 156, 314-316
- flow measurement	73, 75
- liquids, others	31
- oils	24
- pipe losses	53, 156, 303, 335
- pump selection	308, 314-316
- rapid selection table	312-313
- rheology	56-60
- suspensions	56-60
- units	386
- water	19
Volumetric efficiency	
- positive displacement pumps	115, 139
W	
Wash water packages (080)	
- description	101
Water	
- brackish	22
- choice of materials	23, 216-217, 224, 229
- demineralised	19
- fresh water	20, 324
- liquid properties	19-24, 32
- pipe flow	53, 61
- pipe losses, diagrams	62-65
- sea water	22
- sewage	28
Water hammer	
- causes	156, 165
- cavitation	164
- dynamic forces on pump	165
- material removal	137-141
- pressure waves	164, 168-169
- protection equipment	167-169
- reference literature	169
- storage volumes	167
- transient sequence	164-165
- with on-off control	167, 175
Water pump	
- automatic (020)	94
- groundwater	102
- irrigation (035)	96, 102
- marine (045)	96
- standard to prEN 733 (025)	95
Water ring pumps	
see Liquid ring pumps (210)	113
Wave propagation, speed of	
- calculation	15, 164
- definition	164
- liquids	15
- water hammer	164
Wear of metals	
- by abrasion	26, 119, 214, 223
Welder	
- qualifications	273
Welding	
- inspection	210, 273
- procedures	201, 210, 273
- qualifications	273
- review	273
Wet rotor motor pumps	
- installation	286, 337
- non-clogging pumps	286
- pump selection	256, 309-311, 337
Write off time	**298**
Z	
Zero leakage	**105**

19 Useful pump terms translated

A

abrasion	abrasion	Abrieb
acidity	acidité	Azidität
addition	juxtaposition	Anlagerung
aeration	ventilation, aération	Belüftung
ammeter	ampèremètre	Amperemeter
amount	quantité	Menge
analysis	analyse	Analyse
approximate	approché	Angenähert
arc arc,	courbe, coude	Bogen
asbestos	asbeste, amiante	Asbest
asbestos-cement	amiante-ciment	Asbestzement
asbestos fibre	fibre d'amiante	Asbestfaser
atmosphere	atmosphére	Atmosphäre
atomic or nuclear energy	énergie atomique ou nucléaire	Atomenergie, Kernenergie
axial-flow pump	pompe axiale	Axialpumpe
axis	essieu, axe, arbre	Achse, Welle
axis of rotation	axe de rotation	Drehachse
axle	essieu, axe, arbre	Achse, Welle

B

backpressure	contre-pression	Gegendruck
ball bearing	roulement à billes	Kugellager
base	fondation	Fundament
base plate	plaque de fondation, socle	Grundplatte
bearing	palier	Lager
bed	fondation	Fundament
bearing brackets	flasques de palier	lagerschilder
bearing cover	couvercle à palier	lager-deckel
beltdrive	commande à courroie	Riemenantrieb
bend	arc, courbe	Bogen
bid	offre, devis	Angebot
bin	réservoir, bac, caisse	Behälter
blast pipe	porte-vent, buse, tuyau	Düse
boiler feed-water	eau d'alimentation de chaudière	Kesselspeisewasser
bolt	vis	Schraube
bore hole	alésage perçage	Bohrung
bore well	puits de forage	Bohrbrunnen
breakdown	arrêt de service, trouble	Betrebsstörung
building	construction	Konstruktion, Aufbau
building site	terrain à bâtir, chantier	Baustelle
bushing	douille, boite, manchon	Buchse, Mute, Hülse
butterfly valve	robinet modérateur	Drosselventil

C

calibration	étalonnage, tarage	Eichung
canned motor pump	pompe sans presse-étoupe	Spaltrohrmotorpumpe
capacity	capacité de transport	Förderleistung
carbon	charbon, carbone	Kohle
carbonic acid	acide carbonique	Kohlensäure
castillated nut	écrou à créneaux	Kronenmutter
cavitation	cavitation	Kavitation
centrifugal pump	pompe centrifuge	Kreiselpumpe
chain	châine	Kette
chain drive	transmission par châine	Kettengetriebe
chassis	châssis	Fahrgestell
chemistry	chimie	Chemie
chlorine	chlore	Chlor
chlorine dioxide	bioxyde de chlore	Chlordioxid
circulation pump	pompe de circulation	Umwälzpumpe
circuit diagram	schéma de connexions	Schaltbild
circulation	circulation	Umlauf
cleaning	lavage, rinçage, curage	Spülung
clutch	embrayage, accouplement	Kupplung
coal	charbon, carbone	Kohle
coefficient of elongation	coéfficient d'allongement	Dehnungskoeffizient
coefficient of expansion	coéfficient de dilatation	Ausdehnungskoeffizient
companion flange	contre-bride	Gegenflansch
compensation nut	ecrou compensateur	Ausgleichsmutter
compressed air	air comprimé	Druckluft
compression spring	ressort de compression	Druckfeder
compressor	compresseur	Verdichter
connection	assemblage, raccord	Anschluss
connecting rod	bielle	Schubstange
connecting rod bolts	boulon de bielle	Schubstangenschrauben
connecting rod cap	tête de bielle	Schubstangendeckel
connecting rod half brg top	coussinet de bielle supérieur	Oberes Schubstanglager
connecting rod half brg bottom	coussinet de bielle inférieur	Unteres Schubstanglager
connecting rod nuts	écrous de bielle	Schubstangenmuttern
construction	construction	Konstruktion, Aufbau
container	réservoir, bac, caisse	Behälter
contamination	impureté, contamination	Verunreinigung
continuous	continu, à action continue	Kontinuierlich
control valve	soupape de réglage	Regelventil
controller	régulateur	Regler
coolant	produit refroidisseur, réfrigérant	Kühlmittel
corrosion	corrosion	Korrosion
corrosion resistance	résistance à la corrosion	Korrosionsbeständigkeit
counterpressure	contre-pression	Gegendruck
coupling	embrayage, accouplement	Kupplung
cover	couvercle	Deckel
crankcase	bâti	Gestellblock
crankshaft	arbre vilebrequin	Kurbelwelle
crosshead	crosse	Kreuzkopf
crosshead pin	tourillon de crosse	Kreuzkopfzapfen
crosshead pin bearing	coussinet de crosse	Kreuzkopflager
cross section	section transversale	Querschnitt
curve	diagramme	Diagramm
cut-out	interrupteur, commutateur	Schalter
cylinder block	corps du cylindre	Zylinderblock

D

D.C. motor	moteur à courant continu	Gleichstrommotor
dead weight	poids mort	Eigengewicht
deaeration	ventilation	Entlüftung
decarbonisation	décarbonisation	Entcarbonisierung
delivery	capacité de transport	Förderleistung
delivery head	hauteur manométrique, hauteur de refoulement	Förderhöhe
demineralisation	dessalement	Entsalzung
density	densité, poids spécifique	Dichte
depression	dépression, vide partiel	Unterdruck
diagram	diagramme	Diagramm
disc valve	soupape à siége plan	Tellerventil
discharge	issue, embouchure, écoulement, émission, effluance	Ausfluß, Ausflußöffnung, Ausströmung
discharge rate	débit	Förderstrom
dosage	dosage	Dosierung
dosing	dosage	Dosierung
drain	écoulement, effluent	Ablauf
drinking water	eau potable	Trinkwasser
driveshaft	essieu moteur	Antriebsachse
driving clutch	embrayage	Antriebskupplung
driving gear	pignon de commande	Antriebsrad
driving side	côté de la commande	Antriebsseite
ductility	extensibilité, ductilité	Dehnbarkelt

E

efficiency	rendement	Wirkungsgrad
ejector	pompe à jet, injecteur	Strahlpumpe
elasticity	extensibilité, ductilité	Dehnbarkeit
elbow	arc, courbe	Bogen
elimination	élimination, séparation	Ausscheiden
elongation	expansion, allongement, dilatation	Dehnung
end cover	couvercle de fermeture	Abschlußdeckel
energy	énergie	Energie
engaging piece	entraineur	Mitnehmer
engine	machine, engin	Maschine
erection	montage	Montierung

397

example of application	exemple d'application	Anwendungsbeispiel
exchanger	échangeur	Austauscher
exhauster	désaérateur	Entlüfter
expansion	expansion, allongement, dilatation	Dehnung

F

factory	fabrique, usine, ateliers	Fabrik
fault	arrêt de service, trouble	Betriebsstörung
faulty installation	défaut de montage	Einbaufehler
feather	ressort	Feder
feed	arrivée	Zufluß
filter	filtre	Fitter
filter area	surface du filtre	Filterfläche
flange	bride	Flansch
flow	courant	Strömung
flow velocity	vitesse de passage	Durchflußgeschwindigkeit
fluid	liquide	Flüssigkeit
foam	ébullition	Schaumbildung
force	énergie	Energie
forced-feed lubrication	graissage sous pression	Druckschmierung
foundation	fondation	Fundament
friction	friction	Reibung
fulcrum	axe de rotation	Drehachse

G

gasket	bague d'etanchéité	Dichtungsring
gate valve	soupape à coulisse	Schieber
gauge	mesurer	Messen
gauging	étalonnage, tarage	Eichung
gear pump	pompe à engrenages	Zahnradpumpe
gear transmission	engrenage de transmission	Getriebe
gear wheel	roue dentée	Zahnrad
gearbox	carter d'engrenages	Getriebegehäuse
generator	chaudiére à vapeur	Dampfkessel
gland	presse-étoupe	Stopfbuchse
grade	grosseur de grain	Korngröße
graduation	graduation, gamme	Skala
grain	grosseur de grain	Korngröße
granular size	grosseur de grain	Korngröße
graph	diagramme	Diagramm
grey cast iron	fonte grise, fonte moulée	Grauguß

H

hardness	dureté	Härte
heat exchanger	échangeur de chaleur	Wärmeaustauscher
heating	échauffer, réchauffement	Erhitzen
heating jacket	chemise de chauffage	Heizmantel
high pressure	haute pression	Hochdruck
hose	tuyau flexible	Schlauch
hub	moyeu	Nabe
hydrostatic pressure	pression hydraulique de l'eau	Wasserdruck

I

impeller	roue à ailettes	Flügelrad
impermeable	imperméable	Undurchlässig
impurity	impureté, contamination	Verunreinigung
induction motor	moteur asynchrone	Asynchronmotor
inflow	arrivée	Zufluß
initial velocity	vitesse initiale	Anfangsgeschwindigkeit
input side	côté de la commande	Antriebsseite
installation	installation, montage, assemblage, montage	Einbau
installation dimension	dimension de montage	Einbaumaß
insulation	isolation	Isolierung
intermediate stage	étage intermédiaire	Zwischenstufe
internal diameter	diametre interieur	Innendurchmesser
iron	fer	Eisen
issue	éccoulement, effluent, émission, effluence	Ablauf, Auströmung

J

jet	porte-vent, buse, tuyau	Düse
jet pump	pompe à jet, injecteur	Strahlpumpe
journal bearing	palier de l'arbre	Kurbellager
journal bearing	palier lisse	Gleitlager

L

lagging	isolation	Isolierung
laminar	laminaire	Laminar
leakage	coulage, fuite	Leckage
line voltage	tension du résau	Netzspannung
liquid	liquide	Flüssigkeit
loading	charge	Belastung
lube oil pump	pompe de graissage pour paleirs	Lagerölpumpe
lubricant	lubrifiant	Schmiermittel

M

machine	machine	Maschine
magnitude	taille, dimension	Größe
main shaft	arbre principal	Hauptwelle
maintenance	entretien, soin	Instandhaltung, Wartung
manual operation	service manuel	Handbetrieb
manufacturing tolerance	tolérance de fabrication	Fertigungstoleranz
material (of construction)	matériel	Werkstoff
material defect	dégats matériels, defectuosité	Materialfehler
measure	mesurer	Messen
measuring	dosage	Dosierung
measuring error	erreur de mesure	Meßfehler
measuring pump	pompe de dosage	Dosierpumpe
mechanical seal	joint mécanique	Gleitringdichtung
metallic coating	recouvrement métallique	Metallüberzug
method	opération, méthode, procédé	Verfahren
mill	fabrique, usine, ateliers	Fabrik
monitoring	contrôle, surveillance	Überwachung
motor (engine)	moteur	Motor
mounting	installation, montage, assemblage	Einbau, Montage
multi-stage	multicellulaire, à plusieurs étages	Mehrstufig
net positive suction head (NPSH)	nette pression d'aspiration	Gesamthaltedruck
non-return valve	soupape de retenue	Rückschlagventil
nozzle	porte-vent, buse, tuyau	Düse
number of revolutions	nombre de tours	Drehzahl
nut	écrou	Mutter

O

offer	offre, devis	Angebot
oil	huile	Öl
oil level gauge	jauge de niveau d'huile	Ölstandanzeiger
operating instructions	instructions de service	Bedienungsvorschrift
operating temperature	température de fonctionnement	Betriebstemperatur
orifice	orifice	Blende
oscillation	oscillation, vibration	Schwindung
outflow	émission, effluence	Ausströmung
output	débit nominal	Nennleistung
overflow	trop-plein	Überlauf
overload	surcharge	Überlastung
overload protection	protection contre les surcharges	Überlastungsschutz
overpressure	excès de pression	Überdruck
oxidation	oxydation	Oxydation

P

packing	joint, garniture, étoupage	Dichtung
packing fluid	liquide obturant	Sperrflüssigkeit
panel	panneau de distribution	Schalttafel
parallel operation	marche en paralléle	Parallellbetrieb
partial vacuum	dépression, vide partiel	Unterdruck
peak load	charge maximale	Spitzenbelastung
pilot plant	usine de recherche	Versuchsanlage
pinion	pignon	Ritzel
pipe line	conduit de tuyau, tuyauterie	Rohrleitung
pipe socket	tubulure	Rohrstutzen
piping	conduit de tuyau, tuyauterie	Rohrleitung
piston	piston	Kolben
piston displacement	cylindre	Hubraum
piston pump	pompe à piston	Kolbenpumpe
piston rod	tige de piston	Kolbebstange
plain bearing	palier lisse	Gleitlager
plan view	projection horizontale	Grundriß
plant	fabrique, usine, ateliers	Fabrik
plunger	piston, plongeur	Plunger
polishing	rectifier	Schleifen

power	énergie	Energie
power absorption	puissance absorbée	Leistungsaufnahme
power consumption	puissance nécessaire	Kraftbedarf
power demand	puissance nécessaire	Kraftbedarf
power input	puissance absorbée	Leistungsautnahme
pressure	pression	Druck
pressure drop	baisse de pression	Druckabfall
pressure fluctuation	fluctuation de pression	Druckschwankung
pressure gauge	manomètre	Manometer
pressure line	tuyau forcé	Druckleitung
pressure pump	pompe de surpression	Druckerhöhungspumpe
pressure regulator	régulateur de pression	Druckregler
pressure vessel	réservoir de pression	Druckbehälter
priming	aspiration	Ansaugen
process	opération, méthode, procédé	Verfahren
pump	pompe	Pump
pumping head	hauteur manométrique, hauteur de refoulement	Föderhöhe
pumping station	station de pompage	Pumpwerk

Q
| quantity | quantité, debit | Menge |

R
ram	piston, plongeur	Plunger
rate	vitesse, vélocité	Geschwindigkeit
rated capacity	débit nominal	Nennleistung
recrystallize	cristalliser	Auskristallisieren
reducing valve	détendeur	Reduzierventil
regulator	regulateur	Regler
relief valve	soupape d'échappement	Abblaseventil
revolution counter	compte-tours, tachomètre	Drehzahlmesser
rinsing	lavage, rinçage, curage	Spülung
risk of fracture	danger de rupture	Bruchgefahr
roller bearing	roulement à rouleaux	Rollenlager
rotary piston pump	pompe à piston, mouvement de rotation	Drehkolbenpumpe
run-off, discharge	écoulement	Abufluß

S
saddle clip	collier de prise	Anbohrschelle
safety factor	coéfficient de securité	Sicherheitsfaktor
scale	graduation, gamme	Skala
screw	vis	Schraube
scum	ébullition	Schaumbildung
seal	joint, garniture, étoupage	Dichtung
sealing liquid	liquide obturant	Sperrflüssigkeit
sealing properties	properiété d'étanchéité	Dichtungseigenschaft
sealing ring	bague d'étanchéité	Dichtungsring
sealing sleeve	joint d'étanchéité	Dichtungsmanschette
sedimentation	décantation, sédimentation	Absetzen
self-priming pump	pompe à auto-amorçage	Selbstansaugende Pump
separation	élimination, séparation	Ausscheiden
service condition	condition de fonctionnement	Betriebsbedingung
service life	durée de service	Betriebsdauer
servicing	entretien	Instandhaltung
sewage	eaux d'égout, eaux résiduelles	Abwasser
sewage treatment	traitement des eaux	Abwasserbehandlung
shaft	essieu, axe, arbre	Achse, Welle
shaft seal	bague d'étanchéité	Wellendichtung
sight glass	verre de regard	Schauglas
size	taille, dimension	Größe
sleeve	manchon	Muffe, Hülse
slide valve	soupape à coulisse	Schieber
solder	souder	Löten
solidification	solidification	Erstarrung
source	source	Quelle
space required	encombrement	Raumbedarf
spare	pièce de rechange	Ersatzteil
spare part list	liste des pièces de rechange	Ersatzteilliste
specific gravity	densité, poids spécifique	Dichte
speed	vitesse, velocité	Geschwindigkeit
speed range	gamme de tours	Drehzahlbereich
split pins	goupilles	Splinte
spring	ressort	Feder
spring	source	Quelle
stainless	inoxydable	Rostfrei, Nichtrostend
start-up	mise en mouvement	Inbetriebnahme
starting	mise en mouvement	Inbetriebnahme
steam	vapeur	Dampf
steam boiler	chaudière à vapeur	Dampfkessel
steam pressure	pression de vapeur	Dampfdruck
strain	charge	Belastung
strainer	filtre	Filter
strength	solidité	Festigkeit
stress	charge	Belastung
stuffing box	presse-étoupe	Stopfbuchse
stuffing box packing	garniture de presse-étoupe	Stopfbuchspackung
submersible pump	pomp noyée	Tauchpumpe
suction	aspirateur	Absauger
suction head	hauteur d'alimentation	Zulaufhöhe
suction lift	hauteur d'aspiration	Saughöhe
suction line	conduite d'aspiration	Saugleitung
supervision	contrôl, surveillance	Überwachung
supply	assemblage, raccord	Anschluß
supply head	hauteur d'alimentation	Zulaufhöhe
surface	surface	Oberfläche, Fläche
surface finish	état de surface	Oberflägchenbeschaffenheit
switch	interrupteur, commutateur	Schalter
switchboard	panneau de distribution	Schalttafel

T
tachometer	compte-tours, tachomètre	Drehzahlmesser
tank	réservoir, bac, caisse	Behälter
tapping sleeve	collier de prise	Anbohrschelle
temperature	temperature	Temperatur
tender	offre, devis	Angebot
tensile strength	résistance à la traction	Zugestigkeit
tension	tension	Spannung
test pressure	pression d'épreuve	Prüfdruck
testing of materials	essai des matériaux	Materialprüfung
three-phase A.C. motor	moteur à courant triphasé	Drehstrommotor
throttling valve	robinet modérateur	Drosselventil
throughput	debit	Durchflußmenge
time	temps	Zeit
transmission housing	carter de transmission	Antriebsgehäuse
transmission shaft	arbre d'entrâinement	Antriebswelle
tube well	puits de forage	Bohrbrunnen

U
| undercarriage | châssis | Fahrgestell |

V
V-belt	courroie trapézoidale	Keilriemen
V-belt pulley	pouilie à gorge	Keilriemenscheibe
vacuum	vide, vacuum	Vakuum
valve	soupape, clapet, vanne	Ventil
valve body	corps de soupape	Ventilkörper
valve seat	siège de soupape	Ventilsitz
vane pump	pompe à ailettes	Flügelpump
vapour	vapeur	Dampf
velocity	Vitesse, vélocité	Geschwindigkeit
vent	désaérateur	Entlüfter
ventilation	ventilation, aération	Entlüftung, Belüftung
vibration	oscillation, vibration	Schwindung
viscosity	viscosité	Viskosität, Zäghigkeit
volume	quantité, volume	Menge, Volumen

W
washing	lavage, rinçage,curage	Spühlung
waste treatment	traitement des eaux	Abwasserbehandlung
waste water	eaux d'égout, eaux résiduelles	Abwasser
water column	colonne d'eau	Wassersäule
wear	ponçage, abrasion	Abrieb
weight	poids	Gewicht
weld	souder	Schweißen
wheel	roue	Rad
wheel hub	moyeu de roue	Radnabe
working condition	condition de fonctionnement	Betriebsbedingung
works	fabrique, usine, ateliers	Fabrik

Index to Advertisers

Ahlstrom Machinery, Pump Industry	2
Autoclude Ltd	170
BT Tenute Meccaniche Rotanti SpA	Inside Back Cover
Bran + Luebbe (GB) Ltd	306
Clasal N.V.	4
Couplings and Drives (UK) Ltd	Bookmark
DuPont de Nemours	218
Flowguard Ltd	367
Forward Industries Ltd	78
Grundfos A/S	Inside Front Cover
Hughes Wynne Ltd	4
IMO AB	282
ITT Flyg AB	Outside Back Cover
ITT Jabsco	170
Kestner Engineering Co Ltd	262
Liquid Dynamics	78
MDM Pumps Ltd	306
Mackley Pumps, Clarke Chapman Marine	150
Mono Pumps Ltd	282
Production Techniques Ltd	4
The Pump Centre	6
Savino Barbera snc	150
Stork Pumps Ltd	198
T-T Pumps Ltd	322
Pumpenfabrik Ernst Vogel GmbH	198
Wirth Maschinen- und Bohrgeräte-Fabrik GmbH	6

Acknowledgements

The Publishers acknowledge the help and asistance of the following organisations in supplying data, photographs, illustrations and, where appropriate, permission to reproduce material from their own publications:

ABS Pumps Ltd
Alfa-Laval Pumps
Couplings and Drives (UK) Ltd
Dawson Downie Lamont Ltd
Hydraulic Institute, Cleveland, Ohio
Hydraulic Research Centre, University of Bath
IMO AB
Ingersoll-Dresser Pumps (UK) Ltd
Mono Pumps Ltd
Multiphase Pumping Systems Ltd
Sauer-Sundstrand Ltd
Stork Pumps Ltd
Sundstrand International Corporation SA
Tuthill UK Ltd

The Publishers also wish to acknowledge the extensive help and assistance given by their consultant, Brian Nesbitt, in the editing of this edition.